# CHEYENNE BIRD BANTER

*25 Years of Bird and Birdwatching News*

**BARB GORGES**

CHEYENNE, WYOMING
2024

Other books by Barb Gorges:

Quilt Care, Construction and Use Advice

Cheyenne Birds by the Month (with photographer Pete Arnold)

Dear Book: The 1916-1920 Diary of Gertrude Oehler Witte

Cheyenne Garden Gossip: Locals Share Secrets for High Plains Gardening

Copyright © 2024

Except as noted, all essays were previously published from 1999 – 2024 in the Wyoming Tribune Eagle, which has granted permission for reprinting essays 167a and 455, which were written by Ty Stockton and Rachel Girt, respectively.

Published by Yucca Road Press
3417 Yucca Road
Cheyenne, Wyoming 82001

Printed by IngramSpark, USA

Book design by Chris Hoffmeister, Western Sky Design

ISBN 978-0-9992945-7-4

*To Mark, who made me look at the birds again.*

# Table of Contents

Introduction . . . . . . . . . . . . . . . . . . . . . . . . . . . . . . . . . . . . . . **page xv**

## 1999 . . . . . . . . . . . . . . . . . . . . . . . . . . . . . . . . . . . . . . **page 1**

article no.    title . . . . . . . . . . . . . . . . . . . . . . . . . . . . . . . . . . . . . topic

| 1 | Birding the Colorado coast . . . . . . . . . . . . . . . . . . . . . . | *Bird watching* |
|---|---|---|
| 2 | No headline (bird atlas) . . . . . . . . . . . . . . . . . . . . . . | *Bird atlas* |
| 3 | No headline (Cats Indoors) . . . . . . . . . . . . . . . . . . . . | *Cats indoors* |
| 4 | Out-of-towners flocking to southeast Wyoming . . . . . . . . . . . . | *Spring migration* |
| 5 | Cheyenne birders count 153 individual species . . . . . . . . . . . | *Big Day Bird Count* |
| 6 | Backyard building is for birds . . . . . . . . . . . . . . . . . . . | *Bird houses* |
| 7 | Life, death in the backyard . . . . . . . . . . . . . . . . . . . . | *Bird diseases* |
| 8 | Birding know-how a matter of degree . . . . . . . . . . . . . . . . | *Bird i.d.* |
| 9 | Time is here for young to leave nest . . . . . . . . . . . . . . . . | *Birdwatching* |
| 10 | Fall migration can baffle birders . . . . . . . . . . . . . . . . . | *Fall migration* |
| 11 | 'Native species' is convoluted concept . . . . . . . . . . . . . . . | *Bird taxonomy* |
| 12 | Songbirds: Sunflower seed is sure to satisfy . . . . . . . . . . . . | *Bird feeding* |
| 13 | Blustery days challenge birders, send birds packing . . . . . . . . | *Birdwatching* |
| 14 | Juncos add variety to backyard bird feeder visitors . . . . . . . . | *Bird: Juncos* |
| 15 | Take an hour to help count the birds . . . . . . . . . . . . . . . . | *Thanksgiving Bird Count* |
| 16 | Workshops help educate educators on the environment . . . . . . . . | *Environmental education* |
| 17 | Project FeederWatch relies on citizen scientists . . . . . . . . . . | *Project FeederWatch* |
| 18 | Stand up and be counted – by counting . . . . . . . . . . . . . . . | *Christmas Bird Count* |

## 2000 . . . . . . . . . . . . . . . . . . . . . . . . . . . . . . . . . . . . . . **page 18**

| 19 | Refuge offers whooping-good time . . . . . . . . . . . . . . . . . | *Bird: Cranes* |
|---|---|---|
| 20 | Bird count uncovers hidden treasures . . . . . . . . . . . . . . . . | *Christmas Bird Count* |
| 21 | Feeder visitors fewer during mild weather . . . . . . . . . . . . . | *Bird feeding* |
| 22 | Great Backyard Bird Count needs you . . . . . . . . . . . . . . . . | *Great Backyard Bird Count* |
| 23 | Night birding offers unique experience on quiet evenings . . . . . . | *Birdwatching* |
| 24 | Birds are now on the move . . . . . . . . . . . . . . . . . . . . . | *Spring migration* |
| 25 | Birding on a computer screen . . . . . . . . . . . . . . . . . . . . | *Birding software* |
| 26 | Home on the Prairie . . . . . . . . . . . . . . . . . . . . . . . . | *Backyard habitat* |
| 27 | Making birds at home in our 'urban forest' . . . . . . . . . . . . . | *Backyard habitat* |
| 28 | Indoor-dwelling cats can add richness, humor to your life . . . . . | *Cats indoors* |
| 29 | Scouting gets youngsters in touch with outdoors . . . . . . . . . . | *Scouts* |
| 30 | Songs tell of habitat, breeding . . . . . . . . . . . . . . . . . . | *Bird song* |
| 31 | Humans a hazard to navigation . . . . . . . . . . . . . . . . . . . | *Window hazards* |
| 32 | Go Birding! . . . . . . . . . . . . . . . . . . . . . . . . . . . . | *Birdwatching* |
| 33 | Bird identification takes same skills as identifying people . . . . | *Bird i.d.* |
| 34 | Birding Field Guides . . . . . . . . . . . . . . . . . . . . . . . . | *Field guides* |
| 35 | Binocular Considerations . . . . . . . . . . . . . . . . . . . . . . | *Binoculars* |
| 36 | Arrivals, departures mark seasons . . . . . . . . . . . . . . . . . | *Spring migration* |
| 37 | 'Backyard birding' varies by the yard . . . . . . . . . . . . . . . | *Birding Wyoming* |
| 38 | Ears as important as eyes in early morning bird survey . . . . . . . | *Breeding Bird Survey* |
| 39 | Banding volunteers aid research, get up-close contact with birds . . | *Bird banding* |
| 40 | Solitude found off-road . . . . . . . . . . . . . . . . . . . . . . | *Birding Wyoming* |
| 41 | Summer means crowds for both birds and people . . . . . . . . . . . | *Bird: Pelican* |
| 42 | For fun, try cooking outdoors sans pans . . . . . . . . . . . . . . | *Outdoor cooking* |

| article no. | title | topic |
|---|---|---|
| 43 | Hummingbirds making their colorful migrations | *Bird: Hummingbirds* |
| 44 | Outings provide close encounters of the bird kind | *Birdwatching* |
| 45 | Birds congregating, fattening up to prepare for winter | *Fall migration* |
| 46 | Wyoming hosts National Audubon Society board | *Audubon* |
| 47 | Land of open spaces? | *Open space* |
| 48 | Birds make windy hunt bearable | *Hunting* |
| 49 | Birds needn't fear this wolf pack | *Scouts* |
| 50 | A visit to Lucy's garden creek | *Birding Wyoming* |
| 51 | Bird watchers have opportunity to aid research | *Project FeederWatch* |
| 52 | Best Backyard Bird list joins mounting counts | *Bird list* |
| 53 | Clean-up is part of bird feeding | *Bird feeding* |
| 54 | Bird count a time to enjoy winter | *Christmas Bird Count* |

# 2001 . . . . . . . . . . . . . . . . . . . . . . . . . . . . . . . . **page 50**

| | | |
|---|---|---|
| 55 | Calm, cloudy weather makes for unusual count | *Christmas Bird Count* |
| 56 | Lifetime license lends legitimacy to non-native | *Conservation stamp* |
| 57 | Great Backyard Bird Count offers opportunity to aid science | *Great Backyard Bird Count* |
| 58 | Helping to save rain forest has never been so sweet | *Bird-friendly chocolate* |
| 59 | Photographer's identity a mystery | *Birdwatchers* |
| 60 | Mystery photographer identified | *Birdwatchers* |
| 61 | Who's keeping count? Lots of folks | *Bird counts* |
| 62 | Field guide choices now are many | *Field guides* |
| 63 | Birds of a different feather spotted back East | *Travel: East* |
| 64 | Humans, too, flock together during springtime snowstorms | *Spring snow* |
| 65 | Nighttime right time for fauna to frolic | *Bird reports* |
| 66 | Backyard beckons bevy of birds | *Bird reports* |
| 67 | Birders tally 143 species on 'big day' | *Big Day Bird Count* |
| 68 | Putting off yard work helps wildlife | *Bird gardening* |
| 69 | State fair to offer birdhouse, bird feeder contest | *Bird houses & feeders* |
| 70 | Robins fledge, face cruel world | *Fledging* |
| 71 | Some summer ponderings | *Travel: East* |
| 72 | Don't poison your yard to save it | *Bird gardening* |
| 73 | Donations sought for injured birds | *Bird rehab* |
| 74 | Bird flashcards get a technological twist | *Bird education* |
| 75 | Don't rely on color for bird ID | *Bird i.d.* |
| 76 | Local group works to launch exotic bird rescue operation | *Exotic birds* |
| 77 | Riparian areas: Ancient sacred sites still valuable today | *Bird habitat* |
| 78 | Expo is for bird watchers, too | *Bird education* |
| 79 | Fall festivals welcome birds back | *Bird festivals* |
| 80 | Bird travel plans show variety | *Migration* |
| 81 | Did you happen to see the one that got away? | *Bird i.d.* |
| 82 | Feeder watchers keep track of winter birds | *Project FeederWatch* |
| 83 | Pool parties popular with winter birds | *Bird baths* |
| 84 | Count birds while waiting for dinner | *Thanksgiving Bird Count* |
| 85 | Close encounters of the bird kind | *Bird rescue* |
| 86 | Calendar features young artists | *WGFD calendar* |
| 87 | Stop Poaching campaign adds new poster, book, bumper sticker | *WGFD campaign* |
| 88 | Keeping juncos straight isn't for bird-brains | *Bird: Junco* |
| 89 | Count birds, not presents | *Christmas Bird Count* |
| 90 | Apprentices wanted for solitaire sleuthing | *Bird: Townsend's Solitaire* |

| article no. | title | topic |
|---|---|---|

## 2002 . . . . . . . . . . . . . . . . . . . . . . . . . . . . . . . **page 82**

| 91 | Bird count identifies second state record for hermit thrush | *Christmas Bird Count* |
|---|---|---|
| 92 | Looking for birds in all the right places | *Christmas Bird Count* |
| 93 | Some birds aren't crowing about neighbor | *Bird: Owl* |
| 94 | Bird count depends on all bird watchers | *Great Backyard Bird Count* |
| 95 | Immigrant or visitor, red-bellied woodpecker finds food here | *Bird: Red-bellied Woodpecker* |
| 96 | Finding out where the birds are | *Birding hotspot* |
| 97 | Birders use cyberspace to advantage | *Wyobirds* |
| 98 | Racing pigeons | *Bird: Pigeon* |
| 99 | Racing can have lucrative payoff | *Bird: Pigeon* |
| 100 | You don't have to race pigeons to raise them | *Bird: Pigeon* |
| 101 | Neighborhood hosts unknown avian athlete | *Bird: Pigeon* |
| 102 | Winter's over when juncos take wing | *Winter* |
| 103 | Birders flock to observe crane migration | *Bird: Cranes* |
| 104 | Age-old battle between grasslands, forests | *Habitat: prairie* |
| 105 | Cheyenne artist earns conservation stamp honor | *Bird art* |
| 106 | Backyards going to the birds | *Habitat: backyard* |
| 107 | Bluebird housing makes a difference | *Bird: Bluebirds* |
| 108 | Learn to appreciate birds from a distance | *Bird i.d.* |
| 109 | Brown bird mix-up worth unraveling | *Bird i.d.* |
| 110 | Records request ruffles feathers | *Bird records* |
| 111 | Big Count draws flocks of observers | *Big Day Bird Count* |
| 112 | Bird families finding food in many unlikely places | *Birdwatching* |
| 113 | Arizona fire affects family, wildlife | *Wildfire* |
| 114 | Solitary hike yields memorable encounters | *Birdwatching* |
| 115 | Rescuing baby birds not always necessary | *Bird nests* |
| 116 | Bird books cover same turf in different ways | *Bird books* |
| 117 | Park's best-known bird remains elusive | *Travel: Yellowstone* |
| 118 | Some migratory birds more obvious than others | *Migration* |
| 119 | Much of bird watching is bird listening | *Birdsong* |
| 120 | FeederWatch season begins Nov. 9 | *Project FeederWatch* |
| 121 | Local park receives IBA designation | *Important Bird Area* |
| 122 | Birding trips can take watcher to distant lands | *Birding tours* |
| 123 | Do the Thanksgiving Bird Count without leaving home | *Thanksgiving Bird Count* |
| 124 | Make this Christmas a holiday for the birds | *Birdy gifts* |
| 125 | Cardinals top Christmas cards, if not bird count | *Bird: Cardinal* |

## 2003 . . . . . . . . . . . . . . . . . . . . . . . . . . . . . . . **page 117**

| 126 | Arizona fire damage softened by snow | *Wildfire* |
|---|---|---|
| 127 | Bird count picks up four new species | *Christmas Bird Count* |
| 128 | New Web site offers free service to bird watchers | *eBird* |
| 129 | Bald eagles attract commercial interest, festivals | *Bird: Bald Eagle* |
| 130 | Thanks to awareness, birds now visible on range | *Range management* |
| 131 | Bird flashcards evolve into full-service CD-ROM | *Bird flashcards* |
| 132 | Winter bird surveys contrast with backyard experience | *Habitat: Backyard* |
| 133 | River attracts cranes, cranes attract admirers | *Bird: Cranes* |
| 134 | With birds at feeders, cats are sure to follow | *Cats indoors* |
| 135 | Now's the perfect time to identify ducks | *Bird i.d.* |
| 136 | Cheyenne gaining reputation among warbler watchers | *Bird: Warblers* |
| 137 | Birdathon fun for bird watchers, helps raise funds | *Birdathon* |
| 138 | Low numbers, a few surprises at Big Day count | *Big Day Bird Count* |
| 139 | Avian guests make themselves at home | *Backyard birding* |

| article no. | title | topic |
|---|---|---|
| 140 | Knowing birds by ear is helpful on the trail | *Birdsong* |
| 141 | Scout camp reopens after fire | *Wildfire* |
| 142 | Dead blackbird tells no tale about West Nile virus | *West Nile Virus* |
| 143 | Fledglings nearly ready to fly away | *Fledging* |
| 144 | Bird of backcountry moves closer to town | *Bird: Dipper* |
| 145 | Backpack trip to the high country | *Birding Wyoming* |
| 146 | Best backyard bird shows up at dinnertime | *Backyard birding* |
| 147 | A bird by any other name still looks the same | *Taxonomy* |
| 148 | Birding not just a springtime joy anymore | *Fall migration* |
| 149 | Filmmaker catches magic of migration | *Bird film* |
| 150 | Wanted: Project FeederWatch citizen scientists | *Project FeederWatch* |
| 151 | Birding trails increasing in popularity | *Birding trails* |
| 152 | Count on it: This Thanksgiving | *Thanksgiving Bird Count* |
| 153 | Trumpeter performs for crowd at Holliday Park | *Bird: Trumpeter Swan* |
| 154 | How to keep up with birding news | *Birding publications* |
| 155 | Local tradition carried on with 104th bird count | *Christmas Bird Count* |

## 2004 .......................................... page 146

| article no. | title | topic |
|---|---|---|
| 156 | Snow diminishes results of Christmas Bird Count | *Christmas Bird Count* |
| 157 | Winter reading tracks albatross across the Pacific | *Book reviews* |
| 158 | Birders don't always flock together | *Beginning birding* |
| 159 | Backyard bird count migrates back again | *Great Backyard Bird Count* |
| 160 | Starlings aren't the darlings of the bird world | *Bird: Starling* |
| 161 | First sign of spring arrives | *Migration* |
| 162 | Prairie Partners provides for plains birds | *Rocky Mountain Bird Observatory* |
| 163 | Birdhouse site has everything under one roof | *Birdhouses* |
| 164 | Young chickadees may be changing age-old songs | *Bird: Chickadee* |
| 165 | Companions, field guides help with bird identification | *Bird i.d.* |
| 166 | Get gossip on Wyoming birds by phone, Web | *Bird hotline/Wyobirds* |
| 167 | Big Day brings wave of blue-gray gnatcatchers | *Big Day Bird Count* |
| 167a | I'm not much of a birder | *Big Day Bird Count* |
| 168 | West Nile virus frightens letter writer | *West Nile Virus* |
| 169 | Hybrid energy can improve bird watching | *Hybrid car* |
| 170 | Desert travel offers many different birds | *Travel: Albuquerque* |
| 171 | Mountain mud bothers birders, but not birds | *Birding Wyoming* |
| 172 | Brown-capped birds cause cracked ribs | *Birding Wyoming* |
| 173 | Windmills safer for birds than other structures | *Hazard: windmills* |
| 174 | Timing will be everything for birding Alaska | *Travel: Alaska* |
| 175 | Alaska birds add to life list | *Travel: Alaska* |
| 176 | Warblers winging their way through on migration path | *Migration* |
| 177 | Pacific coast down, Atlantic coast yet to go | *Travel: Cape May* |
| 178 | Day at Cape May whets birder's appetite | *Travel: Cape May* |
| 179 | Duck days attract birders to local lakes | *Bird: Ducks* |
| 180 | FeederWatch returns | *Project FeederWatch* |
| 181 | Going where the gulls are adds species to life list | *Bird: Gulls* |
| 182 | Birds are featured in calendar from Game and Fish | *WGFD: Calendar* |
| 183 | Christmas Bird Count debunks stereotypes | *Christmas Bird Count* |

## 2005 .......................................... page 176

| article no. | title | topic |
|---|---|---|
| 184 | Resolution produces list of field trip destinations | *Birding Wyoming* |
| 185 | Bird Seed bandits, How hard can it be to outsmart squirrels? | *Squirrels* |
| 186 | Birder's interest piqued by book a century old book | *Book reviews* |

vii

| article no. | title | topic |
|---|---|---|
| 187 | Fox squirrels pose ethical, biological problems | *Squirrels* |
| 188 | Are condors coming closer to home? | *Bird: Condor* |
| 189 | New doves coming to a feeder near you | *Bird: Doves* |
| 190 | It's time to grab your binoculars and pick a perch | *Birding Wyoming* |
| 191 | Callers: Give a hoot, Birders need to know etiquette of calling | *Bird: Owls* |
| 192 | Working in wildlife biology | *Wildlife careers* |
| 193 | Sharp-tail trip seeks out the other grouse | *Bird: Sharp-tailed Grouse* |
| 194 | Do you know what your hunting machine is up to today? | *Cats indoors* |
| 195 | Studying the science of birdsong | *Book reviews* |
| 196 | "Identify Yourself, The 50 Most Common Birding Challenges" | *Book reviews* |
| 197 | UW researchers seek hummingbird secrets | *Bird: Hummingbirds* |
| 198 | Local bird checklist makes its debut | *Bird checklist* |
| 199 | Backyard hosts lively bird theatre | *Backyard* |
| 200 | Flamm Fest finds record number | *Bird: Flammulated Owl* |
| 201 | New field guide is a useful tool on the road | *Book reviews* |
| 202 | Familiar faces fly by | *Travel: Midwest* |
| 203 | Rescued bird finds his way home | *Bird: Red-breasted Nuthatch* |
| 204 | Letters from a moose hunt | *Moose* |
| 205 | [sidebar about Snowy Range moose] | *Moose* |
| 206 | Time's here for annual bird counting | *Christmas Bird Count, Project FeederWatch, Thanksgiving Bird Count* |
| 207 | The best medium for bird notes | *Bird art* |

## 2006 .......... **page 203**

| | | |
|---|---|---|
| 208 | Woodpeckers find silver lining in devastation done by insects | *Bird: Woodpeckers* |
| 209 | We can have our energy and our sage grouse, too | *Bird: Sage-Grouse* |
| 210 | Bird flu and you, so far, so good | *Bird flu* |
| 211 | Consider it your war paint | *Skin cancer prevention* |
| 212 | Research station is, should be for the birds | *Birding Cheyenne* |
| 213 | Better binoculars make for better birders | *Binoculars* |
| 214 | Spring break trip should be warm | *Travel: Arizona* |
| 215 | Pup retrieves backyard of wonders | *Backyard* |
| 216 | Woodpecker visits might bring home improvements | *Bird: Woodpeckers* |
| 217 | Use a light touch with lawn chemistry | *Backyard* |
| 218 | Shorebird experts reveal their secrets | *Bird: Shorebirds* |
| 219 | A summer of mountain birding | *Birding Wyoming* |
| 220 | Small, flat memories of East Coast visit | *Travel: East Coast* |
| 221 | The last to go: Warblers put on a late season show | *Bird: Warblers* |
| 222 | Grouse losing ground fast | *Bird: Sage-Grouse* |
| 223 | Get creative with gifts for birders | *Birdy gifts* |

## 2007 .......... **page 221**

| | | |
|---|---|---|
| 224 | Out = healthy | *Health outdoors* |
| 225 | This spring watch for 'TVs' soaring above the area | *Bird: Turkey Vulture* |
| 226 | Visiting cedar waxwings learn lesson | *Bird: Cedar Waxwing* |
| 227 | Duck diversity | *Bird i.d.: ducks* |
| 228 | They look helpless, but they probably don't need rescuing | *Baby birds* |
| 229 | Some good guides for days exploring in great outdoors | *Book reviews* |
| 230 | Professor sheds light on crossbills | *Bird: Crossbills* |
| 231 | 'Prairie Ghost' / The 'Where's Waldo' of the wilderness | *Bird: Mountain Plover* |
| 232 | Plenty of bird action despite low count | *Bird count* |
| 233 | The early birder gets the bird in hand | *Bird banding* |

| article no. | title | topic |
|---|---|---|
| 234 | Rosy-finch survey provides bird's eye view | *Bird: Rosy-finches* |
| 235 | A refuge for birds in Alaska | *Travel: Alaska* |
| 236 | Going by the book doesn't prove bird's existence | *Bird identification* |
| 237 | In search of great black-backed gulls | *Travel: Massachusetts* |
| 238 | How to find birds in strange places | *Travel: California* |

## 2008 . . . . . . . . . . . . . . . . . . . . . . . . . . . . . . . . **page 238**

| | | |
|---|---|---|
| 239 | Birds stay warm, despite cold | *Birds in winter* |
| 240 | Doves continue territory expansion, including here | *Bird: Doves* |
| 241 | On tail of secretive goshawk | *Bird: Goshawk* |
| 242 | Authors explore our fine feathered friends | *Book reviews* |
| 243 | How to get energy and save our sage grouse | *Birds & energy* |
| 244 | 12 practical ways you can help keep birds safe | *Bird safety* |
| 245 | Birding naked / It's not nearly as fun as it sounds | *Birding naked/Big Day Bird Count* |
| 246 | Nesting season a time for activity | *Nesting* |
| 247 | Drive the Big Horn loop | *Roadside Attraction* |
| 248 | A splendid book for curious kids | *Book reviews* |
| 249 | Public can chime in on plans for Belvoir Ranch | *Belvoir Ranch* |
| 250 | A fort with stories to tell / Fort Robinson | *Roadside Attraction* |
| 251 | What birds are in your backyard this summer? | *Backyard* |
| 252 | Book review: "Flights Against the Sunset" | *Book reviews* |
| 253 | Ayres Natural Bridge offers cool respite to travelers | *Roadside Attraction* |
| 254 | "A Guide to the Birds of East Africa" | *Book reviews* |
| 255 | Agate Fossil Beds National Monument | *Roadside Attraction* |
| 256 | New Peterson's field guide has it all together | *Book reviews* |
| 257 | Book review / Wild horses won't let you put this book down | *Book reviews* |
| 257a | Hawaii: Hilo-side | *Travel: Hawaii* |
| 258 | Visiting Hawaii helps bird lover add birds to life lists | *Travel: Hawaii* |

## 2009 . . . . . . . . . . . . . . . . . . . . . . . . . . . . . . . . **page 258**

| | | |
|---|---|---|
| 259 | Wolves in Yellowstone | *Birds in Yellowstone* |
| 260 | Hawaii travel guidebook won't disappoint | *Book reviews* |
| 261 | Mountain bluebird is a harbinger of spring | *Bird: Mountain Bluebird* |
| 262 | "Prairie Spring: A Journey into the Heart of a Season" | *Book reviews* |
| 263 | Wyoming has 48 places that are important to birds | *IBAs* |
| 264 | Figure out those chirps with Birdsong CD, book | *Book reviews* |
| 265 | Why can't we encourage the country to "glow locally"? | *Birds & energy* |
| 266 | Q & A with author Rachel Dickinson / Hunts with falcons | *Book reviews* |
| 267 | I-80's Fort Steele isn't just the rest area | *Roadside Attraction* |
| 268 | Alaska's Kenai Peninsula | *Travel: Alaska* |
| 269 | Birders: If you want to see variety, visit Alaska | *Travel: Alaska* |
| 270 | Quilt Wyoming 2009 draws many show entries | *Quilting* |
| 271 | Expedition Island | *Roadside Attraction* |
| 272 | See wild horses on loop drive or in Rock Springs | *Roadside Attraction* |
| 273 | Laramie prison sheds decay / The Territorial Prison | *Roadside Attraction* |
| 274 | Point of Rocks Stage Station State Historic Site | *Roadside Attraction* |
| 275 | Flaming Gorge National Recreation Area | *Roadside Attraction* |
| 276 | Young birds stage a summer drama series | *Young birds* |
| 277 | Overland Trail along Bitter Creek | *Roadside Attraction* |
| 278 | If scary is your thing, go visit former state pen | *Roadside Attraction* |
| 279 | Gilbert's "Flyaway" captivates reader | *Book reviews* |
| 280 | All Aboard! / Get the lowdown on train travel | *Travel: train* |

| article no. | title | topic |
|---|---|---|
| 281 | Fish, float, find wildlife along the North Platte | *Roadside Attraction* |
| 282 | Local birder makes it to "Bird" Mecca | *Cornell Lab of Ornithology* |
| 283 | To feed or not to feed? | *Bird feeding* |

## 2010 ........ page 276

| | | |
|---|---|---|
| 285 | Meditation on pine beetles: Is there life after tree death? | *Pine beetle* |
| 286 | Backyard bird count needs you | *Great Backyard Bird Count* |
| 287 | Rite of Spring: Watch sandhill cranes along the Platte River | *Birds: Sandhill Crane* |
| 288 | Wyoming Quilt Project's quilts now searchable online | *Quilting* |
| 289 | Listening for birds doesn't get easier with age | *Bird i.d.: sound* |
| 290 | Nebraska spring festival is for the birds | *Bird festivals* |
| 291 | "eBirding" our backyards gives science important knowledge | *eBird* |
| 292 | When not to rescue wildlife | *Wildlife safety* |
| 293 | Balloon Fiesta dazzles amateur photographers | *Travel: Balloon Fiesta* |
| 294 | "Birds of Wyoming" is a must have treasure | *Book reviews* |
| 295 | Take a free soak in mineral hot springs | *Roadside Attraction* |
| 296 | Spotting nests isn't easy, but here's an idea of where to look | *Bird nests* |
| 297 | Summer reading list for birders | *Book reviews* |
| 298 | Lake Marie, Snowy Range Scenic Byway | *Roadside Attraction* |
| 299 | New order, new names and new species | *Bird names* |
| 300 | Bird IDs can be tricky, so a photo is always welcomed | *Bird i.d.* |
| 301 | Wyoming Roadside Attraction: Miner's Cabin Trail | *Roadside Attraction* |
| 302 | What the "Bird of the Week" has taught me | *Bird of the Week* |
| 303 | Holliday Park summer bird counts total 43 species | *Birding Cheyenne* |
| 304 | eBird: how to use a scientific database as a vacation planner | *Travel: eBird* |

## 2011 ........ page 291

| | | |
|---|---|---|
| 305 | Birds in fiction need facts, too | *Birds in novels* |
| 306 | Patchwork birding benefits birds | *Birding: local* |
| 307 | Bird feeder quarantine | *Bird feeding* |
| 308 | New field guide is so much more than its title implies | *Book reviews* |
| 309 | Are roadrunners enroute to our residential neighborhoods? | *Bird: Roadrunner* |
| 310 | Butterfly "Big Year" captures heart of one man's passion | *Book reviews* |
| 311 | The Wyoming Dinosaur Center, Thermopolis | *Roadside Attraction* |
| 312 | How to raise a birder: take a child outside | *Birding & kids* |
| 313 | Nici Self Museum | *Roadside Attraction* |
| 314 | Pelicans at Holliday Park: Why do they stop here? | *Bird: Pelican* |
| 315 | South Pass City State Historic Site | *Roadside Attraction* |
| 316 | Planning for serendipity makes for a satisfying birding trip | *Travel: Massachusetts* |
| 317 | Boysen State Park | *Roadside Attraction* |
| 318 | Killer kitchen window adds to national bird death toll | *Bird safety* |
| 319 | Try the Wind River Canyon, by road or river | *Roadside Attraction* |
| 320 | Rendezvous site attracts living history | *Roadside Attraction* |
| 321 | Improve your bird and butterfly eye | *Book reviews* |
| 322 | Crows come home to roost | *Bird: Crow* |
| 323 | Deliberate littering leaves local citizens wondering | *Bird habitat* |
| 324 | Plan to refresh Lake Minnehaha would benefit park visitors | *Bird habitat* |
| 325 | A hawk ate my songbird! | *Bird feeding* |
| 326 | Arctic Autumn: third volume of seasonal quartet | *Book reviews* |

## 2012 ........ page 305

| | | |
|---|---|---|
| 327 | Robins take up year-round residence | *Bird: Robin* |

CHEYENNE BIRD BANTER

| article no. | title | topic |
|---|---|---|
| 328 | Snowy owls' visit a sight to behold | *Bird: Snowy Owl* |
| 329 | It's quite clear – birds losing war on the windows | *Bird safety* |
| 330 | Peregrines back with a little help from friends | *Bird: Peregrine* |
| 331 | Bird Count yields results labeled as "a crazy spring" | *Big Day Bird Count* |
| 332 | New guidebooks take guesswork out of birding | *Book reviews* |
| 333 | Gardener reports from backyard | *Backyard habitat* |
| 334 | Colorado black swift wintering grounds are found in Brazil | *Bird: Swifts* |
| 335 | Mother laid tinder for my "spark bird" | *Spark bird* |
| 336 | Celebrity field guide author visits Cheyenne | *Book reviews* |
| 337 | Goose population success is messy problem for parks | *Bird: Geese* |
| 338 | Nationally known birders have nothing on birds | *Travel: Cape May* |

## 2013 .......... page 316

| | | |
|---|---|---|
| 339 | Winter is good time to spot unusual birds | *Bird irruptions* |
| 340 | Game and Fish needs our help | *Wildlife watching license* |
| 341 | Florida in February is full of fab birds | *Travel: Florida* |
| 342 | Wyoming Birding Bonanza strikes again | *eBird* |
| 343 | Early birds yield clues | *Birds and climate change* |
| 344 | Spring migration surprise delights birdwatchers | *Migration* |
| 345 | Birder learns to look more closely | *Bird behavior* |
| 346 | The generosity of other birders improves travel experience | *Travel: Alaska* |
| 347 | Encourage birding as a lifelong addiction | *Birding & kids* |
| 348 | Fall migration kicks up kites, but not the kind found on strings | *Bird: Kite* |
| 349 | Curiosity, generosity rewarded by UW's Biodiversity Institute | *Bird people* |
| 350 | Project FeederWatch needs you | *Project FeederWatch* |

## 2014 .......... page 328

| | | |
|---|---|---|
| 351 | Owls are among us | *Bird: Owl* |
| 352 | The great migration, sandhill cranes | *Bird: Sandhill Crane* |
| 353 | Let's rethink mega windfarm on behalf of birds, efficiency | *Birds and wind energy* |
| 354 | The bird migration picture gets animation | *Migration* |
| 355 | Trying out Texas birding trail rewards Wyoming birders | *Travel: Texas* |
| 356 | Owl family draws visitors to Lions Park | *Bird: Owl* |
| 357 | Wyoming refuge is a treasure hidden in plain sight | *Travel: Wyoming* |
| 358 | BioBlitz finds birds, butterflies, bees, bats and more | *BioBlitz* |
| 359 | Mind your manners to reduce bird stress | *Birding ethics* |
| 360 | 6 reasons why you should go to "Bird-day" | *Cheyenne Audubon* |
| 361 | Following flock of birders to Sterling was fun | *Travel: Colorado* |
| 362 | Can birds save the world? | *Bird conservation* |
| 363 | Big Bend hosts surprises for local birders | *Travel: Texas* |
| 364 | Feral cat policy will fail | *Cats indoors* |
| 365 | Risking nice Wyoming weather, grebes, loons get caught | *Bird: Loons, Grebes* |

## 2015 .......... page 343

| | | |
|---|---|---|
| 366 | Archiving bird columns shows changes | *Bird Banter archives* |
| 367 | "Habitat Heroes" wanted to grow native plants | *Habitat Hero* |
| 368 | Florida full of great birds and people | *Travel: Florida* |
| 369 | Are you a bird expert? | *Bird quiz* |
| 370 | Birds are always around to fascinate the young | *Birding & kids* |
| 371 | Changes in spring bird count bring up questions | *Migration, bird count* |
| 372 | High-end binoculars, mid-level prices from Wyoming's Maven | *Binoculars* |
| 373 | Hummingbird rescue reveals beauty and mystery | *Bird: Hummingbirds* |

| article no. | title | topic |
|---|---|---|
| 374 | Many mountain birds mean summer of no regrets | *Travel: Mountains* |
| 375 | Virginia is for bird lovers | *Travel: Virginia* |
| 376 | Feed winter birds for fun | *Birdfeeding* |

## 2016 .................................................... page 354

| article no. | title | topic |
|---|---|---|
| 377 | New camera technology can help birders get perfect shot | *Bird photography* |
| 378 | 2 bird books suited for winter reading | *Book reviews* |
| 379 | UW songbird brain studies shed light | *Songbird brains* |
| 380 | Ecotourists enjoy Texas border birds | *Travel: Texas* |
| 381 | Big Day Bird Count results affected by cold weather | *Big Day Bird Count* |
| 382 | Bird count day gives us big picture | *Big Day Bird Count* |
| 383 | Following individual birds brings new insights | *Birdwatching* |
| 384 | New bird singing, maybe breeding | *Bird: Kinglet* |
| 385 | Kids explore nature of the Belvoir Ranch | *BioBlitz* |
| 386 | Pondering how much eagles can take in life | *Bird: Eagles* |
| 387 | Collaboration could keep eagles safe | *Birds and wind energy* |
| 388 | Cranes are a "gateway bird" | *Bird: Sandhill Crane* |
| 389 | Turning citizens into scientists | *Citizen science* |
| 390 | Winter raptor marvels, mystery | *Bird: Raptors* |

## 2017 .................................................... page 367

| article no. | title | topic |
|---|---|---|
| 391 | Eulogy for an indoor cat | *Cats indoors* |
| 392 | Birding by app: New adventures in tech | *eBird* |
| 393 | Bird books worth reading | *Book reviews* |
| 394 | Coast comes through with great birds | *Travel: California* |
| 395 | Citizen science meets the test of making a difference | *Citizen science* |
| 396 | Thrushes take over Cheyenne Big Day Bird Count | *Big Day Bird Count* |
| 397 | Bird by ear to identify the unseen | *Birding by ear* |
| 398 | Bird-finding betters from generation to generation | *Bird finding* |
| 399 | Kitchen window like a TV peering into lives of birds | *Backyard birding* |
| 400 | Project FeederWatch tells us a lot about juncos | *Project FeederWatch* |
| 401 | Wyoming's greater sage-grouse conservation plan | *Bird: Sage-Grouse* |
| 402 | Critics of sage grouse captive breeding doubt it will succeed | *Bird: Sage-Grouse* |

## 2018 .................................................... page 378

| article no. | title | topic |
|---|---|---|
| 403 | Two Christmas Bird Counts – 80 miles apart – compared | *Christmas Bird Count* |
| 404 | Year of the Bird celebrates the Migratory Bird Treaty Act | *Migratory Bird Treaty Act* |
| 405 | Migratory Bird Treaty Act is under attack | *Migratory Bird Treaty Act* |
| 406 | Raptors are popular birds; new book celebrates them | *Bird: Raptors, book review* |
| 407 | How well do birds tolerate people? | *Bird behavior* |
| 408 | World-record-setting birder and author to visit Cheyenne | *Travel: World birding* |
| 409 | Enjoy reading nature writing in three styles | *Book reviews* |
| 410 | Keep birds safe this time of year | *Bird safety* |
| 411 | Bird counting | *Big Day Bird Count* |
| 412 | Burrowing owls materialize on SE Wyoming grasslands | *Bird: Burrowing Owls* |
| 413 | Condor visits Wyoming | *Bird: Condor* |
| 414 | How to prepare for international birdwatching | *Travel: Costa Rica* |
| 415 | Cheyenne bird book coming in late October | *Book review* |
| 416 | Benefit birds (and yourself) with feeders | *Bird feeding* |
| 417 | Try these bird and wildlife books | *Book reviews* |

xii               CHEYENNE BIRD BANTER

article no.   title . . . . . . . . . . . . . . . . . . . . . . . . . . . . . . . . . . . . . . . . . . . . . . . . . topic

## 2019 . . . . . . . . . . . . . . . . . . . . . . . . . . . . . . . . . . . . . . . . . . . page 392

| | | |
|---|---|---|
| 418 | Costa Rica's birds awe Wyoming birders . . . . . . . . . . . . . . | *Travel: Costa Rica* |
| 419 | Wind development on the Belvoir Ranch has its downsides . . . . . . . . . . . . . | *Roundhouse wind* |
| 420 | BirdCast improving birding – and bird safety. . . . . . . . . . . . . | *BirdCast, Roundhouse wind* |
| 421 | Four book reviews: Birds and bears . . . . . . . . . . . . . . . . . . | *Book reviews* |
| 422 | Cheyenne Big Day birders count 112 species . . . . . . . . . . . | *Big Day Bird Count* |
| 423 | What the Roundhouse Wind Energy Project application tells us . . . . . . . . . . . | *Roundhouse wind* |
| 424 | Participating at the Roundhouse hearing . . . . . . . . . . . . . . . | *Roundhouse wind* |
| 425 | Bird families expand in summer. . . . . . . . . . . . . . . . . . | *Bird behavior* |
| 426 | Audubon Photography Awards feature Pinedale photographer . . . . . . . . . . | *Bird photography* |
| 427 | Nestling ID benefits from crowd sourced help . . . . . . . . . . . . | *Bird identification* |
| 428 | How 3 billion breeding birds disappeared in past 48 years . . . . . . . . . . | *Bird conservation* |
| 429 | Alaskan bird behavior intrigues birdwatchers. . . . . . . . . . . | *Travel: Alaska, bird behavior* |
| 430 | Conservation ranching is for the birds – and for the cows . . . . . . . . . | *Bird conservation* |

## 2020 . . . . . . . . . . . . . . . . . . . . . . . . . . . . . . . . . . . . . . . . . . . page 405

| | | |
|---|---|---|
| 431 | Be a Citizen Scientist in your backyard . . . . . . . . . . . | *Great Backyard Bird Count* |
| 432 | Wyobirds and Wyoming Master Naturalists . . . . . . . . | *Wyobirds, Wyoming Master Naturalists* |
| 433 | High capacity water wells . . . . . . . . . . . . . . . . . | B*irds and water wells* |
| 434 | Flock of bird books arrives this spring . . . . . . . . . . . . . | *Book reviews* |
| 435 | How to become a birdwatcher . . . . . . . . . . . . . . . | *Birdwatching* |
| 436 | 2020 Big Day Bird Count best in 18 years. . . . . . . . . . . | *Big Day Bird Count* |
| 437 | Cheyenne Audubon tries a new field trip strategy . . . . . . . . . . . | *Field trips* |
| 438 | Summertime is family time for birds . . . . . . . . . . . . . . | *Bird nesting* |
| 439 | Migratory Bird Treaty Act back in full force . . . . . . . . | *Migratory Bird Treaty Act* |
| 440 | Fall migration . . . . . . . . . . . . . . . . . . . . | *Migration, fall* |
| 441 | Project FeederWatch brightens winter with backyard birds. . . . . . . . . | *Project FeederWatch* |
| 442 | First Cassin's finch visits Gorges backyard . . . . . . . . . . . . . | *Bird: Cassin's Finch* |

## 2021 . . . . . . . . . . . . . . . . . . . . . . . . . . . . . . . . . . . . . . . . . .page 417

| | | |
|---|---|---|
| 443 | Southeastern Wyoming Christmas Bird Counts compared . . . . . . . . . | *Christmas Bird Count* |
| 444 | Great Backyard Bird Count . . . . . . . . . . . . . . . . . | *Great Backyard Bird Count* |
| 445 | Wildlife Conservation license plate . . . . . . . . . . . . . . . . | *Bird conservation* |
| 446 | Salmonella, predator aversion, wind turbines, song i.d.. . . . . . . . . | *Bird science* |
| 447 | Mullen Fire changes forest habitats . . . . . . . . . . . . . . | *Bird habitat, wildfire* |
| 448 | 2021 Big Day brings in birds and birders . . . . . . . . . . . | *Big Day Bird Count* |
| 449 | Close encounters of the robin kind found in backyard . . . . . . . . . . . | *Bird nesting* |
| 450 | Neighborhood Swainson's hawks fledge three . . . . . . . . . . . | *Bird nesting, migration* |
| 451 | Dry Creek restoration to improve hydrology, habitat . . . . . . . . . . . | *Bird habitat* |
| 452 | Cheyenne birders search Pennsylvania and New York for birds . . . . . . . . . . . . |  |
| | . . . . . . . . . . . . . . . . . . . . . . . . . . . | *Travel: Pennsylvania and New York* |
| 453 | Fall reservoir birding is a leisurely affair . . . . . . . . . . . | *Birdwatching* |
| 454 | Bird feeding safety: cleaning, cat fencing, glass obstruction . . . . . . . . | *Bird feeding* |
| 455 | Barb and Mark Gorges, Champions for bird conservation . . . . . . . . . | *Interview* |

## 2022 . . . . . . . . . . . . . . . . . . . . . . . . . . . . . . . . . . . . . . . . . . page 430

| | | |
|---|---|---|
| 456 | Ghosts of Christmas Bird Counts past visit local birdwatcher . . . . . . . . . | *Christmas Bird Count* |
| 457 | How to keep prairie birds, and us, safe . . . . . . . . . . . . . | *Bird safety* |
| 458 | Raptors entice birdwatchers to follow the "The Nunn Guy" . . . . . . . . . . | *Bird trails, raptors* |
| 459 | WGFD bird farm pheasants; sage grouse farming . . . . . . . | *Bird: Ring-necked pheasant, Sage grouse* |
| 460 | How power production underlies bird problems. . . . . . . . . . . | *Bird news* |
| 461 | Cheyenne Big Day Bird Count catches Arctic visitor. . . . . . . . . . | *Big Day Bird Count* |

xiii

| article no. | title | topic |
|---|---|---|
| 462 | Fledge week observations entertain local birdwatcher | *Bird nesting* |
| 463 | Merlin's "Sound ID" uncovers hidden birds | *Birdwatching* |
| 464 | How will the IRA affect birds? | *Bird conservation* |
| 465 | Audubon volunteer reflects on 40 years | *Audubon* |
| 466 | Audubon Rockies' Hutchinson discusses community science | *Citizen science* |
| 467 | Unusual birds "on the road" this fall in southeastern Wyoming | *Migration, fall* |

## 2023 . . . . . . . . . . . . . . . . . . . . . . . . . . . . **page 442**

| | | |
|---|---|---|
| 468 | Several remarkable observations from Christmas Bird Count | *Christmas Bird Count* |
| 469 | Habitat leasing to provide new tool for Wyoming conservation | *Bird conservation* |
| 470 | Birders get look behind the scenes, find more eBird perks | *Bird science* |
| 471 | McLean biography traces passage of Migratory Bird Treaty Act | *Bird conservation* |
| 472 | Longspurs animate local shortgrass prairie | *Bird: Thick-billed Longspur* |
| 473 | Puffin paradise tour held along mid-coast Maine | *Bird: Atlantic Puffin* |
| 474 | House sparrow effect demonstrated in backyard | *Bird: House Sparrow* |
| 475 | Bird Banter includes news from backyard and beyond | *Bird conservation* |
| 476 | Crow loses life but aids researchers for years to come | *Bird: Common Crow, UW Vertebrate Museum* |
| 477 | Not all of Wyoming's birds have been here forever | *Wyoming birds* |
| 478 | Bird strikes, bird movements interest UW students | *Bird science* |
| 479 | Biologist on flyway council protecting migratory birds | *Migration, flyways* |

## 2024 . . . . . . . . . . . . . . . . . . . . . . . . . . . . **page 367**

| | | |
|---|---|---|
| 480 | Christmas Bird Count looks a bit different from 1956's | *Christmas Bird Count* |

## Acknowledgements . . . . . . . . . . . . . . . . . . . . . . **page 456**

## Index of Topics . . . . . . . . . . . . . . . . . . . . . . . **page 458**

## Index of People, Places, Books, Agencies, Organizations, Etc. . . . . . . . . **page 460**

# Introduction

When Bill Gruber, the Wyoming Tribune Eagle's Outdoors editor, proposed in 1999 that I write a column he dubbed "Bird Banter," I was hesitant to accept since there were people in town with more expertise, even academic training, in ornithology.

But I did have a minor in writing along with my degree in natural resource management from the University of Wisconsin – Stevens Point, plus three years on the weekly campus newspaper, first in production, then as a reporter and finally as the Environmental section editor. But most importantly, I was then a stay-at-home mom taking classes at the community college and I had time. The internet was just getting started, putting all sorts of resources and people within reach in addition to my birding friends.

Back then, the Outdoors section was four pages every Wednesday with a lot of information from the Wyoming Game and Fish Department and Bill's weekly column and usually his front page, full-color feature story on his participation in some hearty outdoor recreational activity. Occasionally, my column made the Outdoors front page, too.

Sometimes the columns have been about a bird species, sometimes about a place and its birds. Or about birdwatching or counting birds, bird feeding, birding technology, bird conservation, bird studies, bird books or birdwatchers.

The endless variety of topics mirrors those for the Cheyenne-High Plains Audubon Society's programs. I've been the program chair much of the same time, often writing a column about a program or field trip. In 2014, I started archiving all my Bird Banter columns at https://CheyenneBirdBanter.wordpress.com. It made it easier to search for old columns and other articles, like my favorite about our moose hunt, or the one about Hootie, the rehabilitated, red-breasted nuthatch. And it made it easier to direct callers to the one on what to do about flickers making holes in their homes' siding.

Writing for the WTE means using the WTE version of AP Style. It means no bird names are capitalized except the parts that are a proper name, like Townsend's solitaire. It does make for smoother reading unless you want to talk about how yellow yellow warblers are. Also, when I started, there were "Web sites" and the "Internet." But over time, they morphed into "website" and "internet," the style I decided to use throughout this collection of columns.

There are a lot of word pairs that go through phases of hyphenation and amalgamation, and I've tried to stick with one usage, for instance, "birdwatching" instead of "bird watching" or "bird-watching."

I've sometimes disagreed with editors about using people's first names on second reference. In news stories, John Smith is introduced with his full name and as "Smith" the rest of the article. Bird Banter is classified as commentary or opinion, so most of the editors allowed me to refer to "John" on second reference instead. It feels more like I've been talking to a fellow birdwatcher instead of someone in the news.

Most of the editors did a minimal amount of editing on my columns, but one of my stories that was put on the front page got a complete overhaul of almost every sentence. Sometimes I think that reflects the different ways individuals express the same ideas. When it is about clarity and proper grammar, I appreciate editing.

I recently came across an edition of the WTE that I'd saved from the day after Sept. 11, 2001. The masthead brags that there are 44,000 readers. By 2020, when the state Master Gardener coordinator decided writing a garden column counted toward my annual hours and contact-with-the-public count, managing editor Brian Martin told me there were 10,000 readers.

Now printed only four days, but online seven days a week and print editions delivered by the Post Office instead of carriers (no more seven-days-a-week paperboys like my older son Bryan in our neighborhood 20 years ago, with me as his substitute), the WTE is like many newspapers that have had to adapt to the digital age, where news is breaking every minute.

This collection of columns and other writing for the WTE, like "Roadside Attractions," is in chronological order. In the background, Jeffrey and Bryan grow up and get married and have kids, Mark retires, pets come and go, and family events lead us to travel and find new birds and adventures.

There are recurring themes. There are the annual iterations of the Great Backyard Bird Count, Big Day Bird Count, Thanksgiving Bird Count, Christmas Bird Count and Project FeederWatch. Many birds have their own recurring annual patterns: spring migration, breeding, nesting, fledging, fall migration.

Then there is the latest bird conservation news: declining numbers of sage grouse and other birds continent-wide, the negative impact of wind and solar energy on birds, other wildlife and people, despite the benefits of changing over from fossil fuels. And who knew that windows and loose cats could also have a terrible impact on birds?

The metamorphosis of technology concerning birds,

bird studies and birdwatching in the last 25 years has been tremendous. The Cornell Lab of Ornithology keeps coming up with new ways for eBird to make collecting data enticing for citizen/community scientists. Artificial intelligence like CLO's free Merlin app keeps getting better at identifying birds. Technology for tracking wildlife keeps shrinking and ornithologists quickly adapt it. Social media keeps birders in touch and helps recruit new members to our ranks.

The push to turn lawns into habitat for birds, bats, bees, butterflies and other insects has been gathering momentum over the last ten years, seeping into both my bird and garden columns and publications everywhere.

I don't know how influential I've been in Cheyenne. When the sky at night is still full of unnecessary and harmful outdoor lighting, or people tell me their cat simply won't stand for being kept indoors, or someone else replaces their lawn with gravel instead of low-maintenance native plants, my contribution feels insubstantial.

But then someone tells me they enjoy my columns. Even if I no longer reach 44,000 readers, it is still worthwhile. I appreciate having a role in my community out here on the High Plains of southeastern Wyoming. I appreciate being allowed to educate readers about nature through birds, and to say a word on their behalf.

# 1999

## 1 Birding the Colorado coast
Thursday, February 18, 1999, Outdoors, page C3

There is a section of the Golden Guides' "Birds of North America" I never expected to use unless I became wealthy enough to take ocean cruises.

The section on sea ducks lists species spending winters along the Pacific and Atlantic coasts and summers on the shores of Hudson's Bay. Perhaps by the time I'm old and retired, I thought, I'll have the funds to travel there.

Then I went on a field trip 20 miles south of Cheyenne a couple of winters ago and saw my first oldsquaw [name changed to long-tailed duck in 2000]. This is a sea duck that spends summers on the North American tundra and winters far out in the ocean. But it also has a habit of hanging out on large inland lakes. Our nearest large lake is the reservoir at the Rawhide power plant just off I-25, not far into Colorado. The water remains open at about 65 degrees all winter, unless the plant must go offline temporarily, said Dr. Ron Ryder, Colorado State University wildlife professor emeritus. Ryder has been studying the ecology of the reservoir for 14 years.

The oldsquaw sighting was somewhat unusual and hasn't been repeated yet this season. However, a red-necked grebe, another coastal-wintering waterfowl species was spotted. Birder Gloria Lawrence says the oldsquaw is a visitor nine out of 10 years on the North Platte River and Gray Rocks Reservoir, probably because these waters are farther north and closer to the duck's normal range.

To look for sea ducks, you may accompany Ron and the Cheyenne High Plains Audubon Society to Rawhide Reservoir on Saturday, February 20. The trip is free and open to the public. The group will meet at the Cheyenne Botanic Gardens in Lions Park by 7 a.m. Call Dave Felley at 638-9326 for details.

Ron will be able to take us behind the locked gates, but if you miss the field trip, you can still scope out the bird action from the public observation area. Take I-25 south to Exit 288 (Buckeye) and head toward the mountains for about three miles. You'll need a spotting scope or strong binoculars to appreciate the diverse bird life.

To find out about or to report unusual bird sightings in Wyoming, call the toll-free hotline maintained by the Murie Audubon Society in Casper: 888-265-2473 (the last four digits spell "bird").

When I checked recently, Gloria had listed canvasback, dipper and northern shrike. Last month the hotline had Eurasian wigeon, Lapland longspur and glaucous gull sightings. Many of the birds listed are in the Casper area, but the hotline serves the whole state.

For those of us who like birding best at our kitchen windows, don't forget the Second Annual Great Backyard Bird Count sponsored by Bird Source, a joint venture between National Audubon and the Cornell Lab of Ornithology. It is scheduled for this weekend, Feb. 19-21. Just observe the species occurring in your yard or neighborhood for half an hour or so, and then go on-line to report. You may want to explore the website in advance; last year's data is an interesting snapshot of where birds were wintering. The address: http://birdsource.cornell.edu/.

If you aren't online, you may call me at 634-0463 and leave your name, phone number and species list. I'll pass the information on.

## 2 Bird Banter
Thursday, March 11, 1999, Outdoors, page C3

Atlas was the Greek god whose job for eternity was holding the world up on his shoulders. Putting together the new Colorado Breeding Bird Atlas took 1,295 volunteers a relative eternity. Following 73,000 hours of fieldwork over nine years beginning in 1987, it took another few years to turn all that data into the 636-page book that just came out.

Colorado's atlas is a snapshot of avian breeding activity for 264 species in the state. The protocol called for studying a sampling of 1,760 blocks of land out of a total of 10,000. Each measured 10 square miles and needed to be visited at least four times for at least a half day, perhaps at night if owls or other nocturnal species are involved.

Besides volunteers, the atlas project depended on the good will of hundreds of landowners who allowed fieldworkers access.

Wyoming hasn't done a breeding bird atlas yet. Do we have the dedicated volunteers to accomplish a survey within the recommended 10 years?

One of the authors of the Colorado atlas, Ron Ryder, will speak at the next Cheyenne Audubon meeting about the rigors of undertaking such a project. The public is welcome at the meeting, scheduled for 7 p.m. Tuesday, March 16, in the Exhibition Room at the Laramie County Library.

The Colorado Breeding Bird Atlas is available from bookstores or the Colorado Wildlife Heritage Foundation, PO Box 211512, Denver, CO 80221-0394.

Birders accompanying Ron to look for seabirds at the Rawhide power plant reservoir late last month did find one species. The red-necked grebe normally winters on either the Atlantic or Pacific coasts. More common species such as common merganser, ruddy duck, and lesser scaup were also seen.

The highlight of the field trip was a stop to see a golden-crowned sparrow that's been wintering at a feeder north of Fort Collins. Closely resembling our usual white-crowned sparrow, it was a long way from its usual winter home in California or its summer range in Alaska.

I had a poor showing at our birdfeeder to report for the Second Annual Great Backyard Bird Count in February. Just a few house finches, house sparrows and juncos showed up. People from about 30 localities in Wyoming counted a total of 61 species. Most often seen was the black-capped chickadee (49 reports), followed by the black-billed magpie (35) and the house finch (35).

The species with the most individuals counted was the gray-crowned rosy finch (756), followed by the house sparrow (619) and the house finch (473).

Nationally, out of 39,000 reports, the top three most frequently reported birds were mourning dove, northern cardinal and dark-eyed junco.

The Favorite Bird Survey had 15,000 participants whose top three favorites were black-capped chickadee, northern cardinal and ruby-throated hummingbird. My favorite, the mountain bluebird, didn't make the top 10. It isn't cute or bright red. While I admire its unearthly blue color, the reason I like it is that it's my favorite harbinger of spring. This is the time of year to see it migrating to the mountains. Cheyenne Audubon plans a field trip to Pole Mountain to look for bluebirds March 27. As usual, the field trip is free and open to the public. Double-check with Dave Felley, 638-9326, about plans to leave the Cheyenne Botanic Gardens parking lot at 9 a.m.

Spring migration is underway in Wyoming. Please call in your sightings to Jane Dorn at 634-6328 (evenings). Don't be alarmed if you get the family business answering machine. Please leave a message anyway.

# 3 Bird Banter

Thursday, April 8, 1999, Outdoors, page C3

There are hazards to having no window screens. Eighteen years ago, my open windows let in the fresh spring breezes - and a stray kitten.

I formally adopted the kitten, and in gratitude, Willy brought me treasures that first summer: an earthworm, a sparrow, a mouse and a tailless dead squirrel.

As an active member of Audubon, I identified his bird kills and marveled that he was more likely to kill interesting songbirds visiting our backyard than the abundant resident house sparrows. Willy had a connoisseur's tastes.

I allowed Willy to be an indoor/outdoor cat for eight years, which was longer than the two- to five-year lifespan predicted for the average outdoor cat.

When Willy appeared to be aging, I decided to make him an indoor cat.

This also took care of the hypocrisy of being a bird-killing cat owner and president of an Audubon chapter.

In retrospect, I'm not sure why I let Willy run loose, having myself grown up with two indoor cats. The outdoor life is hazardous – I heard tires squeal and saw cars miss Willy. And loose cats are more likely to be injured in fights and pick up lethal diseases.

Scientific studies show that domestic cats take a huge toll on bird life. Putting bells on cats doesn't save birds because either the cat learns not to make the bell ring or the birds hear it and don't associate it with danger.

A well-fed cat is just as likely to kill birds. Willy was proof of that.

And so are all the uneaten bird bodies neighborhood cats leave in my yard.

I also knew I had no justification for letting Willy run loose on the grounds that a cat needs his freedom. After all, my dog was fenced in.

Livestock are fenced in. At least Willy was neutered and not contributing kittens to the millions that must be euthanized by animal shelters every year.

I wonder now why I put myself through the emotional drain of nights Willy didn't make it home. So many hazards await a cat: nasty dogs, malicious drivers, thugs with guns, cruel children and poisons and traps unintentional and otherwise.

Willy spent his second eight years at home. He became an ardent bird watcher, hiding in the house plants as birds ate sunflower seeds from the windowsill. It was the diseases of old age that claimed Willy's life at age 16.

People who know my Audubon connection have

2                    CHEYENNE BIRD BANTER

been surprised to learn that my family and I adopted two kittens last spring, for reasons only other cat lovers can appreciate.

Henry and Joey are strictly indoor cats. They've discovered the joys of bird and squirrel watching. The boys invent new cat toys and exercise equipment for them, and I've discovered deodorizing, clumping cat litter that really works.

This spring my conscience is clear. My cats will not be killing helpless nestlings of songbird species with declining population problems. How about you?

For information to share with your cat-owning neighbors and tips for keeping your indoor cat content, contact: Cats Indoors!, American Bird Conservancy, 1250 24th Street NW, Suite 400, Washington, DC 20037, phone 202-778-9666 or e-mail, abc@abcbirds.org.

For help identifying spring migrant bird species, attend the next Cheyenne Audubon Society meeting at 7 p.m. Tuesday, April 20 in the Exhibition Room at the Laramie County Public Library. The public is always invited.

# 4 May 6, 1999, Outdoors, page C4
# Out-of-towners flocking to southeast Wyoming

CHEYENNE - Every spring, for several weeks, Cheyenne experiences a flood of visitors that receives no fanfare like Frontier Days, yet the advertising is in place year round.

The influx reaches its peak in mid-May. About that time, overcrowding causes some of the out-of-towners to set up camp in my back yard. One year there was no mistaking for locals the brightly dressed visitors I had: indigo bunting, lazuli bunting, rose-breasted grosbeak, scarlet tanager, green-tailed towhee, rufous-sided towhee, American goldfinch and yellow-rumped warbler.

This big event is, of course, spring migration. Up to 150 species of birds have been counted by members of the Cheyenne High Plains Audubon Society chapter on the annual "Big Day" count.

Just as Frontier Days brings visitors from across the country, so does spring migration. One year the celebrity was a prairie warbler, another year, a prothonotary warbler. Both normally range only east of the Mississippi.

For birds headed further north, Cheyenne is an oasis in a sea of grass.

The mature cottonwoods in our parks and along our creeks are like billboards. The coniferous forests of our older neighborhoods, and the reservoirs and creeks, make this as much a haven for travelling birds as it is for dusty drivers on their way to Yellowstone.

Many Audubon chapters like Cheyenne's, have a traditional Big Day count scheduled for the peak of migration. The actual date depends on the chapter's location because the peak of migration moves north with spring-like weather.

A few years ago, concerned birders set the second Saturday in May as International Migratory Bird Day, to catalyze attention on the decline in migratory bird populations.

Development and other changes in North American breeding areas and wintering grounds in Central and South America have been detrimental to neotropical migratory birds.

For instance, the old way of propagating coffee was to grow it under the protection of shade trees, inadvertently replacing the original forest, to the benefit of many species of birds. Modern technology has been doing away with shade trees, so bird lovers are encouraging coffee drinkers to look for and buy shade-grown coffee.

To try some locally, join Cheyenne Audubon members Friday at Wild Wick's Coffee, 1439 Stillwater (off Dell Range) from 7 to 9 p.m.

Then, the next morning, on International Migratory Bird Day itself, drink some more coffee to wake up and join Audubon members for a beginner's bird walk around Lions Park, starting at the Botanic Gardens at 8 a.m.

The tourism bureau will be happy to know that the natural phenomenon of migration does bring cash to the local economy, if not on the same scale as Frontier Days.

Traditionally, the bird watching class from Casper drives in Friday night before the Big Day and meets us locals Saturday morning to check out our (or should I say the birds') favorite hot spots.

This year's Big Day is scheduled for Saturday, May 15.

Anyone interested may join in the bird watching starting at the Botanic Gardens in Lions Park either at 6 a.m. or 9 a.m.

For more information, call Dave Felley, 638-9326.

During these weeks, scrutinize any movement in the treetops, bushes or unkempt corners of the city. It may be more than the wind.

Thick spots on fences and phone lines may turn into kingbirds or kestrels.

Specks on the far shores of reservoirs could be sandpipers and waterfowl. If you stare at the prairie long enough, you'll start to see shapes like curlews and godwits.

Clouds can turn into white pelicans.

1999

# 5 Cheyenne birders count 153 individual species

Thursday, May 27, 1999, Outdoors, page C4

CHEYENNE - Authors of bird identification guides like to split North America in half. They make two volumes, each close to field-jacket pocket-size.

But if the only field guide in your pocket on May 15 was the western version of Peterson's, Audubon's or Stokes', you would never have identified the most interesting bird to show up in Cheyenne on the Big Day Count.

To find the hooded warbler, you need a field guide for the whole continent, such as National Geographic's or the Golden Guide. Range maps show these small, olive-backed, yellow-bellied, black-hooded birds summering east of the Missouri River, where they are common in swamps and moist woodlands.

And that's where we found the hooded warbler: in a temporary, tree-filled swamp at the Hereford Ranch, where it flitted about catching insects.

These warblers are categorized as rare migrants, according to Jane Dorn, records compiler for the Cheyenne High Plains Audubon Society chapter. This year's sighting was not the first record for Cheyenne; they seem to occur at 5-year intervals.

"We're on a minor flyway that comes up the Front Range," Dorn explained.

"We act a little like a migrant trap."

Birds would rather follow the mountains than cross them, and then Cheyenne appears below like an oasis.

Many species, Dorn said, migrate at night, probably to avoid predators. By sunrise they're ready to spend the day resting and feeding.

Within the Cheyenne area, Lions Park is probably the busiest during migration. Birders from Casper, Riverton and Douglas met the Cheyenne contingent there at 6 a.m. the day of the count, and we discovered 55 species in the following two to three hours.

Our best find at the park was the sora, a small, chicken-like bird from the rail family. I saw it first four days earlier with my class of sixth-graders on a bird watching field trip.

Strutting and pecking in a thin cover of willows next to the lake, only 20 feet from us, the bird obligingly showed his bright yellow bill, chicken-like feet, and short, perpendicular tail.

By Saturday, however, he'd retreated to thicker cover at the other end of the lake. That's when birders were forced to use piecemeal identification tactics: recognizing parts of the bird as they are glimpsed and assembling them like a mental jigsaw puzzle for positive identification.

A first record for Cheyenne was a black-throated

## Species Spotted in the Bird Count

| | | | |
|---|---|---|---|
| Pied-billed Grebe | Cooper's Hawk | Burrowing Owl | Common Raven |
| Eared Grebe | Broad-winged Hawk | Chimney Swift | Black-capped Chickadee |
| Western Grebe | Swainson's Hawk | Belted Kingfisher | Mountain Chickadee |
| American White Pelican | Red-tailed Hawk | Lewis's Woodpecker | Red-breasted Nuthatch |
| Double-crested Cormorant | American Kestrel | Downy Woodpecker | Rock Wren |
| Great Blue Heron | Sora | Hairy Woodpecker | House Wren |
| Snowy Egret | American Coot | Northern Flicker | Ruby-crowned Kinglet |
| Black-crowned Night-Heron | Black-bellied Plover | Olive-sided Flycatcher | Blue-gray Gnatcatcher |
| White-faced Ibis | Semipalmated Plover | Western Wood-Pewee | Eastern Bluebird |
| Snow Goose | Killdeer | Willow Flycatcher | Mountain Bluebird |
| Canada Goose | American Avocet | Least Flycatcher | Townsend's Solitaire |
| Green-winged Teal | Greater Yellowlegs | Dusky Flycatcher | Veery |
| Mallard | Willet | Cordilleran Flycatcher | Swainson's Thrush |
| Northern Pintail | Spotted Sandpiper | Say's Phoebe | Hermit Thrush |
| Blue-winged Teal | Upland Sandpiper | Cassin's Kingbird | American Robin |
| Cinnamon Teal | Marbled Godwit | Western Kingbird | Gray Catbird |
| Northern Shoveler | Western Sandpiper | Eastern Kingbird | Northern Mockingbird |
| Gadwall | Least Sandpiper | Horned Lark | Brown Thrasher |
| American Wigeon | Long-billed Dowitcher | Violet-green Swallow | Cedar Waxwing |
| Redhead | Common Snipe | Northern Rough- | Loggerhead Shrike |
| Ring-necked Duck | Wilson's Phalarope | winged Swallow | European Starling |
| Lesser Scaup | Franklin's Gull | Bank Swallow | Plumbeous Vireo |
| Bufflehead | Ring-billed Gull | Cliff Swallow | Warbling Vireo |
| Common Merganser | California Gull | Barn Swallow | Tennessee Warbler |
| Ruddy Duck | Rock Dove | Blue Jay | Orange-crowned Warbler |
| Turkey Vulture | Mourning Dove | Black-billed Magpie | Virginia's Warbler |
| Northern Harrier | Great Horned Owl | American Crow | Northern Parula |

CHEYENNE BIRD BANTER

| | | | |
|---|---|---|---|
| Yellow Warbler | Western Tanager | Savannah Sparrow | Common Grackle |
| Yellow-rumped Warbler | Rose-breasted Grosbeak | Song Sparrow | Brown-headed Cowbird |
| Blackpoll Warbler | Black-headed Grosbeak | Lincoln's Sparrow | Orchard Oriole |
| Black-and-white Warbler | Green-tailed Towhee | White-throated Sparrow | Bullock's Orile |
| American Redstart | Spotted Towhee | White-crowned Sparrow | Cassin's Finch |
| Ovenbird | Chipping Sparrow | Harris's Sparrow | House Finch |
| Northern Waterthrush | Clay-colored Sparrow | Dark-eyed Junco | Common Redpoll |
| MacGillivray's Warbler | Brewer's Sparrow | McCown's Longspur | Pine Siskin |
| Common Yellowthroat | Vesper Sparrow | Red-winged Blackbird | American Goldfinch |
| Hooded Warbler | Lark Sparrow | Western Meadowlark | House Sparrow |
| Wilson's Warbler | Black-throated Sparrow | Yellow-headed Blackbird | |
| Yellow-breasted Chat | Lark Bunting | Brewer's Blackbird | |

sparrow seen by Bob Dorn. Ironically, the Dorns were just back from a trip to the Arizona desert, where the sparrows are supposed to be. Perhaps this one got a little excited about migration and overshot his destination!

Besides unusual species, the spring bird count documents changes in populations. Crow populations in Cheyenne are exploding. The black-crowned night-heron is pioneering new nesting sites.

How many species are counted on the Big Day really depends on how well the Saturday in the middle of May coincides with the migration wave.

This year the weather was fairly cooperative. The wind from the south may have pushed birds into Cheyenne. On the other hand, as Dorn put it, "Birds don't sit to get looked at when it's windy."

This year's count, at 153 species, was one of Cheyenne's best ever.

Dorn envisions Cheyenne birders someday increasing the count by 20 species, assembling in slightly competitive teams and spreading out at dawn to every known birding hotspot. Who knows what misplaced warblers we'd find then?

# 6 Backyard building is for birds

Thursday, June 24, 1999, Outdoors, page C4

Porch lights seem to be the choice location for robin nest-building endeavors this year, according to calls I've had. One woman wanted to know just how long the parents would continue dive-bombing her. According to "Stokes Field Guide to Birds," robins incubate their eggs about two weeks and then it's another two before the young fledge. But then the robins may raise another brood or two.

The robin's classic bowl of sticks and mud comes to mind when we use the word "nest." And then we think of bird houses. Lately, they've become a decorating motif, quaint little cottages on the coffee table.

There are plenty of books about building bird house folk art. What separates art from utility is the list of house and entrance hole dimensions and mounting heights for different species of cavity-nesting birds. Two books with lists I found at the public library are "The Bird House Book" and "How to Build Collectible Birdhouses."

Bird house dimensions are important. In our backyard right now, we have a Cub Scout special filled with four baby house sparrows, a non-native species that crowds out the natives. To encourage another species, I could change the size of the entrance hole and the mounting height.

When building or buying a bird house, it helps to be realistic about future tenants. If you put up a wood duck box far from their usual habitat, you'll get squirrels moving in. Providing housing for the species already flying through your backyard makes the most sense.

If your yard is forest-like, you probably already attract birds that typically nest in hollows or cavities in mature trees. Here in Cheyenne, look for the house finch, house wren, the nuthatches (white-breasted, red-breasted and pygmy), chickadees (mountain or black-capped), swallows (tree or violet-green), woodpeckers (northern flicker, downy, hairy or red-headed) and screech or barn owls.

Don't forget that part of attracting birds to a bird house means having their preferred food nearby, whether it's seeds or insects. Attracting owls means, of course, providing their favorite food source, rodents "on-the-hoof."

Building houses for bluebirds is a popular pastime. The North American Bluebird Society has tested and developed specific plans and recommendations over the years. They can be contacted at Dept. B, P.O. Box 74, Darlington, WI 53530-0074 or at www.cobleskill.edu/nabs.

Here in the West, bluebird boxes are typically located 100 yards apart on "trails" along fence lines. In southeastern Wyoming we're likely to see mountain bluebirds at elevations a little higher than Cheyenne. Eastern bluebirds have been seen during spring migration. Maybe they're checking for some place that reminds them of the Midwest before they'll stay.

There is plenty of advice out there on building and putting up bird houses. The U.S.

Fish and Wildlife Service has an excellent free pamphlet titled "Homes for Birds" which is available at the field office at 4000 Airport Parkway here in Cheyenne or off the internet at www.fws.gov/r9mbmo/pamphlet/house.

As usual, the internet has zillions of related sites. Try these keywords: birdhouse, bird house, nest boxes. Be sure to try more than one search engine. Some, like www.yahoo.com, cover natural history subjects better than others.

Here is some basic advice I kept coming across.

--Build for a specific species.

--Build out of wood only.

--Don't paint the inside. Leave the area under the entrance hole rough to help young birds climb out.

--Leave perches off. Otherwise, predatory species will perch and reach in and destroy nestlings.

--Include ventilation holes at the top of side walls.

--Include a drainage hole in the floor.

--Make one part of the nest box removeable and clean it out between nestings.

--Protect the nest box from cats and squirrels. Mount it on a pole with a sheet metal cone part way up. Or, what's worked on our bird feeder, slip on a four-foot, 3" diameter PVC pipe, which

is too wide and too slick for squirrels to grab.

--Place the nest box away from heavy cover that may hide agile cats.

--Place away from high traffic areas, human or bird, and not too close to other nest boxes.

--Try orienting the entrance hole north or east, to avoid sun and wind.

--Water in the vicinity will increase chances of attracting swallows. Even a steady drip will attract many species.

--Put up a bird house now and possibly attract a late brood of some sort or at least the house will weather and become more inviting by next spring.

As for those pesky, dive-bombing robins, try installing a wooden shelf somewhere protected, but more convenient – to human habitation that is.

*Cutline for photo by Larry Brinlee*

Jeffrey Gorges, 10, peeks into his backyard birdhouse in Cheyenne recently. The resident fledgling sparrows are now making short flights to neighboring trees and then returning to the bird house. A free pamphlet on building birdhouses, "Homes for Birds," is available at the Fish and Wildlife Service office at 4000 Airport Parkway.

# 7 Life, death in the backyard

Thursday, July 22, 1999, Outdoors, page C4

I love a good murder mystery – in book form – not in my back yard.

I found the first body under a tree in late June, a young blue jay with feathers just beginning to emerge. Looking straight up through the branches above the body, I could see a pile of sticks in a crotch. Falling out of the nest onto hard-packed earth seemed to be a logical cause of death.

Then I remembered the feud our whole family witnessed from the kitchen window. This could be a case of squirrel retribution.

There's some kind of Saturday morning cartoon humor watching a flying blue jay poke a running squirrel in the posterior. But the possibility that both blue jay and squirrel could be responsible for raiding each other's nests reminds the viewer that this

isn't Barney and friends.

This is that old life and death struggle for food and shelter. Blue jays and squirrels compete for nearly the same kinds of foods and trees. And blue jays are known to be omnivorous, eating insects and small rodents as well as the sunflower seeds I put out.

My husband, Mark, found the second blue jay body early in July. Circumstances didn't seem to point to the usual suspects. The body was unmolested by neighborhood cats. The neck didn't seem broken (Bird safety is a good reason for saying you don't do windows – clean ones mislead birds into thinking they can fly through your house.). And there were no outward signs of disease or poisons. Also, a vigorous flock of jays remains in the neighborhood.

It's nice Ziploc makes bird-sized body bags and the morgue is as close as my freezer. Mark and I hope to get friends at the Fish and Wildlife Service to act as coroners and also, for educational use, to act as taxidermists, since they have permits for possessing dead birds.

A couple of years ago, we had a call from a nearly hysterical bird watcher who had found 30 or 40 dead house finches in her yard. Autopsies proved the killer was an outbreak of salmonella passed from finch to finch at infected feeders.

Feeders cause unnatural crowding and need to be cleaned every few weeks, especially if they're up during the summer. Be sure to rinse all the soap off and get them dry before refilling.

This spring another

epidemic hit, leaving bodies on lawns all over Cheyenne. As trees greened up in May, it was with dismay I observed my favorite birches would not be returning from their seasonal somnolence.

Randy Overstreet, assistant city forester, explained to me that the murderers are known but only apprehended with strong chemicals. Bronze birch borers are rice-grain-sized insects laying eggs on trunks and limbs of birches. After hatching, the young bore through the bark and feast on the living tissue just underneath. Younger trees, vigorously growing in the birch's native climate, may be able to grow around the damage, but older trees cannot. Trees here are under stress already, trying to survive in Cheyenne's harsh, alien climate.

I'm a weird person. Even

CHEYENNE BIRD BANTER

if a tree isn't mine and it's the one in the neighbor's yard I observe every day from my window, losing it is a mournful experience for me. What if these boreal skeletons could be thought of as sculpture and left upright? For over half the year living trees are bare, too. In the wild, dead trees provide hollows for cavity nesting birds such as woodpeckers and house wrens.

Birches don't have large limbs that would be dangerous as they decay and drop off. Even cut back to a 12- or 20-foot stump, they could become decorative posts for bird houses or feeders. I like the vine-covered cottonwood trunk at 28th and Evans. It wouldn't be surprising to find a bird nest wedged in the foliage.

The best backyards for birds are full of diversity and seem a little unkempt. I believe every dead tree does not have to be recycled into wood chips or wood smoke. They can continue to serve the avian population – and a few squirrels too.

Not all mysteries are on the bookshelf. You too can be an outdoor detective. Just keep your eyes open. But I can't guarantee you'll find out whodunit at the end of the 29th chapter!

# 8  Birding know-how a matter of degree

Thursday, August 5, 1999, Outdoors, page C2

The phone rings. "Is this the Audubon Society?" I say yes and introduce myself to the caller.

"There's this bird in my yard. It's brown with red on its face."

This is where I offer up my best guess, the house finch. Usually, I can tell by the way callers word the question whether they are, in my mental hierarchy, working on their "first degree" of bird watching or working on accomplishing a higher degree of proficiency.

Some of our bird knowledge seems to be genetic. I have yet to give a talk at a school where the children didn't correctly name the robin. But after that the names seem to be generic categories: "blackbirds," "seagulls" or "ducks."

The ordinary person does not look for birds. He only notices that some bird hits his windshield, the cat dragged in some feathers or some bird has left berry droppings on the front steps.

The first degree of bird watching begins when a person notices some black birds have iridescent heads (grackles), parking lot sparrows come in two styles (male and female house sparrows), and not all birds swimming at Lions Park are ducks (coots and grebes).

To meet the requirement for this first degree, one must find a way to cross paths with birds intentionally. This usually means throwing seed or bread crumbs on the deck or patio. At our house we put up a bird feeder.

This naturally leads to trying to figure out which birds are visiting.

Bird watching isn't just about identification of course. It's also about observing behavior: a flock of goldfinches plays king of the hill on the thistle feeder; the mourning doves have a very peculiar walk; and blue jays grip sunflower seeds with their feet and hammer them with their bills.

Bird watchers attempting the second degree are ready to look beyond their backyards. Birding with other people is the easiest. I started showing up for Audubon field trips. It's so handy to point and ask, "What's that?" And it's even more fun when other people point out a bird and tell me facts not in the field guide.

But perhaps Audubon field trips aren't scheduled as often as the budding birder would like. Here's the first step of the third degree: He decides to plan his own field trip to some of the places he's been before.

However, to really accomplish the third degree in my hierarchy, the birder must intentionally decide to explore a new place. It's finally time to invest in a bird finding book like Oliver Scott's "A Birder's Guide to Wyoming" or, fresh out this spring, the second edition of "Wyoming Birds" by Jane L. Dorn and Robert D. Dorn of Cheyenne. For those of you with the first edition, this one is worth getting. It has easier to read typeface, a water-resistant cover, new introduction with helpful subheadings and more maps and information.

The Dorns have written up 437 Wyoming species, drawing on more than 30 years of personal observation and records going back 150 years. They have charted each species' seasonal occurrence around the state using the latilong system, which divides Wyoming into 28 rectangles and have listed sites where each species has the best chance of being seen.

So, if a birder were to examine her life list for Wyoming and discover she's missing *Amphispiza belli*, the sage sparrow, the entry in "Wyoming Birds" would tell her to look in medium to tall sagebrush between May and September. The best places to look would be 5 to 35 miles west of Baggs, 5 to 10 miles south of Rock Springs, the Fontenelle Dam area in Lincoln County and the Gebo area west of Kirby in Hot Springs County.

The Dorns' book can also be used in reverse. At the back is a list of 124 birding hotspots listed by county. Each entry notes directions for getting there, expected species, best season for visiting and available amenities such as restrooms or campgrounds. Several maps help those of us who do better visualizing directions than reading them.

New to this edition is a section devoted to directions for day tours that link the most notable birding spots.

Just remember to be prepared for Wyoming weather and road conditions so that a day tour doesn't become a week of winter camping.

The Dorns' book is

1999

7

available for $19 from Mountain West Publishing, P.O. Box 1471, Cheyenne, WY 82003. Wyoming folks should add their local sales tax. Shipping is included.

The further degrees of my bird watching hierarchy pertain to how far one travels and how much time is spent birding. Even further up are the birders who volunteer to collect information for scientific studies or get involved in habitat conservation. Somewhere beyond are the people who share their knowledge, leading field trips or writing books. That's where I find the Dorns, helping us all to reach the Nth degree.

# 9 Time is here for young to leave nest

Thursday, August 26, 1999, Outdoors, page C2

August is the time for kids to leave home, avian as well as human, although for college students and kindergartners it's only on a trial basis.

Nevertheless, getting the young to move out of their nest is a perennial parental problem.

Ruth Keto watched a family drama take place in her Sun Valley backyard.

A Swainson's hawk was berating its two young about it being time to hit the road (or the thermals maybe). The young complained loudly, and one continued to cling to the power pole for nearly an hour, as if it were a life raft. As of this writing, the young hawks are still to be seen hanging out around the neighborhood.

Not only do these immature hawks have to leave their nest, they have to flee the country. Swainson's spend the winter in Argentina.

And I worried about sending the boys six blocks to school. Imagine what Swainson's parents think about sending their progeny off to another continent. Of course, birds are supposed to be operating on instinct. If they were capable of agonizing over the perils of a 7,000-mile trip, every winter we'd be finding nests stuffed with the frozen bodies of the timid.

So why don't Swainson's hawks stick around and eat rodents all winter like other hawks? It seems when they don't have to succor nestlings with mice or rabbits, they prefer large, live insects, which are not available here in the winter. A few years ago, farmers in Argentina inadvertently caused thousands of Swainson's hawks to die when they poisoned an outbreak of grasshoppers.

Parents every generation worry they've forgotten to tell the children some essential ingredient for a successful life. I don't suppose all those noisy feeding episodes at the nest are parents imparting words of wisdom such as "Don't forget to preen behind your ears; follow the flock; and don't talk to strange cats."

Despite the desire to empty the nest, it's hard for parents to give up the urge to feed the young. A couple weeks ago we were fishing in the Sierra Madres when I heard a flock's worth of faint but familiar bird song all around me. Having neglected to bring the binoculars (I was using polarized sunglasses to look for fish instead), I had to wait until two birds lighted on a dead branch.

Silhouettes showing crests on their heads reminded me the sounds they made were the same as those from the waxwings raiding the fruit of my neighbor's mountain ash in the winter. Closer inspection proved them to be cedar waxwings rather than Bohemians waxwings.

One waxwing was feeding the other. As it wasn't mating season, I assumed this was a parent feeding a nearly adult-sized juvenile. There were ripe berries everywhere, so why was the parent still feeding it? It might be tough for young hawks to chase down prey but even a youngster unsteady on its wings can bag a raspberry.

Maybe it still wasn't an efficient forager and needed help getting enough calories to put on weight for the winter. Waxwings around here don't migrate so much as spend the winter as itinerant berry-seekers.

Being married to a fisherman takes me places I might not otherwise go, such as the riparian zone where I saw the waxwings. At least, being already mid-August, the mosquitoes on Battle Creek, elevation 7,500 feet, had called it quits. However, the jungle of willow and alder was still cooking in the midday heat.

As we made our way along the banks from fishing spot to fishing spot, I smelled crushed mint and mud. And I felt scratchy. That is, every branch, thistle and nettle tried to leave its mark on my skin. Unidentifiable little brown bird shadows flitted through the dappled leaf light.

Luckily, even this jungle had openings where scattered wild sunflowers grew more than head high, and it was possible to see the sky again and see a belted kingfisher pass by.

In my back yard this past week I heard another familiar bird song. This time it was a goldfinch and one of its young.

The American goldfinch nests late compared to everyone else.

That these seed eaters wait so long makes sense when you realize it takes that long for the new weed and seed crop to mature so they'll have abundant food for their young.

Do goldfinches really drop dry, pointy seeds down the tender gullets of their babies? No. Instead, the experts say they partially digest food and regurgitate. Guess I should have watched those goldfinches more carefully to see if it's true, but I didn't think to wear binoculars while hanging out the wash.

Successful Bird Watching Rule Number 1: Birds are everywhere.

Rule Number 2: Wear binoculars everywhere.

*Cutline for photo of hawk by Larry Brinlee:*

A Swainson's hawk casts a curious eye toward an observer from his utility pole perch Tuesday on Deming Drive in Cheyenne.

# 10

Thursday, September 23, 1999, Outdoors, page C2

## Fall migration can baffle birders

It is possible to watch the fall migration from my basement window. From inside, the sill is at eye-level and from outside, at ground level.

I often gaze up from the computer screen and search my slice of sky, seeking answers to life's persistent questions. Lately, little yellowish, olive-greenish birds, warblers, have come in to inspect the ancient juniper branches and perform pest control.

There it is, the essence of fall bird migration, just three feet away: an ounce or two's worth of bird storing energy for the trip back to the tropics.

Spring migration is easy to get excited about. The big wave of migrants crests about mid-May in Cheyenne, with maybe 150 species observed in one day. Males are in bright breeding plumage and full of their species-specific songs.

In contrast, fall migration lasts for months, with some birds heading south as early as late June and others waiting as late as October or November.

Fall migration slips quietly by except for wedges of noisy waterfowl.

Here in Cheyenne, however, don't be fooled by the local Canada goose flock. It seems to have a diurnal pattern: north to Holliday Park in the morning for handouts, and south in the evening to Crow Creek for safety.

Hummingbird migration is easy to miss. One species, the rufous, doesn't even pass this way in the spring. It leaves Mexico and travels up the Pacific coast to nest in the Northwest. In late June, the males leave first, returning via the Rocky Mountains. By mid-September we see the women and children passing through Cheyenne.

It's hard for me, less than expert, to say whether those giant bees hovering over my hot pink four o'clocks are rufouses or our own broad-tailed hummingbirds coming down from the local mountains. The females and juveniles of the two species look very similar.

According to Fred Lebsack, longtime Cheyenne birder, "Fall migration is fun, if you like a challenge, but kind of tough."

With three quarters of the migrating birds wearing the confusing plumages of juveniles and females, a birder really needs one of those advanced birding field guides.

Adult males, too, are missing some of their diagnostic markings. Visit the mallards at Holliday Park and notice most of the drakes have lost their iridescent green head coloring.

The young of the year are also drab, probably as a result of predator selection pressure. After all, were a goldfinch to pop flightless from the egg in bright yellow packaging like its dad, predators could shop for dinner instead of hunting for it.

To confuse us binocular-wielding types, some bird species take more than one year to mature. Few field guides show them passing through one or more different plumages between nestling and adult.

Adult gulls are already difficult for me to ID because I've never figured out what color their legs really are and whether a spot or a whole ring marks their bills.

Juvenile gulls are impossible because they change gradually from plumage to plumage. In between, they never quite look like their baby pictures in the field guide. Maybe someone could invent time-lapse video field guides.

I met a young bird down at Holliday Park a couple weeks ago. At first glance it was just another brown mallard, except it was standing up instead of slouching like the ducks.

And then the family resemblance struck me: it was a young black-crowned night-heron. Though it wasn't wearing its parents' coloring, it certainly had its parents' attitude of motionless patience, waiting to stab a fish.

Typical of teenagers, it was defying parental wisdom and staying out all day. Night-herons usually fish at night. [Not so much night as dawn and dusk.]

Back at my basement window aviary I can confidently say one of my sightings was a Wilson's warbler. The males wear a small black cap on top of their heads. The immature females are more noticeable for their bright yellow eye-stripe.

Another warbler I saw may have been an orange-crowned. They also prefer low branches and scratching in fallen leaf debris. Fred said he's had Townsend's warblers in his back yard.

Warbler migration will be past us by the end of September, as soon as a freeze puts an end to their favorite food: fresh, not frozen insects.

But the waterfowl will just be getting started.

And did I mention we expect a few Arctic birds to spend the winter down south here in balmy and breezy Wyoming? Stay tuned. The migration saga continues.

# 11

Thursday, October 7, 1999, Outdoors, page C4

## 'Native species' is convoluted concept

A woman researching a fourth-grade statistics and probability question for the WyCAS (Wyoming Comprehensive Assessment System) called me the other day and asked how many native species of birds and fish are in Wyoming.

Math test question writers like to use local color.

At first glance, this question has a straightforward answer, at least for fish, but I soon realized why the Wyoming Game and Fish Department passed the buck to me (besides non-game biologist Andrea Cerovski being out in the field all week).

There are two variables in this simple question: What is meant by native? And what is a species? Biologists are still debating the answers.

First, the word "species." The most accepted definition is, if two organisms can produce viable (fertile) offspring, they are of the same species.

It can take a while to determine this with creatures as elusive as birds.

For instance, the 1961 edition of "Peterson's Field Guide to Western Birds" lists red-shafted, yellow-shafted and gilded flickers living in different parts of the country. However, people living where the species overlapped noticed hybridizations, so the American Ornithologists' Union (AOU) investigated and changed flicker nomenclature.

The 1983 "Golden Guide to Birds of North America" lists all three species under "Common Flicker." But the

1983 National Geographic guide lists all three under "Northern Flicker." Somebody seemed to have missed an AOU update.

However, the 1996 Stokes guide uses the AOU's 6th edition checklist and has the gilded flicker listed separately and the other two as "Northern."

In the 1999 edition of "Wyoming Birds," the Dorns have the good sense to list the information for the red-shafted and yellow-shafted separately (the gildeds are in southern Arizona and Mexico), just in case further investigation changes the AOU's ruling again.

It's important to remember that God did not invent species nomenclature. Linnaeus did, in the 1700s. Like all classification systems, his taxonomy was invented for human convenience, giving us means for naming organisms as well as showing relationships – possibly evolutionary relationships – between them.

For a fascinating perspective on species and evolution, read "Dinosaur Lives," by John R. Horner, curator of paleontology at the Museum of the Rockies, in Bozeman, Mont.

Besides changes in nomenclature, the number of bird or fish species in Wyoming can change because species become extinct, new species are discovered, species are introduced (especially game fish) or species move to Wyoming of their own accord.

Which brings us to the definition of "native." I called

Dr. Ron Ryder, a retired Colorado State University wildlife professor, to refresh my memory. He chuckled and said (the now politically incorrect) definition of native is those species present before the white man came.

So "Wyoming Native" as seen on bumper stickers wouldn't mean the same thing for birds. Being born here isn't enough.

In bird watching circles we quickly dismiss domesticated birds and escaped exotics, though if the Eurasian collared dove begins breeding in the wild, it will be classified as an introduced species, like starlings and house sparrows, which are native to Europe.

What about birds that are native to other parts of North America but move to Wyoming on their own, like the blue jay? It's been years since farmers planting Great Plains shelterbelts inadvertently provided blue jays with stepping stones to the Rockies. For most purposes, they seem to be lumped with our native species.

We can't classify as native the birds migrating through Wyoming. That would be like census takers counting tourists. But what about birds that spend the summer with us? Some spend the majority of the year in Central and South America.

The latest research points to the idea that our migrants are really tropical birds that discovered they could breed more successfully if they made use of the North's seasonally available resources.

Maybe we should ask

Mexican migrants working sugar beets or retired "snowbirds" heading for Arizona where they feel they are native to. And let's not forget the skiers who spend the winters in Jackson. They also have counterparts – birds here only in the winter.

I don't think Lisa Colvin, the researcher in Louisville, expected an essay answer for her math question about how many native bird and fish species we have.

According to my in-house fisheries biologist, Wyoming has 76 species of fish, 54 of which are native.

For birds, I used the Dorns as my authority. Of the 437 species they list as having been observed in Wyoming, one is extinct (the passenger pigeon), 61 are accidentals (off-course migrants like the Arctic tern seen in 1997 or escapees like the Egyptian goose seen in 1962); 18 are visitors (meaning they show up every few years, like your cousin Al the elk hunter); and 86 are migrants (like tourists pausing to refuel at truck stops).

That leaves 271 resident species. Of those, 154 are here only in the summer and nine are here only in the winter. That leaves 108 year-round resident species, but six were introduced: the two mentioned previously, plus other Asian and Eurasian species: the ring-necked pheasant, gray partridge, chukkar (all game species) and rock doves, commonly known as pigeons.

So, after all that math, only 102 species of birds are

CHEYENNE BIRD BANTER

hardy enough to be year-long Wyoming resident natives.

Those fourth graders are in for one heck of a

math problem.

# 12 Songbirds: Sunflower seed is sure to satisfy

Thursday, October 14, 1999, Outdoors, page C4

It's mid-October, and no one seems particularly interested in my bird feeders.

The cats are falling asleep at the window waiting for some feathered action. I wonder, where is everyone?

It turns out the goldfinches and their friends have been having a weed seed bash nearby, where construction left a huge pile of dirt last summer.

The hill sprouted wild sunflowers, mustards and other opportunistic plants now going to seed.

"You birds'll be back," I muse. "As soon as it gets really cold, you'll be back for my premium black-oil sunflower seed instead of this 'cheep' stuff."

My feeders only carry the finest seed, grown over near Carpenter by Jim Dolan.

The Rubbermaid barrel in the garage is just about empty, though, and I'm looking forward to offering this year's vintage: Mycogen Plant Sciences varieties 83-10 and 83-72.

There's a lot of sunflower seed being grown this year, due to low wheat prices and farmers switching to a three-year rotation system. But Jim's seed stands out because of the quaint customs associated with it.

First, seed is planted between mid-May and mid-June. No irrigation is necessary and no cultivation is used since the fields are pre-treated.

About mid-September

the plants mature and begin to dry out.

Ripening sunflowers attract birds. Jim says birds will hit hard if the field is near trees and water. But despite daily predation by birds, farmers have to wait until the plants reach about 10 percent moisture. Otherwise, the seeds will spoil in the storage bins.

A good hard freeze hastens the drying. Then it's time to combine. If conditions are right, just the heads can be cut, and the combine will separate out most of the trash.

Jim funnels the seed through a piece of equipment called a scalper, a whirling metal mesh cylinder that leaves less than two percent trash before shooting seed into the bins.

Now here's the quaint part: Every year since about 1992, Jim's barn has been the scene of the bagging ritual. About mid-November, 20 or 30 Cheyenne Audubon members show up with shovels, scales and sacks. The seed is sent through the scalper once more and Auduboners, standing in a trough reminiscent of grape stompings, shovel it into 25- or 50-pound bags and tie them shut with twine.

Within an hour or two, a couple tons of pre-paid seed orders are bagged and loaded into a convoy of pickups, Toyotas and minivans heading back to the Gorgeses' garage, where the less fortunate pick up their orders.

Those of us able to help bag have an inner sense of harmony with our agricultural ancestors. I'm not sure what my great-grandfather – the Wisconsin dairy farmer who built a round barn – would think of these steel-sided pole barns on the plains, however.

This year's sunflower vintage will be special because Jim plans to retire from farming. Audubon will hardly make a dent in his estimated 240,000-pound harvest, most of which will be commercially bagged for birds or crushed for oil.

In the tradition of Paul Newman giving his salad dressing profits to charities, Audubon profits go to a good cause. They help fund the Audubon Adventures program offered to local fourth through sixth grade classrooms, as well as buying more seed to give to nursing homes and schools with bird feeders.

How much seed should a bird-feeding person buy? Audubon offers its bulk-rate bargain only once a year. Underestimating means buying seed grown who-knows-where, Kansas or Nebraska or someplace. Buying too much means feeding birds into the summer (no problem) or sowing it in the alley.

Storing seed for a year or so is all right if it's kept clean and dry. At our house, a 33-gallon garbage can does the job.

How fast the birds eat seed

depends on how well known the feeding location is, how many feeders are in the neighborhood, whether squirrels and aggressive birds like grackles and blue jays raid feeders, and how many times a day the feeders are filled.

It also matters how prolific the natural seed sources are, how deep the snow gets, how many cats "put birds off their feed" and whether the feeders have been exposed to the contagious disease killing finches.

To avoid this last variable, clean feeders every few weeks with soap and water.

According to a publication put out by "Birder's World" magazine, black-oil sunflower seed attracts the most kinds of birds. It's the best buy, unlike many packaged mixes that have a lot of undesirable seed types the birds ignore and let turn to mush under the feeder. Black-oil is also more nutritious than striped sunflower seeds.

When bird watchers reach the addictive stage of this hobby, they experiment with Niger, corn, proso millet, peanuts, suet, oranges and other fruit. After black-oil sunflower seed, however, the next best thing to offer birds is water.

Yep, our avian pals are just out looking for a good time at a good watering hole.

That weed-seed eating bunch I observed was holding their bash on the banks of a tributary of Crow Creek.

1999

11

# 13

Thursday, October 28, 1999, Outdoors, page C4

## Blustery days challenge birders, send birds packing

Call me a fair-weather birder, but our mid-October snowstorm was not good for binoculars, spotting scopes or field guides.

Of course, the biggest reason for cancelling Audubon's field trip to Glendo Reservoir was poor road conditions.

There were birds to watch at home. Bad weather brought a mob of finches.

The cats sat at the window transfixed as the birds played musical perches on the tube of Niger thistle. Though they show just a slight wash of yellow this time of year, goldfinches still have their distinctive black wings with white wing-bars.

Two days before the storm, Fred Lebsack of the Cheyenne High Plains Audubon Society, Steven Roseberry of the Laramie County Conservation District and I met with Eleanor Grinnell's Community Based Occupational Education (CBOE) science class at the Airport Golf Course.

It was one of those fiercely bright, ferociously windy days. The high school students were to do their monthly water sampling and bird observations at selected points as part of the Audubon Cooperative Sanctuary System program. They document improvements as the golf course adopts environmentally friendly grounds maintenance practices.

Although it too is named for John James Audubon, this program is not related to the National Audubon Society or its chapters. That doesn't stop Fred and me from taking any excuse for bird watching, however, especially for an educational and scientific cause.

But we were disappointed this day. The wind had every bird sitting tight. A few juncos tittered in the brush. Mallards collected along the shore. A pied-billed grebe (not a duck) and a redhead (a duck) didn't let the wind bother them. Both were underwater most of the time, diving frequently after small aquatic animals to eat.

But mostly at each of the eight bird-watching stations, we waited the procedural five minutes, straining in vain for sights or sounds avian, wanting to say "a-ha!" at each falling leaf.

Anyone who's withstood a few falls in Cheyenne recognizes the blustery, sunny weather that precedes a weather front. Lingering migratory birds recognize their flight south is ready for boarding.

Though I will miss them, how can I blame the birds for wanting to double their mileage with a stiff tailwind? Just how many miles does a one-ounce warbler get per insect anyway?

Having struck out birdwise twice within three days, I decided the day after the storm I would visit a reservoir close to home. Early in the morning Jeffrey and I pulled out our winter clothes and the spotting scope and drove out.

It was as cold as any Christmas Bird Count. The brisk breeze and sunshine conspired to make my eyes water. At age ten, Jeffrey is still a good sport, but he soon climbed back in the solar-heated van. He didn't need higher powered optics to appreciate the lone white pelican out on the water.

Scoping the far shore I was able to identify coots and ring-billed gulls. Even with a field guide in hand, I had to let the only two sandpipers remain nameless, which may save their reputation. They can't be too bright sticking around here this late.

The ducks had to remain nameless too. It takes a better birder than I to identify lumps of brown feathers huddled on the far shore or to identify duck butts, which is all there is to see when puddle-type ducks tip to feed.

Had I disclosed my identification difficulties to May Hanesworth, the doyenne of Cheyenne birding, she would have given me unequivocal IDs. Although she had already given up field trips when I first met her in 1989, May had decades of birding experience to draw from.

May was famous in many Cheyenne circles, [including music, music education, Cheyenne Frontier Days and local bridge clubs] but local Audubon members will remember her as charter member and especially as bird report compiler, a job she did with elegance, accuracy and reliability for decades, until she asked to pass it on to someone else three or four years ago.

May died October 5, just five months short of her one-hundredth birthday.

# 14

Thursday, November 11, 1999, Outdoors, page C4

## Juncos add variety to backyard bird feeder visitors

This time of year brings early darkness, candles at the dinner table, Boy Scout popcorn orders in the living room and Audubon sunflower seed orders in our garage (Don't forget to pick up your orders Saturday.).

A more subtle portent of winter is the arrival of the juncos, starting this year with the one I saw at our feeder two weeks ago.

Actually, juncos never appear at feeders as much as they appear under feeders, picking through spilled seed, some of which I spill on purpose.

Now there are half a dozen juncos searching the ground anytime there are any other birds in the yard.

Perhaps it seems like all you get at your feeder are house sparrows and house finches. But if you look closely, you might recognize the varieties of juncos.

In Wisconsin, where I identified my first junco, and the Midwest and most points east, they are what used to be known as the "slate-colored junco." Later, the American Ornithologist's Union changed the name to "dark-eyed junco."

I thought that a picky distinction until I met the western juncos. At one time they were considered three separate species and one variation, but now they are merely races of the dark-eyed junco, versus the yellow-eyed junco of southern Arizona and Mexico.

Any given winter day, my backyard may host the Oregon, pink-sided or gray-headed forms of junco. I might even be lucky enough to see a white-winged or slate-colored.

Differentiating is difficult, with the best delineation given in the National Geographic field guide. Here's the basic breakdown.

All dark-eyed juncos are sparrow-sized, but plain-colored, unlike the streaky looking sparrows. They all have white bellies, "belly" being a technical term to describe part of a bird's topography.

They twitter and flash their white outer tail feathers as they fly away from you.

The slate-colored form is pure gray, though the female is brownish-gray. They are uncommon in the West.

The Oregon male has a very dark gray, almost black hood, brown back and orangish sides. The female is a lighter version. They are common in the West.

The pink-sided variation of the Oregon has a blue-gray hood, pinkish sides and breeds in the central Rockies.

The gray-headed, of the southern Rockies, has a pale gray hood and body, but a bright rufous-colored back.

The white-winged breeds in the Black Hills. It is all blue-gray, with two white wing bars on each wing.

To add to the fun, realize that these races are considered one species because they can crossbreed and produce fertile, if confusingly colored, offspring.

If you look at the field guide range maps for juncos, it shows juncos year round in Wyoming. But they aren't year round in my Cheyenne backyard.

At one inch square, the range map showing the whole North American continent can't show individual Wyoming mountain ranges. If it could, the mountains would be colored to indicate breeding and year round residency, with the plains colored as additional wintering grounds.

Migration is much more complex than the simple maxim we learn as children: south for the winter and north with the spring. After meeting New Jersey ornithologist Paul Kerlinger [director of the Cape May Bird Observatory] in September and reading his book "How Birds Migrate," it seems every species has its own strategy for dealing with the cold season.

For some it's complete migration – all individuals head south, although perhaps the males leave first or don't go as far south as the females.

Many species, like our western juncos, use the partial migration pattern. Not everyone leaves the breeding grounds. Or perhaps some of the individuals breeding northernmost spend the winter in the breeding grounds of the centrally located individuals of the same species. Or the northernmost leapfrog over everyone and winter the farthest south.

After all this hair splitting (or is it feather splitting?), it's a relief to consider the other new seasonal visitor to my backyard: the red-breasted nuthatch.

It looks the same anywhere it is seen on this continent. Its bold black and white striped head reminds me of a miniature badger.

The bird I've been observing swoops up to the feeder in a bossy, efficient way, making everyone else look like they're in slow motion.

No slave to ancestral migration patterns, the red-breasted nuthatch is what is called an "irruptive migrant." It goes where the food is; wherever its favorite food source – coniferous trees – produced the best crop of seed.

Evidently, however, black-oil sunflower will do for right now.

# 15 Take an hour to help count the birds

Thursday, November 25, 1999, Outdoors, page C2

**The bird in your oven isn't the only one worth taking notice of today.**

Today is the only federal holiday with a bird as a mascot.

Cartoon birds sporting pilgrim hats and frozen turkeys from the grocery store hardly resemble the wild bird Ben Franklin thought beautiful and wise enough to propose as our national symbol, however.

If you are reading this Thanksgiving morning, either you aren't the principle cook today or you are taking a breather between candying yams and mashing potatoes.

If so, you have time to take part in the annual Thanksgiving Bird Count.

Of all the bird counting events, this is by far the least strenuous.

Spend a mere hour looking out the window and record the numbers and kinds of birds you see at your feeder. If you look now and then, glance out occasionally while reading all the ads for after-Thanksgiving sales, you'll easily have put in the required time.

When I first heard about the count, I envisioned sated eaters leaning back in their chairs and gazing out the window at the birds while waiting for pumpkin pie.

But in my experience, bird activity this time of year, except for a spate around four o'clock in the afternoon, is highest around midmorning.

Evidently, on a cold day, a bird has to carefully weigh the energy costs of leaving its nice warm bush against the energy benefits of eating breakfast. Many wait for the sun to warm things up, just as my sons wait for someone to turn up the thermostat.

John Hewston's experience also bears out the productivity of bird watching at

1999

mid-morning. A retired professor of natural resources at Humboldt State University in Arcata, California, Hewston has been coordinating the count for the 13 westernmost states since 1992.

The count goes back to 1966. Ernest Edwards and the Lynchburg, Va., Bird Club started it. A friend told Hewston about it, and in 1992 he offered to compile records for the western states.

Last year Hewston compiled reports from 421 people, some of whom submitted more than one count. Perhaps they did their own feeders and then Aunt Lucy's when they showed up at her place for dinner.

Twenty-two counts were submitted from Wyoming, from seven different areas around the state.

In 1998, 157 species of birds were reported, not as good a showing as the 171 species in 1997.

Considering I expect to see only half a dozen myself, those huge numbers reflect the diversity of western habitats. At least one person from tropical Hawaii has been reporting, and the number of reports from arctic Alaska continues to grow.

In between are a variety of types of mountain, desert, plains and coastal habitats.

Hewston, living near the northern California coast, expects to see Stellar's jays, spotted towhees, white-crowned and golden-crowned sparrows as well as juncos, house sparrows and house finches. And Anna's hummingbird, something we don't see even in the summer.

The most common species to be reported on the Thanksgiving count is the dark-eyed junco, followed by the house finch, pine siskin and house sparrow, all birds I expect at my Cheyenne feeder.

As you can see by the accompanying form [not included here], participation is simple. Just imagine a 15-foot-diameter circle on the ground around your bird feeder.

Extend it upward as a cylinder. Keep track of the birds you observe inside this cylinder. You may also keep track of the birds you see outside the cylinder, perhaps species like hawks or gulls that don't use feeders.

If you have an identification question this morning, give me a call.

My husband, Mark, is the one who will be in the kitchen playing with the turkey. Or call me later this weekend.

Please send me your results by next Friday, Dec.

3. Either clip the form, e-mail the essential information to me at the address below, or as a last resort, read your report over the phone to me or to the answering machine.

If you see no birds at all during your designated hour, that is also important information to report.

I will compile our local results before sending the forms on to Hewston. Sometime in late winter you'll receive his newsletter with the western results and interpretation.

So enjoy some armchair bird watching as you contribute to a snapshot look at where birds are this time of year.

And yes, the wild turkey has shown up for the Thanksgiving Bird Count.

Seventy-one were counted last year.

# 16

Thursday, November 25, 1999, Outdoors, page C2

## Workshops help educate educators on the environment

CASPER MOUNTAIN - "Deer, *migrate!*"

On this beautiful September day Anna Wertz, Wyoming Game and Fish Department education specialist and Project WILD facilitator, was playing God. At her command, a herd of human deer rushed from one end of the field, trying to beat each other to limited food, water and shelter at the other end, yet avoiding hungry mountain lions.

Deer that couldn't match up with someone playing the part of necessary habitat would die and become food, water or shelter. This is a lot like real life. Deer that starve to death become

forest fertilizer.

In this rendition of "Oh Deer!" the players were teachers and other educators of children from around Wyoming. We spent the weekend at Girl Scout Camp Sacajawea, taking part in environmental games and activities we'll use with children back home.

Not everyone who took part teaches in the classroom. Caryn Agee, a participant from Worland, works for the local conservation district and frequently teaches visiting school groups at the district's natural area. Some teachers have two identities, like Nancy McFarland, who teaches and ranches up

by Aladdin.

Some of us will have to modify activities for students, like Donna DiPietro of Casper who works with special education students.

Although half of the participants were Casper College education students, they also represented diverse backgrounds and parts of Wyoming.

This chance to play the child and imagine ourselves as deer or even drops of water was part of a joint workshop put on by Project Learning Tree, Project WILD and Project WET.

Some of us came for the teacher recertification credit, but all of us wanted to get

our hands on the amazing activity guides produced by each of the projects. They are only distributed through training workshops.

The idea of holding a joint workshop came to the project coordinators and facilitators while developing Wyoming's natural resource education master plan ("natural resource education" is the common term in Wyoming for environmental education).

A joint workshop is certainly convenient for teachers who then can mix and match activities from all three projects for lesson planning that supports state education standards – and not just in the subject of science.

14

CHEYENNE BIRD BANTER

At the September workshop, the theme is math and language arts.

On the math side, the follow-up to the "Oh Deer!" activity involved graphing the fluctuating deer population.

In "Reaching Your Limits," Project WET facilitator and Casper College adjunct instructor Tammy Brown helped us visualize the concept of parts-per-million as it pertains to safe drinking water.

On the language arts side, Darla West, a rancher from Oshoto and Project Learning Tree coordinator at the time of the workshop, led the "Poet-Tree" activity, delving into formulaic verse such as haiku, cinquain, diamante and picture poetry.

We would-be nature poets were sent outside to be inspired, making this activity part of the first conceptual level: awareness and appreciation of nature.

Other concepts include understanding ecological principles, management and conservation, influence of people, issues and consequences, and finally, taking responsible actions.

At the issues and consequences level, Evert Brown, Casper College biology professor, led the group through Project WET's "Pass the Jug" activity. At first it seemed like a party as we poured water into each other's plastic cups. However, the plot line was really the story of water rights in the West.

One of us was new industry pleading with descendants of the original homesteader for enough water to bring jobs and prosperity to the community. Our personal biases surfaced as we considered who deserves water the most.

With activity guides rated kindergarten through grade 12 (Project Learning Tree has issue-oriented modules for the secondary level), not all activities are suited to every grade, although many can be modified by several levels.

But how does a teacher quickly peruse 1,300 pages for a fourth grade lesson on habitat? Project WET has published an additional guide called "WOW! The Wonders of Wetlands" and Project WILD has added "Project WILD Aquatic."

Each guide is well cross-referenced for grade-level, activity length, topics, concepts, skills, and academic subjects.

As a way to pull together the barrage of experiences at the workshop, at the end of the weekend each participant made a solo hike – a hike through the guides – to find answers to questions such as: "Where can you find tips for bringing nature indoors?" "Where can you find activities that include art?" and my favorite, "How do you make a Blueberry Grunt?" (Start with canned berries and baking mix).

As a final test, each of us researched and presented a synopsis of a series of activities, one from each guide, that will fit a particular theme and grade level.

Thanks to a grant from the Environmental Education Training Assistance Program (and the donation of Project WILD materials by Game and Fish), $35 of the $75 workshop was refunded. Now that's a deal: free food and lodging at a mountain resort, plus enough books and materials to fill eight more inches of bookshelf.

[When it is time to leave at noon Sunday, we all have to brush snow off our windshields. It appears the pseudo-African rainsticks we thought we were making in our free time, pounding nails into cardboard mailing tubes and sealing in a handful of beans and seeds, were really Wyoming snow sticks.]

---

**Project Learning Tree began** in 1973 as a partnership of the American Forest Institute and the Council for Environmental Education (formerly the Western Regional Environmental Education Council).

By using tree and forest concepts, PLT teaches children how to think, not what to think, about local and global environmental issues.

Wyoming PLT's primary sponsor is the Wyoming Timber Industry Association. A 15-member steering committee includes representatives of state and federal natural resource agencies and farming and mining interests.

**Project WET (Water Education for Teachers)** was established in 1984 in North Dakota. In 1990 it went national, cosponsored by the Council for Environmental Education and The Watercourse, a non-profit in Bozeman, Montana, dedicated to unbiased education of children and adults about water and its issues.

In Wyoming, WET is sponsored by the Teton Science School. A network of facilitators around the state can offer training as requested.

**Project WILD** was founded by the Western Association of Fish and Wildlife Agencies and the Council for Environmental Education in 1983. Project WILD uses wildlife concepts to foster awareness of nature and promote discussion of wildlife issues. As with the other projects, the idea is to teach children critical thinking skills, not particular ideology.

# 17 Project FeederWatch relies on citizen scientists

Thursday, December 9, 1999, Outdoors, page C4

This winter I am one of over 13,000 "citizen scientists" across North America.

What that means is, although I hold only a bachelor's of science degree and have never worn a white lab coat, I, too, can contribute to scientific research. So can you – if you learn to identify and count birds at your bird feeder.

Bird watchers have a propensity to quantify their hobby. Some people keep life lists and some people keep backyard lists of the birds they see.

This winter I'll be keeping track of the birds visiting my feeders.

I heard about Project FeederWatch (PFW), which depends on citizen scientists, a few years ago when the National Audubon Society and the famous (in bird circles anyway) Cornell Lab of Ornithology started a new venture called BirdSource.

BirdSource sponsors several kinds of bird studies including PFW. What I didn't know was that PFW started in Ontario in the 1970s and continues as an international effort.

Those long Canadian winters must force people to find exotic entertainment like this.

From November through March, PFW participants document the ebb and flow of bird species that use feeders. I missed the first reporting period, but that's OK. It's even all right to miss some others if I have a schedule conflict.

If I were to submit my data in the traditional data entry, computer-readable booklet at the end of the season, I would be choosing two consecutive days to count in each two-week period. But I decided to go the on-line route, which allows me to report every week by computer.

Picking two consecutive days at least five days apart from the last two leaves, in my case, weekend count days.

To sign up, I could have mailed in my $15 registration fee or sent it via the internet. But I chose to call in with my credit card number.

In return I got a poster, handbook, data entry book and an identification number. The fee may defray costs

as much as it makes people more apt to carry through. Who wants to waste money already invested?

Besides entering information about the birds I see in my backyard, I also describe my feeding setup. At the end of the season I'll describe what and how much I fed the birds. All my data gets compiled with everyone else's.

On the web I can look up animated maps (really!) that show sightings of each feeder species from month to month or year to year starting in 1992.

I begin to realize that in some years the fickle pine siskins weren't personally boycotting me but were wintering in another part of the country.

Ornithologists can't possibly collect as much data by themselves as we citizen scientists can. But they have been able to use our data in published studies about the movement of feeder birds in winter, overall population changes and food and habitat preferences.

An offshoot study was done on the spread of feeder bird diseases. The handbook

and website give descriptions of various diseases, mainly those that afflict house finches. Don't read these right after eating.

Of course, every ardent birder's favorite aspect of their hobby is the chance to report rare birds. It's too bad the tundra swan Jim and Carol Hecker saw at Lions Park two days before Thanksgiving doesn't count. I wonder, if someone put out cracked corn at the lake regularly, could it be claimed as a feeder site? Looks like waterfowl have to be write-in species.

Much of the interesting information about PFW contained in the handbook also is available at the website. The site, though, has bird pictures and descriptions and the maps. If you don't own a computer, use a friend's or one at the library. It isn't too late to sign up to be a Feeder-Watcher, and I hope you will.

If a once-a-year commitment is more your style, read about the 100th annual Christmas Bird Count (Jan. 1 in Cheyenne) in the next Bird Banter on Dec. 23.

## Cheyenne Thanksgiving Bird Count Results

Barb Gorges and the Cheyenne High Plains Audubon Society chapter extend their thanks to all eight parties who participated in the Thanksgiving Bird Count this year.

"We probably doubled the number of reports we had last year," Gorges said.

Here are the local reports. John Hewston will send participants the entire Western compilation around the end of February.

Birds are listed in order of popularity around the designated count circles (number inside/number outside).

| | |
|---|---|
| House Sparrow 141 / 25 | Red-breasted Nuthatch 3 / 1 |
| House Finch 55 / 27 | Black-capped Chickadee 3/0 |
| Dark-eyed Junco 49 / 8 | Rock Dove (pigeon) 3/0 |
| European Starling 15 / 20 | Black-billed Magpie 2 / 8 |
| Gray-crowned Rosy Finch 15/0 | Pine Siskin 2/0 |
| American Goldfinch 11 / 3 | Raven 0/7 |
| Song Sparrow 10 / 0 | American Crow 0/6 |
| Blue Jay 9 / 1 | American Robin 0/5 |
| White-crowned Sparrow 6 / 0 | |

CHEYENNE BIRD BANTER

# 18 Stand up and be counted – by counting

### The 100th Christmas Bird Count

The National Audubon Society wants YOU! -- to count birds for the 100th annual Christmas Bird Count.

The CBC is the annual celebration of a small group's ability to aid bird conservation by changing long-held tradition. Back before 1900 there was a traditional competition on Christmas to go afield to see who could shoot the most birds. Most of the pile of bodies collected weren't eaten or used for scientific purposes.

Around 1900, optics weren't what they are today, and bird watchers frequently shot birds so they could identify them in the hand.

Fortunately, bird watching wasn't as popular back then as it is today. Even so, ornithologist and Audubon Society officer Frank M. Chapman and 26 of his friends thought the Christmas hunt caused unnecessary depletion.

Twenty-five counts were held the first year, mostly in the Northeast – except for five in the Midwest, one in California and one in Pueblo, Colorado. A total of 18,000 birds of 90 species were counted.

The idea of counting birds without guns continued to grow. In 1998, about 50,000 people in 1,800 count areas documented around 58 million birds.

In North America, 659 species were represented. The non-North American counts, in Central and South America and Pacific islands, reached 1,650 species.

CBC data has always been collected for scientific analysis of bird population trends. The annual report of any one of the recent counts is a book larger than the Cheyenne telephone directory.

But if you have web access, www.birdsource.org/cbc, gives you a century's worth of data.

The CBC isn't just for the leisure class anymore, and it isn't always held on Christmas Day. In Cheyenne, we pick a Saturday within the given three-week window. Even if you have to work, just stick your head outside a minute and count how many starlings are congregating on the power line.

If you aren't ready to tramp around in the cold, be a feeder watcher and count the birds in your own backyard.

As a scientific study, the CBC has a couple constraints. Every count area is limited to a 7.5-mile radius. Ours is centered on the State Capitol Building. If you go out on your own, keep track of your mileage by foot or vehicle and note approximately where and what time you see the birds you count.

One major misconception about counting is you have to be an expert ornithologist. Not true. It helps to go with someone knowledgeable, however. Just be sure to wear warm clothes.

On my first CBC, the 83rd count, an Arctic air mass had settled in over the Yellowstone and Tongue River valleys in southeastern Montana. Because I was a novice, Mark, my husband of three months, graciously allowed me to be the recorder. Luckily, birds hunker down in sub-zero temperatures, and there wasn't much besides a tiny flock of horned larks to unsheathe my gloved and mittened fingers for writing the numbers.

The Cheyenne tradition is to get a warm start by meeting at 7:30 a.m. in the lobby of the Post Office on Capitol Ave. and then tour birding "hot spots." It's perfectly all right to drop out when you're ready to go back to your warm and cozy home and check out the birds at your feeder.

With this year's count scheduled for Jan. 1, you may not be ready to join us so early. Remember, there are no rules about what time you have to start or how long you have to be out. If you'd like, pick up the official count form and map from me ahead of time.

You may call your results in to our CBC count compiler, Jane Dorn, at 634-6328.

But it's much more fun to heat up your favorite hot dish or buy your favorite hot salsa and chips and come to the tally party.

After the potluck, we review the winter bird checklist species by species, about 50 or so, and share stories about our day. One year we were able to figure out that two parties of observers counted the same flock of geese because the geese moved between Holliday Park to Lions Park.

Years later, I can look up the first counts I did in Miles City. There in the CBC editions of American Birds are recorded Mark's and my names, names of our hardy bird watching friends, the day's weather and the birds we saw, including all of those ravenous robins we saw in the Russian olives along the rivers.

Note to Prospective Project FeederWatch participants: No, you don't need to count birds at your feeder for all the daylight hours of those two consecutive reporting days every count period. As little as less than an hour total works fine. If you want to take part, send you registration and $15 via internet at www.birdsource.org/pfw, by phone at 607-254-2427, or by mail at Cornell Lab of Ornithology, 159 Sapsucker Woods Road, Ithaca, NY 14850.

# 2000

**19** Thursday, January 6, 2000, Outdoors, page C2
## Refuge offers whooping-good time

Holiday visits with family can easily become a never-ending cycle of cooking, eating and cleaning up. That's why, several weeks before heading to my mother's in Albuquerque for Christmas, I planted the idea of a side trip to Bosque (BOSS-key) del Apache National Wildlife Refuge.

My intentions were to get us out of the house, find grist for this column and avoid the after-Christmas sales.

I've been to "The Woods of the Apache" several times, driving south along the Rio Grande a little past Socorro, New Mexico. The refuge is best known for its wintering flocks of snow geese, sandhill cranes and endangered whooping cranes.

It was originally set up in 1939 for the then-endangered sandhills.

The endangered whoopers have been raised in captivity and trained to migrate to the refuge with the sandhills for the winter.

Seeing whoopers is great, but there are 377 bird species on the refuge checklist and some, like the roadrunner, are equally exotic to us Northerners.

We decided to arrive at the refuge a few hours before sunset, when the geese and cranes start returning for the night from feeding in nearby fields.

The refuge itself includes 57,000 acres. Nine miles of valley include a series of farmed fields, marshes, ponds and woody margins. The Chihuahuan desert uplands on either side are official wilderness.

Examining the ponds, we saw waterfowl common to the Bosque: pintails, northern shovelers, buffleheads, coots and even a few mallards.

As we drove up to the visitor center, my sister, Beth, wondered if her friend still worked for the refuge. In fact, Daniel Perry was working that day and kindly marked out his favorite trails on our copy of the refuge map, as well as the location of the morning's sighting of the two wintering whooping cranes.

At the back of the visitor center a big viewing window had a microphone that brought in the sounds of strutting Gambel's quail.

The busy white-crowned sparrows looked the same as the ones we get in Cheyenne.

In his backyard a few miles away, Daniel said he gets pyrrhuloxia, the southwestern version of cardinals, and black-throated sparrows.

We poked along the 15-mile auto tour loop, playing leapfrog. People passing us as we pulled over to look at birds would themselves be pulled over by the time we continued on. One car, with Albany County, Wyoming, plates belonged to a couple from Laramie who'd recently relocated to Albuquerque.

Just about the time the Chupadera Mountains turned purple in the waning light, we came to the observation deck Daniel recommended.

Thoughtfully equipped by the U.S. Fish and Wildlife Service with both a powerful scope and a Port-a-Potty, it seemed perfect. But no birds were there.

Beth, Jeffrey and I hiked down the road to investigate a small flock of snow geese, including some "blue geese," a dark color phase. A few sandhills accompanied them.

By the time we returned to the deck, Mark and Bryan had two white birds in the scope. It must have been a strong scope, because I couldn't see anything white out there with my naked eye.

Were these snow geese or whoopers? Both are pure white with black-tipped wings that don't show much unless they fly.

Of course, with a way to compare size, identification would be obvious. Snow geese are about 2 feet high, and both sandhill and whoppers stand about 5 feet tall.

When we could make out sandhills standing next to the white birds, we knew we'd found the whooping cranes.

As I looked through the scope, they flapped their huge and wonderfully flexible wings. Just like in the movies. We all got a good look before they moved deeper into the brush.

There were no other people with whom to share the moment with. A steady line of cars lumbered past in the dusk behind us, like elephants, headlights to taillights. It's doubtful anyone not on the deck would have had the angle needed to see the whooping cranes.

We were not entirely alone, however. Occasional sandhills, making their "craa-k" calls, flapped just a few yards over our heads. For one evening, we were privileged to be in just the right place at just the right time.

Conservation note: Whooping crane reintroduction has

18      CHEYENNE BIRD BANTER

not been very successful because the whoopers imprint too well on sandhills and haven't been procreating in the wild. The U.S. Fish and Wildlife Service has decided to put all its crane eggs into the eastern flock instead.

Refuge visitors, as well as locals who enjoy the economic prosperity brought by crane watchers, are petitioning the service to change its mind – and re-evaluate its propagation methods.

Planning a Trip to Bosque del Apache

[The information about visiting the refuge can now be found online.]

# 20 Bird count uncovers hidden treasures

Thursday, January 20, 2000, Outdoors, page C2

Cheyenne's 1999 Christmas Bird Count, actually held Jan. 1, was like an expedition to a dusty attic to unearth avian treasures.

Except the attic was missing the usual white protective sheets of snow covering the landscape furniture.

A total of 42 species of birds was seen, compared to the expected 50-plus on colder, snowier counts when birds have to spend more time out in the open, foraging for food to stay alive.

At 7:30 a.m., Cheyenne High Plains Audubon Society members divided into four groups to cover, before sunset, as much as possible of the 7.5-mile-radius count circle centered on the state Capitol building.

Pete Gardner of Wheatland became our group's "Hawkeye." Without him, we may have missed the red-tailed hawk sitting in a tree where Westland Road crosses over Crow Creek. Or we might have overlooked some of the kestrels perched on telephone wires, looking like slightly oversized berry-eating robins rather than the meat-eating raptors they really are.

According to Jane Dorn, count compiler, we had a lot of northern harriers (formerly known as marsh hawks) "because there isn't any snow and they are able to hunt."

Rough-legged hawk numbers also were high, but not unusual for a raptor that emigrates to Wyoming for the winter.

Speaking of unusual sightings, we thought we might have had two Ross' geese at Lions Park. Conveniently, the two white birds hung around for more than a week after the count. Alas, after closer scrutiny, we determined they were only snow geese.

Although snow geese occasionally visit Cheyenne, I noticed they hadn't been recorded in any of the last 26 Christmas counts posted on the CBC web page.

You too can compare CBC data for past counts anywhere in the country. Go to the website, www.birdsource. org/cbc. Just remember, the count number is one more than the year the count was held. For instance, 1999 [including several days of January 2000] is the year of the 100th CBC.

Two changes in individual species populations of which I've become aware over the 11 years I've lived in Cheyenne show up in the CBC data.

Between 1973, the earliest Cheyenne count shown online, and 1982, Canada geese show up only twice. But since 1988, they appear consistently and in ever-increasing numbers, a phenomenon common in many urban areas, where the geese become fearless when people feed them.

The other species, the American crow, didn't make the CBC until 1986. With hard work, one or two could be found each year after that. By about 1994, we started counting 25 or more each CBC.

Finally, this year, with a count of 109, we didn't have to look hard at all. Crows are everywhere, teasing squirrels and playing with trash. Jane thinks the high numbers mean crows are breeding here now.

Most of the other species counted on this year's CBC are the result of careful examination of Cheyenne's backwaters: brushy habitat along Crow Creek and Dry Creek, abandoned lots, industrial neighborhoods and every other damp and weedy spot visible with binoculars.

Who would have thought a snipe would turn up? I always thought they were mythical creatures of "snipe hunt" shenanigans. But Jane says, "We almost always turn up one or two. They'll stay as long as there's a little open, marshy water."

## Christmas Bird Count Results

| | | | |
|---|---|---|---|
| 2 Snow Goose | 1 Common Snipe | 4 Black-capped Chickadee | 4 white-winged |
| 945 Canada Goose | 194 Rock Dove | 1 Mountain Chickadee | 11 Oregon |
| 2 Green-winged Teal | 1 Great Horned Owl | 25 Red-breasted Nuthatch | 42 race undetermined |
| 725 Mallard | 3 Belted Kingfisher | 4 Brown Creeper | 41 Red-winged Blackbird |
| 1 Canvasback | 7 Downy Woodpecker | 6 Golden-crowned Kinglet | 2 Brewer's Blackbird |
| 15 Common Goldeneye | 6 Northern Flicker | 8 Townsend's Solitaire | 3 Common Grackle |
| 14 Northern Harrier | (red-shafted) | 104 American Robin | 37 House Finch |
| 1 Red-tailed Hawk | 151 Horned Lark | 2 Northern Shrike | 25 Pine Siskin |
| 11 Rough-legged Hawk | 2 Blue Jay | 827 European Starling | 19 American Goldfinch |
| 1 Golden Eagle | 68 Black-billed Magpie | 5 American Tree Sparrow | 249 House Sparrow |
| 4 American Kestrel | 109 American Crow | 62 Dark-eyed Junco including | |
| 2 Merlin | 3 Common Raven | 5 slate-colored | |

2000

# 21 Feeder visitors fewer during mild weather

Thursday, February 3, 2000, Outdoors, page C4

Where have the birds been this winter?

Most of January it was quiet around our feeder. I had a call from a woman concerned that birds weren't coming. Perhaps she wasn't offering the right seed the right way?

My e-mail correspondent up near Chugwater reported a dearth of birds too.

Blame it on mild, snow-free weather. Birds, like us, abandon lunch at home in favor of getting out and about to explore new restaurants when the weather's nice, the weeds are free of snow and they aren't stressed by having to eat constantly as they must during bad weather. We ourselves may consider lunch in Fort Collins if the roads are dry.

But come a good snow-storm, we'd rather hunker down with a bowl of soup at home rather than negotiate messy streets to cross town.

Likewise, the birds flock to our feeders for a reliable source of food.

There are some folks who fear future generations of feeder species will become unable to forage for them-selves, turning into avian welfare recipients. But it appears to me even our finest feeder fare isn't enough to turn house finches into full-time sunflower junkies.

The mild January weather sure has made for some poor Project FeederWatch reports. In one two-day reporting period, the most I had of each species at any one time was one: one house finch and one house sparrow. Granted, I don't watch more than an hour over the two days, but other bad-weather report periods have produced 30 or more house finches, a dozen juncos and almost as many goldfinches.

Let's hope the weather is good and nasty for Feb.

18-21, the third annual Great Backyard Bird Count, spon-sored by BirdSource (a.k.a. Cornell Lab of Ornithology and the National Audu-bon Society).

Check out the count online at http://birdsource.cornell. edu/gbbc. For those of you without web access, I'll have more information in the Feb. 17 column. As submission of data is exclusively elec-tronic, you may call me with your results, and I'll enter them for you.

The count includes one weekday, which isn't a feder-al or school holiday, so even school groups may partici-pate, either by counting birds from the window or on the playground for 15 or 30 min-utes or by taking a short walk around the neighborhood.

Want to know more about our local birds? Pick the brains of the Cheyenne High Plains Audubon Society's ornithological experts.

"Birding for Beginners" is being offered through Lara-mie County Community Col-lege's Continuing Education/ Community Services depart-ment on Tuesday evenings, Feb. 29 through March 28.

Get on a common-name basis with about 100 of Chey-enne's most frequent fliers. Learn some of their unique traits and favorite hangouts. Recognize their voices, looks and attitudes. Also learn to use the quintessential birder's tools: binoculars and field guide. Then you'll fit right in on Audubon's spring field trips.

To register, visit the Con-tinuing Education/Communi-ty Services office on campus (the door by the satellite dish-es) or call 778-1236. You may also visit their home page at www.lcc.whecn.edu/cecs-rm. Note: "lcc" is the web abbre-viation for "LCCC."

# 22 Great Backyard Bird Count needs you

Thursday, February 17, 2000, Outdoors, page C4

Compared to the venerable Christmas Bird Count which celebrated its 100th anniver-sary in December, the Great Backyard Bird Count is just a flash in the pan of data.

Beginning Friday and con-tinuing through Monday, this year's count will be only the third annual Great Backyard Bird Count (GBBC).

Born of a need for data rather than an update of previous traditions, it is a computer project.

The institutional spon-sors, the National Audubon

Society and the Cornell Lab of Ornithology, want to track patterns of bird abundance and distribution. They decid-ed to call on the 60 million bird watchers around the country for help collecting data, whatever their level of expertise in bird identifica-tion and even if they only count for half an hour.

Over time, the data will show if some species are declining in population like the red-headed woodpecker or, for species like the bald eagle, how fast they are

making a comeback.

To take part in this count you need access to the inter-net; at home, work, a friend's, the public library, or by calling me (see the accompa-nying information box).

For this modern count there are no Audubon chapter count compilers, no paper maps of count circles and no potlucks at the end of the day. Instead, it's just you and the birds in your own backyard, and then you and the comput-er as you submit a count for any or each of the four days.

But your bird watching doesn't have to be solitary. The GBBC is so young we can make up our own traditions with family or friends. We could invite our neighbors over to observe our bird feeders over lunch one day and observe their birds the next.

Or we could take a little stroll around the neighbor-hood together. I know there are flickers on my street. They just never show up in my back yard.

Classrooms around the

country are also participating. At the GBBC website, under "Let's Learn About Birds" there are excellent teachers' tips and a six-page bibliography of birding books and media, including juvenile fiction/nonfiction and a glossary of terms.

Other pages on the website – tips on using binoculars and a guide to 50 common backyard birds, including sound clips, range maps and interesting info about each – will improve everyone's level of expertise.

Participation in the GBBC has grown about as fast as personal computer ownership. According to statistics provided by Jackie Cerretani of the Cornell Lab, Wyoming's participation was only 118 counts of the 42,000 submitted nationally last year, but it was up from 19 counts in 1998.

The beauty of computer data collection is that it's compiled so quickly that even the first day of the count you will be able to see results posted on the website.

And then we'll know just where our usual Cheyenne winter birds are hiding out during this mild excuse for a winter.

## To Make Your Count Count, Connect

From a computer at home, work, a friend's house, or the public library, log on to www.birdsource.org and click on "Great Backyard Bird Count."

If you are unable to submit results via computer, leave your information on Barb Gorges' voice mail at 634-0463. You have through Feb. 25 to submit your bird count results.

The Laramie County Public Library has four computers with free internet access available. Ask for help if you haven't been online before. Library hours are M-Th, 10 a.m.-9 p.m., Fri. & Sat., 10 a.m.-6 p.m., Sun., 1-5 p.m. The library will be closed Monday, Feb. 21, for President's Day. Call the library at 635-3561 for more information.

Please include the following information if you leave a voice mail for Gorges.

- Your name and phone number (in case she has questions).
- Your ZIP code.
- Habitat type (suburban, ag land, grassland, forest, water).
- Estimation of your expertise in bird identification (fair, good, excellent).

- Date, start time and duration (hours and minutes) of your observation.
- What kinds of birds you saw.
- The greatest number of each species you could see at one time, to avoid counting the same bird making return trips to your feeder).

# 23 Night birding offers unique experience on quiet evenings

Thursday, March 2, 2000, Outdoors, page C4

[I no longer suggest owling, except for scientific purposes.]

Have you ever tried night birding? The term "bird watching" doesn't really apply, though "bird listening" might.

It wasn't exactly a scene from "Owl Moon," the Caldecott-winning picture book by Jane Yolen, but a nearly full moon was glistening on patches of snow on the mid-February owling expedition.

Members of the Cheyenne High Plains Audubon Society who planned the outing were surprised by the number of people interested in tromping around in the dark and cold on a weeknight.

Kelly Johnson had her Girl Scouts bundled up and

Catherine Symchych, a raptor rehabilitator, came all the way over from Laramie.

Owling means listening for owls. Unfortunately, the wind was still going strong, which meant "low audibility" as well as night's normal low visibility.

During the mating season in late February and March, owls hoot a challenge to territorial trespassers. Researchers improve their odds of hearing owls by speaking up first. And just to make sure the accent is right, they use tape-recorded owl calls.

There are a variety of owls that can occur in Wyoming, so it's important to start with the calls of the smallest owls first. Once the bigger owls, like the great horned, are played, everyone else in the

vicinity stays mute for fear of being eaten.

It's easy enough to copy for your own use specific calls from one of the commercially available bird song tapes.

Get the "Wyoming Bird Checklist" available from Wyoming Game and Fish Department, to decide which of the 15 owl species to use.

When playing the tape, it's important to be in a woodsy area, have warm batteries in the tape player, and allow for quiet between calls so you can listen for responses.

The night we were out on the road by Lummis Reservoir [I probably mean Wyoming Hereford Ranch Reservoir #1], the taped calls were carried far downwind, but responses couldn't be

heard because we were standing upwind by the tape player. At the second stop, outside Lakeview Cemetery, the wind was less boisterous.

Still no owls.

Windless early spring nights are uncommon here, so I'll just have to get up and go owling the next time I notice one.

I did hear birds at night once this winter. Low clouds and a thick fall of snowflakes trapped the streetlights' orange glow and spread it everywhere so bright I could have read the paper.

It was windless and very quiet, no traffic. And then I heard birds twittering in the bushes across the street. Later I heard more somewhere in the back yard.

I've heard that songbirds

2000

21

will huddle together at night to keep warm.

Was I hearing the equivalent of snoring birds, birds trying to get comfortable, or birds confused by the unnatural bright light at 10 p.m.?

If birding in snow and dark doesn't appeal to you, try morning. Check the recording of the night's activities left in the snow.

In the forest I've found rabbit tracks that suddenly stop between two large owl wing imprints and have been surprised by a late awakening grouse exploding from its snow cave.

Early one morning recently, a scant quarter of an inch of snow covered the sidewalks. Whereas in deeper snow footprints are just dark holes, this time even the juncos left prints on the patio so clean, toe joints were visible.

Cottontails left tracks to show where they'd congregated in the middle of the street. A single-minded cat left a single-file string of pawprints down the sidewalk. Down by the ditch, tiny mouse-sized prints emerged from under the concrete barriers and circled back under again.

Every snow print was as if what dogs smell was suddenly made visible.

I'm not sure Lincoln, our dog, is very good at reading smells though. For some reason he got most excited on the way home sniffing the tracks he left when we started out.

Note: Anyone seen their first mountain bluebird yet?

# 24 Birds are now on the move

Thursday, March 16, 2000, Outdoors, page C2

As we turned off Happy Jack Road and onto the gravel road to North Crow Reservoir recently, I asked my husband, Mark, if he'd seen his first bluebird of the season yet.

He had barely answered when, on cue, an azure-colored male mountain bluebird crossed our path, stage left to stage right.

The first bluebird report I heard came last month. Robin Groose, president of Laramie Audubon Society, said his wife, Pat, had seen one down near Walden.

Wintering in southern Colorado and points south, mountain bluebirds come back as early as mid-February. You just won't see them downtown. They are a bird of higher, open grass and shrubland country.

What does an insectivorous species that likes its meals on the wing find to eat in February? Snow fleas?

It appears, according to the bird books, that bluebirds are willing to settle for frozen berries.

Since it was too mushy to ice fish at North Crow, we hiked up a creek into the national forest, into a landscape hardly distinguishable from November's. Same dead-tan grass, same leafless branches, same snowdrifts hidden in the shade of evergreens and rock outcrops.

Same birds too – the resident, year round species – junco, nuthatch, flicker, chickadee.

Then suddenly we were surrounded by unfamiliar bird song from the tops of the ponderosas.

Trying my new skills in Zen and the art of seeing birds with binoculars, I sought out a bird shape on a branch with my naked eyes. Then, without changing my gaze, I lifted the binoculars to my face.

Magic! There she was, a motley, streaky, olivish-yellowish colored red crossbill. As usual, the color part of the name comes from the color of the male. Both sexes, though, have the peculiar upper and lower bills that cross at the tips. All the better for extracting seeds from cones.

Not a predictable species to find in the forest, crossbills are nomadic, pitching their tents wherever pines or spruce have produced a good crop.

Later, my Zen method gave me an excellent look at a white-breasted nuthatch. All the resident species are primarily seed and nut eaters, but less fussy than crossbills. For them, going south or peregrinating to other patches of forest would just put them in competition with the locals.

Back here in town, people are reporting their first robins of the year. I gently break the news that we have robins year round, just not as many in the winter.

No, they don't hide a cache of frozen worms. Instead, robins seek out fruit, hanging out in riparian areas where they can find wrinkled rosehips on wild rose bushes or the produce of other moisture-loving plants.

The redheads are back. The duck species, that is. Take a good look at those ducks down in Lions Park. At first glance it may appear that a drake mallard's green head is not catching the sun right. If it appears a rich auburn color, you're looking at a redhead.

Across the country, birds are on the move. Sandhill cranes, having started out in Texas, are stopping over in Nebraska on the Platte River, taking a breather. Some head as far north as the Siberian tundra.

Other kinds of birds that spent the winter in the Caribbean or as far south as Argentina also are loading up on calories and heading north.

Though the peak for Cheyenne is still two months away, the early birds are arriving daily, in search of the proverbial worm.

# 25   Birding on a computer screen

Thursday, April 6, 2000, Outdoors, page C2

I am not fond of computer games. Maybe because I lack the manual dexterity necessary to land "sim" helicopters or the willingness to commit any of my lifetime supply of brain cells to memorizing volumes of protocol for winning "The Age of Empires," the latest game my fanatic family thinks I should learn.

A new piece of software showed up at our house the other day that may change my mind. Mark is using "Birds of North America" as a study aid before he goes to a workshop on bird inventory methods.

I thought it would be similar to another birding CD-ROM I viewed once, like a digital field guide.

But this disk, produced by Thayer Birding Software, is much more than that.

It includes the contents of the book "A Birder's Handbook," life histories of birds and some 200 articles on bird topics, as well as songs and one or more photos of each of 925 North American birds – and videos of 120 of them.

What the boys latched onto right away was the list of 250 quizzes. By the time

I arrived at the computer, they were in the middle of the "Food, Glorious Food" quiz, 16 ten-second videos of birds eating everything from lizards to carrion.

Next, we asked the software to give us the "Bad Hair Day" quiz, using the multiple-choice format rather than fill-in-the-blank. It gave us 29 birds, all of which had feathered crests at one rakish angle or another.

We chose to play at the easiest level, where we could eliminate really outlandish choices, such as duck or hawk names when the photo was of a sparrow-like bird and have some chance of success.

Other times, if the range map showed the bird wasn't found in Wyoming, Mark and I had to search our memories of formative birding days in the Midwest and on the East Coast or more recent memories of family trips to the Southwest or West Coast.

Some species I'd only seen in passing, while paging through the field guide on my way to identifying Rocky Mountain birds.

As we whipped through

half a dozen quizzes, family of four with faces glued to the screen, I wondered how quickly the boys could have all the answers down, even when they were obscure species like red-faced cormorants or whiskered auklets we're unlikely to see even if we do visit Uncle Peter in Alaska.

The next morning there was still a vestige of interest. Our 11-year-old requested permission to try a few more quizzes before school. No problem.

I've seen similar alacrity for complex detail in kids whether they are hooked on birds or computer games or the intricacies of Pokemon. They memorize tedious minutia in the blink of an eye and understand all the complex qualifiers.

Can my kids make the jump from hours of pointless games to hours of bird study, even progressing to stumping each other by making up their own computer bird quizzes?

Will they finally go outside for the ultimate quiz, identifying birds on the wing? Might they even grow

up to be ornithologists?

Meanwhile, their mother, conserving her precious little gray cells, will make use of her favorite part of the software, the "Identify!" feature. I can plug in what little I know about some bird I can't identify: color, size, habitat, location in North America or type of song, and the computer will generate a list of reasonable identification possibilities without my asking real people stupid questions.

Of course, the software can't give me the same amount of local detail as Bob, Jane, Fred, Dave or other birding friends. And neither can it, though a versatile tool, give me the gossip on who's already come to town for this year's spring migration.

There are three ways to get your own copy of "Birds of North America:" tell your favorite bookstore it's ISBN 1-887148-13-2, go to www.thayerbirding. com or call (800) 865-2473, where you may find, as I did, that you are talking to Pete Thayer himself.

# 26   Home on the Prairie

Thursday, April 13, 2000, Outdoors, page C1

**What can you do to help prairie wildlife thrive in its native home? The best advice: Don't do anything.**

CHEYENNE – On first acquaintance, people from lusher parts of the country believe our shortgrass prairie is in desperate need of improvement.

Not so.

It has developed its own way of dealing with only 15 to 17 inches of annual precipitation from rain and snowfall. Some prairie species won't even thrive in wetter conditions.

The first piece of advice for wildlife lovers contemplating a move to the prairie is, "Don't," said Reg Rothwell, Wyoming Game and Fish Department supervisor of biological services.

Or at least, he said, "Resist the temptation to bring the English countryside with you."

Rothwell finds that many property owners become disillusioned when confronted

2000

23

with badgers hissing warnings, rabbits munching gardens or stripping bark off saplings, ground squirrels tunneling through the lawn and hawks or other predators eyeing their small pets as snacks.

The solution to these wildlife conflicts is for humans to "create as small a footprint as possible," rather like leave-no-trace camping in the back country, Rothwell said.

The benefits of low-impact living mean less work and cost for property owners as well. Unlike standard Kentucky bluegrass lawns, native prairie needs no irrigation, no mowing and no fertilizer or pesticide.

Plus, it makes a much more interesting view compared to monoculture lawn because its variety of grasses and wildflowers attract wildlife.

Don't even bother with the effort of planting trees, said Andrea Cerovski, Game and Fish nongame bird biologist stationed in Lander. Native prairie bird species are ground nesters.

"Trees encourage predation by birds like crows and nest parasitism by cowbirds."

The prairie, a complex ecosystem, doesn't stand up to fragmentation.

Said Cerovski, if a road or other disturbance subdivides the prairie into less than 125- to 250-acre tracts, some native birds will refuse to breed. It's the visibility problem in an area of scarce vegetation. Put yourself in the bird's nest and imagine living in a glass-walled house. How far away would you want

neighbors and passersby?

To nest, most prairie birds scrape a slight depression in the dirt, perhaps lining it with some dried vegetation. Some, like the ferruginous hawk and long-billed curlew, like to incorporate a little cow dung, perhaps for the heat it generates as it decomposes.

Obviously, free roaming cats and dogs are much more of a problem for ground nesting birds than for the tree-nesting species in town.

Grazing, at proper stocking rates, is compatible with most prairie bird species.

Some, like the mountain plover, proposed for listing as a threatened species, demand nearly bare ground before they'll consider nesting.

Others, like sharptailed grouse, prefer a screen of ungrazed vegetation, such as they find in acreage registered in the Conservation Reserve Program.

For the owners of small acreages contemplating grazing horses or other livestock, the best bet is to confer with the experts at the Laramie County Conservation District.

Martin Hicks, wildlife and range specialist with the district, recently advised one landowner how he could safely stock two horses on 11 acres by cross-fencing it into eight pastures and moving the animals every several days. Still, for six months of the year, the horses will require supplemental feeding.

But, Hicks warns, every situation is different. Typically, one horse will need 15 acres of dryland

pasture during the six-month growing period.

An overstocked acre or two in our sea of grass isn't a big problem by itself, but as Rothwell points out, the cumulative effect of everyone in a development confining their horses can cause enough bare ground to make it susceptible to appreciable amounts of wind erosion and adds Hicks, wind-borne weed seeds.

Cerovski, in her other capacity as Wyoming coordinator for Partners in Flight, an international partnership of agencies and organizations devoted to the protection of neotropical migrant songbirds, has been working on the Wyoming Bird Conservation Plan.

The plan, due out soon, will list for landowners and managers best management practices for various Wyoming habitat types, including the prairie.

While many range improvement methods such as mowing or haying or prescribed burning mimic natural events, Cerovski recommends avoiding using them during the nesting season, mid-June through mid-July, and even allowing a two-week buffer on either end to protect early and late nesters.

"June and July are most critical for nest building, egg laying, hatching, growth and fledging the young," Cerovski said.

Insecticides should also be avoided, she said. After all, "those insects are the main food items for prairie birds. If provided with other habitat

components they need, birds will do a really good job of controlling problem insects."

People becoming prairie dwellers have a lot of expertise available to them through the Conservation District and Game and Fish. Those agencies will offer advice on learning how to coexist with the ecosystem.

But it takes personal time and observation to understand the details of what really happens on a particular piece of prairie.

This spring, you can start by training your ear for prairie bird songs. Learn them by using a tape or CD such as those put out by Peterson's Field Guides or the American Birding Association.

Then walk often (with your dog on a leash) and find out who's out there.

As Rothwell put it, "Appreciate the prairie for what it is."

*Cutlines:*

Meadowlarks are among the most common bird species found in Wyoming. Their call provides a familiar backdrop to any prairie outing. Meadowlarks have adapted well to the presence of humans and may thrive in open areas close to town where other prairie species would be crowded out.

Horned larks are common year-round in shrublands and prairies.

Short-eared owls are a "priority" species (see chart at right).

Swainson's hawk (on front page as part of graphic layout).

CHEYENNE BIRD BANTER

## Prairie Birds You'll Want to Meet

Priority* prairie bird species from the draft of the Wyoming Bird Conservation Plan

| | | | |
|---|---|---|---|
| Ferruginous Hawk | R/3, RI, SH, PR, SF (cliffs) | Dickcissel | S/2, PR |
| Mountain Plover | S/3, PR | Lark Bunting | S/4, SH, PR, AG |
| Upland Sandpiper | S/2, PR, AG (crop residue) | Grasshopper Sparrow | S/3, SH, PR, ME, AG |
| Long-billed Curlew | S/2, SH, PR, ME, AG, SF | McCown's Longspur | S/3, PR, SH, AG |
| Burrowing Owl | S/2, SH, PR, AG | Chestnut-collared Longspur | S/2, PR, SH, AG |
| Short-eared Owl | R/3, SH, PR, WL, AG | Bobolink | S/2, SH, PR, AG |

*A Wyoming Priority Species may be rated "3" or common, but over the rest of its range, in other states, it may not be thriving.

Other Bird Species Dependent on Prairie and Shrub Habitat

| | | | |
|---|---|---|---|
| Northern Harrier | S/2, SH, PR, WL | Eastern Kingbird | S/3, SH, PR, AG, FO, RI |
| Swainson's Hawk | S/3, PR, ME, SH, AG, SF (cliff/tree) | Horned Lark | R/4, SH, PR |
| Rough-legged Hawk | W/3, SH, PR, AG | Vesper Sparrow | S/3, SH, PR, AG |
| Sharp-tailed Grouse | R/3, PR, SH, AG | Baird's Sparrow | S/2, PR |
| Say's Phoebe | S/3, SH, PR | Lapland Longspur | W/3, PR, AG |
| Western Kingbird | S/3, SH, PR, AG, FO, RI | Western Meadowlark | S/4, SH, PR, AG |

Species inhabiting Wyoming: S – summer only, W – winter only, R – year-round (resident)

Abundance: 1 – Rare, 2 – Uncommon, 3 – Common, 4 – Abundant

Habitat types, listed for each species in order of where species is most likely to be seen:

PR – Prairie – open expanses of grassland.

SH – Shrubland – dominated by sagebrush, bitterbrush or other shrubs.

AG – Agricultural – pasture, cropland, irrigated meadows.

ME – Meadows – wet to moist meadows and grasslands.

RI – Riparian – bordering streams, lakes or rivers.

WL – Wetlands – water present most of the time.

FO – Forest – dominated by coniferous and/or deciduous trees.

SF – Special Features – unique natural or man-made characteristics.

# 27

Thursday, April 13, 2000, Outdoors, page C1

# Making birds at home in our 'urban forest'

CHEYENNE – Over the last hundred years, part of the prairie in Laramie County has been converted to forest by the residents of Cheyenne.

Either to remind ourselves of homes back East or to block the wind, we've planted trees.

The riparian zones along streams, rivers and lakes, naturally irrigated areas on the prairie grow thick sod, cottonwoods and willows. But here in town, our lawns and trees grow with the help of irrigation water piped over the mountains or pumped from the ground water.

When we want to attract birds to our urban yards, we mimic the amenities of the forest. Sometimes the improvements benefit us as well as the birds.

First, we plant more trees and shrubs, especially the hardy native species requiring less water. Dense plantings, coniferous or deciduous, give birds places to nest and good protection from bad weather and predators such as roaming domestic cats. They may provide food as well for berry eaters like robins and waxwings.

Shane Smith, director of the Cheyenne Botanic Gardens, recommends growing Nanking or sand cherry, woodbine, New Mexico privet and juniper. Other recommendations from the gardens' website include serviceberry, chokecherry, sumac, and varieties of currants.

Besides giving us a little protection from wind, plantings protect us from views of unsightly garbage cans or compost piles.

Birds help us maintain trees. In the spring, warblers can be seen gleaning bugs from the new leaves. Woodpeckers, including flickers, and brown creepers search every inch of the tree trunk year-round.

A plain, ordinary lawn can also attract the native pest patrol. Grackles will patiently pace your sward of green, shoulder to shoulder, their yellow eyes gleaming like searchlights as they delve with their long, sharp bills into the turf for miscreant grubs.

Your end of the lawn maintenance deal is to switch from chemical lawn fertilizers to child/pet/bird-compatible products.

Ken Stevens of Riverbend Nursery recommends "Sustane," a slow-release, balanced fertilizer with microbes that encourage the natural decay and nutrient cycle.

Pesticides? Why use them when you are inviting avian experts? As for weeds, healthy grass will crowd them out.

Occasional dandelion digging is good exercise, and if you miss one, the

2000

seeds will be appreciated by goldfinches.

As testimony to 11 years of organic care, infrequent, deep watering, and grass cutting at the highest blade setting, our lawn has never needed de-thatching or aerating and looks much like the rest of our neighbors' lawns.

But why settle for a boring Kentucky bluegrass lawn? For the same amount of water, or even less if you practice xeriscaping, why not convert to native grasses or the visual diversity of gardens providing seeds and berries for birds?

The birds attracted to our urban forest are often cavity nesters. Trees have to be old and decadent enough to get cavities, but when they reach that point, we usually cut them down because they threaten our safety.

So, we provide bird houses for house wrens, house finches, nuthatches and woodpeckers.

When you are perusing bird house plans, make sure you pick those for species that occur here. For instance, we are not in the purple martin's normal range.

Make sure the entrance dimensions exclude pesky non-native species like house sparrows. Let them build their nests in discount store signs. And erect your bird house so marauding cats, squirrels and starlings can't kill nestlings.

For the cavity nesters as well as the branch and bush nesters like goldfinches, provide fibrous materials like string and hair from people and pets, but not fishing line, or in lengths longer than six inches. And leave some mud for the robins and swallows to plaster their kind of nests.

Forest birds like water for drinking and bathing, though some prefer dust baths. Realistically, during lawn watering season, there's always some water slopping into gutter streams or sidewalk micro ponds. But what about the rest of the year?

If you want to provide five-star accommodations, offer a pool. It doesn't have to be Olympic-sized – I've noticed birds using the dog's water dish. Heated is nice in the winter.

Moving water is especially attractive, whether a fountain or just a milk jug full of water you hang up and then make leak at the rate of a drip every second or so.

Putting food out is always a popular way to attract birds. It is most successful when water and shelter are also present.

Offering sunflower and Niger thistle seed right through the summer gives a nutritional boost to brooding birds and parents feeding nestlings. But in warmer weather, it's doubly important to keep feeders clean and free of deadly bacteria and diseases.

There are lots of selfish reasons for encouraging wild birds to come and live among us: their songs, their colors, their antics and their utilitarian contributions.

Most importantly, birds are still the "canaries in the coal mine" in our age of continuing industry and development. A forest without birds is cause for trepidation.

*Cutlines:*

Hang bird feeders in areas sheltered from the wind.

Robert Dorn puts fresh birdseed into a protected feeder in his back yard recently. Dorn said putting a wire cage over the feeder allows only the smaller birds inside, where they don't have to compete for food with the larger grackles.

Building a birdbath will always attract birds, like this robin in Robert Dorn's backyard. (Larry Brinlee)

---

### Attracting Urban Birds

The list included information for: field guides, local nurseries, Wyoming Game and Fish Department, Birdsource (Cornell Lab of Ornithology and National Audubon Society), Cheyenne Botanic Gardens, the website birding.about.com, Wyoming Wildlife Federation and Laramie County Public Library.

# 28

Thursday, April 20, 2000, Outdoors, page C4

# Indoor-dwelling cats can add richness, humor to your life

**Keeping cats inside promotes a closer relationship with your fascinating pets – and helps wildlife.**

Having cats but leaving them outside is like having children and sending them to boarding school.

If you don't live with them, you won't know them – and you'll miss the whole point of being a cat owner or a parent.

Our cats, Henry and Joey, brother and sister, live indoors where I've been able to closely study their eccentricities, cataloging them in anthropomorphic and animal terms.

*Baby* - Henry enjoys visits by small children, being tucked in with dolls and getting rides in strollers. He finds childish antics fascinating, although Joey abhors them.

*Toddler* - We put childproof latches back on the lower cabinets to keep the kitchen trash and cat food safe.

*Athlete* - I don't know too much about steeple chasing or cross-country track, but the cat versions include hallway dashes, bookcase climbs, tablecloth slides and kitchen counter chasm jumps.

*Parrot* - Joey supervises housework from the vantage point of my shoulder, while humming a sweet purr in my ear.

*Hawk* - Henry prefers higher, more solid perches where he strikes poses as straight-backed as a peregrine, glaring at all us prey moving unaware below him.

*Raccoon* - Perhaps the cats aren't fastidiously cleaning objects when they drop them in the dog's water dish. Perhaps they want to poison their historic adversary. They

CHEYENNE BIRD BANTER

attempted electrocution one night when they knocked the battery recharger into his bowl.

*Botanist/Gardener* - Feline experimentation shows that spider plants are 10 times more edible than geraniums. Sage and rosemary are never harvested, but they require occasional inspections of the soil for bug infestation.

*Pest exterminator* - Explain to me how an 8- to10-pound animal without wings or suction cups on its feet can catch a fly that has the ability to stick to the ceiling.

*Textile expert* - You may think I'm about to tell you how Henry and Joey test textile strength with their claws. Actually, they have refined tastes in fabric: Hand-knit wool and hand-quilted cotton … for sleeping that is.

*Piano player* - Both cats show occasional talent. Rumor has it a composer was inspired by a series of notes his cat played.

*Disco dancer* - Our junk mail is filled with CDs for free internet access. We dangle them in the sunshine over Henry's head, with the reflections causing him to perform feverish dance routines.

*Water witch* - Woe to the family member who leaves out a partly filled watering can or cup. And more woe to the forgetful one who leaves important papers on the counter next to these water sources. So far, the woe is all mine.

*Explorer/escapee* - Every time we reach to open a door, we check around our ankles for the escape artists. Doors leading outside are no more popular than doors to other rooms. Occasionally we'll indulge the cats in their quests for new horizons, even supervising a stroll outside.

*Scientist* - Cats have a healthy amount of curiosity about artifacts in the human home. I can just about hear them thinking, "What causes the stink in the toe of this boot?" "How long can I entertain a human with an empty cardboard box?" "How soon will birds feeding on the windowsill notice I'm here, on the other side of the glass?"

As with adolescent humans, you need to spend a lot of quality time indoors with your cat before it will feel safe and begin to open up to you.

Now, at age 2, Joey even has a nickname for me. "Meow-ma," she says, her amber-colored eyes staring into mine before she launches into a litany of cat remarks. Henry too, has things to tell me, though my translations haven't gotten a whole lot further yet than, "Dinner, please."

I just can't imagine companions such as these being thrown out into the street or field where one of them might fall into dangerous company and be dogged the remainder of his/her life with the reputation of "bird murderer."

*Cutline (Photo of Henry outdoors wearing a harness):*

A harness and leash allow you to take your indoor cat outside and explore nature together. Barb Gorges' cat, Henry, enjoys bird watching. Barb Gorges/correspondent

---

## Author note

Column accompanied by a box of cat facts compiled by Bill Gruber and staff, notably from the Cats Indoors! campaign sponsored by the American Bird Conservancy.

# 29 Scouting gets youngsters in touch with outdoors

Thursday, May 4, 2000, Outdoors, page C1

CHEYENNE – A big part of the word "Scouting," as well as the organizations it represents, is "outing."

Both Boy Scouts and Girl Scouts, in addition to teaching children outdoor skills, also emphasize nature study and the importance of environmental actions such as tree planting, recycling and resource conservation.

Both programs introduce children to the outdoors at a tender age. Local Daisy Girl Scouts, kindergarten-aged, will be going to camp May 20 for a few hours with their parents to play games, sing songs and try nature crafts, said Judy DiRienzo, program specialist for Laramie County Girl Scouts.

Mark Palmer, Tiger Cub leader for Cub Scout Pack 221, said his first-grade boys work on earning "Tiger Paws" with their parents, with some of the activities like "Discover Nature and Energy" or "Go See It!" taking them outside.

At the next level, Brownie (grades 1-3) and Junior (grades 4-6) Girl Scouts and Wolf, Bear and Webelos Cub Scouts (grades 2-5), begin building their outdoor skills under guidance of leaders.

Brownies learn pocketknife safety, knots, safe hiking techniques, fire building, outdoor cooking and map reading, then try sleeping outside.

Cub Scouts learn camping basics as well. Their handbooks include instructions on how to fish and how to identify poisonous plants.

All scouting programs, for whatever age, depend on enthusiastic parents who, with training, well-tested program materials and experienced mentors, can successfully lead a troop, den or pack.

Locally, day camps are available in the summer for elementary-aged Scouts. Girl Scouts offers resident camps in locations around the state.

Cub Scouts can camp overnight with a parent or guardian at special "Partner and Pal" weekend events.

By the time children reach Boy Scouts or Cadet and Senior Girl Scouts, outdoor opportunities – many merit-badge driven – are myriad: backpacking, boating, camping, canoeing, climbing, fishing, orienteering, rifle and shotgun shooting, sailing, snow sports, and wilderness survival camps.

Boy Scout troops routinely camp, even in the winter, when they attempt to earn "frost points," one point for each degree below freezing per night camped out in a tent or snow cave.

2000

Older Girl Scouts aspire to go on "treks." Boy Scouts aspire to "High Adventure"

camps like mountainous Philmont Scout Ranch in New Mexico.

Last summer Troop 102 sent 10 boys and two fathers there on a 100-mile backpacking trip.

**Author Note**

A box included local contact information for both local scouting organizations.

# 30 Songs tell of habitat, breeding

Thursday, May 11, 2000, Outdoors, page C2

[As of 2022, Franz Ingelfinger is a wildlife biologist with Montana Fish, Wildlife and Parks.]

You have no idea how annoying the otherwise sweet "cheer-i-o" of a robin can be until you've had one sing 10 feet from your head at 4 a.m.

I went with the Boy Scouts on my first Ft. Robinson, Nebraska, tree-plant April 8. Although two layers of nylon tent helped keep me warm, it was no insulation from nature's self-appointed alarm clock.

At last month's Cheyenne High Plains Audubon Society meeting, guest speaker Franz Ingelfinger explained the phenomenon of bird song and how it relates to his research.

Birds, male and female, use short call notes for everyday communication.

Usually only the males sing the longer, musical phrases. Simultaneously they broadcast their land claim to keep other males away and invite females, prospective mates, to stop by and check out the quality of their plumage and future nesting spots.

Birds don't sing any louder than necessary, even if their syrinx, or voice box, is capable of louder songs.

Apparently, natural selection has evolved a robin that sings loud enough to be heard in the next territory – but not loud enough to attract the attention of a predator passing through three robin-sized territories away.

Ornithologists have found correlations between the duration and pitch of a particular bird's song and the type of habitat it sings in. Franz mentioned a study of Carolina wrens in which their songs were recorded in one habitat and played back in one with different kinds of vegetation.

The songs didn't carry, or resonate, as well when out of place. For denser brush, the wrens had evolved lower-pitched and purer (less warbling) tones, which don't transmit as well in less-dense brush.

What happens when a bird's habitat is disrupted with human-caused noise?

Franz shared the results of a study in the Netherlands along a busy highway, which showed the density of breeding birds drops dramatically next to the road and gradually recovers with increasing distance away.

The study suggests traffic

noises mask bird song so the birds can't successfully find each other and mate. The study proves your best birding won't be from the highway – you need to get out and walk.

Franz, a graduate student with the Fish and Wildlife Cooperative Research Unit at the University of Wyoming, is studying the effects of natural gas development on birds of the sagebrush-grassland south of Pinedale in an area known as the Mesa, above the confluence of the Green and New Fork rivers.

In his first season last summer, Franz surveyed bird life adjacent to roads in the nearby, nearly established Jonah field (400 vehicles per day) and the Mesa, which was in the initial stages of development (10 vehicles per day).

His preliminary findings for the high-traffic roads show a 50 percent reduction in the number of singing, presumably breeding, birds within the first 100 meters (110 yards).

These natural gas fields are being developed at the rate of eight wells per section (square mile). They are connected by a

spiderweb of roads.

With about 100 square miles affected in the Mesa field alone, what will reduced breeding along the roads mean for the populations Franz is studying – sage thrasher, sage sparrow, Brewer's sparrow, vesper sparrow and horned lark?

Franz's study, sponsored by the US Fish and Wildlife Foundation, Wyoming Game and Fish Department and the US Geological Survey, examines the costs of drilling for only one aspect of the sagebrush-grassland ecosystem. What about impacts on large game, or the invasion of noxious weeds that always follows new roads?

When will the costs add up high enough to make directional drilling cost effective? Drilling in many directions diagonally from one site would reduce the number of sites and roads needed.

Meanwhile, I'm not sure if I envy Franz as he heads out this week for his second field season. He has to be up before the birds and the sunrise, only to be shrouded in dust with each passing vehicle. But the songs of the dawn chorus will be lovely as he walks out into the sagebrush.

# 31 Humans a hazard to navigation

Thursday, May 25, 2000, Outdoors, page C2

The other day, my math students at Johnson Junior High were quick to point out the bird auditing the class from the outside windowsill.

It was a male house finch in his bright-red breeding plumage. His beady stare was making some students uncomfortable as he strutted back and forth.

I laughed and explained that the bird wasn't seeing us at all. The coated window glass acted like a mirror, and the poor bird thought his reflection was an interloper to be challenged.

Occasionally the female came by, and I anticipated the students' next concern.

"Do you think those birds are going to...?"

"Of course!" I beamed, "It's spring!"

Lucky for this particular finch, his run-in with windows was only a territorial problem. Every year windows cause the death of hundreds of thousands of other birds.

The sky and trees reflected in a window make birds think they see a clear flight path. They run into the glass at full-tilt, and if they don't break their necks, a predator may kill them while they lie on the ground, stunned.

Confusing birds is a problem for glass-sided skyscrapers as well as our own homes. Suggestions for protecting birds from this hazard are based on breaking up the reflective surface. Pulling the drapes [might not work] works, of course, but it also negates the reason for having windows.

Netting or window screens in front of the glass helps, and so does the typical dusty coating of most Cheyenne windows.

The popular solution is sticking a hawk silhouette on the glass. This could be more effective if the silhouette flutters in the breeze. Try hanging it from a suction cup.

Windows can be hazardous at night as well. High-rise building managers leaving lights on all night waste energy and reap a fallout of dead, dying and stunned migrating birds on their sidewalks in the morning.

Birds migrating at night use many navigational cues including starlight.

In bad weather, the lights of tall buildings and radio towers attract them, and the birds either smash into the structures or circle hopelessly until falling from exhaustion.

The Fatal Light Awareness Program, FLAP, based in Toronto, has volunteers rescuing stunned birds from around buildings in the financial district. The technique, viable for any stunned bird, is to set it upright in a paper bag, which calms it, insulates it a little and protects it from cats and crows until it recovers and can be released.

FLAP's main goal is to convince managers to turn the lights off during migration season. At www.flap.org, the group lists bird-friendly buildings and manager contact information. FLAP founders hope to see this incentive used in other high-rise cities.

No one used to be aware of the extent of bird migration at night. Radar technicians during World War II referred to the unexplained, ethereal shadows on their screens as "angels." In the early 1960s, Sidney Gauthreaux of Louisiana proved the angels were actually huge flocks of migrating birds.

The latest technology, Next Generation Radar, or NEXRAD, is so sensitive it can track weather by tracking dust particle movement. Birds are detected as so many bags of salt water, just another form of precipitation.

Remember, the human body is 98 percent saltwater.

Using radar to figure out where the birds are is useful to human as well as bird safety when it comes to preventing bird and plane collisions or avoiding poisoning of birds during pesticide applications.

This spring, a cooperative effort called BirdCast, sponsored by the U.S. Environmental Protection Agency, is tracking bird migration in the mid-Atlantic region. Partners in the program are the National Audubon Society, Cornell Lab of Ornithology, the Radar Ornithology Lab at Clemson, the Academy of Natural Science, GeoMarine (a private company) and thousands of bird watchers.

The idea is to match up radar images of bird migration with what bird watchers actually observe, a process known as "ground truthing," and then be able to predict the movement of waves of migration.

For now, we Wyomingites will have to be satisfied with word of mouth for following spring and fall migration across the state.

My bird watching correspondent up in Chugwater would like to start an internet list serve to make it more efficient. Meanwhile, check out www.birdcast.org to see what's going on back East.

Migration is an amazing phenomenon. Millions of birds, some weighing only an ounce, accurately hurtle through the atmosphere for hundreds or thousands of miles using navigation systems we can barely understand. Yet we can mortally confuse them with a few bright lights.

Makes me wonder if we humans are advancing.

# 32 Go Birding!

Thursday, June 8, 2000, Outdoors, page C1

## Mastering bird-watching basics will enhance your time in the outdoors

CHEYENNE – Bird watching is an all-inclusive hobby that has grown immensely in popularity in recent years.

From young children to the elderly or disabled, almost anyone can pick up a pair of binoculars and a field guide and begin to enjoy watching and learning about birds.

Southeast Wyoming lies along the migratory paths of a wide array of bird species and is the year round home to many birds that are as beautiful as they are fascinating.

From identifying birds at the backyard feeder to becoming a full-blown amateur ornithologist, bird watching can be as simple or as all-consuming as you choose to make it.

So, what steps do experienced birders recommend for the casual backyard observer who wants to fan that initial spark of interest in birds into a bigger flame?

"I would suggest you take a class," said Gloria Lawrence. "Or go birding with a group. Go birding every chance you can with people who know birds. Or fumble through the field guide."

Lawrence, who lives near Casper, keeps the Wyoming Bird Hotline up to date. Her interest in birds was sparked by a northern mockingbird that spent a summer singing from the yard light pole when she was a child growing up on a ranch near Chugwater.

She and her husband, Jim, began feeding backyard birds and they learned to identify them, along with those they saw on outdoor trips.

"The spark turned into a roaring fire when Jim and I took a class from Oliver Scott in 1984," Lawrence said. "The fire is burning out of control. I realize in a lifetime I'll barely scratch the surface of what there is to know about birds."

Cheyenne birder Jane Dorn got the tinder for her "spark" – as birders refer to the beginning passion for birds – as a small child growing up near Rawlins, also on a ranch, with a family that hunted and fished.

Dorn could identify game birds and the songbirds her mother fed before she was old enough to go to school.

"I've always watched birds; it's something I grew up doing. I wasn't intensely interested until after taking a college ornithology class," she said.

Jane and her husband, Robert, are co-authors of "Wyoming Birds," a book documenting the occurrence of bird species throughout the state.

"The more you do, the better you get," she said. "Taking a class or going out with a birder is a huge boost to your bird-watching knowledge and shows you what's what locally."

Can a person be too old or too young to take up bird watching?

No, said Lawrence, who helps teach an annual 12-week bird class offered by the Murie Audubon Society at Casper College.

"Many students are middle-aged or older, and many are retired," she said.

"It's a hobby you can pursue for a lifetime," said Dorn, who helped teach a birding class at Laramie County Community College this year.

Birding is ideal for the disabled, and it's easy to add to other outdoor family activities.

One example Lawrence gave of the birding spark flaming at a young age is Joe Scott, whose grandfather, Oliver Scott, wrote the American Birding Association's "A Birder's Guide to Wyoming."

The young Scott, now in high school, and his father, Stacey, like to make the trek from Casper for the Cheyenne High Plains Audubon Society chapter's annual spring bird count.

Scott recently received a grant from the Governor's Youth Initiative for Wildlife. It and other funds he raised will help him build a new flight cage for Casper bird rehabilitators Lois and Frank Layton.

There are just two pieces of equipment needed to enjoy bird watching: A pair of binoculars and a field guide.

"Get the very best equipment you can afford," Lawrence suggested. "I started out with 7x35 Tasco binoculars. When I got my Bausch and Lombs, it opened up a whole new world. Good optics just make birding more enjoyable."

Dorn recommends a minimum power of 7. Go with 8 or 9 if you can afford it. (See the accompanying article on binoculars for a discussion of magnification.)

"Ideally you want to try as many kinds of binoculars as you can," Dorn said.

Choosing binoculars that fit your style of bird watching is as important as fitting them to your hands and eyes.

"If you'll be doing little walking, you can afford heavier binocs with a wider field of view," Dorn said.

She estimated that $200 would buy an acceptable pair of birding binoculars.

Top birders spend as much as $1000. With improvements in quality in recent years, such as lens coatings that improve the brightness of the image, you can get more capability for the same money now.

As a hobby, bird watching doesn't have to be expensive. "You don't need as much (equipment) as golf," Lawrence said.

And, said Dorn, bigger is not always better. "More magnification is not necessarily better. Anything above a 10 you cannot hold steady enough. You buy a scope (with a tripod) when you get serious about shorebirds and waterfowl."

The most important thing about binoculars is to use them, Lawrence said. "Once you get binoculars, use them and use them," until focusing is fast and

---

CHEYENNE BIRD BANTER

automatic. And learn how to use the individual eye focus to adjust for differences between your eyes.

Dorn advises testing binoculars for alignment as well. If the two barrels aren't lined up, you may have a headache by the end of a day of birding.

Field guides are a less expensive tool, running from $15 to $25 apiece.

But, said Dorn, "You'll find you'll want to own more than one."

Lawrence will attest to that. "Jim and I have six bookcases. One is entirely filled with bird reference books, floor to ceiling, probably 250 books," she said.

Both women recommended the newest edition of the National Geographic field guide because it's the most up to date, and it covers bird species for the entire United States as well as exotic species that may show up accidently.

"Peterson's (guides) are still excellent, but you need both the Eastern and the Western guides," said Dorn. "The old Golden (guide) is good, but the nomenclature is sort of out of date."

After the initial investment in binoculars and field guides, you can enjoy bird watching from home.

"You don't have to live any place special to bird watch," Dorn said.

You may enhance home bird watching by making your yard attractive to birds, providing food, water and shelter. On a day too rainy to go out last month, just before the peak of spring migration, Lawrence and her husband counted 38 species from their window.

It is possible to spend a lot of money on the hobby. Birding magazines advertise eco-tours to all kinds of international, bird-rich destinations. And the number of bird festivals around the country, usually celebrating particular species, continues to grow.

There's even one in Wyoming, held in Lander each May.

There's always more to learn about birds, even when you're the teacher. "I learned as much as the students," Lawrence said of her experience. "When you try to describe (an ordinary bird) for someone else, you become more aware of what really looks unique about it."

And there's no limit to how much time some people put into bird watching.

Lawrence, who goes birding all the time, related a typical story. "It's a habit. I was coming up the stairs with a load of laundry when I saw a painted bunting."

This type of bunting shows up accidentally in Wyoming, with only three documented sightings listed in the Dorns' book. After documenting it with photographs, Lawrence added it to the bird hot line report.

The bird hot line has given way to the Wyobirds Google Group.

*Cutlines for Larry Brinlee's photos:*

Bart Rea, Wyoming Audubon president, scans the trees looking for a northern flicker, a type of woodpecker.

Mark Gorges of Cheyenne, left, Jennifer Miller of Basin and Vicki Spencer of Casper look through Golden Guides' "Birds of North America: A Guide to Field Identification," while Neil Miller of Basin consults his "National Geographic Field Guide to the Birds of North America." Members of the Cheyenne High Plains Audubon Society chapter, the Murie Audubon Society chapter of Casper and Meadowlark Audubon chapter of Basin

met at Wyoming Hereford Ranch last month to take part in "The Big Count," one of many outings scheduled by area birders in southeast Wyoming each year.

Wyoming Audubon President Bart Rea of Casper, left, and Dave Felley, secretary of Wyoming Audubon and Cheyenne High Plains chapter president, search for birds on Crow Creek recently.

---

## Bird Watching Stats

According to the Cornell Lab of Ornithology in Ithaca, N.Y., the number of birders in the United States is now estimated at 60 million. No one seems to have kept track of the statistics over the decades, but it's generally accepted that number has grown exponentially in recent years.

According to the lab:

--Bird watching is the fastest-growing form of outdoor recreation in America, second in overall popularity only to gardening.

--By 2050, birding is the only major outdoor recreation that will have grown faster than the national population: It's expected to increase in participation by 53.9 percent.

---

# 33   Bird identification takes same skills as identifying people

Thursday, June 8, 2000, Outdoors, page C1

CHEYENNE – Bird identification is an important part of bird watching.

It's possible to just enjoy watching the birds but knowing who they are allows you to gossip about them with other birders or look them up for more information about their peregrinations and peculiar habits.

Review your field guide in conjunction with the Wyoming Bird Checklist – available from the Wyoming Game and Fish Department – to get an idea of what species you might see.

But otherwise, identifying birds takes the same observational skills and knowledge of habits you use to identify people from a distance when you're out at the mall or if you're a child approaching the playground on the way to school.

As you stand at the edge of the playground, scanning to see who's there, you see the usual kids on the equipment, running endlessly after the soccer ball and hanging back by the school door.

Birds too, have their usual hangouts. Some are to be seen soaring overhead most of the time, and some are always picking around in the underbrush.

2000

31

On a school day you know you won't see the neighbor girl who spends the summers here with her father, just as you wouldn't expect to see hummingbirds or other spring migrant species in Wyoming in January (though unusual events are not impossible).

Time of day is important too. If you were to arrive at school about 7 a.m., you'd catch a glimpse of junior high kids waiting for the bus, and you wouldn't see them again until nearly 3 p.m. Same with birds. Some are very active early in the morning, and some, like robins, can be seen out all day long. And then there are the nocturnal, owl-like children on the playground, hopefully without spray cans in their pockets.

If it's a cold winter day you probably can pick out your best friend because her winter coat is bright pink.

The birds' brightest plumage appears in the spring, however, and usually just on the males. After molting their spring plumage, birds are more difficult to identify.

So, after molting, or on days it's too warm for your friend's pink coat, you may have to go on clues related to behavior and silhouette. Does your friend typically slouch around in baggy pants? Or does she wear leggings and chase the boys? Is she comparatively short, tall, wide or thin?

Birds have distinctive postures and behaviors too. Both mourning doves and rock doves (pigeons) have strange short-legged, head-bobbing gaits that set them apart from many birds, but the mourning dove is a graceful teardrop shape with a long, sharp, pointed tail, and the pigeon is stocky with a squared-off tail.

Maybe your best friend

isn't on the playground yet. You wonder if she's around the corner of the school building. And then behind you, you hear her distinctive giggle. Some birders list birds they didn't see but only heard, while others refuse to list anything they haven't seen. Sometimes, in thick vegetation, bird song is your only clue about where to even look.

When you are bird watching, it's important to identify all the birds you recognize easily. As you sort them out, you may become aware that you see something that looks like a robin but isn't, like a spotted towhee.

On the playground you may see a new kid. Going down your mental checklist, you ask yourself, is the new kid a boy or a girl? Do they look like a 4th grader or a 6th grader? Are they playing with anybody yet? Didn't somebody say a family

finally moved into that house down the block? The quickest way to identify the new kid is to ask a teacher.

The quickest way to identify a bird is to be standing next to a good birder on a field trip. The next best way is to write down all the characteristics you notice: time of day/year, habitat, song, behavior, size, shape, color and markings. Then consult your field guide or another bird watcher.

You'll do fine at identifying birds. You already know how to recognize and name dozens of people, and bird I.D. is nearly the same. Except you can't ask birds their names when you get stuck. And they never wear nametags.

But unlike people, birds don't mind if you stare at them through binoculars.

# 34 Birding Field Guides

Thursday, June 8, 2000, Outdoors, page C4

Each of these general bird identification guides has different strengths to recommend it. Serious birders find themselves owning all of them, plus guides to particular groups of birds such as warblers or hawks, or guides to birds of particular places. Compare guides by checking them out from the public library for a few weeks.

**National Geographic**

"National Geographic Field Guide to the Birds of North America: Revised and Updated (3rd edition)," Jon L. Dunn, 1999.

A little more than pocket-sized, this guide includes

all the North American birds in one book. It has illustrations including immature and female plumage, range maps and descriptions for each species all on the same page, as well as similar species to make comparison easier. This is a favorite field guide of many serious birders.

**Stokes (Western Region)**

"Stokes Field Guide to Birds: Western Region," Donald and Lillian Stokes, 1996.

Each species gets its own page, which can make comparisons difficult.

Not all immature and juvenile plumage is shown

in the photos. But in addition to the usual description and range map is information describing song, feeding and nesting behavior, conservation status and whether it's a species that uses bird houses or bird feeders.

The Quick Index and the Color Tab Index inside both front and back covers are very useful.

**Peterson's**

"A Field Guide to the Birds: A Completely New Guide to All the Birds of Eastern and Central North America," Roger Tory Peterson, 1998.

"A Field Guide to Western

Birds: A Completely New Guide to Field Marks of All Species Found in North America West of the 100th Meridian," Roger Tory Peterson, 1998.

The original field guide concept was invented by the late Roger Tory Peterson in 1934 when he came out with his Eastern bird guide.

He used a system of arrows to point out diagnostic field markings, which birders still depend on. But the newest editions still have the range maps separate from the pictures and description. Plus, you need both the Western and Eastern editions to cover birds in Cheyenne.

**Golden Guide**

"Birds of North America: A Guide to Field Identification," Chandler S. Robbins, et al., 1983, Golden Field Guides.

An old favorite, this was the first to come out after Peterson's. It was the first to put description, range map, picture and even some song sonograms on the same page for each bird species, as well as allowing for comparisons between similar species on the same page.

Sadly, it is not up to date with the latest name changes caused by all the species lumping and splitting adjustments the American Ornithologists' Union has made the last 17 years.

**ABC Guide**

"All the Birds of North America: American Bird Conservancy's Field Guide,' Jack L. Griggs, 1997.

While most of the other guides put out their first editions years ago, this is one of the latest all-new guides.

It takes a radical approach by organizing birds by where they can be found: in the air, on the water, on the shore, on the ground, etc., rather than the usual ornithological order. It also has great introductions to groups of birds. For instance, it discusses birds of prey identification problems and behaviors and compares flight silhouettes of different hawk species. It's not as comprehensive as National Geographic.

**Audubon**

"National Audubon Society Field Guide to North American Birds: Western Region," Miklos D.F. Udvardy, 1994.

When this guide came out in 1977, some birders thought it had poor photos and suffered from the same problem as Peterson's: pictures and descriptions in different parts of the book. It has since been revised with new photos.

**Lone Pine**

"Birds of the Rocky Mountains," Chris C. Fisher, 1997.

Lone Pine Publishers is a relative newcomer on the field guide scene with its 1997 "Birds of the Rocky Mountains." The company also offers "Plants of the Rocky Mountains" and newly released guides to mammals and medicinal herbs of the region.

Because this guide focuses on birds of the Rockies,

it is particularly helpful for outings in our neck of the woods.

More than 320 species are grouped, color coded and shown on a comparative reference chart in this attractive, user-friendly guide.

**Thayer**

"Birds of North America," Peter W. Thayer, 1998, CD-ROM.

Because accessing a CD-ROM in the field still isn't convenient for most people, this technically isn't a field guide. But it includes much more information than any of the other guides, including song recordings, video, a life history for each species and a search mechanism for birds you're having trouble identifying.

# 35 Binocular Considerations

Thursday, June 8, 2000, Outdoors, page C4

--**Magnification:** Binoculars have numbers stamped on them such as "7x35." The first number indicates how many times they magnify what you're looking at. Magnification of 7, 8 or 9 can be held reasonably steady by hand. For anything over 10, you should consider getting a spotting scope and tripod.

--**Brightness:** The second number – the "35" in "7x35" – refers to the size of the objective lens in millimeters. The bigger the number, the brighter the lens and the easier it is to see your bird. For normal daylight conditions, 35 mm is just fine. You would benefit from 40 or 50 mm if you frequently look at birds in poor light conditions such

as dawn and dusk, cloudy climates or dense brush.

--**Field of view:** Somewhere on the binoculars or in the literature packed with them should be information such as "487 ft. at 1000 yds." That means, if you are looking at a bird 1000 yards away, you will be able to see more than 240 feet to each side of it. You'll be able to watch a whole flock or more easily track a bird that flies away. The lower the magnification, the greater the field of view.

--**Depth of field:** This refers to how much area in front and in back of the object you're focused on also will be in focus. The lower the magnification, the greater

the depth of field.

--**Focus:** Binoculars meant for birding have a center focus wheel or lever that focuses both barrels at the same time. However, one barrel should be adjustable independently for a one-time adjustment to your eyes' differences in ability.

--**Coated lenses:** Coatings on all the lenses increases the brightness and contrast of bird images, making them easier to see. Look for advertisements saying "fully coated" or "fully multi-coated" optics.

--**Eye relief:** The measurement from the surface of your eye to the surface of the eye piece should be about 10 mm. Less than

that, and you'll blink excessively. More than that, and you'll miss the optimal focal distance. If you'll be wearing glasses, make sure you can fold back the eyecups and maintain the optimum distance.

--**Fit:** There's nothing like trying binoculars in person. Can you adjust the distance between the barrels to match the distance between your eyes? Can you easily reach the focus wheel with your forefingers without changing your entire grip? Are the binocs a weight you won't mind hanging around your neck for hours?

--**Durability:** Consider cushioning "armor" and water resistance options to protect your investment.

--**Availability:** Local discount and sporting goods stores sell binoculars, although some of what they have to offer may not fall within the 7-9 power or 35-50 mm range preferred by birders. Birding magazines and birding internet sites are rife with binocular ads. If you decide to mail order, find out the details of trial periods and return policies in advance. Uncomfortable binoculars are as big a waste of money as shoes that don't fit.

# 36 Arrivals, departures mark seasons

Thursday, June 15, 2000, Outdoors, page C2

My grandfather, as a child, watched ships dock in his hometown harbor on the North Sea before World War I. When I was little, he still took pleasure in watching arrivals and departures – at Chicago's airports.

The spring bird migration is my version of ship watching. Before the end of March, I heard that the black-crowned night-herons were back at Holliday Park. Also, my newspaper-carrier son reported the sound of a lone killdeer calling in the pre-dawn. As soon as I could, I looked for the birds myself.

By April 1, we had a pair of mourning doves inspecting the patio for spilled seed. A red-naped sapsucker checked the backyard trees for bugs briefly April 21. A week later several white-crowned sparrows came to serenade us. Our whole family heard them because it was warm enough to open the window.

At the beginning of May we spotted the spotted to-whee twice, kicking up dead leaves in the garden. Then I realized the juncos hadn't been around for a while. Just when had they sneaked off to their summer home in the mountains?

Finally, though we'd had goldfinches at the thistle feeder on and off all winter, it was May 3 when the males came in with their brilliant yellow spring feathers. It's been such a pleasure to hear their lilting songs over the cacophony of house finches in lust.

Each spring I welcome familiar birds back, but it's also when I seem to add new birds to my life list, usually two or three on the Big Day count held in mid-May here in Cheyenne.

Considering my modest approach to seeking out new species, I am proud to say I have 286 so far, although it's taken more than 25 years.

This year I had to miss the Big Day. Instead, I was in Lakeside, Ariz., trying not to look at a Lewis' woodpecker overhead while performing the matron of honor duties at my sister's outdoor wedding. According to my records, this was a "life" bird for me.

I added another life bird a week earlier while hanging out the laundry.

A bird I should have seen during my Midwestern years, a chestnut-sided warbler, tumbled out of the tree into the bushes. It is considered way off course when seen in Cheyenne during migration. Luckily, more were seen on the Big Day, so I didn't have to document my sighting with photos or drawings.

On Memorial Day weekend our neighbors described a small bird with yellow markings that died after crashing into their window (even though the shade was pulled. None of our identification suggestions satisfied them.

Finally, Keith took us to the garbage can and pulled out a perfect, but dead magnolia warbler, another bird far from its normal Central Flyway route.

Keith and Mary Ann were very amused when Mark and I insisted on taking the bird home in a bread bag. But it will make a great specimen after taxidermy work by the folks at the university.

I was tempted to list the magnolia warbler even though it isn't ethical to list birds you've only seen dead. I got out my old, mildewed Golden guide and turned to the index where I document the first time I see a species. To my surprise, I'd already listed this warbler in 1975 in Wauwatosa, Wis.

I decided to document this latest encounter anyway, to help me recall this particular spring and to memorialize all the magnolia warblers who don't make it back to Canada to nest another year.

One morning at the end of May I noticed a hermit thrush sitting on the back wall. Thumbing through my old guide showed my first one was in 1979 on Staten Island. Now that's a note that brings back memories enough for a novel titled "Two Young Naturalists in the Big City."

However, now is not the time to write that story. You and I need to get our noses out of the newspaper, get our shoes on and go out and see who's singing.

Spring migration is about over, but the breeding season is in full swing. In a few weeks it will be quieter as parents hunker down over nestlings, unwilling to advertise their whereabouts to marauding squirrels, crows… and cats left outside to roam.

# 37 'Backyard birding' varies by the yard

Thursday, June 29, 2000, Outdoors, page C3

I believe that one's own backyard is sufficient for nature observation and enjoyment. Even the plainest of places has stories when scrutinized over the seasons.

But there is something to be said for travel. Earlier in June our family spent a few days in Lander, where the mountains loom close, and three creeks flow through town.

Our room's door in the unpromising 1950s motel facade opened to reveal a panoramic second-story view of one of the forks of the Popo Agie (Po-po'-zha) River and beyond. What a great backyard.

The treetops appeared to be swarming with grackles. One of many robins swooped straight toward us, disappearing between the roof overhang and the top of our window. By craning our necks, we could see some parental tail feathers sticking over the edge of a nest on a ledge.

The next day my son, Jeffrey, and I went to visit a friend [Marta Amundson]. Her house, sitting on a knoll overlooking Ocean Lake, is that of a poet and artist, where even the lowliest household item or task receives thoughtful consideration.

Over tea on the glassed-in porch, we watched orioles and kingbirds squabble and feed on insects, their bright feathers a welcome contrast to the pedestrian house finches at home.

Later Marta sent Jeffrey and me down the path, through the gate and over the stile to the lake itself. As we carefully tippy-toed over a white alkaline crust covering sticky, stinky mud, Jeffrey said, "You know Mom, you look like, well, you look weird, like a soccer mom or something."

Or something, all right: the epitome of the old lady wearing sneakers and binoculars. I am as old as some grandmothers, but I think it was my floppy old green hat that embarrassed Jeffrey.

Luckily for him there were no other people to see him walking with the lady in the weird hat – or to observe

him building sandcastles like a 6-year-old.

Ocean Lake fills with return irrigation flows from the Riverton Reclamation Project built in the 1920s. Much of its shoreline is managed by the Wyoming Game and Fish Department as the Ocean Lake Wildlife Habitat Area.

On a nearby spit of land I could see several habitat improvements, goose nesting platforms and even an occasional goose, plus a few avocets.

Next to us some kind of shorebird was poking its bill in the wet sand between waves. Shorebirds are difficult to identify, most having brownish backs, whitish bellies, long skinny legs and bills. I looked hard for some notable feature on these birds, finding only a wash of color on either side of their long necks.

On the 14th page of

shorebirds in my field guide, I recognized the shape of Wilson's phalarope, but mine didn't have the bright neck color.

Perhaps they were young and that's why they weren't doing the phalarope thing yet, swimming in tight circles to stir up food from the lake bottom.

A hazy, cloudy sky acted as camouflage for the terns and gulls flying far out over the lake.

Then, from behind a low hill emerged a squadron of large, light-colored birds. Their necks stretched out (not crooked like herons), their long legs following (not short like geese), the sandhill cranes advanced on me until I could see their red crowns. Then they were past, continuing on their mission to banish hunger.

All back yards are not created equal.

# 38 Ears as important as eyes in early morning bird survey

Thursday, June 29, 2000, Outdoors, page C1

LINGLE – It is not quiet before the dawn when it's a summer morning [in June] in Wyoming.

Thirty minutes before sunrise June 8, where a gravel road crosses a canal a few miles west of Lingle, Dave Felley listened intently to distinguish the songs of western kingbirds, western meadowlarks and blue grosbeaks.

It was as if members of an avian orchestra were trying to outdo each other.

Felley's job was to not only recognize the different bird songs but to count how many of each kind.

In the dim light by the side of newly sprouted corn

fields, Felley used his ears more than his eyes to count the breeding birds at the first of 50 three-minute stops he would make on this 24.5-mile Breeding Bird Survey route.

Many species of birds have been well studied and counted, especially game birds and birds that pass through migration bottlenecks, such as shorebirds and hawks, said Felley.

In 1966, ornithologist Chandler Robbins devised the BBS to monitor breeding bird population trends across North America.

"What it's best at is counting territorial adults," Felley said. "Songbirds advertise

their territory, so every singing bird is a breeding bird."

About 2,500 professional biologists and skilled amateur birders across the country drive one or more of the nearly 3,000 BBS routes once every year.

The Lingle route is one of two Felley, a biologist for the U.S. Fish and Wildlife Service's Ecological Services Field Office in Cheyenne, took on four years ago.

This year, for the first time, he joins the 25 percent of participants who conduct a survey as part of their job. Before, Felley did the surveys on his weekends.

He still has to review bird

songs on his own time. "I did that last night," he said. "I went through my CD (of bird songs) for the real tough ones."

More than 90 bird species have been identified on the Lingle route in the 20 years since it was established, according to the BBS database. About 20 of those species show up every year.

How can someone distinguish that many songs? Felley's philosophy is, "If you can remember the melody of a song, you can remember bird song."

National-level BBS coordinators at the U.S. Geologic Survey's Patuxent Wildlife

Research Center in Maryland provided Felley with field data sheets giving start and stop times, a list of birds and a map of the route.

Route 92091 must begin by 4:50 a.m. For Felley, that means leaving Cheyenne at 3:30 a.m.

If the weather is too rainy or windy by BBS standards, Felley has to bag it and try another day. The BBS asks that wind not exceed 12 mph, "except in those prairie states and provinces (like Wyoming) where winds normally exceed Beaufort 3: Leaves and small twigs in constant motion, light flag extended."

Every effort is made to keep survey conditions exactly the same from year to year so that the number of birds is the only variable. The BBS works best if the same observer works the same route from year to year.

Humans and their activities are unpredictable, however. Felley's predawn stops were interrupted by several passing pickup trucks that potentially drowned out some bird songs.

Later, trains, tractors and highway traffic echoed through the North Platte River valley. Bawling calves competed with the less-than-bucolic sounds of center pivot irrigation and screaming domestic peacocks.

From year to year there are other changes: New fields plowed, different crops planted, more houses built, road realigned.

At stop number 34, Felley carefully checked out a lone cottonwood. Last year it held a Lawrence's goldfinch, a vagrant from the Southwest.

Next year, Felley will remember stop number 41, where a northern bobwhite obligingly trotted down a roadside ditch as he pulled up and it stayed to be counted.

The BBS is not a census. Rather, it is an index using breeding birds as indicators of population shifts, declines or increases.

By going online at www.mp2-pwrc.usgs.gov/bbs, researchers and curious bird watchers can find a list of scientific articles written using BBS data, as well as the raw data itself, including data from the 80-plus routes in Wyoming.

Birders interested in running a BBS route may check with Wyoming's coordinator, Andrea Cerovski, Wyoming Game and Fish Department non-game bird biologist at (307) 332-2688 or by e-mailing her at acerov@state,wy,us.

By about the 25th stop, with the sun well up, you realize this is field work, not a bird-watching field day. But for volunteers, the compensation comes in the enjoyment of birding on an early summer morning.

Felley's last stop was east of Torrington where the land is still native prairie, too sandy and hilly to farm. On a fence post sat a grasshopper sparrow, hardly bigger than its namesake. It threw back its head with such abandon for every buzzy trill it gave. It was easy to believe birds too appreciate summer mornings.

### To Learn More

--Folks interested in the Breeding Bird Survey may go online to find scientific articles written using BBS data, as well as the raw data itself, including Wyoming results. The web address is www.mp2-pwrc.usgs.gov/bbs/.

--To find out more about volunteering to run a BBS route, contact the Wyoming program coordinator, Andrea Cerovski....

## 39  Banding volunteers aid research, get up-close contact with birds

Thursday, July 6, 2000, Outdoors, page C1

CHEYENNE – Volunteering to band songbirds supports science. It's also a chance to hold a wild creature in your hand and feel its beating heart.

The volunteers on a recent June morning shared these reasons for gathering on the banks of Deep Creek, but otherwise we were an eclectic bunch:

Donnabelle Leonhardt, retired dairy farmer; Eva Crane, keen birder; J.R. Horton, state parole and probation officer – all from the Lander/Riverton area – and me and my 11-year-old son, Jeffrey from Cheyenne. Two high school students, Joe Scott and Peter Cook, drove over from Casper.

Andrea Cerovski, Lander-based non-game bird biologist for the Wyoming Game and Fish Department, was in charge of the banding station, assisted by Laurie Van Fleet of Game and Fish.

Situated on Red Canyon Ranch, outside Lander, the Deep Creek banding station is one of nearly 500 across the country that are part of the Monitoring Avian Productivity and Survivorship program begun in 1989 by the Institute for Bird Populations.

A cooperative effort of several agencies and organizations, MAPS provides basic data about birds as well as information for land conservation and management planning.

In the predawn darkness, we divided into three teams and set off in different directions. Heading into the willows, we opened 10 mist nets, each 12 meters long, in likely bird thoroughfares.

Forty-five minutes later, we hiked back to check the nets. Irrigation boots were indispensable for getting to nets four and five, which stood in mud next to a beaver pond.

Extracting tangled birds from the fine, hairnet-like threads is difficult. I used a tiny crochet hook to lift a noose-tight filament over the head of a goldfinch.

After being netted, birds are slipped into white cotton bags and carried back to the processing table, where they are banded and information such as wing, tail, culmen (nostril to beak tip) measurements, fat deposits and brood patches are recorded.

Determining age sometimes requires "the Bible," a thick black reference book containing esoteric

36          CHEYENNE BIRD BANTER

data such as how feathers change with age.

Many of the birds of the recent operation were recaptures, most having been banded there within the last two years. One yellow warbler was banded in 1995, the first year Deep Creek was in operation.

After the first two runs, breezes began to blow away horse flies, but they also billowed the nets, making them visible to birds and making the other six runs less productive.

This was Deep Creek's second banding day of the summer, and there will be more every 10 days or so until the beginning of August.

Cerovski doesn't lack volunteers despite the 4 a.m. start time, and she does require prior banding experience or attendance at a training session at the beginning of the season to participate.

Visitors are welcome, however; that day they included home schoolers from down the canyon and third- and fourth-grade students from Fort Washakie who, along with the volunteers, got the chance to hold and release the birds.

*Cutlines for photos by Barb Gorges/correspondent:*

Andrea Cerovski, a Wyoming Game and Fish Department biologist and coordinator of the recent birdbanding operation near Lander, examines a bird that was caught in one of the "mist nets." Birds are studied, banded for later identification and then released.

A young volunteer holds a bird that was captured in a net on the recent bird banding operation along Deep Creek near Lander. Volunteers have fun and contribute to scientific research.

---

### To Find Out More

To visit or volunteer at a MAPS bird banding station, contact one of the following coordinators.

--Andrea Cerovski, Wyoming Game and Fish Department, Lander....

--Jason Bennett, Wyoming Natural Diversity Database, University of Wyoming....

# 40 Solitude found off-road

Thursday, July 13, 2000, Outdoors, page C2

We know from experience the primitive two-track we're following along a ridge top in the Medicine Bow National Forest will soon deteriorate and dead end at an unofficial scenic overlook.

To avoid unnecessary wear and tear on our vehicle, we stop half a mile short, still getting the view of the snowy Sierra Madres.

The rule of thumb for travel in our family is that drive time must be matched or exceeded by time spent "being there."

Being there this time means loading up the day packs and heading off the road. From looking at the topo map we know we only have to follow the steep little valley for about a mile to our destination, the North Gate Canyon of the North Platte.

Either by plan or accident, the sagebrush on the south-facing slope has burned, and the charcoal stumps mark our pant legs.

The antelope bitterbrush is blooming and so are all the colors of the Roy G. Biv rainbow: red/orange of a kind of penstemon and a paintbrush; yellow of wild buckwheats and varieties of DYC's (sunflower types known to some botanists as "darn yellow composites"); green in a thousand shades of leaves and buds; blue through indigo and violet in more penstemons, harebell, bluebells, larkspur and loosestrife.

In the hot pink zone is bitterroot, with its blossoms barely visible above the red gravel, and thickets of wild roses. White is represented too: yarrow, wild geranium and rosettes of evening primrose.

Signs of moose and elk are everywhere as we negotiate the hillside.

Finally, we are walking in a little alley between willows and sagebrush, just across the creek from a north-facing hillside in deep green of fir and spruce.

The creek itself is one long series of beaver ponds. Where one dam has blown out, we see beaver tracks in the silt left behind.

We catch glimpses of snakes and chipmunks, find crickets and dragonflies and watch fish flop, but the birds are easiest to notice.

Mark points out rock wrens singing from the edges of lichen-encrusted boulders. Otherwise, to me most of the bird songs are like hearing a conversation on a New York City subway that is not only indecipherable, but I can't even tell if its Latvian or Estonian.

While we debate the identity of a bird in a tree, another bursts from the grass behind us and disappears into the willows. The size, shape and color of its back reminds me of a meadowlark, but this location is unlikely.

When a second hurtles itself into the brush, Mark decides it's a young grouse, a theory soon validated by an adult blue grouse lifting off. Four more young follow, one by one.

The sky has been gloomy all afternoon, a boon to hiking the treeless side of the canyon. We want to spend more time at the river, but rain feels imminent. Five hundred feet of elevation to climb will be arduous enough without worrying about slipping on wet rocks and grass or being struck by lightning.

We intercept the road near its end and follow it back to our vehicle.

Sometimes it's relaxing not having to plan every footstep, but our boots crunch too loudly on the gravel. Roads are great. Roadless is even better. When I get too old to climb down to the river, I'll still be able to sit here on the ridge and remember what it's like to find my own way without motor noise.

I plan to write a letter asking the U.S. Forest Service to protect its roadless areas from road construction (as

---

2000

proposed in its Roadless Conservation Initiative Draft Environmental Statement), as well as from harmful commercial and recreational activities.

Official roadless areas have been left alone this long. We can continue doing without their meager possible contributions to industry and protect their enormous contributions to healthy wildlife and low-impact recreation.

You too can mail written comments by July 17 to: USDA Forest Service-CAET, attn: Roadless, ....

## 41 Summer means crowds for both birds and people
Thursday, July 27, 2000, Outdoors, page C2

On a recent morning [July 4th weekend] my family drove out to North Crow Reservoir to our favorite access point. We found an encampment at the side of the parking lot numbering half a dozen tents and including girls in pink pajamas, men in lawn chairs and loose dogs lounging.

There are no signs banning camping at North Crow, but there are also no amenities. I know that if I go back now, a few weeks later, I'll find little toilet paper flags behind every bush and tree within 200 yards of the roads, warning me to watch where I step.

A whole book has been written about dealing with toilet issues in the outdoors, including digging holes for deposits, but in a heavily used area that is not a conscientious option.

My preference is to put everything under a dry cow pie or rock for natural decomposition with less unsightliness. I have a friend who takes her used t.p. home in a plastic bag.

Some of the vehicles parked by the encampment may have belonged to people fishing. We're used to seeing up to half a dozen during ice fishing season. But, by nature of their skittish quarry, anglers are a quiet bunch compared to campers.

Passing other smaller encampments, we opted for a parking area with only one vehicle and no loose dogs. The men in my family headed out with fishing rods, and the dog and I, rather than hike around a lake wreathed in tents, opted for higher country and hopped the fence into the national forest.

Summer is a crowded season for birds too. I wonder what the resident species think when the migrants move in for the summer, and everyone has noisy children begging for food.

Last year's young, like the pelican I saw on the lake – the teenagers, so to speak - don't fit in and head for the fringes, the marginal habitat.

That's the story of the American white pelican around here.

The huge, white birds with their famous fish-catching bills may show up in any Wyoming reservoir.

Some have been spotted on F.E. Warren Air Force Base reservoirs this summer, but the breeding records show only a few specific lakes at Yellowstone, near Casper and on the Laramie Plains actually have breeding pelicans.

There's a chance, though, that the young may begin to feel at home in one of the marginal areas. Perhaps the fishing improved, and they'll be the ones to settle new territory.

As long as I brought my own food and water (and toilet paper), I too could explore new territory. Although, with the number of cow pies I saw, it was evident this was merely a pasture some cowboy probably already knows like the back of his hand.

The dog and I hiked up hill after hill to see what was on the other side, and we enjoyed landscape unmarred by empty beverage containers.

The idea of rattlesnakes crossed my mind briefly.

My ancestors, like other immigrants [Wisconsin, 1840s and 50s], must have been considered reckless adventurers by their folks back in the old country. The way Americans crave sport utility vehicles, I'm willing to say this quest for the rugged frontier must be genetic.

Looking back from the top of the last hill I wanted to climb, the pink-pajama encampment was only a dime-sized spot on the edge of the water.

It was getting hotter. Grasshoppers bounced off my shirt. It was time to go back to civilization for a little shade and let the dog get his feet wet.

Down by the creek, the trees were thick with birds and bugs. Before my fishermen were ready to leave, six more vehicles showed up at our parking area, all the occupants heading for the water.

Crowding's OK when the fishing's good.

It was ironic that two evenings later, as I stood in my own city neighborhood, a herd of pelicans passed directly over my head. They've been known to stop over at Holliday Park.

I suppose in the pelican world that would be considered a reckless adventure.

## 42 For fun, try cooking outdoors sans pans
Thursday, August 17, 2000, Outdoors, page C1

### Using natural "utensils" can be an entertaining project for youngsters

CHEYENNE - Mike Randall, Scoutmaster of Troop 102, said the boys would be doing some utensil-less cooking at summer camp, maybe including hamburgers Philmont-style: "Add Cornflakes to the hamburger, slap it on a standing rock by the fire and, when it falls off, it's done."

When my younger Boy

Scout got home, having no slabs of rock available in the back yard, he demonstrated other utensil-less cooking methods for me. If you have a few spare coals after a grilling occasion, you may want to try them too.

He used charcoal briquettes rather than a wood fire because they are easy to start, easy to move and burn uniformly.

At camp the briquettes were started in the campfire. At home he used a charcoal starter chimney. It's like an old-fashioned coffeepot. He poured the charcoal in the upper chamber. In the lower, vented chamber he stuffed a few crumpled sheets of the Wyoming Tribune-Eagle and lit them with a match.

We carefully extracted white, hot coals with long-handled utensils (We could have used sticks, I suppose.), and placed them where we wanted to cook, on bare ground where people wouldn't be walking.

When we were finished, we threw the coals in the grill. At camp they were thrown back in the fire pit.

Here's how to cook if you happen to be outdoors without a pot or pan, but with a few eggs, oranges, onions, a pound of ground hamburger, instant coffee and a couple paper cups.

**--Breakfast:** A chilly morning outdoor breakfast should start with a hot drink. Fill a plain paper cup (unwaxed or uncoated) with water. Pull out three hot charcoal briquettes and push them together, forming a tripod on which to place your cup. Only the bottom rim of the cup will burn. Soon the water will be hot enough for instant coffee, tea or hot chocolate.

For orange juice, cut an orange in half with your trusty pocketknife, either lengthwise or crosswise, and scoop out and eat the insides.

Crack an egg into each of your two small orange bowls, or perhaps use one for sausage. Lay each bowl on three coals. Be sure to turn the eggs and sausage so they cook evenly. If you are truly utensil-less except for your knife, you'll have to make a flipper by cutting and peeling a twig from a non-poisonous source like willow.

If you happen to have along the wooden skewers sold at the grocery stores, you can skewer a raw egg end to end and then fashion forked sticks into a rotisserie to hold the egg over a few coals. My son set his egg baking over coals still in the charcoal starter chimney.

Randall is a proponent of the twizzle-stick method of biscuit making: "Cut open the side of a box of Bisquick (the way individual serving-sized boxes of cold cereal are laid on their sides and cut open to turn them into bowls), stick your finger in and make a hole in the flour. Fill the finger hole with water."

Next, stick the peeled end of a thumb-thick twig in the hole and "twizzle" or roll the other end back and forth between your palms. Dough will collect on the end of the stick. You can then prop the dough end over the coals by resting the stick in a fork of another stick stuck in the ground.

**--Dinner:** Put a cup of water on to heat, this time for mixing your instant soup course. The main entree is meat loaf: ground hamburger mixed with any seasonings at hand, including some of the insides you will be scooping out of the onion you'll cook the meat in.

You can cook meatloaf meatballs in oranges like you did the eggs, but my Scout didn't care for the flavor.

Remember when you're adding seasoning that sagebrush is only edible if you're an antelope. It isn't the sage used in cooking.

Prepare the onion like a jack-o'lantern. Cut the top off and scoop out the insides, leaving two layers intact.

To keep your eyes clear while working on onions, try to breathing through your nose and keep your mouth shut. First, you may want to explain to your fellow cooks why you won't be talking to them for a few minutes.

Fill the onion with hamburger and skewer the lid on with wood skewers from the store, splinters from chopped firewood or whittled twigs. Set the onion on three coals and turn it often so all sides, including the top, heat. The outside layer may burn, but that's all right. It reminds you which sides of the meat loaf are already cooked.

Bigger onions are not better. No matter how often we turned the big 4-inch diameter yellow onion, the meat wouldn't cook in the middle. We finally threw it in the microwave in a fit of hunger. (OK, we could have tried burying it in coals.) The 3-inch diameter onions cooked better.

Ready for dessert? After experiencing all this exotic cooking, your kids may be ready for you to just hand over the marshmallows.

*Cutlines for photos by Kevin Poch:*

Scouts Eric Keto, left, and Jeffrey Gorges, both 11 years old and from Cheyenne, stuff ground beef into hollowed-out onions to make meatloaf meatballs without using pots and pans.

Jeffrey Gorges, 11, prepares an onion for his utensil-less dinner recently. The onion serves as both food and as a cooking pot.

# 43 Hummingbirds making their colorful migrations

Thursday, August 17, 2000, Outdoors, page C2

## The broad-tailed and the rufous are the most common hummingbirds in the Cheyenne area.

Go anywhere in Wyoming's mountains in summer and you're bound to hear the ringing sound of the broad-tailed hummingbird's wings as it skims your hat or nearby flowers. Every mountain cabin seems to be festooned in hummingbird feeders.

A few broad-tailed hummingbirds may be seen in Cheyenne during spring migration, but a hummingbird feeder set out here for the summer is more likely to be emptied by the wind, yellow jackets and house finches.

However, around about August, especially when drought diminishes wildflowers blooming in the mountains as it has this year, we may see hummingbirds hanging out in town around our bountiful, irrigated gardens.

A few years ago, during another dry summer, I saw something hovering over my marigolds which I thought at first was an insect. Now that we're better at spotting them, this August we've had a couple more glimpses of hummingbirds, as have several callers. Maybe it's time to put up our feeder.

To approximate natural nectar for hummers, dissolve one part white table sugar to four parts boiling water. Don't use other kinds of sugar or increase the proportion. Adding red dye is unnecessary. The red on the feeders is enough. If yours has faded, paint the ports with red nail polish.

Be sure to keep your feeders clean. Bill Thompson, author of "Bird Watching for Dummies" recommends cleaning the feeder with hot soapy water before refilling it. During particularly hot weather it may take only a couple days for the solution to sprout mold.

His recommendations of where to put the feeders include where they can be seen easily by you and the hummers, where they are out of direct sunlight, which introduces mold growth, and where they are easy to reach for cleaning and refilling.

There are two species of hummingbirds you may see in Cheyenne. The most likely is the broad-tailed, which is known to summer here but usually breeds in the mountains.

The rufous hummingbird is less common here. It winters in Mexico and migrates north along the Pacific coast as far as Alaska to nest. By June it is heading back southeast through the Rocky Mountains.

According to Jane and Robert Dorn's records in "Wyoming Birds," the rufous has been merely a migrant in the latilong that includes Cheyenne.

There are other hummers seen in other parts of Wyoming: ruby-throated (northeastern counties, rare migrant but the only hummingbird occurring east of the Mississippi), black-chinned (western Wyoming during migration) and calliope (uncommon summer resident in western Wyoming). We are hardly a hummingbird mecca like southeastern Arizona.

Meanwhile, I plant bright tubular types of flowers for the year the broad-taileds decide to spend the summer: bee balm, columbine, four o'clocks, penstemons, petunias, phlox and snapdragons.

I'll have to think about adding bleeding heart, dahlias, nasturtium, zinnias or vines like morning glory. Other plants recommended in books about hummingbirds are less familiar and may not thrive here. Then again, they didn't mention marigolds.

# 44 Outings provide close encounters of the bird kind

Thursday, August 31, 2000, Outdoors, page C2

The day Troop 102 unloaded gear and groceries in front of the lodge at Chimney Park in the Medicine Bow National Forest, our collective subconscious registered the constant "bink, bink, bink" noise over by the flagpoles.

After the boys left to pitch their tents, Mark and I discovered the binking came from a pair of agitated white-crowned sparrows.

Their young were packed into a nest snuggled into the sagebrush.

With most of the daily camp activities away from the lodge, the birds finally settled down and resumed scavenging for food.

With most people away from the lodge, at one point I was alone on the steps taking a breather from merit badge counseling.

One of the white-crowneds landed about 30 feet away and began searching for morsels of bugs for its young. With each of its two-footed hops towards me, I felt more and more invisible.

The fearless bird disappeared under the steps briefly, then moved away, completely unconcerned with quiet human proximity.

The next close encounter was another mystifying example of birds choosing to tolerate humans.

Our family vacations are formed around attending professional functions and visits to friends and family. Recently, the Western Division of the American Fisheries Society met in Telluride, Colorado.

Mark had to take the gondola over a mountain to the conference every morning. The boys and I went to visit my old college friends, Cindy and Mike, near Dolores.

Cindy introduced us to her runner ducks, who as their name implies, kept their distance. The Rhode Island reds were too busy grazing to take much interest in us, and the other chickens, a commercial meat variety, were lying around waiting for the next hand out.

But the six young turkeys formed a gaggle around us (or whatever turkeys form), pecking our toes and taking handouts of grass.

These were an heirloom variety, Bourbon, with feathers of mottled brown and white. The toms were just old enough to begin practicing tail fanning.

Jeffrey decided to spend the afternoon training them to jump through a hoop made of garden hose by throwing little balls made of grass and berries. Mostly the turkeys learned to follow him around the yard, even without food.

Who says bonding only takes place immediately after hatching?

Cindy was a tiny bit jealous, but not having that kind of rapport with her livestock will make culling the flock easier.

We returned to Telluride via the scenic route through Silverton. After experiencing the glut of tourist buses at Molas Pass, we turned off on a dirt road to a Forest Service campground for a walk by Little Molas Lake.

Two female mallards, each with a half-grown duckling following, came out of the rushes, swimming in our direction.

We slowed our pace, watching as they scrambled onto shore. Older son Bryan and I had lagged behind, so it was Jeffrey who, standing stock still, was "sniffed" by the ducks.

It wasn't exactly like a canine greeting, nor was it a quest for food like the geese and ducks in Cheyenne parks display. It was the kind of secure indifference grazing animals show each other as they move about within the herd.

Telluride is no longer the ramshackle town I first visited over 20 years ago. The Victorian-era houses have been refurbished, and nearly all the empty lots have been turned into condos that look like old multi-story mine buildings, complete with rusty corrugated metal siding.

In the evening one can catch glimpses of opulent lifestyles through open windows, but what I found to be truly a sign of wealth was open to everyone: the San Miguel River trail.

The wide gravel path follows the edge of the river for 2 ½ miles through bird-filled willows, skirting back yards and restaurant patios.

Early our last morning we hiked 2 miles up the side canyon to Bear Creek Falls, still impressive in mid-summer.

It's a popular destination – we met a whole parade of people on our way back – so two things amazed me. First, there was no human detritus anywhere, not even a gum wrapper. And the quintessential element of mountain water, in Wyoming or anywhere in the West, was there: The dipper, or ouzel, flew in and out of the water's spray.

# 45 Birds congregating, fattening up to prepare for winter

Thursday, September 14, 2000, Outdoors, page C1

"Birds of a feather flock together," especially this time of year.

Most everyone is finished with quietly raising the young out of earshot of predators and competitors, and the birds are ready to socialize again.

Maybe your yard has been invaded by partiers in black: starlings, grackles and red-winged blackbirds.

Sometimes their ruckus makes me think it's dawn in the heart of the Okefenokee Swamp, like the eco-sounds record I had in the '70s.

Local bird expert Jane Dorn said the birds are fattening up for migration; or for winter if they're year round residents. Jane has noticed that, out in the country, lark sparrows are bunching up as well.

One persistent sound in my yard for the last month has been the red-breasted nuthatch. I haven't seen it yet, but the "ent-ent-ent" is unmistakable.

It's hard to spell bird sounds, so try a recording. Maybe you have one in your yard too and don't know it.

I was worried the nuthatch wouldn't be back after a major pruning job on our two big green ash trees, but it's still around, to be heard at least.

The shaggy branches of the silver maples out front seem possessed as migrating warblers wiggle a leaf here and then a leaf there, looking for scrumptious bugs. I haven't gotten a good look at the yellowish birds yet.

Fred Lebsack spotted Wilson's warblers at the Botanic Gardens. Jane mentioned Townsend's.

I had a call from a reader a couple weeks ago about a bird she thought might be a varied thrush. The range map in my field guide showed it as a species of the Pacific Northwest, no closer to us than western Montana.

Later I mentioned the bird to Jane and she said, "Oh yes, sometimes we get them."

Remember, birds can't read maps. According to Kenn Kaufman in "Lives of North American Birds," varied thrushes have strayed as far as New England.

We took our feeders down in early August when some of the finches looked sick. It's good to break the disease cycle, especially in a season when natural food is available.

Finches and sparrows haven't been seen here since, so when a finch-sized bird perched on the wire the other day, I took a careful look. It was a flycatcher, but which one?

Fred recommended I compare the western wood-pewee, part of the "tyrant" flycatcher group, with the empidonax flycatchers illustrated in the National Geographic field guide. I did.

We're talking about your average little gray birds here, ones that may be molting into even less remarkable plumage. Or they may be the young of the year, wearing some kind of juvenile plumage. And some of these flycatchers, even in peak breeding plumage, can only be told apart by song.

Well, it is not spring – "attract-a-mate-defend-a-territory" – time, so they aren't singing.

Jane mentioned that the reservoirs are full of migrating water birds and shorebirds also wearing confusing feathers. You could be an expert using a high-powered scope and still not be sure of their identification this time of year.

Some, like the Wilson's phalaropes I saw up at Chimney Park at the end of August, at least have an identifiable shape and behavior, sort of like long-necked gourds instead of plastic ducks swirling around a tub of water in some carnival game.

Instead of agonizing over which species of flycatcher I have in my yard, I think

2000

I'll just enjoy observing the bird as it sits so upright and attentive, flicking its head right, then left, up, then down, then flying out to intercept an insect before resuming its perch.

On a recent evening it was so still I think I heard its beak snapping on prey on one of those forays.

Jane said to watch for orioles and western tanagers migrating too. My neighbor recently described seeing in his yard what could have been a tanager: a red/orange-faced bird that was otherwise yellowish, rather than the red-faced but brownish-colored house finch.

Keep your bird water supply full and be sure to call me when you get interesting avian visitors.

# 46 Wyoming hosts National Audubon Society board

Thursday, September 28, 2000, Outdoors, page C2

## The Casper gathering featured field trips to view abundant wildlife

The 35 members of the National Audubon Society's board of directors deal with the same concerns as other organizations large and small: Membership and finances.

As conservation groups, both National Audubon and our local chapter discuss science, advocacy and environmental education.

But here's where Audubon board members differ from those of other groups: After everything is said and done, they go bird watching.

Actually, at the National Audubon board meeting held in Casper two weeks ago, the bird watching came first, about 6 a.m.

The hotel was only a five-minute walk from the North Platte River Parkway. Thick with foliage making it hard to see the river, the parkway was stuffed with migrating warblers.

Like other large nonprofits, National Audubon hires professional staff. But it's probably the only one that expects its employees to acquire binoculars when they start working – if they don't own them already.

It doesn't hurt to have a passport either. The board meets four times a year in locations of concern to bird conservation – even out of the country, as in a recent meeting in Venezuela. A growing number of Audubon chapters are in Central and South America, where some of "our" birds spend the winter.

While Casper probably wasn't very exotic for board members from the West compared to, say, Alaska or the Everglades, it attracted this board meeting for several reasons.

One was the destination for the second morning's field trip, Soda Lake. Just north of town, the reservoir was created by a now defunct refinery and is being studied for the best way to deal with its toxic waste accumulations and still benefit the abundant bird life attracted to it.

The board members, the staff and the rest of us rode out to the lake by bus.

The antelope (all right, pronghorn for you taxonomists) did a fair impression of a filmmaker's Serengeti, with numerous bands crisscrossing the dirt track in front of us.

We had the obligatory predator too: A red fox dashed away along the alkali-covered shore, eliminating our chances to see godwits and other shorebirds.

The national director of marketing had her nose pressed to the window until we stopped and got out to set up the scopes to focus on ruddy ducks, coots, redheads and dowitchers.

It seemed like a wildlife movie with that golden quality of light particular to chilly September mornings. As the sun rises red, perhaps because of dust or smoke in the air, it infuses all the dry, tan grass with colors before fading them out again with increasing brightness.

And for a little audio color on the way back, one of our Wyoming Audubon members who ranches provided commentary on grazing and the strategies for suppressing cheatgrass.

Another attraction for the national board was the Audubon Center at Garden Creek, the centerpiece of cooperation among entities in Natrona County interested in nature, outdoor conservation and environmental education.

At the Western-style barbecue at the center, complete with parking lot attendants directing traffic from horseback, I stood in line with a board member from New York City who wanted to know if he might have seen a white-crowned sparrow.

At this time of year, the white stripes on its head are rather gray. Someone from New York usually sees these birds only in the winter, when their markings are most distinct. Here in the Rockies, the white-crowneds are with us all year, although they move up to the mountains in the summer.

Let's hope that our Audubon guests took away memories of hospitable natives, abundant wildlife, wide-open spaces – as well as plans to visit Wyoming again.

In the meantime, these board members and staffers, in their daily professional lives and especially in their roles as environmental stewards and advocates, may say of an issue, "I wonder what this will mean to those people I met and the wildlife I saw in Wyoming."

# 47 Land of open spaces?

Thursday, October 5, 2000, Outdoors, page C2

**Even Wyoming faces the loss of open space to development, especially in fertile river valleys.**

LARAMIE – Open space is on the agenda in communities across the country as urban and suburban sprawl continues to consume agricultural land.

In Wyoming, however, we still have "Wide Open Spaces," which was the title of a day-long forum held at the University of Wyoming last week.

The conference was sponsored by the Stroock Forum on Wyoming Lands and People and by the university's Institute for Environment and Natural Resources.

The forum presenters and audience members included developers, conservationists, ranchers, farmers and professionals – all interested in land-use planning issues.

What is open space?

Presenter Jay Fetcher, founder of the Colorado Cattleman's Agricultural Land Trust, said he once heard a woman from New York City say it was an apartment with two bedrooms instead of one.

To folks in Wyoming, Governor Jim Geringer said in his opening remarks, "It might be standing on a hill and not seeing any sign of man."

In the forum, discussion was limited to "conserving working landscapes and wildlife habitat in Wyoming and the West," specifically farm and ranchland.

Agricultural producers in the Rocky Mountain region have not been able to keep up with the economy. They subdivide a bit of their ranch or farm and sell it to raise money for their operation, said Ben Alexander of the Sonoran Institute.

And the buyers are affluent people searching for better living conditions in a natural environment with recreational amenities.

Nationwide, said Alexander, 16 million acres were converted from agricultural to residential use between 1992 and 1997.

"We are educated to value open space," said Frieda Knobloch, assistant professor of American Studies at UW. Ever since our country's birth, she said, citizens aspired to live in open spaces and flocked to the frontier.

Converting ag land to rural residential use, said Alexander, is a drain on county finances. Studies show the cost of providing services, such as water, sewage, roads and fire protection exceeds taxes generated by development.

In addition, the land that gets converted tends to be along the valleys – Wyoming's most productive agricultural land.

While the rural human community undergoes a cultural shift away from farming and ranching with the influx of new homeowners, the wildlife community also changes.

Stan Anderson, UW professor of zoology, said fragmentation of landscapes into smaller fenced parcels can interfere with big-game migration routes.

Changes in land use adversely affect the habitat requirements of native wildlife such as sage grouse and songbirds, resulting in the invasion of less desirable species.

Rancher Jim Cole, owner of the Deerwood Ranch outside Centennial, became a developer by default when he decided to sell a few 40-acre lots to subsidize his ranching operation.

But he wrote the subdivision covenants so that he could continue to graze all but the actual 2-acre home sites. In return, homeowners may wander the rest of the ranch.

Jack Turnell, vice president of the Wyoming Stock Growers Association, said the association is looking into establishing a land trust like the Colorado Cattleman's trust.

A rancher could sell his development rights to the nonprofit trust for cash, or he could donate the value of the development rights for a tax write-off.

In return, the present owner holds himself and all future owners to development and management restrictions he and the trust agree upon.

Conservation easements are a similar tool. Local governments can also administer trusts and easements.

Ranchers and farmers would benefit from changes in government policies, said another forum presenter, Saratoga rancher Jim Berger.

He urged repeal of the estate or "death" tax, which frequently forces surviving family members to liquidate all or part of a farm or ranch operation in order to pay the tax. That land often is converted to other uses.

"Ranching is the most irrational act," said Bob Budd, manager of The Nature Conservancy's Red Canyon Ranch near Lander. "People who do it, do it because they love the land."

But, he added, "Where is the market for flycatchers? Where is the market for sage grouse? Where is the market for wild things?"

Budd is a proponent of holistic resource management, an integrated way of making decisions. He proposed that ranchers look at innovative opportunities like grass banking, not only to keep their operations viable but to keep improving them.

Some of the solutions to open space concerns can come from the state legislature, such as tightening lax subdivision laws, said Jean Hocker, founder of the Jackson Hole Land Trust. But saving open spaces "is going to be done with incentives" that are attractive financially and emotionally, she said.

Wyomingite John Turner, president and CEO of the Conservation Fund and former director of the U.S. Fish and Wildlife Service under the Bush administration, said

2000

that although funding land trusts can be difficult, voters around the country have passed hundreds of initiatives supporting open space.

As UW's Institute for Environment and Natural Resources turns its attention to open space issues, Turner said he is certain it will find collaborative, Wyoming-style ways to uphold the Cowboy State's identity as the nation's "open space."

Alan Simpson, a former U.S. Senator from Wyoming and veteran of land-use planning battles, was the forum's luncheon speaker.

He said any effort to shape the look of Wyoming's future will succeed only with "good faith, good science, good sense and good will."

### Want to know more?

For more information on open space issues, contact the University of Wyoming's Institute for the Environment and Natural Resources....

## 48 Birds make windy hunt bearable
Thursday, October 12, 2000, Outdoors, page C2

I made a date with the antelope in Shirley Basin seven months in advance.

When it snowed the week before instead of the week of the hunt, I considered myself luckier than the little warblers caught in the first storm and at least as lucky as the green-tailed towhee that used our patio for a refuge.

By the time Mark and I reached the basin a week later, the snow had melted off to a few streaks secreted in the ripples of the rim, like Styrofoam trash locked in a hedge.

But the wind was relentless, scouring the earth of all loose material.

Most of the vegetation up there is already barely more than boot-heel high.

We hiked for two hours. At the top of the rim, I could raise my foot to take a step and have it pushed sideways by the wind with the strength and noise of a jet-engine's blast. Think blizzard without snow.

Think I'll ask Santa for an anemometer so I can measure the wind speed at which my breath is sucked away.

Such a desolate landscape still has wildlife. A flutter of horned larks greeted us as we turned off the highway. That one and every flock I saw later flew up and then blew away east with the wind.

How exactly do small birds travel against the wind? Do they wait for a calm day and fly back west? Are their bodies stacked up against some fence in Iowa? Is there an undocumented east-west migration phenomenon?

Maybe they hop back along the ground since wind speed decreases the closer you get to ground level. I saw several beetles casually crawling between wind-dwarfed shrubs, and even a few asters blooming within inches of the ground.

You notice these things when looking up can cause your sunglasses to be ripped from your face.

Eagles seem to be strong enough to enjoy the wind.

We saw three immature golden eagles bouncing through the air like kids on carnival rides.

Finally, we decided to seek protection down on the North Platte and arrived at dusk. As I checked the wind-worthiness of a campground cottonwood, I realized I was looking up at the bottom end of an owl. It left, soundlessly, but the rest of the evening was punctuated by weird squawks from an animal in the thicket.

In the morning we discovered the river bottom was decked out in red, orange, gold and green. Accents were provided by red wing linings of a flicker, blue vest of a kingfisher, green heads of male mallards and the rich rust brown feathers of a female northern harrier flying reconnaissance along the riverbank.

Back at our hunting area, herds of antelope blew about in the wind, just out of reach. Unlike birds, you know when antelope are watching you.

They stop, turn their heads toward you, if not their whole bodies, and stare.

I've often wondered why flies, with the ability to cling to the ceiling, get nailed by cats. So too, I wonder how someone on foot in a bare, endless landscape can bring down an animal which can run more than 60 miles per hour and can see a greater distance than a bullet can accurately travel.

Mark was able to compensate for the wind and bring home one antelope. I, with lesser experience, decided not to try.

As I stood on a hill and looked 360 degrees around me, I saw true open space: All the way to the horizon hardly a sign of man beyond the scratch of dirt two-track in the distance.

Selfishly I hoped there won't be too many people coming here in the future who are interested in battling the wind for a look at the play of light on antelope heaven.

## 49 Birds needn't fear this wolf pack
Thursday, October 26, 2000, Outdoors, page C2

A flock in blue and gold juvenile plumage flitted along the sidewalk toward me, darting here and there, smiles sparkling in the sunshine.

It was predictable behavior for a den of second-grade Wolf Cub Scouts just released from school at Lions Park on a warm, crystalline October day.

We had a mission, though: to find at least five kinds of birds.

Luckily, big birds showed up. I had my spotting scope set on two western grebes on the far side of Sloan's Lake and let the boys look one at a time as they arrived.

44          CHEYENNE BIRD BANTER

It isn't unusual to see grebes on the lake about June, floating their chicks around on their backs. But October is considered the tail end of their vacation "up north," as they say where I come from.

It's always fun to point out that grebes are not ducks. Many kinds of birds swim.

The second bird interrupted my mini introduction to bird watching. A black-billed magpie swooped by, raucously reminding me it was time to set the young wolves on the trail.

First stop: Mallards.

They were attracted by all our racket because, in their minds, they've equated humans with handouts. So, they paraded by, green-headed drakes and plain brown hens.

Three down, two to go.

I scanned the ducks, looking for the glowing white beaks and black bodies of their usual companions, the swimming chickens.

There was only one American coot to be found. Must be the one foolhardy one who thinks he'll stay the winter and get first dibs on the best nest site next spring.

We all hurried down the shoreline to get a better look. The coot continued swimming, so the boys had to take it on faith that coots do not have webbed feet like ducks, just lobes on their toes.

As we circled away from the lake, I strained my ears for sounds of birds in trees and was overwhelmed by traffic noise: pedestrian, auto and air. But then a crow did me the favor of proclaiming from the tip-top of a pine tree.

Whew. Five in the bag.

Little birds seemed absent as we infiltrated the Cheyenne Botanic Gardens. Not a house sparrow in sight.

Our own yard's been full since the September snowstorm, and people have been reporting more warblers also. A flock of six or eight pine siskins have been mobbing our thistle feeder early every morning. Some winters we never see them.

A family-like flock of white-crowned sparrows spent the first two weeks of this month with us – two adults and a passel of juveniles, their crowns only dark brown with gray stripes.

A black-headed grosbeak visited our shelf feeder two evenings in a row.

Previously I've only seen them in the yard during spring migration.

Dark-eyed juncos came with the snowstorm, and we're back to our seasonal struggle to differentiate them by race.

The red-breasted nuthatches are boldly flying off with sunflower seeds and hammering against our trees like little vandals.

The neighborhood is besieged by a gang of northern flickers, their calls combining with the occasional blue jay's.

And we've had the privilege of hosting a couple mountain chickadees, although it's easier to hear them than to spot them where they hang out in the neighbor's still leafy crabapples.

The small birds flittering in the crabapples at the Botanic Gardens at the end of our walk were juncos, I think. It was hard to tell.

They were moving as fast and unpredictably as the boys zigzagging back to their family conveyances.

When I finally arrived at my own, I glanced over at the lake one more time.

There, where the mallards had been begging, was one of the western grebes.

I sincerely hope this elegant, black and white, long-necked fish eater isn't turning to breadcrumb handouts for sustenance. It is hard to understand anyway why it chooses to raise its young at Cheyenne's busiest park.

The boys probably won't appreciate having seen grebes for a while. Not until they've seen several decades of starlings and sparrows.

# 50 A visit to Lucy's garden creek

Thursday, November 9, 2000, Outdoors, page C2

I met Lucy nearly 20 years ago when I lived in Casper and worked as an environmental technician, before I cared very much about bird watching.

When I got married and left for a Montana sojourn, we kept in touch by Christmas cards.

On my first spring bird count in Cheyenne 10 years ago, I was surprised to find Lucy among the contingent from Casper. It's a Murie Audubon tradition to push spring a little by coming south to bird with us on our Big Count, because spring migration peaks here about a week earlier.

Since then, I've also had chances to visit Lucy in Casper whenever the state Audubon board meets there. Two weeks ago, we even had time to look for birds together.

For a person who speaks of recently attending her 60th high school reunion, Lucy is remarkably fit and agile, I thought as I puffed a bit following her up to take a look at Garden Creek Falls.

This was where she hung out with her high school friends. It's still popular with the partying set, which is why, she said, she doesn't come up alone.

From the shoulder of Casper Mountain, we had a view of the city and a distant magpie.

But it's a downstream stretch of Garden Creek that Lucy regards as her backyard. Running below her house, the creek is a deeply cut, winding channel stuffed with bushes – just perfect for a second ramble this particular afternoon and just perfect for black-capped chickadees, juncos and pine siskins.

We saw occasional flashes of blue jay and red wing linings of flickers as we hiked down one side of the creek through Nancy English Park and back up the other side.

For 20 years Lucy and two friends have surveyed the creek above and below her house four times a year. The reports are faithfully sent to the Wyoming Game and

Fish Department's non-game bird biologist.

The reports I looked at are a who's who of birds found in Wyoming's riparian areas, although this one is surrounded by a 40-year-old suburban forest planted by adjacent homeowners.

Only one backyard we passed maintained the original flora. What courage these homeowners have in a sidewalks and gutter kind of neighborhood. Imagine the back of a substantial house: sliding glass doors leading to a large patio abutting a well-tended lawn which fades into dry native grasses and rabbitbrush--still within the city limits.

Walking the creek, with houses backing onto it, is a little like walking an alley with a severe erosion problem down the middle.

As we rounded a bend I caught a glimpse of a small log structure, the kind where the mortar makes stripes almost as wide as the logs. I thought it might be some forgotten ranch building now at the edge of someone's yard.

No, said Lucy. Her brother-in-law built it in his back yard for his kids and their cousins, Lucy's children.

I could have sworn the cabin was a contemporary of buildings at South Pass City, except it wasn't sagging. What a great Thoreau house it would make, to live there simply, without stuff, and observe the creek, pen in hand, for a year or so.

Lucy's Garden Creek reports do not mention the species that eats some of the sunflower seed she puts out.

While we enjoyed a cup of tea at the kitchen table after our walks, a doe mule deer nuzzled the few seeds left at the shelf feeder at the windowsill.

The doe's twin sons, nubbins of antlers barely apparent, were still too short to reach. The doe kept her huge, brown eyes on us as she polished the shelf.

The purveyors of bird feeding equipment have all kinds of anti-squirrel devices. Keeping deer out means hanging them higher – and then figuring out how to reach up and fill them.

Biologists talk about corridors wildlife can use to safely travel from one necessary habitat to another, perhaps from wintering grounds to summer range.

Along Garden Creek the wildlife of the mountain and foothills have a way to travel into the city, although we have to be careful when they cross the streets.

But I think this greenway works in reverse for people, taking our imaginations back to the mountain when circumstances only permit a creekside ramble.

# 51 Bird watchers have opportunity to aid research

Thursday, November 9, 2000, Outdoors, page C2

CHEYENNE - Do you find yourself pausing to gaze out at the birds at your feeder? Do you find yourself jotting notes about what you see? Would you like to know what other people are seeing at their feeders?

Then Project Feeder-Watch is for you.

About 14,000 backyard bird watchers from all over the United States and Canada participated in the project during the 1999-2000 season.

This year, the season begins Saturday, although feeder watchers can join at any date.

Project FeederWatch is a scientific study in which ornithologists use the data collected to track the ebb and flow of bird populations across the continent during the winter.

Because it is scientific, participants are asked to choose an observation schedule. You may choose two consecutive days every two weeks if darkening circles on a computer data form to be turned in at the end of the season in April; or you may choose two days as often as each week if reporting via the internet.

This doesn't mean you have to be glued to your window every daylight hour during your observation days. Rather, you record how much time you were able to observe, along with recording weather conditions, ground conditions and the number of birds you counted.

If you have to miss some observation days, that's all right.

Not sure about identifying your birds? Consult the full-color poster of feeder birds given to each new observer. One side features eastern birds, the other side, western birds.

Here in Cheyenne, we're at the interface of east and west, and we make use of both sides.

The project's website has more in-depth information on each feeder species.

Project FeederWatch participants also may take part in the House Finch Disease Survey, which has been expanded to a year-round study.

Feeder watchers' data have been crucial in documenting the spread of disease decimating house finch populations on the East Coast, and new data will follow its transmission pattern in the West.

Project FeederWatch began in the 1970s in Ontario, Canada, at the Long Point Bird Observatory. After 10 years, researchers realized collecting information continent-wide would be more useful, as some bird species travel great distances in the winter looking for food. So Project Feeder-Watch was born.

The project is now a joint research and education project sponsored by the Cornell Laboratory of Ornithology, the National Audubon Society, Bird Studies Canada and the Canadian Nature Federation.

The new Project Feeder-Watch website is slated to debut Saturday. You don't have to be a registered observer to check out data from previous years in an animated graphic format, to read scientific papers based on the data or to learn more about feeding birds.

If you don't have access to the internet, you may log on at your local public library.

Want to Know More? [Box with contact information]

# 52 Best Backyard Bird list joins mounting counts

Thursday, November 23, 2000, Outdoors, page C4

Bird watchers like to count birds almost as much as they like to identify them, it seems to me. And we're entering prime counting season.

Two weeks ago, Project FeederWatch started up its winter count (if you missed the information, call me at the number below); today is the Thanksgiving Count (see the Calendar); the Christmas Bird Count will be Dec. 30; and that will be followed by the Great Backyard Bird Count in February.

There's also a monthly count of sorts compiled by Wayne Tree of Stevensville, Montana. A few days after the end of each month, I get an e-mail request to list my "Best Backyard Bird" seen during that month.

Tree's goal is to get people talking about birds and sharing what they see, he said in his reply to my e-mail question.

He started the BBB list in September 1998. "One day a friend by the name of Jim Brown told me in one of his e-mails to me that his best backyard bird so far for that month was a red-breasted nuthatch. A light lit up in my mind," Tree wrote.

"The toughest part is getting people to share because somehow a lot of people think this is a contest. It most certainly is not. If all you have are house finches or house sparrows or even

mundane black-chinned hummingbirds, then that is what you report."

No hummingbird is mundane in Cheyenne, of course.

About a week after I sent in my best October bird, a black-headed grosbeak, I got an e-mail listing 160 people and their observations, about 70 different bird species.

Most observers are from Montana, but steadily growing numbers are from Wyoming and Idaho, and a smattering of other places.

I like scanning the list as much for the birds as for recognizing Audubon acquaintances from regional meetings.

Tree wrote that the information is also of interest to American Birding Association report compilers and for rare bird hot lines.

One person's best bird is another's pest bird, however. Someone with a sense of humor, or a yard next to a grain elevator, listed the obnoxious European starling.

Someone from California listed a pink-footed shearwater. This is either a case of California boosterism or someone who lives on a yacht.

When I called John Hewston, coordinator of the Thanksgiving Count, at his home in Arcata, California, he said these shearwaters are most likely to be seen on an ocean cruise, rather

than over land.

Project FeederWatch asks detailed questions about your backyard habitat, but the BBB doesn't ask any. We can only speculate on what kind of yard attracts common loons.

The sandhill crane observed by a woman in Gehring, Nebraska, stands to reason, especially if her back yard is along the Platte River or in the grain fields where cranes might stop during migration.

I wonder if the wild turkey, ring-necked pheasant and sharp-tailed grouse weren't seen through shotgun sights, when the concept of back yard was temporarily enlarged.

Some species frequently listed – the pileated woodpecker (6 times) and the Stellar's jay (8) – are no surprise for back yards in forested western Montana.

The other frequently listed species also would be typical for Cheyenne: blue jay (12), red-breasted nuthatch (8), dark-eyed junco (6) and northern flicker (6).

I noticed people around Hamilton, Montana, had a tendency to list blue jay and then specify "eastern." Tree said too many novice bird watchers up there refer to the Steller's jay by its color, referring to it as a "blue" jay also.

Some birds on October's

list are no doubt late migrants: mourning dove, mountain bluebird and Wilson's warbler.

Skimming the monthly list, I play a little game: which bird and back yard would I most like to see?

This time I picked the great tit reported from Edinburgh, Scotland. Tree said the observer is an Idaho teacher living there temporarily.

What is a great tit? Does our library even have a field guide for the British Isles? Is it a seed-eating bird visiting bird feeders? Is it as common as a house finch or as infrequent as an evening grosbeak?

Does a great tit migrate? Where to do Scottish birds migrate? What does it sound like? What do back yards (Don't they call them gardens?) in Edinburgh look like?

Just goes to show, should I ever get bored with the same old birds here in Cheyenne – which is highly unlikely – I just have to look a little farther and start learning all over.

If you'd like to contribute to November's Best Backyard Birds list, e-mail Bruce and Donna Walgren, Wyoming's compilers, before the end of the month at bwalgren@ coffey.com.

# 53 Clean-up is part of bird feeding

Thursday, December 7, 2000, Outdoors, page C2

On one of those days I'd been going brain-cell-to-brain-cell with anonymous software engineers, I finally made a break for daylight, or at least a window overlooking the back yard.

Sunlight revealed dog droppings, inveigling me to go outside to clean up before the next snow and get some fresh air.

Dog droppings only glisten from a certain angle. In our yard, when you wield the long-handled scoop, they blend in with patches of leaves, hide under snow remnants and hunker in shadows of a lawn made lumpy by night crawlers.

One of the lumps this day was the remains of a house finch. Probably dead from house finch disease. Its neck didn't seem broken from a collision, no cats have been seen lately, and the dog's getting too slow to play with birds.

For the sake of future poop patrols, I decided to rake up the clumps of leaves. It was, after all, that balmy, windless afternoon just before Thanksgiving.

And I found another dead house finch. Oh jeez. Time to sterilize.

A week's quarantine is what I tell people. Wash and put away the feeders for a week.

It's hard to sterilize the back yard. I didn't do anything about the branches of the spruce where the house finches line up waiting for their turns at the feeder.

But I finished raking, swept the patio and used an ammonia solution to clean other favorite, whitewashed perches like the TV antenna tower and the railing by the back door.

The wooden shelf feeder I brushed off and wiped with the ammonia. I even threw some on the patio, where wet sunflower hulls and leaves left interesting brown patterns on the concrete. I brought the tube feeder inside to soak in a bucket before scrubbing.

As shadows from the neighbor's garage put the yard in mid-afternoon twilight, I realized my Thanksgiving bird count results were going to be rather poor.

I was able to count a dozen house finches and two gold finches that were picking over the lawn and flower bed.

But there was no sign of the nuthatches and mountain chickadee that have been hanging around.

Birds have no qualms about using the same location for eating and defecating, resulting in disease transmission in crowded feeder situations.

All of us feeding wild birds should have a regular cleaning schedule, perhaps as often as every two to four weeks, instead of waiting for mortalities.

Scientists using Project FeederWatch data from citizen observers across the country have been able to track the spread of disease and the impact on house finch populations. Check http://birds.cornell.edu/pfw for the studies.

The moral is anyone who feeds animals – dogs or birds – is responsible for the resulting byproducts. Just think of it as a chance to go out and get a little sunshine.

Regarding the great tit observed in Scotland (mentioned in my Thanksgiving Day column), my friend Dick Hart here in Cheyenne kindly relayed quotes from his Collins Gem Guide to British birds:

"This common visitor to suburban bird-tables has approximately the same range in Britain as the Blue Tit (all parts of the British Isles, although they are scarce in north-west Scotland); there is also some immigration of both species from Europe."

Its call is described as a "ringing e-hew, e-hew" and as "silvery axe-blows."

My Thayer CD of Birds of North America includes 300 photos of birds around the world. The great tit looks like our chickadee, just a few centimeters larger, with a black cap pulled down over its eyes.

In fact, its genus, *Parus*, is the same and includes 50 chickadee-like species around the world.

Dick wrote, "The various species of tits achieved some notoriety a number of years ago when they learned to pry the foil caps off milk bottles left on people's front steps and drink the milk down as far as they could reach.

"Unfortunately for the birds the Brits have gone to plastic containers like ours."

# 54 Bird count a time to enjoy winter

Thursday, December 21, 2000, Outdoors, page C2

**Volunteers are invited to join local Audubon members on the annual Cheyenne Christmas Bird Count.**

My favorite Christmas Bird Counts are held in absolutely frigid, snow-filled weather.

The birds and those of us counting them are the only creatures abroad. It is so quiet I can hear birds shifting perches inside evergreens.

One of the first CBCs Mark and I did together was in Miles City, in southeastern Montana. It was around zero degrees, no wind and a foot or two of snow. We took the north side of the Yellowstone River, across from town.

Mark graciously gave me the pencil, pocket notebook and job of recorder. I would have been better off with a crayon and a poster-sized tablet so I could write with my mittens on.

48   CHEYENNE BIRD BANTER

Every time we spotted some avian movement, I'd jot down time and place, species and number.

There wasn't a lot of activity. The birds were snuggled down deep in the riverside thickets.

Russian olives are nasty, thorny botanic interlopers, but they do produce edible (for birds) berries and amazing bird observations.

That frozen year we found robins and, another year, a Townsend's solitaire, a gray robin-sized bird.

The Yellowstone doesn't often freeze over completely. In its landscape, twinkling with ice crystals and vapor, there are common goldeneye (ducks that eat small aquatic animals), maybe a bald eagle and, once, a white pelican.

Today I can look at data on the CBC website and tell you precisely what we saw which year. And reading the printed version of the annual reports helps us remember bird watching friends who came out with us, and the weather statistics.

Counts in Cheyenne are a bit different because the city is bigger and takes up more of the regulation 7.5-mile-radius count area centered on the Capitol.

We start out at the downtown post office at sunrise and circumnavigate the state government buildings counting pigeons, starlings and sparrows, and maybe an owl or a brown creeper.

Then we move on to Lions Park for ducks and finches and nuthatches. One year we documented a rare bird, a very cold sage thrasher that should have been wintering in west Texas or southern New Mexico or Arizona before summering again in Wyoming sagebrush.

Counts in Cheyenne in the last 10 years never have been so cold that downtown gets as quiet as the Yellowstone. But there are pockets around the city where a little bit of open water or a weedy patch attracts only us and the birds.

Some years I've been just a feeder watcher, suffering from an untimely bout of the flu. Watching feeders is important, though, because sometimes a species is counted there and nowhere else.

My other favorite part of the CBC tradition is the tally party potluck afterwards. We have to stop counting birds by dark anyway, so we gather at someone's holiday-decorated home to compare notes.

It's a very laid-back, seasonal gathering. We don't get dressed up. In fact, we have to undress some. Long underwear can be uncomfortable indoors.

After supper we gather around the count compiler, who begins reading the bird checklist. Someone from each group of observers volunteers the numbers seen for each species, and somebody else with a calculator adds them up.

For uncommon species like the belted kingfisher, comparing time and location may prove that two observers probably saw the same individual.

Estimating mallards at Holliday Park can be tricky. You think in terms of patches of 10 ducks and count patches instead of individuals.

Even though it's an estimate, it's still useful data when looking for major population shifts over the years.

Professional wildlife biologists can photograph flocks, make an enlargement, apply a grid and count individuals within each grid square.

That may be more accurate, but there aren't enough biologists in North America to cover the territory 50,000 CBC observers did last year.

The Cheyenne count is always [not always] after Christmas, so there's no excuse for not joining local Auduboners.

What better time to review your bird ID knowledge than when there aren't any tree leaves in the way?

Novices are welcome on the CBC. Like other winter activities, you'll be most comfortable if you dress warmly and bring some hot chocolate – and go home before you get too cold. I like cold air. It refreshes me. And it makes me appreciate the miracle of small, feathered beings surviving the depths of winter.

# 2001

**55** Thursday, January 11, 2001, Outdoors, page C2
## Calm, cloudy weather makes for unusual count

The Cheyenne Christmas Bird Count, held Dec. 30, was an unusual day: it was windless. Having skies overcast for a whole winter day was uncommon too.

After lunch the clouds began leaking snow. The flakes fell to the ground instead of being slammed sideways by the wind.

Without sunshine, the birds seemed reluctant to move around, except for what I think of as the "background birds:" pigeons (rock doves, as they are known on the count list), starlings and house sparrows. These non-native interlopers seem unaffected by the cold, which is one reason they interlope so successfully.

Downtown provided some of our group's best birds: a Townsend's solitaire next to the State Museum (attracted by berry-producing junipers) and two brown creepers working tree trunks in front of the Capital.

Unlike the spring count, when we count just the number of species of birds, on the CBC we count individuals. So, all day long every passing starling had to be counted.

Bob and Jane Dorn counted their 600-plus starlings mostly in one huge flock west of F.E. Warren Air Force Base.

While our group was standing on Capitol Avenue, we noticed mallards passing overhead. As one small flock disappeared over the tops of buildings, another would materialize from out of the northwest until we'd counted more than 50.

After some more urban birding at Holliday Park (590 mallards, 557 Canada geese and six goldeneyes); Martin Luther King, Jr. Park (another solitaire, juncos, starlings and crows); and the end of the Greenway by the refinery (Cooper's hawk, kestrel and kingfisher), we headed for the Hereford Ranch.

Getting there via Crow Creek was more interesting than being there: we counted nine rough-legged hawks, two northern harriers and a song sparrow, compared to a few mallards, starlings and pigeons. I couldn't help thinking about the wonderful warblers we see there in the spring.

After lunch, our group dwindled to just our family and Art Anderson, president of the Cheyenne High Plains Audubon Society. Because our van has an entrance sticker, we all piled in and headed for the Air Force base.

Not a bird to be seen out by the lakes. Just a few ice fishermen – one with Hawaiian license plates!

By the time we reached the "FamCamp," the boys opted for naps. Mark, Art and I tromped up one side of Crow Creek and down through the willows on the other side. Not one chirp or flash of feathers as the snow began to fall.

Nearing the van, we finally picked up a little bird talk and followed it to flock of 27 American tree sparrows chattering in the treetops – right over the van.

By the end of the day our group had counted 23 species. At the tally party the other groups' observations gave us a total of 37 species and 5,695 individuals, an average CBC for Cheyenne.

Four additional species were seen during "count week," the three days before and three days after the official count day.

Wyoming has 17 count areas, not very dense coverage compared to, say, the 11 counts on skinny little Long Island or 10 counts around greater Los Angeles. They probably have more counters per area as well.

Each count has its own mix of species. I looked up the Sitka count, done on an island in southeastern Alaska where my brother-in law lives. Their recent counts list at least two kinds of grebes, cormorants, shorebirds, gulls, four kinds of loons and 19 kinds of ducks. But their land birds boil down to mostly crows, ravens, juncos and chestnut-backed chickadees.

The Cheyenne count has, over the span of 40 counts, listed 106 species.

This year's total of 37 doesn't sound like much until you notice that about 20 species have only been seen once, and 45 species have been seen five or fewer times.

Some species, like this year's western meadowlark – seen only once in the previous five counts – are seasonally out of place. Others, like the owls, are not numerous to begin with and are hard to find.

To take a look at count data for Cheyenne or any other count area, go to ….

# 56 Lifetime license lends legitimacy to non-native

Thursday, January 25, 2001, Outdoors, page C2

## Wyoming lifetime licenses and conservation stamps also help non-game species prosper.

I wish I could be a Wyoming native, but some things I just can't help – such as where my mother was when I was born.

I can't even claim any Wyoming ancestors because mine decided to establish a Midwestern dairy farm instead of a Wyoming cattle ranch.

Sometimes it seems that to lobby state legislators effectively I should have a Wyoming surname of several generations' standing. So, how can I prove that Wyoming is where I want to be?

Perhaps I should pin on a list of Wyoming places I've worked or lived: Crook County, Rock Springs, Bitter Creek, Flaming Gorge, a gravel pit west of Green River, Laramie and Casper – besides Cheyenne.

Buying property or financially investing locally won't impress the natives as signs of permanence as both are reversible.

But last week I put my Bird Banter pay into, and my signature on, two irreversible Wyoming investments: a lifetime Wyoming fishing license and a lifetime Wyoming conservation stamp.

Available to anyone who has endured Wyoming for at least 10 years, they are economically sensible.

To be honest, though, my annual fishing licenses have not been economical. Last year, for instance, I caught a total of six nice kokanee – in 30 minutes the last week in December at Granite Reservoir.

Being ready to throw a line when the fishing's hot is part of the cost of being married to a fisherman.

My lifetime fishing license will pay for itself in about 16 years, or less if fees go up. If I move out of state (heaven forbid!) I won't have to buy an expensive nonresident license.

The conservation stamp is required in addition to any kind of annual Wyoming hunting or fishing license. Now that the annual fee is up to $10, the lifetime version will pay for itself in 7½ years.

The real benefit in my mind is that presumably lifetime fees are being invested by the Wyoming Game and Fish Department to benefit wildlife. Whatever benefits game species probably will benefit non-game animals. Like birds. (You were wondering how this discussion would relate to birds, weren't you?)

I visited with Kathy Frank from Game and Fish, and she said this is how things break down:

Lifetime fishing license fees are placed in a special fund invested by the state treasurer. Each year, interest goes to Game and Fish general operations to help even out financial ups and downs. The department otherwise is dependent on annual license fees and is not funded by state government.

General operations such as law enforcement, education and habitat management directly affect game and non-game species. The Game and Fish staff even includes a non-game bird biologist.

The lifetime conservation stamp fees go into the department's Wildlife Trust Fund, established just a few years ago. In addition to the fees, the $14 million principle incorporates the former Conservation Fund and income from Game and Fish products like T-shirts.

The trust fund generates around $1 million a year in interest, which is directed to funding two kinds of grants.

Wildlife Worth the Watching grants totaling $100,000 or more each year fund programs that improve people's appreciation of wildlife. Past grants have paid for projects all over the state such as installing interpretive signs and building nature trails.

The remainder of the interest goes to all kinds of habitat improvement projects.

For you recent immigrants and non-residents, investing in the annual fishing license or any of the other kinds of licenses means you also are investing in the work of Game and Fish. Part of the annual conservation stamp fee goes to improving hunting and fishing access as well.

Of course, you can always make a direct donation. If it's more than $1,000, Kathy said, it can be directed to a grant for a particular project.

You can invest in a conservation stamp without buying a hunting or fishing license. People who enjoy non-consumptive uses of wildlife – for instance, drinking in the view of an elk rather than consuming it – don't pay fees for the privilege otherwise.

The conservation stamp is the perfect way to put your money where you put your camera lens or binoculars.

Meanwhile, I'm wondering just how I can casually flash my new permanent-plastic-lifetime-fishing-license-with-conservation-stamp while leaving messages on the Voter Hotline for my state legislators when I call about wildlife bills.

Perhaps I can figure out how to use it as a name tag next time I visit the Capitol.

Note: for information about the lifetime conservation stamp....

2001

# 57 Great Backyard Bird Count offers opportunity to aid science

Thursday, February 8, 2001, Outdoors, page C1

## Amateur ornithologists in Cheyenne and around America will participate in the annual effort.

CHEYENNE – The fourth annual Great Backyard Bird Count, Feb. 16 through 19, will give local bird aficionados a chance to help document where the birds are before spring migration begins.

The GBBC is one of several citizen-science bird counting projects coordinated by BirdSource, a partnership between Audubon and the Cornell Laboratory of Ornithology. (The National Audubon Society changed its name to simply "Audubon" last month.)

The undertaking makes use of the efficiency of computers. The bird counter, whatever his or her level of expertise, enters data online, and animated maps showing nationwide results will include that input within an hour.

There is no participation fee. If you choose to participate, you can fill out as many checklists as you like – provided your bird checklists from the same day are for different areas.

Last year, 62,475 checklists tabulated more than 4,760,000 individual birds of 419 species across the country.

This year, the GBBC expands to Brazil, Venezuela and Mexico.

Here's what you need to participate:

**--Time:** Watch birds for as little as 15 minutes or as much as an hour on one or more of the official count days, Friday through Monday.

**--Location:** Watch your bird feeders, walk around your neighborhood or visit a park.

**--Some identification skills:** If you can't identify everything you see, that's OK. Just check the box on the report form that says that. To help you out, the GBBC website features its own field guide complete with bird songs. It also has a thorough bibliography describing just about every field guide and bird book available.

**--Internet access:** If you don't have access, see the accompanying information box. If you do, go to the GBBC website to enter your data. You'll find a checklist of Wyoming birds with places to enter numbers counted.

**--Counting protocol:** To avoid counting the same bird twice, keep track of how many of a kind you see at one time, then report the greatest number for that place that day.

For instance, while watching your feeder for half an hour Saturday, you count first three house finches and then four more fly in for a total of seven. Then they all fly off. The rest of your count time you see only two more. You would record seven house finches for Saturday, the most seen at one time. On Sunday, you'd start a new list.

Recording the birds you see in your backyard contributes to a larger picture, helping scientists recognize changes in bird populations and allows them to make connections between weather patterns and bird movements.

"By tracking changes in bird distribution and abundance over time, such a vast database can serve as the SOS signal for species that may be in trouble," said John Fitzpatrick, director of the Cornell Laboratory of Ornithology.

On a smaller scale, you can use GBBC data to track irruptions of pine siskins.

First, check the GBBC website to find out what a pine siskin looks and sounds like. Next, check the Wyoming lists for last year to see how many were counted and in how many different locations. Then check the maps to find out where else they were last winter.

Over the count weekend you will be able to watch the maps, updated hourly, and find out where they are this year.

Analysis by ornithologists after the count may be able to connect weather and food availability with where this species irrupts.

---

## Great Backyard Bird Count 2001

Want to participate as a citizen-scientist in the Great Backyard Bird Count? Here's how.

--Log on to: www.birdsource.org/gbbc.

--If you don't have internet access, call Barb Gorges at 634-0463 and leave your count information on the answering machine. Include your name; phone number; the time, date and how long you counted; where you counted (neighborhood); how many and what kinds of birds you counted; and the snow depth.

--Or you can count with Cheyenne High Plains Audubon Society. Join members Monday, Feb. 19, on an hour-long stroll around Lions Park. Participants will meet at the Cheyenne Botanic Gardens parking lot at 9 a.m. Dress for the weather and bring binoculars if you have them.

# 58 Helping to save rain forest has never been so sweet

Thursday, February 8, 2001, Outdoors, page C4

**Planning to buy your sweetie chocolates for Valentine's Day? Here's a way to help the environment when you do.**

We are at the zenith of the chocolate season, in the middle of the chocolate holidays: Halloween, Christmas, Valentine's Day and Easter.

Chocolate travels better in the cold half of the year.

As a chocoholic known to snack from a bag of chocolate chips when no one is looking – but otherwise concerned with healthy food – I was surprised to learn a few years ago that there is such a thing as organic chocolate.

I choose organic foods often, for my health, for the health of the land and its life forms as well as the health of agricultural workers.

Typically, "organic" means grown without chemical fertilizers and pesticides.

Growing organic chocolate, or rather cocoa trees, is more complicated.

Cocoa has been grown in Central and South America for about 3,000 years, according to the history provided by Ghiradelli's website. The Spanish explorers picked it up and began its distribution to the rest of the world. Cocoa grows best within 20 degrees of the equator, wherever it's hot and rainy.

Traditionally, farmers clear rain forest underbrush and plant cocoa trees in the shade of the native tree canopy. Native birds make their homes there. In fact, scientists have discovered the pink-legged graveteiro prefers traditional cocoa plantations to unaltered rainforest.

Only in the last 50 years or so has chemical-dependent agriculture taken over cocoa production.

The rain forest is now completely cleared away, and new varieties of sun-tolerant cocoa are planted. Unlike their shade-grown counterparts, the new plantations need constant infusions of chemical fertilizers and pesticides. And the monoculture does not attract many birds.

Farming this way means just what it does in the United States: a greater chance of toxic chemicals invading drinking water and the life cycles of native plants and animals.

I won't even begin to pretend to know enough to explain the social justice issues of poor farmers caught in the spiraling costs of chemicals or farmers using banned organochlorines illegally acquired.

Before you completely lose your taste for chocolate, let me tell you the good news.

People in the bird conservation community have been touting shade-grown coffee the last few years and recognition of shade-grown chocolate is growing.

Shade-grown doesn't necessarily mean grown completely without chemicals, so I look for the organic designation.

The purveyors of fine organic chocolate include Cloud Nine, Green & Black's (a United Kingdom company), Newman's Own Organics (run by Paul's daughter Nell) and Sunspire. Those brands are available through the national network of alternative food warehouses, co-ops and stores like Noah's Ark in Cheyenne or Alfalfa's in Fort Collins, Colorado.

I had the pleasure of taste-testing part of a box (the rest of my family is in denial about their chocoholism) of Newman's Own Organics 1.2-ounce Milk Chocolate Bars. The taste difference alone was enough to make me consider giving up my nonorganic chocolate chips.

Isn't it more expensive to buy organic? Yes, it is. But a quick survey of large chocolate bars at a local grocery store showed a range in price per ounce of 28 cents for Hershey's to 68 cents for Lindt, the genuine Swiss article, compared to Newman's at 47 cents and Rapunzel's at 60 cents, as listed in my local co-op's price guide.

So, choosing organic chocolate can become a matter of taste and/or ethics.

Newman's Own Organics developed its own sources for organic cocoa with help from the Organic Commodity Project based in Cambridge, Massachsetts.

The company buys directly from farm cooperatives in Costa Rica and Panama. They offer the farmers financial incentives to comply with organic standards.

Almost all the other chocolate ingredients are organic also, although organic soy lecithin seems to be difficult for everyone except Rapunzel's to find. Newman's lists milk for their milk chocolate coming from a Wisconsin organic milk cooperative, probably the same one I get my organic cheese from.

Perhaps you prefer your chocolate in other forms. Ah!laska makes organic hot chocolate mix, cocoa powder for baking and chocolate syrup. Sunspire makes chocolate chips and chocolate covered nuts and fruit.

The Organic Commodities Project can line you up with bulk quantities if you are a chocolate-based business.

I know I don't do myself any favors giving in to the instant gratification of cheap chocolate every time I'm alone in the kitchen. The idea that I have chips around for baking cookies is just a cover for my addiction.

Maybe it's time to make chocolate eating a communal celebration of the commitment to improve the lives of birds, farmers and other life in the rain forest.

## Organic Chocolate

For more information on organic chocolate, consult the following resources.

### Manufacturers
--Ah!laska, www.ahlaska.com
--Green & Black's, www.earthfoods.co.uk/gbs.home
--Newman's Own Organics, www.newmansownorganics.com
--Rapunzel Pure Organics, www.rapunzel.com
--Sunspire, www.sunspire.com

### Local Source
--Noah's Ark, 1900 Thomes, Cheyenne

### Background info
--International Cocoa Organization, www.icco.org
--Organic Commodity Project, www.ocp.com/ocp-chocolate
--www.healthwell.com (search for "chocolate")
--www.rainforest-alliance.org (check out the "Better Banana Project" too).

# 59

Thursday, February 22, 2001, Outdoors, page C2

## Photographer's identity a mystery

Let us call this "The Case of the Unidentified Photographer" or "Who Shot Those Birds?!"

It was late on a dark and stormy night in January when I finally checked my phone messages. I had an urgent call from Claus Johnson, assistant director of the Cheyenne Botanic Gardens.

Would I, for the sake of bird lovers everywhere, be willing to take on two carousel slide trays labeled "Birds, Vol. 1" and "Birds, Vol. 2"? No one, as far as he knew, had ever investigated the contents. Perhaps the local Audubon chapter could make use of them.

"Certainly," I said when I called back the next day.

The Cheyenne High Plains Audubon Society has a birding class beginning in April at Laramie County Community College. Last year we had to pay $3 apiece for slides to fill out our collection.

More is better. Different lighting conditions or stages of plumage must be compared to help the beginning bird watcher learn identification tricks.

The dark brown cardboard boxes holding the trays sat on my kitchen counter for two

weeks like unopened jewel cases. I wanted to wait to share the riches with someone who could truly appreciate them – someone like expert birder Jane Dorn.

Also, as my sweet spouse pointed out, these slide trays wouldn't fit our Kodak projector. This was my first clue to the origins of the slides.

Kodak-style carousel projection has become standard among institutions, while these other carousel styles persist in homes.

On a cloudy winter afternoon, I pulled slides while Jane peered at them through an antiquated viewer.

It soon became apparent we were looking at the work of a backyard bird photographer. The same platform bird feeder showed up again and again with different birds perched on it each time.

The backdrop was either the backside of the neighbor's house, a modest one-story with grayish-greenish siding and a hipped roof shingled in green, or the alleged photographer's house, with white siding.

Most of the feeder shots were in winter, showing some leafless trees (maybe a crabapple), a juniper, a sturdy clothesline of the type

supported by metal pipe, a wire fence and a metal gate opening to the alley.

Jane and I agreed it could be a neighborhood in Cheyenne, most likely one old enough to have mature trees pictured in slides the most recent of which was labeled with the year 1980.

Some of the slide frames had dates back to 1961. There was no personal notation on any of the slides except for an occasional discrete "LC" in a corner.

Several different slide processors were represented: Agfa, Kodak, GAF, Perutz (German, c. 1963), Ansochrome and RGM Denver. One said, "Arizona Color," which led me to wonder if our photographer was retired and sometimes travelled to Arizona.

The birds represented could be from Cheyenne, even the Rose-breasted Grosbeak, which appears to be caught in a typical late April/early May snowstorm. The summer tanager, though, would be a state record if we knew when and where it was photographed.

After viewing fifty or so slides, we found a silver-haired woman wearing a dress and a red-and-black

plaid wool shirt, hand feeding a gray jay while sitting on a log in the forest, c. 1978. Perhaps she was the photographer's wife?

The shots must have been made with decent equipment – the birds are more than mere specks. Retired men frequently take up hobbies involving lots of technical gadgets.

There were 175 slides between the two bird trays, plus another tray labelled "Animals, Vol. 1" which we didn't have time to investigate. (I've looked since: elk being fed near Jackson, zoo animals, etc.)

The last dozen bird slides proved the most interesting. Part of a man wearing plaid pants appeared in a shot meant to feature a peacock at a picnic area, perhaps the Denver Zoo? One shot featured a pink flamingo walking a manicured lawn.

Then the silver-haired woman reappeared in two shots. This time she was feeding a Clark's nutcracker, possibly at a scenic overlook at Rocky Mountain National Park.

And then – a-ha! The photographer himself, a smiling, silver-haired gentleman dressed as if for a Sunday

CHEYENNE BIRD BANTER

drive, feeding birds at the same overlook. Perhaps he handed the camera to his wife. The slide is dated 1978.

My scenario could be wrong. Not all the bird slides feature that particular backyard. I have to wonder about the shots of a pair of cardinals and some backgrounds that look suspiciously like rural Iowa.

We could be dealing with multiple photographers. We could be dealing with a photographer who took pictures of silver-haired people.

Unlike fictional mysteries, this one doesn't have a solution yet. If you recognize the alleged photographer and his wife, please call me at 634-0463.

I hope it's not too late. I hope their children didn't all move to California. Maybe the photographer and his wife instilled a love of birds in someone younger than themselves who yet survives to read this bird column.

Not only do I want to satisfy my curiosity, but when we use the slides in our bird classes, I'd like to be able to give credit to the photographer(s).

And remember: Label and sign your own work!

*Barb Gorges invites readers to share their bird sightings and stories with her at....*

*Cutline for two photos,*

One an elderly man on a high spot holding out food for a Clark's nutcracker flying away and the other a silver-haired woman in the same location with the bird eating out of her hand, "Do you recognize either of these people? If you do, the writer wants to hear from you at 634-0463."

# 60 Mystery photographer identified

Thursday, March 8, 2001, Outdoors, page C2

*Our story so far: Slides of birds, circa 1960-1980, donated to the Cheyenne Botanic Gardens, cause local Audubon members to seek the identity of the photographer. Photos of two people found with the birds are published with the Feb. 22 Bird Banter column...*

Home-delivered Wyoming Tribune-Eagles are on doorsteps by 6 a.m. weekdays. By 8 a.m. Feb. 22, I had my first call, from Cindy Braden. She identified the woman pictured as her husband's late grandmother, Rhea Clapp.

"She wouldn't be related to my nextdoor neighbor, Bob Clapp, would she?" I asked.

"She's his mother."

I love coincidences.

Cindy, Bob's niece-in-law, was unsure of the man's identity, but she did know the location of the backyard. Bob's daughter, Robin Waterhouse, still lives in the house at 1518 E. 22 St.

The next call was from Jeanette Vandorn. She led a 4-H group years ago that was sponsored by the Cheyenne Garden Club, of which Rhea was an active member.

Vandorn remembers sitting at Rhea's kitchen window bird watching.

Another caller thought the dapper man in the photo had to be Kirk Knox – the man most widely recognized in Cheyenne today for wearing a hat that's not a Stetson or a ball cap.

Marjorie Brink had more information. She was a close friend of Bob's sister, Anna Marie, or Ann as she was called. Their father Leo was a meteorologist in Cheyenne. She thought the Clapps had come from Nebraska originally.

So I went to school Thursday morning pretty sure the unknown woman was Rhea Clapp, garden club member and bird watcher, and still assuming the man pictured was her husband. Until the next call, from Lela Allyn.

"The man in the photo is definitely Cliff Colgin," she said. "I bought my last car from him at Tyrrell's."

Since I couldn't recall the car salesman I met last summer, I was doubtful.

"Oh yes, I remember him," Allyn said. "I bought the car in 1978, and I still have it."

By this time I figured I should call neighbor Bob, but he and his wife Corky appeared to be on one of their out-of-town jaunts.

Saturday I got a call from Helen Colgin. The man in the photo was her husband.

The Clapps and the Colgins were good friends and frequently traveled someplace like Estes Park on Sundays. And yes, they'd wintered in Phoenix. Cliff Colgin always wore a white shirt and jacket, even hunting. They did a lot of fishing, too. And they all took pictures.

Finally, Sunday night Bob called, fresh from a trip to Omaha to visit his 98-year-old aunt, his mother's sister. His parents came from Gordon and Hastings, Nebraska, small towns nearby (OK, so I was off by a bridge-length when I guessed Iowa.). The pictures of the cardinals are probably from there. Perhaps the zoo pictures too.

Leo Clapp began his meteorology career at age 18, flying weather kites, said Bob. He retired from the weather bureau after more than 40 years, in the early '60s. Then he took up photography. So the "LC" on some of the slides is his initials.

We still have to presume he took the other slides as well. He died in 1988 at the age of 89, having lived in Cheyenne for more than 70 years.

Rhea Clapp was a member of Audubon back in the days when there was one chapter for the whole state of Wyoming. She was also known for her expertise in roses, judging shows in the area and once even in Chicago. She died in 1991 at the age of 89. It was then that Bob donated the bird slides and stacks of gardening magazines to the Cheyenne Botanic Gardens.

In 1938 the Clapps built the house on East 22nd, when Bob was 8 years old. They planted the landscape themselves. Rhea brought back an acorn from Chicago and got it to sprout and grow in the backyard. Robin, Bob's daughter, remembers when bird feeding took up the whole back side of the house.

The Colgins lived two

houses away from the Clapps while Bob was growing up, and they became close friends.

Leo and Rhea did travel to Arizona quite a bit, mostly around Phoenix, a destination still popular with Cheyenne's retired population.

And, like many people's children, Ann and Bob left Cheyenne. Ann and her husband lived in California and Oregon before returning. Her widower and three children still live here. Bob spent 23 years in Colorado before coming home. He and Corky sometimes travel to Arizona themselves, to see Corky's sister.

Part of bird identification involves learning a species' preferred habitat and life history. I've been in Cheyenne long enough to recognize a certain human life history pattern that still holds today: parents back in Nebraska, children in Denver or California and winters in Arizona.

Are our migration patterns genetically wired into our brains? Are they the result of environmental influences such as windy winters or overcrowding in the "old country"? Or are they the result of cultural knowledge we pick up from our tribe?

Some juvenile birds make their first migration without any experienced adults along. Do you suppose these youngsters aspire to southern winters after hearing Mom, Dad and the neighbors talk? At this point, late in Wyoming's winter, I myself could stand to hear more.

*Cutline for formal photo of elderly couple:*

"Rhea and Leo Clapp of Cheyenne were avid bird watchers. Bob Clapp/courtesy"

# 61 Who's keeping count? Lots of folks

Thursday, March 22, 2001, Outdoors, page C2

Signs of spring 2001:

March 1- Robin flies over Rossman School.

March 7 - Crocus shoots visible in the school bird garden.

March 8 - Ragged skein of geese seen flying north but so high as to be barely visible to the naked eye. Obviously not the local overwintering flock flying low in and out of city parks.

March 8 - Distinctive killdeer call floats through open car window while driving Avenue C.

March 10 – Worm seen lying on sidewalk in front of B & B Appliance, while snow falls.

Seeing robins is not a very reliable sign of spring, not because we get snow so late but because we have robins that spend the winter.

The proof came while I was comparing results of the Thanksgiving Bird Count, Christmas Bird Count, Great Backyard Bird Count and Project FeederWatch to see if together these surveys could tell me anything about winter bird abundance and movement.

Because the counts don't use the same protocols, each count's data is best compared to its own, from year to year or by location.

The TBC and PFW are strictly bird feeder counts; the CBC surveys 15-mile diameter circles; and the GBBC is a hybrid of feeder and park counts.

Each count reports data differently too. This year the CBC reported how it takes into account the number of observers and the amount of time they spend. For each species counted in a circle, the CBC lists the total number of individual birds as well as the number seen per "party hour" (because birders go out in parties).

As CBC observers are free to count anywhere within the circle, the data don't tell you, for example, that they went straight to a preferred habitat to count geese, such as Holliday Park.

As long as observers go there each year, the numbers of geese can be compared year to year – unless next year there are half as many geese at the park because the other half are in a feeding frenzy over some spilled corn elsewhere. But hopefully local birders know that.

Numbers reported for the GBBC have not been adjusted for the number of observers, and there's been an increase in participation. Last year 98 counts, or checklists, were submitted from 23 locations in Wyoming. This year 202 checklists were submitted from 35 locations.

Cheyenne submitted seven in 2000 and 16 in 2001.

PFW is probably the most reliable way to compare bird movement and abundance over the course of one winter. But again, every observer doesn't always report for every time period.

The key to a well-designed scientific study, as kids participating in science fairs learn, is to have a minimum number of variables. A bird study should be conducted at the same time of day and year, for the same length of time, in the same place, by the same person with an unchanging level of skill, if we truly want to look for changes in the numbers of birds over time. Spring breeding bird surveys performed by ornithologists do that.

But the other hallmark of a good study is replication, or in this case, the number of observers submitting data.

This season the TBC had 449 counts from the 13 westernmost states. The other counts are continent-wide. The latest CBC had more than 45,000 participants counting 63 million birds. About 4½ million birds were documented in 52,000 GBBC checklists last month. Fifteen thousand people participated in PFW.

Besides tracking bird populations in a general way, another benefit of all these counts is that more people get interested in bird conservation. But us backyard bird watchers need a way to get a handle on the tremendous amount of data generated. One way is to track just one species.

So, where has our sign of

spring, the American robin, been all winter?

The TBC, held Nov. 23, 2000, rated the robin as 20th in abundance out of 149 species reported. They were reported in all 13 states except Hawaii, Alaska, Arizona and Wyoming. Montana reported only one.

There were only 23 counts submitted from eight Wyoming locations, so in a strictly feeder-based count, it would be easy to overlook a bird that in winter prefers berries hanging on trees to seed in feeders.

For the CBC, each circle's count is one day long, scheduled sometime between Dec. 14 and Jan. 5. Three robins were seen on the Cheyenne count Dec. 30. That's 0.114 birds per party hour.

Fourteen of 18 Wyoming count circles had robins. The Kane count, up along the Big Horn Reservoir near Lovell, listed 400 robins. Casper counted 99.

The GBBC, a feeder-based count held Feb. 16-19, listed 150 robins for Wyoming. According to the map, none was seen in Cheyenne, five to 10

were spotted in Casper and a bunch in the Big Horn Basin.

PFW will have data from Nov. 11, 2000, through April 6. So far, the animated map for robins in November through February shows concentrations on the West coast and along the Colorado Front Range. Little blips of robins show up in Wyoming from time to time. A lot of robins showed up on Wayne Tree's informal Best Backyard Bird e-mail list for February.

Either robins are flocking to back yards in western Montana (where most of his

contributors live) or birders are picking the robin as their favorite over the winter regulars because the winter's been long and they're ready for a sign of spring.

Robin sightings will increase with the coming of spring, partly because we spend more time out in the increasingly pleasant weather, but mostly because most robins did leave for the winter and are flocking back to southeastern Wyoming for our hospitable trees and tasty – but only seasonally available – worms.

# 62 Field guide choices now are many

Thursday, April 5, 2001, Outdoors, page C2

**Years ago, you pretty much had a choice between Peterson's and Peterson's.**

Field guide: A reference book small enough to be carried outdoors yet comprehensive enough to answer most identification questions for a group such as birds, mushrooms, rocks, etc.

For us bird watchers in the Rocky Mountain West, Roger Tory Peterson's "A Field Guide to Western Birds" has long been a standard. Published in 1941, it was the first book with a systematic approach to bird identification and small enough – at 4½-by-7¼ inches and 300 pages – to drop into a pocket, squeeze into a pack or throw onto a dashboard.

Twenty-five years later the Golden guide came out, about the same size, but including all of North America – very handy for those of us living on the eastern edge of the West.

Golden was an improvement on Peterson's because cheap, modern color printing allowed bird pictures,

descriptions and range maps to be printed on facing pages: one-stop-look-up.

Another 20 or so years later, in 1983, we got the National Geographic guide. More drawings for each species, plus more juveniles, females and obscure species made a bigger book, 5-by-8 inches and 460 pages.

Audubon came out with the first photographic guides for Eastern and Western birds, but no one I know uses them as a primary guide.

In the last five years interest in bird watching has skyrocketed, and so has the number of general field guides. (I haven't room to mention all the specialty guides for groups of species or particular locations.)

First was Stokes in 1996, followed by the American Bird Conservancy's radical guide organized by bird feeding behaviors, National Geographic's third edition and Peterson's third edition

of the Western guide.

Last fall everyone was talking about the advent of "The Sibley Guide to Birds." At 9½-by-6½ inches and 544 pages, it would be huge for a field guide, but doesn't pretend to be a "field" guide.

Advanced birders love the way David Allen Sibley distinguishes details such as the five populations of horned larks and various feather molts of other species. He spent years sketching birds up close while they were being banded.

I like the range maps and the thumbnails comparing similar birds in flight, but for the casual birder, 26 variations on a bird as common as the dark-eyed junco may be overwhelming.

I once identified Kenn Kaufman, author of my favorite "Lives of Birds," at an Audubon conference without being close enough to read his nametag because I noticed the flock of

binocular-wearing females surrounding him.

He's one of the few bird book authors with his picture on the back cover – and it doesn't need digital enhancement. He does, however, use digital technology to improve the bird photos in his new field guide, taking away misleading shadows and cropping distracting backgrounds.

Kaufman's guide has the usual accouterments: quick index by generic name, color-coded pages, introduction to bird watching basics and comments on habitat and voice for each species.

His range maps are exceptional. Not only do they depict summer, winter, year-round and migration ranges in different colors, but where a species rarely occurs or is rare, the colors are paler.

The only disconcerting thing for a veteran field guide user is that Kaufman deviates somewhat from organizing

2001

57

his book in ornithological order. His color-coded table of contents had me stumped when it listed "medium-sized land birds," but immediately following was a photographic table of contents to show which birds he meant.

I still check off my life birds in the index of a first edition Golden Guide my mother gave me in 1973. I bought the next edition in the 1980s, and it's still the one I grab for field trips.

Now a new Golden edition is out. Names are updated, descriptions have been reworked by new authors, and a quick index has been added – although they forgot the check boxes in the regular index.

What I wish they had added are state lines in their range maps. It was one thing to bird in the corner of the country formed by lakes Superior and Michigan back in my youth, but it's pretty hard to eyeball Cheyenne's location in relation to the Canadian and Mexican borders.

So, had I researched the newest field guides sooner, I probably would have chosen copies of Kaufman's as the prizes for the Audubon Award winners from the school district science fair this year.

Oh well. I just hope when I start hinting that I'd like Kaufman for Mother's Day, my family understands it's for the range maps, not the back cover.

## A Field Guide to Field Guides

Following is a "field guide" to the books mentioned in today's Bird Banter.

--*A Field Guide to Western Birds*, Roger Tory Peterson, 1998, Houghton Mifflin.

--*All the Birds of North America*, American Bird Conservancy, 1997, Harper Perennial.

--*Birds of North America*, Golden Field Guide, 2001, St. Martin's Press.

--*Birds of North America*, Kenn Kaufman, 2000, Houghton Mifflin.

--*Field Guide to the Birds of North America*, National Geographic Society, 1999, National Geographic.

--*Lives of Birds*, Kenn Kaufman, 1996, Houghton Mifflin (not a field guide).

--*The National Audubon Society Field Guide to North American Birds* (western volume), 1994, Knopf.

--*The Sibley Guide to Birds*, David Allen Sibley, 2000, Knopf.

--*Stokes Field Guide to Birds, Western Region*, Donald and Lillian Stokes, 1996, Little Brown.

# 63 Birds of a different feather spotted back East

Thursday, April 19, 2001, Outdoors, page C3

Bird watching is a portable hobby.

On a family tour of East Coast relatives over spring break, I was able to add two new species to my life list and introduce the boys to one of my earliest bird acquaintances.

Even binoculars and field guides are dispensable. The days we were on foot (and train and escalator and elevator) in a city, I left them in my suitcase.

It wasn't meant to be a bird watching trip, but when your particular interest is as visible as birds, every trip is a field trip.

As soon as I stepped out of my uncle and aunt's back door to meet their new dog, I could hear bird calls.

The Mount Airy neighborhood of Philadelphia features hundred-year-old houses and hundred-foot, pole-like trees. All of their branches and birds start 80 feet up. Binoculars didn't help me connect the songs with the elusive birds.

Downtown, while we waited for the Independence Hall tour, two hawks swirled around nearby office buildings. Maybe red-tails, maybe red-shouldered.

It wasn't until we were on the steps of the Philadelphia Museum of Art (admiring the owl and other statuary under the eaves) that another hawk buzzed us. This one was definitely a red-tail with its "belly band" of darker plumage.

Out behind the museum we tracked down a very vocal, very cold and plumped-up northern mockingbird. Below, on the Schuylkill River, gulls floated down the current, and teams of rowers sculled against it.

While traversing the New Jersey Turnpike we caught glimpses of great blue herons flying over the grasses and reeds of the Meadowlands marshes, undaunted by jets taking off from Newark.

In New York City, our nephew attends Cardinal Spellman High School. Were the churchmen named for the bird? I'm guessing the other way around, since common names, before the advent of the American Ornithological Union, were more variable than church nomenclature.

My birding goal this trip was to see a cardinal, but I had no such luck in Philly or New York, where we walked through Van Cortland Park.

This is the huge, urban, somewhat natural wilderness on the north end of the Bronx that served as my husband's boyhood backyard. We didn't go into the deep woods, so we saw only a few blue jays and mostly pigeons, crows and starlings.

In the Staten Island ferry terminal two street musicians, twin flamenco guitar players, were asked to leave after only five minutes' worth of performance. The pigeons seemed to be enjoying the music as much as the rest of us and judging by the amount of droppings on the light fixtures, they endure less harassment than musicians.

It being a beautiful day – and we being from a state deprived of ocean views – we opted to ride on the ferry's small open deck. Ring-billed gulls followed our wake so closely we could distinguish first winter, second winter and adults by plumage.

The next day, crossing

CHEYENNE BIRD BANTER

Delaware, we spotted cormorants and a northern harrier. Even at 75 mph you can't mistake the cormorant's weird shape or miss the harrier's white rump.

We skipped the decoy museum in Havre de Grace, Md., in favor of a quick but birdless look at Chesapeake Bay and a visit to an ice cream parlor full of penguins.

As chief navigator, I soon recognized a name on the map, "Patuxent Wildlife Research Center." It's familiar to birders as the home of the nationwide Breeding Bird Survey. So we stopped.

At the visitor center I dragged the binoculars and the family out of the car, picked up a checklist, hiked some of the trails, consulted the front desk staff and added two new species to my life list: the tufted titmouse and the Carolina chickadee.

Other small birds were flitting about the bare beech woods, but who wants too much frustration on vacation?

In Silver Spring, Maryland, our 6-year-old great-nephew got up with the birds. So did we. It's hard not to with a cardinal singing right outside. So, I achieved my trip goal, and the boys added a species to their own memories.

On my first visit to Washington, D.C., I was pleasantly surprised at all the open space, especially along the Mall, which allows sunlight to highlight all the wonderful architecture – and we did have one glorious, sunny day.

I don't know if the American coots paddling the Tidal Basin appreciated the cherry blossoms as much as all of us tourists.

Next trip to D.C. we'll have to plan a hardcore birding trip in May and visit the C&O Canal, joining all the men and women in suits who pursue birds at the height of migration in the well-known hot spot before they head for the office each morning.

At the National Museum of Natural History, we stumbled upon an exhibit of all those birds we didn't see, although Mark thought the specimen of the worm-eating warbler looked like what he saw at Patuxent.

There were other bird encounters in the capital: turkeys in Grandma Moses' paintings at the National Museum of Women in the Arts, birds appliqued on the 150-year-old quilts at the Textile Museum, all kinds of man-made birds at the National Air and Space Museum and great blue herons on many Maryland license plates.

Back at home we heard this spring's first plaintive calls of the mourning dove. It's our turn to play host, I guess. Let's just hope the accommodations are adequate to survive spring snowstorms.

# 64

Thursday, May 3, 2001, Outdoors, page C2

## Humans, too, flock together during springtime snowstorms

Birds that migrate across Wyoming dodging spring snowstorms have my admiration now that I've had similar experience.

The weather was iffy when I headed out for Cody for the state Audubon board meeting two weeks ago this Friday. At Casper I carpooled with another board member, and we began to hit showers, snow or rain – it was hard to tell.

By Saturday morning the hills around Cody had snow, and the mountains did too, although clouds closed off views of the peaks. We slipped over snowy roads to visit the Beartooth Ranch on the Clark's Fork of the Yellowstone River, site of a new Audubon education center.

Other than a few intrepid blackbirds, we saw no little birds at the ranch, just an immature golden eagle and an osprey.

Then we had a five-hour meeting in a windowless room back at the Buffalo Bill Historical Center, on behalf of Wyoming birds.

Eventually it was early evening, and we were threading our way through showers and slush back to Casper.

I avoid highway driving at night, but I had no idea my plan to stay over in Casper also would obligate me to also spend Sunday night away from home.

Birds were flying in the falling snow in Casper when I left Sunday. A flock of Franklin's gulls passed over me by Glenrock. By early afternoon, just as the ice started building up on the windshield wipers, I was stopped by the highway gate at Chugwater, only 45 miles from home.

It was like a fallout of birds during migration, only some of us were trying to migrate south and others north. We all came to roost at Horton's Corner convenience store and gas station, at the back, by the two public phones where we took turns calling home.

While waiting for the road to open again, we huddled over cups of hot chocolate and famous chili while exchanging stories like passengers on a cross-country bus.

No matter how frustrated we felt – college students, grandmothers, mothers, families, turkey hunters and trout fishermen – hanging out in a warm, dry building, however tedious, beat sitting alone in a cold car by the side of the road.

Kinds of birds used to flocking together, huddling deep in bushes for warmth, survive a snowstorm better than those independent meadowlarks whose frozen bodies we find out on the prairie afterwards.

I wonder if some birds that don't normally share communal roosts will do so in times of meteorological adversity. We stranded travelers did.

Over at the Chugwater Community Center, after a spaghetti dinner orchestrated by the director, a video on the big-screen TV and plenty of camaraderie over jigsaw puzzles and card games, 40 or more of us rolled up in the center's blankets or our own and slept on the floor.

Later, through an unshaded window, by a yard light's illumination, I could watch the pantomime of wind harassing tree branches.

Inside I listened to the contented night sounds of the baby belonging to the family sleeping next to me. We were so lucky that what was built to benefit the locals could benefit us wayfarers as well. Thank you, Chugwaterians.

Cheyenne is like that for the birds. We have so many bird feeders and yards planted with protective trees and shrubs. Spring storms bring a lot of interesting migrants here that otherwise might fly over – as travelers have been known to bypass Chugwater.

Besides planting cover and providing food, there is one other favor we can provide: Protect the unwary visitor from local hazards.

May 12 is "Keep Your Cat Indoors Day," sponsored by the American Bird Conservancy.

But every day should be "Take Your Cat Out on a Leash Day," for the peace of mind of us cat owners and the longevity of both cats and birds.

Meanwhile, wipe the dust (or snow) off your binoculars and get outside yourself. The annual migration extravaganza has already begun.

# 65 Nighttime right time for fauna to frolic

Thursday, May 17, 2001, Outdoors, page C2

There's wild night life in Cheyenne in the spring.

The other night, on the way home from quilt club, I had to brake for a fox trotting across Converse at Dell Range. You don't suppose we could get him to use the pedestrian bridge, do you?

Another night, after a Cheyenne Little Theater production, we were driving a friend home along Windmill when I noticed an upright animal shape in the borrow pit.

The object was still there on the way back. Since there was no traffic, we stopped to look. An eye glinted in the streetlight and glared back at us. We were interrupting a fisherman: a black-crowned night-heron.

Reminiscent of its cousin, the great blue heron, but with short neck and short legs, this 2-foot-tall bird was stalking the edge of a puddle, looking for unwary frogs to stab with its long, strong, black bill.

It's predictable that the open space east of the airport would become a red-light district – brake lights, I mean. Maybe that's why that section of Windmill was recently festooned with streetlights; so we won't run over the low life – low profile, that is – such as foxes, birds, mice and frogs.

Even in my respectable neighborhood it's common to see a couple of cottontails loitering on the lawn by the light of the moon.

Later, in the grayness before sunrise, they move from damp lawns to the warmth and dryness of the asphalt streets, where they become nearly invisible - until the dog and I spook them.

If it weren't for rabbits, walking wouldn't give me the upper-body workout I get from hanging onto the leash of my 100-pound dog as he lunges after his instinctual prey.

There's a lot going on during daylight hours too. Here are some reports I received the first ten days of May:

--Velma Simpkins, in Pine Bluffs, has been getting unusual doves, possibly domestic escapees or maybe Eurasian collared doves, which are spreading rapidly since being accidentally released in Florida.

--Wayne Neemann reported a rose-breasted grosbeak in his north-side yard. This grosbeak is an eastern bird normally, but some stray out here during migration.

--Beth Easton wanted to know how to keep a robin from building a nest on top of her wall-mounted porch light. I suggested temporarily hanging chicken wire from the porch ceiling down around the light.

--Belinda and Don Moench were at Twin Buttes Lake on the Laramie Plains when they identified two common loons. The little island in the middle of the lake has a colony of nesting great blue herons. Down at Holliday Park, Belinda saw an eared grebe.

--Eileen Poelma had Canada geese in her yard south of Carpenter.

--Caroline Eggleston described small sparrows with rusty brown caps in her yard in Orchard Valley, probably chipping sparrows.

--Betty Wagner has pine siskins and American goldfinches eating her out of house and home in Sun Valley. Her four feeders have a total of 24 ports, and often each will have a bird.

--Fred Lebsack visited Lions Park May 8 and reported a Tennessee warbler (another eastern stray), a wood duck and a blue-gray gnatcatcher.

--Joanne Mason of Wheatland, one of the students in the "Bird Watching for Fun," class Jane Dorn and I taught this spring at Laramie County Community College, came with the longest list of any of the students of birds she'd seen. The plantings she did about 12 years ago are beginning to really bring in the birds, including a Bullock's oriole, lesser goldfinch, vesper sparrow and clay-colored sparrow.

Things are still slow in my yard. Perhaps too much of our green ash trees got pruned last fall, or maybe there's a hawk roosting on the TV antenna tower, where we can't see it from the window.

I did see a yellow-rumped warbler of the "Myrtle" race (white throat instead of yellow) on May 9, and the next day two red-breasted nuthatches came in. The nuthatches may be the two that spent several winter months debugging our tree trunks.

I'm looking forward to the Big Count on Saturday. It's a chance to renew

60                     CHEYENNE BIRD BANTER

acquaintances with spring migrants – and with bird watching friends who come down every year from Casper for the event.

Every year we find some stray warbler, shorebird or gnatcatcher I've never seen before. It's as if, for us homebodies, migrating birds bring us a bit of more exotic lands for a few days.

# 66 Backyard beckons bevy of birds

Thursday, May 31, 2001, Outdoors, page C2

It's hard to eat breakfast, lunch or dinner these days without picking up the binoculars to admire the birds on my backyard wall.

When I e-mailed my last column to the Wyoming Tribune-Eagle, mentioning the paucity of birds in my yard and all the sightings in everyone else's, a little bird must have been sitting on the wire listening in.

The next day, May 11, my yard was inundated with pine siskins, goldfinches and chipping sparrows, and accented by a black-headed grosbeak, a Bullock's oriole and a rose-breasted grosbeak.

A lazuli bunting showed up, too, and came back with half a dozen friends. They look like small eastern or western bluebirds, with robin-colored, red breasts.

It turns out the buntings like millet, something I usually don't put out because it attracts house sparrows. Dave Felley gave us half a bag when he moved so we've been spreading it out on the top of our concrete-block wall. The buntings have been lining up shoulder-to-shoulder with the mourning doves every day since.

When I observed a house finch drinking out of the dog's water dish, I decided it was time to try the Solar Sipper again. It's like a fancy dog dish with a removable, black plastic bowl inside a red plastic bowl.

It comes with a black lid that's supposed to absorb heat and keep the water from freezing in winter. The lid has a hole in it for birds to stick their heads through and get a drink.

The birds never learned to use it, but the dog learned to knock off the lid and drink.

This time I put it on the back wall without the lid, and birds are using it. The grackles threw in some stale bread and hard raisins and retrieved them when they got soggy. But I still caught a grackle using the real dog dish on the back step.

The green-tailed towhee showed up a week after everyone else.

He's between robin and sparrow size, and he holds his tail up at a right angle. His greenish-gray coloring makes him invisible where he hangs out under the bushes, unless you see the flashy white patch under his chin or his rust-colored cap.

I had three people tell me about western tanagers in their yards before I saw one in my neighborhood. He was drinking water puddled in a crack in the street.

These tanagers are so tropical looking – orange head, yellow body and black and white wings.

You might mistake a black-headed grosbeak for a robin, until you look more closely. They are more orange than robin-red, their heads are blacker, and their wings and tails are spotted

with white. Their thick "gross" – or big – beaks are for cracking seeds rather than drilling for worms.

The rose-breasted grosbeak looks pretty much the same, but instead of orange it has a white belly with a dark pink bib. I've now seen one in the yard four days out of 14. Perhaps I wasn't looking hard enough the other days.

Other than mourning doves and robins, I don't expect any of these spring birds to nest here. Most are on their way to the mountains or farther north.

Some birds get a very early start with nesting and breeding.

We've gone out listening for owls in February because that's when owls set up their territories and hoot at their rivals.

Meriden rancher Dave Hansen was out branding May 19 when he noticed two great horned owls toddling around his home pasture. Had they blown out of the nest?

It isn't unusual for these owls to leave the nest by now, even if they aren't ready to fly yet. They are as big as the adults, just sort of fluffy, sort of chubby, like a two-year old wearing cloth diapers and plastic pants.

As soon as they learn to fly [actually, they are quite good using their beaks and claws to climb trees until they can fly], they will return to obscurity in the treetops

and spend the summer with their parents learning to hunt rodents.

This is the most important time of year to keep all cats indoors – especially if you live out on the prairie with the ground-nesting grassland birds. It's also important not to mow right now.

Cats, unlike native predators, are more numerous and will kill for fun rather than food.

I hope whoever belongs to the gray cat that visits my yard will keep him home. Otherwise, I have to put my dog on guard duty first thing in the morning, and then the green-tailed towhee won't come.

Two bird watchers from California traveling through Cheyenne made arrangements to meet me down at Lions Park on one of the windy days we had whitecaps on the lake.

Other than the western grebes, mallards, a few yellow-rumped warblers and a tree full of goldfinches, there wasn't much to see, and we decided not to walk around the lake.

On the way back to our cars I mentioned the only other sure-fire bird observation we could make would be the yellow-headed blackbirds over by the cattails.

"Yellow-headed blackbirds?" responded the Californians. "We've seen them only three other places!"

So we headed into the

wind and soon were rewarded with a yellow-headed male strolling the path toward us until he was at our feet.

Then he flew up and engaged in aerial shenanigans with a red-winged blackbird a few feet over our heads. The Californians were delighted.

So, one birder's blackbird is another's special species. Should I ever look up the Santa Monica Audubon chapter, I wonder what locally abundant bird they will have that will be my fabulous find?

# 67

Thursday, May 31, 2001, Outdoors, page C2

## Birders tally 143 species on 'big day'

### Some rare birds were among those counted by Cheyenne High Plains Audubon Society members.

CHEYENNE – A scarlet tanager and several black-throated blue warblers were the most exciting birds observed on the Cheyenne Big Count Day, held May 19.

Cheyenne High Plains Audubon Society members and friends counted 143 bird species with the help of birders from Laramie and Casper.

It turned out to be a banner year for warblers in general, with 16 kinds recorded.

At 6 a.m. birders were ready to start simultaneously at Cheyenne's three busiest birding spots: Lions Park, Wyoming Hereford Ranch and the High Plains Grasslands Research Station.

The two rare bird sightings took place at the ranch. Anne and Nels Sostrum spotted the scarlet tanager in bushes along one of the roads.

Scarlet tanagers are red with black wings, unlike the western tanager usually seen here in the spring, which has only an orange-red head.

Scarlet tanagers summer east of the 100th meridian, the line of longitude that bisects the Great Plains.

Black-throated blue warblers, searching for insects around the base of bushes, were easy for everyone in the main group of birders to observe – unlike a lot of warblers, which feed higher up. Warblers migrate at night and feed all day.

Black-throated blue warblers are dark blue with white bellies and black mask and sides. They normally summer in the Appalachian Mountains, the Northeast and the upper Great Lakes region.

Another eastern warbler, the prairie warbler, is also considered rare when seen here, but caused less excitement this year than when first recorded in the state in 1996.

The chestnut-sided, palm and worm-eating warblers also are considered to be outside their usual migration routes.

Several double-crested cormorants have been seen at Lions Park with some regularity this spring, including on the count day. Though common in Wyoming, they're usually less common in Cheyenne.

Steady rain the first two hours of the count made it difficult to find birds because they hide and sit still in bad weather, and raindrops blur binocular lenses.

Because the overcast sky yielded poor lighting, the birds were hard to identify. Their colorful markings looked gray, and the birds often appeared as gray silhouettes.

But by midday conditions improved, and the number of species counted, though not approaching record numbers, was better than expected.

## Big Day Count

Cheyenne High Plains Audubon Society members, along with enthusiasts from Laramie and Casper, counted 143 species of birds May 19.

| | | | |
|---|---|---|---|
| Pied-billed Grebe | Redhead | Wilson's Phalarope | Say's Phoebe |
| Eared Grebe | Ruddy Duck | Red-necked Phalarope | Cassin's Kingbird |
| Clark's Grebe | Northern Harrier | Ring-billed Gull | Western Kingbird |
| American White Pelican | Sharp-shinned Hawk | California Gull | Eastern Kingbird |
| Double-crested Cormorant | Cooper's Hawk | Rock Dove | Loggerhead Shrike |
| White-faced Ibis | Broad-winged Hawk | Mourning Dove | Plumbeous Vireo |
| Turkey Vulture | Swainson's Hawk | Burrowing Owl | Warbling Vireo |
| Canada Goose | Red-tailed Hawk | Chimney Swift | Red-eyed Vireo |
| Green-winged Teal | Golden Eagle | Broad-tailed Hummingbird | Blue Jay |
| Mallard | American Kestrel | Belted Kingfisher | Black-billed Magpie |
| Northern Pintail | American Coot | Downy Woodpecker | American Crow |
| Blue-winged Teal | Killdeer | Northern Flicker | Common Raven |
| Cinnamon Teal | American Avocet | Olive-sided Flycatcher | Horned Lark |
| Northern Shoveler | Solitary Sandpiper | Western Wood-Pewee | Tree Swallow |
| Gadwall | Spotted Sandpiper | Willow Flycatcher | Northern Rough- |
| American Wigeon | Upland Sandpiper | Least Flycatcher | winged Swallow |
| Lesser Scaup | Long-billed Curlew | Cordilleran Flycatcher | Bank Swallow |
| Common Merganser | Common Snipe | Eastern Phoebe | Cliff Swallow |

| | | | |
|---|---|---|---|
| Barn Swallow | Yellow Warbler | Spotted Towhee | Bobolink |
| Black-capped Chickadee | Chestnut-sided Warbler | Chipping Sparrow | Red-winged Blackbird |
| Mountain Chickadee | Black-throated Blue Warbler | Clay-colored Sparrow | Western Meadowlark |
| Red-breasted Nuthatch | Yellow-rumped Warbler | Brewer's Sparrow | Yellow-headed Blackbird |
| Brown Creeper | Townsend's Warbler | Vesper Sparrow | Brewer's Blackbird |
| House Wren | Prairie Warbler | Lark Sparrow | Great-tailed Grackle |
| Ruby-crowned Kinglet | Palm Warbler | Lark Bunting | Common Grackle |
| Blue-gray Gnatcatcher | Blackpoll Warbler | Savannah Sparrow | Brown-headed Cowbird |
| Veery | Black-and-white Warbler | Song Sparrow | Orchard Oriole |
| Swainson's Thrush | American Redstart | Lincoln's Sparrow | Bullock's Oriole |
| Hermit Thrush | Ovenbird | White-crowned Sparrow | House Finch |
| American Robin | Northern Waterthrush | Dark-eyed Junco | Red Crossbill |
| Gray Catbird | MacGillivray's Warbler | (Gray-headed race) | Pine Siskin |
| Northern Mockingbird | Common Yellowthroat | McCown's Longspur | Lesser Goldfinch |
| Brown Thrasher | Wilson's Warbler | Rose-breasted Grosbeak | American Goldfinch |
| European Starling | Scarlet Tanager | Black-headed Grosbeak | House Sparrow |
| Tennessee Warbler | Western Tanager | Lazuli Bunting | |
| Virginia's Warbler | Green-tailed Towhee | Indigo Bunting | |

Thursday, June 14, 2001, Outdoors, page C2

# 68  Putting off yard work helps wildlife

**It's best to delay some chores until young birds have time to hatch and leave the nest as fledglings.**

Procrastination can be a good thing. Spring snowstorms will melt off the driveway by midday if I don't shovel, and fancy computers eventually are available at garage sales.

On the north side of my house is a deep, dark and quiet forest.

Sheltered by the next-door neighbor's house, when the wind gets there, it drops in speed – and drops litter.

The junipers probably were cute little shrubs when they were planted along the foundation 40 years ago. Today they are leviathans, reaching over my head, 8 or 10 feet high and as wide and deep.

I keep thinking I should cut a few branches at Christmas – especially ones shading the window by my computer. The evergreen smell would be nice. But then I forget, and it's May or June before I dig out the pruning saw.

Why do I procrastinate gardening and yard work, which I enjoy?

Perhaps because my other obligations are less forgiving of missed deadlines. Other than the lawn, of which the boys have charge, things grow slowly enough around here there's never a pruning crisis – especially since I cultivate the natural look.

Well, the stars finally lined up right last week, and I found the pruning saw and headed for the woods, intent on bagging a few branches.

Actually, the hunting euphemism doesn't translate here. We don't bag branches. We keep them for yard projects and firewood.

I sawed around the computer window and moved to the next window, but as I grabbed a branch, it squawked.

Mama Robin flew up out of her nest and chastised me from the edge of the neighbor's roof as I hurriedly backed away.

Deep, dark woods may be the epitome of safe bird habitat, but this is the first time the robins have chosen it over the trees out front. In fact, the nest is not deep in the juniper branches, but sort of on top.

By pressing my forehead to the window from inside the house before Mama Robin settled back in, I could see at least three eggs. When she's on the nest, she sits as stoically as an avian Buddha.

A few days later, I had a call from someone concerned because her family cat had slightly mauled a baby bird that fell out of its nest. What should she do?

Here are some suggestions in order of preference.

First, try putting the nestling back in the nest. Some young, however, will just fling themselves out of the nest again, or the nest may be too high for you to reach safely.

Or, if the baby is fairly well feathered and close to being able to fly, let the parents take care of it on the ground. Keep pets and children away.

Once, I tried making a nest out of a bucket, placing it where the parents would visit and feed the baby, but it evidently wasn't cat-proof.

The next option is to buy worms where fish bait is sold and start feeding the baby yourself.

Kelly, who works at the Cheyenne Pet Clinic, said baby birds only need to be fed once a day.

If you're squeamish about worms, try foods from this list she recommends: brown rice (cooked), frozen corn, cooked pinto beans, crushed dog kibble, soaked millet, lean meat, white cheese, fruit (especially oranges), green

2001

vegetables, carrots or squash. For treats, try dabs of yogurt, cottage cheese or dried fruit like raisins – but no nuts.

Kelly also said technicians at The Wildlife Rehabilitation Center, located at the clinic, are happy to feed baby birds for you to get them ready for release.

Let me get on my soapbox here for two ideas.

First, nature doesn't expect every seed to lead to a flower or every bird egg to lead to flight. Some progeny have to become food for others, whether it's baby worms

feeding robins or baby robins feeding hawks.

But on the other hand, bird blood on your cat's paws is not part of the natural balance because domestic cats are not native to our area.

Letting your cat play with baby birds, besides doing damage to individual birds and bird species in general, does nothing for the cat that you and a catnip mouse couldn't do better indoors. And it's safer for your cat, which won't be exposed to bird-borne diseases and other outdoor hazards.

You could build a screened porch-type kennel like a friend of mine has for her cats. They still get to go outside, but everyone is safe.

This is a great time of year to procrastinate the right things.

Put off mowing the prairie, where killdeer and meadowlarks nest on the ground. Save the tree pruning and ditch clearing until the young have cleared their nests by June or July. Let the wild tangle at the back provide escape from predators.

According to Kenn

Kaufman's write-up on robins in "Lives of North American Birds," I may have to wait 12 to 14 days for Mama Robin's eggs to hatch and another 14 to 16 days for the young to fledge.

While I practice procrastinating pruning, if I open the window and let strains of Mozart float down to Mama Robin's nest while the chicks are still in the shell, will they grow up smarter and survive better than other robins? Or will they emerge from the nest chirping the "Piano Sonata in B Flat Major"?

# 69 Thursday, June 21, 2001, Outdoors, page C2
## State fair to offer birdhouse, bird feeder contest

CHEYENNE – A 7-year-old boy gave Wyoming State Fair director Barney Cosner the idea of adding a birdhouse and birdfeeder contest to this year's fair.

The youngster built a birdhouse last year and approached Cosner about hanging it at the fairgrounds.

Cosner said work is ongoing to make State Fair Park in Douglas appealing to family recreation year-round. Playground equipment was installed recently, for example. Some contestants may want to donate their birdhouses to help attract birds to the park.

Birdhouse and birdfeeder entries will be judged in five categories: most functional birdhouse, most functional birdfeeder, "most unique" birdhouse, best craftsmanship and best youth entry.

Contest sponsor Home Depot will provide gift certificates for the winners in each category, and the fair will provide rosettes to each first-place winner and for the People's Choice Award.

Official rules are listed in the Wyoming State Fair 2001 book. County fair offices will provide copies of the rules (page 30) and the Crafts entry form (page 193).

Birdhouses are class K-7-1 and birdfeeders are class K-7-2.

The contest is open to any bird enthusiast, and there is no limit to the number of entries per person, although a $1 entry fee must accompany each entry.

Exhibitors, in addition to permanently marking entries with the entrant's name, address and daytime phone

number, must accompany each entry with a written description.

The description should explain the "inspiration for the design, material made of and/ or used, amount of time to build, unique features, specific birds considered, best habitat location and a few details about the builder."

Exhibits may be sent to the Wyoming State Fair, 400 W. Center Street, Douglas, WY 82633; or they may be delivered in person to the Crafts building Aug. 11 or 12 from 10 a.m. to 6 p.m.

Judging will take place Aug. 13 at 1 p.m.

Exhibits will be released Aug. 19 between 6 a.m. to 1 p.m. To have the exhibits shipped back, exhibitors must include adequate funds with their entry forms.

Cosner said he expects the winning entries will go on tour, possibly with the winners of a new quilt contest.

To make a birdhouse functional, interior and entrance-hole dimensions must be specific to the species of cavity-nesting birds you hope to attract.

The Wyoming Game and Fish Department recommends the book "Woodworking for Wildlife," by Carrol Henderson, which lists dimensions for particular species.

Dimensions are also listed at the U.S. Fish and Wildlife Service website, which also has information on bird feeders. Local USFWS offices may have hard copies of the pamphlets available.

CHEYENNE BIRD BANTER

# 70 Robins fledge, face cruel world

Thursday, June 28, 2001, Outdoors, page C2

Wind, rain and pea-sized hail over the last few weeks did not noticeably affect our resident robin family.

Every day that I checked the nest, it was normal to see one of the parents settled down over the eggs.

Then one day neither parent was there, and I couldn't see the turquoise-colored eggs either. I wondered if the nest had been raided and abandoned.

Turns out it's hard to see into the shadows at the bottom of a nest if you're looking sideways through a window screen. So I removed the screen.

The tiny nestlings were there. Their little bodies quivered with rapid breathing as they slept silently in a heap between feedings, eyes shut tight.

In the gloom of the nest their bright yellow gapes glowed.

When robin babies wake for feeding, these bright-yellow edges of their mouths, plus their red mouth linings, give the parents a target for dropping in food. As they mature, their beaks become totally yellow.

After a week, the three nestlings became pudgy little feather balls overflowing the nest. Their breasts were taking on an orange hue, polka-dotted with brown.

When I opened the window to look at them, their eyes sparkled at me like the jet beads on a cape my ancient relative left behind.

On the 10th day after they hatched, I could count only two beaked faces staring at me. Had one already jumped ship?

But when I checked back a few hours later, the third one had reappeared. Evidently its siblings had been standing on it.

As I watched, one climbed over another and flapped its wings. As soon as I shut the window, Momma Robin returned with more groceries.

She has had the same routine each time – landing on the edge of the neighbor's roof, at the ridge, hopping a few feet, stopping, then hopping again until she reaches the edge of the rain gutter. Then she hops across to the nest.

A worm is neatly looped up and held by her beak, yet she can still make her one-note call to her young.

Now the grackle family across the street is quite raucous. They too are nesting in junipers. I can't see into their nest, but all the hullabaloo every time one of the adults flies into the bush alerts me to their location.

Parents, bird or human, have to deal with the unpleasant chore of disposing of baby poop. Baby birds produce "fecal sacs," which the parents either eat or carry off.

These grackles have been departing the nest with the white blobs in their bills, but as they fly up and over our house, they lose their grips and the sacs splat on our front window.

It would be convenient if I were doing fecal analysis to find out what the young are eating. But since I've been digging alongside the grackles in the garden, I'd guess it is mostly pill bugs and worms.

A commune of mourning doves visits our yard every day. I can't tell who is male and who is female or who is mated to whom. I've seen up to eight birds at a time. They're probably nesting in neighborhood spruce trees.

Between the birds and the bat that zoomed by the other morning before dawn (no – zoomed is not the right word, because bats make no noise, even less than a flit), we should have good insect and weed seed control this summer.

When the young leave the nest, the parent robins will still be taking care of them. There are a lot of predators and dangerous situations out there for which they need parental guidance.

A caller the other day brought to my attention the perils of deep window wells, especially the 6-foot ones at new houses. A baby bird died in hers because it couldn't climb out and was too young to fly.

The caller suggested propping a branch against the inside of the window well or installing covers.

This week I feel a little like Momma Robin will soon. The boys have left home temporarily. Will they be safe out there in the world?

Will they know enough to feed themselves properly?

Will some kind soul assist them when needed, like the driver the other day assisted ducklings across Dell Range?

In the time it has taken to write this, the first young robin has fledged, one is sitting on the rim of the nest, and the other is on a nearby branch. An unhatched egg lies at the bottom of the nest.

Ken Kaufmann writes in "Lives of North American Birds" that once the young are fledged, the male will take over parenting while the female prepares the nest for the second brood. At this rate I can procrastinate pruning the bushes at least another month – maybe even till Christmas.

*Cutline for photo of robin nestlings:*

Tiny nestlings wait patiently for feeding time. Barb Gorges/correspondent

2001

# 71

Thursday, July 12, 2001, Outdoors, page C2

## Some summer ponderings

I think I can pinpoint when summer weather finally reached Cheyenne. One week, shorts seemed too chilly to wear and the trees were trying to leaf out again after the frost.

By the end of the next week, after being away only five days, everything was sun fried, and the hallmark of the high plains summer, the late day thunderstorm, was again a firmly entrenched pattern.

At the beginning of the five-day interlude, when we stepped out of La Guardia Airport, rain had left New York City surprisingly cool and sweet.

But by the time one of my brothers-in-law, Peter, took us on a little expedition two days later, before reporting to the funeral home, it was sizzling again.

Peter showed us the Bartow-Pell Mansion in Pelham Bay Park, in the northeast corner of the Bronx.

Mr. Pell bought the land from the Indians. However, a couple years later, in 1654, the first governor of New York felt it necessary to grant him title. The present house was built 180 years later by a descendant.

Like something out of Masterpiece Theater, the lawn rolled out behind the granite edifice in a series of terraces (one with a fountain) surrounded by a garden wall. A wrought iron gate and arch at the bottom framed a view of woods.

Peter, the brother who migrated farthest from New York, to Alaska, has an exploratory streak. After we perused the herb garden, he said there was this really neat trail from which you could see a swamp.

So we settled Aunt Dorothy in the shade of a huge yew and entered a summer jungle like those I remember from my Midwestern childhood: same nearly impenetrable green humidity, same brambles scratching bare arms and legs, same stealth mosquitoes, whose bites take a few hours to reach and sustain their greatest potency, and same spider webs sticking to my sweaty face.

There was lots of music in the treetops but the singers were all hidden – except a cardinal performing an aria from the top of a large dead tree.

The morning before, we Wyomingites tried to find a sanctuary up near Croton-on-Hudson which was listed on the Saw Mill River Audubon Society website.

We would have missed the small sign buried in the roadside vegetation if we hadn't at that moment pulled over to let traffic pass.

A man driving heavy equipment said the gravel road was washed out further up and we'd have to approach from somewhere else in the trail system.

I was thinking we could maybe negotiate the washout anyway, but then I remembered we had a rental car and not our pseudo-four-wheel drive van.

The recommended trail head turned out to be at the end of a road of exclusive new houses sprawled on a hill overlooking the Hudson River.

The road ended at the entrance to the country club, where we shared the tiny parking space with a limo, the driver snoring in the front seat. Was he waiting for a hiker to return?

But by then it was time to renegotiate Route 9 and get back to the Bronx for the wake.

Our last day in New York was the day of the funeral. We hiked up a sunbaked hillside to the grave site. As we crowded within the shade of a lone cedar tree for a few last words for my late mother-in-law, the cemetery workers waited for us in the shade of the woods at the top of the hill.

I would like to know if those woods are part of the cemetery or whether, when we come back, they will have been cleared for new houses.

Were the woods just the regrowth of land that was farmed 200 years ago? I've forgotten what it's like to live in an ecosystem where if you don't constantly plow or graze or pave, trees grow.

I got the window seat and a bird's view of the landscape from Kansas City to Denver. The grid of fence lines and roads stretched as far as I could see.

Other than the occasional woodlot on the east end of the flight and sections of rangeland on the west, the plow marks were interrupted only by traces of green following streams.

How industrious we are. The cultivation of the Great Plains is so complete. If I were a migrating or grassland bird, I'd be in despair, forced to seek refuge in the feral growth under fences and along ditches, wondering how those blackbirds always seem to adapt.

And in view of having wasted too many too-large servings in New York City restaurants during our visit, I wonder, if we all put on our plates just what we needed to eat to maintain healthy weight, would there be a few more acres left over to restore to native landscapes?

--**Seasonal notes:** You'll be happy to know Mama Robin started her second clutch on the first of July. She built a new nest – away from our original window of observation – but right in view of another window, though I am limiting myself to only one or two looks a day.

Her spotted-breasted teenagers are busy picking ripe sand cherries in the backyard.

Meanwhile, hungry young grackles are still following their parents around, making horrible noises of woe and travail. The goldfinches are in fighting form and loudly proclaiming territorial rights as they finally begin their own breeding.

**72** Thursday, July 26, 2001, Outdoors, page C2

# Don't poison your yard to save it

Which is more hazardous, walking New York City streets or hiking Wyoming's back country? They have muggers; we have bears.

New York got ahead of us last summer with an outbreak of West Nile Virus, which can be fatal to some jays and crows -- and some people.

I was surprised, however, by a news release from the National Audubon Society. According to Ward Stone, chief wildlife pathologist for New York State's Department of Environmental Conservation, among 4,000 dead birds collected and tested for the virus, synthetic chemical pesticides were a contributing factor or cause of death more often than any other agent, including the virus.

Lawn chemicals were among the most common toxins.

About a dozen pesticides approved for backyard use have caused documented die-offs of birds, author Joel Bourne writes in "The Audubon Guide to Home Pesticides."

Nationally, we use three times more chemical pesticides on our lawns at home, school and the golf course than the total amount used by farmers. That statistic comes from David Pimental, a Cornell University scientist.

More than 100,000 cases of pesticide exposure were documented in 1998 at U.S.

regional poison control centers. But the centers do not cover the whole country, and many people do not think to report what just seems like flu symptoms.

What's a conscientious person to do?

Recently Audubon published a poster, "10 Commandments for a Healthy Yard," which helps answer that question. Here are the commandments, with my local interpretation.

**--Go Organic.** For a quick introduction to organic yard care methods – which will save you money as well as make your yard safer – pick up an issue of Organic Gardening magazine or visit www.organicgardening.com.

**--Make Your Turf Tough.** Use grass varieties meant for our climate. Use sharp mower blades and cut high. Water well a couple times a week, rather than watering lightly every day.

**--Go Native!** Plants native to our area will be less susceptible to pests – and take less water. Check with the folks at the Cheyenne Botanic Gardens for suggestions, or visit their website, www. botanic.org.

**--Know Your Enemies.** Figure out what bugs you have, whether you have enough of them to make treatment worthwhile and when in their life cycles is best to strike – and with

what. Call the University of Wyoming Cooperative Extension horticulturist, Liberty Blain, 633-4383, or bring your bugs and diseased leaves to her (in containers) at the Old Courthouse, 310 W. 19th.

**--Treat Only When Necessary.** Use nontoxic methods first, picking off insects, pruning affected areas and hosing down plants. For more remedies, look for books such as "The Encyclopedia of Natural Insect & Disease Control," edited by Roger B. Yepsen or "Rodale's All-New Encyclopedia of Organic Gardening" edited by Fern Marshall Bradley.

**--Pick Your Pesticides.** Don't go for the "shotgun" approach. It will kill beneficial insects as well as pests. Use the least toxic product. The Environmental Protection Agency's rating system is "caution" (least toxic), "warning" and "danger" (most toxic).

**--Use Biological Controls or Biopesticides.** If you can't borrow a goat to eat your thistle, biological pesticides decompose more rapidly and are better at targeting the pest than chemical pesticides. Check the EPA's biopesticides website, .... (although I don't agree that genetically altered plants should be included in EPA's definition of biopesticides).

**--Follow Directions and Protect Yourself.** And don't forget to protect other people, pets and wildlife habitat from exposure. Read the label. Less is best.

**--Respect Your Neighbor's Right to Know.** Ever had your windows open to the light summer breeze – and the chemical drift of whatever your neighbor's lawn care service is spraying? Thank goodness the City of Cheyenne is using modern methods to control mosquitoes instead of spraying malathion everywhere.

**--Teach Tolerance and Be Tolerant.** As the poster explains, "Create natural yards, with a variety of pests, predators, weeds, wildlife and native plant species."

**My favorite:** "Enjoy controlled untidiness, not time-consuming lawn maintenance," and "show by doing."

I don't use pesticides in my yard, so had I been writing these commandments, I would have left out numbers 6, 8 and 9.

But if we were all to subscribe to the organic yard care philosophy, just imagine what the birds would think!

To get your own copy of "10 Commandments for a Healthy Yard," including pesticide information and sources for healthy alternatives, download it from ....

2001

# 73 Donations sought for injured birds

Thursday, August 9, 2001, Page A4

**Juveniles destroyed a dozen cliff swallow nests before throwing 50 hatchlings into Dry Creek.**

CHEYENNE – Donations of food and nesting materials or cash are being requested by Cheyenne bird rehabilitator Karin Skinner.

Last week, three juveniles destroyed about a dozen cliff swallow nests and threw the approximately 50 newly hatched young into Dry Creek where it crosses the Greenway at College Drive.

Thirty of the surviving young were brought to Skinner, an experienced bird rehabilitator who has state and federal permits to care for wild birds.

Fifteen swallows pulled through and are presently being cared for by volunteers at WildKind, a branch of the Humane Society in Fort Collins, Colo., because Skinner could not handle the every-half-hour feeding schedule and still care for her other bird patients.

Donations of nesting materials, such as a heating pad, toilet tissue and paper towels, and food, such as meal worms, wax worms, berries and vegetables, are needed to replenish Skinner's stores depleted by the swallows, and to prepare for their return in a week or so, when their feeding schedule slows.

To arrange for a donation, call her at 778-6177.

Skinner expects the swallows will be released towards the end of the month.

The Wyoming Game and Fish Department is considering how best to handle the case against the perpetrators of this wildlife crime.

# 74 Bird flashcards get a technological twist

Thursday, August 9, 2001, Outdoors, page C2

The Wyoming Game and Fish Commission has approved a "Wildlife Worth the Watching" grant to Audubon Wyoming for a joint educational project titled, "Wyoming Bird Flashcards."

So, yours truly, the (unpaid) project coordinator has to get to work.

The first task is to find photos of 30 more species of birds.

Back in January I perused the Game and Fish's slide bank and found images of the 55 other species we'll be considering as "Wyoming's most noticeable birds."

If you or anyone you know takes pictures of birds, please call or e-mail me for the list. I can use clear slides, prints and maybe even digital images (I'll have to check with the technical guru).

Everyone who donates an image will be listed in the credits.

"Flashcards?" you're thinking. Well, with a twist.

When I first started visiting classrooms to introduce birds before leading field trips, I took a slide projector. But it took almost as long to set up the equipment as it did to give the talk. And it's hard to interact with a class when they are sitting in the dark.

So a few years ago, when someone gave me an Audubon calendar, I started my collection of 8x10 bird flashcards.

Unlike slides, flashcards can be viewed without equipment. They can be passed around, grouped by type, compared side by side, put up on the board or set up in a tree.

As I was blowing up pictures from my field guide one day at the copy shop to fill in some gaps (copies for educational or personal use are legal), I ran into Chris Madson, editor of "Wyoming Wildlife," and I had a flash of inspiration.

Game and Fish probably had lots of bird pictures! Perhaps we could prepare bird flashcards to distribute to educators around the state.

But I found out traditional printing of large color pictures on cardstock is prohibitively expensive. Even the Cornell Lab of Ornithology, when I talked to them, said it was too costly.

Then I had another "aha" moment, derived from my student teaching experience. Modern textbooks now provide worksheet and test masters on CD-ROM so the teacher can print out what they need, and perhaps print out flashcards of just the birds the class will study before a field trip.

CD images can be viewed on computer, of course, so teachers may choose to use them in the computer lab, and community educators can use them with computers hooked up to projectors, as I have at done at Laramie County Community College.

So I started looking into the technical angle of producing CDs. I took a giant shortcut last October when I attended the Wyoming Literacy Conference here in Cheyenne and met Joe LaFleur, the author of the Better Birdwatching CD-ROMs.

He did all the technical work on his CDs, and he is a wildlife biologist. What luck!

Then I asked Audubon Wyoming to be the organizational sponsor when I applied for grants, and so here we are.

The project will be more than just 85 pretty bird pictures. We'll also have information on each species, a list of birding hotspots around the state, species lists for different habitat types (so teachers can figure out what birds to study for their area or field trip) and a list of bird-related resources and how to find them. This is where birders around the state can contribute to the project.

CHEYENNE BIRD BANTER

However, instructional materials are practically useless in Wyoming these days if they haven't been translated as ways to fulfill state educational standards.

So along with flashcard activity ideas and lesson plans will be correlations to standards, not only in science, but geography, math, language arts, and maybe even music. Anyone out there working on their Master's in curriculum design need a project?

Finally, what Game and Fish has learned is don't send valuable instructional materials out to every school or teacher unsolicited. It's like giving away puppies. They'll be perceived as more valuable and more desirable, and treated better, if you ask, say, $25 each.

So Audubon Wyoming and Wyoming Game and Fish plan to show teachers how to use the flashcards in workshops offered for recertification credit next summer, and every participant will get their own Wyoming Bird Flashcards CD.

Each Wyoming school district will get one for their instructional materials center, too, as well as Audubon chapters, bird clubs and any education-minded, non-profit organizations that request one.

Perhaps we'll have to send a CD to Cornell.

I love synergy, when seemingly unrelated things we know suddenly realign and combine to become something new and useful.

**Seasonal Note:** First sign of fall migration: a broad-tailed hummingbird visited the bee balm in my garden July 31.

# 75 Don't rely on color for bird ID

Thursday, August 23, 2001, Outdoors, page C3

One useful way to identify a bird is by silhouette, because us bird watchers are always looking up toward a sky that often backlights our quarry, turning it into a colorless shape.

Also, the color of some birds, like hummingbirds, is produced by iridescence, light refracting within the feather structure, rather than pigment. So, if the sunlight doesn't shine just right, colors won't match the field guide pictures.

Knowing your bird silhouettes can help you distinguish groups of birds, for instance, the ramrod straight posture of a flycatcher perched on a wire, compared to the forward tilt of a finch.

It may even help identify unique-looking individual species. The canvasback can be distinguished from other ducks even in the tricky non-breeding plumage of fall because of its ski slope profile from forehead to bill tip.

I was thinking about silhouettes the other evening on our first-ever Cheyenne High Plains Audubon Society overnight field trip.

The campout had also appealed to Auduboners from Laramie, Casper and Wheatland, and we were sitting companionably by a small campfire along the North Platte.

Above us, the bit of sky between the cottonwoods had been serving as a stage since we'd first arrived late afternoon, when the swallows were performing insect-snatching aerobatics.

During dinner, a common nighthawk, with boomerang-shaped wings, joined in, confusing me with its cry, which I have difficulty distinguishing from the western wood peewee.

Then three American white pelicans drifted by. It was quiet enough and they were low enough that we could hear the sound of their wings creaking as they adjusted to air currents.

Even if the light had not been enough to see their smooth, white bodies, black edged wings and golden orange bills and feet, the pelicans' compact body shape and huge wings make for a prehistoric-looking profile easy to identify.

Later, a skein of Canada geese crossed the sky stage. By then the sunset footlights were too low for markings to be picked out. But the proportions were right, and a few honks dispelled any doubt.

Every summer evening along the North Platte and other rivers there is a magic minute when day gives way to night, when swallows turn into bats.

I can never quite pinpoint it. I look up once and see arrow-shaped swallows darting after airborne food, and the next time I look up, the flying insectivores have scallop-edged bat wings.

The bats, probably little brown bats, swoop much lower than swallows – making silhouettes against the firelight as often as against the starlight. It's a comfort knowing that they may be getting mosquitoes before mosquitoes get me.

In the morning I woke at whatever dark hour it was when the killdeer called. Instead of getting up at the crack of dawn for birdwatching, I laid in our flimsy nylon backpacker's tent "bird listening" – until I heard the sandhill cranes rattle by. Those I wanted to see!

After camp was packed up, Jane Dorn led us back to Saratoga Lake.

It's a hard time of year for bird identification since everyone seems to have molted out of recognizable breeding plumage – or hasn't grown up enough to wear it.

However, the common merganser young were old enough that they had the same strange head shape as their mother, making her difficult to pick out as the eleven youngsters crowded around her.

I could correctly identify the first dark, flying tubular shape I saw as a double-crested cormorant, but on the second dark tubular shape, I missed the difference in bill. When that bird landed on a sandbar, it turned out to be a white-faced ibis (now's a good time to pause and look that up in your field guide!).

The marsh wren doesn't seem to have a silhouette at all. It stays as invisible as its

2001

69

nest, a Baba Yaga affair with legs of bulrush stems – if Baba Yaga were an African native who could weave a conical house of reeds with a hole in the side for a door.

The marsh wren boasts of its secure invisibility by constantly singing, driving to distraction birders trying to see it.

Slow animals are seldom seen against the sky or water as silhouettes. Instead, we need to pick out the pattern of their coloration against the pattern of earth and vegetation.

So, it's surprising that when we field trippers glanced away from the song sparrow we were scoping, we noticed the yellowish, olive brown pattern camouflaging the leopard frog at our feet.

By the end of the day, I was home, back with the familiar silhouettes of doves sitting on the utility lines.

I have to remember though that even the most common birds are worth examining closely, for how else will I notice when I'm looking at the square-tailed silhouette of the rarely seen Eurasian collared-dove instead of the usual pointy-tailed mourning dove?

**Note:** Thanks to everyone who responded to bird rehabilitator Karin Skinner's request for donations. She expects the cliff swallows to be released in the next couple days.

# 76 Local group works to launch exotic bird rescue operation

Thursday, August 30, 2001, Outdoors, page C1

CHEYENNE – When Tina Kozma left Maryland last January, she decided she wanted to volunteer with the local exotic bird rescue group at her new home.

Alas, one did not exist in Cheyenne.

Kozma has owned birds and has helped at a pet shop aviary, handling birds such as canaries, parakeets, finches, cockatoos, parrots and macaws, to keep them accustomed to humans. She soon met Cheyenne resident Jeanine Stallings, who has had a long-time interest in animal welfare.

Together, with wildlife biologist Mary Jennings and former exotic bird breeder Jean Larkin, they formed Prairie Wing Exotic Bird Rescue in June.

"You have ferret rescues and (potbellied) pig rescues," Kozma said of groups found around the country.

Better known are rescue groups for particular dog breeds. Often an expensive purebred is acquired, and then, due to the owner's death or the discovery that the pet is more of a commitment than they can handle, the owners are forced to give them up.

People may buy a bird impulsively, Kozma said. But the new owners may not know birds need lots and lots of social interaction, and without it they may develop obnoxious behavior such as shrieking, or damaging woodwork.

"They don't handle that kind of neglect," Kozma said.

Larkin, who has raised exotic birds for 23 years, said, "I've seen this over and over again. People don't understand the needs of a bird," including proper nutrition, besides seeds, and the need to be socialized.

"Because of their high intellect, they have the intelligence of a 3- to 5-year-old and need constant challenge. When they become part of a human group, they regard you as part of their flock," said Larkin.

Without socialization, she said, the birds may become aggressive, phobic or engage in screaming or self-mutilation.

Larkin didn't let just anyone acquire one of her birds and interviewed prospective owners extensively. "I gave them the third degree," she said.

The larger the bird, the longer it is likely to live. With a lifespan of up to 75 years, a large species like the African grey parrot may have several homes in their lifetime, said Larkin.

While many dog owners can wait out the lifespan of dear Fluffy, and then decide not to acquire another dog, the larger exotic birds are more likely to outlive their owners or their owners' ability or desire to care for them.

Based on ads she's seen in the paper, Kozma said, "I'm thinking there's more (need for rescue work) here than we suspect."

"We have three foster homes available as of now," said Kozma.

Foster birds will get medical attention, but most importantly, volunteers will continue to handle and socialize them. In some cases, the foster family will be working with behavior problems.

Prairie Wing will then work to place the birds in adoptive homes.

Right now, Kozma is advertising for cages, as well as letting the public know the rescue service exists. She's also researching grants and other funding sources to handle medical and operating expenses.

"We're trying to start out small, but my daughter and her friends are interested in fund raising and advertising," said Kozma. They also are interested in organizing bird shows as fundraisers.

CHEYENNE BIRD BANTER

**77** Thursday, September 6, 2001, Outdoors, page C2

# Riparian areas: Ancient sacred sites still valuable today

Nearly 30 years after applying for a "wider opportunity" to spend a week at Girl Scout National Center West, I finally made it to the camp, now known as The Nature Conservancy's Tensleep Preserve, located near Ten Sleep on the western slope of the Big Horn Mountains.

The occasion of my visit was the last hurrah of the Wyoming Riparian Association, of which my husband Mark was a member.

It's not too often a group's mission is accomplished and it formally disbands.

The WRA was formed in 1989 at the request of Gov. Mike Sullivan in order that disparate groups from agriculture and environment, and resource professionals from agencies, would begin discussing what they could agree upon regarding the future of riparian areas.

It was the forerunner of cooperative resource management, now a commonly used strategy for resolving natural resource conflicts.

A riparian area is a type of wetland that is the transition zone between water (rivers, streams, lakes and ponds), and dry upland. It is productive for both wildlife and livestock.

Riparian areas account for only one to two percent of Wyoming's acreage, but if a birder only visited those areas, he'd eventually see a third of the 398 bird species listed in Wyoming Game and Fish Department's bird checklist. Birds whose habitats are listed as wetlands – the actual marshes, lakes and rivers – account for almost another third.

Tensleep Preserve harbors a few wet spots deep in canyons.

Naturalist James "Tray" Davis took us to Canyon Creek. We first dropped down into the canyon the depth of a mere flight of stairs but switched immediately from aridity to humidity.

A huge bush of Rocky Mountain bee plant was humming with butterflies and hummingbird moths. Boxelder and wild clematis formed a screen hiding cliffs rising increasingly higher as we hiked upstream.

We waded the creek several times to get to our destination, the Alcove. Its sandstone overhang had the acoustics of a band shell.

Imagine carrying on a conversation with someone 50 yards away as if they were next to you - provided you faced the rock when talking.

The Alcove is considered to have been sacred to Native Americans for hundreds, if not thousands, of years.

All along the wall we saw pictographs, which experts have recently decided depict images of ancient tobacco seeds, part of a cultural tobacco reverence, perhaps marking growing plots.

Before you spend too much energy considering what archeologists will think of our tobacco advertisements in a thousand years in the future, consider this: a thousand years in the past, a riparian area like the Alcove was receiving special treatment.

The day before, our family unexpectedly visited another ancient riparian landmark, the Medicine Lodge State Archeological Site outside Hyattville.

We were on the way to see the dinosaur tracks between Hyattville and Shell, driving the Red Gulch/Alkali Backcountry Byway through desert as dry as the name of the road.

Around a corner we encountered an old pickup pulling a travel trailer, but it was stalled broadside to the deserted road where the driver had attempted to turn around. He said he and his wife were supposed to meet friends at a campground when their engine apparently vapor locked on the hot, steep, treeless hill.

We determined their destination was not in the forest up ahead, but 20 miles back at the archeological site. So, we took the wife down and found their friends.

Three of the men quickly organized a rescue party while the women stayed behind on the banks of Medicine Lodge Creek, in the shade of cottonwoods, not far from pictographs painted by ancients who had made this riparian area another of their sacred places.

On the way home we drove the Hazelton Road, a primitive scrape along the spine of the southern Big Horns. Our experienced eyes could visualize the treachery that would probably result with snow or rain, even though the nearly treeless slopes were now too dry.

Every other fence post seemed to sport a hawk, and horned larks blew with dust across the road.

The only signs of humans were a few travel trailers and shacks off in the distance now and then, marking summer sheep or cow camps.

The only people we saw were rounding up and loading their livestock – early no doubt, due to the drought. Water is everything.

During its 12-year life, the WRA provided funds for ranchers to improve their riparian areas and for workshops examining riparian values and best management practices.

And now the WRA can be laid to rest because the ancient message has been relearned. The former members will continue to retell it so it will spread like water on parched earth: our green oases are most valuable. They are life.

**Seasonal Note:** House finches are mobbing the sunflowers they planted around our bird feeder. Warblers are fattening up for their trip south by searching out bugs in the trees. First frost may be only days away.

*Ty Stockton becomes Outdoors editor.*

2001

71

**78** Thursday, September 20, 2001, Outdoors, page C2

# Expo is for bird watchers, too

It was a gorgeous day. I haven't used that adjective lightly in years, ever since I took up a last name pronounced the same way.

We had the wind in our faces, the river sparkling in the valley spread out below us, snow sugaring the mountains beyond – and pigeons soaring overhead. We were finding our way through thick grass, yucca and prickly pear cactus.

The boys, Brunton compasses in hand, were taking the bearings and I was providing the pacing. Three years of marching high school band halftime shows with steps measured by 10-yard lines has given me a perfect 5-foot pace.

Before we reached our last point, the wind was blotting our map with drips, and we nearly stumbled straight into a family of black bears.

Luckily, they were only cutouts.

We were at the fourth annual Wyoming Hunting and Fishing Heritage Expo at the Casper Events Center two weeks ago. How about you? Did you go?

If you are a bird watcher who doesn't also hunt and fish, you might say, "Why bother?" But we bird watchers have much in common with people who hunt and fish. For one thing, we should all practice using a map and compass. And Brunton's other product line is binoculars, of which they had a good selection on display.

The Expo filled the Events Center floor as well as the upstairs concourses with booths.

Every hour for three days there were three different presentations in conference rooms and one on the main stage, everything from turkey calling, knots and leaders for fly fishing and Theodore Roosevelt in person, to talks about birds of prey, writing about the outdoors, outdoor survival, making a wildlife video and even an outdoor clothing show.

Friday, thousands of school children piled in to explore the hands-on nature activities, virtual fishing, casting with real poles, canoeing, and mucking about with stream rehabilitation, besides the compass course.

My sons' favorite was the chance to try their accuracy throwing darts with a modern version of the prehistoric atlatl. They were so taken with this that the next day at home they used sticks and duct tape to make their own.

There were still a lot of children with parents, grandparents or scoutmasters in tow during my visit Saturday.

I liked having the chance to meet up with old friends. As much as hunters and fishers like their solitude in the field, I noticed they love to hang out and talk shop almost as much.

Down at the Audubon booth there was a bird quiz featuring donated mounted birds. The walls were decorated with the entries for the contest for their first nature photography calendar.

Down the way, the Wyoming Game and Fish Department's nongame biologists were discussing birds too and sponsoring a drawing for bird clocks.

Lois and Frank Layton of Casper attended their fourth Expo with some of the permanent residents of the Murie Audubon Bird Rehabilitation Center that they run. When I dropped by, the short-eared owl, which lost a wing in a collision with a car three years ago, was on Frank's glove, attentively watching its audience.

The owl, and the other residents of the center, will be temporarily homeless for a couple weeks this month as the old flight barn comes down and the new one, built with donations, goes up.

The Expo is as much about hunting and fishing as it is about wildlife conservation of all kinds, and the associated land use topics.

Bird watchers should be as interested in land management as any of the federal land agencies or the watchdog groups like the Wyoming Wildlife Federation, Wyoming Outdoor Council, Sierra Club and Biodiversity Associates. This was the perfect opportunity to meet them all, under the auspices of a common cause.

Besides organizations representing all the major game animals, there were a few commercial exhibitors like Coleman and Cabela's. But nothing is for sale at the Expo, except raffle tickets and food.

There are no entrance or parking fees for the Expo. There are no suffocating crowds unless you're waiting to try landing a virtual fish.

In the six hours I was there, I didn't get to see and do everything.

Guess I'll have to go back next September. You should come too, unless of course you're out somewhere else bird watching for the whole three days.

**Personal Note:** Five months ago, I contemplated a bird's eye view from New York City's highest rooftop. Now it has disappeared, along with the elevator operator with the eight-minute comedy routine, the gift shop clerk with the snazzy tie, the morning's complement of tourists from all over the world and the invisible mass of office workers.

**79** Thursday, October 4, 2001, Outdoors, page C2

# Fall festivals welcome birds back

"Nearly 40 field trips, including pelagics, shorebirds, neotropical migrants and Florida specialties. Last year's species list – 173!" Florida Birding Festival and Nature Expo, Clearwater, Oct. 4-7.

"Birding and wildlife watching field trips; raptor watch and banding." Florida Keys Birding and

---

CHEYENNE BIRD BANTER

Wildlife Festival, Marathon, Oct. 12-14.

"Pelagic birding trips, field trips, seminars and workshops, boat trips, nature and art exhibits; featuring Debra Shearwater, keynote speaker." Space Coast Birding and Wildlife Festival, Titusville, Florida, Nov. 8-11.

"...featuring Pete Dunne; over 100 workshops and field trips." Festival of the Cranes 2001, Socorro, New Mexico, Nov. 15-18.

"Field trips, seminars, trade show and more!" Rio Grande Valley 8th Annual Birding Festival, Harlingen, Texas, Nov. 14-18.

An informal survey of the September/October 2001 issue of Bird Watcher's Digest shows many bird festivals happening this fall.

Why would someone plan a bird festival in the fall? Then I looked at the locations: Florida, Texas and New Mexico, where birds might spend the winter.

Maybe festival planners think of them as welcome back parties for birds that were away breeding up north all summer.

Maybe folks in the south consider their birds to be away just for the hot summer months, while we northerners consider our birds to be away just for the inclement winter months.

We do have a couple species that return to Wyoming in the fall.

Maybe we should have a rough-legged hawk festival, celebrating their return from their summer sojourn along the Canadian coast of the Arctic Ocean.

However, juncos, instead of flying south, fly down in elevation.

Like people with nearby mountain cabins, they return to town in the fall – if they are the Oregon race. The slate-colored juncos, like American tree sparrows, have to fly back from Canada for the winter.

It is as hard to state a truism about migration that holds true for all birds, even birds of the same species, as it is fall bird festivals. I don't consider Delaware to be very southern, but there's the Snow Goose Festival Oct. 27 and 28. They have a population of snow geese that come back to spend the winter at local national wildlife refuges while most snow geese head for the Gulf or the Pacific coast.

"Congregate....where the birds migrate." The 55th Annual Cape May (New Jersey) Autumn Weekend, Oct. 4-7 is more of a bon voyage party.

Cape May, on the southern tip of the peninsula between Delaware Bay and the Atlantic, is the narrow point of a funnel that collects birds from the Atlantic flyway before spouting them out over the water.

I thought the Great

Louisiana Birdfest was following the southern pattern, but I noticed it's scheduled for April. Spring seems to me to be a logical time to celebrate, when the birds return singing and in colorful breeding plumage.

Then there's the "2nd Annual Wild Bird, Wildlife and Backyard Habitat Expo" scheduled for West Bend, Wisconsin Nov. 2-4. I suppose they are welcoming winter birds back to their feeders. I wonder, what nickname do they have for this event?

Any issue of any bird magazine has dozens of ads for travel in any season to destinations such as Costa Rica, Spain, Venezuela, southeastern Arizona, Alaska, the Amazon, Trinidad, Tobago and the Galapagos.

The ads for Texas are as "booster-ous" as any Texan I've ever met. "Bay City, Texas – come see why we were No. 1 in 1997, 1998, 1999 in the North American Christmas Bird Counts."

"Top 5 reasons to bird in the Texas Hill Country River Region, 5. Painted Bunting, 4. Barred Owl, 3. Curve-billed Thrasher, 2. Black-capped Vireo, 1. Golden-cheeked Warbler."

There's only one problem with travel. You miss what's going on at home. It seems whatever week I take a vacation in the summer, it's at a crucial time for the garden or

when the robins are fledging.

If I were to leave in the fall, I'd miss the warblers passing through in migration. If I were a really good birder, which I'm not yet, I'd miss the migrating shorebirds and waterfowl, but so far I can't tell them apart when they're in their blah non-breeding plumage, so I don't have as much to miss.

Travel, even with a really good bird watching guide, though it may add to your life list, won't give you the familiarity you get with the birds at home. There's something to be said about knowing one place from season to season and year to year.

On the other hand, you can't really appreciate where you live until you've been around. So, if you offer me a pelagic birding trip (pelagic refers to ocean-going birds like shearwaters, frigates and albatrosses), I'll go pack my Dramamine right away – even if I'm gone the same time as the Cheyenne spring bird count!

Meanwhile, you and I can keep our eyes open right here at home. A friend last week reported 30 turkey vultures have been roosting in an evergreen in the Avenues. On a recent Saturday morning, I saw 12 of them circling downtown office buildings. Do you suppose they were using thermals created by the hot air generated all week?

# 80 Bird travel plans show variety

Thursday, October 18, 2001, Outdoors, page C2

I grew up with the simplistic notion that birds fly south for the winter.

But any winter day in

Cheyenne is filled with birds still here: crows, starlings, sparrows, pigeons, as well as juncos, owls, and kingfishers.

Why do some birds change location? The one-word answer: food.

The birds most likely

to leave are insect eaters, like warblers and tanagers, because when it's cold the insects quit reproducing or die.

Birds that eat aquatic animals and vegetation, like ducks and shorebirds, are programmed to leave too, since their food storage units get frozen over in winter.

Birds that can switch over to seeds or eat seeds exclusively (or gleanings from garbage like the crows do) can make it for the winter, as long as they eat fast enough to provide the calories they need to burn to keep warm.

Why do birds bother coming north in the spring? I've heard it's because year round birds can't possibly make use of all the fecundity of northern summers themselves and there's plenty left for the migrants. And migrating birds are looking for a boost in energy for their energy-intensive work of raising young.

Few North American field guides show where migrating birds spend the winter if it's beyond the United States. They may show a little of Mexico, but often stop right at the border.

Imagine my delight when I was told about a website that has range maps including North, Central and South America. Migration is so much more interesting when you can get a bigger picture.

The website, http://wildspace.ec.gc.ca/.html, is sponsored by the Canadian Wildlife Service. You can even choose to read life histories of birds and other animals in French. The western grebe becomes "Grébe élégant."

I found one drawback as I researched the site for the Wyoming Bird Flashcard project. There is no range map for the broadtailed hummingbird because it never goes as far north as Canada.

The range maps themselves have some shortcomings. They don't show migration routes.

The Wilson's phalarope is shown to breed in a swath across Canada and the northern United States, wintering in southern South America, but how does it travel back and forth?

Does it funnel down through Central America, or does it fly out over the Pacific, or maybe the Atlantic?

My Golden field guide, which cuts range maps off mid-Mexico, shows the Wilson's apparently starts over land, while the red phalarope, which breeds in the Arctic, prefers transoceanic routes exclusively.

Wish I could ask a Wilson's phalarope about its precise itinerary next spring when I see one swimming in circles on the pond just down the road from the sewage treatment plant.

These range maps also don't tell you the timing of migration.

Bob Dorn, co-author of "Wyoming Birds," told me that some species of shorebirds meet themselves coming and going.

The adults may hurry to their Arctic tundra breeding ground early in the spring. But as soon as egg laying is accomplished, the males head south again, passing northbound teenagers. Not old enough to breed yet, last year's young are in no hurry and take their time arriving.

The range maps will show an area where for some species, the birds can be found all year round, between the summer-only breeding range and the winter-only range. Wyoming is part of year round range for the belted kingfisher, so it breeds here as well as winters.

But do we have the same kingfisher on Crow Creek all year round, or do the locals migrate further south, leaving their niche open for the kingfishers coming from Canada?

Or do kingfishers that summer in Canada just hop right over Wyoming and winter in Mexico? The inquiring mind needs to research sources more scholarly than field guides.

And then it occurs to me that birds that migrate to South America during our winter are actually enjoying the southern hemisphere's summer. So why don't they breed at the tip of Chile in December the way they breed on the Arctic tundra in June?

Stan Anderson, head of the University of Wyoming's Wildlife Co-op Unit, says it would be too physiologically demanding.

Looking at the range maps though, there seems to be at least one bird that's thought of this strategy. The common snipe's range map shows it breeding in Canada and northern U.S. – and southern South America.

But, according to Stan, the southern birds don't migrate.

Author Scott Weidensaul, in his book about migration, "Living on the Wind," said five billion birds migrate back and forth in our hemisphere.

It seems to me each kind has worked out its own migration strategy. Each has a story for us to decipher.

**Seasonal Note:** The juncos came back to my yard Oct. 10. They thought it was cold enough in the mountains to come down to town. The uncovered sweet peas in my garden had yet to freeze.

# 81

Thursday, Nov. 1, 2001, Outdoors, page C2

## Did you happen to see the one that got away?

Several times lately I've flushed a small brown bird out of the front bushes, but it seems rather late in the season for it to be a warbler searching for bugs, and it's the wrong color, too.

I never think to look for birds out front. We had a feeder full of finches there briefly, before the squirrels destroyed it, but the way this brown bird skulked in the bushes, I didn't think it was one of those constantly exclaiming house finches.

Once, when I was getting the mail, the brown bird flushed and then dropped back in the bushes instantly. So, letters in hand, I decided to stalk it. Shouldn't I be able to see it move again if I stared at that one spot? And I did, for a tenth of a second.

On a bright snowy day in the Medicine Bow Mountains soon after, I was gazing at a vista of coniferous tree trunks, wondering which ones were legs of elk standing still. I've seen so few elk in timber, I wasn't sure I could identify them by just one body part.

I was following Mark, who was slowly and warily following elk tracks in the snow – too slowly for my left big toe, which needed a brisker pace to thaw.

There was no wind, and I could hear nothing but perhaps my own blood circulating.

We didn't talk. We hardly snapped a twig. However, the snow was incredibly loud, making a weird, hollow "skrunk" as our boots compacted it with each step. I was so busy watching where I walked (did the elk really barge through here and not break the branch threatening to poke my eye out?) that my seeing any game was unlikely.

Finally, we headed back, having gotten too far out to want to drag an elk that distance. We loosened up. The squawk of the occasional gray jay, the inquisitive voice of a mountain chickadee and the chatter of squirrels no longer seemed like signals to

our quarry.

And then of course, we caught movement. I could only pick out the curve of its neck and the color of its hair but knew it was only a deer disappearing over the ridge.

The next day the wind was back. The sound of it in the trees, like huge, long trucks rushing past on a wet street, drowned out the noise of our boots on snow and the spitting flurries hitting our hats and coats.

We moved faster and stayed warmer walking logging trails, following game trails and crossing beaver dams, but saw not a single track, though everywhere trees glistened in the dim light where elk had recently rubbed their antlers and removed the bark. Lack of tracks had us mystified.

The coniferous forest is a colorless place on a dull, snowy day: dark gray tree trunks, dark green, nearly black, needles, gray snow and gray sky. But the snow wasn't deep enough mid-October to cover the scattered bunches of fresh green grass everywhere. Since elk are grass grazers, rather than browsers of twigs like deer, their absence seemed even

more mysterious.

Finally, we began to parallel Fox Creek, bright with tawny dried grass and red shrub stems. A handful of ducks whistled away and suddenly short-lived sunshine shimmered silver on the open water and fluttered gold on remaining willow leaves.

But of course, we weren't over there along the creek. We were back in the dark timber, negotiating a giant's game played with unpainted pick-up sticks left from a forest thinning project.

We saw only one party of hunters while walking, but driving again, we were soon blocked by two pickups parked side by side in the narrow road. About 20 yards into the timber, munching leavings of a fresh forest thinning job, was a moose. Then another materialized to make a pair, a bull and a cow.

It isn't moose season in the Medicine Bows very often, so we all enjoyed just watching, as if we were in our own private Yellowstone. Moose are so huge. I think their bellies would come even with our four-wheel drive windows. And talk about legs looking like tree trunks!

All things considered,

I think I prefer antelope country, where I don't have to check my bearings with Mark every ten minutes. Besides, it's tough looking for game while trying to run a forest obstacle course at the same time. And there are other hazards in the woods.

After Mark indulged me in a walk into the forest in the dark after dinner, he reminded me of how many mountain lions were radio-collared for a study around here a few years ago. It's creepy, suddenly changing from the stalker to the stalked.

Well, maybe someday you'll see the ones that got away from me. The little brown bird will show an identifying mark and the tree trunks will turn into elk legs.

**Seasonal Note:** Things are picking up at my backyard bird feeders: many goldfinches (now pale), pine siskins and juncos (Oregon race), a few grackles and red-winged blackbirds, and one each so far of blue jay, red-breasted nuthatch, mountain chickadee, white-crowned sparrow, white-winged junco (migrating from its summer home in the Black Hills) – and someone's junco-eating gray cat. How are things in your yard?

# 82 Feeder watchers keep track of winter birds

Thursday, November 15, 2001, Outdoors, page C4

CHEYENNE – Do you like to watch birds at your feeder? Do you sometimes jot notes about which species have visited? Do you like to compare notes with your neighbors, or even friends and family far away?

Then joining Project FeederWatch is your next step.

Project FeederWatch

is one of several "citizen science" projects sponsored by Birdsource, a collaboration of Cornell Lab of Ornithology and the National Audubon Society.

Citizen scientists become the eyes and ears of researchers, collecting data across North America. This will be the 15th season feeder

watchers report the birds they see in their backyards.

The data collected can show long-term trends in bird distribution and abundance. Results are published in scientific journals, bird newsletters and magazines, as well as newspapers.

The more that is understood about birds, the more

likely conservation measures will benefit them.

The results are also tabulated and available at the Project FeederWatch website, http://birds.cornell.edu/pfw.

At the Data Retrieval page, you can ask for the Wyoming summary for last season and find that American goldfinches were at 42

2001

75

percent of reporting locations in November, but none of the 19 Wyoming Feeder watchers reported a white-crowned sparrow until the end of December.

Over in the map room section, you can watch an animated map showing the fluctuation of goldfinch numbers month to month across the continent.

If you need help identifying birds, go to Cornell's Bird of the Week page, at http://birds.cornell.edu/bow, where birds featured in the past will probably include species you're looking for, with illustrations, identification tips, life history information and even songs.

Feeder watchers choose either a weekly or biweekly reporting schedule. If you're away from home for part of the winter, that's all right (it appears three of Wyoming's feeder watchers left for a couple months mid-winter).

And it isn't necessary to observe your feeders all day long either. A quick count in the morning or just before dusk, when feeder activity is at its greatest, is quite sufficient.

After all, even if you can only devote a few minutes, remember the researchers can't make it out to your yard at all.

Prospective feeder watchers can find out how to sign up by calling (800) 843-2473 or visiting the website.

The $15 sign-up fee gives participants a research kit that includes a full-color feeder bird poster and calendar and the FeederWatcher's Handbook, as well as data summaries and findings published in Birdscope, Cornell Lab of Ornithology's quarterly newsletter.

# 83 Pool parties popular with winter birds

Thursday, Nov. 15, 2001, Outdoors, page C2

Pigeons and doves are among the few birds with the ability to suck liquids, states author Kenn Kaufman in "Lives of North American Birds."

This is because the parent birds produce a milky substance from a gland in their mouths, "pigeon milk," that the young suck by inserting their bills into the corner of the parents' mouth.

The ability to suck carries over into getting a drink of water.

Other birds can fill their mouths with water, but then have to tip their heads back to let it run down their throats.

I'll have to wait until spring to observe mourning doves again, though I suppose I could go down to one of the bridges over Crow Creek and observe the pigeons.

However, I can observe a house finch getting a drink in my backyard. It sits on the edge of the bowl, leans forward to dip its bill in the water and then leans back before dipping again, see-sawing until it's had enough, or until too many other birds crowd in.

Whether or not you plan to compare bird drinking habits, providing water for your backyard birds is a good idea, especially in the winter when natural water sources may freeze.

My present method uses an overpriced, black plastic bowl, about the size of a large dog dish, which was supposed to absorb solar radiation and keep the water from freezing in cold weather.

The bowl is too small to qualify as a bird bath, though a few grackles splashed around in it this summer. But it's easy to pick up and bring in any morning I need to run hot water over it to release the ice before refilling it.

The system we used years ago in southeastern Montana, where winter temperatures were often subzero, involved a shallow plastic garbage can lid used upside down as a liner for a traditional, pedestal-styled bird bath.

The lid being flexible, we could just pop out the circular ice chunk every morning and refill the lid with a kettleful of boiling water. Sometimes the water froze over in an hour, so I'd repeat the procedure later in the day.

The good thing about these two systems is that they cost nearly nothing and are easy to disassemble and bring in for a soap and water cleaning. However, they don't provide a constant supply of unfrozen water.

There are plenty of bird bath options (water for bathing is as important to birds as water for drinking). Some come with electric heating elements built in, or you can buy the element separately.

The luxury models of heating elements are thermostatically controlled so you don't boil any bird feet or waste electricity unnecessarily. They will also shut off automatically if the bird bath goes dry.

I checked locally and both A & C Feed stores have heating elements with or without the bird bath, as do McIntyre's Garden Center and Oasis Market.

Whatever you use for a bird bath, it should be shallow, with the water not more than 2 inches deep. It's better if the bottom slopes down toward the center.

If your bird bath is just a plastic bowl, you may want to hold it down with a large rock in the middle or mount it somehow to keep the wind from tipping it over. Then again, if your site is too windy all the time, the birds won't visit.

Cornell Lab of Ornithology recommends that the bird bath be out in the open, away from hiding places for predators (i.e. house cats), yet with perches not too far away so the birds have some place to sit while they preen.

Roger Tory Peterson, in his video about creating backyard bird habitat, expounded on how attracted birds are to moving water. Although elaborate spigots are available, one easy method is to hang a plastic jug full of water over the bird bath, with a small hole pricked in the bottom to provide a constant drip.

I wonder, if I painted the jug black, would it drip most winter days?

Meanwhile, we had such

warm weather clear through the first week of November, that I had to water my still-blooming snapdragons and rudbeckia.

I let the hose run in the garden, and next thing I knew, the siskins, finches and juncos had abandoned the bird feeders for a raucous water party.

I was extremely lucky last week and had a Townsend's solitaire land in the top of one of our big trees. This time of year they make a one note "bink" call over and over, instead of their spring melody.

Unfortunately, our mountain ash is too young to provide a berry crop this year for a fruit eater like the solitaire.

Or was it looking for one of my wild garden water parties it heard about?

# 84 Count birds while waiting for dinner

Thursday, November 22, 2001, Outdoors, page C3

CHEYENNE – So, you're taking a little break from turkey preparation or consumption and taking a look at today's paper.

How about taking a look at the birds at your feeder and, with just a little effort, participating in the 10th (western) Thanksgiving Bird Count?

All you need to do is glance out the window a few times over the course of an hour and note how many and what kinds of birds you see at your feeder.

Back in Virginia, in 1966, Ernest Edwards started the tradition of counting birds on Thanksgiving and it has grown in popularity.

However, by 1992, when John Hewston, a professor of wildlife at Humboldt State University in Arcata, California, was asked to take over the count in the western states, there were only about a dozen western participants.

Since then, Hewston has recruited quite a few more bird watchers. Last year, in the 13 western states, including Hawaii and Alaska, 449 people made 462 counts, tallying 149 species of birds.

After the first of the year, Hewston sends out a report listing the bird species in order of most counted. In 2000, the house sparrow came in first place, with 3,137 counted, followed by the dark-eyed junco, house finch, black-capped chickadee, pine siskin, American goldfinch, European starling, California quail, mourning dove and white-crowned sparrow.

The quail, which ranges from California up through Oregon and into Washington state, is a clue that over half the count participants come from those three states.

Wyoming had 23 participants, above average for the other states.

Our list of top 10 species for 2000 began with the house sparrow, house finch and dark-eyed junco, three birds very common at Cheyenne feeders.

It then continues with gray-crowned rosy finch, evening grosbeak and western scrub-jay. The first two are unpredictable and the third prefers juniper woodlands.

Our top 10 continued with black-capped chickadee, American goldfinch, rock dove (pigeon) and Cassin's finch, the last another species hard to find at local feeders.

How will the 2001 count shape up? You can get your own report from Hewston, the count compiler, by simply taking part in the count and sending in your results.

And, if you are motivated enough to make a little outing after dinner, consider making a second count at a local park.

[The 2001 Thanksgiving Bird Count Form was formatted to be clipped from the paper. It included instructions, a short list of potential bird species and blanks for count information.]

# 85 Close encounters of the bird kind

Thursday, Nov. 29, 2001, Outdoors, page C2

It was cold enough to freeze water in the birds' water dish (my friend Marta recommends I get one of those heated dog water dishes), but not so cold a morning I thought I needed shoes just to grab the dish to bring it in when I let the dog out.

However, Lincoln made a beeline for a spot on the patio and I forgot all about the water dish.

Huddled on the concrete was a stunned American goldfinch, a male in winter plumage, just the lightest wash of yellow on his feathers.

Birds thumping into our kitchen window rarely happens anymore.

Between the pair of feline faces nearly always present in the window when the birds are active, and the dust on the glass, few birds think our window reflects a continuation of our backyard.

My first impulse was to rescue the goldfinch from the onslaught of the breath of a 100-pound dog, so I scooped him up and cupped him in my hands, leaving his face free.

His toes were just a tickle on my skin, his black eye, alert and unblinking.

"Oh, I am so sorry about the window," I thought. The bird was so light it was as if I held my imagination.

But my bare feet were standing on cold reality.

Bird feet are engineered differently. A goldfinch weighing less than half an ounce feels less cold. First, bird feet are mostly bone, sinew and scale, with few nerves.

Also, as Bill Thompson, author of "Bird Watching for Dummies" explains it, the arteries carrying warm blood from the heart to the toes are interwoven in the legs and feet with the veins, so the arteries warm the cold returning blood. And because bird feet don't have sweat glands, they don't stick to metal perches when it's cold.

After a few minutes, even after moving to the doormat, I realized it might take longer for the bird to recover than to frostbite my feet.

By this time, the rising sun was eye level and shining on the bird feeder.

Our sunflower seed tube feeder has a saucer attached to the bottom and a wire cage around it that keeps squirrels out and lets small birds in. So I carefully placed the goldfinch on the feeder, safe from stray cats, close to food, facing the sun.

Twenty or thirty minutes later he was gone, though he may have been back later as one of the flock emptying the thistle feeder.

Of the disciplines available to me in the College of Natural Resources at the University of Wisconsin at Stevens Point where I was getting my degree 25 years ago, studying wildlife was far more popular than measuring trees, digging soil pits or analyzing water samples.

It seems to me most people have an urge to touch an animal (especially furry ones) or interact with it. For some people pets or livestock are good enough. For others, only wild animals will do.

For some of us, observation is fine, but there are always crazy tourists trying to pet the buffalo.

Some of the legitimate ways to handle live wildlife are catching and releasing fish, helping band birds, getting certified for rehabilitation work or becoming a wildlife field biologist.

But sometimes wildlife comes to you. I heard of two incidences this fall where birds chose to interact with people.

The beginning of October I got a call from a friend who'd been sitting on her deck with out-of-town visitors when a blue jay convinced them to feed it by hand.

A few weeks later another friend told me about having a blue jay light on her shoulder as she was carrying seed out to fill the feeder.

Our consensus was that these must have been young birds who didn't know better than to trust humans, but being from a relatively smart species, they'd learned it was possible to manipulate food sources.

I once read how to train chickadees to eat from your hand. It involves sitting as still as a bird feeder until they get used to you.

But I don't get chickadees in my yard very often, and I don't think the most prevalent birds, house finches, are as smart.

We have some sort of relationship anyway. They know when the backdoor opens, they don't need to fly far. Within a few minutes the disturbance is over, more seed's been spread and it's time to get back to eating.

How much more trust could a birdwatcher ask for?

# 86 Calendar features young artists

Thursday, November 29, 2001, Outdoors, page C4

## The youngsters' works will help publicize the Stop Poaching program.

CHEYENNE – Blair Edholm's prize-winning entry for the 2001 Wyoming Hunting and Fishing Heritage Expo poster contest, sponsored by the Wyoming Game and Fish Department, will be getting additional publicity as the cover for a new calendar put out by the department and the Wyoming Wildlife Protector's Association.

"When we saw the artwork from the poster competition, we decided to use some of the posters for the calendar," said Jay Lawson, chief game warden for Game and Fish.

Besides providing recognition to the school children who did the artwork shown for each month, Lawson said the calendar's other purpose is to publicize the Stop Poaching campaign.

The toll-free number to report poachers, 1-800-442-4331, is listed on each page.

And to promote law-abiding hunting, license application deadlines are printed in red.

Printing of the calendars was paid for with federal restitution funds from people convicted of poaching.

The calendars are available free from the department, though quantities are limited. Call 777-4600 to inquire.

Edholm was a third grader at Rossman School last year when her art teacher, Virginia Allshouse, made entering the poster contest an assignment for all the third graders.

Edholm's portrayal of the theme, "Exploring Our Wildlife Legacy," chose to use torn paper collage for trees, water, fish and the fisherman.

Classmate Christina Dewey, whose entry illustrates October, used the same technique to portray a family of bears in the mountains.

Drawings by two Dildline School students, Brittany Huie and Tyler Davidson, illustrate March and April. The other months are illustrated by students from Buffalo, Casper, Lyman and Sundance.

The theme for next year's contest is "Discover Wyoming's Hidden Treasures." The poster contest is open to students in grades 3-6. A poetry contest is open for students in grades 7-9. For students grades 10-12, there's an essay contest.

Entry forms will be available mid-December.

Contact Traci Sasser at Game and Fish, 777-4600, for more information.

CHEYENNE BIRD BANTER

# 87 Stop Poaching campaign adds new poster, book, bumper sticker

*Thursday, November 29, 2001, Outdoors, page C4*

CHEYENNE – If you enjoy Dragnet, a good "who dunit" or tales of good winning out over evil set in Wyoming, there's a new book out this year you might enjoy.

The third edition of "The Quest to Safeguard Wyoming's Wildlife Resource" is a collection of short stories – 50 case histories of poachers convicted of all kinds of transgressions against Wyoming's wildlife, to be exact.

It also gives an inside look at what Wyoming's wildlife law enforcement officers do and how the wildlife forensics laboratory contributes to solving cases.

The 144-page book is compiled by Russ Pollard, Wyoming Game and Fish Department wildlife law enforcement coordinator, and published by the department in cooperation with the Wyoming Wildlife Protectors Association.

Available free from Game and Fish, the book is part of an ongoing campaign to publicize the department's Stop Poaching Hotline, a joint effort of the department and the association since 1980 for reporting suspicious activities related to wildlife.

The toll-free number, 1-800-442-4331, is also appearing on newly redesigned bumper stickers and posters.

"We have seen a fairly sharp increase in calls," said Jay Lawson, Game and Fish's chief game warden, but not necessarily an increase in violations.

The increase in the number of people carrying cell phones is responsible, Lawson thinks. "It's almost like having this giant Neighborhood Watch," he said.

Besides the use of cell phones, wildlife law enforcement has benefited from a change in public perception.

Back in the old days, poaching was sometimes considered to be a way for a poor man to feed his family. However, a survey of Wyoming residents conducted for Game and Fish published this spring shows that food is considered by respondents to be only a minor reason for poachers' violations today.

The violation that irked survey respondents the most was the wasteful taking of game and fish when an animal was killed and the meat allowed to go to waste.

"The Quest to Safeguard Wyoming's Wildlife Resource" documents case after case of violators killing for the thrill, killing for trophy antlers or abandoning their kills when they felt threatened by law enforcement.

In many cases, a tip from an observant citizen, whether they be hunter, angler or outdoor recreationist, leads to an investigation. The cases, written with professional understatement, don't mention the bravery necessary for officers confronting alleged poachers who may or may not be carrying weapons.

The various authors sometimes let the villains speak, "I wondered when you guys would catch up with me," said one.

"Goodness me, I don't know anything about elk," said another, eventually confessing to an overlimit of elk because he was so entranced with using his new gun.

Not only are poachers shown to be stupid and/or greedy, they often have histories of other kinds of violence and lawbreaking, making wildlife law enforcement even more dangerous.

Since 1980, 25 percent of calls to the "Stop Poaching" hotline have led to arrests.

Callers are not required to reveal their names, testify in court or sign a deposition, but they can be rewarded when an arrest is made or a citation issued. More than $130,000 has been paid so far.

Stop Poaching books, posters and bumper stickers are available at the Game and Fish office located at Central Avenue and Bishop Boulevard.

# 88 Keeping juncos straight isn't for bird-brains

*Thursday, December 13, 2001, Outdoors, page C2*

It's that Christmas Bird Count time of year again. Jane Dorn, our local count compiler, is going to ask me what kind of juncos I counted.

Sometime between the 1966 and 1985 editions of my favorite field guide, four species of juncos were lumped into one, the "Dark-eyed Junco."

On the off chance that the American Ornithologists' Union could someday split them out again, we should keep track of all the different kinds. But it isn't easy.

Let Roger Tory Peterson refresh your memory. "Juncos are unstriped, gray, sparrow-shaped birds with conspicuous white outer tail feathers, gray or black heads, pale bills.

"Species with gray sides: White-winged Junco (white wing bars); Slate-colored Junco (fairly uniform gray); Gray-headed Junco (rusty back).

"Species with rusty or 'pinkish' sides: Oregon Junco (rusty or brown back)." The pink-sided junco, a paler version of the Oregon, was lumped with it prior to 1961.

But even if you can remember all these side and back color combinations, consider that the Oregon female, paler than the male, looks an awful lot like the pink-sided, except the pink-sided has an even paler throat.

One junco species didn't get lumped, the former Arizona or Mexican junco. It was renamed the "Yellow-eyed Junco." You have to go to Mexico or southeastern

Arizona to find one.

Cheyenne in winter, however, is a great crossroads for the other juncos.

The white-winged breeds in the Black Hills and winters in southeastern Wyoming and eastern Colorado.

The slate-colored breeds in Canada and the northeastern states, but winters just about everywhere between our borders with Canada and Mexico.

The Oregon breeds in the Pacific Northwest and may travel in winter as far as the western Great Plains. The pink-sided sub population breeds in Montana and northwestern Wyoming and winters in the southern Rockies.

However, according to Sibley's range maps, the gray-headed breeds only as far north as Colorado and Utah and winters south into Mexico.

But birds don't read field guides so I can understand why Jane is interested in knowing exactly which juncos we see. Recorded observations are what continually redefine a bird's actual range.

What makes it tough to identify the juncos feeding under my kitchen window is that they hybridize. Peterson said, "There is frequent hybridization or integration; therefore it is impossible to name all individuals."

When Dave McDonald, University of Wyoming zoology professor, spoke at the Cheyenne High Plains Audubon Society meeting last month, he described his work with rosy finch species: black, brown-capped and gray-crowned.

To my untrained eye, they look like another group ready to be lumped, and in the past they sometimes have been. But genetic comparisons of blood drawn from birds captured and released shows they are quite distinct.

Dave thinks it might have to do with their preferred breeding habitat being high mountains that act like isolating islands, or maybe a predilection for breeding in the same place they were hatched.

On the other hand, one of Dave's graduate students did extensive genetic comparisons of burrowing owls and found those living as far apart as California and eastern Wyoming were genetically indistinguishable.

I wonder, are their wintering grounds like Club Med and they leave in spring with new friends for locations other than their birthplace?

In the discussion after Dave's presentation, he mused on what genetic testing for juncos might show. I hope, should he ever get funding, he remembers I volunteered the flock in my backyard.

So Jane, until I get the blood tests back, I'm afraid many of the little gray birds I count on the CBC will merely be listed as "Dark-eyed Junco, subspecies unknown."

Seasonal note: Bird watchers of all skill levels, including feeder watchers, are invited to participate in the Cheyenne CBC Dec. 29. Call Jane at 634-6328 for details.

# 89 Count birds, not presents

Thursday, December 20, 2001, Outdoors, page C4

## Christmas bird counts scheduled across Wyoming

CHEYENNE – The 102nd Christmas Bird Count started last weekend, but the counts for many areas are yet to come. The counts for Yellowstone National Park, Green River, Laramie and Buffalo, among others, have already taken place.

Before the CBC season ends Jan. 5, 13 more counts will be held statewide, including Cheyenne's on Dec. 29.

Each count is a census of the birds in a 15-mile diameter circle. How thorough the census is depends on the number of observers who participate, whether they are expert birders of novice bird watchers.

Anyone who enjoys birdwatching is invited to participate in one of the counts listed, as either a field observer accompanying a group, or as a feeder observer at home, but the person listed for the count should be contacted in advance.

What began in 1900 with 27 observers counting in a limited area of the East Coast has spread to over 50,000 birders counting 54 million birds in 1880 count circles spread over the Western Hemisphere from the Arctic Circle to southern South America last year.

This year all count compilers will be sending their results in online. Data will be available much more quickly than in the past, when it took nearly a year to compile it in a book the size of Cheyenne's telephone directory.

Historical data is available at the CBC website, though some is still being entered into the data base.

The data will soon be available in a query format that will allow anyone to more easily mine it for information about the range expansion, decline or irruptive patterns of various bird species.

[Accompanying the article was "Christmas Bird Count Schedule," a list of the remaining Wyoming CBCs and their compilers and their phone numbers: Bates Hole, Cheyenne, Clark, Cody, Crowheart, Dubois, Evanston, Gillette, Sheridan, Star Valley, Storey/Big Horn, Teton Valley. A map showing the extent of the Cheyenne CBC count circle was also included.]

# 90 Apprentices wanted for solitaire sleuthing

Thursday, December 27, 2001, Outdoors, page C2

I've been hoping that all the avian activity across the street in my neighbor's junipers would turn out to be waxwings someday, but it always seems to be starlings.

Then Dec. 12, looking out the window while discussing food coop finances on the phone, I saw a robin and another sort of colorless bird there.

As soon as I hung up I grabbed the binocs, but the bird had left the branch. Missed it! No, wait ... there by the sidewalk ... all gray, with white and black lines etched over its back where the wings lay.

Now its head is turning towards me ... white eye-ring ... Townsend's solitaire! Wouldn't it be great if it stuck around for the Christmas Bird Count?

Because solitaires would rather pluck berries in winter than visit feeders, it seems like an unusual bird to me. And some years, looking at the CBC data, there are fewer solitaires than robins, another species with preference for fruit in the winter.

There are 28 years of CBC data posted for Cheyenne at www.audubon.org/bird/cbc, that show the Townsend's solitaire was always reported. In the '70s it averaged five individuals per count and in the last five years, 14.

This alone doesn't tell us anything about population trends because results may vary due to numbers of observers and the amount of time they put in.

So the data also include number of solitaires per "party hour" (birders travel in parties of one or more people). That way we can compare results from year to year and location to location.

In the case of the solitaire, average frequency back in the '70s was 0.208, and in the last five years, 0.509. I would guess the number of berry-producing junipers has more than doubled in 20 years as well.

Two years ago, the Yellowstone National Park count came in first with 4.436 solitaires per party hour (54 birds), compared to our 0.351 (eight birds) in 47th place.

That same year the count in Bend, Oregon, reported the most solitaires, 134. Our meager tally came in 73rd. But we must keep things in perspective.

Of the approximately 1,880 count circles all over the Western Hemisphere, only 207 reported any solitaires last year.

Only us westerners in North America can see them on a regular basis – and only if we know where the junipers and other berry bushes are within our 15-mile diameter count circles.

There's a mistaken impression that only expert birders can take part in the CBC. If that were the case, then no one would ever go on their first one. Serving sort of an apprenticeship for a few years not only helps you with bird identification, but with learning where the birds are.

In 1989, Mark, the boys and I met the grand dame of Cheyenne birds, May Hanesworth, now deceased, who had been compiling count data for over 40 years.

In her 80s by then, she no longer went out, but at the tally party afterward (when the "parties" get together to party), she did interrogate us about every nook and cranny in town where she expected particular birds to be.

Even experienced birders have to apprentice on a count new to them. Last year a longtime friend and resident of Miles City, Montana, who moved to Helena, Montana, wrote, lamenting, "I wasn't sure I wanted to count birds here anyway. I loved it in Miles City where I knew where to go and what to look for."

Novice birders have another role besides perpetuating count circle knowledge. When nearly all pairs of binoculars are focused on some nondescript sparrow, someone has to turn 180 degrees and say, "What's that?"

And because we're counting the total number of birds seen, everyone will love you for asking the question, even if "that" is another flock of starlings.

While folks who are willing and able will meet at 7:30 a.m. in the lobby of the Capitol Avenue Post Office this Saturday to tromp around downtown and Lions Park for a few hours, others are needed at home to count birds at their feeders.

I hope you'll participate. It's a great holiday tradition, great winter recreation and a chance to add to scientific knowledge.

If you can't join the main party, then keep track of the number of birds you see within the count circle, the number of miles you travel, and the number of hours you're out and then join us for the Saturday evening tally party.

For details or to report your data, call the count compiler, Jane Dorn, 634-6328.

# 2002

Thursday, January 10, 2002, Outdoors, C1

## Bird count identifies second state record for hermit thrush

CHEYENNE – A hermit thrush, aptly named for its shy and retiring ways, was the star of the Cheyenne Christmas Bird Count held Dec. 29.

Birder Bob Luce, new to the Cheyenne count, pointed it out, and with help from experienced Cheyenne birders, was able to identify the robin-like bird with the spotted breast.

Jane Dorn, count compiler, said there has been only one other winter record of the thrush in Wyoming, which is otherwise a somewhat common bird in summer in the state's coniferous forests.

Normally the hermit thrush winters no farther north than central Arizona and New Mexico.

Total number of birds counted, 4,138, and total number of species counted, 36, were down from last year's 5,686 birds and 43 species despite 24 observers, which was nearly double the number for previous counts.

The weather was colder than last year, with a high of 29 degrees, and a low of 11 degrees.

There was no precipitation and only a trace of snow on the ground.

The wind was out of the northeast and fairly calm, with gusts reaching only 18 mph.

Though the 1,536 Canada geese counted were not as many as last year, only one other CBC has recorded over 1,000 geese. In the 1980s, geese were sometimes not observed on the CBC at all.

Crows also seem to be following the same pattern. While 97 were seen this year compared to 109 last year, crows used to be scarce. Only by the 1994 CBC did numbers counted exceed 10.

The pine siskin was reported only during the week of the count (the three days either before or after the count) and not on the count day. However, their winter populations are irruptive, meaning they go where the food is. So apparently some other location had a better seed crop.

Neither white-winged nor red crossbills, other irruptive species, were seen on the count this year.

Rough-legged hawks, as expected, continued to be the most numerous raptors in winter.

To compare this year's results with previous years or other locations, go to www.audubon.org/bird/cbc.

### Cheyenne Christmas Bird Count, Dec. 29, 2001

36 species, plus one other species seen week of the count.
4138 birds counted.
24 observers.

Canada Goose 1536
Mallard 574
Green-winged Teal 3
Common Goldeneye 12
Northern Harrier 2
Red-tailed Hawk 2
Rough-legged Hawk 18
Golden Eagle 2
American Kestrel 1
Common Snipe 1
Rock Dove (pigeon) 376

Great Horned Owl 1
Belted Kingfisher 5
Downy Woodpecker 12
Northern Flicker 11
Blue Jay 4
Black-billed Magpie 41
American Crow 97
Horned Lark 21
Black-capped Chickadee 3
Mountain Chickadee 10
Red-breasted Nuthatch 6

Brown Creeper 10
Golden-crowned Kinglet 14
Townsend's Solitaire 29
American Robin 24
Hermit Thrush 1
European Starling 792
American Tree Sparrow 14
Song Sparrow 5
White-crowned Sparrow 1
Dark-eyed Junco (total: 78)
  Gray-headed 3

Oregon 14
Pink-sided 20
Slate-colored 8
unspecified 33
Red-winged Blackbird 42
House Finch 41
Pine Siskin (week of
  the count only)
American Goldfinch 19
House Sparrow 330

# 92   Looking for birds in all the right places

Thursday, January 10, 2002, Outdoors, page C2

Turnout for the Cheyenne Christmas Bird Count Dec. 29 was better than any other year we can remember ... for number of observers participating.

Despite the bank's temperature proclamation of 17 degrees at 7:30 a.m., 20-odd people (and maybe we did look odd to early post office patrons) were gathered in the downtown post office lobby, ready to beat the bushes.

We had enough people to choose three teams so we could hit our three favorite birding spots simultaneously. But unlike choosing up grade school kickball teams, team members chose which team they wanted to join.

The downtown Capitol complex buildings were not very exciting this year. I can tell you where we've seen a great horned owl, brown creepers and Townsend's solitaires in previous years, but this time we didn't have much to show the folks new to our count except for the three species federal law allows to be "controlled" without special permits: pigeons, starlings and house sparrows.

By 8:30 a.m. we were at Lions Park, and luckily it was early enough or cold enough the mallards and Canada geese were still on the ice sleeping, their heads tucked under their wings, instead of swimming in the open water.

Counting sleeping birds is so much easier.

Ten common goldeneyes dropped in, their white breasts and sides making them easy to pick out from the 700 mallards and geese.

A tree full of goldfinches delighted us, along with a downy woodpecker, over by the old greenhouse. On the west side of the park, the cottonwoods gave us eight brown creepers, flickers and both kinds of chickadees.

The highlight was six golden-crowned kinglets playing a fast-paced game of hide and seek among the branch tips of a spruce, frequently hanging upside down for a moment. Their constant frenetic movement in search of bugs and their size of about three and a half inches long combined to make trying to glass their best field mark, golden head stripes, quite a challenge.

As our group straggled around Discovery Pond on our way back to the parking lot, I found myself alone – and attracting squirrels. Did you know the fox squirrels in the park have gotten so tame they will approach and sniff a stick you hold out to them?

As one squirrel and I re-enacted Columbus meeting a New World native (no shared spoken language, just lots of eye contact and gestures), I was aware the tribe was gathering around us. Not just six or eight more squirrels, but a dozen mallards were quietly moving in.

Just call me St. Francis of Assisi – courtesy of the folks who feed wildlife in the park.

Luckily, I rejoined the birders in time to have a solitary Townsend's solitaire pointed out to me before it flew over the Botanic Gardens' greenhouse roof.

As I traversed our traditional routes, I was remembering birds we've seen other years and bird watchers who've died or moved away. I was even thinking fondly of trees and bushes the parks department has removed.

The bush where we unexpectedly found the sage thrasher several years ago was taken out when the sunken garden was filled in. I know the park people think shrubbery can hide people with nefarious agendas and is not safe, but I hope they will replant some.

The trip out to the base in the afternoon was a little different this year, due to security concerns. My family, including my visiting sister and brother-in-law, had to show photo identification, even though we had a vehicle sticker.

I'm not sure all the fuss was worthwhile. Where Crow Creek runs through the family camping area, it looks

like good, brushy bird habitat with great big cottonwoods overhead. But all we counted were about a dozen magpies and a flock of pigeons in the distance.

As we tromped through the fresh skiff of snow all the way down to the bridge and then back up and around to the nature trail, I wondered if we were just visiting at the wrong time.

Maybe we need a fourth group first thing in the morning to check the base when birds are most active.

Finally, just after 4 p.m., as we skirted the backside of the mall in the van, I saw black birds in the cattails. Mark obligingly backtracked through the parking lot, and we were able to tell that they were four red-winged blackbirds.

I wonder if those birds decided belatedly to head south to join the rest of their species after we saw them on a day so much colder than any up until then. I always think of the red-winged blackbird's song as an element that proves it's spring, even if we're due for a few more snow showers. Then again, along with the robins one of the other groups counted, perhaps red-wings are only a sign of spring to those of us who stay inside too much all winter.

# 93   Some birds aren't crowing about neighbor

Thursday, January 24, 2002, Outdoors, page C2

As I turned off the hairdryer a little after 7 a.m. one morning recently, I heard

the end of a ring. Hoping it wasn't the last, I grabbed the phone. It was my neighbor

across the alley, Sue.

"That owl is in the tree again, just west of you and

the crows are picking on it."

Naturally, I immediately abandoned my comb,

2002

83

grabbed binoculars and headed for the alley.

Sue was there and coached me until I was able to see the great horned owl myself, ensconced in spruce branches.

One cawing crow flew at the owl, waggling its claws in its face, but the owl didn't budge.

The crow returned to a safe perch on the powerline, flaring its fan-shaped tail. Ravens have wedge-shaped tails and haven't, apparently, moved into our neighborhood yet.

Sue's neighbor across the street thinks this might be the owl they had hanging around for a couple years. Sue thought maybe it liked our alley because there's a yard light that can illuminate scurrying rodents, though a nocturnal hunter like the owl is well adapted for working in the dark.

Great horned owls prey on wildlife as large as Sue's small dog, but she was more concerned about the owl's welfare and us disturbing it. So after another good look at the avian Buddha, I returned to my yard and morning chores.

Meanwhile, the lone crow had succeeded in attracting at least five others to its cause

(get the pun?).

Two were in my tree, heckling from the back row. Two swayed on the cable TV line, trying to catch their balance and dignity without missing their timing for hurling invectives.

I couldn't see the spruce anymore, but it sounded like two more crows were in there with the owl. They carried on for at least another half hour.

A few weeks before, before Christmas, Sue had left an owl message for me about 7:15 a.m., which I didn't pick up until much later, but I could remember hearing a mob of crows right about then.

The best part of this owl experience has been to find someone happily excited about having a natural predator in the neighborhood, though the crows are not.

Often enough I get calls from people concerned that hawks are eating the birds at their feeder. Isn't that what sharp-shinned hawks are supposed to do? Isn't a hawk a bird too?

I just figure, when I put out seed, I'm feeding herbivores directly and indirectly feeding carnivores, whether they come to my yard or not.

Great horned owls prefer

bigger prey than finches and sparrows. Cornell Lab of Ornithology's on-line field guide mentions they especially like hares and rabbits.

I know we have plenty of cottontails hopping around the neighborhood at 5 a.m. The dog is always trying to drag me along after one whenever we get to do the paper route.

Squirrels are on the list too. We have plenty of those. Five of them come by every morning to sample our sunflower seed.

"....and the occasional domestic cat," reports the CLO. With my luck, it would be my cat on her annual accidental outdoor foray whose bones and hair get turned into owl pellets, instead of the loose cats that defile neighborhood gardens and terrorize wildlife.

Mammals make up three-quarters of the average great horned owl's diet, though 50 species of birds have been recorded as prey, from songbirds to grouse, herons, ducks, geese, hawks, and even other owls.

I wonder if the owl I saw was house-hunting as well. Mid-winter is when owls announce their territories and some may begin nesting in

February. They have to start early because incubation takes a month and getting the young airborne takes another two and a half to three months.

However, great horned owls are lazy. They prefer to use old hawk nests in big trees, and I haven't noticed any around here. Otherwise, they are comfortable in a greater variety of habitats than any other owl.

Wouldn't it be fun to have owls for neighbors? It would mean our 50-year-old suburbanized neighborhood has an original piece of the natural mosaic, even though the prairie and its creek-side cottonwood fringe have been swapped for lawns and evergreens.

**Bird Alert:** Two sightings of the red-bellied woodpecker have been reported in the Pioneer Park neighborhood.

This woodpecker, which is normally seen in eastern Nebraska and further east, has a wide red patch covering the entire back of its neck but has barely any red on its belly.

Please call 634-0463 to report additional sightings and take photos if possible.

# 94 Thursday, February 7, 2002, Outdoors, page C1
# Bird count depends on all bird watchers

Join in for the 5th Annual Great Backyard Bird Count Feb. 15-18, and you won't even have to put your coat on if you can count your backyard birds from the window.

Curious ornithologists at the National Audubon Society and the Cornell Lab of Ornithology wanted to know

where birds were between Christmas bird counts and spring migration counts, and so they organized the GBBC, tapping into a huge reservoir of birdwatching volunteers.

Last year bird watchers across North America counted birds in their backyards and local parks and sent in

53,000 checklists.

There is no fee, no advance sign up and no particular level of bird watching expertise necessary to participate.

The first step is to watch the birds in your backyard or a local park for at least 15 minutes. Be sure to put your

coat on if you go out!

Make a list of all the birds you can identify and the greatest number of each you can see at one time.

Then submit your checklist online at www.birdsource.org. If you don't have your own computer, use one of those available at the

84          CHEYENNE BIRD BANTER

public library.

At the website you fill in location information and the number and kinds of birds you saw.

You can submit a different checklist for each day of the count you wish to participate. You can also count birds at more than one location if you submit separate checklists for each.

Be sure to visit the map room and find out what birders across the continent are seeing over the weekend. Find out where the pine siskins are this winter, as well as some of the other irruptive species such as redpolls and crossbills, which also travel to wherever food supplies are best.

The GBBC website has a wealth of information about winter birds and bird watching, so don't wait, visit before the count begins.

## 95 Immigrant or visitor, red-bellied woodpecker finds food here

Thursday, February 7, 2002, Outdoors, page C2

Birds are illiterate, at least in the usual sense. However, the most successful, longest-lived birds are very good at reading signs in their environment to avoid danger and locate food, shelter and the opposite sex.

Birds do not read field guides.

A red-bellied woodpecker was seen in January in Cheyenne several times by three different people.

I was a little skeptical when I got the first call. I've never seen a red-bellied woodpecker, which, despite its name, is recognized by its black and white striped back and the red on the top of its head (male only) and back of its neck.

Jane and Bob Dorn, authors of "Wyoming Birds," list only two records of the species in the state. One was January 1993 in the latilong that contains Douglas and the other May 1992 in the Cheyenne latilong.

For the purposes of bird records, the state is divided into 28 latilongs, each measuring one degree of latitude by one degree of longitude.

Red-bellieds are birds of the southeastern United States that have gradually increased their range to the north, and now, apparently, to the west.

In the 1961 edition of his western bird guide, Roger Tory Peterson mentions red-bellieds are casual to Colorado, meaning a few records, but they don't merit an illustration. The Stokes' 1996 western edition doesn't mention them at all.

"The Sibley Guide to Birds," 2000 edition, shows the westernmost boundary of the red-bellied's range approximately at the 100th Meridian, that magical line of longitude marking the difference between eastern and western species of biota in North America.

The 100th Meridian slices vertically through the middle of Nebraska, a mere 250 miles east of Cheyenne. What's that distance to an eastern bird with a decent set of wings?

This winter's visitor could be here by some accident of weather – and that would have to be some accident to get the wind to blow out of the east long enough.

It's more likely the intervening Great Plains, thanks to all the mature windbreaks, can now host a species dependent on large trees full of bugs and seeds and fruit.

How many other red-bellieds have visited Cheyenne birdfeeders without being recognized as unusual? How many have met disaster shortly after arriving, such as plate glass windows, storms, loose cats and natural predators, and are never seen by bird watchers?

Chances are we'll have more reports of red-bellied woodpeckers, if only because the number of bird watchers continues to increase.

In this month's issue, National Geographic used the estimate of 63 million bird watchers in this country alone to justify launching its own birding magazine.

What will happen to our red-bellied visitor? We must assume, until proven otherwise, that there's only one, since only one female has been seen each time.

It could survive the winter quite well using the three well-stocked backyard feeding stations it has already found.

It's not a seasonal migratory species, and it may not be inclined to move in the spring, so it could become a resident.

And, compared to its stronghold in the southeastern U.S., it doesn't have as many species of woodpecker competitors out here.

However, a few observations of red-bellied woodpeckers in Wyoming won't change the "accidental" status of the species until there are breeding records.

If the conditions that allowed one member of the species to find its way here stay constant, chances are more will follow and then breeding could happen.

Birds are opportunists. Short of being dropped here by the wind, a bird wouldn't travel to Cheyenne if it hadn't read signs along the way for favorable conditions for survival.

Whether it becomes a resident depends on finding enough of the habitat it is used to, or adapting to what is available.

It's about the same for the rest of us coming to Wyoming from elsewhere. Except, we people have the ability to make things more like our old homes, so we tend to plant trees, diminishing the grasslands and their species.

# 96
Thursday, February 21, 2002, Outdoors, page C2

## Finding out where the birds are

Where can teachers take children bird watching, and what birds could they expect to see? I asked that question a couple months ago while preparing a section on field trips for the Wyoming Bird Flashcards CD project.

Next time I write a grant, remind me to figure in field research. It would have been fun traveling the state looking up good bird watching locations.

But even had I visited extensively, there's nothing like a report from someone on site, someone who has intimate knowledge from observations made over time.

I was wondering how I could find people willing to help when I realized the list of Wyoming Christmas Bird Count coordinators was the key. Since then, over a dozen coordinators have generously shared information.

So, where do you take children, or any novice bird watcher? Responses included the Laramie River Greenbelt in Laramie, Burlington Lake in Gillette, Dry Lake outside Lander, Clear Creek

in Buffalo, several parks in Casper along Garden Creek and the North Platte River, Gray Reef Reservoir south of Casper, Clarks Fork River up by Cody, Hume Draw in Sheridan and the elk feeding grounds along the Snake River up by Jackson.

Do you see a pattern – river, lake, draw, creek, reservoir? With the exception of recommendations to visit the cemeteries in Sheridan and Casper and a few notes to look for horned larks on the uplands, water is where the birds are, in Wyoming, or anywhere else.

Some birds, such as waterfowl, depend on aquatic animals and vegetation for food. Others like the protection from land predators while they float in a lake. Some like to hide or nest in bushes and trees which grow thanks to their proximity to water.

Some birds eat the other animals attracted to aquatic environments, and of course, nearly all birds need water for drinking and bathing.

Wet spots are not equally rich in bird species. Should

a field trip leader or teacher take children to the local park, or to some really spectacular hotspot a couple hours away? Luckily for us, some of the best birding around Cheyenne is Sloans Lake and the area around it in Lions Park.

If you are a teacher trying to spark an interest in birds in your students, do you give them a long term, close-up look at birds at the feeder outside the classroom window, or do you take them for a day to where multi-colored birds are flitting from branch to branch and blanketing the surface of the lake?

If you are a birder intent on adding species to your life list and choose to dash from place to place, you'll never know what a hotspot your own backyard might be. A friend of mine is discovering bird by bird how interesting her own backyard is.

She works at her computer at a window overlooking a well-developed feeding station with all the usual bird foods, plus dog chow. It seems the most unusual

visitor this winter, the red-bellied woodpecker, is attracted to the dog's food (see Feb. 7's Bird Banter for more about the red-bellied woodpecker).

Already she's developing some sense of the time of day the birds prefer to visit and which bird doesn't seem to be just another house finch. Her backyard list has over 30 species so far.

It is this kind of intimate knowledge of a place I was after when I contacted the CBC coordinators. I, a mere reader of lists, might predict certain species in certain habitats, but every spot on earth is unique.

Only the local observer knows which lake has trumpeter swans and what time of year, unless of course the local observer is Terry McEneaney and he has written the book, "Birds of Yellowstone."

Meanwhile, it's my hope Wyoming's teachers and children will go out and watch birds, and through their observations, write books for their own local hotspots.

---

### Bird Books

You can locate a lot of birds with the help of the following books:

--"Wyoming Birds," by Jane Dorn and Robert Dorn, available through the Rocky Mountain Herbarium, University of Wyoming, Box 3165, Laramie, WY 82071.

--"A Birder's Guide to Wyoming," by Oliver K. Scott, available through bookstores or the American Birding Association, 1-800-634-7736.

--"Access to Wyoming's Wildlife" and "Wyoming Wildlife Viewing Tour Guide," both available at the Wyoming Game and Fish Department, 5400 Bishop Blvd.

---

# 97
Thursday, March 7, 2002, Outdoors, page C2

## Birders use cyberspace to advantage

Ever have the feeling you live in a parallel universe? Once in a while you find a

metaphorical door open and discover, for instance, that people you talk to regularly

are having a whole other conversation among themselves via e-mail.

Some months ago, a birding friend held a door open for me, but I put off stepping through until a couple weeks ago when I finally signed up for the Wyoming Bird Discussion Group and Bird Alert listserv known as Wyobirds.

Some of you savvy internet users know all about listservs, and now I'm convinced they are the greatest thing since spotting scopes.

Serious birders have always had a communications system. When Gloria Lawrence saw a pale-phase gyrfalcon in her backyard on the banks of the North Platte River west of Casper Feb. 26, she knew exactly which birding friends would want a phone call. She also posted her sighting for all the Wyobirders.

Gloria maintains a toll-free, state-wide bird hotline, 307-265-2473, sponsored by Murie Audubon Society. She can receive and post rare bird reports and migration information. Too bad my phone doesn't blink when something exciting happens.

Now that I've subscribed to the Wyobirds listserv, reports from a network of nearly 50 members around the state (plus a few northern Coloradoans who occasionally bird Wyoming) come right to my computer as e-mail messages.

Some of you will cringe at the idea of even more e-mail. One option is not to subscribe, but to just go to the website and peruse the archives at your leisure. However, if you want to post any replies or reports for the edification of the group, you need to subscribe, which costs nothing except the time it takes to send an initial e-mail.

Wyobirds was started last May by Will Cornell of Rock Springs. A recent transplant from Kansas, Cornell modeled his listserv after one for that state, managed by his friend and birding mentor, Chuck Otte.

Starting with only five members, the first few postings were Will's reports from birding trips. Then other birders from Rock Springs, and then Green River and Casper, began sharing their observations and answering each other's questions about where to find birds.

The summer months were slow, except for an announcement in July from a member in Laramie that the fall shorebird migration was underway. By September the first sighting of a rough-legged hawk was reported. That's the Arctic-nesting hawk that thinks Wyoming is a balmy place to spend the winter.

Migration is a good time for finding birds rare for Wyoming. Last fall there was a red knot near Casper, a surf scoter near Lake Hattie and a little gull (that's its official name) at the sewage ponds in Green River.

Spring migration is already astir, with reports of mountain bluebirds north of Cheyenne Feb. 19 and eastern bluebirds Feb. 21 at Bessemer Bend.

For serious birders able to chase after rarities, the listserv makes an excellent, low-cost alert system. Meanwhile, the rest of us enjoy knowing there's more out there than the house sparrows in our backyards.

But some discussions and reports are of interest to backyard birders too, such as where blue jays are nesting, or that a Townsend's solitaire was heard singing somewhat prematurely Feb. 21 in Green River, or that someone has rosy finches at their feeders.

I saw one example of political lobbying in the archives, but it was quite forgivable. It was against a proposed law allowing falconers to remove wild peregrine falcon chicks from nests and raise them for their sport.

Wyobirds is a good place to pose a bird question. What is the name for a female swan? The young are cygnets, the male is a cob and the female is called a pen.

If you are a grad student studying the mountain plover, this is the group to ask if they've sighted any. Or if you are travelling across Wyoming on Christmas break, Wyobirders will gladly help you find rosy finches.

Late fall there was a flurry of messages about dates and contacts for various Christmas Bird Counts, and already there's increasing discussion of spring birding trips. But unlike members of listservs for indoor hobbies, there's a chance when Wyobirders go outside we'll see each other.

Sign up information for Wyobird, which is now a Google Group, was included.

# 98 Racing pigeons

Thursday, March 14, 2002, Outdoors, page C1

Though they don't have the prestige of horses, these birds are the thoroughbreds of the sky.

CHEYENNE – It looked like a normal gathering of men on a snowy Saturday morning in February up at the Laramie County Sheriff's Posse building on Yellowstone Road. They were wearing the usual jeans, jackets and embroidered ball caps, except they had white smudges on their shirt fronts. And they were attending a basket show. Baskets full of racing pigeons.

You may have thought pigeon racing was as archaic as an Andy Capp cartoon, but it is alive and well all over the world, though less visible in Cheyenne.

This morning, Cheyenne was host to the winter basket show and auction for the Northern Colorado Flyers, a group with its roots in Nunn, Colo. Along with sacks of pigeon feed for raffle prizes, shallow crates, or baskets, were stacked up around the room. Most baskets were made of metal or plastic and a few were actually wicker.

All were full of pigeons waiting for the judges visiting from Oklahoma, Colorado and Utah.

Bill Hill of Tulsa, Oklahoma, was the senior judge for this event. He explained, "There's 28 classes, starting with 100-mile young cock and young hen (birds that have flown at least one 100-mile race). We look for

2002

conformation, muscle tone, feather quality. We try not to let beauty sway us – until the end."

Steve Buehler of Lakewood, Colorado, was the eye sign judge for this show. He and many in the pigeon business swear the way a bird's eye looks is predictive of racing success. He looks at color, the circle of correlation around the pupil and other attributes, though he said some pigeon breeders think it's all foolishness.

Other pigeons were on display, waiting to be auctioned. Their admirers were examining their pedigrees and making their own judgements. Experienced flyers were taking the birds in hand and feeling the heft of well-developed pectoral muscles that power the wings, making a racing pigeon capable of flying 40 miles an hour for hundreds of miles without stopping.

Rick Brown, who flies his birds east of Greeley, Colorado, said his list of attributes includes "conformation, condition, including health, how soft the feathering, how complete the feathering, racing attributes, balance, fullness of the wing and how feathers are formed."

Pigeon breeders learn genetics, said Stan Freeman of North Platte, Nebraska. Although the emphasis is on breeding for racing ability, some breeders play with color. "It's a given; a blue cock and a red hen, of the young, the blues are hens and the reds are cocks." White is not a very common color, said Freeman, due somewhat to natural selection. White birds are more often taken by

hawks when flying because they contrast with the landscape so well.

A basket show, though a nice chance to visit with other flyers, win ribbons and buy new birds at auction, is not the proving ground that a race is. And just as in horse racing, there's prize money to be won.

Everyone at the Cheyenne show pointed out the most successful flyer present.

Francisco Hernandez of Denver, Colorado, has been an extremely successful breeder, even though he considers his loft of about 75 birds to be small by most standards. "I send for special races, 40 birds a year. I don't buy birds. I fly ones that do good in 300s (miles)." He sends birds to races in Mexico and all over the U.S.

The way winnings are paid out is rather complicated, but recently three of Hernandez's birds brought in $131,000.

Pigeon racing, unlike horse racing, is not limited to the rich. Doug Donner of North Glen, Colo., and Phil Calerich of Brighton, Colorado, remember that as boys in the 1950s, they could afford to fly their birds from as far away as Glasgow, Montana, or any train station in between.

"We'd fly out of Cheyenne, Chugwater, Wheatland, Glendive or Glasgow," said Donner. The boys would put their crate of birds on the train in Denver with $5 or $10 in an envelope for the stationmaster at a particular station, who would then take the crate off the train, feed and water the birds and release them in the morning. He'd ship the empty crate

back to Denver, where the boys would be waiting for their birds to fly in.

Wally Sabell of Arvada, Colorado, got into flying pigeons at a young age and used his hobby to drive his ambitions. "I started at 9 and I'm 72 now. It helped me go into business. My stepfather said I had to pay for the feed, so I went out and got a job and became an entrepreneur to pay for the pigeons. Money helps. It can be a rich man's sport, it can be a poor man's sport. I know every pigeon I've got. I've got 500. (I decided) when I have all the money I want, I can have all the pigeons I want. So far, my wife's let me have all the pigeons I want."

Summer is the usual racing season, but the balmier U.S. climes allow for winter racing. Sabell was just back from the Million Dollar Race World Flight in Sun City, Arizona. He also enjoys entering pigeons in a 400-mile race in Florida. Recently he entered races in China and Africa.

Pigeon racing is a peculiar sport. The owners frequently send their birds to the starting point of a race without attending themselves. Club members will meet the night before to ship the birds out on specially designed trucks and check and synchronize their clocks. Then they each take their clock home and wait for their own birds to arrive so they can clock them in.

The home loft of each pigeon serves as a finish line. The carefully measured and calculated distance from the start of the race to the loft is divided by the time it takes the bird to fly home. This

allows comparison of each bird's speed. The fastest is the winner.

But the fastest bird won't win until it actually enters the loft and someone can take off the special rubber band called a counter mark put on its leg at the beginning of the race, and clock it in and seal it for inspection by race officials.

Modern technology has benefitted pigeon racing. The latest is snap-on scan bands, explained Freeman. "It means it doesn't take two people to train pigeons," one to release and one to be back at the loft. The bird just walks over a plate and is scanned right into the clock and computer. "When your bird gets home, you don't have to be there." He said it works well for a friend who works for the railroad. Now when he's on a run his wife doesn't have to wait all day looking for birds.

Only two club members at the Cheyenne show were women. Donna Case, from Nunn, Colorado, and her husband Steve mark their baskets with the name of their loft, "Flying Nunn." She explained her presence at the registration desk. "I write the best, so I have to take care of the money and the concessions and go for anything they need. I go with Steve when they drop the birds. The last four years we've gotten involved in it quite a bit."

Dixie Rapelje and her husband Dan have the Double D loft, also in Nunn, Colorado. She was pouring coffee in the kitchen. "I do a lot of helping out. My husband, he inherited his loft and the rest just kind of evolved. His father had pigeons for years."

In October, Denver will host the 2002 convention of the American Racing Pigeon Union, including races. The AU, as it is commonly referred to, is one of several national clubs. How do flyers enter their birds in national races when their loft may be on the other side of the continent? How did Sabell enter his birds in races in China and Africa?

For the Denver convention, three of the flyers attending the Wyoming show, Sabell, Calerich and Buehler, will be loft handlers. For an entry fee of $100 per bird, pigeon flyers from all over the world will send teams of six young birds sometime in March, April or May.

Birds only 30 days old have not yet imprinted on their natal lofts. Denver will become their point of reference as they are trained over the summer. One bird from each team will be in each of the six lofts. Three thousand birds from all over the world are expected.

Their breeders will arrive in October to witness the 300-mile race. The five best birds from each of the six lofts will be auctioned to pay race expenses.

A winning breeder could buy his bird back. But what do you do with a homing

pigeon trained to home in on someone else's loft? Sometimes birds come home when they aren't supposed to. Jay Welden of North Platte, Nebraska, said, "I sold birds to Dallas, and two of the three of them came back (to North Platte)" because someone inadvertently let them loose.

The record, said Welden, is a bird in India that flew 7,000 miles. He himself has entered a bird in an Orlando, Florida, race which takes the birds three or four days to fly 1,500 miles. Birds won't fly at night, he explained. On long races the clock stops half an hour after sunset and doesn't start until half an hour before sunrise.

"Nothing (is) better or more pleasing than to send a bird 600 miles and then see it come in the loft," said Welden, especially considering the obstacles: shooters, wires, storms and hawks.

Experienced birds rarely get lost. Most losses occur in training, allowing the survival of the most fit.

And racing pigeons are fit. They have as much in common with plain old street pigeons as marathon runners do with couch potatoes. Winning birds owe their success to their breeding as well as to their training. Their owners think of them as athletes.

Calerich said, "They can fly 6-8 years, depending on

the season," – how battered they are by weather and obstacles. "Common pigeons live one to three years, racing pigeons live as long as 20 years in breeding capacity."

Good training, good health and good ventilation and any loft can win, he said.

Homing pigeons have a long history of working for mankind. Records of their domestication go back to 5000 BC. The first Olympic athletes brought pigeons with them to the games and released them when they won so the folks back home would get the good news.

The famous Rothschild fortune was built on getting information on the stock market ahead of everyone else – by pigeon. With the ability to travel 40-50 miles an hour, pigeons were far speedier messengers than horses. The telegraph ended this type of airmail, but there are situations in which homing pigeons are still of use.

Pigeons were very successful in carrying messages during World Wars I and II. Ed Eaton of Colorado said the American military doesn't use homing pigeons anymore, but pigeons were found in enemy trenches during Desert Storm.

Pigeons are used in rescue work, said Eaton. "They train birds to see orange in the ocean." The birds fly

with a search and rescue pilot, but with their superior eyesight they can see the life raft first and will peck at a release mechanism.

Eaton said a Fort Collins rafting company uses pigeons released from remote locations to return clients' rolls of film for processing.

Steve Dermer, a Cheyenne flyer, was responsible for local arrangements for the show and auction. The day's responsibilities had him moving almost as fast as a prize-winning pigeon, which some of his birds were at this event. Out of the 205 birds present, three of his won blue ribbons.

The auction was a success. The donated birds brought in $5000, said Dermer, covering the cost of bringing in the judges and other club expenses. One bird went for a high of $550. The average was $158, a tidy sum, but not as much as the thousands of dollars paid for birds in some circles.

Now about those smudged shirt fronts. Pigeon feathers are coated in a very fine powdery white talc-like substance called plume that gets on everything. It looks as though when these pigeon lovers handle their birds, they use a firm, reassuring grip and hold them close, physically – and mentally.

# 99 Racing can have lucrative payoff

Thursday, March 14, 2002, Outdoors, page C1

CHEYENNE – Steve Dermer has 150 pigeons for breeding or racing in his lofts just outside Cheyenne. It doesn't sound like he's ever had a day without them.

"My dad and grandparents and great uncles flew pigeons. I have had them all my life," he said.

How much time do pigeons take? "I pretty much

breathe and live pigeons – you get out of it what you put in," replied Dermer.

Back in 1984, Dermer's work paid off. Breeding, training and flying conditions

all came together and made one of his birds the winner of the big Topeka Midwest Classic held annually in June. That meant that 5,332 birds from 429 other lofts in nine states didn't fly as fast. Dermer won $5000.

He considers the Topeka, Kansas, race, which draws participants from as far away as Chicago and Minneapolis, to be on a parallel to the big race in Barcelona, Spain, which is located at about the same latitude, runs about the same time and draws similar numbers of birds, but from all over Europe, especially Belgium. Nearly one out of every hundred Belgians flies pigeons.

For Dermer, an official survey had already calculated the distance from the Topeka drop to his loft at 504.685 miles. His bird made it in 12 hours and 4 minutes, at the rate of 1242.583 yards per minute, roughly 42 miles per hour. Because everyone's loft is a different distance from the drop point, winners are the fastest flyers rather than the first to get home. How far a loft is from the drop point determines whether a pigeon is entered in a 100-, 300- or 500-mile race.

Earlier, Dermer had lent this speedy bird to a friend in Kansas City, Misouri, for breeding. When it escaped and flew back to Cheyenne, Dermer decided it must be special enough to enter in the big race.

Winning birds start with good breeding. "All my breeders, I've got pedigrees on them back five generations," explained Dermer. Many of his birds are Meulemans, a strain originally from Holland.

Dermer believes in eye sign as a predictor of a bird's racing success. He likes birds with constantly moving eyes because they are constantly surveying their surroundings, a trait they need for getting home quickly and safely.

He also looks inside their pupils to see how strong their eye muscles are, believing that is also an indicator of success.

Next Dermer plays matchmaker, "Keep them alone a few days and then they'll usually mate right up," he said. The birds form such tight bonds, there is rarely any misbreeding, even though they all range freely inside the loft.

The loft itself is a breezy place, but ventilation is important to the birds' health. Cold temperatures don't bother the birds or Dermer. He's more worried about loose cats climbing into the loft.

By the end of February, nesting bowls are usually in place in the cubbyholes, or rather, pigeonholes, that line one side of Dermer's breeding loft from floor to ceiling. It takes only 11 days for eggs to be laid, and about two weeks to hatch.

A week later the young are feathering out, and they've been banded with a seamless ring on their leg that identifies the year they were born, the club they fly with and their own unique number. Dermer's birds are also identified as belonging to the American Racing Pigeon Union, designated as the AU, so anyone finding a bird can go on the internet and use the numbers to track down and contact the local racing club's secretary.

By the time the birds are two to two and a half months old, they are ready for their first toss, said Dermer. No one has completely figured out what makes a homing pigeon home in so well. They may be keying in on their home range with their acute sense of smell. With each training flight, they learn the local topography too.

Dermer trains his birds by taking them to work with him in Denver and releasing them there. But at first that backfired when they kept coming in an hour late from races. The next year Dermer tried broadening their horizons to the east by having his mother release them on a trip to Nebraska. A little training to the east every year has fixed the problem.

Dermer sends only the healthiest birds to races. "I believe in natural health. I haven't medicated my birds in two years." Instead, he uses supplements mixed with water.

Pigeon athletes can have some of the same problems as their human counterparts. "We've got a team that tests the birds for steroids and amphetamines at the big races," said Dermer.

Birds in their first season are categorized as "young birds," and Dermer will begin flying them during the racing season that starts in May and continues for about 11 weekends. After the first year they are considered "old birds" and entered in old bird races.

On average, Dermer said, pigeons have five or six years of racing before retiring, with their life span at 12-14 years. He had a bird once that flew 500-mile races 19 times in nine years. The bird's last race was through brutal humidity. When he came back in poor shape, Dermer decided it was best to retire him to breeding instead of trying for an even 20 races.

What happens to old pigeon flyers? Florida. "All the best guys move down there," said Dermer. He shook his head at the thought of the super competitiveness he's heard about.

A competitive bird not only has to have speed, but said Dermer, "They got to have a compass." Besides the smell and sight of home, and maybe even the sound of the wind over the Rockies, or the pattern of stars visible to their eyes even in daylight, it's been shown homing pigeons can use the earth's electromagnetic field somehow.

Flyers in the region, said Dermer, think they are experiencing higher than normal losses of birds. They think it's due to all the new communications towers springing up. Certainly the guy wires are responsible for killing birds, but the thinking is that radio waves may be confusing them.

There's one other important trait for a racing pigeon: heart. "That bird that will hurt itself to get home is the bird to race," said Dermer.

In Dermer's lofts are sharp-eyed, alert birds in perfect, iridescent plumage gently cooing and milling about at the interruption by humans.

Dermer knows each one, although he's never named them. Outside it's cold and sunny and windy and Dermer's teenagers, the birds hatched last spring, are out enjoying a frolic. Twenty or so disappear over the roof of the house and then zoom back, banking sharply for a turn over the field, like a squadron of Blue Angels.

It isn't hard to see why earth-bound people get attached to these pigeons. They are birds that can soar the skies yet will still come home to roost.

## 100 You don't have to race pigeons to raise them

Thursday, March 14, 2002, Outdoors, page C4

CHEYENNE – Pigeons make great pets. They don't bark – they only coo. They don't scratch up the furniture – they live outside. And old pigeon droppings make great garden fertilizer.

How could your parents complain?

Pigeons lead a simple life. All they need is water, feed, grit for digestion and a clean, dry, airy place to live. The more time you spend with them, the more attached they will get to you.

The pigeon fanciers interviewed for this week's stories have racing pigeons, but there are other kinds. One example: Russian tumblers do somersaults when they fly.

If you are interested in finding out more about racing pigeons, contact the American Racing Pigeon Union at www.pigeon.org, or call 1-800-755-2778, or write to them at PO Box 18465, Oklahoma City, OK 73154-0465. Ask for their free information.

Locally, Cheyenne pigeon racer Steve Dermer invites anyone interested in getting started to call him at 778-7424. Before going crazy over pigeons, check your town or city ordinances for restrictions on the number of pigeons you can keep. It appears Cheyenne ordinances don't mention any restrictions on the number of birds you can keep or any restrictions relating to pigeons.

## 101 Neighborhood hosts unknown avian athlete

Thursday, March 14, 2002, Outdoors, page C2

One summer evening years ago, Mark and I were finishing up some yard work. The wind was blowing dark clouds in and throwing small birds about. It felt like we were battening down the hatches for a storm at sea.

Our younger son, who was just learning to talk, had been out too, entertaining himself, when he came to find us, announcing, "Mom, druck! Dad, druck!"

Parents of toddlers have to have vivid imaginations to understand early childhood language, and failing that, have to ask for a reinterpretation, but all our son would say was, "Druck!"

He indicated he wanted us to come in the garage where the pickup was parked and the garage door was still up. Oh, all right. Was he trying to tell us something about the truck? But then he bent down to look underneath, and I did too, and there was a pigeon looking back at us.

I guess a pigeon may look like a duck to you if you're only 2 years old and your parents, card carrying members of the National Audubon Society, have been too snobbish to point out pigeons to you. This pigeon had bands on its legs. We'd heard of homing pigeons. Maybe this was one. Did people still have homing pigeons? Perhaps the storm had confused it.

I called someone who I thought might know how to track down the owner of a pigeon with a band, but they didn't think anyone would want a homing pigeon that wasn't smart enough to find its way home.

Our next-door neighbor, Deb, owned a parrot, so we asked her for help. She had an extra cage and took the pigeon home and fed it bird seed and watered it.

Every morning Deb left the cage open on her back porch and each day the pigeon flew out – and came back. This went on for over a week. Deb must have been getting attached because the day the pigeon didn't come back, she was a little sad. We hoped it had avoided the loose cats in the neighborhood and found its way home.

More than anything, we wished we'd known where the bird had been coming from and where it was going to and if it was far off course.

Now that I know how to look up the owner of a racing pigeon, I wish I had copied down those band letters and numbers. The letters stand for national pigeon organizations which all have websites with band listings and people to contact.

If you can remember AU stands for American Pigeon Racing Union, you can search for them on the internet, or use their address, www.pigeon.org. There they list the other major organizations, the International Federation of American Homing Pigeon Fanciers (IF) and the Canadian Pigeon Racing Union (CU). The National Pigeon Association (NPA) represents non-homing pigeons.

The AU has a whole page on how to take care of a lost pigeon. We, however, had relied on common sense and Deb's parrot experience, but we managed to do the right things.

First, a lost pigeon is thirsty. Provide a shallow dish or margarine container with about an inch of water in it, not a bottle.

They will also be hungry. They are raw seed eaters, so anything around the house, such as unpopped popcorn, uncooked rice, split peas, barley or bird seed is acceptable.

And most importantly, keep them safe by putting them in a container a dog or cat can't get into.

After 24 or 48 hours, the AU experts say most pigeons have recovered and are ready to go home. Don't bog them down by trying to attach any notes to their legs. However, if you can track down the owner, they may prefer to pay to have you ship the bird home. A pigeon temporarily lost to severe weather is still wanted.

Pigeon racers are eager to share their hobby and help beginners get started. Steve Dermer, the Cheyenne flyer interviewed for this week's story, offered me some young birds this spring. I wonder what my wild backyard birds would think?

Though many of the pigeon fanciers I interviewed have been involved for decades, younger generations are represented as well. I called Glen Gleason, past president of the Northern Colorado Flyers club, for clarification of a point, he told me about a new junior member, 13 years old, whose family is just establishing their own loft west of Cheyenne.

It's incredible to think that when that modern American boy breeds and races pigeons, he is sharing the experiences of flyers all over the world – and all through history.

# 102 Winter's over when juncos take wing

Thursday, March 28, 2002, Outdoors, page C2

How can you tell it's spring in Wyoming? The birds' water dish isn't frozen over every morning. House finches are proclaiming breeding season. Tiny blades of green show under the lawn's brown thatch. And it has started snowing more often.

For me, this winter will be remembered for the Townsend's solitaire flitting about the neighbor's junipers, eating berries.

For everybody else in the Cheyenne birding community, the bird of the winter has been the red-bellied woodpecker, our unexpected visitor from the east. I've heard reports now from half a dozen people who've seen it. Will it stay in Cheyenne year round because it is a year-round resident in its normal range, or will it get wanderlust again, which is what brought it here in the first place?

I knew it was nearly spring when John Hewston sent me the results of the 2001 Thanksgiving Bird Count. He said once he doesn't use computers.

This year 451 counts were reported from the 12 westernmost states (no reports from Hawaii this time), from 255 different cities and towns.

The 10 most numerous birds to be reported in those 15-foot diameter count circles around feeders were, in descending order: house sparrow, house finch, pine siskin, dark-eyed junco, mourning dove, black-capped chickadee, California quail, American goldfinch, European starling and white-crowned sparrow.

The quail making the list can probably be attributed to almost a quarter of the reports coming from Hewston's home state, California.

In Wyoming, 23 reports were submitted from seven different cities and towns. We had eight counts from Cheyenne.

Our state's top 10 list was a little different, with the last two entries reflecting our location in the Rocky Mountains: house sparrow (610), house finch (174), pine siskin (73), American goldfinch (48), rock dove (pigeon) (39), dark-eyed junco (29), black-capped chickadee (28), European starling (22), black-billed magpie (21) and mountain chickadee (18).

Twenty other species made up the rest of the Wyoming list. All were typical for winter – sparrows, woodpeckers, jays, nuthatches – except for three. Six sage sparrows, three mourning doves and a yellow-rumped warbler were pushing their luck sticking around Wyoming in late November.

No crows were on the list. Evidently no one feeding crows in Thermopolis was counting. The city now estimates there are 3,200 of these corvids causing civic problems.

Let me know if you'd like to see a copy of the Thanksgiving count report.

The Project Feederwatch season, November through March, is almost over. Four of us in Cheyenne have been submitting reports online.

Four is a good number of participants, compared to six in New York City, as reported by a map of Feederwatcher locations. It's also surprising. The image of the Manhattan skyline notwithstanding, much of the city has plenty of yards and places for bird feeders.

This winter, house finches, dark-eyed juncos and house sparrows have been my regular visitors, with the occasional starling or crow.

I've had other species visit only once or twice: sharp-shinned hawk, northern flicker, blue jay, Townsend's solitaire (I don't count seeing it in the neighbor's yard) and American goldfinch. No chickadees or nuthatches this year, a local aberration, since they have been visiting other people's feeders.

I've appreciated the variety of subspecies and hybrids of juncos because it felt like I had more kinds of birds. But when warm weather comes, the juncos will head out, north or up to the mountains.

And when they head out for good, that's when I know winter is truly finished, again.

92                                              CHEYENNE BIRD BANTER

# 103 Birders flock to observe crane migration

Thursday, April 4, 2002, Outdoors, page C2

Common wisdom has it birds fly north in the spring and people on spring break either head south to warm up or head west to ski. More than a dozen Cheyenne bird watchers recently headed east instead to see a bit of the sandhill crane migration.

Lots of birds migrate, but few are as dramatic about it as the sandhill cranes. They come from their various wintering grounds in New Mexico and Texas to a 100-mile stretch of the Platte River where they all lay over, eat waste grain and fatten up for the rest of their trip and the breeding season. Some continue north as far as Alaska and Siberia.

In March and early April, it isn't hard to find sandhills between North Platte and Grand Island, Neb., even at 75 miles per hour. Cranes, their nearly four-foot lengths bent double to feed, show as gray lumps moving through old cornstalks.

An individual crane is a graceful, elegant bird to study, small flocks are interesting, but the spectacular part comes at twilight. Thousands of cranes gather in the shallow reaches of the river to spend the night.

They like best where the river has been scoured clean of vegetation so no predators can hide. Before all the dams, the river scoured itself with frequent floods. Now crane conservationists take heavy equipment to the brush.

Cheyenne High Plains Audubon Society president Art Anderson organized our 300 mile expedition east. Friends of his have blinds just downriver from the Rowe Sanctuary run by Audubon Nebraska near Gibbon, where Mark and I visited eight or nine years before.

In that time, Kearney, Nebraska, has become the crane capital: more motels, crane information at the rest area and messages on business marquees welcoming crane watchers. I didn't stop to shop, but undoubtedly there's crane stuff for sale.

At the sanctuary, a whole building is now dedicated to crane souvenirs and several people were available to answer questions. I overheard visitors discussing the likely location of a lone whooping crane.

There was a happy delay when we arrived at the gate to our destination. Right alongside the long dirt driveway was a flock of sandhills—and the whooper.

There are few whooping cranes worldwide, a few hundred, but at least they are easy to pick out because they stand taller than the sandhills and they gleam bright white. We set up spotting scopes and studied the whooper's bright yellow eye and red and black facial markings. The evening breeze was ruffling its plumy feathers.

By show time, we were ensconced in a really swank duck blind. Line after line of sandhills came in from upriver. Some from nearby fields flew right over our heads, close enough we saw their bills open as they made their creaky calls to each other. In near dark their slow wingbeats and the way their long legs extend well past their bodies make them easy to sort out from ducks and geese, small flocks of which were flapping furiously in staccato counterpoint.

A lot of the cranes headed for the big flock around the bend from us. The noise of those thousands of voices reminded me of the roar of fans at Oakie Blanchard Stadium I can hear from my house on a fall evening.

When we left, tardy geese were still coming in, crossing in front of the nearly full moon as if to recreate the artwork hanging in a hunting lodge.

The other show is when the cranes lift off around sunrise. However, the next morning a stolid sky, promising precipitation, squelched the exact moment of dawn. As the day brightened by imperceptible degrees, I became aware that the gray-colored sandbar was really a mass of cranes.

Though we'd pried ourselves out of our sleeping bags before 6 a.m., the cranes in front of the blind were reluctant risers. A lot of other cranes passed by, enticing a few strings to lift off and peel away, but most appeared to still have their heads under their wings.

Finally we gave up and went indoors where a picture window gave us a view of a really nice bird feeding station. There were gobs of goldfinches, several cardinals, nuthatches, chickadees, downy woodpeckers, juncos and, drumroll please: a pair of red-bellied woodpeckers. So now I don't have to wait until I see the one that's been reported in Cheyenne in order to add this species to my life list.

The red-bellieds and the whooper were the perfect souvenirs of my chance to escape routine for a couple days outside in a different landscape – my chance to take a spring break.

Note: Reservations for Rowe Sanctuary's blinds are available by calling 308-468-5282. Visit their website at www.rowesanctuary.org. Contact the Kearney Visitors Bureau, 1-800-652-9435, for lodging information and other viewing opportunities.

# 104 Human interference changes age-old battle between grasslands, forests

Thursday, April 18, 2002, Outdoors, page C2

## Bird Banter

Which is better? Trees or grass? Here on the Northern Great Plains there's a line of skirmish wherever the grasslands meet ponderosa pine.

Historically, whenever a company of pines marches down the hill and attempts to take the valley, the grasslands fight back, first with dense root mats, then with the occasional fire.

In the thickest coniferous forest, where sunlight hardly reaches the bare forest floor, there isn't much bird song. The few kinds of birds found there specialize in seeds from the trees and whatever bugs live among them. Even fewer kinds specialize in preying upon seed-eating mammals and other birds.

The grasslands by contrast are swarming with birds. The first weekend in April, our family camped out with the Boy Scouts at Fort Robinson State Park near Crawford, Neb. It was a great luxury to wake up to the sounds of multiple meadowlarks and killdeer echoing in the valley.

People have taken sides in this battle between the graminids and the conifers. Up in the Black Hills, where we spent a few days before the campout, miners cut a lot of trees for lumber during the gold mining frenzy of the last quarter of the 1800s. But then people also took away fire. The results of changing the rules of engagement are hard

to see unless you've done photo documentation.

In 1974, a book titled "Yellow Ore, Yellow Hair, Yellow Pine" by Donald R. Progulske was published by South Dakota State University, Brookings (Agricultural Experimentation Station Bulletin 616). In it, photos from General George Armstrong Custer's 1874 expedition to the Black Hills are compared with photos taken by Richard H. Sowell at the same locations nearly 100 years later.

The trees appear to be winning. In many cases Sowell finds the same pile of rocks seen in the foreground of an 1874 photo, but instead of a view across a valley to a distant peak, only tree trunks can be seen.

At Wind Cave National Park, people have taken the side of the grasslands. Fire has been brought back to the battle, with strict restraints, to benefit herds of large grass eaters, buffalo.

In contrast, where fire broke out about 13 years ago at Fort Robinson, burning away ponderosa pine on steep hills, people have taken the trees' side. This year 23,000 ponderosa seedlings were planted by about 1000 Scouts, similar to previous efforts in other parts of the park since the fire.

The grasslands are getting the last laugh. This year we planted on the same hillsides

as the last time I came. There were very few survivors of the earlier planting. Judging by the porcupine damage on the few mature trees that survived the fire, the continued dry conditions and the healthy stands of grass, even well-intentioned intervention by people isn't going to give trees the upper hand soon.

But like so many other localized battles, grassland and forest have more in common with each other than their mutual enemies, the outside forces that threaten their battle ground.

Twenty-four years ago, I became intimately familiar with the northern Black Hills as a surveyor's aid helping to set Forest Service boundary monuments. The Hills are a patchwork of privately owned valleys and publicly owned hills.

Perhaps I should retake all the photos I have from that summer to compare and document the progress of the war between civilization and nature, though some changes, such as non-native grasses grown for hay substituting for native species in the valleys, may not be apparent to the casual observer.

I got lost trying to find the house in Spearfish, South Dakota, where I rented an apartment. The view from the front door had been fields, the Interstate and Lookout Peak. Now

the fields are full of streets and buildings.

Deadwood and Lead are landlocked, but if ownership of the surrounding, privately-owned hills ever changes, more casinos, resorts and tourist attractions will be carved out.

Mountain carving continues at Crazy Horse Monument. The chief's face is finished and the horse's head has been etched in the rock with dynamite. Associated visitor facilities continue to expand.

Over at the famous faces, the Mount Rushmore parking garage has been artfully inserted and a granite plaza and several monumental gateways lead to a larger visitor center.

Sadly, where battles have raged or are raging, one can't go home again. Neither can the meadowlark, whose grassland is taken over by forest.

Presumably, along with increasing hordes that need to be artificially entertained, there will be visitors who want the natural experience and the tourism industry will provide for them. The birds are there. I glimpsed a kingfisher, wood ducks, hawks and plenty of juncos (I couldn't see if they were the endemic white-winged). Meanwhile, for the hurried or undiscerning, there's always Bear Country USA.

# 105 Cheyenne artist earns conservation stamp honor

Thursday, April 25, 2002, Outdoors, page C1

For the first time since its inception in 1984, a Cheyenne artist has won the Wyoming Conservation Stamp Art Competition sponsored by Wyoming Game and Fish Department.

Artist Renee Piskorski's winning portrayal of mountain bluebirds will be printed on 300,000 stamps for the 2003 hunting and fishing season, and the original will be framed and added to the gallery of previous winners at the department's Cheyenne headquarters.

Beginning Jan. 1, anyone can own a print of the winning painting by purchasing the conservation stamp for $10 through the department or any outlet that carries hunting and fishing licenses, or by purchasing a full-size, limited edition print.

In addition to being collectibles, the conservation stamps must accompany any Wyoming hunting or fishing license. The fees collected go into the Wildlife Trust Fund for habitat acquisition and improvement, non-consumptive use of wildlife and nongame projects.

Competition coordinator Mary Link said 97 artists from 29 states and Mexico entered and that more Wyoming artists than usual participated, 42 percent.

Of the other entries that placed or received honorable mention, three were from Wyoming, two from Utah, plus one each from Nebraska, Ohio and Connecticut. The judges included art and bird experts.

Piskorski said wildlife artists consider the Wyoming contest to be second in prestige only to the federal duck stamp competition, due to the quality of the competition and the prize money offered ($2,500 for first place).

"It's important for me to get the correct anatomy and habitat," said Piskorski, discussing her winning technique for painting wildlife. "I need to go to their environment to view them. I take lots of photographs and study videos. Then I do thumbnails, sketches.

"For the stamp you have to remember it's very small. It was difficult to paint them (mountain bluebirds)

larger than life. It felt like a Hitchcock movie.

"I'm always trying to keep in mind the mood I'm trying to create, using the light and the weather. And I keep a color palette in my mind while I'm sketching."

Piskorski said she has never had formal art training, but has been painting most of her life, and seriously for 15 years, beginning with a request for her paintings from the gallery owner she worked for. Her career snowballed from there. Her entries in the competition in past years received third, fourth and sixth place.

As a professional artist, she can justify trips to Yellowstone and the Tetons two or three times a year for research, continuing an outdoor lifestyle that began with hunting and fishing trips with her dad, who was an artist himself, an engineering draftsman with a bent for drawing political cartoons.

"In the end, I hope the viewer will feel the same emotion that I felt while painting it," Piskorski said of her work, "and maybe have

an even greater appreciation for the natural world and want to preserve it."

More of Piskorski's work can be seen at Deselm's Fine Art gallery, 303 E. 17th St., where she will be featured tonight at the weekly Artists' Hangout, beginning at 5 p.m. Artists and art appreciators are welcome to stop by.

Her winning oil painting, "Sagebrush Outlook," will be on display at the Wyoming State Museum, 2301 Central, upstairs with all the other entries until May 25. Then it will travel with the "Top 40," to Cody, Thermopolis, Dubois and Pinedale. All of the entries, except the winner, are for sale, with commissions also benefitting the Wildlife Trust Fund.

The State Museum hours in April are 9 a.m. – 4:30 p.m. Tuesday – Friday and 10 a.m. – 2 p.m. Saturday, and in May, 9 a.m. – 4:30 p.m. Tuesday – Saturday.

Rule books for next year's competition, featuring the mountain lion, will be out in the fall.

# 106 Backyards going to the birds

Thursday, May 2, 2002, Outdoors, page C1

## Habitat can also be pleasant for people

When Sue and Chuck Seniawski moved to the Monterey Heights neighborhood about 13 years ago, their backyard was not fit for man or beast.

"The backyard was absolutely bare when I got started – just grass, with a couple

trees in front of the house," Chuck said.

The Seniawskis worked out a landscape plan through Tom's Garden Spot, a nursery no longer in business, and now those trees and shrubs provide a sanctuary for them and a variety of birds.

In one hour on an April afternoon, about 10 species were observed.

Any grade-school child can point out the three major needs of wildlife the Seniawskis have provided: food, water and shelter. As it turns out, what's good for wildlife

is good for people.

Reg Rothwell, author of Wyoming Game and Fish Department's free publication, "Wildscape," said good landscaping will increase property value, "but it will also provide auditory and visual screening, protection from wind and excess solar energy and give privacy for the home. Wildlife habitat comes with it."

**Shelter**

Though the term "birdhouse" implies birds may seek shelter from weather in them, only a few species use natural or man-made cavities, and then usually only for nesting. Most look for shelter in vegetation.

Publications about creating backyard bird or wildlife habitat start with planning for and planting trees. However, most are written with the eastern U.S. in mind and recommend kinds of trees that cannot live long in Cheyenne's environment or need a lot of water to survive.

Rothwell champions native species for their suitability. "If I can't get natives, I want something like natives," he said.

At the Cheyenne Botanic Gardens, director Shane Smith estimates 80 percent of the plant species recommended for Cheyenne on their website are identified and growing in Lions Park, so people can visit and find out what they look like.

The website lists both local and area nurseries, but Smith recommends checking local nurseries first.

"Take the compass into account," said Smith, giving his general planting rule of thumb. Plant coniferous trees on the north and west side of the house to insulate it from winter wind.

Plant deciduous trees on the south side so that their leaves shade the house in summer, but when their leaves drop in the fall, solar rays will warm the house.

Rothwell, Smith and University of Wyoming Laramie County Cooperative Extension horticultural agent Catherine Wissner warn against planting aspen because it is short-lived, and a longer-lived tree would be a better investment in time and money. Also, one aspen will send out suckers all over the yard, attempting to turn it into a forest.

Wissner said landscaping advice is also available through her office, especially through the master gardener program. Two of the current master gardeners specialize in trees and may be available to come out and look at potential planting sites.

Shrubs are perhaps more valuable than trees for providing shelter for some birds, said Smith. However, one book on gardening for birds pointed out that rigorous pruning may cause growth too dense for birds to navigate easily.

**Food**

Trees and shrubs can be selected to do double duty as both shelter and food sources if they produce flowers, berries, cones, seeds or other kinds of fruit.

Fruits of chokecherry and Nanking cherry make good syrup and jelly, but the birds will want their share. If your goal is backyard wildlife habitat though, there will be plenty for everyone.

Flowers, whether in the garden or on trees and shrubs, will attract birds. It's the flower nectar attracting hummingbirds and orioles, flower petals for evening grosbeaks, and the insects attracted to the flowers for insect-eating birds.

Bird feeders are not an essential element of a backyard habitat, but they do add to enjoyment. A sunflower seed or niger thistle feeder like the Seniawskis have, covered with cheery-voiced goldfinches, is hard to resist.

**Water**

Birds visiting the Seniawskis' yard drink and bathe all winter in the heated bird bath located up on the deck. Down below is a pedestal-style bird bath. Birds will appreciate a simple pan of water on the ground as much as an elaborate waterfall or pond, especially if you clean it regularly to avoid the spread of disease.

**Nest boxes**

Only certain bird species are interested in nesting in a structure or cavity. Some of the swallows prefer to build their own with mud.

Backyard birds in our area that might be interested in your handiwork include downy woodpecker, northern flicker, black-capped and mountain chickadees, red-breasted nuthatch, house wren and house finch. The mountain bluebird, wood duck, common merganser and American kestrel will also use nest boxes but have habitat requirements beyond Cheyenne's average backyard.

The size of the entrance hole determines if the intended species will be able to use it without aggressive species not native to our area, starlings and house sparrows, taking it over. Nest box specifications are available from the U.S. Fish and Wildlife Service's free pamphlet, "For the Birds," and from The

Birdhouse Network.

**Hazards**

Just as visitors to your home should be protected from injury, so should your avian visitors.

Keep your own cat indoors, or build a "cat haven" as Pat and Paul Becker have done. Make sure shrubbery that might hide a loose cat is far enough away from water and feeders so that birds, especially ground feeders like juncos, have a chance to see the cat coming and to escape. A dog installed in the yard makes a great cat repellant.

Pesticides poison insects and seed-producing plants, the very things that attract birds to your yard. If a bird eats enough poisoned insects, it will die.

High amounts of lawn care chemicals were found in birds succumbing to West Nile virus on the East Coast.

The National Audubon Society website offers alternatives, though you can consider the birds themselves as part of your pest management strategy.

**Housekeeping**

Nature is not tidy.

RockShe doesn't rake up dead leaves and bag them. Instead, decomposing leaves offer sustenance for insects, slugs and worms – and the birds that eat them, before completely breaking down and nourishing the soil. Chuck Seniawski allows leaves to remain under shrubs because leaf litter and its denizens attracts green-tailed and spotted towhees.

*Cutline for photo of cat in a cat kennel:*

A cat kennel, like the one keeping Pat and Paul

Becker's cat contained, lets cats enjoy the outdoors without providing them a chance to kill avian visitors. The tube on the right allows the cats to go in and out of the house. Kevin Poch/staff

*Cutline for photos of Sue and Chuck's backyard:*

Top – A downy woodpecker takes time out to feed on some black oil sunflower seeds in the Seniawskis' backyard. Above – Sue and Chuck Seniawski enjoy their backyard that they have slowly turned into a great bird habitat in north Cheyenne. Larry Brinlee/staff

## Resources

**Cheyenne Botan ic Gardens, Lions Park, 637-6458, www.botanic.org/treelist**
A list of trees, shrubs and wildflowers recommended for Cheyenne and coded for growing attributes (drought tolerance, etc.) is available at the solar greenhouse or online. A library of gardening books is available to members.

**University of Wyoming Laramie County Cooperative Extension Horticulturist, 310 W. 19th St., 633-4383**
Horticulturist Catherine Wissner can give you advice or may send out a trained master gardener for onsite evaluations.

**Laramie County Conservation District, 11221 US Highway 30, 772-2600**
Ask for the free "Backyard Conservation" booklet. Free pamphlets address building a pond, attracting butterflies, bees, bats as well as reptiles and amphibians, plus other topics. The district also sponsors tree planting programs.

**National Audubon Society, Audubon at Home, www.audubon.org/bird/at_home**
Topics include why and how to reduce home pesticide use, how to increase backyard biodiversity and how to garden for birds and other wildlife.

**Laramie County Library, 2800 Central, 634-3561**
Look for books in either the gardening section or the bird section (598s), but when plants or animals are mentioned, check them against a list of species native to our area. For an especially enjoyable and philosophical outlook, read Gene Logsdon's "Wildlife in the Garden." For particular information for the Rocky Mountain region, read "Backyard Birdwatcher's Home Companion" by Donald S. Heintzelman.

**Wyoming Game and Fish Department, 5400 Bishop Blvd., 777-4600, gr.state.wy.us**
Ask for the free publications "Wildscape – landscaping for wildlife and the homeowner" and "Wyoming Bird Checklist." A series of bulletins on specific wildlife topics is also available. Ask at the office or check the website for a complete list.

**The Birdhouse Network, Cornell Lab of Ornithology, PO Box 11, Ithaca, NY 14851-0011 http://birds.cornell.edu/birdhouse**
Another citizen science project like Project Feederwatch, TBN is set up to accept reports about nest box success from member observers. Anyone can access the site, which has an incredible amount of information about building, buying or placing bird houses, plus a nesting success data base, and for live looks inside, nest box cams.

*Additional resources not included when this article was published:*

--U.S. Fish and Wildlife Service, Ecological Field Services, 4000 Morrie Ave., 772-2374
A limited number of pamphlets on attracting and providing for birds in your yard is available at the office. The same pamphlets are available on line, ttp://migratorybirds.fws.gov/pamphlet/pamphlets.

Up to three copies of "For the Birds," which includes nest box specifications, can be ordered free by e-mail, IMBD@fws.gov or by writing Office of Migratory Bird Management, U.S. Fish and Wildlife Service, 4401 N. Fairfax Drive, Suite 634, Arlington, VA 22203.

--Hummingbirds! www.hummingbirds.net
The website provides information about attracting, watching, feeding and studying North American hummingbirds and includes photos and migration maps. Our most common species are broad-tailed and rufous hummingbirds.

--National Wildlife Federation, www.nwf.org/backyardwildlifehabitat
Even if you aren't interested in certifying your backyard habitat through NWF's program, check out the how-to section, tips, photo gallery and online planner. "Explore Nature in Your Neighborhood" is this year's theme for National Wildlife Week.

--Project FeederWatch, Cornell Lab of Ornithology, 159 Sapsucker Woods Road, Ithaca, NY 14850, http://birds.cornell.edu/pfw
Members participate in a citizen science project by recording the birds that visit their feeders November through March. The website is available to everyone and is chuck full of information and data about feeding birds.

## The Seniawskis' Backyard Plant List

Remember, trees in neighbors' yards contribute to bird habitat.

**Broadleaf trees**
Aspen
Flowering Almond
Flowering Crabapple
Locust
Narrowleaf Cottonwood, N
Seedless Mountain Ash

**Broadleaf shrubs**
Alpine Currant, N
Canada Red Cherry
Cotoneaster
Saskatoon Serviceberry
Spirea (white, pink, blue)
Sumac

**Evergreens**
Austrian Pine, D
Bristlecone Pine, D, N
Colorado Blue Spruce, D, N
Ponderosa Pine, D, N
Juniper shrubs, D, N

Designation from Cheyenne Botanic Gardens list:
    N – native
    D – drought resistant after establishment

**Tree Planting Disclaimer**

Not all birds appreciate trees. Birds, such as the western meadowlark, grasshopper sparrow, killdeer and bobolink nest on the ground in wide-open spaces.

If wide open describes your property, consider allowing it to continue as grassland bird habitat rather than transforming it into forest.

Avoid mowing during nesting season, now through July. Keep dogs and cats confined or on a leash so they won't harm eggs and young.

Be aware that a pole or tree may provide avian predators such as crows with a watch tower and launching pad to use in their quest for prey.

2002

## Birds Observed in the Seniawskis' Backyard, 1990-2001 [not included]

| | | | |
|---|---|---|---|
| Bluebird, Mountain | Grosbeak, Rose-breasted | Peewee, Western Wood | Tanager, Western |
| Chickadee, Mountain | Hawk, Sharp-shinned | Redpoll, Common | Thrush, Swainson's |
| Crossbill, Red | Jay, Blue | Robin | Towhee, Green-tailed |
| Crow, American | Jay, Steller's | Siskin, Pine | Towhee, Spotted |
| Dove, Mourning | Junco, Gray-headed | Solitaire, Townsend's | Vireo, Red-eyed |
| Falcon, Prairie | Junco, Oregon | Sparrow, Chipping | Warbler, Nashville |
| Finch, House | Junco, Pink-sided | Sparrow, Harris' | Warbler, Orange-crowned |
| Flicker, Northern (Red-shafted) | Junco, Slate-colored | Sparrow, House | Warbler, Townsend's |
| Flycatcher, Hammond's | Junco, White-winged | Sparrow, Lincoln's | Warbler, Wilson's |
| Flycatcher, Western | Kinglet, Ruby-crowned | Sparrow, Song | Warbler, Yellow |
| Goldfinch, American | Longspur, Lapland | Sparrow, Tree | Waxwing, Cedar |
| Goldfinch, Lesser | Merlin | Sparrow, White-crowned | Woodpecker, Downy |
| Grackle, Bronzed | Nuthatch, Red-breasted | Sparrow, White-throated | Wren, Rock |
| Grosbeak, Black-headed | Owl, Great Horned | Starling, European | |

Thursday, May 2, 2002, Outdoors, page C2

# 107 Bluebird housing makes a difference

I didn't grow up with bluebirds. First, because I was a suburban kid. Second, I didn't look much at birds; and mostly, it was a period of time during which bluebird populations in the Midwest had decreased due to pesticides and loss of cavity nesting places like hollow trees and wooden fence posts.

Notes in my old field guide indicate my first eastern bluebird was July 1975, when I rode my bike out into the countryside between split shifts for the food service at the University of Wisconsin-Stevens Point.

Then, in 1978, I reported for the campus paper on a talk by Vincent Bauldry of Green Bay, Wisconsin, who had built a better bluebird nest box. He had 21 years of experience showing that a hole in the roof (screened to keep out predators) had greatly improved nesting success. By the end of the year, Bauldry's design got national coverage when I sold the story to Organic Gardening magazine.

Bauldry claimed his design imitated rotting wooden fence posts where eastern bluebirds liked to nest. Added moisture from rain helped the eggs stay hydrated and excluded competing species, such as starlings.

When I mention this "skylight" concept to bluebird box experts out here, they think I'm crazy, maybe because few cedar fence posts on western rangeland rot in the dry climate.

There are, however, other modifications Bauldry used that can be seen in modern bluebird nest box plans today.

A traditional bird house, the kind people now paint decoratively and display in their living rooms, is often cube-shaped with a peaked roof. Bauldry made his with a flat roof and made it more than twice as deep so eggs or nestlings were beyond reach of a marauding racoon's arm.

The increased depth, he said, would also keep the young in the nest longer, so they would be stronger when they fledged. He added horizontal saw cuts on the inside of the front wall to help the babies climb the greater distance to the entrance hole.

Bauldry eliminated any kind of twig-like perch sticking out by the entrance, making it more difficult for nuisance birds to find a vantage point from which to harass the bluebirds.

His nest box design is clearly utilitarian, with one side swinging open so old nesting material can be cleaned out between broods, a feature of most modern nest boxes for any species.

I don't know if it's Bauldry's innovation, but his box and several modern boxes usually have the back wall extend either below the floor and/or above the roof so there's something to nail to the fence post or other support. Bluebirds evidently don't care for the rock-a-bye-baby effect of hanging bird houses in trees.

When Alison Lyon [Holloran] of Audubon Wyoming gave a presentation on mountain bluebirds at the Wyoming State Museum in conjunction with the opening of the Wyoming

Conservation Stamp Art Competition, Show and Sale last month, someone in the audience wanted to know how to attract mountain bluebirds to Cheyenne.

Unless the city expands its limits halfway to Laramie and dedicates the land to open range, and also raises the elevation from the present 6,100 feet to something over 7,000, the mountain bluebird's preference, it's unlikely they will ever do more than pass through during migration.

The most dependable place to see mountain bluebirds close to town is Curt Gowdy State Park, where several Eagle Scout candidates have installed nest boxes and a little further up Happy Jack Road, at North Crow Reservoir.

Reports of mountain bluebirds begin in the last half of February, making them a harbinger of spring for me. I once made a quilt and re-colored the Flying Swallows pattern to represent them and commemorate this annual event.

CHEYENNE BIRD BANTER

It turns out my favorite sign of spring, the spot of sky blue in a gloomy landscape, can be found all winter as close as southern Colorado, though many more head south into Mexico.

By summer, mountain bluebirds can be found from Arizona and New Mexico north to eastern Alaska.

Catching a glimpse of a mountain bluebird is always a treat. I double check to make sure what I'm looking at isn't a western or eastern bluebird. Much less abundant in Wyoming than the mountain, their ranges extend into western and eastern Wyoming, respectively, overlapping the mountain's range.

With red on their breasts, I always think of these two as the blue-coated versions of the closely related robin, since they are the same general shape, though smaller.

Mountain bluebird males are blue all over, and the females are less bright, more gray.

Besides dressing up the landscape, bluebirds eat insects. That is one reason why so many smart people encourage bluebird nesting around their property by installing a series of nest boxes along a "bluebird trail." Providing nest boxes makes a difference to the birds too, especially for the eastern bluebird, whose population has made progress in recovering.

If you like bluebirds or are serious about building a bluebird nest box or trail, the ultimate resource for information and specifications, such as using a 1 and 9/16 inch entrance hole, is the North American Bluebird Society website, www.nabluebirdsociety.org. You can also write to them at The Wilderness Center, P.O. Box 224, Wilmot, OH 44689-0244.

# 108 Learn to appreciate birds from a distance

Thursday, May 16, 2002, Outdoors, page C2

**Most birdwatchers never have a close encounter with their quarry.**

Identification of birds in the field is based on patterns: color, shape, behavior, location, and voice, because even with binoculars, an impression is all we get.

The skillful birder knows just what to key in on for each kind of bird. To distinguish two of the large, soaring birds around Cheyenne this summer, key in on the black and white pattern on the underside of the wings.

Swainson's hawks have the forward half or leading edge of the wing and the body light and the trailing edge of the wing dark. Turkey vultures, which seem to be sticking around instead of migrating through Cheyenne, are just the opposite. In addition, their wings tilt upward in a shallow "v."

The sparrows I like best are the ones with distinct markings. Half a dozen white-crowned sparrows, their heads marked with alternating black and white stripes, were in our backyard the last week of April and first week of May, singing non-stop all day until they continued their migration to spend summer in the mountains.

Some birds are so distinct, a silhouette is enough. Such is the case for the white-faced ibis, several of which have been reported the last couple weeks at reservoirs around southeastern Wyoming. I glimpsed one wading in Crystal Reservoir May 5.

Certainly the white on this ibis species's face is hardly enough to justify its name. You'd think it would be named for that incredible bill. Look up a picture of this exotic-looking shorebird to understand how amazed I am when I see one.

Short of putting bird food on the windowsill or getting trained or licensed to handle live wild birds for research or rehabilitation, birders don't usually get very close to birds, especially to owls.

If you are very lucky, you may notice one imitating a lump of tree trunk as it naps during the day. Otherwise, all you have to go on are signs and sounds: a hoot in the dark, a rabbit leg on the lawn, wing marks in the snow or pellets under the tree.

Andrea Cerovski, Wyoming Game and Fish Department nongame bird biologist, is on tour this year with Jupiter, a great horned owl.

During their appearance at the April Cheyenne High Plains Audubon meeting, Jupiter viewed the room from his position on Andrea's glove.

His eyes were huge and black, and he stared back at the audience. He is new to the lecture circuit and was finding everything to be of interest, especially the Cub Scout den that came in. Perhaps the high voices of young children reminded him of squeaking prey.

Jupiter's feet, with their huge talons, restlessly renewed their grip on the glove. His beak, though short, has to be sharp enough to shred flesh. His soft-edged feathers would, were he to hunt, break the sound of his flight so prey couldn't hear his approach until too late.

Hissy, the resident great horned owl at the Wyoming Children's Museum in Laramie, has had a lot more public exposure and seemed rather bored when introduced at the Audubon Wyoming reception there a few days later.

Because their mottled coloring camouflages the contours of their closed wings, it takes a close look to see why these two birds are not in the wild. Each has a mangled, unusable wing, due to collisions with man-made obstacles.

I'm like many people who wish they could communicate with wild animals. I want them to respond to me, yet when I sat still enough that a mountain bluebird foraged three feet away, as happened recently and unexpectedly, I was as happy to be accepted as nothing but a nonthreatening part of the landscape.

But I'm glad the law bans people from making pets of wild birds. We've interfered enough already, especially

2002

99

by changing many of their habitats.

When identification is especially vexing, I wish I could hold the bird in my hand. But if you've ever taken ornithology and studied bird skins, or "birds on a stick," you know that a bird in hand, out of context, can be just as vexing.

Then there's the occasional dead bird. For any dead bird, anytime of year, bag it. Pull a plastic bread bag over your hand like a glove. Pick up the bird and pull the bag inside out over the bird and tie the bag shut. Then double-bag it and take a close look through the bag at the bird if you like, before disposing of it.

After June 1, the county environmental health department's system will be in place for examining dead birds, especially black birds, for West Nile virus.

Meanwhile, if you'd like to see live wild birds, some preparing to nest and some just passing through in migration, join Audubon members on the Big Count this Saturday. You'll never believe how many kinds of birds there are here unless you see them for yourself.

# 109 Brown bird mix-up worth unraveling

Thursday, May 30, 2002, Outdoors, page C2

Lazuli bunting, indigo bunting, rose-breasted grosbeak, green-tailed towhee, yellow-rumped warbler, yellow warbler: the birds of spring migration are a colorful bunch.

Less obvious are the "LBJs" – "Little Brown Jobs," as they are sometimes called.

To appreciate the visiting LBJs, we must first distinguish them from the usual brown riff-raff, so let's review.

First and most prominent in the cityscape is the house sparrow. They build messy nests in three-dimensional signs or any other cavity, pick miller moths out of car radiators in parking lots and are seen wherever there's urban detritus to peck.

This time of year, the male house sparrow is a rich chestnut brown pattern on the back and wings, with a black goatee and bib and gray crown (top of head). The female has no black markings. Both have pale gray breasts and bellies.

The second most common brown bird in Cheyenne is the house finch. They like big trees and bird feeders stocked with black oil sunflower seed.

Both male and female are about the same size as the house sparrow but appear more slender and have breasts streaked with grayish brown lines.

The male in prime breeding plumage right now has a bright red forehead, red breast and red rump. Lesser males and males at other times of the year have faded pink or even yellowish hues.

As members of the finch family, house finches have big, thick bills for cracking seeds.

House finches and house sparrows are here year round. Once you can reliably distinguish them, male and female, you'll discover there are other brown birds eating at your feeder or flitting through the parks.

The pine siskin is a streaky brown bird easy to mistake for a female house finch. When it flies, a little yellow shows in its wings and tail.

Siskins can show up in Cheyenne any time of year. They are smaller than a house finch by three quarters of an inch and have a smaller bill suited to eating the tiny niger thistle seeds they love.

Some people in town had siskins all winter, but our tube-style thistle feeder was untouched until the beginning of May. Now it's crowded with siskins and the closely related goldfinches, one for each of the eight perches and a dozen waiting for openings. Their cheerful songs get a bit strident as they retreat to the trees while I refill the feeder.

Occasionally I will see one kind of sparrow often enough that I begin to distinguish and remember its unique field markings. Such is the case with the clay-colored sparrows showing up in our backyard this month.

As small as a siskin, the clay-colored sparrow still has to head at least as far north as Montana to nest. It has a very pale gray breast that sort of glows from a distance. The rest of it is a non-descript brown pattern with darker streaks over the head, through the eye and in front of the eye. Luckily it has a unique song, four buzzes in each phrase, like an insect.

There's another petite, nondescript sparrow in my yard, the chipping sparrow.

Chippies have one notable field mark, a rusty brown crown. They will spend the summer here, especially in weedy places, so there is plenty of time to get to know them well.

Sometimes a birdwatcher has to be able to identify parts of a bird because it won't come out for a clear view.

This spring I think Mark and I identified a new (for us) brown bird. As it skulked in the bushes, there was a flash of brown, then an eye and a leg. The pale breast had markings.

The best identification we could make using our field guide was the veery. But there are three or four other thrushes that look similar. I'll have to learn more about them before my next encounter.

So far I've been discussing backyard brown birds. Don't forget the bigger migration picture and the bigger brown birds.

Shorebirds are on their way north to breed and stop briefly at our local reservoirs. Identifying the different species is a real art. For me, it's like trying to distinguish identical triplets.

One must find the slight differences in brown plumage, body shape, length and tilt of bill, length and color of legs. Heaven forbid their legs or bills are discolored by mud from probing the shoreline

100    CHEYENNE BIRD BANTER

for edibles.

The problem is, the sandpipers, dowitchers, yellowlegs and the other shorebirds that nest in the far north are never here long enough in one season to become as familiar to me as their relations that nest here, such as the killdeer.

I guess that means at my present rate I'll never run out of obscure brown birds to learn to identify.

# 110 Records request ruffles feathers

Thursday, June 13, 2002, Outdoors, page C2

One of the fringe benefits of writing this column is hearing from other people watching birds.

In the last month there have been reports of robins building a nest on a ladder and nesting on a porch light fixture plus one report of a robin attacking what it thought was its rival – its own reflection in a clean window.

The rest of the year the most common calls are requests for help identifying birds.

Luckily there aren't many calls from people with rare bird sightings. If someone were to insist, for instance, that they have a pink flamingo in their garden – and it's not plastic – I would refer them to Jane Dorn, whose training and expertise in birds extends far beyond our local backyards.

Jane compiles the reports for our local Christmas Bird Count, the Big Day spring count and local reports for American Birds, a quarterly journal.

If a species is unusual for Wyoming or for the time of year, Jane will ask the observer for more information because she is also a member of the Wyoming Bird Records Committee.

A photograph of the bird in question is extremely useful, or verification by one or more knowledgeable birders. It boils down to credibility

and the honor system, unless Jane gets a chance to run out and see the bird herself.

An observer can send a report directly to the committee, in care of the Wyoming Game and Fish Department nongame bird biologist, but there are advantages to working with Jane for rare bird sightings in our area.

First, she is intimately knowledgeable about the birds here and second, she may be able to vouch for your credibility and birding ability when your report is being reviewed for inclusion in official state records.

Eventually the accepted reports are used to revise new editions of the "Atlas of Birds, Mammals, Reptiles and Amphibians in Wyoming," published by Game and Fish.

Some rare bird observations indicate nothing more than a migrant blown off course, while for other species, reports begin to accumulate, showing they are changing their migration patterns or expanding breeding ranges.

Imagine my surprise and dismay when a request for documentation after last month's Big Day count was met with hostility and suspicion.

The best birding on the Big Day count starts at sunrise, so it helps to have birders in our hottest bird spots simultaneously.

Each year, while Jane and her husband Robert, also an expert birder, are scoping the Grasslands Research Station first thing, birders from Casper and our local Audubon chapter start at Lions Park. We can't wait to see what unusual migrants will turn up.

This year, two out-of-state birders met up with us for a little while and were the first to spot a Connecticut warbler, a first record for the Cheyenne latilong (a latilong is an area one degree of latitude by one degree of longitude). The two enthusiastic birders helped many people get a chance to see it.

Each year, before the Casper birders leave town, I try to record their observations, but with so many people participating this year, I didn't get a chance to check with everyone.

Knowing most of the Casperites and the two out-of-state birders subscribe to the Wyobirds listserv, Jane and I compiled a preliminary list and sent it over the internet with a request for documentation of two rare species, including the warbler (I only saw a few of its feathers).

The responses of the two out-of-state birders, who have evidently birded the Cheyenne area frequently on their own, appeared quickly. One asked, who is Jane Dorn and why should he report anything to her? The other

complained that the state records committee had never acknowledged other reports he'd sent in and he wasn't going to send in any more.

Since then, two Casper birders have sent Jane excellent documentation for the warbler, and one of our local birders may be able to do so for the other bird, a glossy ibis.

This whole episode brings up two points. One, if visitors have the ability to identify rare birds and they take the time to befriend and share their talent with the locals, it is time and expertise that is greatly appreciated. Otherwise, they appear to be roving rare bird baggers.

Second, the all-volunteer records committee needs to figure out how to deal with its backlog of reports. Modern communications technology would benefit the scattered members who find it difficult to meet in person.

There's also a third point to make. As willing as birders are to serve as citizen scientists, there is an increasing amount of data organization and processing needed for wildlife planning and management purposes. The state wildlife agency is the logical institution to handle it.

Game and Fish should consider increasing its nongame bird staff so data can be prepared in a useful and timely way.

# 111 Big Count draws flocks of observers

*Not found in my newspaper clippings.*

Members of Cheyenne High Plains Audubon Society broke out their hats and gloves to brave cold and foggy conditions for the beginning of the annual Big Day bird count May 18. By afternoon the skies cleared and temperatures reached 67 degrees.

Local birders were joined by others from Rock Springs, Laramie, Casper, Wheatland, Colorado, Nebraska and Mississippi for a total of more than 40 observers, probably a record.

The high number of species counted despite unfavorable weather, though not record breaking, was due to the large number of observers getting an early start at 6 a.m. in several locations simultaneously.

Two species recorded on this count, the glossy ibis and Connecticut warbler, will be the first records for Cheyenne if their identification can be officially confirmed. Both species are considered to be accidentals, having previously been recorded in Wyoming only once each.

The glossy ibis, native to Florida and the coasts of southeastern and eastern U. S., and the Connecticut warbler, which migrates through the eastern U.S. and breeds in Canada, are difficult to distinguish from their western counterparts, the white-faced ibis and MacGillivray's warbler, respectively.

## Final Results Cheyenne Big Day Count, May 18, 2002

146 species

1 - Lions Park, including view of Kiwanis Lake

2 - Wyoming Hereford Ranch #1 Reservoir

3 - Wyoming Hereford Ranch

4 - High Plains Grasslands Research Station

5 - Other city locations, including various backyards

| | | |
|---|---|---|
| 1 -- -- -- -- -- | Pied-billed Grebe | |
| -- 2 -- -- -- -- | Eared Grebe | |
| 1 2 -- -- -- -- | Western Grebe | |
| -- -- 3 -- -- -- | Clark's Grebe | |
| -- 2 -- -- -- -- | American White Pelican | |
| 1 2 -- -- -- 5 | Double-crested Cormorant (Holliday Park) | |
| -- 2 -- -- -- -- | Great Blue Heron | |
| 1 2 -- -- -- 5 | Black-crowned Night-Heron (Holliday Park) | |
| -- -- 3 -- -- -- | Glossy Ibis* | |
| -- -- 3 -- -- -- | White-faced Ibis | |
| -- -- -- -- 4 -- | Turkey Vulture | |
| 1 2 -- 4 5 | Canada Goose (Holliday Park) | |
| -- 2 -- 4 -- | Green-winged Teal | |
| 1 2 3 -- -- -- | Mallard | |
| -- 2 -- -- -- -- | Northern Pintail | |
| 1 2 -- -- -- -- | Blue-winged Teal | |
| -- 2 -- -- -- -- | Cinnamon Teal | |
| -- 2 -- -- -- -- | Northern Shoveler | |
| -- 2 -- -- -- -- | Gadwall | |
| -- 2 -- 4 -- | Redhead | |
| -- 2 -- 4 -- | Lesser Scaup | |
| -- -- 3 -- -- -- | Bufflehead | |
| -- 2 -- -- -- -- | Hooded Merganser | |
| -- -- -- 4 -- -- | Common Merganser | |
| -- 2 -- -- -- -- | Ruddy Duck | |

| | | |
|---|---|---|
| -- 2 -- -- -- -- | Northern Harrier | |
| -- -- 3 -- -- -- | Sharp-shinned Hawk | |
| -- -- 3 -- -- -- | Cooper's Hawk | |
| -- -- 3 4 -- | Swainson's Hawk | |
| -- -- 3 4 -- | Red-tailed Hawk | |
| -- -- -- 4 -- | Ferruginous Hawk | |
| -- -- 3 4 -- | Golden Eagle | |
| -- -- -- 4 -- | American Kestrel | |
| -- 2 -- -- -- -- | Sora | |
| 1 -- 3 -- -- | American Coot | |
| -- -- 3 -- -- -- | Black-bellied Plover | |
| -- -- 3 -- -- -- | Semi-palmated Plover | |
| -- 2 3 4 -- | Killdeer | |
| -- 2 -- -- -- -- | American Avocet | |
| -- 2 -- -- -- -- | Lesser Yellowlegs | |
| -- -- 3 -- -- -- | Willet | |
| 1 2 -- 4 -- | Spotted Sandpiper | |
| -- -- 3 -- -- -- | Whimbrel | |
| -- -- 3 -- -- -- | Sanderling | |
| -- 2 -- -- -- -- | Long-billed Dowitcher | |
| -- -- -- 4 -- | Common Snipe | |
| -- 2 -- -- -- -- | Wilson's Phalarope | |
| -- -- -- -- 5 | Franklin's Gull (South Industrial Road) | |
| 1 -- 3 -- 5 | Ring-billed Gull (South Industrial Road) | |
| -- -- 3 -- -- -- | California Gull | |
| 1 -- 3 -- 5 | Rock Dove | |
| 1 -- 3 4 5 | Mourning Dove | |
| -- -- -- 4 -- | Great Horned Owl | |
| -- -- -- -- 5 | Burrowing Owl (NE Cheyenne) | |
| -- -- -- -- 5 | Chimney Swift (St. Mary's Cathedral tower) | |
| -- 2 3 -- 5 | Belted Kingfisher (Greenway on Crow Creek) | |
| 1 -- -- -- -- -- | Red-headed Woodpecker | |

| | | |
|---|---|---|
| -- -- 3 4 -- | Downy Woodpecker | |
| 1 -- 3 -- -- | Northern Flicker | |
| 1 -- -- 4 -- | Olive-sided Flycatcher | |
| -- -- -- -- 4 -- | Western Wood-Pewee | |
| 1 -- -- -- -- -- | Willow Flycatcher | |
| 1 -- -- -- -- -- | Least Flycatcher | |
| -- -- -- 4 -- | Dusky Flycatcher | |
| -- -- 3 -- -- | Cordilleran Flycatcher | |
| -- -- -- 4 -- | Say's Phoebe | |
| -- -- -- 4 -- | Cassin's Kingbird | |
| 1 2 -- 4 -- | Western Kingbird | |
| -- 2 -- 4 -- | Eastern Kingbird | |
| -- -- -- 4 -- | Loggerhead Shrike | |
| 1 -- -- -- -- -- | Plumbeous Vireo | |
| 1 -- 3 4 -- | Warbling Vireo | |
| -- -- 3 -- -- | Red-eyed Vireo | |
| 1 -- 3 -- -- | Blue Jay | |
| -- -- -- 4 -- | Black-billed Magpie | |
| 1 -- -- -- -- -- | American Crow | |
| -- -- -- -- 5 | Common Raven (east end Happy Jack Road) | |
| -- -- -- 4 -- | Horned Lark | |
| -- -- 3 -- -- | Northern Rough-winged Swallow | |
| -- 2 -- -- -- -- | Bank Swallow | |
| 1 -- 3 4 -- | Cliff Swallow | |
| -- -- -- -- 5 | Barn Swallow (Wyoming Information Center) | |
| -- -- -- 4 -- | Black-capped Chickadee | |
| -- -- -- 4 -- | Mountain Chickadee | |
| 1 -- 3 4 5 | Red-breasted Nuthatch | |
| -- 2 3 4 -- | House Wren | |
| -- 2 -- -- -- -- | Marsh Wren | |
| -- -- 3 -- -- | Ruby-crowned Kinglet | |
| 1 -- 3 4 -- | Blue-gray Gnatcatcher | |
| -- -- -- 4 -- | Eastern Bluebird | |

| Code | Bird | Code | Bird | Code | Bird |
|---|---|---|---|---|---|
| --- 3 – 5 | Mountain Bluebird (north of Riding Club Road) | 1 -- 4 -- | Northern Waterthrush | --- 3 -- -- | Rose-breasted Grosbeak |
| 1 2 3 4 -- | Swainson's Thrush | --- 3 -- -- | Connecticut Warbler* | --- 3 – 5 | Black-headed Grosbeak (backyard) |
| 1 -- 3 -- -- | Hermit Thrush | --- 3 -- -- | MacGillivray's Warbler | -- -- -- 4 -- | Blue Grosbeak |
| 1 2 3 4 -- | American Robin | 1 2 3 4 -- | Common Yellowthroat | --- 3 4 5 | Lazuli Bunting (backyard) |
| 1 2 3 4 -- | Gray Catbird | 1 – 3 -- -- | Wilson's Warbler | -- -- 3 -- -- | Indigo Bunting |
| -- -- -- 4 -- | Northern Mockingbird | 1 -- -- -- -- | Yellow-breasted Chat | 1 2 3 4 -- | Red-winged Blackbird |
| --- 3 4 -- | Brown Thrasher | --- 3 – 5 | Western Tanager (Holliday Park and backyard) | -- 2 3 4 -- | Western Meadowlark |
| 1 -- -- 4 5 | Cedar Waxwing | 1 – 3 4 -- | Green-tailed Towhee | 1 2 – 4 -- | Yellow-headed Blackbird |
| 1 -- -- -- -- | European Starling | 1 – 3 4 -- | Chipping Sparrow | -- -- -- 4 -- | Brewer's Blackbird |
| 1 – 3 -- -- | Tennessee Warbler | 1 -- -- 4 -- | Clay-colored Sparrow | 1 – 3 4 -- | Common Grackle |
| -- -- -- 4 -- | Orange-crowned Warbler | 1 – 3 -- -- | Brewer's Sparrow | --- 3 4 -- | Brown-headed Cowbird |
| 1 – 3 4 -- | Virginia's Warbler | --- 3 4 -- | Vesper Sparrow | -- 2 -- -- -- | Orchard Oriole |
| 1 -- -- 4 -- | Northern Parula | 1 -- -- 4 -- | Lark Sparrow | 1 -- -- 4 5 | Bullock's Oriole |
| 1 2 3 4 -- | Yellow Warbler | --- 3 4 -- | Lark Bunting | 1 -- -- -- 5 | House Finch (backyard) |
| 1 – 3 -- -- | Chestnut-sided Warbler | -- -- -- 4 -- | Savannah Sparrow | 1 -- -- 4 5 | Pine Siskin (backyard) |
| 1 – 3 4 -- | Yellow-rumped Warbler | 1 – 3 -- -- | Song Sparrow | 1 2 3 4 5 | American Goldfinch (backyard) |
| -- -- 3 -- -- | Townsend's Warbler | --- 3 4 -- | Lincoln's Sparrow | 1 -- -- -- 5 | House Sparrow (backyard) |
| 1 -- -- -- -- | Palm Warbler | --- 3 -- -- | White-crowned Sparrow | | |
| 1 – 3 4 -- | Blackpoll Warbler | -- -- -- -- 5 | McCown's Longspur (NE Cheyenne) | | |
| --- 3 -- -- | American Redstart | | | | |

\* First record for Cheyenne if identification is verified.

# 112

Thursday, June 27, 2002, Outdoors, page C2

## Bird families finding food in many unlikely places

It was easy to see where the fish were biting Father's Day at Glendo Reservoir. Six or eight boats radiated from a point of land like magnets around a pole.

So we broke out our rods, hiked down to the water's edge, impaled leeches and began casting for walleye. Besides the boaters and a couple other families on shore, we joined the avian fish-wishers.

Two or three western grebes, just white dots bobbing on the edge of boat wakes, could be heard chattering between dives for fish. When they moved in close to shore, I wondered: Have they learned that a flock of people means fish, or are both humans and birds good at recognizing fruitful fishing spots?

Soon after we arrived, a female common merganser surfaced nearby, another bird that dives for fish. As I cast unsuccessfully for the umpteenth time, the male came in as if he was following my bait, flying so low over the water he barely kept his wing tips dry.

After a while, son Bryan and I, the non-fanatic half of our fishing family, decided to try our luck birding over by the power plant and hiked an interpretive trail provided by the Bureau of Reclamation.

We seemed to be alone in the canyon except for noisy red-winged blackbirds in the marsh and less identifiable birds singing in the junipers. Even at 11 a.m. on this warm day, there was a terrific amount of bird flitting and song: orioles, kingbirds, flycatchers, sparrows, nuthatches, and flickers with yellowish wing linings, possibly those red-shafted-yellow-shafted hybrids I hear about.

Though we were away from the hum of boat motors, we still found avid fishers.

First, a great blue heron rose up from the cattails and perched in a dead tree (yes, I know, herons have the wrong kind of feet to be officially classified as perching birds). It and another heron patrolled the canyon the whole time we were there.

Then, we heard the familiar rattle of a belted kingfisher overhead. It was so odd to see it against the backdrop of dry juniper hills.

As Bryan and I sat on a bench on a little observation platform in the middle of the marsh, we heard a terrific splashing and thrashing nearby. It was as if invisible men were wrestling in the shallow water. Then we saw a fin. Whew. It was just carp spawning.

If "carpe diem" means "seize the day," did carp get their name because they would be easy to seize in the shallow water? Too bad we didn't have nets. However, the fishing half of the family brought in both walleye and carp for dinner.

Earlier in the week, Mark and I had a chance to stroll around Bear Lake up at Rocky Mountain National Park. Mark said seeing so many visitors reminded him of the Bronx Zoo, except for the lack of families with strollers. But two minutes later we saw one.

On my first visit, 40 years ago to the week, snow made the trail to Nymph Lake impassable. This year there was only a tiny patch for visitors to marvel over. But there was no change in the locals trying to separate tourists from their dough—or rather, anything edible.

As soon as we seated ourselves on a bench with a view, an enterprising golden mantled ground squirrel (cousin to the smaller chipmunk) quickly appeared, willing to pose for peanuts. But we declined.

Soon a drake mallard paraded by us and swam all around the edge of the small lake, but no one seemed interested in rewarding his enterprise. Down at the other end, visitors refused to fall victim to a couple of camp robbers, otherwise known as gray jays.

It is good news to me that national park visitors are showing signs of accepting the mantra "Do Not Feed the Animals."

Drought conditions may be changing the foraging habits of some birds. Many recent postings on the Wyobirds listserv are about an influx of lesser goldfinch sightings along the Colorado-Wyoming border.

The field marks for the lesser goldfinch male are a black cap that extends to the back of the head (further than the American's), greenish rather than yellow back and a large white patch on both

the top side and underside of the wing at the base of the primary feathers, most noticeable in flight.

The American goldfinch, the regular species in Cheyenne, has large white patches only in its "wing pits." What's confusing is some males might not yet have attained full yellow coloring on their backs and look sort of grayish.

The tree swallows on the Laramie Greenbelt are busy catching flying insects to feed their young, however, I saw one taking a breather of sorts. It was sitting on top of a tall post, its small, weak

feet hooked into the rough end-grain as the wind buffeted it like a tethered kite.

A robin family whose nest site is somewhere nearby has been coming to our yard for the millers. And the grackles have been climbing the trunks of our trees like overgrown nuthatches, gleaning something edible from bark crevices.

Though sometimes we might be in competition, lucky for us, more often one species' pest is another species' manna.

# 113 Thursday, July 11, 2002, Outdoors, page C2
## Arizona fire affects family, wildlife

Once, in just a few days, I managed to add 20 birds to my life list. We were visiting Arizona, where even the most common birds were species I'd never seen before, such as yellow-eyed juncos and acorn woodpeckers.

I would not have expected to travel to Arizona before reaching retirement age, but we were visiting my sister Beth who was working as a recreation planner for the Apache-Sitgreaves National Forest and lived in Pinetop-Lakeside.

Many of the new birds were seen on a short trip further south, to the Arizona-Sonora Desert Museum and Madera Canyon near Tucson, but even Pinetop, up in the White Mountains, had a new southwestern species for me, the black phoebe, and one I had yet to identify in Wyoming, the pygmy nuthatch.

At first, Beth rented a

cabin so buried in the ponderosa pine she shivered even in summer. When she bought a house, she made sure it didn't have a single pine on the lot. Though she's worked a lot of fires in her career, I think she was considering solar heat more than fire safety.

Last year, Beth relocated to the Forest Service office in Heber where her husband is a wildlife biologist. Their present house is in a piney neighborhood, but on a lot open except for a few small trees next to the house and a couple larger ones out by the road.

The small trees succumbed to some quick chainsaw work a couple weeks ago. After my brother-in-law left to work the Rodeo fire, Beth had to evacuate. Luckily, her office also had to evacuate, to an old field camp where Beth could leave the family pets.

Except for the day the

Chediski (think "Cheddar Sky" when you pronounce it) fire singed the Heber office's front lawn, Beth and her few remaining coworkers came back every day to man the phones.

Most of what I know about the fires I learned from the Wyoming Tribune-Eagle's coverage and secondhand from my mother who has been acting as family dispatcher. Beth is home now, however, the fire season is not over.

Many of the homes lost belonged to seasonal residents who can stay home in Phoenix while they decide what to do about rebuilding, but what about Beth's coworkers and other year round residents who lost their homes?

What happens to wildlife? If Bambi's neighborhood burns and he escapes to the next valley, will the deer there share their scarce resources during this drought

year, or will Bambi have to make do with the most marginal habitat? Or does he change the pecking order in the new valley and force some other deer to relocate, causing a ripple effect?

What about birds? I wouldn't think your average bird could out-fly a crown fire racing through treetops. Do birds smell smoke and evacuate an area? Or, if there are any birds left after a fire, is it because their territory happened to be one of the many patches the fire skipped?

Coincidently, during the first week of the Rodeo-Chediski fires, I happened to read the May/June 2002 issue of the U.S. Department of Interior's "People, Land and Water" magazine, which was devoted to wildland fire and the new national fire management plan.

The article that caught my eye was "Birds and Burns

in Ponderosa Pine Forests," written by Natasha Kotliar from the Midcontinent Ecological Science Center. The MESC is one of 16 science centers in the Biological Resources Division of the U.S. Geological Survey and is headquartered in Fort Collins, Colo.

She described a plan to study the ecological consequences of three fire conditions: unburned forests, prescribed understory fire and wildfire, in three locations including Arizona.

The article didn't list where the no burn areas were going to be, but there will be an abundance of burn areas to choose from.

Beth and her husband have worked plenty of fires before, but never so close to home that they needed to evacuate. This time, they will have front row seats for studying the recovery of the land and wildlife – and the people.

Even though the fire isn't completely contained as I write this, already both Beth and her husband are beginning rehabilitation work. Wish for them not so much rain that the soil washes away, but only enough that the seeds sprout.

# 114 Thursday, July 25, 2002, Outdoors, page C2
# Solitary hike yields memorable encounters

I didn't do my homework in early July. I expected to be able to look for birds while exploring another riparian area below the dam at Grayrocks Reservoir on the Laramie River east of Wheatland.

Had I studied the entry for the Grayrocks Wildlife Habitat Management Area in the Wyoming Game and Fish Department's publication "Access to Wyoming's Wildlife," I would have already known there is no public access below the dam.

The dog and I resigned ourselves to sitting in the shade at the side of the car, waiting for Mark to limit out. We can't both fish when the dog's with us. Periodically I sent Lincoln down the rocky cliff, caused by the low water level, to cool his paws.

The fishing was good, so we returned the next week. This time I studied the map by the restroom and discovered Cottonwood Draw was part of the management area. It looked interesting, but since everyone else wanted to fish and we'd left the dog home in the shade, I was on my own.

"Lions and tigers and bears, oh my!" I thought, as I left the car to bake in the parking area. I seldom

hike alone. "Mountain lions and ticks and rattlesnakes!" I revised.

As I followed an overgrown two-track, each of my steps caused grasshoppers to ricochet off the dry grass – and my bare legs. At least they don't bite or sting. They might, however, chew on me if I sat still long enough.

When the track petered out, bushwhacking was as easy as stepping around multitudes of prickly pear cactus until I came to a side draw stuffed with juniper. Deep in the shade was a trickle of water and long green grass, and bits of bird song.

Later, I decided to walk in the dry main creek bed, even though it seemed circuitous. Someone had left tracks from either a bicycle with wide tires or a motor bike, though the area is closed to motor vehicles.

There was no breeze and no shade except where water has cut high, steep banks, leaving several trees dangling overhead with half their roots exposed, like cut-away models explaining arboreal hydrologic systems.

The further up the creek I went, the more anonymous bird singing I heard (still need time to learn those

signature tunes).

One unseen bird, probably a northern mockingbird, had a repertoire of half a dozen whistled phrases. It whistled one and then I whistled it back and then it whistled the next one – sort of a dry country "Dueling Banjos" duet including fair imitations of jays and crows.

A Cooper's hawk (I think) swished close over my head, diving evasively among the trees with a kestrel on his tail in hot pursuit, all the action occurring in a flash like a one-second clip from some old war movie.

Then, practically under my next step, the sand exploded and a toad or frog landed a foot away. We stared at each other for several minutes while I tried to figure out what markings I should memorize so I could look it up when I got home.

Unfortunately, my amphibian and reptile field guide is for the east and central United States and uses terms I am unfamiliar with, making me as frustrated as any novice birder consulting their first bird guide. The white line down its back appeared to make the animal I saw some kind of toad.

Though a couple spots in

the dry creek bed appeared damp, the first standing water was at least a mile upstream from the reservoir. Just above that, water was flowing. I wished I'd been hiking in sandals and could easily slosh a bit to cool off. Then I remembered all the cactus I'd come through.

A few wildflowers bloomed on the low banks. Upstream, the creek glittered invitingly in a haze of green. Too bad I told the fishermen I'd be back by a certain time and had to turn around to keep the appointment.

Fishing was not so good this second trip, so I don't know when we'll get back. Fall would be nice, but then it will be "Lions and tigers and hunters, oh my!" Hunting season is not impossible, just more people.

Solitude is a luxury that is sometimes a result of timing – not everyone is willing to walk away from a lake on a hot day in July – as much as it is space.

Hiking with a dog is not hiking alone either. But though one gains in personal observation and reflection in solitude, having companions along adds their viewpoints, their observations, their knowledge, and later, the strength of their shared memories to the richness of the experience.

As I retraced the old track past the pair of lark sparrows feeding on insects in a juniper, I wondered if the creek would have enough water next time to interest my three favorite fishermen in hiking further up the way with me.

# 115 Rescuing baby birds not always necessary

Thursday, August 8, 2002, Outdoors, page C2

Mid-July I got a call from a member of the staff at the Cheyenne Pet Clinic. Would I know where to find killdeer? Someone had brought in a chick and the staff wanted to release it near other killdeer in hopes they would foster it.

There was no information about where the chick had come from. Why was it brought in?

"They said it fell out of the nest," replied the staffer. We had to laugh. Killdeer nest on the ground.

In fact, here in the grasslands, most birds nest on the ground, including the western meadowlark, and even hawks such as the northern harrier (formerly named marsh hawk) and ferruginous hawk.

Some grassland birds may nest in the few available bushes, but otherwise, it's an entirely different group of birds adapted to building nests in trees in our yards, along riparian areas (streams and creeks) or in forests.

In other treeless habitats birds also nest on the ground. Think of all the shorebirds, penguins and seabirds.

Just yesterday I was reading one of Christopher (Robin) Milne's autobiographical books in which he describes finding an owlet on the ground near his home in Dartmouth, England, and how he thought he needed to take it home and raise it himself.

Well-informed people in this country know that they need a permit to raise wildlife and besides, owlets walk around on the ground before they learn to fly as a normal part of their development. Burrowing owls even nest underground.

I think there is a default setting in our brains when the phrase "baby bird" is uttered. We automatically envision a tree with a cozy nest of tiny, featherless robins. Their parents take turns perching on the rim, stuffing worms and insects into their gaping mouths.

For many birds, this picture is accurate. When their eggs hatch, the young are helpless, naked and blind creatures that spend a week or two in the nest. They are classified as altricial young and are called nestlings.

Ground nesting bird species tend to have precocial young called chicks. Shortly after hatching, the chicks, covered in downy feathers, are running around after their parents. Think about domestic ducklings and chickens. Though they can't fly right away, falling out of the nest is not one of their problems.

When do baby birds need rescuing? My rule of thumb is when a life-threatening catastrophe is human caused, such as last summer's incident when children tore down swallow nests on the Greenway. Or there are loose pets that may cause injury. I

hope you've been too smart and soft-hearted to let your pets roam, especially during the May, June and July nesting season.

There will always be a baby robin that leaves its nest prematurely. Even if you put it back, whatever defect in the nest construction or in its baby brain that caused it to fall or jump the first time will usually cause it to do it again.

It's important to remember that for every year's crop of young animals, a high percentage is meant to be food for other young. Even People for the Ethical Treatment of Animals can't turn carnivores into herbivores. Just make sure the balance of nature isn't upset because your Fluffy or Fido is pretending to be one of the native predators.

Even if people mistakenly rescue a bird, it's still a good sign that they care about the welfare of wildlife. Perhaps they are ready to take other, less direct actions, on behalf of wildlife, such as using organic lawn care products, recycling and supporting organic farming, pollution control and native landscape reclamation.

Though this year's nesting season is nearly finished, except for the goldfinches and a few birds trying to get a second brood in, it's not too soon to make your own small contribution and work on turning your cat into a house pet.

If the part you dislike about house cats is the litter box, let me put in a plug for Arm and Hammer's "Super Scoop" clumping kitty litter. It is more expensive per pound than regular clay, but it lasts longer and works better.

I take a minute a day to scoop tidy, nearly odor-free litter box lumps into an empty produce bag, bread bag or cereal box liner, and maybe add a little fresh litter to the box if the level is getting low. However, I only dump the entire litter box contents once or twice a year. A 14-pound box of litter lasts my two cats about three weeks.

Think of it, something as minor as fresh-smelling, easy to use cat litter could improve the chances of survival for birds in your neighborhood.

A couple days after receiving the killdeer call, I was at the clinic for my menagerie's annual visit, so I was able to meet the young women who had been on the hunt for a foster home for the chick.

They spent hours hiking over hill and dale before finding likely killdeer parents. Because the chick was not part of a study, it was not banded or fitted with a radio transmitter. No one will ever know its fate.

In the natural world, sometimes it's better to leave well enough alone.

# 116 Bird books cover same turf in different ways

Thursday, August 22, 2002, Outdoors, page C2

One of the pleasures of travel is visiting used bookstores, especially in university towns and large cities.

Buying secondhand books saves money, but it also gives me a chance to acquire books sometimes too obscure for the local bookstore or library to carry, such as one I found in Albuquerque, N.M., last month.

"The Birder's Miscellany, A Fascinating Collection of Facts, Figures and Folklore from the World of Birds," by Scott Weidensaul (Simon and Schuster, 1991) is a slim 135 pages.

I was attracted to the title as well as the name of the author. Last winter I read Weidensaul's "Living on the Wind, Across the Hemisphere with Migratory Birds" (North Point Press, 1999). I started out with the public library's copy, but soon determined I needed my own, both because I enjoyed the writing style and because it's a good reference.

Unfortunately, "The Birder's Miscellany" is out of print, but you may be able to find a copy through Barnes and Noble or Amazon's out-of-print sections of their websites, or through an online catalog such as bookfinder.com.

It's worth finding this compendium of odd facts and figures to answer questions such as, "What's the biggest bird?" That answer,

says Weidensaul, needs to be qualified.

The heaviest and tallest living bird is the ostrich (350 pounds, six feet). The heaviest bird that can fly is the mute swan (up to 50 pounds). The bird with the longest wingspan is either the marabou stork of Africa or the wandering albatross of the Southern Hemisphere (12 feet).

Did you know the domestic turkey's heart rate at rest is 93 beats per minute, compared to 480 for the blue-throated hummingbird? In flight, that hummer from Mexico has 1,200 beats a minute.

Besides exploring the range of physical attributes, Weidensaul also explores bird behavior and birds in folklore and history in a style that invites reading his book cover to cover.

In contrast, the 600 pages of "The Sibley Guide to Bird Life and Behavior" (Knopf, 2001) could be used to press a few wildflowers, and though just released last year, it was already available at a Boulder, Colo., used book shop in June.

It is a companion to "The Sibley Guide to Birds" (Knopf, 2000), which, compared to all other field guides, has excellent illustrations of each species' various plumages but no information about behavior or habitat, two things which sometimes

help clinch identification.

Evidently David Allen Sibley was saving that stuff for the second book.

I thought "Bird Life and Behavior" might be similar to another book I have, Kenn Kaufman's "Lives of North American Birds," but it isn't.

For this second volume, David Allen Sibley is the illustrator and one of the three editors. The other two editors, Chris Elphick and John B. Dunning, Jr., along with 46 other expert birders and biologists, contributed articles for the text.

Where Kaufman systematically provides a photo and an account for each species, this book starts with a 120-page introduction to basic ornithology, including biology, behavior patterns and bird conservation issues.

This first section is more technical and thorough than Weidensaul's book, but not as much fun to read.

The remainder of "Bird Life and Behavior" is divided into 78 chapters, one for each North American family of birds, from "Loons" to "Old World Sparrows."

Any birds pictured are meant to illustrate a particular bit of information, so when the text refers to a species you aren't familiar with, you may have to grab a field guide.

If you want to know more about mountain bluebirds, for instance, you look in the

table of contents for "Thrushes." Otherwise, back in the index, under "Bluebird, Mountain," you are referred to 459-60, 461 and 464.

The first reference compares mountain bluebirds to the other bluebirds, stating that they "occur at high elevations throughout the western mountains, often in recently burned areas." We also learn that they like to winter in open, arid grasslands, and that their populations have benefited from the increasing numbers of nest boxes provided.

After finding the specific references to mountain bluebirds, you can read the whole chapter for general and comparative information about thrush species (including the robin) under various subheadings: taxonomy, habitats, food and foraging, breeding, vocalizations, movements, conservation and accidental species. Each chapter is set up the same way.

The thrush chapter is written by John Kricher, and in the "Author Biographies" section you can read his list of credentials.

As with any encyclopedic tome, I'll be reading this new Sibley guide, bit by bit, as questions come up. And bit by bit, I hope its overwhelming amount of information will seep in and stick to my brain.

# 117 Park's best-known bird remains elusive

Thursday, September 5, 2002, Outdoors, page C2

**A copy of the book "Birds of Yellowstone" wasn't as handy for this trip as it could have been.**

There wasn't a hummingbird to be seen on our family's mid-August trip through Yellowstone.

No familiar rattle of feathers to indicate our examination by a broad-tailed hummingbird, and no sign of the other species common to the park, calliope and rufous.

Their absence shouldn't have been a surprise to me since we're used to seeing hummingbirds in Cheyenne as early as the end of July, when they begin to leave the high country.

The checklist and ecological chart at the back of Terry McEneaney's definitive book, "Birds of Yellowstone" confirmed my observations. Plus McEneaney, the park's bird biologist, rates hummingbirds as difficult to see even at the peak of summer.

I checked out a copy of the "Birds of Yellowstone" from the public library last winter while doing research for the "Wyoming Bird Flashcards" CD-ROM. I decided a trip to Yellowstone merited buying my own.

After finding a copy in the park, I looked up the trumpeter swan, the species I most wanted to see and have never seen in the park. Mary Bay on Yellowstone Lake and the Seven Mile Bridge on the road to West Yellowstone were recommended, but the time of year wasn't perfect. Every potential swan turned out to be a white pelican.

My first trip to Yellowstone was 30 years ago and most of the half dozen trips since then have been quick drive-throughs. This time we stayed one night in a six-plex "cabin," but we still traveled only the front country.

Driving from one scenic icon to the next must be what it's like to do the Stations of the Cross. Instead of stopping for prayers, we stopped for photographs. The Yellowstone pilgrimage quickly becomes a litany of boardwalks and blacktop, cameras and crowds, as well as mud pots and moose, rivers and ravens, geysers and gray jays.

After dinner, on our way back to our Canyon cabin, we made one more stop at the Norris Geyser Basin. Our rule is "waste no daylight hour" since our trips to Yellowstone are infrequent.

We came upon an audience waiting for a geyser show and decided to wait too. However, as the landscape marked with plumes of steam darkened, we decided to head back, but still finish the rest of the loop trail.

It's not easy to find solitude on a boardwalk in the Yellowstone caldera basin, but twilight is a good bet. We were alone, except for killdeer skittering on the thin sheets of water spreading across mineral deposits. Only the occasional pop or hiss of a mud pot added to the normal outdoor sounds.

Early morning, especially on a Sunday, turned out to be another good time to find the park alone. Our cabin on P Loop was only a 20-minute walk through forest to Grand View Point which overlooks the Grand Canyon of the Yellowstone.

We hiked down the Red Rock trail and waited, alone, for the clouds to roll back so we could take pictures of the Lower Falls. An osprey swooped picturesquely in front of the spray, chased by two kestrels. It's that black dot in the middle of my picture.

Later, our family was the only human life to be seen on a long section of black beach near Gull Point. And we even picked an unpopular, though not unpopulated, picnic area for lunch.

Shortly after sitting down at the table I realized we weren't alone. First one gray jay and then another lighted in the tree branches over our heads. What beady eyes!

Then two Clark's nutcrackers came on the scene. The four birds shuffled from perch to perch, but never shifted their gaze from us for long, in hopes we'd leave a crumb.

The two species are in the same bird family, of similar size and gray color, and both have been referred to as "camp robbers." When they are next to each other, it's easy to see the nutcracker has a bill twice the length of the gray jay and has black wings.

My list of additional birds for this trip is not long: kingfisher, kinglet, redpoll, merganser, swallow, woodpecker, unidentifiable ducks, various blackbirds, and even a species as pedestrian as the robin.

Erik Blom, columnist for Birdwatcher's Digest in an article in the September/October issue, lists Yellowstone as one of the top 25 places in the country every bird watcher needs to visit at least once.

Once is not enough, I've decided, unless the trip is just a little earlier in the summer, on some kind of excursion with a local expert like McEneaney, and most importantly, less by car and more by foot.

# 118 Some migratory birds more obvious than others

Thursday, September 19, 2002, Outdoors, page C2

It isn't a good idea to park in the shade in our driveway this time of year. Splatters of orange fruit are augmented with crunchy seeds. The robins are fattening up for migration.

The neighbors across the street have a lovely old mountain ash full of orange berries, and the robins seem to know better than to defile a tree that provides their food source, so they come across to ours to perch and defecate.

It's really not a problem. We park in the garage, and the fruit stains disappear with a snowfall or two. The seeds get swept away with each pass of the snow shovel.

We've actually benefited because mountain ash trees have sprouted in our garden, and last year one was big enough to transplant.

The robins are very obvious as they swoop back and forth across the street. If we're lucky, they won't eat all the berries right away and there will still be some for the Townsend's solitaire if it spends the winter in our neighborhood again.

Just when the leaves begin turning yellowish is the right time of year to keep an eye open for leaf-sized yellowish birds flitting among them. I've already seen a couple Wilson's warblers (black spot on top of the head) inspecting the bushes for insects.

Many migrating birds merely infiltrate the local landscape, the way warblers do. Others, such as the shorebirds, stop over in wet places that are only on the regular routes of committed bird watchers.

Doug Faulkner of Denver is one of those birders. Here's the list he reported on the Wyobirds listserv for Cheyenne, September 8. It includes local wet areas such as Lions Park.

The sightings include:
Wilson's Warbler
Townsend's Warbler
MacGillivray's Warbler
Blue-gray Gnatcatcher
Cassin's Vireo
Empidonax sp. (flycatcher species)
Hermit Thrush
Swainson's Thrush
Black-headed Grosbeak
Short-billed Dowitcher
American Avocet
Pectoral Sandpiper
Stilt Sandpiper
Solitary Sandpiper
Least Sandpiper
Baird's Sandpiper
Red-necked Phalarope
Wilson's Phalarope
Franklin's Gull
Ring-billed Gull
California Gull
ducks, mostly Mallard and Northern Shoveler

I'm sure Doug saw other, more common species, including the Canada geese at Lions Park, but because they are common, they didn't catch his interest.

I'm impressed by the list of sandpipers. These are the little brown birds with long legs that skitter at the edge of the water, probing the muck with their long bills, looking for invertebrate animals to eat.

Spotted sandpipers, which breed here in the summer, are not on Doug's list and may have migrated already. However, the pectoral, stilt and Baird's sandpipers are on their way back from nesting above the Arctic Circle.

When those three species migrate, they bustle right through here to spend the winter in southern South America.

The least and solitary sandpipers also breed in Alaska and Canada, but not quite as far north.

The least winters from the southern U.S. into the northern half of South America. The solitary prefers to winter further south, from the tip of Texas into Argentina.

"Our" sandpiper, the spotted, breeds all across the U.S., except for the southeast and far southwest and doesn't winter nearly as far south as the others mentioned above.

It's really a pity that none of my six bird watching field guides have range maps that extend further than central Mexico.

Instead, I depend on the Canadian Wildlife Service's Ontario website, http://wildspace.ec.gc.ca/, to find out the rest of the story.

This oversight on the part of the field guides is either because the information wasn't available at the time they were written or because they are, after all, merely North American field guides.

But it leads to this provincial feeling that migratory birds are "our" birds and they merely visit lands to the south during inclement winter weather.

In truth, some species spend more time away than here, especially migrants passing through.

We don't have an international airport in Cheyenne, but if you know where to hang out, where the travelers come to roost, this is a good time of year to catch a glimpse of a few fascinating foreigners.

Our berries, our insects and our muck are our gifts of hospitality.

# Much of bird watching is bird listening

Thursday, October 3, 2002, Outdoors, page C2

My grandmother had a conch shell by her front door. We kids liked to put it to our ears and listen to the ocean. Later, we learned we were hearing our own sounds – our blood rushing through our heads. When my ears get stuffed up with a cold, the outside world recedes and all I can hear is the sigh of those inner tides.

I had my ears plugged up the other day for our family's annual expedition to Pole Mountain to sight in our rifles. Little foam ear plugs aren't as good as the headphone types, but they still block out the natural sounds I enjoy hearing.

Mark said, "See if there's water in the creek," and I was only 10 feet away before I finally heard it, full of water from recent rain.

High pitched twittering was missing, too. Perhaps the shooting startled the birds, though I saw several, including a flock of Clark's nutcrackers.

Even if they do protect my ears from the wind, I just can't wait to pull those ear plugs out. I'm not looking forward to a decline in hearing as I age. Much of bird watching is actually bird listening.

However, deafness is no hindrance for one of Casper's sharp-eyed birders. His biggest challenge is communicating the finer points of bird identification with all of us who can't sign.

There are dozens of companies providing binoculars and scopes to help birders see better, but only one company in recent birding magazines offers a product to help hearing.

The Orbitor resembles a small version of the equipment scientists use in the field to record bird song. It is an 8-inch parabolic dish with a microphone in the middle, hooked up to headphones, though it can also accommodate a recording device. It also has a scope built in to help you focus on the bird you are listening to.

One satisfied customer, quoted in the advertising, called it "binoculars for the ears." You can read more promotional material at ramphastos.com.

For those whose houses are too well insulated, there is another device to bring the sound of birds at the feeder inside. The Nokida Naturescout (www.nokida.com) is billed as a "high fidelity stereo nature monitoring system for your home." My feeder birds are close enough to the window that what I need sometimes is a way to turn down the volume on their incessant chatter.

Every spring, as bird song fills the air, I realize I've let another winter pass by without studying bird song. If this is the year I finally get around to it, I have plenty of options.

First there's the three-CD set from Cornell Lab of Ornithology, "Bird Songs of the Rocky Mountain States and Provinces" (http://birds.cornell.edu). Then there's Thayer Birding Software's "Birds of North America" CD (www.thayerbirding.com) with which you can see and hear your chosen bird at the same time.

Thayer and Cornell are advertising their new joint CD-ROM, "Guide to Birds of North America," which includes 710 species' songs as well as other identification information.

The Peterson field guide series' "Birding by Ear" tape has been around for a long time, and Dover Books (store.doverpublications.com) has Donald J. Borror's "Songs of Western Birds," also on tape.

The "Birdsong Identiflyer" (that's not a typo) advertises in all the birding magazines and online, www.identiflyer.com. It is a hand-held machine that uses cards with pictures of 10 birds each. You insert a card and then push the button corresponding to the bird you want to hear.

There are several birding websites that include bird songs. To play sound files, most rely on RealPlayer software and explain how to download a free version of it. Some of those sites are:

--www.naturesongs.com/birds – more birds than most sites, but no pictures.

--www.naturesound.com – pictures and sounds for a few birds.

--www.birdwatchersdigest.com/audio_index.html – sponsored by the magazine.

--birds.cornell.edu/bow – Cornell Lab of Ornithology's Bird of the Week archives with songs from their Library of Natural Sounds, plus pictures and information.

--birding.about.com/cs/onlinebirdsong/index.htm – appalling amount of advertising, but has links to sites featuring North American and foreign bird songs.

One of my favorite sites, which I've mentioned before, the Canadian Wildlife Service's Ontario website, wildspace.ec.gc.ca/, has bird songs to go with photos, but it isn't easy to get from bird song to bird song.

Sometimes there's the problem of too much noise. Last week Art Anderson and Chuck Seniawski birded with Eleanor Grinnell's science class from the Community Based Occupational Education high school program to survey birds at Kiwanis Lake at the Airport Golf Course and found 22 species.

The very next day, Jim Hecker and I went out with another CBOE class, but that thumper truck was breaking up concrete right by the parking lot. We didn't see as many individual birds as there were species seen the day before.

Even the Orbitor can't help identify birds if all the birds have flown.

CHEYENNE BIRD BANTER

# 120 FeederWatch season begins Nov. 9

Thursday, October 17, 2002, Outdoors, page C2

A small flock of white-crowned sparrows blew into our backyard with the bad weather at the beginning of the month for a two-day stay.

My first clue was a sweet phrase of familiar, but out of place, song. Common near timberline in the summer, to hear it at the beginning of October in Cheyenne, on the plains, must mean migration – and that it's an individual bird which hasn't heard birds are supposed to sing mostly to establish breeding territories.

Once I heard the song, it wasn't difficult to catch a glimpse of the brown sparrow with the black-and-white-striped head. Two adults appeared to be traveling with half a dozen young. Although they might be mistaken for house sparrows, the young have conspicuous brown stripes on their heads and a much perkier attitude.

They also weren't interested in our new tube-style sunflower seed feeder, preferring to feed on the ground.

It must have been the inclement weather that made me think, "Gosh, this would be a good species to report for Project FeederWatch."

The Project FeederWatch season doesn't begin until Nov. 9, when birds supposedly have settled down for the winter, and it ends in April with the onset of migration.

Project FeederWatch has been going for 15 years (longer in Canada) and now includes nearly 17,000 "citizen scientists" in the U.S. and Canada.

The Cornell Lab of Ornithology has already mailed me my research kit.

This year the poster-sized calendar is illustrated with bird photos taken by FeederWatchers. As a renewing member, I didn't get another copy of the handbook, and because I enter my counts online, I didn't get the data entry booklet either. I just have to remember or ask what my password is.

I'll probably continue to make every Saturday and Sunday my count days each week (FeederWatchers who use data sheets report every two weeks).

I leave a sheet of paper and a pencil by our best bird watching window and whenever I walk by, I check and see how many of what species I can see, adding tick marks for a species if I can see more of them at once than in my previous observations.

Some FeederWatchers glue themselves to the window all day long, but that isn't necessary. It's even all right to skip your pre-determined count dates for other obligations. When I submit my data, the form only asks how much time I spent, what the weather conditions were like and how many and what kinds of birds I saw.

Anyone who has watched the birds at their feeder knows that the presence of some species can be erratic. Maps made from data sent in by FeederWatchers can help determine where those white-crowned sparrows are when they aren't here. One or two were actually seen in the Casper area several weeks in mid-winter last year.

Besides tracking irruptive migration (when flocks move major distances to better food sources), data has tracked the effect of disease on house finch populations back east. House finches have not been eliminated, but their population is now maintaining itself at lower numbers.

Project FeederWatch data will also show the effects of West Nile Virus on bird populations. Though the corvids, crows and jays, are the most likely victims, 110 species of birds have been documented with the disease.

Here in Wyoming, for the winter of 2000-2001, crows were the 17th most frequently reported bird, visiting 26 percent of feeders in groups averaging 3.62 individuals. We'll have to wait and see what effect the virus' summer activity will have on crow populations we see this winter.

No scientific experience is necessary to become a FeederWatcher. The handbook and online resources help with explaining protocol, and new members get a full-color poster for help with bird identification.

There are three ways to become a FeederWatcher. Send $15 and your name and address to Project Feeder-Watch, Cornell Lab of Ornithology, P.O. Box 11, Ithaca, NY 14851-0011.

Or call 1-800-843-2473 during Eastern Standard Time business hours and use your credit card.

Or register online at birds.cornell.edu/pfw. Even if you don't sign up, there's a lot of good information about bird feeding at the website.

It takes about three weeks for your kit to be shipped, including your identification number. If you sign up now, you won't miss any count periods. Signing up later is fine, as long as you do it before renewals start March 1.

No two winters are the same. I'm looking forward to comparing my data from last year, available online, with this year.

Maybe this year the evening grosbeaks will come by. Maybe redpolls will find my house. Maybe more red-bellied woodpeckers will be seen in Cheyenne. And maybe by April, before this Project FeederWatch season is over, the white-crowned sparrows, the young with their new black and white head feathers, will make it back to my yard to be counted.

2002

# 121 Local park receives IBA designation

Thursday, October 31, 2002, Outdoors, page C2

## Sloans Lake and the many birds that frequent the area have helped Lions Park become an Important Bird Area

Lions Park is finally an official state Important Bird Area. Those of us who start our annual spring bird count there thought it deserved recognition as soon as we heard the definition of an IBA.

However, it was not easy to convince the technical review committee.

The idea of identifying places important to birds, publicly or privately owned, was started in Europe in the mid-1980s by Birdlife International. The National Audubon Society translated it for the U.S. in 1995.

Audubon Wyoming began soliciting for nominations a few years later and hired an IBA director, Alison Lyon, in 2001 with help from Partners in Flight and other grantors.

Alison, who earned her Master's at the University of Wyoming studying sage grouse, is developing a program that can directly improve the welfare of birds in Wyoming.

When Alison asked Cheyenne High Plains Audubon Society members if we had a site to nominate, we immediately thought of Lions Park.

Art Anderson, chapter president and retired U.S. Fish and Wildlife Service biologist, took charge of the nomination, setting up a meeting in the spring of 2001 with Dave Romero, head of the parks and recreation department, now retired.

Dave and his staff were very enthusiastic about the nomination and provided maps. IBA designations can be touted in civic and tourism advertising, and funding may be available for conservation improvements. There is no regulatory component.

An IBA must meet at least one of four criteria and Lions Park meets numbers one, two and four.

The first criterion includes importance for a species of concern in Wyoming, which in this case would be the western grebe that nests at the lake.

The second criterion, a site important to species of high conservation priority, is met by several of the species on that list that have been seen at the park.

The fourth is the park's strongest suit: a site where significant numbers of birds concentrate for breeding, during migration or in the winter.

Lions Park has a reputation during spring migration for diversity and numbers of birds. I once recorded 60 species in two hours. Migration was also Gloria Lawrence's arguing point in getting the nomination accepted.

Gloria and her husband Jim drive down from Casper every spring to join the chapter in birding the park. She is one of the seven members of the Wyoming IBA technical review group made up of a cross-section of the state's ornithological experts.

The members are Stan Anderson, University of Wyoming, Laramie; Tim Byer, Thunder Basin National Grasslands, Douglas; Andrea Cerovski, Wyoming Game and Fish Department, Lander; John Dahlke, consultant, Pinedale; the Lawrences, Murie Audubon Society, Casper; and Terry McEneaney, Yellowstone National Park.

The nomination originally included all Cheyenne's city parks and the Greenway. However, some technical review group members argued that a city park's intense human use couldn't possibly be compatible with bird use. Maybe, they suggested, Lions Park is more important to bird watchers than to birds.

In reality, the parks are microcosms of the city. Cheyenne is an oasis for migrating birds that funnel along the Front Range.

While Lions Park has gotten a close inspection every year for one day mid-May, turning up all sorts of warblers thought to be unusual for this area, undoubtedly these same warblers can be found in any neighborhood with large trees and many bushes. That was true this spring when a chestnut-sided warbler visited my backyard and a magnolia warbler visited the neighbors'.

Lions Park does have one characteristic that our backyards don't have – a lake. So in addition to neotropical migrants like warblers, it gets a variety of shorebirds and waterbirds.

Funding from Partners-in-Flight, passed through Audubon Wyoming, has become available to our local Audubon chapter for monitoring work.

Many of the previous records are from Christmas Bird Counts and spring Big Day Counts which lump observations from around the city.

Now the chapter needs to plan for making more detailed surveys, training volunteers in survey protocol and compiling databases useful to science. Then we can figure out what conservation projects might be of benefit to both birds and the park.

While Lions Park will never achieve global status like Yellowstone National Park may, the information we collect at least gives us more understanding about where we live.

The Lions Park IBA is too new to be listed in the new publication, "Important Bird Areas of Wyoming." However, if you are interested in reading about the first 20 state IBAs, contact either me (see below) or Alison Lyon in Laramie, alyon@audubon.org, 307-721-4886.

# 122 Birding trips can take watcher to distant lands

Thursday, November 14, 2002, Outdoors, page C2

Who would have dreamed we'd celebrate a white Halloween in a drought year – and that the snow would be around for another week?

The wintery cold and gloom had birds swarming our feeders the way they never do during our otherwise mild winters. Wyoming birders reported mixed flocks of Canada and snow geese heading south.

And remember those icy streets the week before? That weather was foreshadowed back in August when Jose (pronounced Joe-say) Goncalves, owner of J-G Travel in Denver, called me.

Since I'm listed as a chapter contact on the National Audubon website, it isn't unusual to have the occasional traveler call and ask where to bird while visiting Cheyenne.

However, this time I got an offer to travel. Jose has been in the travel business for over 25 years, organizing custom tours for various groups including museums and universities. Where would I like to take a birding trip? Central or South America? Africa? Asia?

The intriguing part of the offer was I could go for free if I were the trip leader. This was enticing. I haven't been out of the country since 1975, except to walk over to Juarez, Mexico and to fish in Saskatchewan.

"But I don't know anything about foreign birds," I protested. No problem. Jose hires local experts. My main job would be to recruit the other travelers.

I am familiar with this way of organizing trips. My brother-in-law has recruited fellow travelers for several. He brought his digital photos with him when he visited this summer, so we got to see Spain and China by hooking up his camera to our TV.

Peter has two essential attributes that make these adventures successful. He knows people who can afford to travel, and people automatically trust him because he's a retired priest.

My mother always warned me about strangers, so to take Jose out of that category, I invited him to be Cheyenne High Plains Audubon Society's guest speaker for the October meeting. He hesitated and joked about the weather that time of year.

Oh pshaw, I said, you can always stay over with one of the members after the meeting if it gets too nasty.

The streets did ice over. Jose e-mailed me later to say he got back home to Denver that night just fine, though a bit slowly.

It proved Jose not only knows the weather patterns where he travels, but he is intrepid. Unfortunately, many of our chapter's members and friends weren't and missed a fine presentation.

Jose is a trim, silver-haired gentleman with an accent that hints of not just world traveling, but world living. He was born in Portugal (thus, the Portuguese pronunciation of his name), and then his father moved the family to a ranch in Mozambique when Jose was young. He went to school in South Africa before attending Denver University.

Jose's slides were all his own work from a variety of African wildlife-watching safaris and included wonderful portraits of lions and elephants and birds.

While other wildlife photographers would have brought animal photos exclusively, Jose slipped in a few of the tourist accommodations. I liked the thatched-roofed tree house myself.

Anyone paging through the back of a birding magazine can choose from any number of pre-planned trips for which I would probably never meet the owner of the company.

Ads in the November issue of Birder's World name-drop exotic geography: Costa Rica, the Dry Tortugas, San Blas, Trinidad, Tobago, Belize, Tikal, Amazon, Galapagos, Machu Picchu, Yucatan, Panama, Ecuador, Australia, New Zealand, United Kingdom (well, it would be exotic to me since I've never been there), Borneo, Vietnam, Cambodia, Sri Lanka, New Guinea, China, Indonesia, Thailand, India, Bhutan, Malaysia, Himalayas, Scottish Isles, Iceland, Pribilof Islands, Venezuela, Peru, the Arctic, Newfoundland, Patagonia, Morocco and Spain.

Some of these locations are mentioned in half a dozen ads, making me wonder if the bird watchers stand elbow to elbow.

Many destinations are to developing countries where eco-tourism is a major industry.

Jose said his African guides go to "guide school" and are knowledgeable interpreters of their local environment.

For the tropical birds, I wonder if it matters what time of year one looks for them. Does their plumage change seasonally in a place when there are only two seasons? When are the wet and dry seasons anyway? Guess I better study a little world bird geography.

From my family's point of view, this would not be a good year for me to fly off to foreign parts for a few weeks, free ticket notwithstanding.

Meanwhile, I told Jose to keep in touch. I'll let you know if he tells me about some enterprising Denver birder organizing a tour.

And meanwhile, I'm envious of those migrating geese whose sole preparation for foreign travel seems to be gorging on food.

# 123 Do the Thanksgiving Bird Count without leaving home

Thursday, November 28, 2002, Outdoors, page C2

If you are reading this Thanksgiving Day and you have an hour of daylight to spare, you can take part in the Thanksgiving Bird Count. It's as easy as counting the birds at your feeder, which means you probably don't even have to go outside.

Last year, 448 people in 12 western states participated. John Hewston of Arcata, California, has been compiling the results since 1992.

It is OK to surreptitiously peer over Aunt Edith's shoulder at dinner and watch the birds out the dining room window. If you have relatives that have been featured in Dear Abby, maybe birds can provide a better topic of dinner conversation than your cousin's off-color jokes or medical concerns.

Perhaps you are merely a visitor to Cheyenne. There is no residency requirement to take part. Just count where you are and e-mail or call in your results today.

Some people look forward to a little excursion before or after dinner. It would be a great time to visit one of our local parks and select a site for counting.

Here are the directions:

1. Select a circular area on the ground 15 feet in diameter, to include feeders, bird baths, shrubs, etc. – even a body of water. Imagine the circle extending upward as a cylinder.

2. For one hour, count the numbers of birds of each species that come into this circle or cylinder. Count the maximum number of individuals of a species seen at one time. Otherwise, you may be counting the same chickadee every time it comes for another helping.

3. Record the number of birds of each species seen inside the count circle. You can also record birds seen outside the circle, but keep that list separate.

4. Record conditions and location information: location of count circle (address), habitat type (kind of vegetation and amount), number and kinds of feeders and baths, weather, temperature and beginning and ending times.

5. Send your report, your information from steps 3 and 4, as well as your name, mailing address and phone number to me by e-mail, bgorges2@juno.com, or phone, 634-0463, by Dec. 6. You can call today and leave a message.

In a few months John will send you the compiled results.

If you have an identification question, I'll return your call tomorrow.

Here's a list of the birds I've had at my feeders lately:

--House sparrow – chestnut brown back with plain, pale gray breast.

--House finch – brown bird with streaky brown breast. Males have red head, chest and rump markings.

--American goldfinch – slightly smaller than house sparrow, but unstreaked, grayish-yellowish body and black wings with white wingbars.

--Dark-eyed junco – smaller than house sparrow, various colorations of gray, white and pale reddish brown, but all have white outer tail feathers visible when they fly.

--Red-breasted nuthatch – even smaller than a junco, blue-gray back, pale red breast, black stripe through eye.

--Robin – some of them think Cheyenne is south enough. They are berry eaters this time of year.

--Blue jay – robin-sized, but blue and white with a crest on the top of the head. Stellar's jays have been reported in town, but they are blue and black.

--European starling – robin-sized, brownish black with white speckles (stars) and short, stubby tail.

--Downy woodpecker – slightly larger than a house sparrow, black and white.

--Northern flicker – brownish robin-sized bird with spotted breast, black crescent-shaped "bib" at top of breast, and very long bill. When it flies, look for the white rump patch and red wing linings.

This must have been a good year for flickers. I've had a lot of calls about them. Technically, they are in the woodpecker family, but they are as likely to peck out grubs in lawns as bugs in tree bark.

If you are really lucky, a hawk may visit your feeder too. Usually it's the little sharp-shinned hawk, a forest bird that is willing to negotiate backyard branches.

However, a couple weeks ago, while on our daily morning walk, my neighbor and I saw crows harassing a rough-legged hawk. After nesting on the Arctic tundra, this large raptor winters across the U.S., but usually in open country, not in town.

It's that chance of unpredictability that makes watching even my bird feeder interesting. I can't wait to see what shows up today – in your yard and in mine.

CHEYENNE BIRD BANTER

# 124 Make this Christmas a holiday for the birds

Thursday, December 12, 2002, Outdoors, page C2

Satisfying the wild bird lover on your Christmas gift list can be as easy as buying a sweatshirt decorated with chickadees, a clock with bird song chimes or chirping plush toys, not to mention fine bird art in all kinds of media.

However, none of these gifts do much for the birds themselves unless part of the profits benefit bird conservation.

Consider turning the wild bird lover into a knowledgeable bird watcher who can contribute to citizen science bird counting efforts such as the Christmas Bird Count, the Great Backyard Bird Count or Project Feederwatch.

You could pick up the basic field guide, "Birds of North America" by Kenn Kaufman, for about $15 at a local bookstore, and a pair of 7 x 35 Bushnell binoculars at Kmart for about $25.

If your bird watcher is more advanced, you'll have to do some sleuthing. Do they already have a copy of "The Sibley Guide to Bird Life and Behavior" or "A Field Guide to the Birds of the West Indies" by James Bond?

Don't try to pick out binoculars for the advanced birder. Pricey models have too many variables that must fit the individual user's eyes.

Does your bird watcher subscribe to Bird Watcher's Digest (800)

879-2473 or Birder's World (800) 533-6644?

Both magazines are filled with advertising for all kinds of bird identification and observation gizmos, even special clothing such as field vests with pockets designed to fit field guides.

However, all the latest bird watching accoutrements advertised in those magazines are merely trappings of a personal hobby and won't help the birds if the bird watcher doesn't share their observations and knowledge.

Feeding wild birds can be a hobby that benefits some kinds of birds directly. The gift ideas range from a simple shelf and a bag of black oil sunflower seed to elaborate spring-loaded, squirrel-proof dispensers and custom seed blends.

Don't forget water. A large plastic dog dish filled less than 2 inches deep is easy to bring in and thaw under the kitchen tap if you aren't ready to finance a heated bird bath.

There are other gifts that delight the bird lover/watcher and benefit birds. Three major bird conservation organizations provide informative and colorful magazines as part of membership: National Audubon Society (sign up through the local chapter, 634-0463), Cornell Lab of Ornithology (607) 254-2425 and American Birding Association (800) 850-2473.

Perhaps the person on your gift list is already a member and is ready for a more altruistic gift. You can make a donation in their name to that organization or pick one of the many others such as the American Bird Conservancy (540) 253-5780.

Maybe you are the kind person who remembers your pets at Christmas and would like to do something for the birds too. Here are suggestions.

--Avoid planting trees in grassland bird habitat. Plant more fruiting trees in town.

--Keep your cat indoors or on a leash or in a kennel at all times.

--Lobby for bird-friendly legislation and policies. It isn't as much fun as counting birds for scientific study, but protecting habitat is the most efficient way to help wild birds.

--Conserve resources, "reduce, recycle, reuse." Owning too much stuff wastes energy and resources which require mining, drilling, timbering, spraying – all activities usually detrimental to birds. Besides, the simple life will give you more time to enjoy bird watching.

Actually, these suggestions would all make good New Year's resolutions.

When all your shopping is done and you can finally put your feet up, you'll be happy to know there are things you can consume, of which every

ounce helps birds.

Shade-grown coffee and organic chocolate are grown in the shade of forest trees, the time-honored family farmer's method, in Central and South America, where our neotropical birds spend the winter. The mega-farms use new varieties that require sun, which requires cutting the forests and spraying the crops, leaving no place for birds.

Jane Dorn was telling me last week that she read that the particular bee that pollinates coffee plants prefers shade, so shade-grown plants are also much more productive than those receiving chemical fertilizers.

Locally, organic coffee is offered by Coffee Express, Starbucks and sometimes City News.

If you do an internet search, the key phrases are "organic chocolate," which will give you mouth-watering sites like Dagoba Organic Chocolate (541) 664-9030, and "shade-grown coffee," where I found gourmet blends offered by Andeano Gold (888) 213-5059.

Finally, one of the best gifts you can give someone is your time. Arrange to take your friend or family on a little bird watching field trip, either your own itinerary or with a group. The memories of real birds will be more valuable than any flock printed on a sweatshirt.

# 125 Cardinals top Christmas cards, if not bird count

Thursday, December 26, 2002, Outdoors, page C2

There's the Christmas Bird Count and then there's the Christmas card bird count. As I write this Dec. 19, the tally is two chickadees, six cardinals, a cinnamon teal, five birds of undeterminable species – and two penguins.

Last year, I identified Canada geese, blue jays and a junco plus the popular chickadees and cardinals.

The jackpot was provided by an Audubon card sent by Audubon friends featuring a red-bellied woodpecker and a white-breasted nuthatch.

John Hewston, compiler of the Thanksgiving count, has also noticed the northern cardinal seems to be a favorite on Christmas cards, "or of people who select them."

I think cardinals are so popular because their bright red feathers fit the seasonal color scheme when they are depicted perched in an evergreen.

However, I was surprised to find a cardinal on the cover of this month's issue of the Wyoming Game and Fish Department's "Wyoming Wildlife" magazine.

What was Editor Chris Madson thinking? He's a pretty astute student of nature and I would expect he'd be aware that cardinals are considered to be rare in Wyoming.

"Rare" is the technical term used in the Wyoming Game and Fish Department's "Wyoming Bird Checklist" and rates "1" on a scale of abundance from one to four.

The checklist also identifies the cardinal as a species seen in Wyoming only during spring and/or fall migration.

Of the 28 latilongs formed by the gridwork of degrees of latitude and longitude that biologists use to locate animal observations in this state, the cardinal has only been seen, and without any signs of breeding activity, in seven latilongs, as shown in the "Atlas of Birds, Mammals, Reptiles and Amphibians in Wyoming" also published by Game and Fish.

One of those seven latilongs contains Cheyenne. However there is no asterisk to indicate the observation has been scrutinized and accepted yet by the Wyoming Bird Records Committee.

And in the list published by the Cheyenne High Plains Audubon Society, no cardinals have been seen in any of 40 years' worth of data for the Cheyenne Christmas Bird Count.

Cardinals are most abundant in southeastern United States. Thirty or forty years ago, I would see them at my grandparents' feeder south of Chicago, but not at home a mere 100 miles to the north. Since then, they have extended their range north through most of Wisconsin.

Cardinals are classified as permanent residents within their range, so the few observed in Wyoming were more likely to be juveniles on a road trip, now that their range extends as far west

as the Wyoming-Nebraska border, than birds lost during migration

It appears Chris is another victim of a pretty passerine face. He explained to me that he'd had this particular cardinal in the photo file for several years and kept passing it over for December issues because he knew cardinals are not typical Wyoming wildlife.

He took as a sign the submission by a Wyoming photographer of another cardinal that he could finally justify using it on the cover.

Cardinals are also just over our southern border, in Colorado. It's only a matter of time before they become common residents of eastern Wyoming too, like the blue jay, another formerly eastern U.S.-only species. Chris's cover choice serves as a heads-up. When you hear that distinctive cardinal whistle, look up.

It would be neat if a cardinal made an appearance for the Cheyenne Christmas Bird Count Jan. 4. Why not plan to join in the fun and look for cardinals – and maybe be part of a historic moment?

Other winter birds are drab by comparison, though close examination shows the beauty of their sophisticated, subtle coloration. Why is the chickadee motif nearly as popular as the cardinal? Maybe it's because their black and white heads make them easy to depict. Or maybe it's because in cold

weather they fluff up into little round balls, multiplying their cuteness factor.

But there's something even more appealing about a red bird in winter. When snow makes the landscape monochromatic, or as is the case most of the winter in Cheyenne, the snowless landscape is dull, red is a desirable accent. Our eyes are attracted to red-stemmed shrubs, red sumac, red berries and red bows.

How did red become a symbolic color for this time of year? There's probably an anthropologic answer published somewhere explaining why people have a yen for red in winter. I suspect both our hunter-gatherer ancestors and animals today roaming the land had/have an eye out for the color that could mean dried rosehips or other fruit. Marketers of packaged foods certainly understand the use of the color red.

Christmas card designers no doubt have their own statistics showing the appeal of cardinals. So when you go out today to buy next year's Christmas greetings on sale, don't be surprised if the cardinal cards have already flown.

If you missed them, you could still buy cards with a nice winter landscape and ink in a small red dot on a distant tree branch. Everyone will know it could only be a cardinal.

---

116

CHEYENNE BIRD BANTER

# 2003

## 126    Arizona fire damage softened by snow

Thursday, January 9, 2003, Outdoors, page C2

**The Chediski Fire left areas of the Apache-Sitgreaves National Forest looking like a New England winter scene.**

The view from my sister and brother-in-law's house two weeks ago was of ponderosa pine forest decorated for Christmas in several inches of fresh snow.

Beth and Brian Dykstra live in Heber, which, like other small towns in the White Mountains of central Arizona, has been reduced to a vacation destination. Once one of the area's logging communities, founded before the turn of the century, it is now a string of businesses along the highway backed by rural subdivisions built in second growth timber.

Last June, the Chediski fire caused Beth and Brian and all their neighbors to evacuate for more than a week. Brian, a U.S. Forest Service wildlife biologist, had already headed to the Rodeo fire near Show Low, 40 miles east. He worked as a "dozer boss," clearing fire lines and saving homes.

Beth, a Forest Service recreation planner, left the family pets with other evacuees before reporting back to the district office in Overgaard, Heber's twin community, to field phone calls from distressed residents and deliver daily updates to evacuees.

Fire came within three-quarters of a mile of Beth and Brian's house and within 500 yards of the office but burned homes of several friends and coworkers.

Six months later it was only natural that while waiting for the Christmas turkey to roast we should tour the nearby burned areas.

The layer of snow prevented us from seeing ground damage. Beth said most of the ash disappeared during the annual "monsoon" rains later in the summer. In many places, without any assistance, tree seedlings were soon sprouting as thick as grass.

Volunteers wanting to plant trees have been turned away. What's really needed, Beth said, only partly in jest, are people to stomp on some of the seedlings and keep the forest from growing back so thick again.

However, in other places where the fire burned too hot, the soil has become "hydrophobic" and won't absorb moisture at all, putting off regrowth indefinitely.

Where we first turned off Highway 260 into the Black Canyon, some of the ponderosa had blackened trunks and their lower limbs held needles singed an orange color.

Beth said if a tree retains at least 30 percent of its needles undamaged, it may recover. However, the drought responsible for the severity of the fire may continue, adversely affecting survival.

For a stretch the road seemed to have contained the fire to one side of the canyon, but then there was a whole expanse devoid of green, orange or any other color except the blue sky silhouetting each charcoaled tree.

In another part of the country, New England for instance, this could be a typical winter landscape – except maples don't have a ponderosa's shape.

During an hour's drive we saw over and over the mosaic pieces of unburned, singed and burned forest. Coming out on the highway again by Overgaard, Beth pointed out the mounds of snow marking foundations of houses unlucky enough to catch stray embers while neighboring houses survived.

Both Beth and Brian were, after the fire, immediately assigned to the Burned Area Emergency Rehab team to assess areas most in need of erosion control before the rainy season began. Brian said 50,000 acres (of the 176,000 acres of Forest Service land burned) were seeded aerially with a mix of native grass species and then mulched with hay.

"Bale bombing," dropping 1000-pound bales from airplane cargo nets, tended to spread the hay unevenly. In other places, members of a four-wheeler club from Phoenix were happy to use their off-road vehicles to deliver 70-pound bales for other volunteers to spread.

Brian hopes the state game and fish department will take measures to reduce the elk herd before their grazing undoes the rehabilitation effort.

We saw mourning doves while on our tour. They too see the seeding effort as a food source when usually, by this time of year, they would have already migrated to the desert.

Beth has since been assigned to the salvage team. "Categorical exclusions," actions excluded from analysis as in-depth as an environmental impact statement, are allowing the clearing of dead trees within 100 feet of roads, buildings, recreation areas and power lines. These are places where the trees could cause damage when they eventually topple. Burned forests are hazardous places to be on a windy day.

Many of the burned trees I saw were too small to be valuable timber. Beth said they would be knocked down to become soil amendments. The larger trees are salvageable for lumber up to 18 months after a fire.

Burned trees are generally regarded as too messy to be harvested for firewood, although someone is working to set up a power plant that could use them for fuel.

Damaged trees will attract insects and insects will attract birds. A researcher from Northern Arizona University has contacted the Forest Service about studying the effects of the fire on the hairy woodpecker, an insectivorous species.

Someday managers will figure out how to achieve natural ponderosa forest, those grassy parks dotted with mature "yellow pine," yet still satisfy the nation's lumber needs.

Meanwhile, months from now, the spring thaw will bring flooding to fishless streams, but it will also show what healing has taken place as it melts the gauze bandage of snow.

*Cutline:*

Burned trees poke out of the snow cover in the Apache-Sitgreaves National Forest in Arizona. Barb Gorges/staff [No, I wasn't staff!]

# 127 Thursday, January 19, 2003, Outdoors, page C3
# Bird count picks up four new species

CHEYENNE – No snow, no wind to speak of and temperatures ranging from 32 to 50 degrees made comfortable conditions for 20 observers participating in the annual Cheyenne Christmas Bird Count Jan. 4.

The tally was 5,648 individual birds of 47 species, plus one species observed during the week of the count but not count day.

Mild weather this winter may be responsible for the number of robins still here and the absence of northern or high-altitude species such as the rosy finches.

In the eastern U.S., West Nile Virus has decimated crow populations, but in the 15-mile diameter count circle centered on Cheyenne, 250 crows were counted, up from 97 last year.

Crows were not observed on counts before 1987.

Four species appeared on the count for the first time: Eurasian collared-doves have been showing up regularly at a south-side feeder; the northern bobwhite was observed feeding on food scraps thrown by crows from a trash container; the white-throated sparrow, considered an eastern U.S. species, was visiting a north-side feeder; and the wood duck has been observed at Lions Park for several months.

Five species have been seen on this and the 40 previous counts: northern flicker, horned lark, Townsend's solitaire, house finch and house sparrow.

Ten other species have been seen on this and at least 35 other counts: mallard, rough-legged hawk, great horned owl, downy woodpecker, blue jay, black-billed magpie, mountain chickadee, American robin, European starling and dark-eyed junco (slate-colored and Oregon races).

The redhead (duck) has been recorded only once before. Ruby-crowned kinglets have been observed on two other counts and Harris' sparrow on three.

Christmas Bird Count data for previous years and other locations is available online, www.audubon.org/bird/cbc

## Christmas Bird Count results

| | | | |
|---|---|---|---|
| Canada Goose 1451 | Merlin 1 | Mountain Chickadee 22 | race unknown 48 |
| Green-winged Teal 2 | Northern Bobwhite 1 | Red-breasted Nuthatch 34 | Dark-eyed Junco, |
| Mallard 1776 | Wilson's Snipe 2 | White-breasted Nuthatch 7 | slate-colored 12 |
| Northern Shoveler 9 | Rock Dove (pigeon) 265 | Brown Creeper 10 | Dark-eyed Junco, |
| American Wigeon 2 | Eurasian Collared-Dove 3 | Golden-crowned Kinglet 6 | white-winged 1 |
| Redhead 1 | Great Horned Owl 3 | Ruby-crowned Kinglet 3 | Dark-eyed Junco, Oregon 8 |
| Common Goldeneye 20 | Belted Kingfisher 2 | Townsend's Solitaire 7 | Dark-eyed Junco, pink-sided 10 |
| Wood Duck 1 | Downy Woodpecker 11 | American Robin 11 | Dark-eyed Junco, |
| Northern Harrier 2 | Northern Flicker, red-shafted 9 | European Starling 687 | gray-headed 4 |
| Sharp-shinned Hawk 4 | Horned Lark 74 | American Tree Sparrow 4 | Red-winged Blackbird 138 |
| Northern Goshawk 1 | Stellar's Jay 2 | Song Sparrow 5 | House Finch 129 |
| Red-tailed Hawk 3 | Blue Jay 5 | White-crowned Sparrow 1 | American Goldfinch 8 |
| Ferruginous Hawk 1 | Black-billed Magpie 10 | Harris' Sparrow 3 | House Sparrow 515 |
| Rough-legged Hawk 7 | American Crow 250 | White-throated Sparrow 1 | *Species observed week of |
| American Kestrel * | Black-capped Chickadee 4 | Dark-eyed Junco, | the count, but not count day. |

118 CHEYENNE BIRD BANTER

# 128 New website offers free service to bird watchers

Thursday, January 23, 2003, Outdoors, page C2

With a chirp-chirp here and a chirp-chirp there, here a chirp, there a chirp, everywhere a chirp-chirp, Old MacDonald had a farm, e-i-e-i-o.

The pursuit of bird counting has entered a new era. The National Audubon Society and Cornell Lab of Ornithology have introduced www.eBird.org, a Web site where any bird watcher can keep track of his own data from any time of year and have it compiled with everyone else's in North America.

I've always looked forward to the annual tradition of the Christmas Bird Count, that 103-year-old institution. And I've adapted to Project FeederWatch and the Great Backyard Bird Count, recent additions to the citizen science field.

However, eBird sounds a little compulsive.

The only other list I keep is my life list. I check off names of birds in the index of my first field guide, now moldy and broken, when I see the species for the first time, and also record the date and place to make a kind of album of memories.

Unlike me, a lot of birders like to keep records. They keep a life list, a year list, maybe even a daily list, a backyard list, a trip list and lists for their favorite places to watch birds. There are several computer record-keeping programs just for birders.

eBird is free and allows you to enter your records from 1960 to the present.

You will always be able to view them at the Web site, at "My eBird," but you can also download them to your own computer.

Everyone can look at the maps and lists of compiled data, but personal information about participants is not publicly available.

Being a new site, there isn't a lot of data to look at yet, so I chose to look up Ithaca, NY, home of Cornell. The data maps went back to 1984 – for goldfinches anyway. I was impressed with the choice of 10 degrees of "zoom" available, from the whole continent down to a particular street.

Bird counts, such as the Breeding Bird Survey and Monitoring for Avian Productivity Success, done by professional wildlife biologists, ornithologists and other trained people, have strict protocol to eliminate certain biases so that data can be accurately compared from year to year and site to site.

The Christmas Bird Count, a volunteer effort, has fairly strict protocol as well, with each count in a predetermined area and survey forms that ask for number of observers plus time and mileage spent observing. Project FeederWatch and the Great Backyard Bird Count have protocols too, though much simpler.

Scientists have found that even though PFW and GBBC participants are of all skill levels, the sheer quantity of data collected overcomes

biases well enough to make data useful in monitoring population, mid-winter migration and disease trends.

Just what have eBird designers done to eliminate bias from the deluge of data they are asking for?

For instance, many people on a field trip will list the birds they've seen, but not count the number of each kind. Often the names of "trash" species, very common non-native birds, are never even mentioned.

If eBird scientists aren't careful, they could be led to believe that some of us see only one of a kind of each species – and no starlings and house sparrows!

Here's how eBird works. It has four observation types. "Casual sighting" means there was no measure of time spent observing, distance traveled, or area covered. You just wanted to note the Townsend's solitaire you saw while out walking the dog Tuesday.

"Stationary count" means you recorded how much time you spent bird watching in one place, such as observing birds at your feeder.

"Traveling count" means you measured distance and time while on a bird walk or a trip by any other mode of transportation.

"Exhaustive area count" would be physically exhausting because you would be beating the bushes to count all the birds in an area of a known size.

And if you didn't count

how many there were of a particular species, you can just mark "x" to mean a species was present.

I guess I'll add eBird stationary counts to the others I do. When the Project FeederWatch season ends in April, I can just continue counting birds in my backyard through the summer.

I tried entering the backyard birds I saw yesterday. Of several methods offered for locating my yard, I chose the aerial photo map. It is between 4 and 10 years out of date, based on building construction in my neighborhood.

It was a little difficult finding the birds on the checklist, but with time I'll be more familiar with the process and I'm sure the site will improve.

eBird is being established at a critical time. When West Nile Virus got started in this country, it was the corvid species, crows and jays, which people observed dead or dying. Scientists think other birds may be just as affected but are not as noticeable. Information from eBird might show if this hypothesis is right.

Counting is work. So sometimes we need to just relax and watch birds without scientific motives. I hope the eBird folks understand if I enjoy the sandhill crane mass migration along the Platte this spring and merely submit "Sandhill Crane – x."

2003

119

# 129 Bald eagles attract commercial interest, festivals

Thursday, February 6, 2003, Outdoors, page C2

This time of year, between mid-December and the end of February, interior-nesting (as opposed to coastal-nesting) bald eagles are attracted to large lakes and rivers with open water where they feed on fish and waterfowl, etc.

This time of year bird watchers are attracted to large lakes and rivers with open water where they feast their eyes on bald eagles. Purveyors of goods and services have learned how to feast on the bird watchers.

Business often has some natural resource at its root, though a hole in the ice doesn't usually bring to mind profits.

However, a plethora of bird festivals has now come into being, capitalizing on the highlights of bird behavior. Bird Watcher's Digest (www.birdwatchersdigest. com/festivals/festivals) lists 24 eagle festivals in January and February alone.

By March, festivals concentrate on welcoming migrating birds, or in the case of West Palm Beach, Fla., wishing them farewell. I'm a little alarmed at the thought of the Bethel (Alaska) Bunting Bash. I think organizers were intent on alliteration rather than whether potential

visitors might envision attacks on flocks of rare McKay's buntings.

Bald eagles make a good focus for a festival. They are large, easy to see, distinctive looking and they gather in particular places in the winter.

A good reason for celebration is the bald eagle's tremendous comeback from the days when DDT poisoning was about to cause its extinction. It has moved from endangered species status to merely threatened and is presently proposed for delisting altogether.

Also, why not celebrate our national bird? Ben Franklin, who proposed the turkey instead, was never happy with the bald eagle as our national symbol.

Countries may have chosen eagles for their reputation for using formidable beaks and talons to defend its nests and to hunt vermin.

However, our founding fathers picked the wrong eagle. Ours is as fond of carrion as fresh meat. So maybe a national symbol willing to clean up the environment is not so bad, Ben.

Bald eagle festivals I looked up were sponsored by chambers of commerce,

Audubon societies and other conservation groups, park departments, state conservation departments and the U.S. Army Corps of Engineers.

The Corps, with a reputation for flooding prime farmland and wildlife habitat, has, with its dams, ironically and inadvertently provided many instances of open water coveted by the eagles.

What does one do at an eagle festival? You look at eagles. Take part in guided tours by land or boat or look through spotting scopes set up by sponsors.

There are talks about eagles, maybe a parade, photo contest, art exhibit, poster contest, poetry contest, storytelling, souvenirs, book signings, music, theater, conservation displays, games, crafts and Volkswalking. At one festival there's the quintessential Methodist pancake breakfast billed as "Breakfast with the Eagle," because a live eagle named Emi presides.

A well-advertised bird festival can mean a lot to a small town like Concrete, Wash., on the banks of the Skagit River. Visitors infuse the town with money spent for food and lodging, etc.

But the real value is the

new value local residents now place on eagles. I don't care whether it's because they have their own interests or the eagles' at heart. Either way, eagles will be protected.

Here in Wyoming, we don't have an eagle festival, but we do have the long-running Lander Bird Festival in May. The second annual Platte Valley Festival of Birds is slated for the same month in Saratoga.

Though we have eagles spending the winter in the state, I don't think they are in high enough concentrations yet to warrant the tourism industry making up full-color brochures.

Well, maybe there are other options: the Crow Counting Convention, Flicker Festival, Siskin Celebration, Blue Jay Jubilee, Flycatcher Fair, Gull Extravaganza, Pelican Party, Swan Symposium, Hummingbird Holiday, Swallow Soiree, Warbler Wingding, Gadwall Gala, Junco Jamboree, Falcon Fiesta, Goldfinch Gathering, House Finch Forum, Plover Powwow or Shoveler Shindig. But please, no Bluebird Bash.

# 130 Thanks to awareness, birds now visible on range

Thursday, February 20, 2003, Outdoors, page C2

The Society for Range Management held its six-day, annual meeting in Casper earlier this month. In the 23 years I've been a member there have been noticeable changes.

First, a majority of presentations were made with PowerPoint rather than slide or overhead projector. The abstracts were on CD rather than in a book. Fewer people wore ties. More

women attended in a professional capacity.

Most amazing to me was the inclusion of three sessions of papers on rangeland birds.

The society began in about

1948 as a group of western public land managers who seceded from the Society of American Foresters to concentrate on rangeland, usually considered to be the naturally treeless regions.

The new society soon attracted producers (ranchers), wildlife biologists (primarily big game) and more recently, folks from the minerals industry concerned with mined land reclamation.

This year a symposium titled "Rangeland Birds and Ecosystems" sponsored by Audubon Wyoming, broadened SRM's perspective even more. It was divided into three sessions: "Birds as Environmental Indicators in Rangeland Ecosystems," "Resources and Management Practices for Healthy Rangeland Ecosystems" and "Partnerships and Funding Opportunities."

For most ranchers and others working on rangeland, birds have merely provided background music, but interest is building. As many as 150 people at one time attended these talks.

How did birds get on the agenda of a society which uses a cowboy, "The Trail Boss," for its logo?

The invitation came from Bob Budd, an extraordinary man who has been building bridges for some time now between livestock producers and the environmental/biology community.

Years ago, he was in Cheyenne as the executive director of the Wyoming Stock Growers Association. Even though I represented Audubon, I found him easy to visit with and open-minded.

I wasn't nearly as shocked as the ranching community was when he took a position with The Nature Conservancy to manage their Red Canyon Ranch property near Lander. He now also directs science stewardship and planning.

Bob has been able to demonstrate ranch management practices that benefit livestock and wildlife. He won't allow ranchers to make "environment" into a four-letter word, or on the other hand, let environmentalist keep old stereotypes of ranching.

Bob's credibility continues to remain high, high enough to be the newly elected president of SRM.

At the bird symposium I was reminded one can never generalize about rangeland.

Fire as a management tool may work wonders in one place and create a long-lived disaster in another.

One can't generalize about rangeland bird species either. One prefers bare ground, another prefers a jungle of sagebrush. One species is happy with a couple hundred acres of unfragmented grassland, another needs 50,000 acres including three distinct types of habitat for breeding, nesting and raising young.

One of few generalizations to be made is that bird species present in a rangeland ecosystem can indicate its health – usually taken to mean approximating historical conditions.

Managers have learned to create perfect and uniform pastures with plants livestock need to graze for maximum weight gain. But just a little untidiness, a little bare ground here, a little ungrazed patch of shrubs there, can produce the full historical spectrum of birds.

This sounds suspiciously like advice given to birdwatching homeowners to lay off the pesticides and

pruning shears and let at least part of the yard go natural.

Wildlife biologists at the symposium presented best management practices for birds for several types of habitat. I thought, gosh, I've heard this before – 20-some years ago from my range management professors at the University of Wyoming. Maybe what's good for cows can be good for birds.

Rangeland has perpetuated itself for millennia, and we are still figuring out the intricate relationships between climate, soil, plant and animal, and how to take advantage of them.

No one can predict when a tiny facet of scientific understanding will catch the light and shine it on matters of human importance.

Three of my former UW classmates, also still SRM members, study things as obscure as the way mesquite beans weather in Texas. Seems to me the study of rangeland birds has as much beneficial potential.

# 131 Bird flashcards evolve into full-service CD-ROM

Thursday, March 6, 2003, Outdoors, page C2

Beauford Thompson, formerly a Davis sixth-grade teacher, is at the root of the origination of the Wyoming Bird Flashcards CD-ROM, produced jointly last year by Audubon Wyoming and Wyoming Game and Fish Department with a Wyoming's Wildlife Worth the Watching grant.

About 10 years ago Beauford invited Cheyenne-High Plains Audubon Society

members to go bird watching with his class. Several of us were available for the ensuing field trips with him and other teachers, but I was the one to visit classes the day before to introduce birds we might see.

It's hard to make eye contact with students sitting in the dark while using a slide projector, so I switched to 8½ x 11-inch flashcards. They are more versatile, allowing

for comparison and individual study.

The next evolutionary step came at Kinko's, where I was making another flashcard when Chris Madson, editor of WGFD's Wyoming Wildlife magazine, walked in. I asked him if the Game and Fish had any bird photos that could be used for flashcards. What if we had sets printed and distributed to teachers?

Color printing is

prohibitively expensive, but then I remembered my recent student-teaching experience with Kathryn Valido at Afflerbach, and how many workbooks are now on CD. The teacher selects worksheets to print out and duplicate.

With a western meadowlark image borrowed from Chris, Rainbow Photo was able to put it and accompanying text on CD. I showed that

2003

121

and the flashcard printout to the other Audubon Wyoming board members when I asked them to sponsor the project in order to search for grant money.

The Wyoming's Wildlife Worth the Watching grant became the first and only grant applied for.

I soon realized the CD format, not only cheap to produce and distribute, lent itself to being used by students and teachers like other software on school computers and in labs.

In addition to the bird images and information, I added:

--A guide to using the CD

--Suggestions for introducing birds and bird watching

--A model for planning a field trip

--135 places around the state to watch birds

--Bird checklists to identify what birds may be seen when and where

--300-term glossary

--Standards-based activity

ideas for students K-12 in all content areas

--Additional resources including Wyoming organizations and agencies, books, CDs and internet sites.

It's taken me awhile to realize I've essentially compiled a book – with the help of about 45 other people. I hesitate to single out contributors, but without Dave Lockman, now retired from the Game and Fish, and Mike Randall, the tech consultant, no one else's contributions would have made it to disk.

This first run of the CD is serving as a beta copy. There are a couple technical gremlins, mostly for users who need to update their internet browsers (the CD operates like a web page, though it isn't necessary to be online to use it unless you want to click on the links in the resources section).

Also, some of the photo credits got scrambled; new bird books and websites are coming out; teachers have

great ideas to share; and the state is updating its science education standards. I'm working on a website to hold this additional information until we come out with WBF 2.0.

Meanwhile, to demonstrate the CD, third-grade teacher Kathy Hill and I are offering "Birds in Your Classroom," a workshop for K-12 teachers at Jessup School March 22, which has been approved for recertification credit (see below to contact me for more information before March 14).

Several years ago, Kathy built a school bird habitat at Jessup, also with a Wyoming's Wildlife Worth the Watching grant.

The Flashcards CD is available free to Wyoming educators through Audubon Wyoming or the Game and Fish. To get a copy, give me a call or send an e-mail (see below) including your mailing address, email address, school and grade level/

subject area.

The CD is available to non-educators who send $12.50 to Audubon Wyoming, 400 East 1st, Suite 308, Casper, WY 82601.

Because the intent of the CD is to educate people about birds in Wyoming, there are no restrictions for loading it on multiple computers or school computer networks or making additional free copies for teachers or students.

Unlike other authors, in this instance I'm receiving no royalties or pay for my work, but I am eagerly anticipating feedback from readers – or should I say, users: more resources to post, useful comments for improvement, great ideas for lesson plans – and maybe even a noticeable increase in everyone's understanding of the natural history of this great state we call home.

# 132 Winter bird surveys contrast with backyard experience

Thursday, March 20, 2003, Outdoors, page C2

*"....For, lo, the winter is past, the rain is over and gone; The flowers appear on the earth; the time of the singing of birds is come...."*

*- Song of Solomon*

Though the spring equinox is today, we Wyomingites hope to have a few snowstorms yet. In our climate, our definition of spring as an improvement over winter does not include the cessation of precipitation, especially in a drought year.

However, a sure sign of spring is always the house finches' increased singing – males advertising for mates and defending territories.

But before rhapsodizing about spring, let's review the winter.

My own backyard has been abysmal. I count birds every Saturday and Sunday November through March for Project FeederWatch. Though I don't watch the feeders every minute, I've been hard pressed to count more than one or two birds per weekend the last couple months.

Last weekend I thought I

would finally have a chance to check the "no birds observed" box on the report form, but then about 5:30 p.m. Sunday there was a flutter around the sunflower seed feeder for a few minutes. Five house finches, one house sparrow plus a crow flying overhead seemed like a bonanza.

I think back to other winters and wonder if the lack of birds is due to mild weather, an unseen predator (there was a sharp-shinned hawk in the backyard tree last week), or some avian complaint

about our seed and feeders.

In desperation, we've finally put out millet, hoping to attract house sparrows so that their loud chatter will advertise our yard.

Last week's mail brought the Thanksgiving Bird Count report for 2002. Twenty reports were submitted from Wyoming listing a total of 27 species.

Many were birds common to Cheyenne backyards (each species name is followed by the number of individuals observed statewide): house sparrow (240), house finch

(94), American goldfinch (46), pine siskin (29), black-capped chickadee (18), rock dove (17), mountain chickadee (11), dark-eyed junco (9), blue jay (8), European starling (7), red-breasted nuthatch (4), northern flicker (3), evening grosbeak (3), downy woodpecker (2).

Obviously, some Wyoming backyards are more rural and were able to add these species to the report: Canada goose (45), western scrub jay (32), gray-crowned rosy finch (21), black-billed magpie (9), song sparrow (8), Cassin's finch (8), horned lark (7), American tree sparrow (7), Steller's jay (3), ring-necked pheasant (1) and Clark's nutcracker (1).

The sage sparrow (11) and spotted towhee (1) appeared to have put off migration past their typical departure dates.

The Christmas Bird Count isn't very indicative of backyard birds since we go out along the creeks and lakes looking for birds, although the flock of Eurasian-collared doves, first timers on the Cheyenne CBC, is still showing up regularly at the same feeder.

The Great Backyard Bird Count, held Feb. 14-17, does not restrict observers to their backyards as evidenced by 69 species reported for Wyoming including bald eagles and trumpeter swans.

The top 10 species reported in Wyoming (each followed by the number of reports it appeared in and the total number of individuals reported) were: house sparrow (64, 1590), house finch (50, 580), black-capped chickadee (49, 171), European starling (35, 1240), pine siskin (33, 775), northern flicker (31, 47), downy woodpecker (31, 50), black-billed magpie (28, 98), American goldfinch (26, 294) and common raven (10, 70).

The four reports submitted for Cheyenne list only Canada goose, gadwall, mallard, rock dove, downy woodpecker, American crow, white-breasted nuthatch, dark-eyed junco and house sparrow.

No house finches. Four reports are not a good statistical sampling, but in view of my own backyard experience this winter, maybe it means something. You can check the data at www.birdsource. org/gbbc/results.htm.

Project FeederWatch data is also interesting, however, it needs to be examined species by species. The reports submitted online are already available, but it will be months before the results from observers using paper forms will show.

I find it fascinating to look at the animated maps and see how observations for a species change over the course of the winter. Check for yourself at http://birds.cornell. edu/PFWMaproom.html.

Meanwhile, the millet we put out is attracting squirrels. House finches are singing in the neighborhood, and I hope to hear them soon in my own yard.

## 133 River attracts cranes, cranes attract admirers

Thursday, April 3, 2003, Outdoors, page C2

The morning sun was about to wash the Platte River in gold and rose-colored light. The shallow water gleamed silver wherever it wasn't full of thousands of sandhill cranes agitating for lift-off. Their craa-acking calls filled the air.

It's a spectacle that can be witnessed every year in March and early April along an 80-mile stretch of the Platte in central Nebraska. Mass gatherings of the 500,000 cranes don't happen on any other river between the cranes' main wintering grounds in southern Texas and New Mexico and their breeding grounds, which can be as far north as Alaska and eastern Siberia.

Every evening the cranes leave the fields and wetlands, where they've been feasting on waste corn, invertebrates, worms and snails. In squadrons they arrive at the river to stand together, resting in the wide and shallow channels to avoid predators such as coyotes.

Before sunrise, small groups of the leggy 3- to 4-foot-tall gray birds begin to fly out to the fields, their bodies streamlined and the flapping of their six-foot span of wings propelling them at a majestic speed.

Sometimes, with a roar of wings and a clamoring of voices, a whole section of river seems to lift off in response to some seen or unseen disruption.

A couple weeks ago, members and friends of the Audubon chapter in Cheyenne watched cranes from behind a screen of tall weeds on the river bank a respectful distance away. Some of us wondered who the first non-natives were to remark on this natural extravaganza.

We agreed it wouldn't have been the wagon train pioneers in the 1840s. They didn't even leave Missouri until April each year. Early trappers on their seasonal peregrinations may not have arrived at the right stretch of river at the right time either.

But by 1974, the crane phenomenon was well enough known that the National Audubon Society, with funds from Lillian Annette Rowe, bought 2½ miles of river channel to establish a bird sanctuary.

Dams along the Platte have almost eliminated the spring flooding that controlled vegetation growth, so Rowe Sanctuary staff and volunteers use mechanical means to maintain the wide channels and open sandbars the cranes prefer. The river through the sanctuary looks bulldozed because it is.

There's been a lot of progress since my first visit several years ago. And in the year since my last visit, the effort to build a visitor center was begun and completed, producing the second largest straw bale-constructed building in the United States.

What was once a natural event known only to locals and bird watchers is now a well-advertised tourist attraction. Crane viewing opportunities, either free or for a

2003

123

fee, are marked on a special map available from local businesses from Kearney to Grand Island. Passengers from a tour bus filled Rowe Sanctuary's gift shop by the time we arrived.

I'm not sure how I feel about commercializing bird watching opportunities. On one hand, some of us would rather discover nature on our own – not an easy task when you need to know somebody who knows somebody who will allow you to find a crane viewing spot on their property.

Instead of interpretive signs at eye-level, we must, as the early settlers did, bring our previous experience and our future research to bear on our ability to understand what we've seen. It's called learning by discovery.

On the other hand, a non-profit organization like Rowe Sanctuary is funding its conservation efforts by charging for observation blind reservations and by selling the best selection of crane-related items to be found.

People often don't value

an experience unless they pay for it. The more people who come to value cranes, the better chance necessary habitat management will be supported politically and financially.

As long as people promoting eco-tourism keep the welfare of the wildlife and natural resources their first priority, they will be assured of having the basis for their business continue indefinitely.

In that respect, eco-tourism benefits from good stewardship in the same way

as other uses of renewable resources such as timbering, grazing, farming, hunting and fishing.

Whatever we think of the politics of river and wildlife management, there is still the soul-pleasing aspect of sharing the sunrise with thousands of birds who have figured out how to travel hundreds of miles on nothing but waste corn and creepy-crawlies. Don't ever underestimate a birdbrain.

# 134

Thursday, April 17, 2003, Outdoors, page C2

## With birds at feeders, cats are sure to follow

Last month I complained my backyard feeders were attracting very few birds.

Then a flock of 15 juncos showed up the day before the big storm and some of them are still here, more than a week into April. Evidently, they aren't ready yet to return to the mountains, their summer home.

Birds with more normal migration patterns have been observed. Wilson Selner called last week about a spotted towhee in his yard (formerly named the rufous-sided towhee). This is about a month earlier than I've seen these robin-colored birds in my yard.

The turkey vultures are back too. We'll wait and see whether they are passing through or staying to nest in town.

Twice in the last week I've caught the wispy sound of cedar waxwings while out walking the dog. They can be year-round residents – if they find enough berries

and blossoms.

Along with the return of birds to my feeders is the return of an unwanted visitor, a black cat – definitely bad luck for a bird crossing its path.

I am not too fussy about the demise of a few non-native birds like house sparrows – birds that crowd out the native species. However, it's usually the native birds, especially those passing through on migration, which become victims. Or ground-feeding birds like juncos. Or ground-nesting birds like meadowlarks and other grassland species.

Is it possible to be a cat owner and a bird watcher at the same time? Yes, if you keep your cat indoors.

If you are interested in the conservation of wildlife, remember domestic cats are not predators native to North America. The native fauna have not evolved skills for evading domestic felines. They aren't fair game.

Experts estimate the loss of hundreds of millions of birds each year, not to mention small mammals, amphibians and reptiles. Well-fed cats and belled cats are still successful hunters.

If you value your cat's well-being, the American Veterinary Medical Association reminds you to keep your cat indoors because loose cats are more likely to contract fatal cat diseases and rabies.

Outdoor cats also transmit diseases to humans. Almost all human cases of pneumonic plague have been linked to cats. Cat-scratch fever infects 20,000 people a year and is particularly dangerous to children and people with compromised immune systems. Toxoplasmosis is a problem for pregnant women and their babies.

My bird-watching, cat-owning friends who still allow their cats outdoors unsupervised say their felines are incapable of adapting to

the indoors.

As the owner of two indoor cats, I can imagine outdoor cats mean less hair, less furniture scratching and less kitty litter. But on the other hand, I like knowing that my cats are not in danger of being hit by a car, swallowing poison, being abused by people, killed by other animals or caught in a trap.

Besides, what point is there in having a pet cat if it isn't around to pet?

Pet ownership means pet-proofing your house, but it is possible to convert a cat to the indoor lifestyle with minimal impact by following these tips from the American Bird Conservancy's website, www.abcbirds.org/cats.

--Make the change gradually. Slowly limit the time your cat is allowed outdoors.

--When the cat is indoors, pay more attention to it. Invent cat games and toys. Play with your cat instead of watching TV.

--Provide scratching posts

124                    CHEYENNE BIRD BANTER

and trim your cat's claws every week or two.

--Provide interesting places to lounge, such as by the window overlooking your bird feeder.

--Provide quality, clumping litter. It won't be so hard to make yourself clean it once a day.

--If you are gone a lot, your cat would appreciate a companion. ABC recommends a dog, or another cat, of the opposite sex. My cats are brother and sister and regularly nap and play together.

--Provide fresh greens. The pet stores have kits for growing catnip, etc.

--Take your cat outside once in a while, either on a leash or in a catproof enclosure.

Whatever it takes to make your cat an indoor cat, know that bird watchers will thank you.

As for the neighbor's cat in your yard, try explaining to the neighbor the rewards of an indoor cat, keep the bird feeders away from the bushes and let the dog out frequently. If all else fails, borrow a trap from the Cheyenne Animal Shelter, 632-6655.

# 135 Now's the perfect time to identify ducks

Thursday, May 1, 2003, Outdoors, page C2

A cold, nasty spring day is a good day for ducks. My husband Mark and I laid the spotting scope in the backseat of the car and headed out to check the reservoirs on Crow Creek.

Spring is the easiest time of year to identify ducks: the males have complete breeding plumage; the drab and confusing females are swimming close to their associated males; and there aren't any half-grown young with half adult plumage.

There are difficulties though. The day we went out, cold wind made my eyes water when I tried to look through the scope. Some ducks dive, so when Mark got the scope centered on an individual, by the time I took a look-see, there was only empty water, with the duck reappearing somewhere outside the scope's field of vision.

At our city park lakes, at first glance every duck seems to be a mallard, the males sporting those distinctive green heads. But don't shortchange yourself by assuming the ordinary.

Mallards are puddle ducks. When they tip over to feed, only their tails are visible. When one comes back up,

look to see if it's a drake mallard with bright green head and yellow bill, or does it have a green head and a big black bill like a spatula (male northern shoveler)?

Or is it pointy-tailed and brownish gray all over except for a white breast ending in a streak up the side of its neck (male northern pintail)?

Or is it completely blah brown and gray, but with a very black butt (male gadwall)?

Or maybe the most noticeable field mark is a white crescent on its cheek (male blue-winged teal). Or the whole duck is a reddish, cinnamon brown (male cinnamon teal).

Or maybe the duck you're looking at keeps diving, completely submerging to feed. Does it have a dark red head and a light-colored bill (male redhead)?

Or are the duck's head and breast black followed by a pale gray back, white sides and black tail (male lesser scaup)? Or perhaps it is patterned black and white with a big white splotch on both sides of its head (male bufflehead).

Or maybe the duck has a white cheek, black cap, redbrown body with a tail held

at a jaunty, nearly vertical angle (male ruddy duck).

Or maybe the head seems small for a duck and there are ragged feathers sticking out from the back of the head (some kind of merganser) like the birds we saw at Lions Park and couldn't narrow down to species.

Or maybe the bird floating on the lake isn't a duck at all. The American coot is really a swimming chicken – all black with a white bill and lobed toes instead of webbed feet.

While we saw the above-mentioned birds Easter weekend, this is not a complete list of duck and duck-like possibilities.

It was evidently too early for grebes, non-duck waterbirds easy to spot at Lions Park. The pied-billed (black and white bill) grebe is half the size of a mallard. Even though it's light brown, it reminds me of a rubber duckie the way it bobs and dives.

In contrast, the western grebes that show up every year are large gray birds with long, elegant, white-fronted necks.

While some of the ducks I've listed might have spent the winter here or may spend the summer and nest, others

are passing through to higher latitudes.

But even for the ducks that summer here, don't wait too long to try your hand at identification.

By July some ducks begin to molt or the young, soon adult-sized but not adult-colored, start paddling around and then you're stuck trying to use much more subtle field marks such as the shape of the head or the shadow of a facial marking. It's less frustrating to spend time with a field guide and in the field studying the birds while markings are clear.

Ducks, because they are often in the middle or on the other side of a lake, are often out of binocular range and difficult to identify. With a little help from other birders and a spotting scope, you should be able to figure out most of the ducks – males in breeding plumage anyway.

Then you can start on the next challenging group of birds. Where Crow Creek was flooding the day we were out, Mark and I saw the quintessential long-legged, long-billed, mottled gray shorebird.

Oh gosh. Thirty-eight pages of the field guide to pick from. Was it a kind of plover,

2003

yellow-legs, willet, sandpiper or dowitcher? Maybe

I should take up a different challenge next instead, like

warblers or flycatchers.

# 136 Cheyenne gaining reputation among warbler watchers

Thursday, May 15, 2003, Outdoors, page C2

After a winter of watching seed-eating birds from the windows, it's time to get out and do some weekend warbler watching.

Cheyenne has a reputation for attracting all sorts of unlikely warblers. In the past 10 years the Big Day bird count has produced a cumulative total of 26 species. There are only 55 species in all North America.

The warblers, in turn, attract birders—even from Casper.

The Big Day count is not as formal as the Christmas count and only counts the number of all species observed, not the number of individual birds seen. Cheyenne – High Plains Audubon members choose the date, usually the closest Saturday to the middle of May. We expect that to be the height of spring migration.

This year you can participate by meeting us between 6 and 6:30 a.m., Saturday in the parking lot of the Cheyenne

### Cheyenne Big Day Count -- Warbler species observed, 1993-2002

| Species name | Status | '93 | '94 | '95 | '96 | '97 | '98 | '99 | '00 | '01 | '02 |
|---|---|---|---|---|---|---|---|---|---|---|---|
| Tennessee Warbler | rm | | | | | | | x | | x | x |
| Orange-crowned Warbler | s, unc | x | x | x | x | x | x | x | | | x |
| Nashville Warbler | rm | | | | | x | | | | | |
| Virginia's Warbler | s, rare | x | | | | x | x | x | x | x | x |
| Northern Parula | rm | | | | | x | | x | x | | x |
| Yellow Warbler | s, com | x | x | x | x | x | x | | | | |
| Chestnut-sided Warbler | rm | | | | | | | | x | x | x |
| Magnolia Warbler | rm | | | | | | | | x | | |
| Black-throated Blue Warbler | rm | | | | | | | | | x | |
| Yellow-rumped Warbler | s, com | x | x | x | x | x | x | x | x | x | x |
| Black-throated Green Warbler | rm | | | | | | | | x | | |
| Townsend's Warbler | s, rare | | | x | | | | | | x | x |
| Prairie Warbler | rm | | | | x | | | | x | x | |
| Palm Warbler | rm | | | x | | x | | | | x | x |
| Blackpoll Warbler | rm | x | x | x | x | x | x | x | x | x | x |
| Black-and-white Warbler | rm | x | | | x | x | x | x | | x | |
| American Redstart | s, unc | x | x | x | x | x | x | x | | x | x |
| Prothonotary Warbler | rm | | | | | x | | | | | |
| Ovenbird | s, unc | | | x | x | | | x | x | x | |
| Northern Waterthrush | s, rare | x | | x | x | x | x | x | x | x | x |
| Connecticut Warbler | rm | | | | | | | | | | x |
| MacGillivray's Warbler | s, sc | x | | x | x | x | x | x | | x | x |
| Common Yellowthroat | s, sc | x | x | x | x | x | x | x | x | x | x |
| Hooded Warbler | rm | | x | x | | | | x | | | |
| Wilson's Warbler | s, sc | x | x | x | x | x | x | x | x | x | x |
| Yellow-breasted Chat | s, unc | | x | | | x | | x | x | | x |
| yearly total | | 11 | 9 | 13 | 12 | 16 | 11 | 16 | 15 | 17 | 17 |

Compiled by Barb Gorges from Cheyenne - High Plains Audubon Society records.

Wyoming seasonal status as listed in "Wyoming Birds" by Jane Dorn and Robert Dorn

**rm** = rare migrant

**s** = summer resident

Abundance: **r** = rare, **unc** = uncommon, **sc** = somewhat common, **com** = common

Botanic Gardens in Lions Park. Bring your lunch if you want to stay out with us that long, while we trek to two other local hotspots.

Lions Park is a great place to look for warblers because it has the big trees and the bushes, but it really is just a microcosm of all of Cheyenne, a wooded island on the Great Plains for a group of birds technically known as "wood-warblers."

To me, the warbler family resemblance is that all are small busy birds that are, if not all yellow, partly yellow or yellowish, with the exception of a few such as the northern waterthrush - which ornithologists may decide to put in its own family anyway.

The beginning of the warbler wave is the end of April, when yellow-rumped warblers first show up. They are one of the few warbler species that are willing to subsist on berries until it's warm enough to hatch insects – the only food for most warblers.

Last year the wave peaked at 17 species on the Big Day.

The warblers that visit Cheyenne aren't always the ones predicted from studying range maps in field guides – birds don't read books – so that's what makes warbler watching so much fun.

I have a vivid mental picture from 1996 when Oliver Scott, grand old man of Wyoming birding, heard something in a thicket on the outskirts of Cheyenne and despite his 80 years, jumped a fence and "bagged" the first prairie warbler ever recorded in Wyoming. It was only about 600 miles off its official course.

The prairie warbler was observed again in 2000 and 2001. This might indicate a trend, a range expansion, or maybe we have increasing numbers of better prepared warbler watchers.

Over half the warblers seen in Cheyenne the past 10 years are considered rare migrants, birds thought not to regularly migrate through Cheyenne.

However, in the case of the chestnut-sided warbler, the range map shows its migration route staying east of a line nearly the same as the 100th Meridian, through Kansas, Nebraska and the Dakotas, but its breeding range stretches west across Canada into Alberta. Why would this species not cut kitty-corner through Wyoming?

Twelve of the warbler species observed on Big Days spend the summer and breed in Wyoming, but typically not in Cheyenne. These birds prefer the mountain forest. Some, such as the yellow warbler, provide the background music along willow-choked mountain streams.

Of course, warblers migrate in the fall too, but they aren't as much fun then. They don't come as a wave as they do in the spring. Instead, they slip south over several months.

Plus, the males have lost their distinctive breeding plumage, the young of the year have indistinct juvenile plumage and the females are still confusing. You must be a birder who has run out of other identification challenges to want to take on fall warblers.

I have seen a few warblers from inside. Just last week we had a yellow-rumped flitting 10 feet from the living room window. Another time I saw a Wilson's from a basement window, but generally, most of us wait for the weekend to get outside and do our warbler watching.

The wave will be tapering off over the next few weeks. If you want to try warbler watching, find someplace like Lions Park that has big old cottonwoods. Stand still, watch the treetops and wait for movement to indicate a bird. Then, still with your eyes on the bird, quickly raise your binoculars. Some warblers prefer lower strata, so check smaller trees and bushes too.

You know, nothing's flitted by the window over my computer this morning. Maybe it's time to take a break and hang out some laundry.

# 137

Thursday, May 29, 2003, Outdoors, page C2

## Birdathon fun for bird watchers, helps raise funds

"Did you get the blue jay?"

"Over here, Barb, you can get a good view of the worm-eating warbler, right there by that white patch on the tree."

"Listen, Barb, hear the ruby-crowned kinglet?"

Lucky for me, I was doing my Birdathon on our Audubon chapter's Big Day bird count May 17 and I was getting all kinds of help finding birds – even from my sponsors, for whom each additional bird species I saw was going to cost them more.

It was a great day. I ended up counting 87 species, more than I had predicted, probably due to having practiced on several preliminary field trips. Some of my sponsors had opted for certainty though and pledged a lump sum.

Audubon hit on a great fundraising idea when it came up with Birdathon more than 15 years ago. What better activity than birdwatching could convince a bunch of birders to raise funds?

Eating maybe. It works for the local chapter of the Rocky Mountain Elk Foundation which recently hosted 640 people for a dinner and auction and netted $70,000. Blake Henning, southeast Wyoming director, said the funds will be used for habitat improvement and education.

I think sportsmen are used to paying for the upkeep of the wildlife they enjoy. They buy hunting and fishing licenses and pay excise taxes on their gear so parting with more money at an event is unremarkable. Maybe it is easier for hunters and fishers to justify paying for their wildlife recreation because they often bring home tangible products.

Getting bird watchers to raise funds is sometimes like pulling hen's teeth. Maybe some birders feel justified in not contributing because bird watching is a non-consumptive use of wildlife.

Birdwatchers like to eat

2003

127

too, though. Murie Audubon members in Casper hold a successful annual fundraising banquet that provides what they need for their programs, including education and raptor rehabilitation.

Half the battle in fundraising is to provide a really good reason for people to part with their money. This spring Audubon Wyoming asked each board member to participate in a Birdathon.

I decided my potential sponsors would be more likely to contribute to the state Important Bird Area monitoring and conservation work rather than, say, utilities or staff health insurance; though, arguably, funding office and staff should allow progress towards organizational goals in a more efficient manner than volunteers can usually accomplish alone.

Because the welfare of wildlife is the responsibility of each state and sometimes the responsibility of the federal government for species that are migratory or threatened and endangered, people might think government money is enough to take care of everything.

However, I haven't heard of any government wildlife agency that isn't under-funded or that hasn't any ideas on how to use gifts of money or volunteer time.

But the bottleneck in most wildlife conservation work is habitat, where the animals eat, sleep and breed, which has been adversely affected by human activity directly or indirectly.

Habitat is on public as well as private land. While government programs have offered incentives to private landowners such as the Conservation Reserve Program, or can punish land abusers through environmental regulations, there are gaps to be filled by non-profit organizations.

The Wyoming IBA program not only identifies areas important to birds, but it evaluates sites for habitat improvements and offers assistance to public and private landowners alike.

My Birdathon sponsors this year were recruited from the list of distant friends and family who request Bird Banter by e-mail.

Interestingly, most of my sponsors don't even get to visit Wyoming often. Either they believe in the intrinsic value of wildlife conservation, or I did a great job of arm twisting. Thank you, Art, Anonymous, Cindy, Dale, Mom, Eric, Iren, Jeff, Warren, Pat, John, Lynne, Mark and Beth. Hope your arms didn't feel a thing.

Fundraising is not easy. There's always the possibility that approaching a friend for a contribution will cause that person to relegate you to the same category of acquaintance as telemarketers.

But if you are a better friend than that, let me tell you about these Duck Derby tickets Audubon Wyoming wants me to sell....

# 138 Low numbers, a few surprises at Big Day count

Thursday, May 29, 2003, Outdoors, page C3

At 130, the total number of bird species counted on Cheyenne's Big Day on May 17 was lower than most years but included a few special birds.

Thirteen species of warblers were counted, down from 17 the last two years. However, three of the warbler species had not been seen before on any of the Big Days in the last 10 years. The first, the black-throated gray warbler, seen at Lions Park, is a western species, but commonly found in Wyoming only in the western half of the state. The worm-eating warbler seen along Crow Creek usually breeds in southeastern U.S., east of the Missouri. The third, also an easterner, the golden-winged warbler, was found in the arboretum at the High Plains Grasslands Research Station.

Other unusual birds this year were a great egret and great-tailed grackle south of the Hereford Ranch, and at Lions Park, a broad-winged hawk and a northern parula, another warbler species rare for this area.

The weather was predicted to be first windy and then stormy, but instead it was relatively calm with a high of 79 degrees. Thirty birders started out at Lions Park, including 10 from Casper. Later in the morning half a dozen Laramie birders added to the count.

## Cheyenne Big Day Count, May 17, 2003

| 130 species | | | |
| --- | --- | --- | --- |
| Pied-billed Grebe | American Wigeon | Common Merganser | Willet |
| Eared Grebe | Green-winged Teal | Ruddy Duck | Spotted Sandpiper |
| Western Grebe | Mallard | Osprey | Long-billed Dowitcher |
| American White Pelican | Northern Pintail | Broad-winged Hawk | Common Snipe |
| Double-crested Cormorant | Blue-winged Teal | Swainson's Hawk | Wilson's Phalarope |
| Great Blue Heron | Cinnamon Teal | Red-tailed Hawk | Franklin's Gull |
| Great Egret | Northern Shoveler | American Kestrel | Ring-billed Gull |
| Cattle Egret | Gadwall | Sora | California Gull |
| Black-crowned Night-Heron | Redhead | American Coot | Rock Dove |
| Turkey Vulture | Ring-necked Duck | Black-bellied Plover | Mourning Dove |
| Canada Goose | Lesser Scaup | Killdeer | Great Horned Owl |
| | Bufflehead | American Avocet | Burrowing Owl |

| | | | |
|---|---|---|---|
| Chimney Swift | Horned Lark | European Starling | Savannah Sparrow |
| Broad-tailed Hummingbird | Tree Swallow | Golden-winged Warbler | Song Sparrow |
| Belted Kingfisher | Northern Rough- | Orange-crowned Warbler | Lincoln's Sparrow |
| Red-headed Woodpecker | winged Swallow | Northern Parula | White-crowned Sparrow |
| Downy Woodpecker | Bank Swallow | Yellow Warbler | Dark-eyed Junco |
| Northern Flicker | Cliff Swallow | Yellow-rumped Warbler | McCown's Longspur |
| Western Wood-Pewee | Barn Swallow | Black-throated Gray Warbler | Black-headed Grosbeak |
| Willow Flycatcher | Black-capped Chickadee | Blackpoll Warbler | Lazuli Bunting |
| Least Flycatcher | Red-breasted Nuthatch | American Redstart | Red-winged Blackbird |
| Cordilleran Flycatcher | White-breasted Nuthatch | Worm-eating Warbler | Western Meadowlark |
| Say's Phoebe | House Wren | MacGillivray's Warbler | Yellow-headed Blackbird |
| Western Kingbird | Marsh Wren | Common Yellowthroat | Brewer's Blackbird |
| Eastern Kingbird | Ruby-crowned Kinglet | Wilson's Warbler | Common Grackle |
| Loggerhead Shrike | Blue-gray Gnatcatcher | Yellow-breasted Chat | Great-tailed Grackle |
| Plumbeous Vireo | Townsend's Solitaire | Western Tanager | Brown-headed Cowbird |
| Cassin's Vireo | Veery | Spotted Towhee | Orchard Oriole |
| Warbling Vireo | Swainson's Thrush | Chipping Sparrow | Bullock's Oriole |
| Blue Jay | American Robin | Clay-colored Sparrow | House Finch |
| Black-billed Magpie | Gray Catbird | Vesper Sparrow | Pine Siskin |
| American Crow | Brown Thrasher | Lark Sparrow | American Goldfinch |
| Common Raven | Cedar Waxwing | Lark Bunting | House Sparrow |

# 139 Avian guests make themselves at home

Thursday, June 12, 2003, Outdoors, page C2

In our family, the months of May and June are traditional for family gatherings: weddings, anniversaries, new babies, birthdays and graduations. So, the avian open house in my backyard fits right in.

First, at the beginning of May, the white-crowned sparrows always visit – unobtrusive guests blending in with the usual brown birds hanging out on the ground picking up spilled seed under the feeders – except that they sing.

I could listen all day to white-crowned sparrow singing. After a couple of weeks, in the midst of the commotion of other arrivals, I suddenly realized they'd slipped out without a chance for me to thank them for all the songs. No matter – I'll catch them at their next gig in the mountains this summer.

After a whole winter without so much as stopping by for Sunday dinner, the entire goldfinch contingent descended upon us and have been eating us out of house and home ever since. They gossip constantly and bicker over who gets which of the eight perches on their favorite thistle feeder.

There's a new thistle feeder too, except it's designed with the seed ports below the perches. The goldfinches hang upside down and reach for the seed the way they would on a ripe sunflower head.

The new feeder takes a little getting used to. The neophytes flutter a lot at first before learning to relax and hang by their toes. Two of the goldfinches got so good at it, they tried hanging on the old feeder one day. However, the feeder wasn't designed for this, with the seed ports under the perches further away than on the new feeder, so the birds really had to stretch. They looked like little yellow bats.

Goldfinches are always cheerful and entertaining company, so as far as I'm concerned, they can stay as long as they like.

On the other hand, two blue jays have recently discovered our yard. Like haughty, well-dressed, querulous relations (but not like any one in my human family!) they scare all the other guests into silence. They mostly stay away from the feeders by the house and join the mourning doves on the back wall where we spread millet. Luckily, mourning doves are too pea-brained to realize if they are being snubbed.

One year we had a whole parade of western tanagers on the back wall. However, this year I only saw one flash through the yard. I've been getting reports from other backyard birders though, so this must be the year the tanagers are obligated to visit the other side of the family.

I've also gotten a lot of calls about orioles in Cheyenne's residential neighborhoods. The Bullock's oriole, western counterpart of the Baltimore, prefers to hang its pouch-shaped nest in cottonwoods by a creek. Whether they'll nest in town is something we can only surmise if we observe their unique nests.

The other orange bird has been by, the black-headed grosbeak. The striped-looking female visits most often. She doesn't mind sharing the shelf feeder with the house finches.

I always look forward to a visit from her cousin, the rose-breasted grosbeak. I heard reports and even received a digital photo from another backyard birder, but it was the third week in May before one came here. All black except for a white underside and a hot pink bib or ascot, one male shyly joined the finches. He must have liked the fare we offered because later he was happily singing from a treetop.

Chipping sparrows, spotted towhees, green-tailed towhees and several warblers put in appearances as expected, but why is it some of my favorite guests visit least often? Absence makes the heart grow fonder? I know my annual spring avian open house has reached its zenith when the buntings come. This year it was two lazuli bunting males, glimpsed only briefly three mornings. They truly are a piece of sky, as their name implies.

Now, at the beginning of June, most of these house guests have moved on to summer homes in the country or the mountains. The goldfinches and doves will stick around, but things will get quiet when everyone goes on nesting duty. And then it won't be long before the robins bring their fledglings out for a worm hunt on the lawn.

Having company is work. We have to fill the feeders every day, sweep the seed hulls and wash off the whitewash, but what's a little housework when my favorite company is visiting? I'm already looking forward to next year.

# 140 Thursday, June 26, 2003, Outdoors, page C2
## Knowing birds by ear is helpful on the trail

Our recent Saturday excursion was not meant to be a bird-watching outing. Instead, hiking Laramie Peak with seven teenagers was an endurance test for us four adults.

Who in their right mind would burden themselves with binoculars and field guides for a 10-mile round trip including an ascent (and then descent) of 2,800 feet in elevation?

I wouldn't have made use of binoculars anyway – I was too busy watching my step on the rock-strewn path. But I could hear distinct bird songs in the trees. However, even with binoculars, I doubt I would have found the singers in the foliage.

Was there any chance I could remember the songs until I could get home and compare them with the "Bird Songs of the Rocky Mountain States and Provinces," the three-CD set from the Cornell Lab of Ornithology?

Although I can read music for at least one instrument, I'm no good at notating avian glissandos. Only one bird song from the trip stuck with me because it was like the sound made by squeezing the inflatable seahorse beach toy I had as a small child.

Distinguishing bird songs shouldn't be a big deal. We can distinguish hundreds of human voices. The trick is to hear them often enough – with an associated name and face.

Back at home I was faced with picking from recordings for 259 species. How do super birders do this?

Besides being people with good memories, good birders get lots of field experience and they study. They study books, magazines, tapes, videos and CDs, and they study with people who know things they don't know yet.

Because I already recognize a few bird songs, I knew what species my mystery bird wasn't – robin, blue jay, crow, starling, red-winged blackbird, etc.

Because I heard the song in the forest, I perused the play list for forest birds – all those little gray jobs I don't know, such as flycatchers and vireos.

Putting one of the CDs on the computer meant I could click on any of about 80 tracks, at any time, to make comparisons. The track number corresponded with the bird's name in an accompanying booklet and a narrator announced the name of the bird at the beginning of each track. My mystery bird turned out to be a veery – a little brown bird.

Since my next excursion will be specifically for bird watching, I decided to study in advance. First, I needed a list of birds. My destination was listed in the book, "Wyoming Birds," by Jane and Robert Dorn, and the entry for it lists about 25 possible species.

I could also have come up with a list by matching the section of Wyoming, habitat type and time of year of my trip with the information from the "Wyoming Bird Checklist," put out by the Wyoming Game and Fish Department.

Next, I used a field guide to discover, for instance, that the difference in looks between rock wrens and canyon wrens is minimal.

I read up on their life histories in "Lives of North American Birds," by Kenn Kaufman, which told me more about their habits and behaviors. They both like rocky places.

I compared their songs on the Cornell CDs. Luckily, the canyon wren was very distinct. The CD booklet described its song as "clear series of descending, down-slurred whistles ending with a bzzz, bzzz, bzzz." The recording was made in Arizona, and allowing for regional dialects, I think I'll be able to match that song in the field.

Listening to each song over and over, while visualizing the bird that sings it will help me learn the rest. Another CD, Thayer Birding Software's "Birds of North America," has photos and bird songs combined. I created a list of the birds to study so I won't have to always navigate the entire list of 900-plus species.

Although these research materials are the distillation of the experts' knowledge, visiting with a local expert gave me a shortcut to information after our Laramie Peak hike.

Because the ordeal took longer than expected, our dinner picnic at the trailhead at Friend Park was late and we found ourselves driving out of the foothills on dirt roads at 10 p.m.

What looked like rocks in the road would suddenly explode and careen up and over the hood of our vehicle with the beat of wings.

We thought it might be nighthawks, but when I

talked to Jane Dorn later, she said they were common poorwills. I'm not sure I would have thought to consider poorwills, but when I looked them up, the description fit perfectly.

Kenn Kaufman wrote, "In dry hills of the west....Drivers may spot the Poorwill itself sitting on a dirt road, its eyes reflecting orange in the headlights, before it flits off into the darkness."

The only orange I remember was the rising moon. These birds must have had their eyes closed.

At any rate, exploding rockets of feathers work well for keeping hike-weary drivers awake. And they aren't as scary as finding deer in the headlights.

# 141 Scout camp reopens after fire
Thursday, June 26, 2003, Outdoors, page C3

CHEYENNE – In June of last year, as Camp Laramie Peak filled with smoke from the Hensel fire, Boy Scout officials made the decision to cancel the first week-long camp session.

A few days later, the U.S. Forest Service evacuated the area. The camp, located west of Wheatland, remained closed the rest of the summer. Staff and campers were sent to other Boy Scout camps.

Camp Laramie Peak, one of several camps run by the Longs Peak Council, serves scouts from northeastern Colorado, southeastern Wyoming and southwestern Nebraska, although it also hosts troops from outside the council.

Camp ranger Steve

Sunderman, of Cheyenne, estimated that one third of the camp, or about 65 acres, burned. The hardest hit area was the Mountain Man campsite where two cabins were lost. Losses in other parts of camp included black plastic surface waterlines, one new latrine and several old latrines scheduled to come down anyway.

The dining hall and other buildings received metal roof replacements five years ago and the cook's cabin and the chapel will get metal roofs soon.

"We did some work last fall," Sunderman said. "We cleared out two construction Dumpsters worth from the Mountain Man area and replaced the majority of

waterlines."

The scouts also extended an invitation to local residents to cut burned trees for firewood. Logging of burned areas is anticipated this fall.

Sunderman hopes a burned area will be set aside and that a fire interpretation program will be established. In addition to the melted Dutch oven he salvaged from the Mountain Man camp, he hopes to get a copy of a video of the fire from one of the neighbors. He also sees potential for service projects trimming and thinning trees.

"A lot of our trails aren't burned," Sunderman said. However, the Harris Park trail used for the 50-Miler program now has large dead trees that need to be removed

to make it safe, so rock climbing has been substituted for the long-distance hiking program this summer. The Mountain Man campsite has been relocated, but otherwise, things are back to normal, with 220 people expected in camp for the first session.

Last summer was Sunderman's first year as camp ranger. He recalls volunteers and fire trucks filling the adjacent meadow, assembling to protect structures, and Forest Service firefighters marching down from burn areas. Conditions are wetter this year, so far.

"This year I tried to burn slash and it wouldn't light," Sunderman said.

# 142 Dead blackbird tells no tale about West Nile virus
Thursday, July 10, 2003, Outdoors, page C2

While I was away over a weekend, my friend Ruth left two phone messages. The first, left Saturday, was, "Call me."

By the time she left the second call Sunday morning, she knew she wouldn't reach me before leaving town herself.

Apparently, she had found a dead blackbird while mowing her neighbor's lawn. Would I find out if it had West Nile virus? Instructions for

retrieving the double-bagged body followed.

Well, what are friends for? However, I hoped the neighbors weren't watching as I pulled up in Ruth's driveway late Sunday afternoon, got out of the car, walked to the big black plastic trash can provided by the city, lifted the lid, removed the carefully wrapped package and left with it.

The bird had been gently roasting all day in the solar-heated garbage container,

but it only gave off a faint smell. At home I put it in the refrigerator until appropriate offices would be open on a weekday.

A few weeks before, a neighbor had called about dead blackbirds in her yard, concerned about West Nile virus. I had given her my best guess on who to call. Now it was my turn to find my way through bureaucratic channels.

Turns out the state veterinary lab has a special

toll-free West Nile virus hotline, 877-996-2483.

Among other information in its extensive message, it listed the only species of birds accepted for testing: crow, raven, blue jay and magpie.

Based on Ruth's description and its heft, I knew without having to open the package, this dead bird was not one of the four species mentioned. So, I unceremoniously removed it from the fridge and dumped it in my own big black garbage container.

We will never know if Ruth's bird died of West Nile or something else.

Before the coming of the virus, people reported dead birds in their yards for which the cause of death apparently was not a predator or an accidental collision with a window.

There are a host of bird diseases and sometimes the avian amenities we provide in our yards lead to an unnatural concentration of birds and the setting for a mini epidemic.

A few years ago, a woman called me, reporting 30-40 dead blackbirds in her yard. If memory serves, autopsies diagnosed salmonella.

Disease around feeders and bird baths is more likely in summer. So if you notice sick or dead birds in your yard – and you haven't sprayed chemicals lately – put away your feeders for a while.

It breaks up the party and encourages the birds to go after all those annoying bugs hanging around the garden and patio. Otherwise, wash your feeders frequently.

Birds and other animals can be carriers of West Nile virus, and when a mosquito bites one of them before biting humans, there's a slim chance for serious medical problems for the humans.

However, Cheyenne and Wyoming, except for mountain snow melt areas and some streams, are the most mosquito-free places I've ever lived.

Wyoming has other insect diseases of concern besides West Nile. No one this time of year spends the day afield without checking for ticks afterwards to avoid Rocky Mountain spotted fever.

Diseases carried by insects have always been around and have had major impact on civilizations.

It would be nice if we could do away with all creepy-crawlies. Of course, that would mean no more songbirds since so many are dependent on insects for food. Even a lot of the seed-eating species depend on insects and other arthropods to feed their young.

There is worry that West Nile will itself decimate bird populations. Chickadee numbers seem to be declining in the northeast, where West Nile was first diagnosed on this continent (American Birding – The 103rd Christmas Bird Count, National Audubon Society).

And animals recovering from West Nile virus appear to have permanent brain scarring (Smithsonian magazine, July 2003).

My advice is to take recommended precautions. Empty standing water every couple days, use mosquito larvae-eating fish in ponds, dress appropriately for areas with lots of insects and monitor your own patch of terra firma for changes.

A couple blue jays showed up in my yard this spring for a few days and then disappeared. West Nile crossed my mind. But the quiet was, apparently, due to nesting. This morning the racket in the alley turned out to be a young blue jay.

Oh good. Come on over, son – help yourself to some of the bugs in my yard!

# 143 Fledglings nearly ready to fly away

Thursday, July 24, 2003, Outdoors, page C2

Summertime and the kids are out of school and everywhere. Summertime and the chicks and nestlings are out of the nest and flopping or flying somewhere.

Last week a speckle-breasted young robin fell in my neighbor's window well and needed a helping hand to get out.

Lately, three mourning doves are perching on the utility lines out back and two of them, noticeably smaller than the third, are always pestering it for attention.

Down at Lions Park there's a nearly indistinguishable mix of immature blackbirds. It would be easy to shrug and say, "Ah yes, the young of the year," except that six of us from the Cheyenne High Plains Audubon Society needed an exact count by species for a survey now that the park is a state Important Bird Area.

At the times of year of the Christmas count and spring migration count, most birds are full grown and definitely easier to distinguish. At least as these blackbirds swarm across the lawn, the yellow-headed young appear lighter than the red-winged.

While we stood at the edge of Sloans Lake, a flotilla of teenage mallards, distinguishable as such because they are not quite as bulky as adults, headed our way. So young and already they've been corrupted by their parents' point of view that humans are a food source.

The goslings at Kiwanis Lake, on the other hand, looked a lot younger, still fluffy and sort of yellowish and about three-quarters the size of the Canada goose elders closely guarding them.

Afterwards, while walking along the edge of Sloans Lake back to our vehicles, I realized I'd seen a belted kingfisher fly by three times in the same direction in the space of a couple minutes.

I don't think it was one bird circling. Instead, it might indicate a whole family. Somewhere nearby they've found a hole for nesting, in a tree or in a bank.

Over at Mylar Park, though not yet an IBA site, we attempted to get a handle on over 70 swallows as they and a chimney swift or two darted by.

The swallow choices in our area are: barn, cliff, violet-green, tree, bank or northern rough-winged. But only the cliff swallow has a head so dark combined with a tail not forked – like most of the swallows we were seeing.

The white spot between the eyes is also a good field mark for the cliff swallow, but a lot of these aerialists were spotless youngsters. For a few minutes they perched shoulder to shoulder on utility lines and then later on a wire fence, jostling around a bit like excited elementary students waiting for a lesson in pizza eating to begin.

On the south side, along the Greenway near Morrie Avenue, we counted an uncommon bird, a lazuli bunting. I use the term "uncommon" loosely, meaning uncommon for me, not necessarily the official abundance level.

It was a spot of blue perched on a prominent dead twig at the top of a tree, singing away. Though every spring migration a few lazulis visit our yard, it's just not as thick and bushy and inviting for nesting as Crow Creek, I guess.

Here at home, the last week or so I've been hearing and observing two chickadees hanging out in our backyard spruce. I think they might be mountain rather than black-capped. If the former, what are they doing here? They should be nesting in the mountains.

Maybe our 6,100-foot elevation and all the conifers we residents plant makes Cheyenne appear montane to them.

The mountain chickadee's best field mark is white supercilium, according to "The Sibley Guide to Birds," in other words, white eyebrows. Except this time of year those white feathers have worn down and are hard to see. However, after molting in late summer, they will be nice and bright again. Maybe these two chickadees will stick around that long so I can tell what species they are for sure.

At the tail-end of June we were out looking for fish at Wheatland Reservoir No. 3, on the plains north of Laramie. Two young American white pelicans were gadding about on their own. They could almost be mistaken for adults except for the smudge of gray on the tops of their heads. One can always hope that these young birds will someday choose to come back as pioneers rather than just wanderers at these lakes where no pelicans nest.

Wouldn't it be fun if pelicans moved into Lions Park? We'd probably have to build an island in Sloans Lake to give them protection from loose dogs. I'm not sure what they'd make of the paddleboats. If you are an angler, pardon me while I dream without considering the cost of these voracious fish eaters.

At a puddle adjacent to the reservoir, young avocet chicks were poking about with more serious intent than human children probably would. At the water's edge, they were studiously learning to feed rather than noisily splashing for fun.

But then again, if play is practice, the architects of sandcastles may become architects of cathedrals and excavators of interesting rocks may excavate the cure for cancer. Ah, summer. Each frolicking offspring is the germ of infinite possibilities.

# 144

Thursday, August 7, 2003, Outdoors, page C2

## Bird of backcountry moves closer to town

Two friends and I have a favorite hiking destination in the Pole Mountain area. We're beginning to hike it often enough to notice seasonal and yearly changes in vegetation, stream flow – and our physical fitness.

Our destination is a shaded spot on the bank of a creek from which we have a view down a narrow little canyon [now named Hidden Falls at Curt Gowdy State Park]. We always enjoy a bit of rest there before striking back up the trail to a solar-heated vehicle.

This summer the falling water can be heard, but not seen. When my teenage son, Jeffrey, accompanied us one morning, he scampered up and around the canyon walls and reported that apparently natural debris had shifted most of the flow out of sight.

Whether the change has anything to do with it or not, this summer we've noticed a new bird.

As swallows whipped around and robins called, a quick gray shape determinedly skimmed the surface of the creek, following it to the dark recess made by the rock wall of the formerly visible waterfall.

The first time we saw it, we looked at each other and said, "Dipper?" The second hike it happened we were sure the stubby-winged bird was an American dipper.

Pole Mountain, in my mind, is not high country. I've seen dippers as they dart in and out of waterfalls high in the mountains or dive into rapid mountain rivers, but I wouldn't expect to find one along a tame little stream so close to the plains.

The dippers (there are other species in Europe and South America) are aquatic songbirds. They dive for insect larvae in streams, propelling themselves underwater with their wings or they simply walk the stream bottom. They can stay submerged up to 15 seconds before popping up on a rock midstream where they stand and bob a bit before plunging back in.

Formerly called the water ouzel, this species of the Rocky Mountains builds a domed nest just above streams or behind waterfalls where it will receive continuous spray, according to the field guides. They've also adapted to human construction by building nests under bridges that span mountain streams.

I've never seen a dipper nest. Even if the creek wasn't too cold to wade, I'd be reluctant to follow the Pole Mountain bird for fear of disturbing it.

With some birds you might wait until after nesting season to take a look, but dippers are year-round residents. They stay all winter, as long as the water stays open, and move downstream only if it freezes.

How do dippers react to people? Our hiking destination is in the middle of this dipper's territory, judging from the way it passes us as we sit at the edge of the creek. I don't know how many other people visit. If this is the dipper's first year in this canyon, it may find

2003

133

it to be too populous and have to move on.

However, indirect pressure from people may be more detrimental. Kenn Kaufman states in "Lives of North American Birds" that the dipper has declined or disappeared in some of its former haunts because of declines in water quality affecting its food source. This makes dippers a good indicator of the presence or absence of pollution.

The cause for decline in water quality is usually people, directly or indirectly. All kinds of human activities can pollute mountain streams with additional sediment or unnatural chemicals.

It's ironic that the more we help people to connect with wildlife and the outdoors, the more likely they are to want to work, live and recreate in wild places and the more often wildlife is pushed out.

Maybe, tongue-in-cheek,

we proponents of wildlife conservation could be more helpful if we chose to live in urban apartments, spend our vacation at home and enjoy wildlife programs on TV, leaving the outdoors to those sometimes dangerous wildlife species.

Rather than the classic motto of opponents of industrial sitings, "Not in my backyard," our new motto could be "Staying in my backyard!"

The Pole Mountain dipper is now practically in our backyard. We must be doing something right. However, we need to respect our new resident's privacy if we want to encourage it to stay.

I don't suppose we could salt the creek with caddis fly larvae as sort of a neighborhood welcoming committee's plate of cookies, could we?

# 145

Thursday, August 21, 2003, Outdoors, page C2

## Backpack trip to the high country a chance to do some great birding

A shadow passing over the rocks and turf where I sat made me look up. A falcon!

Even without binoculars, those boomerang-shaped wings are a good field mark.

Mark saw the falcon from where he was fishing and thought it had the "dirty wingpits" of a prairie falcon. I, of course, had once again left the binoculars at home in the interests of packing light.

If this was a prairie falcon circling over Brown's Peak, 11,700 feet in elevation, in the Snowy Range where we adult and teenage members of Boy Scout Venture Crew 102 were catching our breaths, maybe its name contains only part of its identity.

Luckily, my ability to identify alpine avians wasn't as severely tested by the other birds we saw on our backpacking trip.

Our campsite was filled with the unmistakable song of mountain chickadees. Gray jays should change their nicknames from "camp robber" to "camp cleaner" since they seemed to be making

a thorough search for the soup that slid off the stove, even after we thought we'd cleaned it up.

No hummingbirds in camp. But while plodding up a trail, one buzzed me and my new red pack. From what I've heard since though, many hummers had already migrated down to feeders in Cheyenne at the time of our trip the first weekend in August.

All along the trail between Lewis Lake and Deep Lake, white-crowned sparrows kept up constant chirps and whistled songs, but otherwise kept hidden – except for one that insisted on singing from an exposed branch over our heads as we all took a standing break in the meager shade of tree line.

A species of extremely blah-colored bird could be seen moving in the boulder fields at the edges of the North and South Gap lakes. It wasn't until I got home I remembered that tail wagging could indicate an American pipit, a bird that

nests above tree line.

Robins, one of my favorite natural lawn ornaments, seem so out of place at high elevations. But then I realize those wet meadows crowded with yellow and purple daisy-like flowers and other plants fading or about to flower are more like lawns, lawns far superior to what we offer in town.

Domesticated lawns tend to be pesticided monocultural deserts with only a few sprinkler heads making the soil moist enough to probe.

Deep Lake is ensconced in the middle of a "non-motorized zone" in the Medicine Bow National Forest. Cars are all crammed into trailhead parking lots at least four miles away, but there are still sounds of traffic— jets and puddle jumpers.

Recreationists traveling by foot are quieter than the birds (well, except for one evening when the Venture Crew felt moved to sing for a little while). It's a shame motors in the air can't be restricted as well.

Two days after the crew's return, our family took off for another trip, packing in to Rock Creek from the Sand Lake road by way of Crater Lake.

No planes could be heard over the crushing sound of the creek next to our first night's campsite. Not many birds could be heard either, except one that seemed to be following me around, pip-pipping. This trip I had the binocs, but the trees were too thick to find and identify birds.

The second night we returned to Crater Lake. There's a comfortable rock seat from which early the next morning I had a view of the water and Mark fishing below.

The boys were still snoozing in the tent, and I was busy scribbling notes when I became aware of little bird voices all around me. Some were definitely chickadee songs, but unfamiliar tunes were coming from tiny birds flitting in the branches directly in front of me.

Thumbing my mental field guide (none of the books at home were worth dragging around the horrible hairpins of the Crater Lake trail), I decided on kinglets.

The ruby-crowneds always give Cheyenne a wave of their wings during spring migration, and the golden-crowneds appear on the Christmas count. But having them pointed out to me only once in a while makes me less than a nodding acquaintance.

When I was a teenager, my aunt and uncle introduced me to the alpine country through backpacking. Though the physical exertion and accomplishment is a good feeling, I mainly continued all these years for the vistas.

I still love the views, such as getting a closeup of the Snowy Range or looking down into Rock Creek from the trail. But I'm more interested in the details than I used to be.

So it's a good thing now that I myself travel with teenagers who, bored with my slow pace, notice the American dipper in the stream below us – and are still willing to point it out, knowing we have to come to a complete stop to take a look.

# 146

Thursday, September 4, 2003, Outdoors, page C2

## Best backyard bird shows up at dinnertime

The most amazing bird stopped by at dinnertime last week. My family and I were remarking on three yellow jackets hovering behind the screen of the open window next to the table, hovering like hungry urchins, attracted to the smell of the roast chicken, when Bryan, our older son, said, "Hey guys, there's a kingfisher!"

I flew around the end of the table and squatted down beside Bryan to replicate the angle of his gaze and quickly spotted the belted kingfisher after it let loose a string of rattling cries, its characteristic sound.

Sitting on a limb high in one of our big green ash trees, the kingfisher appeared much larger than the usual perching robin and its silhouette displayed the craggy crest of feathers on its head. It let out another raucous call and flew off.

I was dumbfounded by the idea of a kingfisher in our backyard. They are usually in proximity to one of the local creeks or lakes. The ditch a block away from our house doesn't appear to support any fish, a kingfisher's chief prey.

Perhaps the reputation of our backyard avian eating establishment has spread to the kingfisher community. Though we haven't put any sunflower seed out this summer, every day we have a herd of grackles, starlings, robins and flickers probing the lawn and fluffing the garden mulch in search of munchable critters.

But this kingfisher may not have been quite as disappointed as a steak eater in a New Age café. The reference book, "Lives of North American Birds," states kingfishers have been known to eat berries, and we have a bumper crop of chokecherries this year.

It was later that I realized "Belted Kingfisher" would make a terrific entry for my Best Backyard Bird report for August.

Wayne Tree of Stevensville, Montana, started the Best Backyard Bird reports in 1998 and now nearly 200 people all over the western U.S. and a few points beyond contribute their bird sightings by e-mail.

Bruce and Donna Walgren of Casper help by compiling the Wyoming data, and the whole report comes back by e-mail to each observer a week or two later.

Best Backyard Bird reports are meant to be fun rather than scientific, because each person participating can list only one species, and there are no criteria for selection except that the species was in the observer's yard.

Predictably, many birders, like most people, being just teensy part braggarts, want to list the most unusual bird to show up in their yard. However, after a long winter, the best birds sometimes are common signs of spring such as robins or bluebirds.

Last May when our yard hosted all kinds of flashy migrants, I chose to report "American Goldfinch" because we were getting so many at the feeders and I enjoyed watching them so much.

In July's list a couple in Casper listed "European Starling." I wonder if that's the only kind of bird in their yard because they live next to a grain elevator, or if they just enjoy the antics of pesky, but intelligent birds.

When a really wild species appears on the list, such as some kind of backwoods owl, I wonder if the observer lives so deep in the woods the town they list as their location is the closest civilization, but actually miles away.

From years of visiting with people about the birds they see in their backyards, I know that each entry on the list represents a whole story. The robin listed in June could represent the nestlings installed on top of someone's back porch light fixture.

So, I find the Best Backyard Bird list just a little frustrating when all that's mentioned is the observer's name, town and species of the month. I want to know more.

How many feeders do they have? What kind of food do they put out? How do they provide water? Do they watch their yard all day long, or give it the occasional glance? Have they been birding a long time? Is the species they listed a first for their yard? Is it common in the area, or is it as unusual for the area as for their yard?

I assume a lot of the reports are accompanied by commentary and I could find out more from Wayne or Donna and Bruce about any particular report but including remarks regularly would be too huge a task.

However, the real value of Best Backyard Birds is knowing such a wide variety of birds are unafraid of human habitats, so many people are taking note of the birds in their own backyards – and so many are interested in communicating their birding experience.

You can also participate.

Within a couple days of the end of this month, send your name, town and best backyard bird for September to bwalgren@coffey.com.

# 147 Thursday, September 18, 2003, Outdoors, page C2
## A bird by any other name still looks the same

It's a bird book author's nightmare and a marketer's dream. It's the American Ornithologists' Union's third supplement to their 1998 checklist of official bird names and taxonomic order as recently published in The Auk (and available online at www.Birding.com).

All bird publications, including the Wyoming Bird Flashcards CD I finished making last year, are seriously out of date.

Most of the AOU's new changes don't affect us in Wyoming, but this time there are some doozies.

The AOU has three goals when making changes to its official checklist. It wants each bird's scientific name to reflect its relationship to closely related birds. It wants to comply with standardized common names for birds around the world. And it wants to list all birds in ornithological order, based on evolutionary development.

Because people in different locations have had different names for the same plants and animals – not to mention different languages, about 150 years ago scientists started using Latin, the historical universal language of scholars, to give them each a unique name.

The Latin, or scientific, names are also part of the taxonomic system developed by Linnaeus to categorize living things. Each scientific

name starts with the genus, which is shared with a plant or animal's closest relatives. The second part is the species name, the plant or animal's individual name.

On the other hand, common names for birds were originally whatever people observing them wanted to call them. Naturally, there was some confusion as to which small yellow bird the name "yellow canary" referred to.

Evidently ornithologists gave up on the idea of all of us learning the Latin names because now organizations like the AOU are working hard to standardize common bird names – even from continent to continent. That is why our robins are listed by the AOU as "American Robin" because lurking out there are "European Robin" and "Rufous-tailed Scrub Robin."

Ornithologists try to find the oldest common name, and since ornithological study is comparatively young here in North America, we frequently must give up our own bird names, such as sparrow hawk, pigeon hawk and chicken hawk, for names based on older European terms: "American Kestrel," "Merlin" and "Peregrine Falcon."

So now the bird almost every English speaker anywhere would call a pigeon, and for which the AOU's former common name was

"Rock Dove," will now be known as "Rock Pigeon." I guess we commoners were right all along. "Rock" just distinguishes it from other pigeon species.

Three-toed woodpeckers, also found in Wyoming, used to be split into two subspecies, but those subspecies have been elevated to genus level. So now we'll have the American three-toed woodpecker and across the Atlantic they will have the Eurasian three-toed woodpecker.

Genetic studies drive most AOU changes. Comparing DNA is more precise than examining the expression of genes, such as a bird's internal structure and external looks, as was done previously.

Since new information comes to light constantly, conclusions often have to change. So in this supplement the AOU has also done additional shuffling, but luckily, most associated common names have stayed the same.

Finally, ornithologists worldwide have decided loons, long the first group of birds listed in North American field guides arranged in ornithological order, are no longer the most primitive. Geese, ducks, swans, quail and grouse will now come before loons.

I wonder how field guide authors feel about this – especially David Allen Sibley, who just came out this spring

with his first eastern and western field guides. As usual, we'll just have to remember all the previous names of each bird and look them up in the index.

So now "90 of Wyoming's Most Noticeable Birds," as listed on the Wyoming Bird Flashcards CD-ROM, are no longer in correct ornithological order either.

But the CD will need revision before its next printing anyway since the Wyoming Department of Education has rewritten its educational standards for which I wrote bird-related activity ideas.

Meanwhile, first editions are available free to Wyoming K-12 teachers by sending a request to Audubon Wyoming, dwalter@ audubon.org, 307-235-3485, 400 East 1st Street, # 308, Casper, WY 82601.

Other Wyoming educators, from preschools, home school associations, church schools, community colleges, the university, 4-H, scouts, non-profits or government agencies, etc. involved in education can get a free copy by calling Wyoming Game and Fish Department, 777-4600, and requesting a CD from Judi Lemons. Outside Cheyenne, call 1-800-842-1934.

For everyone else, copies are available from Audubon Wyoming at $12.50 each.

CHEYENNE BIRD BANTER

# 148 Birding not just a springtime joy anymore

Thursday, October 2, 2003, Outdoors, page C2

The next time Doug Faulkner plans to come up from Colorado to bird Cheyenne, I hope to tag along again.

He's one of those people who, after scanning acres of ducks, can look around and say, "Gee, it'd be nice to see a peregrine," and wham, something nails a duck and seconds later we all get a chance to see a peregrine falcon standing on its prize on a sandbar in the middle of a drought-stricken reservoir, only a mile south of Cheyenne's city limits.

By the way, the colloquial name for the peregrine was duck hawk. Chicken hawk, a name I mentioned in my last column, referred to red-tailed hawks.

For whatever reason, perhaps years of attending children's soccer games on Saturday mornings, I've never done much purposeful birding in the fall. Besides, it didn't seem appealing because many birds are more difficult to identify than in the spring. They've molted out of their distinctive breeding plumage or they are the young of the year and haven't acquired adult feathering.

Fall birding for me has always been just a matter of what crosses my path. So it was interesting to revisit spring birding haunts and see what was flitting. Technically, this excursion was during fall migration, even though it was the last weekend of summer.

Doug, who is a bird specialist for the Rocky Mountain Bird Observatory located at Barr Lake State Park outside Brighton, Colo., gathered up a group of six other birders for a second annual fall foray to Cheyenne.

First stop, where I met the group, was at the Wyoming Hereford Ranch by the horse barn, overlooking the riparian thicket of Crow Creek. I arrived earliest, but the vista was pretty quiet. Two big bird lumps were sitting in the treetops, one a turkey vulture and the other an unidentifiable hawk showing me only a speckled shoulder.

A lone car pulled into the avenue of cottonwoods and then stopped – a birder, of course. It was Gary Lefko, part of Doug's group. He was studying a small bird lump in one of the trees, which in turn studied us. It had a faded red breast, white belly and a face like a bluebird. It hunched like a bluebird, but had its wings tight across its back where we couldn't examine them for blueness.

Was it an eastern or a western bluebird? Mountain bluebirds have no red markings. When Doug came along at last, he pointed out the obvious field mark. Easterns have a red breast that comes up to their chins like a turtleneck sweater while westerns have the equivalent of a v-neck. So, we had an eastern.

"O.K., we can go home now!" Doug said. Eastern bluebirds are rare enough here to be celebrated as the find of the day.

Back at the creek overlook, the turkey vulture took off, the hawk had gone and small birds were jumping. "Western tanager, western wood peewee, Townsend's solitaire, ruby-crowned kinglet, Wilson's warbler!" Everyone was calling something.

Some of these species, such as the tanager and later, the green-tailed towhee we saw by the office, come through my yard in the spring on route to the mountains, but I had never seen them in the fall before.

The Wilson's warblers were the most numerous. At Lions Park, they seemed as thick as butterflies in the garden. Over the course of the morning, we also saw yellow-rumped, orange-crowned, Townsend's and MacGillivray's warblers plus a chestnut-sided warbler which had none of its chestnut-colored field markings this time of year.

Undoubtedly, any neighborhood in Cheyenne with mature vegetation is hosting these travelers. The week before I'd glimpsed a Townsend's warbler in my own bushes as it fueled up on bugs to continue its trip from breeding grounds somewhere between southeast Alaska and Washington state to wintering grounds stretching from California into Mexico and Central America.

At the reservoir, the coots were easily identifiable, same all-black plumage. Pintails still had pointy tails and gadwalls were still black behind. We'd seen blue-winged and green-winged teal in the creek.

The birds that had lost the most coloring were the phalaropes, those sandpipers that swim in circles to churn up food. In the spring, the Wilson's phalaropes are marked with red and black, but winter plumage is gray and white.

Then it was pointed out that these particular little whirling dervishes were red-necked phalaropes instead. They were just passing through from a summer spent high in the Arctic.

Since my North American bird field guides don't show where these phalaropes winter, I had to do a little more research to discover that they prefer the open ocean, south of the Equator, off western South America. It's amazing the endurance of a 1.2-ounce bird with a wingspan of only 15 inches.

I'm glad the visiting Colorado birders took me along for a bit. Birding in the fall, though challenging, turns out to be just as exciting as in the spring.

# 149 Filmmaker catches magic of migration

Thursday, October 16, 2003, Outdoors, page C2

Last week, "Winged Migration" made a one-night stopover at the Lincoln Movie Palace.

The film, by French director Jacques Perrin, released in 2001 and then in the United States in 2002, features a cast of thousands of ducks, geese, swans and cranes filmed up close as they make their way through spring and fall migration. Cameras were so close, viewers can overhear the birds' every utterance, as though listening in on airline pilots' communications.

This was not a documentary explaining migration. In fact, with its occasional subtitles and the ambiance of its cinema club showing here in Cheyenne, it reminded me of watching other foreign films, especially since many of the birds featured were not North American.

You know how it is when you watch a foreign film. The subtitles at least give you the gist of the story line, but you know that if the characters were from your neighborhood, speaking your language, you'd get all the inside jokes.

Some of the jokes in this film, for instance the elegant crane that kept slipping and landing on its chin, were broad enough humor to transcend cultural – and species – differences.

One of the disconcerting things about foreign films though, is when they portray Americans. The accent never seems to ring quite right. In this case, the flock of Canada geese traversing what looked like Monument Valley desert left us birders in the audience scratching our heads. However, the footage of sandhill cranes along the Platte River was exactly my experience there.

Though I knew I wanted to review the movie, pencil and paper were left at home on purpose. And, as much as possible, I shut down the analytical part of my brain that was looking for information about unfamiliar birds. This was a movie to enjoy for its ballet of avian performance. One difference between this and TV wildlife documentaries was the relief from constant narration. It was almost all close-ups of beautiful birds in flight moving over (mostly) beautiful landscapes, with occasional touches of ethereal music.

There was a plot. Birds fly north in spring. Then they nest. Later, the narrator, in his thick French accent, explained simply that birds fly south for the winter, to where food sources are still available.

There were the antagonists, the obstacles to success: the hunters, the predators, the fatal accidents and the tar pit in the nightmarish industrial zone. But there was also the deus ex machina, the boy who frees the trapped goose.

"Winged Migration" is by no means a complete exploration of the phenomenon. The movie focused entirely on large species, ducks, geese, cranes and large seabirds, probably because they are easier to find with the camera than a four-inch kinglet.

It did not explain that not all bird species migrate nor even all individuals in migratory species (witness the comfortable geese on our local ponds). The details of migration are not cut and dried, north and south.

If watching the movie has piqued your interest in migration, let me recommend one of my favorite books, "Living on the Wind, Across the Hemisphere with Migratory Birds," by Scott Weidensaul, a satisfying piece of literature.

Paul Kerlinger's book, "How Birds Migrate," is a very useful and readable reference. You may also enjoy the lengthy review of "Winged Migration" in the October 2003 issue of "Birding," the American Birding Association's publication.

What is needed is one of those accompanying coffee table books that give the background on how the film was made and the life history and conservation status of the avian stars. I'm sure, as this becomes a cult classic for birdwatchers, someone will publish one.

Then we'll learn more about bar-headed and red-breasted geese, and about the gliders, balloons and Ultra-Light Motorized aircraft that were used over all seven continents, and the hundreds of people listed in the credits. If you want to film a particular migration between Europe and Greenland, you have to contact the locals to know when and where to find the birds.

Perhaps the DVD version has all that extra information. Sony Pictures has also made "Winged Migration" available on VHS, but really, you can't fully enjoy the amazing cinematography on any small screen, including the usual cinema multi-plexes. The Lincoln's huge screen was perfect.

Whichever way you can find it, this is a fabulous 89 minutes of birdwatching. Forget the field guides. Just fly along for the ride.

# 150 Wanted: Project FeederWatch citizen scientists

Thursday, October 30, 2003, Outdoors, page C2

Nov. 6 is the opening of the Project FeederWatch season. It brings with it images of juncos scrabbling in the snow over spilled sunflower seed and little puffs of frozen chickadee breath blowing away in the wind.

It's hard to believe winter when it's 80 degrees in the third week of October and I can shuffle through the fallen leaves barefoot.

Some summer birds have been loath to leave because of the balmy weather and the continuing natural food supply. Though the recent invasion by a flock of American white pelicans was unusual for Holliday Park, and they generally vacate in September, there have been observations of lone stragglers in past winters.

A lone American coot seen down at Sloan's Lake in Lions Park may hang out all winter also. This would be considered rare, but the coot has through November to accomplish normal migration. That lone western grebe, if it leaves this week, won't be too late either.

In the cottonwoods overhanging the lake I noticed a few yellow-rumped warblers Oct. 19, when other warbler species are long gone. But yellow-rumpeds are a special case. They've learned to diversify with the seasons, to change from an insect diet to a fruit diet when the weather cools. At least one has been seen in Wyoming as late as the first week in December.

I'd love to have yellow-rumpeds on my first Project FeederWatch report, but, as berry eaters, they are hardly likely to visit my sunflower and thistle seed feeders.

The Project FeederWatch website shows there were about 20 FeederWatchers in Wyoming last year, with four of us in Cheyenne strung out evenly in an unmeditated transect of the city from west to east.

Over in Ithaca, New York, home of Cornell Lab of Ornithology, a Project FeederWatch sponsoring organization, only four sites were mapped in the city proper. Wouldn't it be neat if we could beat that number this year?

What's keeping you from becoming a citizen scientist?

Do you think you need a scientific background? Project FeederWatch will give you instructions. You need to be able to count, to identify species and to record data. Many classrooms of school children have mastered these requirements.

Do you have trouble identifying birds? The Project FeederWatch research kit includes a full-color poster plus more help online. Set your feeder close to your window and study birds at your leisure.

Are you an expert birder who finds house finches passé? One option is to report observations of various races of the dark-eyed junco. In Cheyenne, that's a challenge. Jane Dorn, co-author of "Wyoming Birds" once looked out at my flock of juncos and said, "I think you have some hybridization."

Is your time at a premium? Project FeederWatch protocol asks for two consecutive days every two weeks until early April, if you're filling out the data booklet, or you can choose to report as often as every week if submitting data online.

The research kit includes a full-color wall calendar for you to mark your schedule. But if you can't make every date, it isn't a problem. Just record your amount of effort, whether it's 8 hours or half an hour. I usually figure about an hour's worth of accumulated glances out the window each weekend I'm home.

Is the $15 registration fee bothering you? Although Project FeederWatch is sponsored by the Cornell Lab of Ornithology, National Audubon Society, Bird Studies Canada and Canadian Nature Federation, registration fees pay just about all its $200,000 annual budget. This includes mailing research kits and the quarterly Birdscope newsletter, keeping up the website and processing and analyzing data, which are reported in scientific journals and birding magazines.

Are you wondering how reporting your few house sparrows can possibly be of interest? After all, there were more than 15,000 count sites last year, all over North America. But sites are sparse here in the west. We need you!

The accumulation of data since 1986 shows population trends and will be able to show just what kind of impact West Nile Virus is having. The optional House Finch Disease survey has been documenting the spread of several other diseases.

The accumulation of data from week to week shows, for instance, some birds don't commit to a particular location for the whole winter and that certain finch species seem to irrupt south on a two-year cycle, tied to the seed crop success in the northern boreal forests. This year we're due to see more common redpolls and evening grosbeaks.

Whether you use paper or computer, you have a record of the activity at your own feeder. You can analyze your data for the effects of local conditions. You are the local expert. You are the citizen scientist.

# 151 Birding trails increasing in popularity

Thursday, November 13, 2003, Outdoors, page C2

There was quite a display of birding trail maps and brochures from all over the country on the table at the recent Cheyenne-High Plains Audubon Society meeting at the YMCA.

Guest speaker, as well as founder and director of the Great Pikes Peak Birding Trail, Gary Lefko of Nunn, Colo., had quite a collection to share. One was for the Oak Leaf Birding Trail. In smaller letters it said, "Milwaukee County Parks, The Great Wisconsin Birding Trail." That county includes Milwaukee and many suburbs, including my hometown.

Excitedly, I unfolded a big map showing the trail, a brown line connecting green park and parkway spots.

As a teenager, I discovered my closest county parks by bicycle and observed, without benefit of binoculars, my first two interesting birds, rose-breasted grosbeak and indigo bunting.

My exploration was also done without benefit of a map. In a city of nearly perfectly gridded streets, wherever a meandering path beside the Menomonee River led me, the nearest corner street sign instantly plotted my location.

So it was with some surprise that I studied the map and realized how close I grew up to an extensive parkway, now identified as part of a birding trail.

This concept seems to have originated in Texas as The Great Texas Coastal Birding Trail. There are many birding hotspots along the Gulf Coast and someone finally put them all together on a map for the benefit of visitors.

Someone also understood that these visitors might be of interest to the local tourism industry and thus was born a partnership model, combining nature and economics, used by most subsequent birding trails. Many also make use of Federal highway enhancement funds from ISTEA (Intermodal Surface Transportation Efficiency Act).

Nothing says a visitor has to stop at all the sites along a particular trail. A birder might spend all of a morning's prime time at just one site, especially if it's a state park with hiking trails.

Sites are not limited to public lands. Anyone can nominate any site for the Great Pikes Peak Birding Trail. Gary uses a nomination form and a set of criteria to determine listing worthiness, but no sites are listed without permission of the landowner. Many have access restrictions, for instance, The Nature Conservancy requests birders arrange and pay for guided tours. In northwest Minnesota, there are sites you might not expect, such as three wastewater treatment plants.

Gary started his fascination with birds years ago, and while living in Colorado Springs, Colo., became intrigued with the birding trail concept. He enlisted the aid of his local chapter, Aiken Audubon Society, and now Audubon Colorado, under contract with the state game and fish agency, is poised to expand Gary's work into a statewide system.

A techie by interest and occupation, Gary has developed a very deep website to support the trail, www.GreatPikesPeakBirdingTrail.org. For each trail site, he has a list of birds of interest, a checklist, bird photos, photos of the area and all the directions and contact information for planning a visit, including the nearest food, accommodations and services.

Colorado, being a destination for outdoor recreation pursuits, will be able to use this information to entice people to extend their visits, i.e., spend more money. So, the Great Pikes Peak Birding Trail is funded in part by the Colorado Springs Convention and Visitors Bureau and local businesses and towns.

I think Cheyenne and southeastern Wyoming have the same potential. We have bird species that out-of-state birders salivate over, such as burrowing owls and sharp-tailed grouse. Even those obnoxious yellow-headed blackbirds at the south end of Sloans Lake in Lions Park are on the wanted lists of many ardent birders. And fables of our spring and fall warbler migrations are spreading and attracting the attention of folks beyond our city limits.

Talk is stirring at the Audubon Wyoming office of developing a statewide birding trail. We can wait and see what form that will take before setting up something for Cheyenne. Meanwhile, check out some of these other resources:

"Wyoming Wildlife Viewing Tour Guide" available through Wyoming Game and Fish Department.

"Access to Wyoming's Wildlife" also available from the Game and Fish.

"Wyoming Birds," by Robert and Jane Dorn, available through the University of Wyoming's Herbarium.

"A Birder's Guide to Wyoming" by Oliver Scott and available through bookstores or the American Birding Association, 1-800-634-7736.

While we wait for direction from Audubon Wyoming, our local chapter is working on a Cheyenne-area bird checklist that could mention some of our favorite hotspots and be something for area businesses, the library, the visitor's bureau, Game and Fish, etc., to hand out.

Wish I'd had a checklist for Milwaukee County 30 years ago. The Oak Leaf map has one now, but funny thing, the indigo bunting isn't listed. Maybe it's too uncommon to be mentioned and I was very lucky to see one there, next to the golf course at Hansen Park.

# 152 Count on it

Thursday, November 20, 2003, Outdoors, page C1

## This Thanksgiving, take time out to keep track of the birds

Birds are a Thanksgiving tradition for most of us in the United States, usually a roasting turkey in the oven. However, for John Hewston, professor of wildlife at Humboldt State University in Arcata, California, the holiday means flocks of wild birds. He is the compiler for the Thanksgiving Bird Count.

Many people are familiar with the longer running, much larger and more intensive annual Christmas Bird Count sponsored by the National Audubon Society, but this count is much simpler to take part in for the average admirer of backyard birds.

Parties of observers sally forth in all kinds of weather for the Christmas Bird Count, rambling over hundreds of 176-square mile, designated count areas, but each Thanksgiving Bird Count observer watches a 15-foot diameter cylinder around their own bird feeders, from a window in the warmth of their own home, or the home of the relatives with whom they are sharing that roast turkey.

And while dedicated Christmas Bird Count birders may brave the elements from dawn to dark, Thanksgiving Bird Count participants are limited to one hour, although, if they wish to make things more challenging, they can spend another hour at another location. Last year, Hewston received data from 472 counters in the 11 western continental states and Alaska for a total of 486 counts.

This was the highest number of participants in the 11 years of the count. Hewston hopes to top 500 this year.

The first Thanksgiving Count was instituted by Ernest Edwards and the Lynchburg Bird Club in Virginia in 1966. Over the years it migrated west until Hewston took over compiling the western count in 1992.

Although count results may vary depending on who participates, such as the year reports from Hawaii documented only spotted doves and mallards, the data gives an overall picture of bird population trends from one year to the next and population differences from one place to another.

Hewston ranks the species counted. Last year there were more house sparrows (2,905) than any other species, but the house finch was seen at a greater percentage of count locations, 54.9 percent, closely followed by dark-eyed junco at 54 percent, but with the house sparrow trailing at 43 percent. A total of 176 species were reported inside count circles.

Hewston also shows results by state. California, of course, beats us all, with 119 counts recording over 100 species because it's the winter destination for so many migrating birds. In Wyoming we had 20 counts recording 27 species. The house sparrow, at 240 individuals, was most populous, followed by the house finch at 94.

Hewston is always recruiting new counters. If you are already feeding birds and will be at home for at least an hour Thanksgiving Day, save the accompanying count form and take part. Sometime in late winter you will receive the results in the mail.

**Feeding stations**

Don't have a feeder yet? Get a sack of black oil sunflower seed from local businesses such as Big R or A&C Feed. Black oil sunflower is enjoyed by more bird species in Cheyenne than any other kind of seed. Spread some on your patio, windowsill or on top of your back wall.

The squirrels may find your seed first and if you object to squirrels, you may want to invest in a tube-type feeder with the wire mesh protecting it, or the hopper-type feeder that closes when heavy mammals or large winged marauders such as crows, land on the perches. Shelf and table-style feeders work well for bird species uncomfortable clinging to tubes or reaching into hoppers. Always leave a little seed on the ground for the ground-feeding birds.

Of course, a variety of food will increase the variety of birds. Suet appeals to woodpeckers, white proso millet to ground feeders and the fine, niger thistle stocked in special thistle feeders attracts goldfinches and pine siskins. Fruit, especially if you have any berry-producing bushes, will attract robins and Townsend's solitaires. If you put out peanuts for the blue jays, be prepared for a squirrel attack.

However, the big drawing card may be water, especially if you can keep it from freezing on cold days. Specialty bird stores carry bird baths with heating elements, but even if you use an old plastic garbage can lid upside down, flexing it to break the ice out each morning and pouring in a kettle of hot water, the birds will appreciate it.

Cover is the other important aspect to attracting birds to your yard. Feeders themselves should not be placed too close to bushes that might hide predators such as loose cats. But having a bushy place within diving distance is appreciated by seed-eating birds when meat-eating hawks glide by, and also when weather is wild.

It sometimes takes a while for birds to discover your new feeding station. You may spend the entire hour for the Thanksgiving Bird Count watching and no bird comes. That can also happen at a well-established feeder, so just report "no birds seen." You might make a note on your report if you have just set up your first feeder.

**Identify the birds**

If you do have birds come, you'll have to be able to identify them for the Thanksgiving Bird Count. With feeders close to a window, or with binoculars, most of the typical Cheyenne seed eaters are easy to distinguish.

Most likely to show up

2003

141

first at your feeder, especially if you use one of those seed mixes with a lot of millet, and the red milo that most local birds ignore, is the house sparrow. It is brown, but its breast and belly are a clear gray. Later in the winter the male will develop a black bib under its chin.

The next most common visitor is the house finch. It is grayish brown, but its breast and belly are streaked rather than plain like the house sparrow. The male has bits of red showing about its head, breast and rump and by spring these areas will be brighter.

Dark-eyed juncos are in the sparrow family, but they appear to be gray birds, rather than brown. We have several races that show up in Cheyenne. Some are gray and white, some appear to have rust-colored backs, some appear to have a black hood, and some appear to have pink-colored sides, but the key identification marks are that these patches of color appear plain, not streaked like most sparrows, and when a junco flies, it flashes its white outer tail feathers.

Black-capped chickadees are less common, but due to their popularity as decoration on cards, mugs, etc., they shouldn't be too difficult to identify. Be aware that we also have mountain chickadees. Their field mark is white eyebrows. Red-breasted nuthatches are about the same size, but more of a bluish color, more of a stream-lined shape, and of course, their breast feathers are a shade of reddish brown. Their most notable characteristic is that they examine trees by walking down the trunks headfirst.

You may be lucky enough to attract goldfinches, but don't expect them to be bright yellow this time of year. The males don't even have black caps now, but you'll notice their black wings with white wingbars.

There are many good field guides available at the Laramie County Library or local bookstores to help you identify your avian visitors. But be aware before you purchase one that "Western" field guides may not cover all of the birds flying through Cheyenne. Look for "Blue Jay" in the index of western guides. The more up-to-date guides will list "Eurasian Collared-dove," an immigrant species which has been seen here the last few years. The very latest guides will have changed the official name of pigeons from "Rock Dove" to "Rock Pigeon." One of the best guides for first timers is Kenn Kaufman's "Birds of North America."

**Beyond the Count**

There are plenty of books on wild bird feeding, as well as websites. Be sure, however, to check if the information is geared for our part of the country. Check the range maps in a field guide to find if the bird species mentioned would even be here in the winter.

Bird watching, especially at the window, can be a lot more than counting numbers of birds and species. Observing bird behavior year round can be an intellectual as well as entertaining pursuit.

Once you get a Thanksgiving Bird Count under your belt, maybe you'll wonder about all the winter birds that don't eat seed or come to feeders. You'll be ready to take part in the Cheyenne Christmas Bird Count Jan. 3. Meanwhile, enjoy the birds in your backyard – and send in your Thanksgiving Bird Count results.

Information on how to conduct a count and a list of species to look for was included.

*Cutlines for photos by Ty Stockton:*

A red-breasted nuthatch, left, gathers seeds from under Chuck and Sue Seniawski's feeder as an Oregon junco moves in from the upper right to do the same.

Sue Seniawski watches a robin perched in a tree above her birdfeeder in her north Cheyenne backyard.

A male house finch surveys Chuck and Sue Seniawski's backyard from its perch in a tree above the Seniawskis' feeder. The bird would be counted in the Thanksgiving Bird Count if it were to sit in the tree on Thanksgiving Day, because it is within the 15-foot diameter imaginary cylinder around the feeder.

# 153 Trumpeter performs for crowd at Holliday Park

Thursday, November 27, 2003, Outdoors, page C2

The man walking the circumference of Lake Minnehaha was approaching us when he noticed our binoculars and paused to make a remark about the trumpeter swan we were watching.

"Yeah, there was a flock of them here last week."

"You mean the pelicans?" I asked.

"Oh, no, swans."

We had seen pelicans about then, but we hadn't been at the lake when he was. However, it's unlikely a whole flock of swans would escape notice by the local birding community.

We shrugged, and the man continued on. At least he wasn't mistaking the white domestic geese that sometimes show up on this small lake at Holliday Park in the middle of Cheyenne.

This particular trumpeter swan visited between about Nov. 8 and 12, feeding on submerged vegetation growing from the lake bottom. It was still using the lake as of press time. It may not have attracted much attention from motorists on busy Lincolnway since, with its head and elegant neck underwater much of the time, it mostly resembled a large white, inflated, plastic bag stranded in a raft of Canada geese, dwarfing them and the other waterfowl.

American white pelicans have thick necks about equal the length of their long, orange bills. By contrast, trumpeter swans have long, slender necks and much shorter black bills. Males and females are indistinguishable unless you have them in hand and check under their tails. The more common tundra swans are smaller and their bills are also black, but with

142             CHEYENNE BIRD BANTER

large yellow markings up by the eyes.

The mute swan, star of European fairy tales such as "The Ugly Duckling" (my favorite version is told by Marianna Mayer and illustrated by landscape artist Thomas Locker), has a black and orange bill. Like so many other birds introduced to North America for their entertainment value, the mute swan has escaped domesticity and, in the northeast, is becoming a nuisance to people and out competes native waterfowl.

Tundra swans have been common enough to support hunting seasons in some states, however, numbers for the trumpeter continue to improve more slowly.

Originally found across most of North America, by the early 1900s, the trumpeter was nearly completely decimated due to over-hunting for its meat and feathers, loss of breeding and wintering habitats and other effects of increasing human population. Remnant flocks held out in remote places like Yellowstone National Park, where hot springs keep water open in the winter. Red Rock Lakes National Wildlife Refuge in southwestern Montana was devoted to its recovery in 1931.

Ruth Shea, executive director of The Trumpeter Swan Society, told me that today there are about 17,000 trumpeters summering in Alaska, 3,700 in western Canada, 3,000 in the Midwest and about 300 in the greater Yellowstone area, for a total of about 25,000.

In winter the Canadians crowd into the Yellowstone area. Biologists attempt to disperse the birds to keep them from damaging their food source. Also, having all your swans in one basket makes the flock susceptible to die-off from any passing disease.

The problem is that when there were less than 100 trumpeters south of Canada in the thirties, the birds lost a lot of their species memory. With swans, migration routes seem to be less imbedded in their DNA and more likely passed down by example. They still know how to get as far south as Yellowstone, but old routes beyond seem to be lost.

Ruth expects our trumpeter is like the few other lone birds that show up in central and eastern Wyoming during fall migration. It could be a subadult from Canada exploring new wintering areas, but no one knows for sure.

David Allen Sibley's range maps show rare observations in states across the country, except for the eastern seaboard where there are none.

How did this one swan find our island of city trees and this speck of a lake? Maybe we will see it again now that we are on its mental map.

But swans have a lot of challenges when they migrate, much like the Ugly Duckling's. First, when flying with a wingspan of six to eight feet at speeds of 40-80 miles per hour, collisions with power lines become deadly. Next, a trumpeter has to find open, shallow water for feeding – all winter long.

Then, in states where the similar-looking tundra swan is hunted, a few trumpeters are always mistakenly shot. Anywhere waterfowl hunting has been permitted, some swans will die from poisoning by picking up old lead shot as they graze submerged vegetation. And, though the trumpeter is our largest native waterfowl species, standing an imposing four feet tall, coyotes are often successful in bringing it down.

Comparing range maps in old field guides with those published since the 1990s, it is easy to see the results of

restoration work. Breeding and wintering ranges in Alaska and Canada have expanded, as has the year-round range in greater Yellowstone.

There's even a dot of purple over western Nebraska and South Dakota indicating the year-round restoration flock there. Ruth mentioned that in nearby northeastern Wyoming, one nesting pair has been documented several years. Other restoration flocks are in Minnesota, Michigan, Wisconsin, Iowa and Ontario, Canada.

Trumpeter swans live 15-25 years in the wild, if we can give them what they need: the right space, natural food and fewer obstacles.

Wouldn't it be great if trumpeters began to visit Lake Minnehaha so regularly that every casual observer became a swan expert?

---

### Trumpeter Swan Society

The Trumpeter Swan Society was established in 1968 to coordinate restoration and research efforts among multiple agencies and organizations. To learn more about trumpeters or to report observations, visit the TTWS website, www.trumpeterswansociety.org.

## 154 How to keep up with birding news

Thursday, December 11, 2003, Outdoors, page C2

I am an amateur watcher of birds. Other than a college ornithology class, the bird knowledge I have has been gained informally, by observation, by talking to people and by reading.

I've picked up a lot from Audubon chapter members,

many of whom are experts on local birds. Some are even formally educated and employed bird biologists.

My library has expanded from a single field guide to about two dozen reference books plus the whole internet – sometimes very

useful when local experts aren't available to answer my questions or the questions I get from readers.

But no science is static, so it's important to read the periodicals. My husband, Mark, and I have been reading Audubon magazine for years,

but it deals with conservation issues affecting birds more so than birdwatching, which is of high interest to local chapter members.

So, a few years ago I responded to subscription offers from Bird Watcher's Digest and Birder's World. Both

2003

143

magazines are informative as well as entertaining, written so even novice birdwatchers can enjoy articles about attracting birds to backyards or anecdotes from the field.

Then, after several years of participation in Project FeederWatch, I finally joined the Cornell Lab of Ornithology. Now in addition to the quarterly newsletter, Birdscope, I get the quarterly magazine, Living Bird. Both focus on the Lab and its far-ranging research.

And then there's the American Birding Association. Because people who are my fonts of local birding wisdom belong to the ABA, I always figured it was over my head. But when it sent me a membership offer this fall, I reconsidered. After nearly five years of exploring birdwatching topics through this column, I decided I needed to expand my horizons.

As I suspected, the ABA is geared for the serious birdwatcher, though it is still accessible for us aspiring to higher expertise. Shortly after I joined, however, I had an encounter that personified the elitist stereotype I feared. It started with a phone call from an impatient visitor from the Midwest who'd left his directory of ABA contacts at home but got my number from someone at the Cheyenne Botanic Gardens.

I have had many nice people with bird questions referred to me, but this man was in a hurry. He was sure he'd seen a kingbird at Lions Park that had a bill too big to be just a western. Could it be a Couch's or a tropical kingbird?

Having never heard of either of these species, I quickly scrambled through my Sibley's and found that they range from Mexico a little way into Arizona and Texas, and they are almost indistinguishable from our western kingbird, except they lack white outer tail feathers.

My very apparent ignorance made the caller even more snappish. Wasn't there anyone else who could come down immediately and verify his rarity? I gladly passed him off to a more knowledgeable birding friend who went to the park but didn't see the bird.

Later, my friend, who also belongs to the ABA, told me that our western kingbird sometimes loses the white color of the outer tail feathers in the fall before migrating. And he agreed that this particular specimen of ardent birder came off as rather unpleasant.

Luckily for the ABA, the members I know are much kinder and more patient with those of us of lesser experience. Half the members, according to a 1999 ABA survey, can identify over 300 species by sight and 75 species by sound. About 40 percent bird more than 50 times a year, and for half the 20,000 members, birding is their main leisure activity.

The ABA, in addition to promoting birding skills and ornithological knowledge, even for those under 18, also provides volunteer opportunities using birders' expertise. Among its programs is support of a conservation project at the location of each annual convention, plus involvement in issues directly affecting birds.

Too often birdwatchers are reluctant to get involved in the politics of conserving birds. They would rather run out to get a last glimpse of the endangered spotted owl than ask for an alternative to cutting the whole forest.

Birdwatchers can also be consumers of products of which the collection or manufacturing can have negative impacts on birds. How can one lament the effects on wildlife from drilling for oil in the Arctic National Wildlife Refuge, yet purchase a new SUV that gets less than 10 miles to the gallon?

It isn't possible to live without any impact on the world's resources, but it is irresponsible to race after elusive life list birds and ignore the health of those birds and their environment. So I'm glad to find that an organization like the ABA caters to listers, but reminds them of their responsibilities.

Audubon and Cornell, with their partnership on the Christmas Bird Count, Project FeederWatch and other BirdSource programs, are also making the connection between birdwatching and bird conservation.

But before Ted Williams' latest piece in Audubon magazine can cause too much heartache, or the latest article in the ABA's Birding magazine, describing the feather-length difference between longspurs, gives me a headache, it's not a bad idea to step outdoors and hear the twitter of the plainest juncos and remember why I was attracted to birdwatching in the first place.

## Birding publications and organizations:

www.americanbirding.org (1-800-850-2473)
www.audubon.org (local chapter, 634-0463)
www.birdersworld.com (1-800-533-6644)

www.birdwatchersdigest.com (1-800-879-2473)
www.birds.cornell.edu (1-800-843-2473)

# 155 Local tradition carried on with 104th bird count

Thursday, December 25, 2003, Outdoors, page C2

## Christmas Bird Count

Jan. 3, beginning at 7:30 a.m. at the Post Office, 2100 block of Capitol Avenue.

Tally party potluck: 6 p.m. Westgate community building, Gateway Drive.

Call Barb at 634-0463 for more information or visit www. audubon.org/bird/cbc.

Two family Christmas traditions at our house are baking stollen like my German grandmother's and using my grandfather's icebox cookie recipe. To those, I have added participation in a tradition that stretches back further than their immigration to America, the Christmas Bird Count.

The original bird count participants burned off holiday calories by hiking through the woods and fields to shoot as many birds as possible. In 1900, an informal network of 27 birdwatchers, mostly on the East Coast, decided to compete with each other by counting live birds instead of dead ones. From 25 locations they toted up 18,500 birds of 90 species.

Now the count is a little more formal and much more widespread. To make competition fair, each count area is a 15-mile diameter circle. In Cheyenne, ours is centered on the Capitol building, and this year's count will be on Jan. 3.

There are more than 1,900 counts now, involving 50,000 people, mostly throughout North America. This year, the usual CBC partnership of Cornell Lab of Ornithology,

National Audubon Society and Bird Studies Canada welcomes the National Network Bird Observers and Instituto Humboldt, both of Columbia, and the Gulf Coast Bird Observatory of Mexico.

Though there remains some competitive spirit – who saw the most species, the most individual birds or the most unusual birds, the collected information is becoming more interesting as data to track trends such as population abundance and range and migratory behavior. Data for Cheyenne is available online back to 1974.

I've been helping the Cheyenne count compilers for 14 years, but this will be the first as compiler myself. However, I expect lots of help from the members of Cheyenne-High Plains Audubon, which sponsors our count.

A lot has changed since May Hanesworth used to write the results longhand and I would type them for her to send to the Wyoming Tribune-Eagle and CBC headquarters. During Jane Dorn's years as compiler, the CBC went online, and we used my computer to fill in the check list of birds.

This year, count compilers had the choice of allowing participants to sign up online and pay the $5 per person fee for field observers or closing their count to new observers or, what I chose, potential participants can contact the compiler for more information. Our chapter traditionally pays everyone's fee but

asks for donations.

The Cheyenne count is a little less structured in other ways also. Many compilers will hold an organizational meeting a few weeks in advance. They assign people to parties and each party is assigned to a particular route. I suppose this is necessary if, like the Point Reyes, Calif., count last year, you have 184 people showing up. But they had more species to count, 205 compared to our 48, since they host a lot of wintering migratory birds.

Here in Cheyenne, we don't worry about bumping into each other. Traditionally, we've met at 7:30 a.m. at the Post Office lobby on Capitol Avenue and then walked the Capitol district before heading to Lions Park and other hotspots.

The last few years we've had enough folks to simultaneously hit the park, Wyoming Hereford Ranch, High Plains Grasslands Research Station and F.E. Warren Air Force Base, dividing into groups based on who has permits for the station and the base. This makes sense because birds are more active and easier to find at the beginning of the day.

After the first few hours, the groups break down. People have toes to warm and other obligations. A few folks continue to nose around the creeks, Little America and other spots that may have open water.

It's important to stop at home and check the bird

feeder. There is no fee for people who only participate by watching their feeders. The protocol is to report the highest number of individuals of a species seen at one time. Otherwise, a very busy chickadee becomes a flock of fifty.

No birdwatching experience is necessary for becoming an observer with one of our field parties. Each needs someone to record numbers, someone to help count the starlings on the wire, and someone to turn around and say, "What's that over there?"

Perusing the CBC website ahead of time to see what birds are typical for Cheyenne counts (our code is WYCH) is good preparation. But nothing beats being at the elbows of good birdwatchers.

If you go out to count birds on your own, also keep track and report how long, how far and by what means you traveled.

Although you can call or email your observations into me, it's more fun to come to the tally party and potluck afterwards. That's when you find out who saw what and where. The final results archived on the CBC website won't include stories.

If the birds stick to their own traditions, we'll find the belted kingfisher on Crow Creek and rough-legged hawks on the prairie. As for me, I'll pack along the traditional hot chocolate and left-over Christmas baking to share.

2003

# 2004

## 156 Snow diminishes results of Christmas Bird Count

Thursday, January 8, 2004, Outdoors, page C2

If someone was counting the human population of Cheyenne last Saturday, based on the number of pedestrians observed, they might have come up with only 14 of the 53,011 reported by the Census Bureau – those of us foolish enough to be outside on the annual Christmas Bird Count.

Birds visible in the blowing snow underrepresented actual numbers as well. Most were hunkered down, waiting out the storm. Where one might expect the twitter and movement of juncos and other sparrows in tangles of shrubs, or the rhank-rhank of nuthatches in trees, most often there was only the steady tisp-tisp of tiny snow pellets hitting Gore-tex outerwear. Some years we see more than 50 species. This year it was 35, plus three observed during "week of the count" (the three days before and three days after count day, Jan. 3).

Canada geese, however, were easy to find, bunched up in open water, unwilling to fly out to snow-covered fields to feed as usual. Water in this dry country is easy to pinpoint. Between Hereford Reservoir No. 1, Lake Minnehaha and Sloans Lake, 2,092 geese were counted, up from 1,451 last year.

House sparrows were in great abundance if you knew where to look. At Avenue C-1 and Jefferson Street, a couple hundred swarmed between feed at one house and cozy bushes at another.

Over at the South Fork subdivision west of South Greeley Highway, what at first looked like another flock of house sparrows feeding on the ground between homes turned out to be 40 horned larks. The presence of grassland birds wasn't too surprising since the subdivision was recently carved out of the surrounding prairie.

Lapland longspurs are not found often, but a birder joining us from Ovid, Colo., who has lots of longspur experience from living in Kansas, was able to identify their peculiar call as they flew over with flocks of horned larks.

House sparrows and European starlings don't seem to limit activity during snowy weather, and I would think American crows wouldn't either, but on this count, we were hard put to find them and their close relations, the black-billed magpies. Last year, we counted 250 crows and 48 magpies. This year, we were down to 41 crows and six magpies. Since crows and magpies are among the most noticeable birds to be affected by West Nile virus, this decrease isn't too unexpected after a summer when the first human cases occurred here.

Warblers don't normally show up on our Christmas count. In 29 years of available data, only twice have they been observed. The yellow-rumpeds seen Saturday are the most likely to winter here since they are one of the few warbler species that can change from a summer insect diet to an after-frost berry diet.

In great contrast was the Guernsey – Ft. Laramie count held Dec. 20. This is a new count designed by the Cheyenne count's former compiler, Jane Dorn, who, with her husband Bob, has retired to the Lingle area.

The center point of this count circle is the Platte-Goshen county line where the railroad tracks cross it. The 7.5-mile radius stretches from the east end of Guernsey State Park to the west side of Ft. Laramie National Historic Site. A map shows no mountain ranges on this far eastern edge of Wyoming, but the land is a wonderful jumble of geology and habitats.

Ten of us met at the main entrance to Guernsey State Park, drove along the reservoir edge and hiked up Fish Canyon. There was snow in the old road tracks in the shade, but otherwise, we were shedding layers as we went. The high for the day was 61 degrees.

After lunch at the Oregon Trail Ruts State Historic Site, we explored Hartville, an old mining town set in a narrow, winding canyon. We parked by the churches for a better look at a downy and a hairy woodpecker in the same tree and were greeted by two locals – two inquisitive black dogs. Further up, we were entranced by a front-yard feeder full of goldfinches.

Lucky for us, the open water at Grayrocks Reservoir was at the lower end, within the count circle. A thousand mallards attracted 31 adult bald eagles and three immatures. Most of the eagles merely stood around on the ice, but one aerialist performed, stooping to slam into, then eat, a duck.

We ended the count at Fort Laramie, the historic site, not the town, hiking the Laramie

River in two groups in opposite directions and finding great blue herons.

While the group I was with waited back at the cars for the other, the sunset turned the hills pink, and two bald eagles flew low overhead, along with skeins of geese so high they could have been mistaken for wisps of cloud.

We missed the companionship of Barbara Costopolous of Guernsey, whose husband's funeral and burial was that day. We counted 31 species this year (plus seven week of the count), but with her help next year, who knows?

## Cheyenne Christmas Bird Count

Tally taken Jan. 3, 2004
35 species and 4,579 individuals count day
cw – count week only, 3 species
Canada Goose 2092
Mallard 796
Northern Shoveler cw
Green-winged Teal 3
Common Goldeneye 10
Common Merganser 1
Sharp-shinned Hawk 1
Rough-legged Hawk 4

American Kestrel 1
Wilson's Snipe 2
Rock Pigeon 133
Belted Kingfisher 1
Downy Woodpecker 4
Northern Flicker 8
Blue Jay 3
Black-billed Magpie 6
American Crow 41
Horned Lark 305
Red-breasted Nuthatch 9
White-breasted Nuthatch cw
Brown Creeper 5

Golden-crowned Kinglet 6
Ruby-crowned Kinglet 1
Townsend's Solitaire cw
American Robin 60
Brown Thrasher 1
European Starling 369
Yellow-rumped Warbler 2
American Tree Sparrow 16
Song Sparrow 7
Harris's Sparrow 1
White-crowned Sparrow 2
Dark-eyed Junco, race unknown 50

White-winged Junco cw
Slate-colored Junco 35
Gray-headed Junco 8
Oregon Junco 10
Pink-sided Junco 54
Lapland Longspur 2
Common Grackle 1
House Finch 74
Pine Siskin 1
House Sparrow 454

## Guernsey – Ft. Laramie Christmas Bird Count

Dec. 20, 2003
31 species and 2,907 individuals count day
cw – count week only, 7 species
Canada Goose 938
Mallard 1528
Green-winged Teal 2
Common Goldeneye 4
Common Merganser 1
Hooded Merganser cw

Wild Turkey 12
Great Blue Heron 2
Bald Eagle, adult 31, imm. 3
Sharp-shinned Hawk 1
Rough-legged Hawk 1
American Kestrel cw
Merlin cw
Killdeer 1
Ring-billed Gull cw
Herring Gull cw
Rock Pigeon 2

Belted Kingfisher 3
Downy Woodpecker 3
Hairy Woodpecker 1
Northern Flicker 3
Northern Shrike 2
Blue Jay 8
Black-billed Magpie 9
American Crow cw
Horned Lark 6
Black-capped Chickadee 9
Townsend's Solitaire 35

American Robin 70
European Starling 131
American Tree Sparrow 20
Song Sparrow cw
Dark-eyed Junco 26
Red-winged Blackbird 13
House Finch 4
Pine Siskin 5
American Goldfinch 26
House Sparrow 7

# 157

Thursday, January 22, 2004, Outdoors, page C2

# Winter reading tracks albatross across the Pacific

A media tradition, apparently with East Coast origins, is to offer a summer reading list. I imagine normally harried commuters spending a week's vacation at the shore, snoozing in the sun on short-legged beach chairs stuck in the sand, thick best sellers shading their midriffs.

Here in Wyoming, the land of more active, year-round recreational pursuits, long winter evenings are a better time to peruse a reading list.

My book nomination, however, is full of beaches.

"Eye of the Albatross – Visions of Hope and Survival" by Carl Safina was a book I read rave reviews for when it first came out in 2002. But it wasn't until my brother-in-law, Peter, picked up the paperback edition at a bird festival in Alaska where he lives, and sent it on to us, that I finally read it.

You might read this book if you merely want life history information on several sea animals (it has an index), especially albatrosses, and one Laysan albatross in particular which Safina dubs Amelia for her role in opening new avian aviation frontiers.

Amelia builds a nest for her one egg just steps away from the entrance to derelict military barracks, now serving as a National Wildlife Service research facility on Tern Island in the French Frigate Shoals of the Northwest Hawaiian Islands, where Safina is a guest.

One of the scientists outfits Amelia with a transmitter that will allow satellite tracking to find out for the first time how many hundreds of miles an albatross will travel in search of food. During the course of a year, Amelia also carries the structure of the book.

Albatrosses are amazing. Perfectly adapted for gliding effortlessly in solitude for days on end in search of prey, such as jellyfish swimming just below the surface of the water, they also live long lives of social intricacy.

Safina has ferreted out a lot of scientific and historical information, but it's the way he puts it together with his experiences that makes this book a good read as well as educational.

2004

Himself a scientist with a Ph.D., earned by studying seabirds as well as the founder of National Audubon's Living Oceans Program and now president of the Blue Ocean Institute, Safina doesn't let science extinguish the spiritual. In fact, the more he immerses himself in seabird knowledge, the more spiritual his contemplations.

The way living things adapt to climate and each other is always awe inspiring no matter where it happens. However, Safina explains three major human-induced upsets seabirds cannot adapt to quickly enough to avoid extinction. One is global warming weather changes, causing prey species to become less available and causing more flooding of the few islands where seabirds now nest.

Another is the proliferation of plastic debris that has been dumped in the oceans and with which the birds get tangled up or fill their stomachs and starve to death. Even the invasion and establishment of biological debris, such as a non-native grass species, can have a deleterious effect on nesting.

And finally, for albatrosses, there is the danger of long-line fishing. The baited hooks being let out behind boats, just under the surface of the water, are too tempting. Too many albatrosses died before someone figured out a way to scare the birds off and get the lines to sink more quickly.

While every detail of albatross life history fits together to form an incredible whole, all of the threats Safina identifies could become raveling threads of tragedy. Doubtlessly, the world will find a new balance if a species becomes extinct, but before it does, there are usually other, even human, victims of the same circumstances.

Luckily, for the sake of his readers, Safina is an optimist. Luckily, Amelia surmounts obstacles and her chick fledges to wander toward the coasts of Russia and Japan.

Luckily, for the sake of the future of the oceans, there are the people on Tern Island and elsewhere whom Safina introduces who are working hard to understand the lives of sea creatures and what they need from us to survive.

Nearly a thousand miles from an ocean beach, it's easy to enjoy a book like this, but then feel helpless about making any personal decisions that will improve the ocean situation. However, if you look up www.blueoceaninstitute.org, you'll find the "From Sea to Table Seafood Miniguide" which lists seafood by species, explains present circumstances and compliments those fisheries with good management. All the information is in an easy-to-print-and-pocket document you can use to determine ethical eating choices.

There you'll also find more of Safina's writing and his explanation of the Sea Ethic, the counterpart to Aldo Leopold's Land Ethic.

It wasn't until I finished reading the book and fanned back through it, that I realized the handwritten note addressed to us on the flyleaf was written and signed by Safina, rather than my brother-in-law. So, it felt even more as if I had just finished reading a long, but interesting, personal letter instead of a book.

# 158 Birders don't always flock together

Thursday, February 5, 2004, Outdoors, page C2

I meet people who, when learning I watch birds, say, "Oh, that's something I'd like to get into someday." Or, "I watch birds at my feeder." Getting them to come out on an organized field trip is another matter.

Having read a lot of novels, I would know what to expect of a weekend at an English country house: dinners, fox hunting, murder, etc. But not a lot of novels are based on bird watching, so if you've never been on a field trip, let me shed some light, using for example, the recent mid-January joint field trip sponsored by Cheyenne-High Plains and Fort Collins Audubon societies featuring a tour of Cheyenne's birding hotspots.

First, you have to find a field trip. This one was listed in the Wyoming Tribune-Eagle and other media as free and open to the public. You can also call the local Chamber of Commerce, visitor's bureau, library or local bird columnist to find a contact or look up the nearest Audubon chapter through www.audubon.org.

CHPAS welcomes novice birders. There are just a few things novices should know to better blend in with the group. Dress for the weather and for walking. Avoid wearing white or bright colors that can alarm the birds.

Leave pets at home, although well-behaved children are welcome. Bring food and water, and if traveling to the boonies, bring a bit of toilet paper and a bag to bring it home in.

Make sure your vehicle is gassed up and ready for the kinds of roads to be driven. The bird watchers' vehicle of choice lately seems to be the small all-wheel-drive wagon – rugged, roomy enough and economical on fuel. If you are new to the group or a novice birder, see if you can carpool – it's much more fun, less polluting and makes it easier for the group to park. Be sure to offer the driver a little something for gas for long trips.

Arrive at the meeting place early or on time. Many field trips peregrinate, moving from place to place like a flock of birds, and unless you know the route, you may not be able to catch up.

Don't be afraid to introduce yourself. This trip, we had 14 folks from Fort Collins, six from Cheyenne and one from Laramie. Next time I think I'll bring nametags for everyone.

We started at the Wyoming Hereford Ranch with admonitions to not cross fences, open gates or point binoculars at windows, since we were guests on private property. With more and

148      CHEYENNE BIRD BANTER

more birders visiting, we should also stay out of the brushy side of Crow Creek in the bunk house area.

I was worried that the serious birders might be bored because the reservoirs were mostly frozen and the beginning of spring migration still far off. But they were quite satisfied with identifying blue-winged teal in the creek – unusual, but not unheard of this time of year. A white-winged junco was another nice find.

Over at Wyoming Hereford Ranch Reservoir No. 2, we parked on the road's shoulder, and several folks brought out spotting scopes. Everyone had a chance to see the common goldeneye close up.

At Hereford No. 1, the geese on the far side, near the only open water, turned out to be decoys tended by several hunters. At Holliday Park, the ducks and geese were all crowded around a gentleman ignoring the "do not feed" sign. The visiting birders were impressed by the black-crowned night heron rookery, the nests in the tops of the trees clearly visible without the leaves.

Lions Park produced an American wigeon, several black-capped chickadees and one of the top finds of the day, golden-crowned kinglets playing hide and seek among the cones at the top of an evergreen.

As satisfying as being the first one to identify a bird is being able to help other people see it. Most birders are very generous this way. As a novice, it's instructive to watch which way the binoculars are turning and ask for help to see what everyone's looking at.

But don't hesitate to bring to attention a bird no one else is observing, even if you don't know its name (but be sure to commit to memory early the identification of the too common starling and house sparrow).

One of the less experienced birders was the one to point out the Townsend's solitaire at the Cheyenne Botanic Gardens.

After lunch at the gardens' picnic tables (it was sunny, windless and in the 40's), we found red-breasted nuthatches and a brown creeper in the spruces by the parking lot at Little America.

By mid-afternoon, most of the local birders had dropped out, but the Fort Collins crew proclaimed that they like to bird till they drop.

I wonder how many additional stops they made on their way home.

While it was, with the exception of the birds I've already mentioned, a slow day, much of the fun was meeting other birders. One man from Fort Collins asked if I knew a couple from Casper he met in Alaska at a birding convention. Well of course – I was visiting them at their country house the next weekend (minus foxes and murder).

Many of the Fort Collins birders said they are looking forward to coming up again on their own, now that they know where to go and where to find the public restrooms.

And we've been invited to visit them.

Watch for announcements about the next adventure or call the columnist.

# 159 Backyard bird count migrates back again

Thursday, February 12, 2003, Outdoors, page C4

"One junco, two sparrows, three house finches...."

It must be time again for the annual Great Backyard Bird Count. According to scientists at Birdsource, the cooperative effort of the Cornell Lab of Ornithology and the National Audubon Society, which sponsors it, Friday through Monday is the time to document birds at the end of winter, to count survivors before migration.

The tally dates may be a little premature this far north, but Wyoming birders have jumped right into their roles as citizen scientists. Last year 115 reports, or checklists, from 32 towns and cities were turned in. Casper turned in the most reports, 10, and Powell recorded the most species, 29. Cheyenne submitted four reports for a total of nine species of birds.

The GBBC has taken full advantage of computer technology. No need for count compilers or for coordinated outings. Anyone can take part and enter their own data on their own computer or on a computer at the library, and there is no charge. The GBBC website, www.birdsource.org/gbbc, even promotes it as a family-friendly activity.

All the directions and help needed to participate in the seventh annual count are on the website.

First, print out a checklist of the birds to be seen in your area. One option is a list of backyard birds. For Cheyenne, it is a list of about 20 species. The other option is a complete list of 138 species that have been reported in Wyoming on the GBBC over the last six years.

Not all counting is done in the backyard, as illustrated by the inclusion of two dozen waterfowl species, unless, as one Wyoming report indicated, the backyard is in Yellowstone National Park.

If some of the birds are not familiar, look them up in the "Learn about birds" section.

Count in one or more locations over one or more of the four days but keep separate records for each count. Each time, watch the birds for at least 15 minutes (preferably 30).

Watch at a bird feeder or walk the neighborhood or park but walk less than one mile.

Record the species seen and the largest number of individuals of a species seen at any one time. Don't count the same chickadee each time it flies in to pick up a seed.

Submit this checklist report online. Then, as the weekend progresses, check out the Map Room for the results across the continent.

Last year, 47,740 checklists were submitted, documenting over four million individual birds of 573 species. The house sparrows in one backyard may not seem

2004

149

important, but as part of this massive amount of data, Birdsource scientists say they contribute to an overall picture and understanding of North American birds.

To improve checklists from all backyards, the GBBC website also includes links to information about creating bird friendly yards and feeding wild birds. Then, with a little work, the count next year might be, "One sharp-shinned, two waxwings, three downy woodpeckers, four blue jays, five creepers, six goldfinches...."

# 160 Starlings aren't the darlings of the bird world

Thursday, February 19, 2004, Outdoors, page C2

The European starling has done as well as any immigrant left to fend for itself in the middle of New York City. It was introduced in 1890 by Shakespeare aficionado Eugene Scheiffelin (also spelled Scheffland) who, as head of the American Acclimatization Society for settlers from Europe, wanted to bring to his new home all the birds mentioned by the bard.

The starling's legendary success, nearly 200 million individuals in North America and spreading worldwide, has come at the expense of native birds such as bluebirds, nuthatches, swallows and woodpeckers – all species that nest in cavities coveted by aggressive starlings.

Though in Europe, their native land, starlings might be prized as caged pets because of their ability to mimic, or as canned pate de sansonnet, meat pie of starling, I was reminded by recent phone calls that here they can be pests.

Callers want to know how to keep large flocks of starlings from roosting in their trees and buildings and leaving all that whitewash. They have reason to be concerned as starlings can transmit diseases to people and livestock, and their droppings encourage the growth of pathogens. Starlings also eat crops and contaminate livestock feed.

The first step is to determine whether the birds observed are starlings because all birds in the United States are protected by law except for pigeons, house sparrows and starlings.

Starlings are chunky little birds, short-tailed and short-winged, which appear black, somewhat iridescent, though in the fall new feathers are tipped in white, like little stars. By spring, the stars wear off and their bills become bright yellow.

A quick look at a field guide shows that other black-colored birds have easily distinguished differences.

Starlings showed up in Wyoming in the 1930s and are now found in human-modified areas throughout the state. They are classified as predacious birds that can be taken in any manner except as restrictions apply to methods. Catherine Wissner, at the University of Wyoming Cooperative Extension office in Cheyenne, reminded me that it's illegal to use firearms in the city.

Stan McNamee, director of the Laramie County Weed and Pest District, advises against chemicals except where starlings have invaded a contained area such as a barn. It's highly likely that without confinement, other kinds of birds would be poisoned as well. He said the most effective chemical requires licensing.

Non-lethal methods of dealing with starlings are based on making them unwelcome. Wyoming Game and Fish Department biologist Steve Tessmann recommends a publication available at the Cheyenne office, "Homeowner's guide to resolving wildlife conflicts." He said a few nights of sporadically flashing floodlights could disrupt roosting starlings and run them off for good.

Fluttering mylar strips as another deterrent were recommended in a University of Florida Cooperative Extension bulletin. I suppose I should be grateful for the remnants of a plastic bag left by the wind in the top of one of our backyard trees.

Bob Lee, director of Environmental Management for the City of Cheyenne, recommends trying a modern-day scarecrow, a garden hose sprinkler with an electric eye that shoots water whenever it detects movement.

A Nebraska Cooperative Extension publication I found through the website www.invasivespecies.gov, recommends limiting food and water available to starlings. In a livestock operation, that means cleaning up grain spills and feeding pellets larger than a half inch in diameter, which is larger than starlings can swallow.

In your backyard, limiting food means avoiding the starling favorites, bread crusts, milo, millet and platform-style feeders in favor of black oil sunflower seed and tube or hopper feeders that are screened or balanced to keep out bigger birds.

Starlings are always squeezing their messy nests into tight, human-made places in addition to natural cavities, so screening those openings can help. Make sure nest boxes for other birds have no perches in front of the entrances and the entrance holes are sizes that exclude starlings.

It even matters how you prune your trees. Starlings prefer large, densely branched trees compared to more natural, open branching.

You can also find companies that make obstacles to roosting, spikes and coils and generally barbaric looking items, which can be installed on roofs and other problem areas.

Though the Columbia University web page on *Sturnus vulgaris* (Latin for starlike and common) is quite scientific, the author proposed, possibly tongue in cheek, "... maybe combining trapping with a pate production plant would make it (trapping) cost effective."

A weed or pest is only a plant or animal out of place. Isn't someone in Wyoming

Late winter in Wyoming means only a few more snowstorms until summer. However, it's still worthwhile looking for signs of spring.

On Washington's real birthday, I saw my favorite sign, my first mountain bluebird, up at Curt Gowdy State Park. At almost 1,500 feet higher than Cheyenne, the advent of spring should be two weeks behind there. But skimming over the brown grassland at about 7,500 feet in elevation in late February is normal for mountain bluebirds. The northern edge of their winter range is only southern Colorado, so they don't have to travel far to be here.

Bluebirds are insect eaters, seen typically perched on a fence post or the tip of a shrub before launching themselves after a flying delicacy, or just hovering like a hawk, waiting to pounce on an unsuspecting caterpillar. Are there any live insects so early? A fly-fisherman, standing on ice on the edge of open water at Granite Reservoir that same day, seemed to think so.

Bluebirds, like their close relatives the Townsend's solitaire and American robin, will eat berries, but as we rambled the nearly snowless open country dotted with pine, the mountain bluebird's favorite terrain, the berries seemed few and far between. The bluebirds were also, so maybe it all works out. By the time the rest of the crew, wintering as far south as Mexico, heads back, insects should be hatching.

Other people may count the robin as their sign of spring. I used to, until I started getting outside more in the winter and realized there are always a few around, especially in riparian thickets with berries. I don't think they really count for spring until they show up on our lawns.

In lower country, too low for mountain bluebirds, I look for western meadowlarks. Maybe it's more precise to say I listen for them, though most of the time they are singing in plain view on a fence post. Perhaps they make a good first sound of spring because it means it's warm enough to have windows open.

You'd think that migrating birds would have some inside information on weather so they wouldn't start back until conditions were perfect. This isn't so. I remember an April a few years ago when we took John Flicker, newly appointed president of National Audubon, on a field trip around Cheyenne. Fresh snow glinted on everything that bright morning, and there, like a drop of frozen sunshine, was a dead meadowlark, yellow belly visible in a snowdrift.

Birds here year round make changes in honor of spring. Some, like the male house finches and goldfinches, get brighter plumage. Others start hanging around in pairs. On our February ramble, up above a rocky outcrop, I saw two ravens fly looping patterns in perfect tandem, as if performing a three-dimensional skating routine. A pair of mallards has been swimming quite cozily in the ditch by my house since mid-February.

The noise level in the neighborhood has changed too, or maybe I'm just not bundling up my ears as much. The house sparrows and house finches are really making a racket. It seems a little early to be defending nesting territories but having experienced spots of warm weather over the last couple months, maybe they are itching for spring as much as anyone.

I know I shouldn't tempt fate by mentioning this, but does it seem to you that in town the ground has been even more brown than white this winter? Though it is nice not to have to shovel often or watch snowbanks turn black and then turn into slushy reservoirs at every curb, are we missing anything, besides recharging the aquifer, by not having a blanket of snow to protect the prairie for the winter?

I hope you were careful about what you wished for on Groundhog Day, especially in the Arctic where aerial photography over 50 years has documented how a warming climate is thawing the permafrost and increasing woody vegetation.

Audubon Alaska executive director Stan Senner (you may remember his visit to Cheyenne years ago when he was director of Birds in the Balance) is quoted in the December 2003 issue of Audubon magazine, "Almost every Arctic nesting bird will be affected in some way by climate change. The northward march of woody vegetation may extend the ranges of birds like the Arctic warbler. But birds that nest in open situations, like the long-tailed jaeger, may be limited by more woody vegetation."

Who knows how a warming climate will play out here? Maybe we'll have mountain bluebirds regularly on the Christmas Bird Count. There is already a report for one in Wyoming as early as February 3 and one as late as January 1. If the bluebird season extends any further, what will I do for a sign of spring?

My cousin (once removed, I think) sent pictures from northern Wisconsin today of snow hip deep. I grew a little wistful – until I remember spring in snow country is mud season. I appreciate again living in what was once known, perhaps only prematurely, as the Great American Desert.

# 162 Prairie Partners provides for plains birds

Thursday, March 18, 2004, Outdoors, page C2

Say "observatory" and we think of astronomy.

Say "bird observatory" and first thing to come to a birder's mind is Point Reyes Bird Observatory, established in California in 1965. However, an internet search last week gave me 27 more bird observatories in the first 50 hits.

The one closest to home is the Rocky Mountain Bird Observatory (formerly the Colorado Bird Observatory), founded in 1988 and headquartered in Brighton, Colorado, at Barr Lake State Park.

The purpose of a bird observatory is the conservation of birds. Last month, Rocky Mountain Bird Observatory biologist Tammy VerCauteren gave a presentation in Cheyenne about her work as coordinator of the Prairie Partners program which exemplifies the observatory's mission of research, monitoring, partnership, education and outreach.

The shortgrass prairie, the western part of the Great Plains stretching from Canada to Mexico, including eastern Wyoming, was overlooked when concern was raised over the decline of bird populations nationwide – until recently, when it was discovered that prairie species are declining the most rapidly.

Research documenting the decline doesn't in itself help birds. Research that shows what is causing declines still won't help unless the information is passed on to the people who make land use decisions.

In this case, 70 percent of the shortgrass prairie is privately owned, so Prairie Partners works not only with state and federal land agencies, but must work to reach farmers, ranchers and other landowners and managers.

Of course, land management suggestions need to be economically feasible to be taken seriously. With funding from various agencies and private foundations, Rocky Mountain Bird Observatory was able to publish "Sharing Your Land with Shortgrass Prairie Birds," a 36-page manual that describes the region's ecology, birds and management recommendations.

Some suggestions are as simple as not mowing at night during the two or three months ground-nesting prairie birds are resting on their nests. Others are more elaborate instructions for grazing strategies depending on whether the birds to be benefited prefer taller grass or no grass.

Mountain plover, a species once petitioned for listing as threatened or endangered, prefers to nest in heavily grazed, nearly bare situations and even in plowed fields. Rocky Mountain Bird Observatory offers to survey and flag plover nests two to three days before farmers cultivate. Advertising the Mountain Plover Number, 877-475-6837 (April 12- July 4), brought a good response

last year. Tammy expects even more calls as word gets out. Just lifting machinery or avoiding the nest by a few inches is all that is necessary.

My favorite win-win recommendation is directions for building an escape ladder that allows birds that have fallen into stock tanks to climb out instead of drowning and contaminating the water.

In addition to consulting on bird-friendly practices, Rocky Mountain Bird Observatory knows where the assistance and money is for habitat improvements. While most farmers and ranchers are familiar with the 20 or so Farm Bill programs and working with the Natural Resources Conservation Service, there are also private lands programs through the U.S. Fish and Wildlife Service as well as cost-sharing assistance from the Prairie Partners program itself.

Of course, the most convincing information comes from peers. Tammy has organized workshops hosted by ranchers in which friends and resource professionals meet on the land. In addition to grazing and farming operation suggestions, one might hear about economic diversification, such as tapping into the cultural and wildlife resources.

For instance, getting listed as a site on the Colorado Birding Trail helps make more people aware of the benefit of maintaining land in agricultural production and

brings revenue to rural communities offering services to travelers.

Rocky Mountain Bird Observatory's urban workshops, which bring people out to farms and ranches so they will understand where food comes from, have been immensely popular.

Also very popular is another Prairie Partners publication, "Pocket Guide to Prairie Birds." Measuring about three by four inches, it truly is a pocket field guide. Nearly all of 23,000 copies printed so far have been distributed for free, and Tammy is looking for funding to print more. Each of 86 prairie species has a clear photo, a range map covering the prairie states, a few of the most diagnostic markings needed for identification, and most importantly, a description of the species' favorite habitat and feeding practices.

A quick glance at the food icons on the bottom of each page shows that prairie birds are big on insects and rodents – the bane of farmers and ranchers. Perhaps we will be rewriting that song from Oklahoma about farmers and cowmen to read "Oh, the farmer and the plover (or harrier and the cowman) should be friends."

The folks at the Rocky Mountain Bird Observatory are definitely the friends to make when it comes to doing something for birds on the prairie.

152     CHEYENNE BIRD BANTER

**Rocky Mountain Bird Observatory [Current name: Bird Conservancy of the Rockies]**

The Rocky Mountain Bird Observatory, in addition to its conservation work in prairie, wetland and forest habitats, offers educational programs such as Women Afield (girls ages 12-18), Birds Beyond Borders (pairing schools in the US and Mexico) and field experiences for groups of all ages. Check the website at www.rmbo.org or contact them at 14500 Lark Bunting Lane, Brighton, CO 80603, 303-659-4348 (Education Hotline: 303-637-9220). For more information about the Prairie Partners program, contact Tammy VerCauteren at the Fort Collins field office, 970-482-1707 or tammy.vercauteren@rmbo.org.

# 163

Thursday, April 1, 2004, Outdoors, page C2

# Birdhouse site has everything under one roof

Ah, Spring! It's house-hunting season for couples seeking a place to raise their offspring, looking for a safe place to lay an egg, searching for the perfect birdhouse. The most comprehensive avian real estate listings are to be found at The Birdhouse Network website, www.birds.cornell. edu/birdhouse.

The birds that use birdhouses don't usually build their own homes, though they may make some minor alterations. If there's a shortage of hollow trees or former woodpecker holes because local humans have been too quick to tidy their yards and woods, a birdhouse will do.

Birds that insist on placing nesting materials within a structure, natural or unnatural, are known to ornithologists as cavity nesters. Birds such as robins prefer to build on ledges and branches and the grassland birds nest right on the ground, at the mercy of loose dogs and cats.

Folks at The Birdhouse Network have taken pains to distinguish which birds like what kinds of birdhouses and which locations will attract them and allow them to nest successfully.

Birdhouse Network range maps show the most likely species to be found in our area, though they may prefer various habitats: American kestrel, wood duck, eastern screech owl, red- and white-breasted nuthatches, northern flicker, tree swallow, violet-green swallow, black-capped and mountain chickadees, mountain bluebirds and house wrens. House sparrows and European starlings, nonnative species, are cavity nesters too, but do not need encouragement as they frequently steal housing from the other species anyway.

Because birds are particular about what they are looking for in a house, builders have learned to accommodate them, and so the network's website features woodworking plans for each species.

Some of the recommendations for building birdhouses are like building safety codes:

--Use untreated wood at least ¾-inch thick so the nestlings aren't poisoned and are insulated a bit from heat.

--Build an extended, sloped roof so that starlings or cats can't perch over the entrance hole and attack emerging birds.

--Leave inside walls rough or grooved horizontally so young birds can get a toehold when ready to climb out.

--Recess the floor and make sure it has drainage holes.

--Drill ventilation holes at the top of the sides of the birdhouse so nestlings don't cook on warm days.

--Dispense with the idea of an outside perch in front of the entrance hole. It gives predators a place to perch and reach in and grab young.

--Provide a means for opening a side of the birdhouse for monitoring or clearing out old material.

While all this wonderful information is available free to the public, The Birdhouse Network is actually another citizen science program of the Cornell Lab of Ornithology, like Project FeederWatch. Network participants monitor one or more birdhouses or nest boxes, as they are also known, and send the data in online.

Joining The Birdhouse Network requires a yearly participation fee of $15 (or $12 if you already belong to the Cornell Lab of Ornithology), and in return, in addition to the already freely available website information, you get a subscription to the quarterly Cornell Lab of Ornithology newsletter, The Birdhouse Network's newsletter and access to the continent-wide online database, which in the four years of the program, has accumulated over 40,000 nesting records, helping scientists learn more about birds whose populations seem to be declining.

Cornell advertises this program as perfect for families, so it has made taking part achievable by almost anyone interested in birds. The website provides step by step directions, a glossary of bird terms and even photos of nesting materials and eggs so you can figure out what species chose to nest in your birdhouse. If you aren't a carpenter, there's a list of sources of readymade shelters.

But most importantly, Cornell has made the network's website fun. Do download "Big Bluebird Movie," a short claymation feature created by elementary students in Caldwell County, Kentucky. It really is cool. Then there are the nest box cams, which are not very different from the photos excited new parents might send. And if you join the network, you can compare notes on nestlings with other members through an e-list.

The idea that wild birds might condescend to use a house provided by us is, I think, a bit of fulfillment of the need we have to feel connected to other animals. Sometimes the urge to attract wildlife, especially by feeding elk and deer, is detrimental to them, but in

2004

153

the case of birdhouses, with the decline of natural cavities due directly and indirectly to human activities, it's the least we can do.

Historically, Cheyenne, built on the nearly treeless prairie, would have had cavity nesters only in the cottonwoods along the creeks.

Now we have a backyard forest already decayed enough to provide for a few chickadees and nuthatches, but birdhouse developments

are a kind of urban growth hard to disagree with.

# 164 Thursday, April 15, 2004, Outdoors, page C2
## Young chickadees may be changing age-old songs

Black-capped chickadees were the stars of the storytelling at the Cheyenne-High Plains Audubon lecture last month.

Dave Gammon, a doctoral candidate in the biology department at Colorado State University, was the storyteller. As all graduate students do, Dave had had the opportunity to ask a question and investigate possible answers and was now ready to tell his story.

To begin with, in deference to his major professor's expertise, he chose to study chickadees. While Dave observed a captive specimen in the lab, it proceeded to sing a variation on the standard "fee-bee" tune that they are known for.

Reading the literature regarding black-capped chickadees, Dave discovered a study that showed that all across the country, they have one song that sounds pretty much the same everywhere, except in certain pockets.

Dave discovered in Fort Collins, Colorado, the males have three songs (only males sing). Where others sing the standard fee-bee, these birds have added an introductory syllable for a second song Dave describes as "fa-fee-bee," or sometimes a third, "chick-a-fee-bee."

All up and down the Poudre River corridor, full of trees essential to chickadee habitat, the songs are similar, though there is a noticeable variation from northwest to southeast. However, out on the prairie, in isolated islands of trees in small towns and on ranches, chickadees have added additional introductory notes to the "fa-fee-bee" song.

A lone bird on a ranch near the Wyoming border, which Dave recorded and nicknamed Ivan, was singing half a dozen introductory notes. To some extent, Dave recorded something similar among the chickadees at Guernsey in eastern Wyoming. Why does this happen and how does it happen? These were the questions Dave set out to answer.

He employed about 50 volunteers, who helped capture songs with dish microphones, and the good will of more than 20 landowners. He was able to incorporate and replay samples of those songs for us in his PowerPoint presentation as well as depicting them graphically as sonograms – lines representing the pitch, depth and length of sounds.

In chickadee culture, Dave said, the males are the first to rise. They sing without much notice of other males until they realize the females are awake. Then they stop abruptly and get directly to the mating business. Later in the day, singing is more a matter of declaring territorial boundaries.

Perhaps having a repertoire of more than one song type helps these chickadees communicate better. Many songbirds have more than one song and scientists seem to think it's an advantage.

Dave tested to see if different songs were reserved for females, the males' way of showing off, but could find no statistical evidence.

Perhaps defending males would match particular songs of aggressors or vice versa, sort of a "Your Mama" insult competition, escalating until fisticuffs – or at least wing beating – occurred. But unlike other bird species, there was no significant statistical difference.

Perhaps, thought Dave, these changes in chickadee song are merely accidental, the result of young birds making mistakes and never being corrected. Does humanizing that idea make parents responsible for the beginning of heavy metal music?

A chickadee nestling, Dave said, is born in a cavity of a tree, insulated from noise. During incubation and after hatching, he is unlikely to hear his father sing near the nest because it would attract predators. After a few weeks, the youngster leaps from the nest, never to return, and moves one or two kilometers away where he

stays the rest of his life.

Normally, in prime chickadee habitat, where the woods stretch for miles, wherever the young chickadee lands, he will be surrounded by chickadee mentors. If he makes singing mistakes, and he will – Dave has recorded juveniles really jazzing things up – he'll learn to conform.

But if any young chickadees ever disperse as far as old Ivan's lonely place, they'll probably wind up sounding much like him. What would their mothers think if they knew!

Apparently, there are other pockets of subversive chickadee song in an example of convergent evolution: Martha's Vineyard, Massachsetts; Puget Sound, Washington; Fort Lupton, Colorado; besides Guernsey. The only Cheyenne chickadees Dave found were mixed pairs of mountains and black-caps – another interesting conundrum.

And then one of the audience members, visiting from Casper, thought maybe her backyard chickadees might also sing "fa-fee-bee." Dave's eyes lit up.

The new questions are: How widespread is this phenomenon? How long ago did these breaks from the standard "fee-bee" occur? Will a multiple song repertoire eventually prove to be advantageous to chickadee survival and population growth? What new variations will this year's hatchlings come up with?

Dave would like to squeeze in one more chickadee field season. We hope wherever he lands his first job after earning his degree, it's in black-capped chickadee habitat.

# 165

Thursday, April 29, 2004, Outdoors, page C2

## Companions, field guides help with bird identification

I am a social bird watcher, either out on an Audubon field trip or with family and friends. If I see a bird by myself, it's usually in relation to some other activity, like the opportunity I had to hike the nature trail at Hynd's Lodge up at Curt Gowdy State Park when I gave my son a ride for a Scout event a couple weekends ago.

With husband Mark also away, I found myself on my own. So, having heard various glowing reports of waterfowl sightings at Wyoming Hereford Ranch Reservoir Number One, I decided to see for myself, by myself. I didn't even bring the dog who accompanies me on so many rambles.

My first stop at the west end of the reservoir gave me three American white pelicans on the island, surrounded by a flock of small indistinguishable birds standing in the shallow water.

The lighting was bad because the sun isn't very high at 8 a.m. mid-April and was shining in the spotting scope. The wind joggled it too and made my eyes tear. Apparently, the birds were sleeping with their heads tucked in, making identification impossible – for me anyway.

Normally, I would turn to other birders with better optics and more experience and say, "So, what do you think they are?" And then I realized, I hadn't even remembered my field guide. If you identify something fairly unusual on your own and want to report it, everyone else will examine your credibility before accepting it. That's why it's nice to bird in a group – plenty of backup.

It will be too bad if that flock with the pelicans turns out to have been some unusual shorebird. Most likely they were Franklin's gulls, including black-headed adults and a lot of pale-headed immatures – but I wouldn't bet my meager reputation on it. Don't even ask me about the duck silhouettes I saw.

Along the north side of the reservoir, at a better angle to the sun, field marks became useful identification tools again. There were northern shovelers (green heads and bright chestnut brown sides), northern pintail (white neck streak), American wigeon (wide white stripe from bill to crown), green-winged teal (red head with green over the eyes like a pair of fancy shades) and blue-winged teal (white crescent on each side of the face).

The passel of double-crested cormorants flying looked as streamlined as double-barreled shotguns with wings. One American coot flashed its white bill, but with its unique, chicken-like shape, I didn't need that field mark. An American avocet, white with pink and black markings, was stalking prey in the shallows on its long, thin legs.

By the time I got to the dam, more wind was making the water increasingly choppy. I was able to pick out several western grebes even though the gray water and whitecaps nearly camouflaged them. Between glassing the water with binoculars from inside the car and getting out with the scope, I lost sight of a pair of what probably were common mergansers. A window mount for the scope would have been handy, except the car was vibrating too much in the wind.

There were other birds I could confidently identify. Killdeer and red-winged blackbirds were providing background music. Mourning doves were decorating the fence. A belted kingfisher flew a sortie overhead, as did a northern harrier.

Below the dam, more ducks flew up from the creek and Canada geese struck statue poses, only their long black necks visible, appearing like iron pipes jutting up from the thick grass. Several turkey vultures coasted low while two Swainson's hawks held onto their cottonwood perches as the wind buffeted them. No warblers, probably too windy or too early in the season, but most likely because I didn't have the 20 extra pairs of eyeballs we'll have for our Big Count on May 15.

Farther up the road, the pastures were being flooded.

Mallards dotted the short grass like rocks. I noticed one, and then another, though also of mallard hen brown color and size, they had extremely long legs – and very long, down-curved bills. Hmm, godwits, willets, yellow legs? Nope. Long-billed curlews.

I checked my book when I got home. I'm certain what I saw wasn't any outlandish species of curlew. The cinnamon brown of the underwings when one of the birds flapped clinched the i.d. If the meadow remains somewhat undisturbed this spring, a curlew nest wouldn't be surprising.

I survived my solo field trip. It was only five miles from home. Of course, my most amazing sighting was while pulling into my own driveway. Two Eurasian collared-doves flitted from the top of one of our front yard trees and over the roof of our house.

Still considered uncommon new immigrants to Cheyenne, I won't tell you how many hours we drove around town looking for them for the last Christmas Bird Count. There's a possibility these birds were similar looking, ringed turtle doves, someone's escaped pets, so I quickly put out some millet on the back wall to see if they would come back for more study.

I'm glad I went out so I could see the doves and all

2004

155

the other birds. Next time though, I'll bring the book.

# 166
Thursday, May 13, 2004, Outdoors, page C2

## Get gossip on Wyoming birds by phone, web

Much of spring bird migration is invisible, especially the species that fly at night – and especially if your bird-watching is confined to your windshield as you commute back and forth to work.

Obvious migrants like robins are visible flocking on lawns. Grackles, shiny black with iridescent heads and long tails, have been clustering in blooming crabapples and raiding bird feeders. Mourning doves are pacing the pavement or clutching powerlines.

An astute birdwatcher can find Swainson's hawks cruising Cheyenne's residential neighborhoods now that they've returned from wintering in Argentina. Two flew over our house today. I wonder if they've found a spruce tree to nest in nearby, or if the harassing blue jays and crows will be too much to put up with.

If you don't have time to scope out local reservoirs or glass the treetops carefully, much of the parade will pass you by unless you have a source for avian gossip. Traditionally, birders have used the phone to spread the word about rare and unusual bird sightings. Gloria Lawrence in Casper maintains the Wyoming Bird Hotline sponsored by Murie Audubon Society.

If you see something worth crowing about, say white-faced ibis or marbled godwit, you can leave a message including your name, phone number and when and where you saw the bird. Gloria then adds your bird and its location to the message she leaves for callers who check in periodically to see what's up.

A few years ago, mention of a snowy owl just north of Cheyenne brought birders in from over 100 miles away. It was a life bird for several people, the first time they had ever seen the species.

Now we also have the Wyobirds e-list for spreading the bird word. Internet-savvy folks groan when they think about subscribing to another e-list. If the membership is large and vocal, it can mean a lot of e-mail messages every day. But this is Wyoming, so there are fewer of us, and we're a bit taciturn.

Currently there are about 70 members. Messages average less than a dozen a week, with a few more during migration, so it's obvious there are a lot of "lurkers" – people who read but never write. Shyness may have to do with a subscriber's confidence in identifying birds correctly, but Wyobirds is not about snobbery.

Recently, a subscriber received kindly worded help identifying a confusing-looking bird in a photo he had taken and attached to his query. Two possibilities, both gray birds with white eye-rings, were deftly sorted out by list owner Will Cornell.

Will, a birder in Rock Springs, began the Wyobirds e-list in 2001, modeled after his experience in another state. An internet company provides the service for free (a small charge Will pays eliminates the advertising) so subscription is also free.

Will's description of the list is, "Wyobirds is a medium by which birders can post birds seen in Wyoming, upcoming bird related events, keep in contact with other birders, post photographs of birds in Wyoming, ask questions and keep abreast of bird related news. Basically, if it relates to birds in the state of Wyoming, it is welcome for posting to Wyobirds."

Over the last few weeks, postings have included American pipets and broad-winged hawks near Casper, a peregrine falcon south of Cheyenne, a common loon near Sundance, a snowy egret near Rock Springs, mountain blue birds nesting north of Cody and American white pelicans at Saratoga Lake.

The e-list can also work in reverse. Travelers post messages asking where to find a particular species they want to add to their life list or where to find good birding. Researchers ask for help and Audubon chapters post field trips.

Because Wyobirds is a public list, you can choose to subscribe and receive each e-mail as it is posted by other members, with each message identified in the subject line by "Wyobirds" in brackets, or you can go to the website and read the archives at your convenience without subscribing.

You can post a message using the Wyobirds e-mail address whether or not you subscribe and you can reply to other people's messages. However, your reply will only go to the one person. To make your reply part of the general discussion, you need to send it to the Wyobirds e-mail address.

Since Will is the owner of the Wyobirds e-list, he is the person who deals with the hosting company, arbitrates good taste and good manners, and who encourages the timid to post. However, it is all the subscribers who participate who give the e-list content and voice.

That means 70 potential reporters can help you stay connected and "see" migration this year – even if you are trapped at your computer too many of the daylight hours.

---

### Bird Info

Wyoming Bird Hotline: no longer exists

Wyobirds e-list: now a Google Group

# 167 Big Day brings wave of blue-gray gnatcatchers

Thursday, May 27, 2004, Outdoors, page C1

CHEYENNE – The blue-gray gnatcatcher was the bird of the day for members and friends of Cheyenne-High Plains Audubon Society last Saturday who were out on their annual Big Day bird count.

The 4-and-a-half-inch bird was abundant in the four major locations searched: Lions Park, Wyoming Hereford Ranch, High Plains Grasslands Research Station and F.E. Warren Air Force Base.

Blue-gray describes most of the gnatcatcher's feathers; however, it is a white ring around the eye and a white-edged black tail, held at a jaunty angle, that are the determining field marks. A bird of the eastern and southwestern United States, this species is considered uncommon in Wyoming, normally summering in the southwestern part of the state.

The date of the Big Day is selected to coincide with the predicted height of bird migration, or at least the closest Saturday to allow for the availability of observers. This year, Cheyenne chapter members were aided by birders from Casper, Laramie and Colorado. Nearly 30 people were either part of the main group or reported birds they saw on their own.

Despite the healthy turnout of birders, the total number of species was only 125, compared to 130 last year and 150 typical in the 1990s.

There were close to 30 species missing this year that were on last year's list. However, around 20 listed this time were not seen last time.

Black-capped chickadees were missing, as were several of the regular warblers – blackpoll, MacGillivray's, American redstart and yellow-breasted chat – making for a total of 10, rather than the 13 seen last year, out of a total of 27 warbler species that have been seen in Cheyenne over the last decade.

Speculation on reasons for lower species numbers include West Nile Virus, changes in observers and areas covered, timing not coinciding with the peak of migration and, of course, weather.

Three days before the count, Cheyenne had three inches of snow after several rainy days. The week before, temperatures were in the 80s following previous unseasonably warm weeks, which exacerbated drought conditions and speeded up some of the local phenology – the advancement of spring. However, the day of the count was mostly sunny with a high of 62 degrees.

The already fully leafed trees made it harder to find birds, though some types of trees were still not beyond the budding stage.

One of the highlights of the day was finding a palm warbler in the pines at the research station. Considered a vagrant in the western U.S. and rated rare in Wyoming (on a scale from abundant through common, uncommon, rare to casual visitor), it evidently doesn't always take the most direct route from its wintering grounds in the far southeastern U.S. to its breeding grounds across Canada.

Two other vagrant species counted were the worm-eating warbler, also an eastern species, and the great-tailed grackle, a bird the size of a magpie.

Another rare species observed was a dunlin, one of the smaller shorebirds, which has records of observations across the state during spring and fall migration.

Other species of note for this year are the uncommon migrants: greater scaup, broad-winged hawk, black-bellied plover, red-necked phalarope, northern waterthrush and white-throated sparrow.

Other uncommon birds, which have records as summer residents, include Virginia rail, long-billed curlew, red-eyed vireo, chimney swift, plumbeous vireo and clay-colored sparrow.

Not all observations were of birds. A red fox was at the entrance of the Cheyenne Botanic Gardens greenhouse at 6 a.m.; a skunk trundled out from underfoot at the Wyoming Hereford Ranch – and didn't stop to spray; and half a dozen mule deer greeted birders at the gate to the research station.

So downy they appeared to be wooly, two great horned owl youngsters, also at the research station, scrutinized bird watchers from a low tree limb after the two adults flew.

Results from the Big Day bird count usually find their way into various reports compiled by organizations such as the American Birding Association and the Wyoming Rare Bird Committee. The more often unusual species are reported, the more likely range maps will be revised in future North American field guides and bird atlases.

For local Wyoming Important Bird Areas, the Wyoming Hereford Ranch and Lions Park, Big Day observations confirm their importance to migrating birds. As data is collected from year to year, understanding of sometimes elusive avian animals continues to improve.

## Big Day Bird Count species list

Cheyenne Big Day Count, May 15, 2004

125 species

l – Lions Park, Wyoming Important Bird Area

h – Wyoming Hereford Ranch (reservoirs and home ranch), Wyoming Important Bird Area

r – High Plains Grasslands Research Station

f – F. E. Warren Air Force Base

o – species seen only outside of the four locations listed above

|  |  |
|---|---|
| l h f | Canada Goose |
| h f | Gadwall |
| h | American Wigeon |
| l h r f | Mallard |
| l h f | Blue-winged Teal |
| h | Cinnamon Teal |
| l h f | Northern Shoveler |
| h | Northern Pintail |
| h f | Green-winged Teal |
| h f | Redhead |
| f | Ring-necked Duck |
| h | Greater Scaup |
| h f | Lesser Scaup |
| f | Bufflehead |
| l h f | Ruddy Duck |
| l | Pied-billed Grebe |
| h | Eared Grebe |
| l h f | Western Grebe |
| h f | American White Pelican |
| l h | Double-crested Cormorant |
| l h f | Great Blue Heron |
| l f | Black-crowned Night-Heron |
| h f | White-faced Ibis |
| l | Turkey Vulture |
| r | Cooper's Hawk |
| r | Broad-winged Hawk |
| h f | Swainson's Hawk |
| h | Red-tailed Hawk |
| r | American Kestrel |
| r | Merlin |
| h | Virginia Rail |
| f | Sora |
| l h f | American Coot |
| h | Black-bellied Plover |
| l h f | Killdeer |
| h | American Avocet |
| h | Lesser Yellowlegs |
| h | Willet |
| l h f | Spotted Sandpiper |
| h | Long-billed Curlew |
| h | Marbled Godwit |
| h | Baird's Sandpiper |
| h | Dunlin |
| h r | Wilson's Snipe |
| h f | Wilson's Phalarope |
| h | Red-necked Phalarope |
| l h | Franklin's Gull |
| h | Bonaparte's Gull |
| l h f | Ring-billed Gull |
| h | Forster's Tern |
| h f | Rock Pigeon |
| l h r f | Mourning Dove |
| h r | Great Horned Owl |
| o | Burrowing Owl |
| r | Common Poorwill |
| l | Chimney Swift |
| o | Broad-tailed Hummingbird |
| h f | Belted Kingfisher |
| l h | Downy Woodpecker |
| l h f | Northern Flicker |
| l | Western Wood-Pewee |
| o | Least Flycatcher |
| o | Cordilleran Flycatcher |
| h | Say's Phoebe |
| h | Western Kingbird |
| h | Eastern Kingbird |
| l | Plumbeous Vireo |
| l | Warbling Vireo |
| h r | Red-eyed Vireo |
| f | Blue Jay |
| r f | Black-billed Magpie |
| l f | American Crow |
| r f | Horned Lark |
| l | Tree Swallow |
| l h | Violet-green Swallow |
| l | Northern Rough-winged Swallow |
| h | Bank Swallow |
| l h f | Cliff Swallow |
| l h f | Barn Swallow |
| r f | Red-breasted Nuthatch |
| r | White-breasted Nuthatch |
| h r f | House Wren |
| l | Ruby-crowned Kinglet |
| l h r f | Blue-gray Gnatcatcher |
| l h r f | Swainson's Thrush |
| l h f | American Robin |
| l h | Gray Catbird |
| l h f | European Starling |
| l | Orange-crowned Warbler |
| l h f | Yellow Warbler |
| h f | Yellow-rumped Warbler |
| l | Black-throated Gray Warbler |
| r | Palm Warbler |
| h | Worm-eating Warbler |
| h | Ovenbird |
| l | Northern Waterthrush |
| l h | Common Yellowthroat |
| l | Wilson's Warbler |
| l | Western Tanager |
| l h | Green-tailed Towhee |
| l h r f | Chipping Sparrow |
| l h | Clay-colored Sparrow |
| l | Brewer's Sparrow |
| r | Vesper Sparrow |
| l h r | Lark Sparrow |
| r | Lark Bunting |
| r | Savannah Sparrow |
| l h f | Song Sparrow |
| h | Lincoln's Sparrow |
| f | White-throated Sparrow |
| h f | White-crowned Sparrow |
| l | Black-headed Grosbeak |
| o | Lazuli Bunting |
| l h f | Red-winged Blackbird |
| h r f | Western Meadowlark |
| l f | Yellow-headed Blackbird |
| o | Brewer's Blackbird |
| l h f | Common Grackle |
| f | Great-tailed Grackle |
| h f | Brown-headed Cowbird |
| l h | Bullock's Oriole |
| l h f | House Finch |
| l h r f | American Goldfinch |
| o | Evening Grosbeak |
| l h f | House Sparrow |

Ty Stockton His regular column, "The Great Outdoors" page C2

# I'm not much of a birder

## Saturday's Big Day Bird Count was more of a leaf count

As we walked through Lions Park Saturday looking for birds for the Big Day Bird Count, Barb Gorges or one of the other members of the Cheyenne High Plains chapter of the Audubon Society would point up into the trees and call out excitedly.

"Ooh, a purple-spotted grackle-flicker!" one would say. Sometimes someone would exclaim, "Wow, there's a yellow striated gnarly-beak!" Once, I'm pretty sure another birdwatcher pointed out a bunting-bellied warblefoot.

Each time someone noticed a bird, I'd peer up into the branches looking for the object of interest. All I could ever see were leaves. Sometimes, I'd look up into the trees, then realize I had no idea what I was looking for, so I would begin frantically flipping through the pages of my field guide. The names in the book would confuse me, so I couldn't remember if I was looking for a black-capped grosbeak or a gross-capped blackbeak.

"Yeah," I'd say each time. "What a neat bird. There's another one I can add to my life list.

For a beginning birdwatcher, knowing the terminology is the first step to acceptance by the other birders. After only a couple of minutes with the group, I realized a life list was a collection of names of the birds the birdwatchers had seen at one time or another. A bird doesn't have to be rare or spectacular to go on a life list.

On the other hand, a "life bird" is a rare specimen. It's one a person can't expect to see more than once or twice in a lifetime, unless the birdwatcher travels to far-off, exotic lands, like southern Colorado.

I made my share of mistakes early in the day. Once, after misinterpreting what the other birders were saying, I pointed to what turned out to be a very common house finch and said, "Hey, there's a live bird over there."

Barb and her husband Mark were very patient with me. Barb explained, in a low voice, that the bird was indeed alive. Then she told me how to tell that the bird was a house finch and not a more exotic specimen.

"That's good, though," she said. "Point out all the birds you see, even the common ones. We might not have counted them yet."

I thought I was doing a pretty good job of disguising my ignorance, but when a black-crowned night-heron flew over, I realized I still needed to work on my credibility.

"Black-crowned night-heron," I said, pointing to the passing bird.

Before looking up, another birder said, "No, that's a … Well I'll be darned. It is a black-crowned night-heron."

The pats on the back and the looks of incredulity on the birdwatchers' faces caused me mixed emotions. On one hand, I could tell they were amazed I actually knew what a night-heron was. But on the other hand, they were genuinely pleased with me and my identification skills. I finally started to feel that I was adding something to the expedition

Our next stop on the journey was Holliday Park. We only paused a few minutes there, but it was long enough for me to make another correct identification

"Redheads," I said as a few ducks swam past.

Again, I was rewarded with 20 amazed faces.

After Holliday Park, we went to the Wyoming Hereford Ranch. From a long way out, I could tell that a few of the wading birds were American avocet, and I said as much.

"Way to go, Ty," Barb said. "I think you're getting the hang of this.

But it turned out I wasn't.

There were more birds on the water I couldn't identify than those I could. I recognized pelicans and mallards, but I couldn't make out the coloring on a flock of teal.

"Cinnamon teal!" four of the birders chorused in unison. I looked at my binoculars, realizing this kind of outing requires a better magnification than a 10-power set of lenses.

As we stood on the bank of the lake, more and more shouts of identification rose up. Like at Lions Park, they all blended into each other. I found myself searching the

water and my field guide for red-breasted phalaropes, pen-striped chukar-dumpers and black-headed yellow birds.

Barb, bless her heart, noticed my confusion and wandered over.

"It takes some practice to learn all these birds," she said. "You can't become an expert birdwatcher overnight."

I guess not. But being surrounded by people who all knew more about birds than I'll likely ever know, I didn't feel like a complete idiot. They were all helpful, and they all tried to point out birds to me at one point or another. Despite my ignorance, they didn't make fun of me or try to avoid me

Now if you'll excuse me, I need to go study up for my next birdwatching excursion. I need to figure out what the heck an indigo-collared rufus-bloke is.

Ty Stockton is the outdoors editor at the Wyoming Tribune-Eagle. He can be reached by ….

# 168 West Nile virus frightens letter writer

Thursday, May 17, 2004, Outdoors, page C2

My bird phone calls this time of year usually concern identification of migrants or the inconvenient nesting of robins. Two weeks ago, however, a caller said she had received an unsigned letter from a neighbor asking her to stop feeding birds. The anonymous writer worried that birds concentrated around a feeder will contribute to the spread of disease, including West Nile virus.

While I abhor unsigned criticism, the letter writer did bring up points worth examining.

First, various bird diseases have been transmitted at feeders, including salmonella, and some of those are deadly for birds and or transmissible to people. That is why it is important to keep feeders clean and to quit feeding for several weeks if any sick birds are

observed – whether lethargic or with facial tumors or other growths or sores.

My caller estimated she feeds five pounds of seed a day, so I said she should probably clean once a week. Scrubbing feeders with a solution of one part bleach to nine parts hot water is the usual advice.

Raking up seed hulls and any seed that has been rejected is important too.

Buying sunflower seeds already hulled or making or selecting mixes minus unpopular seed such as milo can reduce debris.

Nectar feeders need to be cleaned every few days so that hummingbirds don't succumb to the lethal molds that can grow so fast in sun-warmed sugar water.

Summer bird feeding, as Francis Bergquist points out in an article in the May issue of Wyoming Wildlife magazine, is more for the pleasure of the bird watcher than the needs of the bird. Because birds have plenty of natural food sources available, stopping feeding for the summer is not a problem.

Providing a bird bath instead of food could be beneficial and entertaining, but may further antagonize my caller's neighbor because shallow, stagnant water is where Culex tarsalis, the mosquito responsible for Wyoming's West Nile virus woes, leaves its larvae to hatch. Dumping out the bird bath water every three or four days will kill the larvae before they hatch, but other scummy organisms may need scrubbing.

At last week's Cheyenne-High Plains Audubon Society meeting, guest speaker Terry Creekmore, vector-borne disease coordinator for Wyoming Department of Health and head of the state's West Nile virus surveillance program, shed additional light on birds and the virus.

He said infected mosquitoes can infect birds, people and other animals when they bite. Birds are bitten on bare skin around their eyes and feet or under their wings. Rarely do human West Nile virus cases come from

anything but mosquitoes.

Terry said certain birds, for the few days the virus is active in their bodies, can infect the mosquitoes that bite them. Laboratory studies of 30 bird species show several that can harbor enough of the virus to pass it on to mosquitoes: crow, blue jay, grackle, house sparrow, house finch, robin, red-winged blackbird, mallard and starling. These are some of our most common city birds.

Gus Lopez, director of the Laramie County Health Department, who also attended the meeting, was quick to point out that human infection rates are dependent on numbers of mosquitoes. With Laramie County's BTI-based larvicide spraying program and education efforts, the prevalence of the virus can be kept low, though not eliminated.

There is a slim possibility that West Nile virus, along with other diseases, can infect a person handling a sick or dead bird. To dispose of a dead bird, use a plastic bag like a glove, pick up the bird and then pull the bag over it, tie it shut, seal it in another bag and put it outside with your other refuse.

To find out if the state wants to analyze your dead bird and how to submit it, call 1-877-WYO-BITE. At

Wyoming's West Nile virus website, www.badskeeter. org, check out all the statistics and advice.

Basically, the risk of getting seriously ill or dying from West Nile virus decreases with the increase in funds available for spraying in the area, with increasing dryness of the weather and with increasing distance from rivers, creeks and irrigation water. Young and healthy people are least susceptible to the fever and encephalitis that may result from infection. Bird feeding is not listed as a hazard.

In Wyoming, people are at more risk of death or injury from a traffic accident than from death or illness from West Nile virus. In 2003, 165 people died and 6,248 were injured in more than 16,000 traffic accidents. In that same year, only nine people died of West Nile virus out of 392 cases reported.

Just as you buy a car with safety features, wear your seat belt and drive safely, you can look for and eliminate standing water in such places as clogged rain gutters and avoid spending the evening mosquito hours outdoors or wear protective clothing and DEET insect repellent.

What about the birds themselves? Crows, ravens, magpies and the different

species of jays seem to have the lowest survival rates. However, in the wild, any evidence of dead birds is quickly eaten, so the total effect of the virus is not usually directly measurable.

However, last summer, radio-collared sage grouse being studied for the impact on their populations of coalbed methane drilling inadvertently became subjects of an impromptu West Nile virus study when researchers found them dead or found their empty radio-collars. This summer, $1 million has been granted to begin a three-year study of the relationship between sage grouse, West Nile virus and the stagnant waters of coalbed methane discharge ponds.

The future looks brighter. The virus first appeared in New York City in 1999 and in Wyoming in 2002, but already there is a horse vaccine, and introduction of a human vaccine is only a couple years away.

Hopefully, birds that survive West Nile virus will be able to pass their antibodies on to their young so that eventually equilibrium will be established as it has in Africa, where the virus was first isolated in 1937. I predict a plethora of wildlife study topics for graduate students for quite some time.

# 169 Hybrid energy can improve bird watching

Thursday, June 19, 2004, Outdoors, page C2

**An electric engine makes it easier to sneak up on wily wildlife.**

My husband, Mark, and I have found the perfect car for birdwatching, the 2004 Prius, Toyota's second-generation hybrid energy car.

After observing one on a field trip in January, coincidental to realizing the advancing age of our family fleet, we ordered in March and were able to take possession of our own Prius just before the Big Day bird count mid-May.

The five-passenger, mid-sized, moderately priced Prius has aerodynamic body design, front-wheel drive, four-cylinder gas engine and an electric motor which kicks in whenever necessary to provide more power. You don't plug the car in to recharge the battery for the electric motor. Instead, energy captured from coasting and braking does that. A computer coordinates everything.

When the car stops, so does the gas engine. This is disconcerting at first, but ideal for using a window-mounted spotting scope – no engine vibration. When only the electric motor runs, you could sneak up on wildlife – or pedestrians if you aren't careful.

Mark and I haven't owned a sedan-styled vehicle in 20 years, so we were pleased to find the folding split rear seat and hatchback made it easy to stow our spotting scope when it is still attached to the tripod.

The window glass is solar filtering but doesn't appear to interfere with using binoculars when it's too cold and windy to open the windows.

The top-of-the-line Prius features a navigation system, but we're pretty good with a map and the odometer, so we didn't opt for that. What will be handy is the $1500 IRS tax credit for alternative energy vehicles.

I've ridden in the back seats a few times now, and they seem comfortable. With four doors, carpooling is easy.

The birdwatching hobby, once expanded beyond the backyard, can include a lot of driving, so the Prius improves the number of miles we can drive per gas dollar, compared to our eight-year-old Ford Explorer SUV which gets 20 miles per gallon on the highway with the wind, and our 12-year-old Plymouth Voyager minivan, which averages 22 mpg in town and, perversely, hits 30 only when it hauls six people and their luggage over mountain ranges.

While the EPA rates the Prius at 60 mpg in the city and 50 on the highway (city driving charges the battery better), real life may be less. Turning on the heat or the air conditioning will also lower mileage. The other day, traversing Cheyenne from the southeast to the northwest corner, the in-dash computer clocked me at 47 mpg, but 55 mpg on the return, evidently downhill, trip.

On the Big Day, a couple of birds gave the Prius a splat of approval. Certainly, any vehicle that can cut gas usage will also gain the approval of birdwatchers and other wildlife enthusiasts who deplore the rampant drilling in prime wildlife habitat. The Prius also has a near zero rate of emissions.

Hybrid vehicles are not going to be the ultimate solution for breaking our dependence on oil which, in addition to habitat destruction, is partly responsible for wars. Hydrogen fuel cells may be an answer, except I've heard the source of hydrogen is expected to be natural gas.

Consumer preference for alternative energy can force companies to proceed with innovation and forestall the collapse of our freedom of transportation. Mass transit is all well and good, until you want to check out waterfowl at a not so local reservoir at 6 a.m. I haven't heard of any regularly scheduled bus or train going to Gray Rocks Reservoir.

If consumers can "drive" the preferred shade of green car paint from avocado to imperial jade mica, turn airbags into standard safety features and make so many other safety improvements available, certainly we can make the change to safer energy – safer for birds, other wildlife and people.

A type of hybrid energy is also in the Honda FCX, and the technology is spreading to several other vehicles, the Ford Escape and the Toyota Highlander, both small SUVs. General Motors has hybrid city buses, and even the U.S. Army is working on hybrid tanks, according to a story in this paper last month.

I predict that in 10 years, hybrid energy will be made passé by even better alternatives.

The most useful characteristic of the Prius will have its greatest effect in big cities. This is its ability to shut off its gas engine when other vehicles would idle and resume movement at the touch of the accelerator. Imagine if 10 lanes of LA freeway were jammed with quietly waiting electric motors. There would be heard the sigh of a clean breeze, the song of birds – and probably the blare of car stereos.

If you see me still driving our SUV, it's only because I'm hauling something too grubby for the new car's upholstery, or the kids took the van to Scout camp or Mark has further to drive that day. It will take time to completely update our fleet.

Call me if you want the 50-cent Prius tour, since our local dealership sells them too fast to have one on their lot very long. Or look for an elegantly aerodynamic – and quiet – car at local birding hotspots.

2004

# 170 Desert travel offers many different birds

Thursday, June 24, 2004, Outdoors, page C2

*"....so it is in traveling, a man must carry knowledge with him if he would bring home knowledge."*

Samuel Johnson, April 17, 1778.

Travel broadens my knowledge, but it's also useful to know something about where I'm going, so I don't mistake a new bird for one I know at home or fail to appreciate new species.

Earlier this month, my family and I hiked the trail at the top of Sandia Crest, elevation 10,600 feet, overlooking Albuquerque, New Mexico.

We were surrounded by familiar birds and their songs and calls even though we were almost 600 miles south of home in Cheyenne: gray-headed junco, yellow-rumped warbler, mountain chickadee, brown creeper, red-breasted nuthatch, white-breasted nuthatch and Steller's jay. White-throated swifts and violet-green swallows reeled over the cliff edge.

The 800-acre bit of Rocky Mountain spruce-fir forest in the middle of the southwest desert is similar to the same forest in Wyoming, but at a higher elevation. Conversely, the spruce-fir forest occurs at a lower elevation in Montana than in Wyoming. The formula is this: a climb in elevation of 1,000 feet ecologically equals traveling north 600 miles and equals a drop in temperature of 3 degrees Fahrenheit.

If there are any birds peculiar to the crest (specialty birds as they are known to avid birders), I didn't know what they were. A bird checklist for the area or having a local birder along would have been helpful.

However, I did discover a peculiar tree, identified by an interpretive sign as "corkbark fir." I couldn't find it listed in the index of my Peterson field guide to Western trees. Does the scientific community change the names of trees as often as it does bird names? Luckily, in my 1979 edition of Ruth Ashton Nelson's "Handbook of Rocky Mountain Plants," she mentions corkbark fir is a southwestern variation of subalpine fir, the usual sidekick to Engelmann spruce.

The next day, in Albuquerque, at a mere 5,000-foot elevation along the bosque, or riparian woods, of the Rio Grande, we were in desert summer heat, and we finally found birds different from home. No sooner did we emerge from our car in the parking lot at the Rio Grande Nature Center State Park, than we saw a greater roadrunner snatch up a lizard and stride away, clasping the small, pale body in its long, strong bill.

There are no records of roadrunners in Wyoming. According to my newest field guide, they are not found north of southeastern Colorado.

While the nature center's pond was full of whole families of the familiar – Canada goose, mallard, cinnamon teal, wood duck and even a pied-billed grebe on a floating nest – we were able to identify a green heron, a rare migrant in Wyoming, but a nesting species at the pond.

Black-chinned hummingbirds, considered rare summer residents in Wyoming and only in the western part of the state, were swarming feeders, completely oblivious to people a couple yards away.

We saw great-tailed grackles, a species confined to southern Texas and farther south in 1900, but which has expanded its range to 19 states, according to an article in the June issue of Birding magazine. This spring we had a report of three of them at F.E. Warren Air Force Base. Now that I've seen them in the flesh, or feather, it's obvious that they are much bigger than common grackles (18 versus 12 inches) and though the length of crows, they are quite slender.

Half of the doves flapping around Albuquerque are white-winged, another species apparently heading to Wyoming. The white-winged dove's tail is blunt compared to the mourning dove's. Its folded wings show white along the lower edge but are otherwise plain gray, whereas the mourning dove has black spots.

A traveling birder, Paul Lehman of Cape May, New Jersey, reported a white-winged dove in Burns last month. He also reported a mourning warbler southeast of Cheyenne, another bird with few records in Wyoming.

I looked up this new (to me) warbler and discovered it is nearly identical to its western counterpart, MacGillivray's warbler, except it has a white eye-ring. Is this a case of us Westerners neglecting the study of eastern birds and failing to identify them properly when they appear, or is it the traveler putting a familiar name to a nearly familiar face in a strange land?

As it turns out, the warbler was also identified by several local birders this spring, and Paul Lehman has impeccable credentials. The fourth edition of the National Geographic Field Guide to the Birds of North America lists him as chief map consultant. It shows the mourning warbler's migration range skirting Wyoming a couple hundred miles to the east. I suppose that's one range map to be amended in the fifth edition.

At any rate, our visitor from New Jersey is a perfect example of Samuel Johnson's observation about carrying knowledge when traveling. I should probably invest in a copy of the "New Mexico Birdfinding Guide" for all the best places to bird and their specialties before visiting my mom again.

Why not find out what makes a travel destination unique? Otherwise, except for family reasons, one might as well stay home.

# 171 Mountain mud bothers birders, but not birds

Thursday, July 8, 2004, Outdoors, page C2

Ever notice how often precipitation in Wyoming is dangerous? When temperatures are cold it forms ice and drifts, stranding people and animals.

When it's warmer, precipitation comes as fog, thunderstorms, hail, tornados and floods. This summer, we've been treated to the unusual – days of gentle drizzle. Drizzle, however, can make a malicious, muddy mess of roads, which it did in the latter part of June, on the eve of our Audubon chapter's fourth annual birdwatching camp out.

Just before Mark and I were due to leave home Friday afternoon for Friend Park Campground on the west side of Laramie Peak in the Medicine Bow National Forest, our boys drove in from the Boy Scout camp on the east side. "Slick roads, Mom!" they reported.

A quick call to the Forest Service district office in Douglas confirmed what we suspected, Friend Park was out of reach, but the Esterbrook campground was possible. That's where the Wyoming Native Plant Society was meeting Saturday morning, and we had already planned to join them later for dinner.

So I called everyone I had a number for and gave them the change in plans and prayed anyone else would be too much of a fair-weather birder or have too much common sense to chance the mud on the road into Friend Park.

Our birdwatching goal was to check out several types of habitat with Bill Munro, the district wildlife biologist, as our guide. We were particularly interested in burn areas that might attract uncommon woodpeckers. Two years ago, when we first tried to schedule the annual camp out at Friend Park, we got smoked out by the Hensel fire.

Our route along Horseshoe Creek was full of birds, including the spotted towhee, green-tailed towhee and lazuli bunting. These are birds I see in my Cheyenne backyard only during spring migration, and they made for bright and enjoyable identification practice for the novice birder in our group.

Two families of pygmy nuthatches scampered around the ponderosa pine branch tips. At only four inches long, it's a wonder these sociable gossips manage to be so voluble while stuffing their bills with insects for their young.

Almost all large soaring birds overhead were turkey vultures – the few others remained unidentified. Flitting in the willows along the creek were yellow warblers, robins, goldfinches, broad-tailed hummingbirds and cedar waxwings.

At a bridge, we found an American dipper working the stream. Scrambling around, we were able to find both its old and new nests stuck to the underside of the bridge. The other bird wading the creek, doing an imitation of a dipper, turned out to be a spotted sandpiper.

Sunshine and a veneer of mud on the roads dogged us all the way to the scout camp at Harris Park where we found a bird-full place for picnic lunch.

Our older son, Bryan, the camp's ecology director this year, showed us a recently deceased bird. Jane Dorn identified it as a western wood-pewee. Yesterday it had been lethargic, Bryan said, and today it was dead. Its lack of fat reserves led us to think that while the previous week's cold weather had kept flying insect levels down, it may also have contributed to the starvation of this particular flycatcher.

However, another kind of flycatcher, the Say's phoebe, its nest stashed under the eaves of the dining hall, seemed no worse for wear. House wrens were acting suspiciously like they had a nest in the tool shed, and the western tanagers, which Bryan said were nesting over the nature lodge, could be seen busily feeding, making up for lost time.

The Hensel fire had crept into the upper end of the camp in 2002, so we followed the Black Mountain road on foot through camp and out into the forest to explore the burn.

Someone earlier in the day had complained that there was no good, leafless, time of year to bird the coniferous forest, but here there was nothing but black trunks and a carpet of green splashed with wildflowers. It still was not easy, especially when the sought-after woodpecker species are mostly black and their white markings would be shining like sunshine on charcoaled bark. We found hairy woodpeckers, but no three-toed or black-backed.

For three of us, the Black Mountain fire lookout became our goal and we concentrated on the ascent rather than birdwatching, but we didn't miss the blue grouse unconcernedly grazing just off the trail. On the way down, it was much easier to observe the treetops and notice a Swainson's thrush singing the cascading melody we'd heard all day.

Evening showers while we shared potluck back at the campground made the trip down to pavement afterwards a bit of a nailbiter. So for next year, we've set our sights on another good birding location but with more gravel on the road: the Sierra Madres, on whatever weekend follows the July 4th holiday.

For those of you managing water supplies during this multi-year drought, we will entertain suggestions to plan a camp out in your area, but we can't guarantee rain, only good birds, good food and good company.

# 172 Brown-capped birds cause cracked ribs

Thursday, July 22, 2004, Outdoors, page C2

Bird research can be hazardous to the researcher.

Just ask University of Wyoming zoology professor Dave McDonald's graduate student. She was poised high in the Snowy Range to ascend a crack in a rock with technical climbing gear to photograph a nest of brown-capped rosy-finches. It could possibly be the first such nest documented this far north.

Instead, a variety of emergency medical technicians and sheriff's deputies became intimately familiar with the nest location because one of the student's pieces of climbing protection came out of the crack and she was dumped onto the jagged boulders several feet below.

Hours later at the hospital, she was diagnosed and treated for a dislocated hip and two cracked ribs.

The trip out was excruciating for her and her rescuers. Getting to the ambulance involved negotiating a snowfield and about a mile of narrow trail, which meant the litter, centered on the trail, left the bearers to scramble over rocky or boggy terrain on either side, huffing and puffing in thin air at over 10,000 feet in elevation.

The adventure actually began the day before, July 10, when the nest was first found.

Dave sent out an invitation to the Wyoming birding community to join him in a search for brown-capped rosy-finches.

To his surprise, about 20 people between the ages of three months and 70 years old showed up that morning at the Forest Service's Centennial visitor center.

About half were wildlife biologists on their days off, but everyone was anxious to see this bird, even if it meant hiking steep trails and terrain.

For many of us, the brown-capped rosy-finch would be a life bird, one we'd never seen before.

The Snowy Range, remnant of higher mountains embedded within the lower but more extensive Medicine Bow Range, has been designated a Wyoming Important Bird Area primarily because of its importance to the brown-capped rosy-finch, which is listed as a declining and rare species in the IBA site description:

"The site is the only area in the state of Wyoming where Brown-capped Rosy-Finches occur and breed. In addition, the species is considered a 'species of local concern' within the Forest Service."

Alison Lyon, IBA coordinator for Audubon Wyoming, devised a survey form which was handed out to all the observers.

Dave made suggestions for areas to search based on his previous observations.

This species of rosy-finch locates its nests in cliffs near snowfields. The melting snow attracts an abundance of hatching insects that make good food for nestlings. Other times of the year, rosy-finches are seed eaters.

Mark and I were lucky enough to observe one brown-capped rosy-finch for 20 minutes as it foraged among the rocks high above Lake Marie.

Imagine a milk-chocolate brown bird with raspberry-flavored sides. It was a life bird for both of us.

One group saw American pipits skylarking but no rosy-finches. Two others had several sightings and a fourth group had only rosy-finch fly-overs, but nearly stepped on a pipit's nest full of eggs.

It was a fifth group lucky enough to come upon the rosy-finch nest that would become infamous. First, they observed several birds fly and twice saw one disappear near a rock face. Getting closer, they were able to determine the bird was flying into a crack, and they were able to hear the chicks peeping. But then a snow cornice above them broke, hurling ice and rocks, and they had to make a run for it.

Brown-capped rosy-finch

breeding populations center on the peaks of the Colorado Rockies, ranging north only a little way into Wyoming and south in winter only as far as northern New Mexico.

The black rosy-finch and the gray-crowned rosy-finch have much more extensive ranges across the west. By August, when their nestlings are on their own, all three species begin to gather in mixed flocks, eventually moving south or to lower elevations for the winter.

Banding data is just beginning to unravel the extent and timing of rosy finch travels. For instance, a gray-crowned banded in a yard above Lander one March showed up three days later at a banding station in Jeffrey City. Abandoned swallow nests in highway underpass tunnels near Laramie provide roosts after breeding season.

In addition to another survey next July, Dave is thinking about making a third attempt yet this summer to document the jinxed nest, but it won't be his grad student climbing up there.

She has plenty of rosy-finch literature to read and data to analyze from her field work with the other two species earlier this year, giving her time to heal completely.

CHEYENNE BIRD BANTER

# 173 Windmills safer for birds than other structures

Thursday, August 5, 2004, Outdoors, page C2

Birds and wind farmers think alike when it comes to maximizing the power of wind. Both are attracted to places like rims and passes where wind speed accelerates. Raptors (hawks and eagles) riding thermals and migrating passerines (songbirds) are the most likely species.

Unfortunately, birds may collide with the wind turbines and their towers. However, of the estimated 100 million to one billion birds that die in collisions with man-made objects every year, only one or two of every 10,000 are attributable to wind turbines, as of 2001. Communications towers and windows are much deadlier.

Greg Johnson, a wildlife biologist for Western EcoSystems Technology, Inc., an environmental consulting firm headquartered in Cheyenne, pointed me towards www.west-inc.com, where the studies he and his company have done on wind power and avian mortality are posted.

WEST has done studies in 11 states, but the one most interesting to me was for the Foote Creek Rim Wind Power Project built near Rock River.

In the first building phase, completed in 1998, 69 turbines were installed on 2000 acres of grass and shrub lands. Only 26 acres were disturbed by construction and five miles of road were built. Five meteorological towers were also installed.

Each 600-kilowatt wind turbine tower is 131 feet tall, and the blades have a diameter of 138 feet. Study plots were established at the base of each turbine and meteorological tower. Carcass searches were done every four weeks for all plots and every two weeks for half.

Results had to be adjusted for "removal bias" – how many carcasses were eaten or taken away by scavenging animals, and "searcher efficiency bias" – how proficient people were at finding dead birds or piles of feathers.

The study plots were salted with carcasses of non-protected birds such as pigeons, house sparrows and game farm birds, identified unobtrusively with duct tape, to see how many were found by searchers and scavengers. The searchers were 80 percent proficient. Scavenger success depended on the season.

Over the study's three years, carcasses of 122 birds of 37 species were found. The 69 turbines were responsible for 83, and the five meteorological towers killed 36 (three were unknown). Passerine species accounted for 112 birds (92 percent) and 36 of those were horned larks. Half the birds were probably migrants. Few mortalities occurred in the winter.

Overestimating bias, the average number of avian mortalities calculated was 1.5 per turbine per year. On the other hand, the meteorological towers each averaged 8.09 birds per year.

Mortality rates will obviously vary from site to site because of differences in bird abundance, species composition, landscape features and wind plant features. Right away though, one can see that making meteorological towers free-standing like turbine towers, rather than supported by guy wires, probably would reduce mortality.

In another study at Foote Creek Rim, researchers examined the avian mortality differences between blades manufactured with paint of high ultra-violet reflectivity, 60 percent, and the usual 10 percent of normal paint, but though birds can see UV light, that didn't seem to make a significant difference in this study. What did reduce mortality was larger, slower-moving blades.

One surprising finding of the first study was the number of bat deaths. One would think that bats, equipped with echolocation abilities, would be able to avoid towers and blades.

At Foote Creek Rim, 47 dead bats were found the first year, 18 the second and 14 the third year. Perhaps there is a learning curve. All were found during the summer, most in August. Considering bat hibernation and migration, that timing isn't surprising. But it is surprising that the turbines were responsible for all the deaths—none of the carcasses were found under the meteorological towers.

There's high bat mortality at other wind plants also. One hypothesis is bats send out echolocation signals less frequently while migrating. So how did they avoid the meteorological towers at Foote Creek Rim? Perhaps the explanation will be related to how the rotation of wind turbines befuddles radar.

Another question needing research: as wind plants begin to be developed in sage grouse habitat, will the increase in traffic have the same negative effect on them as it does in oil and gas developments?

Personally, I've been in love with windmills, with their symmetry and kinetics, ever since I spotted my first Chicago Aermotor in a pasture. Wind power has to be better for us and wildlife than bulldozed habitat, tailings piles, evaporation ponds and plumes of particulates.

To look at it another way, one of the Foote Creek Rim turbines can power at least 150 of our houses, according to the American Wind Energy Association calculations (www.awea.org), and probably more with Wyoming's notorious wind. How many more birds are killed by thumping into the windows of those same houses? But that's a topic for another day.

# 174 Timing will be everything for birding Alaska

Thursday, August 19, 2004, Outdoors, page C2

The last week of July, two reports came over the Wyobirds e-list, both indicating the first wave of fall migration – shorebirds – had already started passing through.

If plovers and sandpipers make it this far south by then, will there be any left in southeast Alaska when we visit Mark's brother, Peter, in Sitka a month later?

It's time to take the advice I give to other birdwatching travelers and study up on where we'll be going.

My first step is to query my Thayer Birding Software "Birds of North America" CD-ROM. I ask it to make lists of all the common birds in Alaska for two habitat types, Ocean Shore and Open Ocean. I don't expect the forest birds to be too different from those of Wyoming's mountains, but the seabirds are going to be all new to me, and I'm not very good at shorebird identification.

As I suspected, the Open Ocean list includes over a dozen birds I've never seen in person, only paged past in field guides: red-throated and Pacific loons, sooty shearwater, pelagic cormorant, common eider, three scoter species, three new gull species and the common murre. However, though they can all be observed off the Alexander Archipelago that makes up much of southeast Alaska, only a few are summer birds.

Our plans are to spend a couple days in a boat fishing, so if the Dramamine works, I may get to see another seabird, the auklet with the descriptive first name, "rhinoceros," which translates as "nose horn," not "large African mammal."

The Ocean Shore list includes about a dozen shorebirds from the Wyobirds reports. Looking in "The Sibley Field Guide to Birds of Western North America," I note that according to the range maps, some, such as the whimbrel, semipalmated plover and semipalmated sandpiper, migrate from northern Alaska through the Sitka area, while others, such as the greater yellowlegs, have summer or breeding ranges that include southeast Alaska.

It's difficult to tell by looking at the range map in Sibley if some species ever occur around Sitka. The entire North American continent is shown only as a one-inch square, and the jumble of island boundary lines makes it difficult to pick out narrow bands of color indicating season of residence.

The good old American avocet, an easy to identify shorebird here in Wyoming, was listed by Thayer as common on Alaska shores, but even after I put on my glasses, I couldn't see any color of occurrence anywhere in the state. At least the map for the wandering tattler shows a distinct band of yellow indicating a migration route from its breeding range in interior Alaska and the Yukon, through Sitka, to its wintering grounds along the California coast.

When exactly is migration for these species? Undoubtedly, it starts as early as July for the migrating shorebirds we see in Wyoming. Will that make late August late enough to see a winter species such as the black-legged kittiwake?

What I really need is a book like Terry McEneaney's "Birds of Yellowstone" for which he made charts showing in which habitats and which months a species might be seen.

From the descriptions available in the American Birding Association catalog and Barnes and Noble's book list, I don't see a book precisely answering my needs or with timely shipping.

What would procrastinators do without the internet? So, I go in search of a checklist for Sitka that will have the birds I might see in August.

Lonely Planet's guidebook to Alaska mentions Saint Lazaria Island near Sitka and its huge number of birds. Several ocean species nest in colonies on the cliffs. Online I find that it is part of the Alaska Maritime National Wildlife Refuge. No bird list on the website though.

At dogpile.com, a conglomerate of search engine results for "Alaska+bird" gives me the Alaska Raptor Center in Sitka and the Alaska Bird Observatory in Fairbanks, but no lists.

However, the Juneau Audubon Society has a checklist which I can print out. Juneau has similar habitat to Sitka and is only about 100 miles away. Unfortunately, the list gives no migration information.

The most common species listed for summer include birds I know well: mallard, bald eagle, blue grouse, Steller's jay, common raven, barn swallow, American robin, dark-eyed junco and pine siskin. In another category of common species are those also found in Wyoming, but which I'm not good at identifying – various shorebirds, warblers and sparrows, etc.

But happily, some of the distinctive common summer birds will be new to me: mew gull, glaucous-winged gull, Arctic tern, marbled murrelet, northwestern crow, chestnut-backed chickadee, winter wren and varied thrush. All of these are species I'm unlikely to see in Wyoming. I just hope it won't be too late in the season.

At least when it comes to identifying fish, I'll be traveling with my personal, professional fish biologist. As for bears, I'd rather not have to identify any. A few whales would be nice. Maybe even a pigeon guillemot. Wish me luck.

CHEYENNE BIRD BANTER

# 175 Alaska birds add to life list

Thursday, September 2, 2004, Outdoors, page C2

"It's easy to find. Stay to the right of that cruise ship until the entrance of the bay and then stay to the left. It's cream-colored with green trim," said the woman renting us her float house for two nights. We stood on the dock in Sitka, Alaska, our rented 18-foot skiff packed with provisions and fishing gear for our family of four plus Peter, Mark's brother.

Luckily, Peter lives in Sitka and has fished the sound quite a bit. Also, from topo maps, we knew Camp Coogan Bay was only six miles away.

We found the small, isolated frame house on a barge moored to a dock which was anchored to the shore of a cove bordered by the deep green of Tongass National Forest.

While fishing and kayaking the mile-long bay, belted kingfishers and bald eagles were our most common companions. Small birds floating on the calm waters were marbled murrelets, my introduction to the alcid family, a group of ocean bird species that only come on land when they nest, in colonies, on cliff ledges or in burrows or rock crevices.

On shore, the forest floor was steep and deep in wood at different stages of spongy decay, with little undergrowth except for a thick layer of moss. Walking the trail-less ground was like climbing a pile of well-upholstered couches. Invisible birds twittered at the tops of 150 to 200-foot-tall spruce and hemlock.

The day we chartered a fishing trip, we came in behind St. Lazarius Island, famous for its bird nesting colonies. I was able to glimpse two more members of the alcid family, pigeon guillemots and puffins.

Later, I realized there may be two species of puffins in the area, and I think what I saw (though I couldn't stomach binoculars) were the all-black bodies and colorful heads of the tufted puffins as they skimmed the waves, rather than the white bellies of the horned puffins.

Between tours of historical Russian and Tlingit landmarks, we hiked forest trails and I finally caught up with the chestnut-backed chickadee, another new species for me. It was easier to find in the muskeg bogs where stunted tree growth put the tops of lodgepole pines nearly eye-level.

On a particularly grueling hike up Gaven Hill we frequently paused for breath. It was one staircase after another, literally. Sitka-area trails seem to be either boardwalks across the bogs or wooden steps up mountains. Some tiny movement or sound caused us to examine shadowy tree branches and discover a robin shape which was actually a varied thrush, another new species for me, as was a winter wren we saw later.

I did spot one lonely robin in town after nearly a week – the same day I finally noticed starlings. The Northwestern crows were much more abundant. They

are a distinct species and caw with a discernable accent. I noticed them and at least one raven stockpiling food and treasures in rain gutters.

There are no house sparrows or house finches in Sitka. Peter said he gets fox sparrows at his feeder. I think I saw some of them down at the waterfront. In southeast Alaska, they are a very dark variety.

Gulls are an enigma to me. My pre-trip research had shown a possibility of glaucous-winged gulls, so I gave every gull a good look, but so many appeared to be murky-colored immatures or too far away.

Finally, I found a flock close in on a gravel bar and memorized the distinct markings of one bird: pink legs and red spot on lower mandible. These turned out to be perfect field marks for the herring gull – common on either coast. I should have been looking for an absence of black on the wing tips. However, I felt better after reading the note in the field guide which said herring gulls hybridize extensively with glaucous-winged gulls where ranges overlap, making them difficult to distinguish.

My shorebird identification skills are about on par with my gull abilities. One day, we saw a handful at the water's edge that Peter said were turnstones. Black turnstones have about the most distinguishable plumage of any shorebird that might be in Sitka, even in late summer, so I added them to my list of

new life birds.

My complete Sitka bird list is not very long, even though I had help from Peter. But I'm happy about every bird I found while on what was essentially a family vacation rather than a birding trip.

For those of you interested, the fish species we caught included pink, silver and king salmon, halibut, flounder, rockfish, Irish lords and lingcod, plus crabs too small to keep. We also found starfish, mussels, squid and scallops and sighted humpback whales, sea lions, lots of red squirrels, but, thank goodness, no bears.

If we're lucky enough to visit Sitka again, I hope we go earlier, maybe June. I'd sign up for a boat tour of St. Lazarius Island and look up local contacts in the American Birding Association directory.

Another trip might also give us a better feel for the normally cool and wet temperate rainforest climate. As it was, every day was sunny except one, and our first day the temperature hit a record-breaking 89 degrees.

Since our return home, Peter reports that it has started raining again, which, he says, makes the salmon happy.

*Cutline*

Barb and Bryan Gorges explore the Estuary Nature Trail at Starrigaven Bay in Alaska while on a family vacation to the southeast thumb of the 50th state [Hawaii was the 50th state. Ty must have written the cutline.] The boardwalk on the nature

2004

167

trail is much more intricate than the other boardwalks

the Gorges family encountered on their trip. Jeffrey

Gorges/courtesy

# 176 Warblers winging their way through on migration path

Thursday, August 5, 2004, Outdoors, page C2

Early mornings mid-August get a chill snap to them that foreshadows September and indicates warbler weather – warbler migration weather.

A trickle slowly builds through the last week of August. By then, you can stare at almost any deciduous tree and see the flutter of the leaves, branch by branch, as these small passerine birds hunt for insects and other arthropods.

The migration will continue into October. The last warblers to leave will probably be the yellow-rumpeds. They don't mind eating berries when the insects die off. The other warbler species are stricter insectivores.

The best time to look for warblers is early morning. They migrate at night and come to earth by dawn quite ravenous. They flit frantically, as if they've had three cups of coffee on an empty stomach. It makes them hard to track with binoculars, especially since they all seem to be shades of yellow, greenish yellow or olive green – the same colors as leaves losing chlorophyll.

Identifying fall warblers can be tough since the adult males are no longer in their distinctive breeding plumage, the young don't have all their adult feathers, and the females are so subtly marked, they tend to look all alike. But if you identify them as Wilson's warblers around here, you could be

right as often as fifty percent of the time.

The last weekend in August, our family attended the Rocky Mountain Bird Observatory member's picnic at their headquarters in Barr Lake State Park near Brighton, Colorado. One of the activities was visiting a bird banding station in the park.

With newly banded birds in hand, RMBO staff member Arvind Panjabi was able to compare male Wilson's of different ages. The younger the bird, the more yellow-green feathers are interspersed with the cap of black feathers on the top of its head. For the females, the cap is just a gray-green smudge.

Arvind didn't think the Wilson's warblers being caught in the mist nets that day were the ones that spent the summer in the mountains. He suggested that these were the birds that nested in Alaska and Canada, and the mountain populations migrate later. No one will know for sure until more banded birds begin to be recaptured at other banding stations.

The different populations of Wilson's probably winter in different areas as well. Some go only as far south as the Gulf Coast and Florida, and some are found throughout Central America. Other warbler species spread out into South America.

Another activity at the picnic was a talk by RMBO volunteer Bill Schmoker about learning to recognize bird

songs and calls. He claims bird songs aren't any harder to remember than snippets of popular songs, even bird calls of just one note.

It helps to see the bird which is singing or calling when learning new vocalizations. I had that opportunity to make a connection at the banding station when some of the Wilson's chipped loudly while being held. When released, they didn't fly far and continued their one-note chips from cottonwood branches overhead.

Back at home, with a window open one morning, my subconscious identified the same chip and sure enough, there was a Wilson's in the tree outside. However, even if it had never made a sound, I could have found it by following the stares of my two indoor housecats.

Warblers weren't the only species to be caught in the mist net while we visited. A young western wood-peewee modeled its cream-colored wingbars which will turn whiter with age. We were also afforded the treat (well, maybe you have to be a birder to enjoy it) of watching a Hammond's flycatcher, a petite bird, work to swallow a moth.

We have spring migration records for about 25 warbler species in Cheyenne from our Big Day bird counts, but we don't have a comprehensive count like that in the fall, and because there's no guarantee that what flies north

will fly the same route south, we can't suppose all the same species will be here now.

However, we do have observations accumulating through e-mail postings to Wyobirds. One e-mail posted last week by Ted Floyd, editor of the American Birding Association's magazine, after a visit to Cheyenne, listed yellow, yellow-rumped, Townsend's and MacGillivray's warblers and the common yellowthroat. Also, Vicki Herren, a Cheyenne-High Plains Audubon Society member, identified a Nashville warbler, considered a rare migrant here.

September is the height of the warbler season, so it isn't too late to get out and look for activity in the tree branches. By October, the show will be over except for a few stragglers and some of those berry-eating yellow-rumpeds hanging out as late as November.

The Sibley bird books are probably the best at elucidating the different species at this time of year. There's also the Stokes Field Guide to Warblers, though I haven't seen it yet myself.

But like not knowing the name of the driver on a country road who gives you a happy wave of the hand in passing, it isn't necessary to know a bird's name to enjoy that brief moment when it examines you with its bright black eyes before turning to clean another beetle from the branch.

CHEYENNE BIRD BANTER

# 177 Pacific coast down, Atlantic coast yet to go

Thursday, September 30, 2004, Outdoors, page C2

By the time you read this, I'll be back from the East Coast.

Only once before have I made it ocean to ocean in one year. That year [1979] I started out in spring as a naturalist-in-training on Staten Island, New York, then spent the summer in Wyoming as a soils tech in Rock Springs, and when the field season finished, drove up to Seattle to visit my aunt and uncle.

These same relatives now live in Philadelphia, the destination of this latest trip. My sister and I had asked our mother where we could take her to celebrate her 70th birthday, and she chose her brother's.

Since I travel these days with an eye for the birds, it didn't take long for the realization (perhaps it was with help from Aunt Pat) that Cape May, New Jersey, is only two hours south of Philly. Don't you just love the miniature geography back east? People measure travel by hours rather than miles because of the traffic congestion. If the East Coast had Wyoming's unimpeded highways, most of it would be within a day's drive of Philadelphia.

Cape May is a Mecca for observing migration. Located on the southern tip of New Jersey, it is a natural stopping place for birds during both spring and fall. Capitalizing on this, the Cape May Bird Observatory was established in 1975 to count the birds. In addition to research and conservation projects, it has extensive education and recreational birding programs.

Judging by the CMBO's advertising in birding magazines for its seasonal festivals, I would gather the height of spring migration is mid-May, and in the fall it's Oct. 29-31. Rats.

Our trip to Southeast Alaska in August was on the late side of migration up there, and September at Cape May will be on the early side. Someday I may be lucky enough to travel to bird and slip in visits to nearby relatives, rather than the other way around.

As far as I'm concerned, except for Cape May, the itinerary is totally up to Mom. From the looks of my internet search, the 16 antique stores should provide her plenty of entertainment anyway.

Cape May is apparently full of quaint Victorian-era architecture and is preserved intact as a National Historic Landmark City. Of course, local businesses capitalize on the fact. In addition, the local entrepreneurs have always encouraged their reputation as a seaside resort – for the last couple hundred years.

Though Southeast Alaska has been a destination for adventurous vacationers since the late 19th century, it seems only since the closing of its pulp mill more than 10 years ago has Sitka gotten more serious about the tourism industry.

When we visited the Alaska Raptor Center, we were astounded by the $12 per person entrance fee, and the new million-dollar visitor-office-rehabilitation facility with permanent staff of five and hundreds of volunteers. It's a far cry from Lois and Frank Layton's new, but just as effective, pole building flight barn outside Casper.

By the way, congratulations to Lois and Frank for being in the first group, including Curt Gowdy and Olaus and Mardy Murie, inducted into the Wildlife Heritage Foundation of Wyoming's Outdoor Hall of Fame earlier this month. The Laytons were honored for their 45-year commitment to bird rehabilitation.

At the ARC we joined a group of 50 or 60 cruise ship passengers bused in for a tour of the facility and a presentation with a resident, permanently injured, bald eagle. Then we were shepherded towards the gift shop.

Busload after busload all summer long gives the place a commercial air. At least the entrance fees help pay for the grand facility to accommodate so many visitors as well as public education and rehab of birds.

For a family like ours, familiar with facilities in Cheyenne, Casper, Laramie and Fort Collins, it seemed like overkill. But perhaps for the majority of the visitors, it was their first exposure. One should never discount the influence of 30-minute tours on the future welfare of wildlife.

The Cape May Bird Observatory, conveniently located in another tourist destination, may be similar. Operated by New Jersey Audubon (not affiliated with and predating establishment of the National Audubon Society), it does have a gift shop, according to the website www.njaudubon. org/centers. I just hope that at the end of the summer season they still have t-shirts my size.

In the next column you'll find out if I did actually get to Cape May – after all, it is hurricane season. However, Philadelphia itself has a good reputation for bird watching.

I don't expect to add a lot of birds to my life list as I did in Alaska. The birds of my Midwestern youth are pretty much the same species as in the east, except for the seabirds that might overfly the cape, though I still may pick up species I wasn't paying attention to 30 years ago.

Checking out the Cape May Rare Bird Alert website, I was surprised to see some familiar, but normally western, birds listed – American white pelican, American avocet and lark sparrow. Just visiting, like me. Just a long way from home for a short while.

# 178 Day at Cape May whets birder's appetite

Thursday, October 14, 2004, Outdoors, page C2

I have been to Mecca. But it's Cape May, where bird watchers raise their binoculars to the sky in exaltation of migration.

Midday Sept. 22 did not have a wave of raptors flying over the Hawk Watch platform that overlooks the Atlantic at Cape May Point State Park. I saw only a couple osprey and kestrels.

The best numbers at this southern tip of New Jersey come from passing cold fronts and this day was in the 80s and sunny. The trip, begun with my mother and sister to visit my aunt and uncle in Philadelphia, miraculously coincided with a lull in damaging storms generated by hurricanes.

Pete Dunne, in 1976 the first to start counting hawks officially for the Cape May Bird Observatory operated by New Jersey Audubon Society, relates the legendary day a year later, when unique weather conditions produced 21,800 hawks, almost all sharp-shinneds and broadwings, giving Cape May the nickname "Raptor Capitol of North America."

The cape, or peninsula, is formed by Delaware Bay on the west and the Atlantic on the east, making a funnel that gathers the birds on their southward trip into a narrow stream at Cape May Point before they reluctantly cross open water.

Clay Sutton, author of an article in this month's Birding magazine, published by the American Birding Association, relates another legend. In 1970, 24,875

American kestrels were counted in a single day! However, in 2001 and 2002, kestrels barely broke 5,000 for the entire count season, September 1-November 30, though they normally average about 10,000.

While kestrel numbers have declined overall, peregrine falcon counts have improved, from 60 in 1977 to a high of 1,791 twenty years later. Sharp-shinned hawks are the most common, averaging 28,512 per year.

Swainson's hawks average three, with never more than 10 counted in a year, so far. Being a western species (the hawk most commonly seen over Cheyenne in the summer), one has to wonder how they get so far off course.

A CMBO intern was on hand at the Hawk Watch platform to document any hawks and answer questions, but I was wondering about the white birds in the pond below and approached another birder for help instead.

Turns out he was also a CMBO intern, hired to count five days a week at the Avalon Sea Watch, a few miles up the beach. It was his day off.

The white birds were a mute swan (an invasive, non-native species that has somehow achieved protection under the Migratory Bird Act, I recently learned), great egret, snowy egret and a mixed flock of laughing gulls (no longer sporting their black head feathers of the breeding season), Forster's terns and herring gulls, the ubiquitous gull of my

Alaskan trip in August.

The Sea Watch was started in 1993 and runs mid-September to mid-December, with mid-October to mid-November being the migration peak. The most abundant species are red-throated loon, northern gannet, double-crested cormorant, brant, surf scoter, black scoter, white-winged scoter and parasitic jaeger.

My relatives and I did not make it to Avalon. But now I understand which migrations CMBO's late October bird festival targets. It would be a great time to visit again.

There's also the option of hiring a guide through the observatory. The price is reasonable if you have a small group to share it. It would be worth the help sorting over a dozen other birding hotspots in Cape May County, including national wildlife refuges, Nature Conservancy holdings, and state and private lands open to birders.

The CMBO has a long list of educational and recreational opportunities also, some of which are free.

The daily "Monarch Tagging Demo" coincided with our visit.

Staff and volunteers demonstrated how a small patch of orange scales on the forewing is brushed away and the tiny sticker is placed on the clear membrane, then folded over the edge. The sticker has a unique number plus contact information for the Monarch Monitoring Project.

Monarchs produce several generations over the summer,

but only the last one is of the individuals that will head south and hopefully make it to the mountains of central Mexico. Of thousands of tagged butterflies, about 30 have been caught there, and others have been documented in places in between.

Monarch spring migration is not as direct. The first generation to leave Mexico stops to mate along the way where milkweed is growing, and it is their progeny or later generations that make it the rest of the way north.

Afterwards, we stopped at CMBO's Northwood Center, buried in the trees, where we saw an incredible display of bird watching and feeding equipment. The woman running the gift shop cleaned and adjusted my binoculars for free – as advertised – but told me I really should get new ones. The pay from about a year's worth of Bird Banter columns would buy a nice pair I saw on display.

We were only in Cape May a few hours, long enough to get my feet wet in the Atlantic (and my shoes and socks—thanks, Sis!) and vow to come back someday. However, it is good to be home again in the land of broad landscapes and straight-forward highways – and where wet things dry out in a quarter of the time.

*Cutlines:*

Business is slow on a balmy September afternoon at the Hawk Watch platform staffed by the Cape May Bird Observatory. Under the right weather conditions,

thousands of raptors can be seen in one day as they funnel over this southernmost tip of New Jersey.
Barb Gorges/WTE

A young herring gull wanders the beach at Cape May Point State Park, located at the southernmost tip of New Jersey. Barb Gorges/WTE

# 179 Thursday, October 28, 2004, Outdoors, page C2
# Duck days attract birders to local lakes

Great minds think alike, certainly the minds of the field trip chairs for Cheyenne-High Plains and Laramie Audubon Societies, Art Anderson and Rhett Good, respectively. Fall is a great time for duck watching and Hutton Lake National Wildlife Refuge is a great place to go. Thus, members of both chapters converged there Oct. 9.

It was a wonderful day to be outside. The dirt track of a road was almost completely dry. The wind was hardly enough to ripple the water and didn't tear up my eyes when I peered through the spotting scope. Sunlight glinted off the Jelm Mountain observatory.

The refuge, southwest of Laramie, includes several lakes besides its namesake, all within practically a stone's throw of each other. They lie out on the Laramie Plains with hardly a tree in sight.

The refuge lakes are part of a collection spread between the Medicine Bow Mountains to the west and the Laramie Range to the east. They are hollows blown out by the wind and filled with water naturally or with a little human assistance.

All birds that swim are not ducks, of course. The most abundant bird on the water was the American coot, more closely related to chickens than ducks. Completely black except for white bills, coots are easy to winnow out while

searching for more interesting birds, but they can be fun to watch. Several were playing king of the hill on a pile of debris.

Ducks get easier to identify by October and November because they resume the brighter colors of the breeding plumage they discarded in May and June. Their summer feathers, called eclipse plumage, are drab.

Females are never easy to identify. Going by body shape is almost better than sorting through particular arrangements of brown feathers.

The easiest duck for me to pick out on this trip was the American wigeon. The light streak from the top of its bill up and over the top of its head was quite distinctive.

Someone pointed out a ruddy duck. If it hadn't been holding its tail in that diagnostic, peculiar upright position, I'm sure I would have given up. Unlike the other duck species males, its breeding plumage season runs much later, March through August. Right now it is gray instead of ruddy colored.

The male green-winged teal were easier to pick out, sporting a wing-shaped patch of green over each eye. Unless they fly, you may not see the speculum, or patch, of green on each wing.

Other ducks observed included northern shoveler, ring-necked duck, canvasback, gadwall, lesser scaup, redhead and only a couple

mallards. Another non-duck waterbird, eared grebe, also made an appearance.

The first birders to arrive at the first lake were able to identify black-bellied plover and a bald eagle before they flew off. When the rest of us arrived, the long-billed dowitchers and American avocets were still probing the shoreline unconcernedly. Someone identified a California gull.

At a second lake, a sharp-eyed birder noticed that the motionless lump sitting on the hillside opposite was a ferruginous hawk. As we moved on, a prairie falcon crossed our path. Later, a red-tailed hawk gave a nice performance. Northern harriers, however, were the most common hawk of the day.

Canada geese aren't hard to identify, but every time a small flock flew over, we had to make sure they weren't the sandhill cranes we kept hearing off in the distance. Finally, when we stopped on a high spot overlooking the last refuge lake, someone was able to scope out four tall, pale gray beings in a distant pasture feeding on insects, rodents, seeds, etc.

While all optic equipment was focused on the lakes, other terrestrial birds were also noted: horned lark, song sparrow and marsh wren.

We left the refuge to check out other Laramie Plains lakes, but one was back lit and by the time we got to Twin Buttes Lake, wind

was chopping at the water and only gulls were circling round.

When I got home, the new issue of Wyoming Wildlife had arrived, and perusing it, I came across mention of an online map of duck migration based on reports by hunters, www.ducks.org/migrationmap, on the Ducks Unlimited website.

Now that would be handy. I was thinking it might track migration so that I could find out how much earlier pintails and blue-winged teal migrate. But it isn't very specific, has no archives, and it needs a lot more participants.

So, we'll just have to get out and look for ourselves on local lakes, which I did the very next day, at Sloans Lake in Lions Park. With no one to turn to for authoritative identification, I scrutinized what I determined was a female ruddy. Three days later, birding with Cub Scouts, we scoped a bufflehead and a redhead, both ducks, though they don't use that noun in their official common names.

A week later, Mark and I identified eared grebe, pied-billed grebe and more ruddy ducks.

Lions Park was designated a Wyoming Important Bird Area for a reason. It's not just for mallards begging handouts. I'm looking forward to more trips around Sloans Lake to see what the season brings in.

2004

171

# 180 FeederWatch returns

Thursday, November 11, 2004, Outdoors, page C2

## Website is a nest of information

The new Project FeederWatch season begins this Saturday, so if you've considered joining, sign up now and make the most of the yearly $15 membership fee.

This will be my sixth season reporting on the birds in my backyard (and front yard now too). In preparation, I visited the Project FeederWatch website last week, www.birds.cornell.edu/pfw, to see what was new and to log on and update my information.

The Project FeederWatch home page is the portal to rich resources, even if you don't sign up to submit data. First, there are the headings across the top of the page, "About FeederWatch, Instructions, Data Entry, Explore Data, News, About Birds and Bird Feeding."

I skipped the first two, knowing already that this is a joint research and educational project of Cornell Lab of Ornithology, National Audubon Society, Bird Studies Canada and Nature Canada, attracting 16,000 participants last year who reported, either online or on paper forms, weekly or bi-weekly, over 3.7 million birds at their feeders. Some watch less than an hour each time, like me, and some spend hours at their window.

Project FeederWatch makes it easy for the rankest beginner to take part by sending a poster to help with identification, plus a handbook explaining bird watching, bird feeding and bird counting.

Online, the heading "All About Birds and Bird Feeding" is an even richer source of information. If all of you consulted this section, I may never get another phone call asking about bird feeding, making bird columnist a lonelier job. Topics include feed, feeders, tricky IDs, diseased birds, strange-looking birds, Bird of the Week and a link to an online field guide.

This last is a whole other wonderful website, www.birds.cornell.edu/programs/AllAboutBirds. While this is no substitute for flipping book pages when comparing birds for identification purposes, it has detailed information on each species (not just feeder birds), from sound and video recordings to egg descriptions.

Besides the species accounts are these headings: "Birding 1-2-3, Bird Guide, Gear Guide, Attracting Birds, Conservation, and Studying Birds." This last is information on how to sign up for Cornell's famous Home Study Course.

I found myself exploring the field guide quite a while before backtracking to the Project FeederWatch home page. Then I went to the "News" heading where vast amounts of data have been distilled into scientific reports and feature articles. But if you are feeling adventurous, try "Explore Data." That's where, among other things, I can access my own data submitted since 1999, neatly charted by year.

Again I got sidetracked. First, I looked up the map showing FeederWatcher locations last year – 21 in Wyoming including five of us in Cheyenne. Our state top 25 species list begins with house finch, followed by house sparrow, American goldfinch, dark-eyed junco, northern flicker and starling. Seventy species were reported, but some were one-time wonders such as mountain bluebird late winter. And some, like the bald eagle, hardly fit the definition of feeder bird. The reports of flocks of turkeys and chukars seem more like something from a gamebird farm.

While I was in the Map Room, I checked out mountain chickadees on animated maps of North America showing observation locations month by month and year by year. I also looked up the population trend graph and found fairly consistently over time that these chickadees visit 50 percent of participating feeders in the Northern Rockies, averaging three birds per count.

Back at the Project FeederWatch home page is a series of links down the left side. "FeederCam" will give me a live view of the feeders at Cornell in Ithaca, NY, when my computer decides to be more compatible.

"Participants' Corner" has a lot to offer, including photos and stories. My favorite anecdote was from a woman in Maine covered by a swarm of 30-50 chickadees on her

shoulders and arms, and nuthatches on her head, when she stepped outside during an ice storm.

From the "Feeder-bird Quiz" I learned that the most commonly reported bird is the dark-eyed junco. The "Young FeederWatchers" link has charming artwork plus extensive coverage of Monty the (stuffed) Moose's trip to Cornell. His classmates in British Columbia, who participate in the classroom version, were able to observe him filling feeders through the FeederCam.

Along with various news stories on the home page, I noticed there will be a prize, including binoculars, for the FeederWatcher who submits the millionth checklist, expected to occur this season. Let's see, if I'm home for all my weekly counts, that will be 21 chances between now and the end of the season April 8.

Finally, I logged on with my personal password and number and updated my feeder site description so I'm all set for my first count day on Saturday. I'll do the usual, leaving paper and pencil on the table under the window so that any of us in the family can jot down what we see whenever we walk by – if it is a new species or a greater number of any species recorded earlier in the day. Later, I'll enter the data online.

I'm ready, and I know the birds are too. They ignored our feeders until that six-weeks-late frost/snow on Halloween. Maybe I'll put out millet for those Eurasian collared-doves still hanging around our neighborhood.

**Author note**

To sign up for Project FeederWatch, either call 607-254-2427 during business hours, Eastern Standard Time, or go online to www.birds.cornell.edu/pfw.

## 181 Thursday, November 25, 2004, Outdoors, page C2
# Going where the gulls are adds species to life list

It was obvious, based on time of which day and the location – early morning Saturday near a wetland in Fort Collins, Colorado, – that the flock of four Subaru Outbacks and five other fuel-efficient vehicles gathering belonged to birders, especially since one bore the plate "Skuas," referring to a type of oceanic bird.

Another clue was that about half the vehicles were then left behind in the parking lot.

Birders carpool not only to lessen the necessity of drilling for oil in the Arctic National Wildlife Refuge, but because we're sociable, and it's easier to share sightings while enroute. It's also easier to park where there isn't a lot of room to pull off the road, which was the case at our first stop at the edge of Long Pond.

A local resident stopped to inquire what we were looking for. With nearly a dozen scopes on tripods set up, we were either peering through the windows of the waterfront homes on the far side or checking out the birds on the water.

We were on a gull trip this mid-November day. Our leaders, Doug Faulkner and Tony Leukering, had the expertise and the optics to find something beyond the most common species of the plains, the ring-billed gull.

My expertise lags far behind, but at least I don't refer to them as seagulls. However, I wished I'd studied up the night before. Instead, I had to juggle my notes and my field guide with frozen fingers while gray chill also found my toes.

Squinting through my scope made my eyes water, increasing the difficulty of picking out how much black and gray marked the inside of a wingtip of a floating gull surrounded by a flock of common mergansers.

Birders, sharing observations of particular feathers seen from about 400 yards away, had it slightly easier because of the landmarks on the opposite shore, such as the green canoe, the overturned red canoe and the collection of chaise lounges.

Three gull species, herring, California and Thayer's, were identified. The Thayer's, normally an Arctic breeder wintering on the west coast, is considered rare in Colorado and has not yet made the records in Wyoming. I could add it to my life list, but not to my list of birds I can identify by myself.

Someone also picked out a large gull, white head with marbled brown body, and determined it to be a young great black-backed gull. It certainly was larger than the other gulls, and also far from home, the Atlantic seaboard.

As we wandered from lake to lake, we found a very pale gull normally seen along the northeastern and northwestern coasts of North America. The back of the glaucous gull lives up to its name which is Latin for a silvery, bluish color. I think I can add that species to my self-identifiable list, unless someday I have to compare it to the glaucous-winged gull, which strays much less often to Colorado and Wyoming from the west coast.

If you've only buzzed by Fort Collins on the Interstate admiring the snow on the peaks, or only shopped College Avenue, you might find it incredible that it's a hot spot for rare gulls. However, one look at a map more detailed than a road atlas shows you are in lake country. The area at the foot of the foothills is pockmarked with ponds. All are manmade. Some reservoirs cover almost a square mile, making lakefront developments common.

Luckily, lakefront has been set aside in the Open Space system. One of our stops, Fossil Creek Reservoir, just west of the Windsor exit, has recently been developed for wildlife viewing.

No rare gulls at Fossil Creek, but I did pick up a new life species. The American Ornithologists' Union has very recently determined that the four smallest of the 11 races of Canada goose are now to be known as a separate species, the cackling goose.

Without DNA testing equipment, birders will have to depend on relative size, color and location for identification. Doug said the geese we were seeing were cackling, migrating through from their tundra breeding grounds. For a long goose discussion, go to www.sibleyguides.com.

I also saw a species that I thought was a genuine life bird for me, only to discover once home that I saw it in New Mexico 10 years ago. The greater white-fronted goose's name refers to its white face. Otherwise, it is blah gray-brown. But it's the orange bill, and orange legs if you can see them, which stand out in a crowd of cackling/Canadas. We counted six of them swimming in a line like the ducks on that pull-along toy I had as a toddler.

Like so many other field trips, this one was open-ended. A couple folks from Casper turned back around noon and a couple more of us from Cheyenne headed home around 2 p.m., the rest disbursing later. No new gulls were added without us, but we missed the trumpeter swan.

Tony said an increase in the sightings of rare gulls is partly due to increasing population and range thanks to people inadvertently providing more food sources, but also because more people are looking for gulls, and more people are capable of picking out the rare species.

To become a gull expert, I should probably invest in that huge book, Gulls of North America, Europe and Asia by Olsen and Larsson. But nothing takes the place of field observation and the patient mentors I've met so far.

# 182

Thursday, December 8, 2004, Outdoors, page C2

## Birds are featured in calendar from Game and Fish

The bright goldfinch on the front told me something interesting was afoot when Wyoming Department of Game and Fish's 2005 Wyoming Wildlife magazine calendar showed up in the mail recently. I was right. Seven of the 12 months feature birds, and three of those are songbirds.

I also like the pithy comments by the magazine's senior editor, Tom Reed, which tie each featured species to that particular month, that pose and that species' status in Wyoming, all with entertaining informality.

Judith Hosafros, assistant magazine editor, is in charge of the calendar project. She said it has mainly been a subscription promotion – give a gift subscription and get a gift – but it's so popular now, the calendar has been designed to be offered to the public as well. In the future it should be available wherever the magazine is sold. Profits support wildlife habitat acquisition.

Judith had some design help this year, but format decisions, making it 14 by 22 inches when open, using slick paper and including moon phases and the previous and next month on each page, were hers. The calendars were printed by Pioneer Printing of Cheyenne,

Always a very elegant publication, previous calendars have placed in the top four at the Association for Conservation Communication competitions. I think this next year's will do well also, especially since the photos are less stereotypical. The bald eagle is standing over its (mostly hidden) dinner, with commentary pointing out how roadkill benefits our national symbol. The obligatory elk picture is of a cow and nursing calf rather than a trophy bull.

The idea of featuring a particular photographer each year started a few calendars ago. This time it's a team, F.C. and Janice Bergquist, of Saratoga. I recognized them not only because they frequently have work published in Wyoming Wildlife and national birding magazines, but because a few years ago, editor Chris Madson introduced me to the Bergquists and they generously donated use of many bird images for the Wyoming Bird Flashcard CD project.

Francis began his fascination with wildlife photography around 1976. Janice blames herself for that because a friend helped her pick out his first camera as a present. It was a hobby until Francis retired three years ago – but a hobby that has paid for itself.

The first step was to research magazines and send for their want lists.

"The more you send, the more you get published," said Francis in a phone interview. The more familiar your name, the more likely publications will call you for particular photos, he added.

Janice has always been involved in the business end and recently has begun taking photos herself, her specialty being butterflies and flowers.

"It's just a lot of fun," she said.

Their son Greg is also getting published, including the cover of the latest Birds in Bloom magazine and the cover of Wildbird sometime this spring.

Wildlife photography becomes an obsession, Janice said, especially in spring with migrating birds. She and Francis are up at daylight and out until 10 or 11 a.m. when the light becomes too harsh. When I called, Francis was out in their yard attempting to shoot a flock of hundreds of Bohemian waxwings.

He hasn't gone digital yet, he said, since so many publications still want slides. However, they can be scanned if digital is required. When he does convert, Francis expects all his Canon lenses will fit the Canon digital body. His largest is a 500 mm, onto which he can add a converter to make 700 mm, but Francis thinks anything larger is too heavy. His portable, tent-like blind works well to bring birds close. Within minutes of setting it up, the birds forget all about him.

Most of the Bergquists' photography is within 100 miles of home, plus a few trips to Arizona, New Mexico and Nebraska. Francis considers himself to be a birder but doesn't keep a life list. However, he does keep track of great birding locations, such as the Tucson, Ariz., water treatment plant. He says it looks like a pristine marsh – with viewing platforms.

OK, here are three ways to get one or more of these great calendars.

Give a gift subscription to Wyoming Wildlife ($12.95) between now and January 31 and get a free calendar sent to you. Call the subscription service at 1-800-710-8345.

Or call 1-800-LIV-WILD (1-800-548-9453), Game and Fish's Alternative Enterprise office, and order calendars at $9.95 each, plus the sales tax for your Wyoming county, plus $5 each for shipping to you or someone on your holiday gift list. You can even ask for a free gift card.

Or send your calendar order to the Cheyenne office, 5400 Bishop Blvd., 82006, or visit weekdays between 8 a.m. and 5 p.m. Enter the visitor parking lot just opposite the Wyoming Department of Transportation complex on the west side of the I-25 exit at Central Avenue. If you don't see calendars among the other Game and Fish gift items when you walk in, just ask at the front desk.

There's nothing like sending a bit of Wyoming

when I think my friends and relatives have everything else. And there's nothing like a bit of Wyoming's wildlife in the kitchen to improve my daily view.

# 183 Christmas Bird Count debunks stereotypes

Thursday, December 22, 2004, Outdoors, page C2

What comes to mind when you hear the words "Christmas Bird Count"? A snowy, Grandma Moses, New England winter scene peopled by eccentrics bundled in wooly scarves, craning their necks to study treetops at the break of day?

There have been changes since the first count held in 1900, especially in the distribution of individual counts. Even that first year, though most of the 25 were in the northeast, there were counts in California – so much for the snow part of the stereotype – and Canada.

The publication American Birds, analyzing last year's count, shows nearly 2,000 counts were held across North America, and new ones are showing up further south.

Counts in Brazil, Chili, Columbia, Costa Rica, Ecuador, Mexico, Panama and Peru as well as Bahamas, Bermuda, Dominican Republic, Puerto Rico and Virgin Islands help us discover more about where "our" summer birds migrate to for the winter.

We also still have a lot to learn about species that don't migrate.

What began as a replacement for the Christmas "side hunt," a traditional competition to see who could kill the greatest number of birds and small mammals, has become an increasingly valuable tool for monitoring bird populations, even though it is conducted by non-scientists, or citizen scientists, as we are

called by the National Audubon Society, sponsor of the Christmas Bird Count.

A panel of scientists has concluded that the count data is scientifically useful. Though they would like to see the protocol be more standardized, for instance, observers of equal ability putting in equivalent effort on equal length routes by the same means of transportation on the same day each year within 15-mile-diameter count circles that are representative of local habitat, they sensibly realized that we birders are too tied to tradition to go for that. Plus, the new data wouldn't have any statistical relation to the previous hundred years' worth collected in its own haphazardly consistent way.

So, scientists will adjust our results with statistics to give them important information on population trends and movements. The panel recommends more research results be posted on the official website.

Now, about that perception that people taking part in a Christmas Bird Count are somewhat eccentric, a government survey showed bird watching is one of the most popular outdoor pursuits. No longer only the domain of scholarly gentlemen and little old ladies in tennis shoes (I admit, I have a gray hair), birding is an easy way for those of us no longer making a living on the land to connect with the outdoors.

The great thing about birds is they are everywhere year round. However, comparing last year's results, all locations are not equal. The Chesterfield Inlet count in Nunavet, northwestern Canada, recorded only two species compared to 231 species in Corpus Christi, Texas. And then there was an Ecuadorian count with 423 species.

There will always be a few eccentrics such as Kelly McKay of Hampton, Ill., who determined last year to participate in one count during each day of the season, December 14 through January 5. The problem is most counts are scheduled for weekends, when birders are available. So, he had to drive 7,100 miles and slept only a total of 32 hours. What a highway menace.

There are some things about the count here that don't change. We still start early, when the birds are most active. We still wear warm clothes, and we still warm up in the evening at the tally party where we share anecdotes and totals.

You are welcome to take part in the Cheyenne Christmas Bird Count on January 1.

There are three ways to participate. The most fun is to meet with Cheyenne-High Plains Audubon Society members in the lobby of the downtown Post Office, 22nd and Capitol, at 7:30 a.m. There you can join a group that is headed for one of the local hotspots and you can stay with them as long as you

can stay warm. Even if you aren't a birding whiz, extra eyes are always helpful.

If you want to bird on your own within the 7.5-mile radius count circle based on the Capitol Building, contact compiler Greg Johnson, 634-1056, gjohnson@west-inc.com, for maps, data sheets and instructions.

Counting birds at feeders is also important. If you don't have the time or the inclination to drive or walk around town, observing your own feeder as little as a few minutes may be the perfect way to participate.

You can call or e-mail results to Greg, but it's more fun to come to the tally party potluck. It will be at 6 p.m. at the Westgate community building on Gateway Drive, first building on the left after you turn off Carlson. Call Mark Gorges, 634-0463, if you have questions.

If you won't be in Cheyenne on Jan. 1, go to www.audubon.org/cbc for the date and location of other counts, as well as past years' results.

The Christmas Bird Count is a fun way for the appreciative person to make a scientific contribution to understanding the natural world, that world on which human health and well-being is dependent.

Whether wearing wooly scarves or not, we birders get a start on personal well-being resolutions when we get up and out of the house to follow in the footsteps of those first counters.

# 2005

## 184 Resolution produces list of field trip destinations

Wednesday, January 5, 2005, Outdoors, page C2

Here we are at the top of the 2005 calendar, with a total of 53 Saturdays for field trips. This year has a bonus because it starts and ends on Saturdays.

My resolution is to get to know birds better by getting out more often. One of the best ways to do this is on organized field trips.

A week or so ago I was compiling a record of Cheyenne – High Plains Audubon Society field trips for the past 17 years. There is a noticeable, yearly pattern.

Unlike scheduling monthly chapter programs for variety, field trips thrive on return engagements. In bird watching, no matter how many days you visit the same place, any one of them could be the day you see an interesting bird behavior, a bird that's new for you, or rare for the whole birding community.

The field trip year for Cheyenne birders is anchored by two major events, the Christmas Bird Count, usually held the Saturday after Christmas, and the Big Day bird count held on, or the first Saturday after, May 15. Both events concentrate on Cheyenne, especially the two designated state Important Bird Areas, Lions Park and Wyoming Hereford Ranch. Both sites are representative of the city in general, a forested island on the plains, attractive to avian life.

What also attracts birds and makes a good field trip location is water, the centerpiece of both of those IBAs and most of the past destinations.

Time of year is also important. With the exception of the Christmas Count and excursions around town in January, mostly to combat cabin fever, admire chickadees and to see if there is any open water where a lost duck has unexpectedly dropped in, migration is the big draw.

Mountain bluebirds cruise in as early as February, and after that it's a steady stream of visitors. Things settle down briefly in June, but then in July, Arctic-nesting shorebirds have finished their parental duties and start the parade through Wyoming in reverse.

By November, birders are watching for stragglers, wondering if they'll stick around to be counted at Christmas and wondering also if later and later dates for the last observation of a migrating species reflects global warming.

With the advent of spring migration, and again in the fall, the chapter's constellation of field trip destinations is broader. To the west are Hutton Lake National Wildlife Refuge and all the other Laramie Plains Lakes.

To the east are sharp-tailed grouse dancing grounds and further east is the area referred to as Goshen Hole, a collection of public access areas in the vicinity of Hawk Springs Reservoir, such as Wyoming Game and Fish Department's Table Mountain and Springer-Bump Sullivan Wildlife Habitat Management Areas.

To the south are Pawnee National Grassland and the reservoirs along the Colorado Front Range.

The big reservoirs to the north, along the North Platte, Alcova, Pathfinder and Seminoe, are a little far for a day trip, but Murie Audubon members from the Casper area keep close tabs on them.

Though farther, Cheyenne birders are much more likely to make an overnight trek to Nebraska to see the sandhill crane migration sometime during the height of the phenomenon, between mid-March and mid-April. We're there more to enjoy the mass of birdlife rather than the diversity of species, but also cherish the hope we'll glimpse a rare whooping crane.

Come summer, water is still an attraction, but Cheyenne – High Plains Audubon members also begin to head for the mountains, just like the juncos. It looks like the Snowy Range survey for brown-capped rosy-finches will be repeated after last summer's success.

Then there's the annual chapter camp out which over the years has met more weather-induced obstacles than the Christmas Count. We've tried twice to hold it at Friend Park, at the foot of Laramie Peak, but the first time we got smoked out by a forest fire and last year the mud was too deep.

This year, the plan is to schedule the camp out for July 8-10 and headquarter it at Battle Creek in the Sierra Madres. The gathering of birders will be put to work looking for nesting flammulated owls and purple martins.

One of the enjoyable past camp outs was to the Saratoga area. Several Wyoming Game and Fish Department public access areas, Treasure Island, Foote, and Saratoga Lake, are in the North Platte River valley, featured in the annual Platte Valley Festival of the Birds June 5-6.

Other areas with public access administered by Game and Fish are cataloged in their publication, "Access to Wyoming's Wildlife." Reviewing the table of contents is like reading the names of old friends, stirring up memories of many family outings, with or without Audubon.

Bird watching is a classic example of what can be a solo recreational pursuit. But the advantage to an organized field trip is that someone is bound to know something more about birds than I do, which is a much better way to learn than by reading, especially since local knowledge of local birds may best that of a book written for all North America.

I don't know yet how many return engagements will be scheduled by the chapter this year. Each will be a welcome reunion, if not an adventure to some place new.

# 185 Bird seed bandits

Wednesday, January 19, 2005, Outdoors, page C1

## How hard can it be to outsmart squirrels?

Fox squirrels are a by-product of bird feeding in Cheyenne. While they are cute and fuzzy and entertaining, the ones attracted to my yard have also been destructive, crashing bird feeders and stripping tree bark, not to mention stealing food meant for birds.

Originally, Cheyenne had hardly any trees and no tree-type squirrels. Birds had no competition at the feeder until, the story goes, somebody imported a few squirrels from Nebraska.

Much thought by people who feed birds has gone into outwitting squirrels. The problem is they seem to adapt to all our strategies to exclude them. Fighting them off is a bit like fighting an infection with antibiotics. Do you use the lowest level of technology that will do the job for now, or do you use a well-fortified feeder to begin with? It all depends on your means and patience.

Feeding birds in Cheyenne is as simple as throwing black oil sunflower seed on the ground. It's everybody's favorite, and you'll get a wide assortment of seed-eaters including sparrows, juncos, finches, chickadees and nuthatches – and eventually, squirrels.

The first level of advice often given is to offer squirrels their own feeding station stocked with favorite foods, such as dried corn. Many companies offering bird feeders also offer a platform on which to spike a whole ear.

### Baffling the wee beasties

However, with five furry and frisky eaters now gnawing on my trees, I'd rather not attract them to my yard at all. Putting sunflower seed in a tube, hopper or platform feeder protects it only somewhat from squirrels.

These kinds of feeders can be set on a pole, especially if you live where the wind tends to dump seed out of hanging feeders, but sooner or later the squirrels learn to shimmy up the pole.

Commercially made baffles are available that mount on the pole below the feeder. Some look like large, upside down, plastic salad bowls, so perhaps you can drill a hole in the bottom of that extra one you got for a wedding present.

Ruth Keto said greasing her feeder pole with canola oil has worked well so far in her Sun Valley neighborhood. It's not certain yet how often the oil needs to be reapplied to keep it slippery, or if it's actually a matter of fastidiousness which the squirrels will eventually overcome and finally get their paws dirty.

In our yard, we tried slipping a 6-inch diameter plastic pipe over our feeder pole before setting it in the ground. The same length as the pole was above ground, it worked because the pipe is too big around for the squirrels to get a grip – until the plastic weathers and the surface becomes rougher.

Lela Allyn has a solution that recycles two-liter pop bottles. She cuts a hole in the bottom of a bottle the diameter of the pole, then slits it all the way up the side. She slips the bottle around the pole and tapes up the slit. It takes several pop bottles, starting at ground level, to bypass the distance squirrels in her Cheyenne backyard have learned to jump.

Pop bottles applied to Lela's clothesline in the same way have protected feeders hanging from it. Any squirrel stepping on a pop bottle will cause it to spin and the little seed burglar will lose its footing.

Feeders hanging from the arm of a pole or tree branch are usually invaded from above. Once again, a dome-shaped baffle, this time hung above your feeder, could solve your problem, whether commercially produced or of your own invention. These also serve a secondary purpose in partially protecting the feeder from snow and wind.

### Caging the consumables

Putting your feeder in a cage is another way to keep out squirrels. It also has the benefit of keeping out large birds, such as grackles and blackbirds, which may monopolize feeders.

Our family bought a Duncraft sunflower seed tube feeder in 1993, and it is still in good shape. It came with a plastic-coated wire mesh fence around it, capped by a plastic roof and a plastic tray at the bottom. The wire mesh had big enough openings for a small bird to reach the seed ports, but not a squirrel.

After years on the pole protected by the plastic pipe, we moved it to a tree branch in the front yard. In only a couple weeks, we caught a squirrel wedging itself under the roof and between the tube and the cage.

An inspection of new

2005

177

Duncraft products at a local store showed we could buy the new version with a presumably squirrel-proof locking mechanism on the cap of the tube, plus metal roof and tray securely attached to the mesh.

Instead, we bought a new cage. This is complete with a wire top and bottom, and it will fit most tube feeders. The top opens with a presumably squirrel-proof latch so that you can fill the feeder. The handle of the feeder fits through a slot when the cage is closed. So far, so good. Of course, it's only been a few months.

Small wire cages are sold for holding blocks of suet. Woodpeckers and chickadees, which normally like to eat insects, are attracted, but so are squirrels. We had one of these suet feeders, but the birds never had a chance at it. The squirrels hung from it and nibbled. Finally, they unlatched it so the whole block fell out. I see in a catalog there's now a big cage just for hanging a suet feeder inside.

Platform feeders attract ground-feeding birds that will not tackle a tube feeder. Dark-eyed juncos are ground feeders, though they will use a platform four feet in the air. Cage adaptations are available commercially, but I'm thinking I could fix something over the top of our shelf feeder. It has to be removable so the feeder, like all feeders, can be cleaned every few weeks to avoid spreading bird diseases.

Duncraft has come out with a platform feeder guaranteed squirrel proof, based on the theory that squirrels need both paws to grasp a seed. They claim they have a metal grid with spacing too close together for two paws in one opening, but large enough for bird beaks. The platform is entirely metal so the squirrels won't chew their way in to the booty. How long will it take them to learn to use their paws to scoop seed instead?

For about as long as we've had that sunflower tube feeder, we've had the same brand of tube for niger (also spelled nyger) thistle seed. This seed is very fine and needs ports, or tube openings, that are very small. Luckily, they automatically exclude squirrels and large birds in favor of the thistle-eating species with thin bills such as goldfinches and pine siskins. That's good, because thistle seed is quite a bit more expensive than sunflower seed, and I'd hate to waste it on squirrels.

On the other hand, if you enjoy feeding the increasing numbers of Eurasian collared-doves, and the mourning doves when they come back in the spring, you are out of luck. Cage methods probably won't work well because the doves are about the same size as the squirrels, and the squirrels like the doves' favorite food, white millet.

## Springing surprises

One obvious solution to the squirrel problem is to decimate the population. However, without the proper licensing, this may be against the law in the ordinary backyard. Instead, members of the bird feeding community have become quite inventive and several have patented their anti-squirrel technology.

First, there's the Twirl-a-Squirrel Electronic Baffle I saw in a catalog. The weight of the squirrel activates a motor that starts twirling your tube feeder until the squirrel falls off. I think it's only a question of time before one of them figures out how long it has to hold on before the batteries die.

Another battery-operated feeder, by Duncraft, actually zaps squirrels with electric current they say birds can't feel.

Then there's the Yankee Flipper by Droll Yankees. This operates on batteries also, but it flips the squirrel off. For $10 you can buy the action-packed video that shows how effective this feeder is. Recently, the company added the Yankee Dipper, Yankee Tipper and Yankee Whipper, which all use the principle of perches that collapse when a large enough animal lands on them.

Then there is spring technology. Hopper feeders are roofed containers filled with seed that spills out a crack at the bottom where it is caught on a tray, or perhaps the seed is available through a series of ports along the bottom while birds perch on a bar. Barbara Costopoulos of Guernsey loves her spring-loaded hopper feeder. She has it adjusted so that the weight of a squirrel will close the ports.

Another of her feeders is by the Perky-Pet company. It looks like a square tube feeder wrapped in metal fencing and decorative metal leaves. When a squirrel lands on a perch, the metal fencing, attached by springs, is pulled down and a leaf blocks each seed port, like the portcullis on the entrance to a castle.

## Quality counts

If no one has been feeding birds or squirrels in your neighborhood for a long time, you may be able to get away with a lightweight feeder – for a while.

The first time we hung a feeder in our front tree, it was a Mother's Day gift from the boys, bought with their meager allowance. First the squirrels took the cap off the tube and reached in for the seed. Next, when the seed level got too low, they began breaking off chunks of the thin and brittle plastic tube so they could reach farther in. Finally, the feeder was knocked to the ground. Destruction was complete in about two weeks.

Paying for quality is cheaper in the long run. But don't forget to protect your investment. Use eyebolts and snapping clips so your hanging feeders can't be swung loose by squirrels or wind. Save your money for bird seed.

Bird feeding information:

Check out Cornell Lab of Ornithology's Project Feeder-Watch Web site, http://birds. cornell.edu/pfw, and also the book, The FeederWatcher's Guide to Bird Feeding.

## Feeder sources:

While there are many bird feeders available, local sources of bird feeder brands mentioned are:

--A & C Feed, 721 W. 22nd, 634-7391 or 4509 Driftwood Dr., 638-1770, including Duncraft.

--Murdoch's, 3773 E. Lincolnway, 632-7888, including

Duncraft and Perky-Pet.
--Riverbend Nursery,

8908 Yellowstone, 638-0147, including Droll Yankees.

--For the Bird Quest tm Twirl-a-Squirrel

electronic baffle, see www. BirdQuest.com.

# 186

Wednesday, January 19, 2005, Outdoors, page C2

## Birder's interest piqued by book a century old

My Garden Neighbors, True Stories of Nature's Children by L. A. Reed, B.S., M.S., with illustrations by the author, Review and Herald Publishing Association, Washington, D.C., 1911, copyright 1905 by L.A. Reed.

A couple months ago, Edna Hudson called me to see about donating bird books. I found a home for them at the Rocky Mountain Bird Observatory, which will either put them in its library or sell them at a fundraising auction in April.

But one book, the subject of this review, caught my eye, and I will either mail it or a check to Rocky Mountain Bird Observatory later. Edna said she bought it for resale at her shop long ago but had no other information.

"My Garden Neighbors," by L.A. Reed, is apparently written for children. One of the last pages is in smaller type and is addressed to teachers, giving them study ideas and informing them outdoor study aids body as well as mind.

Then there's the cover's patina of hard use. Another clue was stuck between the pages, a certificate for junior membership in the Seventh-Day Adventist Young People's Society of Missionary Volunteers, dated 1929, for William Archer, Boulder, Colo.

A search of the online used book seller, Alibris, shows two copies in much better shape available for $16 and $37, but no background

information available.

With skepticism, I Googled the publisher's name. What I found is no doubt familiar to some folks because Review and Herald has been the publishing house of the Seventh Day Adventists since the 1850s. They still publish children's nature literature.

Too close to deadline, I can't find out more about the author, whom I suspect is a man [Lucas Albert Reed]. For one, it would have been highly unusual in 1905, the date of the first copyright, for a woman to have a Master of Science degree. Plus, even though the stories in each chapter are told in the third person, they are told from the point of view of "the man."

And while the animals are referred to in that quaintly polite, somewhat anthropomorphic style characteristic of the age, there's no softening of the reality of nature.

For instance, Mr. Sparrow, after suffering the loss of a leg by slingshot, nearly loses his mate to another male before adapting.

Reed draws parallels to the way animals handle life and the way people could learn from them, but by no means is he as sentimental as other writers for children of the same era. In fact, when the man's adopted stray cat begins "going to the bad," killing several birds a day, the man chloroforms it without a sugar-coated euphemism.

In this modern age of effective kitty litter, the man

may have been able to keep his cat indoors. However, on other topics he shares a modern birder's viewpoint. For instance, house sparrows (he calls them English sparrows) are not to be encouraged since they are an invasive species that competes with the natives.

After the twelve chapters of stories about birds, spiders, and other garden neighbors, Reed provides "An Invitation to the Birds." His admonition against loose cats, red squirrels and house sparrows, and his prescription of tangles of bushes and shrubs, watering places, nesting places and various grains to feed is hardly different from that of modern experts, although finding hemp seed at the feed store today may be difficult.

Also at the end of the book are individual species descriptions, including range. This is where I thought I might learn what part of the country Reed gardened. First, I had to translate some of the old-fashioned names. I think "The Snowflake" should be revived for the snow bunting.

Reed mentions some other former names such as Summer Yellowbird (yellow warbler), Myrtlebird (yellow-rumped warbler), Cherry Bird (cedar waxwing), Thistlebird (American goldfinch), Chewink (eastern towhee), Firebird (Baltimore oriole), Blue Canary (indigo bunting) and Yellow-Hammer (northern flicker, yellow-shafted race).

Did you know that Lewis and Clark first collected for science what was originally known as the "Louisiana Tanager," named for the Louisiana Purchase? Now we call it the western tanager.

Reed lists a couple other western species, but I don't think he did more than travel through the west because he lists the red-breasted nuthatch, white-breasted nuthatch, brown creeper, yellow-breasted chat, and chipping sparrow as eastern-only species. I'm pretty sure they've been breeding in Wyoming and the west more than 100 years.

On the other hand, Reed lists the blue jay as being found in North America in general and we know it is a species still expanding its range into the west, though there are other blue-colored jays already here.

More than the change of bird names and distribution in the last hundred years, what is notably different in Reed's book from our modern lives is the amount of time "the man" has for nature observation. It is explained, "The man's health had failed, and the doctors had advised him to live more out of doors. That is how he came to have a garden."

Perhaps we too should take the doctors' advice and learn to take time to observe such events as construction of a spider's web – from start to finish. It may be healthful as well as instructive.

2005

179

# 187 Fox squirrels pose ethical, biological problems

Wednesday, February 2, 2005, Outdoors, page C2

Just minutes after e-mailing the Jan. 19 Birdseed Bandits story to the Wyoming Tribune-Eagle Outdoors editor, I watched as a squirrel extracted sunflower seed from my tube feeder while it was hanging inside its supposedly squirrel-proof cage.

The squirrel hung from the outside of the top of the cage by its hind feet, stretched full length and reached through with its front paws for one of the tube's lowest seed ports. The squirrel's weight caused the cage to tilt sideways, but the free-hanging tube within remained vertical and the end of it touched the side of the cage. That the squirrel found a weak spot in my defenses does not surprise me.

I'm also not surprised a story about keeping squirrels out of birdfeeders generated several kinds of responses. One reader, in Billings, Montana, reminded me of the red pepper cure. Squirrels hate it and birds can't taste it. However, she said keeping enough on the bird seed is a lot of work.

A Casper reader said he has a 95-percent effective system. He hangs his feeders from a horizontal steel cable with baffles between, above and below feeders.

A reader from Greeley, Colorado, wanted to know why I didn't mention live-trapping. He said he's released problem squirrels as far away as Casper and Cheyenne. Gee, thanks a lot! Obviously, this solution only changes the location of the problem, not to mention moving live wildlife across state lines is illegal.

Wyoming Game and Fish Department warden Mark Nelson, stationed in Cheyenne, encourages anyone wanting to borrow a trap and legally move a squirrel to call him at 638-8354.

Outside city limits, where hunting fox squirrels is legal, one needs a state small game license. Gun and bow season is Sept. 1 through Dec. 31, though for falconers it is year long.

Then a local caller who feeds 12 squirrels a day, without apparent harm to trees, asked why I was so intent on decimating them. Personally, I like the little imps. I just hate seeing the many places on our trees where they've stripped the bark. Lisa Olson, city forester, said squirrels feed on the cambium layer, the layer responsible for tree growth.

According to the American Society of Mammologists, fox squirrels also eat other tree parts: seeds, fruits, buds and some flowers. They are particularly fond of acorns, walnuts, pecans, etc., but since Cheyenne doesn't offer a lot of nut trees, I'm thinking they've latched on to bark instead. On the plus side, they prune my trees nicely while gnawing off twigs for nest building, meaning fewer visits required by an arborist.

The Billings reader reminded me squirrels will also eat baby birds and bird eggs, a fact documented by the mammologists society's paper, and an additional reason to consider the effects on the balance of nature of a species that was brought by people to our city.

The problem with feeding squirrels is that if we supplement their diet, they are likely to produce more than the average three pups per litter and even nest twice a year. Good nutrition equals increased fecundity which can mean increased population and increased tree and bird damage.

There are some people who might point out that Cheyenne's historic vegetation, except along the creeks, was treeless. But for those of us who appreciate trees and are too kind-hearted not to feed the squirrels and won't or can't hunt them, we need to look at their natural predators. Dogs and cats kill them too, but kindhearted people don't allow that to happen.

In the simplified version of the perfect predator-prey relationship, as the prey species population increases, predators move in and/or finding more to eat, are able to produce more young.

Eventually, the effects of their higher numbers outstrip the prey population's ability to reproduce. Normally the predators do not get every last squirrel. Instead, they starve, leave or produce fewer young than average or don't reproduce at all. With less hunting pressure, once again prey numbers begin to climb, and the cycle begins again. Wildlife managers have learned that a healthy population fluctuates.

Hunting, collisions with cars, power line electrocutions and disease also limit squirrel populations, but at least with natural predators, we have more watchable wildlife. Those species mentioned by the mammologist society I've seen within Cheyenne's city limits are red-tailed, ferruginous and rough-legged hawks, great-horned owls and red foxes.

There's a chance that my squirrel problems are localized. If I keep spilled seed cleaned up and keep squirrels out of the feeders, they might move on. If, however, it is a city-wide problem, we need to look at how to attract more of the natural predators. Besides food, they want a place to nest or den safely. How about adding a crossing guard or tunnel for those foxes that insist on navigating Dell Range Boulevard?

It is unlikely Cheyenne will ever be without fox squirrels. As another of many species affected by human action, intentional and unintentional, will nature find a balance for squirrels which people can live with? Some say bird feeding also presents ethical wildlife problems, but that's a discussion for another day.

*End of Ty Stockton's editing.*

# 188 Are condors coming closer to home?

Wednesday, February 16, 2005, Outdoors, page C2

If seeing is believing, can you believe what someone else is seeing?

In January someone reported seeing a California condor in the Alcova area. The news was reported second hand on the Wyobirds e-list with a follow up of another second-hand report about a second person's observation. As I usually do when confronted with obscure or unusual species, I checked records in Wyoming Birds by Jane and Robert Dorn.

No condors are listed in the Dorns' second edition, or in the 2004 edition of the Atlas of Birds, Mammals, Amphibians and Reptiles in Wyoming available from the Wyoming Game and Fish Department. The Wyobirds posting refers to two Wyoming records, and although I've sent an inquiry about them to the author of the posting, I have yet to receive a reply. When I asked Jane, she thought one of the records referred to was one so very old that it was impossible to establish the credibility of the witness.

Greg Johnson, a local wildlife biologist with a good network of professional contacts, remembered hearing about a condor seen at Flaming Gorge awhile back.

The California condor is an endangered species. Ten thousand years ago it ranged across the southern United States, but by the time of European settlement, it was found only in a coastal strip from British Columbia to Nuevo Leon, Mexico.

By the 1930s it was confined to central California. The last wild condor was trapped in 1987 and joined 26 others in a captive breeding and reintroduction program in California. Some of the condor's survival problems still exist: collisions with power lines, shootings, lead poisoning from feeding on carcasses of animals that have been killed with lead shot as well as the killing of young condors by eagles and coyotes.

However, a reintroduction effort in northern Arizona back in 1996 has been very successful. Condors released there are courting, nest building and in the last two years, have fledged three young birds.

Apparently, one of the Arizona birds, Condor 19, took a two-week trip north in August 1998 and followed the Green River. In the Peregrine Fund's archives of field notes, researcher Shawn Farry reports being contacted by Loren Casterline, an adult who was with Varsity Scout Troop 1834 at Kingfisher Island in Flaming Gorge Reservoir Aug. 6. The site is in Utah, five miles south of the Wyoming line.

While the scouts swam, the condor left a perch 100 meters away and landed within several meters of Casterline (condors are known to be curious about humans), but then his movement caused the bird to take off over the lake. A Fish and Wildlife Service news release mentions "Flaming Gorge, Wyoming," so perhaps there was another observation.

Condor 19 returned to the Arizona release site seven days later, making a total round trip of at least 600 miles. The U.S. Fish and Wildlife Service agent in Casper, Dominic Domenici, contacted Arizona about the new reports and learned that the whereabouts of two immature condors was unknown.

Condors are fitted with two radio transmitters – one on each wing – but ordinary radio telemetry used for locating wildlife doesn't work for a species capable of ranging as far as the condor. As funding becomes available, one of the radio transmitters on each bird is replaced by a solar-powered, satellite-based one that reports Global Positioning System fixes, or locations.

Locations are transmitted every hour through the day and then the data is e-mailed every evening to researchers, conveniently plotted on a topographic map. Apparently, the missing condors are not wearing the improved technology. How likely is it a condor really has visited the middle of our state? Unlikely, say the Arizona researchers, but not impossible.

The second observer is said to have been able to make a direct comparison with a nearby eagle, but depth of field can play tricks. None of the other Casper birders combing the area saw any condors. Turkey vultures have a wingspan of 5½ feet, eagles are at 6½, while the condors are at 9½ feet. Though condors have a red-skinned head devoid of feathers like a vulture, their bodies are 46 inches long – 20 inches longer than the vulture's.

Adult condors have a black and white pattern under their wings that is the reverse of the silver and black pattern for vultures. Immature golden and bald eagles have white under-wing markings, but an immature condor has none.

The way to make a rare bird report more creditable is to have more people see the bird and to photograph it. At this point, the sightings remain unconfirmed. If a formal report were to be submitted to the Wyoming Bird Records Committee and accepted, what would it mean? It would mean a condor has scouted the neighborhood. Whether or not it leads to return trips and breeding records for a future edition of the Atlas, only the condors can tell.

2005

# 189 New doves coming to a feeder near you

Wednesday, March 2, 2005, Outdoors, page C2

Pine Bluffs has that clean, scrubbed look of a small town on the treeless Great Plains – the look that comes from the cleansing effect of strong wind.

But on the side of the Lincoln Highway, 40 miles east of Cheyenne, is a small house cozily set in a thicket – a haven for birds.

Velma Simkins said she's lived there since 1933, when she and her husband bought the house as a young married couple. They planted the trees and shrubs.

She called me the other day for help identifying strange doves feeding on the seed she puts out. She's familiar with the Johnny-come-lately species, the Eurasian collared-dove, but thought these were different. Since a traveling birder reported white-winged doves in nearby Burns last year, my husband Mark and I decided it was worth looking into, but by the time we could check a week later, the strange doves were gone. We decided to make the drive anyway and meet Velma.

It wasn't hard to find her house, across from huge steel grain elevators. Sixteen Eurasian collared-doves were perched on nearby utility lines. A few blocks down another dozen of the pale gray doves with black marks across the backs of their necks were perched near another grain storage facility. Not a pigeon in sight. Have the doves run the pigeons off?

Velma said the only pigeon she sees is a pet belonging to the neighbors. There haven't been any pigeons since the old elevators were replaced. I imagine new, tighter facilities probably give pigeons less access to food and means may have been taken to eliminate them entirely, since they are one of three introduced bird species that can be controlled (the others are house sparrow and European starling).

So why are there Eurasian collared-doves in the vicinity? Are they hardier or smarter? We are still learning about this Middle East species that successfully colonized Europe, was later brought to the Bahamas as a caged bird where it escaped in 1974 and soon made its way to Florida and beyond. The similar looking ringed turtle-dove has escaped frequently but has not prospered or expanded its range beyond a few urban areas.

The first Cheyenne record for the Eurasian collared-dove was the Big Day Count in May 1998. Now they can be found regularly. I counted a flock of 20 in my neighborhood last summer, but there seemed to be fewer mourning doves. Perhaps the collared-doves, which are year-round residents, are getting the jump on the native migrating mourning doves when it comes to claiming good nesting habitat.

Eurasian collared-doves first show up on the 87th (1986-87) Christmas Bird

Count, but in only two count circles. By the 103rd count they were observed in 263 circles. The earliest observation for Project FeederWatch, an annual, winter-long count, was 1995, but another was not recorded again until 1998, and then every year after that.

The Great Backyard Bird Count held in February 1999 recorded collared-doves in eight southeastern states. This year's map shows the western frontier as a curve stretching from southern California up through Idaho and points east, though it seems to have skipped Nevada.

Kenn Kaufman notes in his book, "Lives of North American Birds," that in Europe, in warm climates, this species raises up to six broods a year. That's only one or two young each, but the young have a habit of dispersing long distances. They feel at home wherever there are large trees and open ground.

What about that other dove, the white-winged dove? It is a bulkier, darker gray bird with large white wing patches that become crescents when the wings are folded. It is a native species expanding its northern most limits, traditionally the southern edge of the southwestern states. Kaufman mentions it as an important pollinator of giant saguaro cactus.

There are a few records for Wyoming, one going back to 1954. But if the observations

in northern New Mexico and Colorado are any indication, we'll probably be seeing more of them.

I visited with Ron Ryder, retired Colorado State University wildlife professor, and he said scientists are looking into this explosion of doves. Eurasian collared-doves have been observed in just about every county in his state, even in inhospitably cold places like Gunnison and the San Luis Valley.

They are flocking to grain elevators like pigeons, he said, noting a recent posting on Cobirds, the Colorado bird watchers' e-mail list. It mentioned a flock of 200 Eurasian collared-doves, with a few white-wings in Flagler, Colo. He also mentioned wildlife managers are discussing listing the Eurasian collared-dove as a game bird. As an introduced species, it is not protected by the federal Migratory Bird Treaty Act.

The word "invasive" conjures up rapacious, economically debilitating species like kudzu, zebra mussels and leafy spurge. But other than apparently displacing a few mourning doves and pigeons, no one is sure yet what this wave of immigration will mean if it continues.

Kaufman said in his 1996 book, "If it spreads in North America as it did in Europe, the Eurasian Collared-Dove may soon be among our most familiar backyard birds."

Gosh, what will the robins think?

# 190 I spy with my little eye...

Wednesday, March 16, 2005, Outdoors, page C1

## It's time to grab your binoculars and pick a perch

Spring puts birds on the move – as well as birdwatchers.

While winter birding can be interesting – feeder birds, straggling migratory species or irruptions of more northern species such as the flocks of Bohemian waxwings seen this winter – we're talking maybe 50 species or so at the most on the Cheyenne Christmas Bird Count.

But come February, the first mountain bluebirds flash through town on their way to higher elevations. New species of waterfowl begin to show up on Sloans Lake in Lions Park. You can stay right here and let spring migration wash over you.

At the peak mid-May, as many as 150 species have been counted on the Big Day Count. Some of them, such as the black-bellied plover, would take elaborate travel plans to visit later on their Arctic breeding grounds.

A trip to Mexico would be necessary to find other species, such as Bullock's orioles or lazuli buntings, on their wintering grounds.

But it's nearly spring and the roads are mostly clear of ice and snow and birdwatchers are particularly eager to travel. While it's possible to find a bird almost any time or place, birders make the best use of their resources by hitting the hot spots.

A hot spot by definition is where a lot of species of birds can be found, or a lot of few species. Practically speaking, it is also publicly accessible.

Because birds spend so much time eating, all you have to do is look for bird food – aquatic creatures and plants, small birds and mammals, insects and other arthropods, fruits, seeds and other vegetable matter. These are usually found in a wet place.

Even though Wyoming's dry climate makes it a simple task to catalog them – lakes, reservoirs, streams and rivers – you can shorten your search by consulting local birders.

You can join the American Birding Association and get the directory of members. Or look for the closest Audubon chapter by going to the National Audubon Society website for a list of contacts.

There are also e-mail lists for birders. The Wyobirds list frequently posts queries from travelers, especially specific questions such as "Where can I find brown-capped rosy-finches?"

In response to my posted question, "What local places would you recommend to Cheyenne readers?" I received several replies.

### Sundance

I heard first from Jean Adams of Sundance reminding me of Sand Creek, east of Sundance. It is mentioned by all the state guides. U.S. Forest Service Road 863, beginning at Beulah, runs south through the steep-sided valley into the Black Hills National Forest.

In his book, "A Birder's Guide to Wyoming," Oliver

Scott says to look for lazuli buntings, canyon wrens, vireos and white-winged juncos, a locally common subspecies of the dark-eyed junco.

Published in 1993 by the American Birding Association as part of its series of state "birdfinding guides," the book is organized into three major routes across the state that include many hot spots. Its charm is in the personal narrative.

Oliver says of Sand Creek, "Occasionally, Canyon Wrens have been found here. If they are here, their remarkable song is easily heard. In fact, I have heard them on most of the canyon walls in this valley, but they can't be counted on – which is true of this bird over most of Wyoming."

Adams also mentioned Whitelaw Creek in the Warren Peak area, north out of Sundance on Forest Service Road 838 and then east along Forest Service Road 851.

### Riverton

Riverton-area birders flock to Ocean Lake, a Wyoming Game and Fish Department Wildlife Habitat Management Area, created by a reclamation project in the 1920s. It is on local birder Bob Hargis's map for exceptional migrating birds such as the parasitic jaeger, Sabine's gull, dunlin and other great shorebirds.

There's also a bit of a migrant trap at Long Point. Hargis said it has "yearly produced warblers galore including Virginia's,

chestnut-sided, and blackpoll, also tanagers, orioles and three types of vireos."

Ocean Lake is 17 miles northwest of Riverton on U.S. Hwy. 26. Turn north onto Gabe's Road, then east on Long Point Road.

This and other Game and Fish public access areas are mapped in the book, "Access to Wyoming's Wildlife," available from the department.

Frequently, what attracts fishermen also attracts birds, but be aware of hunting seasons.

### Casper

Members of Murie Audubon Society were quick to list the multitude of opportunities in the Casper area. First on their list is Edness Kimball-Wilkins State Park, a few miles east of town. It is also listed in the "Wyoming Wildlife Viewing Tour Guide" available from the Wyoming Game and Fish Department.

Murie member Rose-Mary King wrote, "From the first parking lot on the east side of the main road, take the path going east and you will walk about three miles around the entire park. This will take you through varied habitats along the North Platte River. Dogs must be on leashes."

Ann Hines added, "If you leave the paved path after you pass the gazebo and follow the human paths into the woods on the east side, there is some very good birding all the way back to the fence and the end of the park. We often

2005

183

find good warblers in the hedges in the spring."

Hines also recommends the two miles of path along the river in Casper which can be accessed from Crossroads Park. From I-25, exit at Poplar Street and head north a few blocks to the entrance.

Should you head south on Poplar instead, toward Casper Mountain, you can see about birding at the Audubon Center at Garden Creek. Call 473-1987 for permission. Further up is the sign for Rotary Park, the location of Garden Creek Falls.

"Hummingbirds come here early in the spring. Spotted towhees (formerly known as rufous-sided towhees) and green-tailed towhees can be seen singing in the top of the shrubs as you go up to the falls," Hines said.

Eventually, Poplar Street joins up with Casper Mountain Road. "Mountain bluebird, black-throated blue warbler, Townsend's warbler and blue-gray gnatcatchers are among species found on this road," said Hines about the climb up and over.

King said Grey Reef and Alcova reservoirs are good birding spots for water and shore birds. In their book, "Wyoming Birds," Jane and Robert Dorn include the reservoirs in one of their suggested state birding routes. They say best birding is April through November – weekdays especially.

The book is available through the Laramie Audubon Society. Call Deb Paulson, 307-742-5623.

### Jackson

Bert Raynes wrote, "I don't consider anywhere in Jackson Hole meets the definition of "hotspot" but (the book) "Birds of Grand Teton National Park" is still useful, despite its age."

Perhaps Bert is unfairly comparing Wyoming to Costa Rica. He wrote the book in 1984. He also coauthored the book, "Finding the Birds of Jackson Hole," with Darwin Wile in 1994. The earlier book has color photos of 60 bird species taken by famous wildlife photographers and describes the different habitats. The second book is a collection of explicit directions for bird walks, hikes and drives.

### Lander

Lander is only 25 miles from Riverton, but it is at the base of the Wind River Mountains and local birders like Jim Danzenbaker are likely to head up Sinks Canyon, so named because the Middle Popo Agie River suddenly disappears underground at the Sinks and re-emerges at the Rise. This geologic phenomenon is protected within Sinks Canyon State Park.

Danzenbaker wrote, "During both spring and fall migrations, the area around the Sinks and the Rise can yield Virginia's warblers (more than 50 percent of visits), chestnut-sided warblers (fall), many buntings, tanagers, orioles, vireos and other warblers. Summer produces many breeders including dusky flycatcher,

red-naped sapsucker, canyon wren, blue and ruffed grouse and prairie falcon."

He adds that should you arrive before winter leaves the canyon, you might also see chukar, goshawk, Bohemian waxwing, Townsend's solitaire and golden eagles.

Danzenbaker also said Dry Lake is a good place for shorebirds and ducks. "I've had 15 shorebird species there on a single visit."

From Lander, head eight miles south on U.S. 287. Look for a pullout on the east side of the highway, just before the Rawlins turnoff. You'll need a spotting scope since the lake itself is not publicly accessible.

Another place Danzenbaker likes is Ray Lake and Marsh, visible to the west on U.S. 287 about eight miles north of Lander.

### Laramie

Closer to home, Deb Paulson of Laramie recommended the Laramie River Greenbelt Trail along the Laramie River on the west side of town, as well as the University of Wyoming campus with its mature trees which attract ruby-crowned kinglets and other small forest birds.

Fall migration is also good on campus – lots of warblers and pine siskins.

### Yellowstone

While no one responding mentioned Yellowstone National Park, "Birds of Yellowstone," a book by park bird biologist Terry McEneaney, is essential. He lists 75 "suggested intensive birding

areas," but he also treats 20 species in depth.

One of the most useful parts of the book is the checklist that graphically shows when and how likely and in what habitat a species may be seen.

McEneaney will be in Cheyenne for Cheyenne – High Plains Audubon Society's Big Day Count May 21 and will give a public talk about Yellowstone birds in the evening.

### Cheyenne

You can enjoy a perfectly wonderful spring migration right here in Cheyenne. With all the visitors passing through, it's almost as good as visiting a whole new place every time you take a look.

Lions Park has been designated a state Important Bird Area through the program administered by the National Audubon Society. Some 40 IBAs are described in the publication, "Important Bird Areas of Wyoming," available from Audubon Wyoming's Casper office, 307-234-1795.

Sloans Lake in the middle of the park, has big trees and thick shrubs that are all attractive to birds which have been traveling over the treeless plains.

The Wyoming Hereford Ranch, on Campstool Road, is also an IBA and has the same attributes. However, it is private property so stay along the roads and do not cross the fences.

So many birds, so little time!

# 191 Callers: Give a hoot

Wednesday, March 16, 2005, Outdoors, page C2

## Birders need to know etiquette of calling

A windless winter night is a good time to call for owls. Their response is only muffled by traffic, barking dogs, shifting feet and blood rushing in your eardrums.

The February evening the Cheyenne – High Plains Audubon Society sponsored an owl outing attracted a field trip record of 31 local listeners. I'm guessing popularity was due partly to owl mystique and partly because the evening followed an unseasonably warm day.

Calling owls has been immortalized by the children's picture book, "Owl Moon," the story of father and child crunching over the snow on a moonlit night. On our outing, starlight was soon washed out by a huge orange moon just past full, however, snow was hard to come by.

We didn't have anyone willing to imitate an owl, so we used recordings. First, we played the northern saw-whet owl several times, with pauses in between to listen for responses. We then moved on to long-eared, eastern screech and great horned owls. Starting with the largest would have inhibited smaller owls.

The chapter had another owl outing five years ago, but this time I heard concerns about its effect on the owls.

Calling owls or any other bird is actually a form of avian harassment. It works because birds respond to an intruder on their territory. Songbirds tucked away in leafy shrubbery will respond to the vocal sound birders make, "pish, pish." It sounds like the avian alert signal, so they come out to see what's going on.

You can understand why frequently antagonizing birds into alarm mode is not good. The American Birding Association states in its Code of Birding Ethics, "Limit the use of recordings and other methods of attracting birds, and never use such methods in heavily birded areas or for attracting any species that is threatened, endangered or of special concern, or rare in your local area."

Cheyenne birders hardly ever resort to pishing or recordings, and I was surprised once when a birder from out of state on one of our field trips started playing calls. As a birder friend said, "It seems like a cheesy way to bird," in a recreational situation.

Calling owls is a time-honored way of surveying for them, however. Jennifer Bowers, a local wildlife biologist, told me about her experience listening for Mexican spotted owls in Apache-Sitgreaves National Forest in Arizona.

From the end of April to the end of July, her workday started at 4:30 p.m. with a hike into the backcountry, arriving at the beginning of one of many survey routes about 7 p.m. One kind of survey involved stopping at predetermined intervals and hooting. I can just imagine the interview for this job. Bowers said it's difficult to demonstrate hooting over the phone – it's rather loud.

The other type of survey was to hoot where a pair of owls was known to nest. By the end of the season, the young were noticeable.

The surveys took about an hour or two, depending on the amount of paperwork generated by responding owls. The biologists were usually back at the field camp around midnight, only to get up and do a willow flycatcher survey at 4:30 a.m.

Inconveniencing Cheyenne owls once every few years for the sake of education is worthwhile, I think. But this time we decided to limit our calling to areas we visually survey on the Christmas and spring counts.

Apparently, we weren't disturbing very many owls. We tried calling among the big trees at Lions Park. We tried out Crow Creek where we'd seen great horned owls last year. Nothing. By then some of the neophyte owlers must have thought we were crazy.

We had one more stop to make, the High Plains Grasslands Research Station, where we had written permission to visit that evening. We parked along the road near the entrance and before the last car door closed, people with better hearing than me heard a saw-whet.

We walked along the road until the calls became more distinct. While great horned owls are often seen on our counts on the piney island of the station, Jane Dorn, who has surveyed it for about 25 years, recalls only one other saw-whet. She said this time of year they are migrating through to their preferred nesting habitat in the mountains, so that's why we wouldn't find them on the Christmas count or the spring count in May.

Saw-whets are tiny, only eight inches long and not even three ounces. The great horned owl is 22 inches and three pounds. I can only surmise that an owl that calls constantly as this saw-whet is advertising for a mate and unconcerned with a larger owl making a midnight snack of it.

We had a beautiful moonlit evening outdoors and we gained new information. Also, it will be a long time before any of us forget the northern saw-whet owl's call. Now that we know it, we may start hearing what we would have otherwise missed before we could distinguish it from all the other night sounds.

*Cara Eastwood begins as Outdoors Editor.*

# 192 Wednesday, March 30, 2005, Outdoors, page C2
## Working in wildlife biology

Three elementary students won Audubon awards at Laramie County School District 1's science fair earlier this month: Bailey Pawling, Rossman; Colby Styskal, Hobbs; and Marcela Means, Pioneer Park. Each had a topic fitting the criteria of birds, wildlife and or environmental science.

A few days later, proud parents saw their children making presentations to 45 people at the Cheyenne – High Plains Audubon Society meeting. I saw future wildlife biologists.

I'm guessing that most parents, like mine when I was young, have little idea of career opportunities in wildlife.

At my suburban high school career day, the closest my friend Jackie and I could come was a presentation by the National Park Service where we were told it needed secretarial and maintenance workers. Years later I realized that even the park ranger's job involves acting as public information officer, an educator and a law enforcement agent.

A wildlife biologist is someone who manages wildlife and wildlife habitat, especially where wildlife and people have conflicting needs. Wildlife habitat is the space where animals find food, water, shelter and a place to raise young.

Sometimes the wildlife biologist gets to play with animals, banding birds, wrestling elk or shocking fish, but a lot of the time they must crunch data, write reports and look for funding.

Having a degree and experience in a closely related field, as well as being closely related to a couple of wildlife biologists and friend of many, I'll tell you parents what I've observed of the career possibilities.

First, a college degree in wildlife is a necessity in this very competitive field, but even college students have opportunities for short term and seasonal field work in trapping, tracking, observation, etc.

Most wildlife biologists would prefer to work year round so they get a master's degree to qualify for permanent positions but find themselves doing more and more paperwork.

Some deal with planning to influence the size of species populations and others research how particular species are affected by changes. Usually, people who want to study the pure biology of animals get their degrees in zoology.

Where do wildlife biologists find jobs?

Probably the largest sector is government. Every state has some form of game and fish department to manage the wildlife within its boundaries. Each state has ownership of its wildlife. The U.S. Fish and Wildlife Service deals with the big picture – especially for migratory species.

Federal land agencies such as the Forest Service, Park Service, and Bureau of Land Management manage wildlife habitat. Other federal agencies, the Bureau of Reclamation, the Bureau of Indian Affairs, and the Geological Survey may also hire wildlife biologists, as might local agencies like the county conservation district.

In the commercial sector, companies that have an impact on the environment, such as mining, drilling or timber harvesting, hire biologists full time or as consultants to help them comply with environmental protection regulations. Consultants can also be hired to work for the government. Privately owned ranches, resorts and hunting preserves might also hire a wildlife biologist to manage habitat for the wildlife that use their property.

A degree in wildlife adds to the credentials of hunting and fishing guides, people leading bird watching field trips or writing about or photographing wildlife.

Non-profits employ a sizeable share of wildlife biologists. Audubon Wyoming hired someone specializing in birds to lead the state Important Bird Area program and habitat enhancement work.

A peculiar kind of institution, the bird observatory, has evolved over the last 40 years. It raises funds or receives grants or contracts to investigate bird populations and issues, but it may also have a membership component and lead educational activities. There is a host of groups for specific species, like elk, trout, bighorn sheep and even some that focus on specific wildlife diseases.

Then there's the academic world where instructors and professors split their time between teaching, research and writing. Presenting papers at The Wildlife Society, an organization of professionals, is one way to spread new information that will improve wildlife management.

There aren't enough paying jobs for all of us who appreciate and would like to work for the benefit of wildlife, but there are a lot of opportunities for "citizen science" involvement.

The Christmas Bird Count is one example of laypeople contributing observations to the scientific record. Or you might volunteer to help a group improve habitat by planting trees or removing weeds.

With the amount of information now available, it is possible for anyone to be knowledgeable enough to write lawmakers and decision makers on behalf of wildlife. Even if this seems a lot like throwing pebbles in the lake, eventually enough collect to be visible above the water's surface and to be taken into account.

Every time you choose alternatives that are energy efficient, recyclable, renewable, sustainable, clean or organic, you are helping wildlife.

If you can't afford to be the typical American consumer, just think of it as doing your part to consume fewer resources that require disturbing wildlife habitat.

OK, you Audubon Award kids, whether you become famous ornithologists who specialize in avian oology (bird biologist studying bird eggs) or an occupational therapist with a special place in your heart for oystercatchers, thanks for your enthusiasm.

The future looks so much brighter.

*Marcela Means, as of November 2022, is a wildlife biologist for the Bureau of Land Management in Idaho.*

# 193 Sharp-tail trip seeks out the other grouse

Wednesday, April 27, 2005, Outdoors, page C2

As the driver of the last car in the caravan, I expected whatever the attraction was that had caused lead driver Bill Gerhart to pull over to the side of the gravel road would be even with or ahead of him.

The goal of our Audubon chapter field trip was finding sharp-tailed grouse north of Hillsdale. In the dimness before sunrise on this calm, mid-April morning, I searched the pasture ahead for any movement on the lek.

Leks are dancing grounds, where males come back every year to congregate, display and compete for females as the females hang out at the periphery. When I finally caught a flash of movement, it was even with my car. How nice for my passengers who had never seen sharp-tails before.

Mention grouse around here lately and most people think immediately of the sage grouse. Technically known as "Greater Sage-Grouse," it is in the news as a declining species in the way of oil and gas drilling.

The image of the sage grouse male in full display appears in wildlife publications regularly. He has two large, yellow-skinned, inflated air sacs embedded in a drooping white neck ruff and a fan of spikey tail feathers.

The sharp-tailed male, on the other hand, has just one spikey point to his tail, and no white ruff, though he does have small purple air sacs on either side of his neck.

Because sharp-tails don't make the news as often as their relatives, I asked Kathleen Erwin, a wildlife biologist with the U.S. Fish and Wildlife Service office in Cheyenne, about them.

Though their population has declined, they are not in as much trouble as sage grouse, she said, except for the Columbian sharp-tail, a sub-species that prefers mountain shrub habitat found in south-central Wyoming and other western states. It is being petitioned for addition to the list of threatened or endangered species.

As so often is the case, the changes to native habitat are a problem. However, sharp-tails will adapt to using cropland more so than sage grouse. We saw two fly over fresh green shoots of winter wheat on our field trip.

Bill, a Wyoming Game and Fish Department wildlife biologist as well as one of our trip leaders, said the Conservation Reserve Program started in 1985 encouraged the reseeding of cropland with grass species for erosion control and provided a good base for sharp-tails to rebound.

Drought is a factor right now, Kathleen said. The grouse need enough vegetation to hide their eggs from predators. Their nests are mere scrapes on the ground. Drought also cuts down on the number of insects available to feed the young right after hatching, before they grow into the adult diet of mainly seeds, buds and leaves.

The development of native prairie habitat also brings new predator species that the birds aren't used to, said Kathleen. Converting prairie to houses brings domestic cats and more skunks.

And then there's the competition. Ring-necked pheasants brought in by game bird farms will push sharp-tails out, said Kathleen. Later on our trip we had an excellent view of two cocks of this Asian species fighting on the side of the road.

Sharp-tails are a resident species in southeastern Wyoming as well as grasslands extending north into Canada, which means you should be able to see them any time of year. But they are a lot easier to find in spring on leks.

The six or seven sharp-tails we found were completely oblivious to us as we watched from our vehicles.

The males held their pointy tails erect, stretched their stubby wings horizontally and with head down, stepped rapidly in little circles, advanced on their rivals or retreated. The white of the undersides of their tails was the only contrast to the color of the dry grass landscape or the rest of their feathers that are also the color of dry grass.

With our windows open we heard tail feathers rattling and the weird cooing sound as the males deflated their air sacs. When the orange globe of sun slipped over the uncluttered line of the horizon, all the photographers were happy.

This particular morning it was we who left first, rather than the birds, to search out other leks. Often, the shadow of a passing hawk sends all grouse airborne. While a hawk in flight is often favorably compared to a fighter jet, flying grouse are the epitome of short-winged, big-bellied bombers. They prefer to flap and glide, and never far from the ground.

In late April, we are now part way through the pageant of spring migration. The snow geese have come and gone, and the warblers and shorebirds are just now showing up. If we miss the ducks, we'll see most of them again in fall migration. But the grouse show is mostly finished. Many people view sage and sharp-tail leks every year and some, wildlife biologists as well as volunteers, perform surveys.

Every year we hope they find good news. Otherwise, it could forecast the future demise of something more important than just another roadside attraction.

2005

For a copy of Habitat Extension Bulletin 25, "Habitat Needs and Development for Sharp-tailed Grouse," call Wyoming Game and Fish Department, 777-4600.

# 194 The lion inside

Wednesday, May 11, 2005, Outdoors, page C1

## Do you know what your hunting machine is up to today?

When you see a cat walking along the top of your backyard fence or concealed under a bush ready to spring, what comes to mind?

-- "Another pesky predator after my chickens!"

-- "I wonder if that's the cat that left nothing but a pile of feathers under my bird feeder?"

-- "Fluffy! Where have you been the last five days? I've been worried sick."

All of these responses illustrate aspects of the issue of domestic cats roaming outdoors.

But the impacts of house and feral cats on wildlife is a growing problem – especially as felines eclipse dogs as the most popular pets in the United States.

If more cat owners knew how harmful their pets can be to birds and ground-nesting mammals, experts say, they might be more careful about letting these hunting machines loose in the yard.

### Cats – predators or prey?

The issue of cats on the loose heated up last month when Wisconsin sportsmen voted to recommend feral cats be classified as an unprotected species, allowing them to be hunted.

The reasoning behind the vote was that [nonnative] domestic cats compete with native species like hawks and owls for prey.

Cats' impact is estimated at [hundreds of] millions of bird deaths every year in this country and probably more than a billion deaths of small mammals. Also, feral cats carry diseases that endanger domestic and wild animals as well as people.

While the Wisconsin sportsmen want their recommendation to go to the legislature, the governor has said he will not sign any legislation allowing cat hunting.

But Wyoming statutes classify stray cats with the red fox, coyote, porcupine, raccoon and skunk as predators that can be hunted without a license all year [where hunting is allowed].

While Wisconsin won't be implementing hunting policies like Wyoming's anytime soon, the discussion has been successful in raising awareness of the feral cat problem.

The impact on birds

Where do roaming cats come from? A surprising number are companion animals of people who regularly let their cats out.

A study by Carol Fiore and Karen Brown Sullivan at Wichita State University examined the hunting habits of 41 pet cats allowed outdoors.

They found that 83 percent of the cats killed birds. They also verified the deaths of 4.2 birds per cat per year but added that kills undoubtedly were under-reported.

[Few cats brought kills home. However, fecal analysis for less than half of the cats (less than half the owners provided indoor litter boxes) showed that they were eating far more birds than their owners were aware of.]

Surprisingly, the researchers found that the most prolific cat hunters had been declawed. That dispelled the myth that a well-fed or clawless cat doesn't hunt.

Fiore and Brown also found that 43 percent of the bird kills happened in May and June. That coincides with nesting season.

Non-game bird biologist Andrea Cerovski of the Wyoming Game and Fish Department said ground-nesting birds and those that nest low in shrubs are the most susceptible.

A fledgling can't sustain flight for long, and even young raptors can have problems with cats. She said waterfowl are also susceptible when they molt and are waiting for new flight feathers.

Since cats are not a part of the natural ecosystem, Cerovski said, they are particularly hard not only on birds but also on amphibians, reptiles and small mammals.

Each natural predator species fills a niche in the ecosystem in balance with its prey, Cerovski said. But cats kill additional prey. That means the populations of prey species might not reproduce enough to keep up, so the natural predators have less to eat and their populations drop as well.

Susceptible

ground-nesting species in this area include vesper, lark, savannah and grasshopper sparrows; lark bunting; western meadowlark and killdeer. Cats also can disrupt nesting ducks and geese as well as game birds since most of them nest on the ground.

[Cerovski noted that new housing developments on the prairie are a problem. "It's the cumulative effects – it's not just your cat," she said.]

A paper published by the Journal of Land Use and Environmental Law says the Migratory Bird Treaty Act makes it unlawful to kill migratory birds. The owner of a cat that kills them could conceivably be charged.

Few birds survive a cat encounter. Veterinarian Robert Farr cares for injured wild birds.

"I don't see a lot of injured songbirds," he said. "Most of the time, if a cat gets a bird, that's it."

### Colonies of feline hunters

In a study published by The Wildlife Society in 1999, two California parks were compared for their populations of feral cats in relation to the number of birds found at the parks.

One park had no feral cats; the other had a colony of 25 cats that were fed daily. There were twice as many birds seen in the park without the cats. Two ground-nesting species were not seen at all in the park with cats.

Cat colonies begin when owners fail to find lost cats or owners dump unwanted pets. The homeless cats then gather where there is garbage or someone puts out food.

Sue Castaneda, director of the Cheyenne Animal Shelter, said cats are not allowed to run loose in Cheyenne. Although no licensing is required, cats must be tagged for rabies.

But according to Castaneda, few stray cats wear tags.

Of the 2,431 cats picked up or turned in to the shelter last year, only 57 were reclaimed. [Another 1,100, many already dying of respiratory disease, were put down (compared to only 890 dogs).]

Veterinarian Karen Parks, owner of the Cat Clinic of Cheyenne, agreed that in this area, cats don't get the same respect as dogs.

"This is a human-caused problem," she said. "(Cats) are looked at as disposable property."

While she said she knows cat colonies are unpopular with wildlife biologists, the managed cat colony is a compromise that suits her, Parks says.

She said such colonies can be found everywhere in town, including behind her clinic.

Three years ago, she trapped, neutered and vaccinated 13 cats and began feeding them. Since then there have been no kittens. The colony is down to nine cats, and although two new felines showed up, they appeared tame and Parks was able to trap them and find them homes.

Parks' colony is unusual. Other managers suffer from burnout or depletion of funds trying to keep up with neutering and vaccinating the new cats moving in.

Vaccination is an equally important part of Parks' program.

The American Bird Conservancy says cats are the domestic animals most frequently reported as rabid. Feral cats can transmit diseases to native wild cats like mountain lions, bobcats and the Florida panther. They also carry diseases that are transmissible to humans.

Currently, Parks sends six or seven feral cats a week to Colorado State University for spaying, neutering and vaccination before returning them to where they were trapped. She said it gives veterinary students experience and, if no other cats ever were abandoned, that eventually would be the end of cat colonies.

Roaming is risky for cats Parks said 85 percent of her clients keep their cats indoors or supervise them outdoors.

[Compared to Maine where she practiced previously and where most cats were outdoors, she said emergencies are way down here.]

The idea that cats are smart and can fend for themselves is inaccurate: The life span of the average stray cat is only three years, said Tara Knight, Parks' assistant.

According to the Humane Society, "Cats kept exclusively indoors often live to 17 or more years of age."

Farr said 95 percent of feline injuries he sees are caused by outdoor hazards. Cats are hit by cars, preyed on by coyotes, get in fights, develop wound infections or are poisoned by antifreeze. They also are attacked by dogs and abused by people.

**Tips for keeping both your cat – and wild birds – safe (front page)**

**Keeping cats safe**

**Spay or neuter your cat**

Veterinarian Karen Parks of the Cat Clinic of Cheyenne said she has no tolerance for clients' desires to let their cats produce litters. Cheyenne Animal Shelter statistics show there are plenty of cats that need homes without adding more kittens.

**Keep your cat indoors**

The American Bird Conservancy's "Cats Indoors!" campaign has many ideas for turning outdoor cats into happy indoor cats. Play with them!

**Outdoor enclosures**

Build or buy an outdoor enclosure where your cat can safely be left. Check with the SafeCat Outdoor Enclosure (www.just4cats.com), The Cat Enclosure Kit (www.cdpets.com) or Kitty Walk (www.midnightpass.com).

**Adopt a friend for your cat**

Consider adopting a companion cat or dog of the opposite sex.

**Grooming and cleaning**

Trim claws every week or two. Scoop the litter box daily. If you use clumping litter, it needs to be changed only every two to four weeks.

**Don't feed stray cats**

Take them to the animal shelter. They may be someone's lost pet. If not, adopt them, get them spayed or neutered, vaccinate them and make them an indoor cat.

**Keeping birds safe**

Putting a bell on a cat doesn't work because birds don't associate its tinkling noise with danger. Keeping cats indoors or under control is the only solution.

**But what about your neighbor's cat?**

--Ask your neighbor to keep the cat indoors or in an outdoor enclosure. If the neighbor does not, humanely trap the cat and take it to the shelter and explain to whom it belongs.

--Keep bird feeders away from places where cats may hide. Try placing poultry or rabbit wire fencing around bird feeders and bird baths. The fence need only be 2 feet high and 4 feet in diameter. If a cat tries to jump over it, it gives birds a chance to fly away.

--Keep your cat in your yard, or keep other cats out, by installing cat-proof fencing. Two brands are the Cat Fence-In System, www.catfencein.com, and Affordable Cat Fence, www.catfence.com. Both are mesh netting systems that attach to the top of the fence. Another, Kitty Klips, www.corporatevideo.com/klips, offers free directions for installing PVC pipe so cats cannot clear the fence.

# 195 Studying the science of birdsong

Wednesday, May 18, 2005, Outdoors, page C1

## "The Singing Life of Birds, The Art and Science of Listening to Birdsong"

Donald Kroodsma's "The Singing Life of Birds, The Art and Science of Listening to Birdsong," has been released at this most appropriate season.

More birds sing in the spring than any other time of year, the drawback is birdwatchers will be out in the midst of migration rather than reading a book.

And it's a big book – 480 pages. The good news is that the author is a storyteller as well as a scientist.

Of the nearly 10,000 bird species worldwide, some have their songs encoded in DNA and some learn their songs.

"Of those that learn, some do so early in life, some throughout life; some from fathers, some from eventual neighbors after leaving home; some only from their own kind, some mimicking other species," Kroodsma writes.

"Some species sing in dialects, others not. It is mostly he who sings, but she sometimes does, too. Some birds have thousands of different songs, some only one, and some even none. Some sing all day, some all night. Some are pleasing to our ears, and some are not. It is this diversity that I celebrate."

Kroodsma, professor emeritus at the University

of Massachusetts, Amherst, and a visiting fellow at the Cornell Lab of Ornithology, uses thirty bird species to illustrate the different aspects of "avian bioacoustics" as the field is known.

His research is often opportunistic, taking advantage of travel over the years to compare the repertoire of different populations of the same species or similar species.

In 1968, as a senior at the University of Michigan, Kroodsma was introduced by famed ornithologist Olin Sewall Pettingill to the bird then known as the rufous-sided towhee. It has since been split into two species – the spotted towhee, found here in the West, and the eastern towhee that Kroodsma heard in Michigan singing its distinctive "drink-your-te-te-te-te-te-te-te-te."

In grad school at Oregon State, Kroodsma found that the western bird, the spotted towhee, sang the tune differently. This bird leaves off syllables at the beginning and performs twice as many variations.

While you might think determining whether a towhee has eight or four different songs would be an exercise in frustration – it isn't for Kroodsma.

After 30 years of experience, he can practically see the sonogram printout of what he records with his parabolic dish microphone before the computer spits it out. The sonogram is like the musical staff showing the pitch and duration of sounds.

But why do the Oregon towhees have more songs, why do they share songs with their neighbors and even match them as if in reply, while the New England towhees don't?

Kroodsma attributes this difference to the Oregonians being on their territories year-round while the others are migratory and may not have the same neighbors from year to year.

Kroodsma tested his hypothesis in 1987 while visiting Florida where the white-eyed variation of the eastern towhee lives year round. As he had predicted, the bird had a large repertoire shared by neighbors. Someday Kroodsma hopes to get to the Great Plains where both spotted and eastern towhees are migratory and see if they both sing like the birds in the northeast.

Kroodsma also hopes we will all learn to listen to birdsong more closely.

In Appendix II he describes the equipment needed

to make his recordings. For every fascinating story he tells, there is at least one sonogram printed with an extensive caption at the back of the book. To help us train our ears, he includes a track of the recorded bird song on the CD included with the book.

This might limit where you sit and read to places within reach of a CD player.

You can read the author's commentary while listening to each track and find the page number of the corresponding sonogram.

Buried in the back of the book, I discovered the Notes and Bibliography where Kroodsma has included footnotes, citations for studies he mentioned and more commentary.

While many of the references cited are from professional journals that are probably incomprehensible to anyone but ornithologists, Kroodsma does offer reading recommendations for the rest of us.

But first I plan to digest what he's written.

By following his instructions, I might be able to recognize individual songs when the spotted towhees visit my yard next week on their annual spring tour.

# 196
Wednesday, May 18, 2005, Outdoors, page C4

## "Identify Yourself, The 50 Most Common Birding Challenges"

"Identify Yourself, The 50 Most Common Birding Challenges" by Bill Thompson III (2005, Houghton Mifflin, $19.95)

Where birding field guides to birds leave off, "Identify Yourself" takes over. It contains a collection of columns written for Bird Watcher's Digest by the author and the magazine's other editors.

The book tackles all the problematic groups – sparrows, shorebirds, gulls, hawks, swallows and warblers for beginning to intermediate birders.

I found it handy this spring for an unusual sparrow in our backyard. Looking up the visitor was slightly complicated by the author's idea that I should first learn the song sparrow thoroughly.

I don't see song sparrows often, but the other comparisons helped me select the Lincoln sparrow. This surprising identification was poignant since we'd just begun to mourn the passing of our elderly dog of the same name.

"Identify Yourself" is not a stand-alone. One still needs Kaufman, Sibley or Peterson to get the full story on a species. But the book is profusely illustrated by Julie Zickfoose and in my experience, it is always helpful to see a questionable bird from more than one artist's point of view. The book's most helpful advice is in an introductory chapter's 20 simple rules for bird identification. While all field guides have something similar, this one is worth reading.

I like Rule 1, "Look at the bird, not at the book." I can't tell you how many times I've stared at a bird, reached for the book and frantically flipped pages only to look up and discover the bird has flown.

I also like Rule 21, the bonus rule, "See more birds. Have more fun." You can't go wrong with a book that while parsing chickadees cheerfully admits, "We will only be discussing sitting birds. Trying to separate the two in flight is insane."

# 197
Wednesday, May 25, 2005, Outdoors, page C1

## UW researchers seek hummingbird secrets

Hummingbirds are captivating creatures.

Bradley Hartman Bakken, a Ph.D. candidate at the University of Wyoming, can entertain an audience for an hour with his PowerPoint show of fascinating hummingbird facts.

For instance, hummingbirds can flap their wings 30 to 80 times per second and their tiny hearts beat 500 times per minute at rest and 1,200 when active. They are the only birds or vertebrates that can fly backwards.

But Bakken didn't come to the study of hummingbirds through the pursuit of trivia. His interest is in the physiology of kidneys, and he began working with hummingbirds as an undergraduate under the guidance of Carlos Martinez del Rio, a professor in the Zoology and Physiology Department.

Bakken discovered that hummingbirds use – or actually don't use – their kidneys in a unique way.

They live almost entirely on flower nectar, except for the occasional bit of protein from a passing insect.

Because nectar is mostly water and hummingbirds need a lot of sugar, they must expel a lot of water. Bakken said if humans drank as much water as hummingbirds in proportion to their body size, they would die – their kidneys would be overwhelmed.

But hummingbirds and humans are both able to lose water through breathing and evaporation through their skin.

Hummingbirds are just much more efficient. Because they are so small, their proportion of skin surface to body volume is very high.

Humans, being bigger, have a much lower ratio so sweating doesn't help us as much.

In fact, hummingbirds give off so much sweat, another hummingbird researcher, Ken Welch, a visitor to UW from the University of California Santa Barbara, said they smell like wet dogs.

The only time hummingbird kidneys kick in is when the birds are feeding, Bakken said. They don't use their kidneys overnight since they can lose as much as 11 percent of their body weight just by evaporation during that time anyway.

If humans lost as much, they'd be in a coma.

Bakken has been working with captive broad-tailed hummingbirds. His next step will be to determine the effect of a lower surface to volume ratio in a larger hummingbird. He travels this fall

to Santiago, Chile, for three months to study the giant hummingbird. This larger bird weighs 21 grams and is similar in size to a sparrow – dwarfing Wyoming's broad-tailed hummingbirds that only weigh about 3.5 grams. [Later, he'll look into nectar-feeding bats in Mexico.]

Scott Carleton, another UW doctoral candidate, also works with hummingbirds to explore the physiology of energy use.

"They have the highest mass specific metabolic rate of birds," he said, "and they're easy to study – it's all nectar."

Carleton wanted to know whether hummingbirds operate more on stored energy or on the energy that comes directly from nectar.

Because the sugar from sugar beets has a different carbon isotope signature than cane sugar, the breath of a hummingbird can be analyzed with a mass spectrometer to determine how much of which sugar it is burning.

The hummingbird is fed one kind of sugar and then switched to the other. The first sugar, stored as fat, produces one isotope and the other, burned as it is consumed, shows the other.

Welsh traveled to Wyoming to spend three weeks here this spring to answer a similar question. He demonstrated how his research subject takes a sip of nectar from within a mask that analyzes the carbon dioxide and oxygen in its breath. The results appear instantly on a computer graph.

What fuels a hummingbird when it takes its first sip of the day?

It appears that within five minutes the bird fuels hovering flight totally on the sugar it is ingesting. Humans also make use of sugar quickly when exercising intensely, but still have to get 50 percent from stored energy.

The researchers also found that caged hummingbirds, which are less active than those in the wild, are able to maintain their weight when tempted with a constant source of sugar water. Except twice a year, Bakken said.

Even though day length is controlled and never changes for his hummingbirds in the laboratory, they tend to put on extra weight in the spring and fall as if in preparation for migration.

Bakken said this could be in their genetics and not just a response to day length or temperature.

### Hummingbird facts

--331 species of hummingbird are found in North, Central and South America, but fossils of modern-type hummingbirds from 30 million years ago have recently been found in Germany.

--Ecuador has the most species, 163, and the U.S. has 19. Wyoming has black-chinned, calliope, broad-tailed and rufous hummingbirds.

--The rufous migrates farther than any animal when comparing distance to body weight. In spring they migrate from Mexico along the coast of California to Alaska, then return south via the Rocky Mountains in July and August. If a 6-foot man took as long a journey proportionate to his size, he could make 13 trips to the moon and back.

### What to feed hummingbirds

Wyoming wildflower nectar is 80 percent water, so make yours four parts water to one part sugar. Use red feeders, but don't dye the nectar. Use only regular table sugar – either cane or beet sugar but not honey since it's too waxy and could carry mold spores.

# 198 Local bird checklist makes its debut

Wednesday, May 25, 2005, Outdoors, page C2

Trumpets please!

The long-awaited "Checklist of the Birds of Cheyenne, Wyoming and Vicinity" is now available at the Cheyenne Audubon website and at local Audubon events.

The checklist has the names of 324 species seen in the Cheyenne area, including season and abundance information for each.

It's a handy guide for boasting. For instance, in our backyard this spring we've had white-crowned sparrows for several weeks. The checklist shows the species as "C," or common, in spring and fall migration. But it gets an "R" for rare in winter and not observed at all in summer.

A field guide's 1-inch square range map of North America for this species makes it hard to tell specifics for our location. The description may mention that white-crowns spend the summer in the mountains, which explains why they aren't here then. But there's no wow factor for white-crowns in Cheyenne in spring.

On the other hand, the lesser goldfinch that spent an hour at our thistle feeder last week gets a wow. It's marked as rare during spring and fall migration and has not been reported any other season.

A local checklist is an invaluable aid to any birders new to an area or traveling through. It keeps them from calling the state bird hotline for sightings of the yellow-headed blackbird, for instance. It's a common bird here, but rare on either coast.

For us locals, the checklist gives us guidance on which warblers to study before spring migration or which sparrows to pick from when faced with an identification challenge.

It's also a useful way to check off – hence the name 'checklist' – which birds you've seen in Cheyenne. You could have one copy of the list for each year, one for each field trip and one just for your backyard.

If you don't write down every bird you see while peddling the Greenway, reading the list when you get home will jog your memory.

Where, you may wonder, does the information for this checklist come from?

This index to local birding was compiled by three members of the Cheyenne-High Plains Audubon Society. Jane and Robert Dorn, co-authors of "Wyoming Birds," have 25 years of Cheyenne birding experience documented by extensive notes. Co-compiler Greg Johnson is close on their heels in expertise.

Whether to list a bird as abundant, common, uncommon or rare is a judgment call and every checklist has its own definitions.

For this one, expect to observe an "Abundant" species almost every outing with little effort, in appropriate season and habitat. Think house sparrow, house finch, western meadowlark, red-winged blackbird, European starling and American crow.

"Common" means some effort may be required to locate the species – like the American goldfinch, Swainson's hawk, great blue heron, American coot.

"Uncommon" means considerable effort is usually required – or serendipity. These include the ruddy duck, ferruginous hawk, Eurasian collared-dove and belted kingfisher.

"Rare" means that the species is difficult to find because of low numbers and/or secretive habits. I hope that the Laramie County Community College birding class members realized how lucky they were to see a black-necked stilt a couple weeks ago.

There's also the category "I" for introduced species such as northern bobwhite, an escapee from game bird farms. The category "V" stands for vagrant – birds blown off course during migration – like the painted bunting for which we've had one or two sightings in the last 25 years.

In addition to his avian expertise, Greg digitally formatted the checklist. He is also the one who has his phone number on it for receiving updates and additions.

There's a distinct advantage to modern technology when compiling and publishing a checklist.

First, it's easy to update. Greg will never have to re-type any of the 324 bird names unless the American Ornithologists' Union changes them.

The biggest advantage is that no one has to fund the printing of hundreds of copies – and then get stuck with them when they need updating.

While the chapter will have a limited number of printed checklists available, anyone can go to the website to print their own. Portable document files, PDFs, print up cleanly, without any Internet tag lines. And the checklist is formatted for standard paper size.

I was not happy when I tried to piece together a legal-sized checklist for Juneau, Alaska, I found on the Internet last year.

Of the checklists I've collected from around the country, some are printed on cardstock, some on slick paper with color and some cost a dollar.

But for a sponsoring organization like Audubon for which one of the goals is education, it doesn't make sense to restrict access to a great learning tool like the checklist.

Last year Greg agreed to be chapter compiler and was the one to finally bring the checklist to life. I'll have to ask him if he has any great ideas for the next step. That would be taking all the random bird sightings and getting them into a publicly accessible internet database.

But right now is the height of migration. Time to get out there, checklist in hand.

# 199 Backyard hosts lively bird theatre

Wednesday, June 22, 2005, Outdoors, page C2

I could spend all day at the kitchen table this time of year. It has a great view of the backyard where avian dramas unfold by the minute.

The neighbor's row of crabapple trees and our shrubs form the stage backdrop. Dappled sunlight shines through boughs overhead like spotlights. We've scattered props, most recently adding a birdbath.

Within minutes of filling it the first time, a house finch perched on the rim. It dipped to scoop up a drink and tilted its head back to swallow. I haven't caught the mourning doves drinking yet. Doves and pigeons are the only birds that can suck water. However, they prefer their water at ground level, not on a pedestal.

Since it is a birdbath, a lot of splashing goes on. But most of the action is center and front, where we have the sunflower tube feeder. We go through more seed in May and June than we do the rest of the year.

It's an energy intensive season. Laying eggs and feeding young takes a lot of calories. By the first of June I have a hard time telling streaky brown females from streaky brown juveniles because both indulge in begging behavior.

I researched the differences through my new subscription to The Birds of North America Online: www.bna. birds.cornell.edu/BNA. It's the definitive modern compilation of bird information. A few years ago, the 18,000-page hard copy was offered for $2,000. Now all the information is available, searchable and constantly updated for $40 a year.

Compare "Begging often so intense that to a casual observer, fledglings and parent may appear to be fighting," with "Courtship feeding – female assumes a distinctive begging posture by fluttering her drooped wings, tilting her head up and giving excited call notes."

When house finch parents feed young or males feed females, it's normally regurgitated stuff. One day I observed another, more equal feeding strategy.

A female sat on a perch at a seed port feeding. The activity around her was like Grand Central Station. Often birds would maybe get one seed before being replaced by an aggressor. Since the feeder is encased in grill work, birds frequently land on it before grabbing a perch.

This time I watched a male land above the feeding female but instead of scaring her off, he reached past her and took a seed, then she took one. They alternated several times.

Our thistle feeder didn't get much action over the winter. It's one of those that attempts to exclude house finches by placing the ports below the perches. One of the target species, goldfinches, didn't arrive in our yard until May. It took a little practice for them to feel comfortable feeding while hanging upside down like bats.

One observant male house finch decided to give it a whirl. He actually was able to peck at the seed before losing his grip. Other house finches were watching and soon mobbed the feeder, tying up all six perches.

2005

193

But none of them got further than a fluttering of wings. Finally, they all took off, leaving the thistle to the goldfinches. None have given the feeder another try.

Some of the seed-eating players prefer to wait for the chow to hit the ground. Lazuli buntings and mourning doves are fond of millet scattered on the patio. It gets to be quite a mess though.

Of course the day it was swept clean, two Eurasian collared-doves stopped in and didn't find anything to eat. Since they live in the neighborhood year round now, maybe they'll give us another chance.

A migrating black-headed grosbeak also found no food. They always enjoy our shelf feeder, but with all the rain, I had decided to clean it and let it dry out.

Not all yard visitors are attracted to seed and feeders. A brown thrasher picked over our lawn and tree debris one day, looking for insects, as has a Swainson's thrush we've seen several times.

A drawback to feeding a flock of house finches is that their constant squabbling drowns out other birdsong.

By concentrating, I've been able to pick out the melody of a vireo in the background a couple times. One morning a green-tailed towhee was in full throat right outside the bathroom window, which also overlooks the backyard.

Lately, our dinner-time companion has been a western wood-pewee. It perches very uprightly on a wire, making finch posture look poor by comparison. Its head constantly turns watching for flying insects. Then it swoops out to nab one and returns to its perch.

Over the course of the summer the dinner theater productions will change. The crows have been quiet, but they'll be back when their young fledge. Another month or so and those won't all be large insects hovering over flowers. Some will be hummingbirds stopping by on their journey south.

Gazing at an aquarium is, I heard once, a way to lower one's stress level. The antics of birds aren't nearly as smooth and quiet as swimming fish but gazing into our backyard aviary works for me.

# 200 Flamm Fest finds record number

Wednesday, July 20, 2005, Outdoors, page C2

Kim Potter undeniably deserved to be crowned Queen of Flamm Fest earlier this month. Like other queens, she displayed talent – a talent for finding flammulated owl nests.

Having honed her skills in Colorado, Potter was able to find a large aspen with a hole 20 feet above ground. By lightly scratching the bark she got a female flammulated owl to come to the entrance. Because it was the second weekend in July, we were certain that Kim had found the first documented nest in Wyoming.

Flamms are tiny – less than 7 inches long and weigh just over 2 ounces. They are 5 inches shorter and weigh less than half the amount of the northern flickers that make many of the holes the owls nest in. Flamms prey on insects, especially moths, by inspecting infested trees.

Their name probably comes from an old word that means "with flame" as some appear to have a reddish brown color. Flamms are a western mountain species, although they are seen at low elevations during migration. The U.S. Forest Service considers them a sensitive species.

Several years ago, Rocky Mountain Bird Observatory biologists Doug Faulkner and Rich Levad made a list of bird species that had not been documented in Wyoming but which they felt should be here because of the similar habitat the birds use in neighboring states.

With their knowledge of preferred flammulated owl nesting habitat in Colorado, Rich and Doug made an educated guess that other RMBO biologists confirmed when they found a flamm in the Battle Creek area three years ago.

Historically, these owls have been considered rare, but most likely their camouflage coloration, small size and quiet hoots made them easy to overlook.

Thus, we created Flamm Fest, the nickname given to the fifth annual Cheyenne-High Plains Audubon Society campout. Our mission was to spread out and see how many more flamms we could find.

Just about every one of the 31 participants, ages 11 and up, got a good look at one, either the female or, on Friday night, a male responding to Kim's tape. She was demonstrating the survey techniques we would be using the following evening.

We divided into nine teams and each was assigned a route to drive. At half mile intervals the recording was played and surveyors waited for an answering hoot.

The road our group was to travel was closed to vehicles so we set off on foot at twilight, only to discover a culvert was missing over a wide stretch of icy water. Everyone crossed with different degrees of dryness.

On the way back we walked without turning on flashlights and stopped every 500 paces to call for owls. We did have a response from a saw-whet owl, but no flamms.

Five of the teams were luckier and counted a total of 10 flammulated owls. At a lot of the survey points it was too windy or too close to running water to hear return hoots. At some points the habitat was very different. But it is just as important to know where the owls are not as it is to know where they are.

Other owls that responded or were seen were long-eared, eastern screech, great horned and possibly a pygmy.

During daylight on Saturday, we checked out the only known colony of purple martins in Wyoming. They also like old flicker holes in

194                                                    CHEYENNE BIRD BANTER

old aspen trees.

The whole grove was aflutter with several other cavity-nesting species: mountain bluebird, red-naped sapsucker, house wren and tree swallow.

Purple martins in the west are a different subspecies than those in the eastern part of the country. The westerners don't use manmade apartment-style bird house complexes – but then no one has ever put one up near where they live in the forest. We looked for other colonies but didn't find any.

One unexpected bird was a bushtit down along the shrubby lowlands of the Little Snake River valley. Both the tiny round bird and the spruce tree it nested in were completely out of their normal forest habitat.

We were also very close to the state line. A GPS reading may show the nest is a latilong breeding record for Colorado. But the bird itself, since it flew over the fence marked "Wyoming State Line," will at least be a Wyoming observation record.

Our Flamm Fest campers

were from an unexpected diversity of locations. From Wyoming, 19 people represented Cheyenne, Casper, Lingle, Riverton and Saratoga. We also had birders from the Denver area, western Colorado, Salt Lake City, Rapid City, New York City, Washington, DC and Chicago.

If a simple Cheyenne chapter outing and the lure of flammulated owls can draw this group, who knows whom we'll find on next year's campout to the Bear Lodge in the northeast corner of the state.

Also, what species might we find?

Broad-winged hawk, golden-winged warbler, yellow or black-billed cuckoos and black-backed woodpecker are some of the Black Hills specialties not found elsewhere in Wyoming.

We've got to find another catchy title – and maybe a trophy if Kim joins us again and proves to be Most Valuable Birder.

Mark your calendars for June 23-25, 2006.

# 201 New field guide is a useful tool on the road

Wednesday, August 17, 2005, Outdoors, page C2

It's been about a dozen years since our last summer family trip across the ocean by car. Whether you call it the sea of grass or the Great American Desert, crossing the Great Plains is a test of endurance even in the modern age.

Speeding along at 75 mph keeps the tedium to a minimum. Books on tape keep us awake and air conditioning keeps the exhaustion from hot weather at bay, but we are never truly on the Plains, only passing over them.

Now here's a way to help us slow down and enjoy. Published just in time for the planning of our excursion was "Birds of the Great Plains," published by Lone Pine Publishing International.

Though there is a plethora of bird field guides for North America, a regional guide is useful. The bird descriptions can concentrate on the local plumage variations and the range maps have more detail.

The book defines the Great Plains as a large vertical rectangle including the eastern parts of Montana, Wyoming, Colorado and northeastern New Mexico plus all of North Dakota, South Dakota, Nebraska, Kansas, Oklahoma, the panhandle of Texas and the western edges of Minnesota, Iowa and Missouri.

Although the Great Plains extend into Canada, Lone Pine, a Canadian company, does not include them here because they are covered in their provincial guides.

Within the given geographic boundaries, the introduction maps three major ecosystems from west to east, short-grass prairie, tall-grass prairie and tall-grass savanna. There is also the island that is the Black Hills and oak-pine woodland in the southeast corner of Oklahoma. Unfortunately, there are no descriptions of these areas.

The range maps accompanying each species only show

the Great Plains, so I have the same complaint I have for even the general North American bird guides – I want to know everywhere a species goes. [Some warblers spend more time in South America than here.]

But minor matters aside, "Birds of the Great Plains" has features other guides would do well to emulate. It begins with a color-coded reference guide with thumbnail pictures of each species covered, sorted by taxonomic group. On the back cover are thumbnails representing each group, again color-coded and with page numbers.

The color for each group also marks the top edge of the pages where the pictures are repeated in a large, easy-to-see format, one species per page.

At the bottom of each page are the finer points of identification – size, habitat, nesting and feeding behavior, voice and comparison to similar species.

What I like about this guide is the long introductory paragraph for each species, the story teller's approach that includes gee whiz facts, literary references, even how the bird was named – everything that distinguishes it from others and makes you care about it.

These mini essays would make a book in themselves. They are unique in the world of field guides, but not surprising when the major author, Bob Jennings, had a 30-year career as a naturalist and interpreter in addition to being an avid birder. The other two authors, Ted T. Cable and Roger Burrows, also have extensive credentials in nature interpretation.

"Birds of the Great Plains" would make a good first field guide for our area. It isn't cluttered with east and west coast species we'll never see here. It acknowledges using less than scientific terms when describing birds, but it has a glossary and the requisite diagram of a bird's parts.

It also has, separate from the index, a checklist of all 457 birds that have been officially recorded in the Great Plains states and indicates which are introduced, threatened or endangered.

But how useful is this field guide here in Cheyenne, on the west edge of the plains?

I compared it to the Cheyenne – High Plains Audubon chapter checklist. Only 24 Cheyenne species, out of 324, were missing in the guide,

and all of those are rare or uncommon, except for the mountain chickadee. But at least the descriptions of other chickadees mention it as a similar species.    On the other hand, the book has 50 eastern species not on the Cheyenne list so you'll be set for future immigrating species and that visit back to the grandparents in Nebraska.

In case you don't have "relative" destinations on the Great Plains, this field guide gives you 91 top birding sites, highlighting

20 with descriptions of the landscapes and the species to be seen.

In Wyoming, Devils Tower National Monument, Thunder Basin National Grassland, Keyhole State Park, Glendo Reservoir and Table Mountain Wildlife Habitat Management Area make the general list.

If we weren't also visiting relatives in the Great Lakes states (and Lone Pine has bird guides for Chicago, Michigan and Wisconsin), "Birds of the Great Plains"

could easily be the only field guide we'd throw in the car.

Lone Pine publications, with their tough and flexible, plastic-coated covers, are good for throwing around. My copy of "Plants of the Rocky Mountains" survived last summer as a well-thumbed reference book at Camp Laramie Peak.

Whether or not we get time to bird, Bob Jenning's essays will make good reading in whatever spare minutes come along.

# 202 Familiar faces fly by

Wednesday, September 14, 2005, Outdoors, page C2

A bald eagle before breakfast. Where else but Eagle River, Wis.? We were waiting on my cousin Dick's porch for everyone to assemble for pancakes.

He and his wife Judi built their retirement home on the Deerskin River. Because landowners are no longer allowed to clear trees down to the water's edge, the eagle followed the course of the river behind a screen of maples and alders.

Where else but northern Wisconsin would we see eagles? Back in Wyoming and many other places, thanks to the Endangered Species Act. Though we were three days' drive from home, other birds were also familiar in the northern forest at mid-August.

After breakfast, we took two canoes to a put-in upstream on the Nicolet National Forest. We paddled through a sea of rushes growing in a recently drained reservoir where the river was trying to establish its course

again. It left barely submerged sandbars to catch us by the keel. Once past the site of the defunct dam we had to dodge rocks instead.

The first familiar bird spotted was a backlit hawk standing on a snag. We have lots of those unidentifiable hawks at home, but in Wisconsin the choices also include red-shouldered hawk.

Dick and Judi (and Buddy, the dog who hyperventilates at the sight of a paddle) were in the lead when they spotted a river otter. We were able to catch a glimpse of its face before it retreated. For a while, a great blue heron scouted the river ahead of us, letting us approach and then flapping ahead. Was it watching for the fish we spooked and sent downstream?

We flustered a family of common mergansers, four young and their parents. As we approached, they too would take off, but instead of flying clear, they raced over the water six abreast like synchronized swimmers, feet

paddling furiously and wings beating, leaving a wake the width of the river.

This was repeated time after time. Finally, the river was wide enough that they allowed us to slip by. Then they paddled furiously again – upstream this time. These fish-eating, diving ducks can be seen not far from Cheyenne at Curt Gowdy State Park. Another familiar fish eater, the belted kingfisher, crossed the Deerskin a couple times.

Back at the house, Dick and Judi's feeders were busy with black-capped chickadees, a brown creeper, red-breasted and white-breasted nuthatches, and goldfinches – pretty much like home except we get mountain chickadees too. However, the ruby-throated hummingbirds working the nectar feeders probably will never show up in our garden – we have several western species instead.

The new addition to my life list was only a flash seen

through the windshield as we drove a forest road – and I was sitting in the back seat. It was a pileated woodpecker.

Lately, in the frenzy over the comeback of the ivory-billed woodpecker, bird magazines are at great pains to point out the differences between the two big black and white birds with red crests. At this point no one wishing to keep their credibility intact will report an ivory-billed outside the Deep South. But there are a couple records for pileateds in Wyoming.

While we did pack a pair of binoculars and a bird book, this trip was primarily about visiting lots of relatives. Even so, birds were everywhere. Nighthawks dived after insects over Susie's barn near North Platte, Nebraska. Cardinals whistled above Hertha's house in Skokie, Illinois and also at the Argos' in Crown Point, Indiana.

Cousin Bill, the horticulturist, has a garden in south central Michigan to delight nectar and seed eaters alike. In the oak woods at Cousin Dean's near Grand Rapids, Michigan, and in the woods at the edge of the hay meadow at the Dykstra farmstead not far away, chickadees had the leading song.

Chickadees also had the distinguishable melody in the Schmeeckle Reserve at the University of Wisconsin - Stevens Point, where I spent so much time scrambling around 30 years ago.

So many leaves. You really would have to learn the local bird songs when the birds can hide so well. I suspect there were a lot of interesting species if we'd known what to listen for.

But shining like beacons along the Rock River, south of Rockford, Illinois, were great egrets, their all-white plumage flashing in the sunlight as they waded, hunting for breakfast.

The birds we saw from the S.S. Badger, which ferried us across Lake Michigan, were also completely visible, though I wondered why we saw a great blue heron flying five miles from shore since they also wade for food.

Everywhere on this trip we saw monarch butterflies – even in the middle of our lake voyage. Can they successfully migrate over 50 or more miles of open water?

Returning home, my e-mail was full of Wyobird postings heralding the first migrating warblers. Right now you can keep a sharp eye on the leafy trees and find several species of these small, yellowish birds looking for insects.

For two weeks we were the ones passing through everyone else's habitat, but now that we're home, it is the birds in fall migration passing through ours.

# 203 Rescued bird finds his way home

Wednesday, October 12, 2005, Outdoors, page C2

Hootie's mother once told him she met his father at a backyard feeder in Cheyenne early last spring. She said as soon as the weather warmed they went up to the forest to look for a building site, eventually settling on a ponderosa pine with a broken top and a rotten spot.

Hootie's mother did most of the tree excavation and stuffed the new home place with shredded bark, grass, feathers and fur. It became a cozy fit as the six red-breasted nuthatch youngsters grew larger.

"Stay away from the doorway!" was his parents' constant refrain as they darted in with different kinds of beetles, spiders and caterpillars up to 18 times an hour.

"The squirrels will get you if you don't watch out!" they said as they left with another fecal sack as smelly as any diaper.

One of the furry monsters climbed within a few feet of the nest entrance but was met by Hootie's mother. Perching above, her wings outstretched, she swayed slowly from side to side, mesmerizing the would-be baby eater until it woke with a start and fled.

Keeping Stellar's jays away was more difficult. Hootie's parents spread pine pitch around the edges of the entrance. None of the vain big birds would risk dirtying its feathers by poking its head in. The pitch kept out pesky ants too.

Finally, in late July, three weeks after hatching, the big day arrived. The children were dressed in garb nearly identical to their parents'. Each had a blue-gray jacket, pale red vest and flat black hat. Their white faces were marked with a black stripe through the eye.

Hootie's parents laid down squirrel fur across the sticky doorsill and began to encourage their children with a new song, "Come on out, the weather's fine. Flying is a wonderful thing. You'll love it. And walking up and down tree trunks is a hoot."

As Hootie tottered at the threshold of the bright new world, the rush of air triggered his genetic reminder to flap. He clumsily made it to the neighboring tree. Soon all his siblings were enjoying flying and finding insect treats hiding in tree bark, though their parents planned to feed them for two more weeks.

Two bicyclists were also enjoying the sunny day, following a narrow trail through the forest. But they didn't look like any of the dangers Hootie's mother had described. He never even saw what he hit.

On the ground stunned, he thought maybe there was something about using wings his parents had forgotten to mention.

But he could still flap them. It was his leg that wouldn't hold up.

"Mommy!" he cried. "Daddy!" And then he was scooped up and put in a small dark place, just like his old nest.

Six weeks of recovery in a bird cage was like returning to the nest. Someone brought him turkey scratch and then mealworms every day and someone cleaned up after him. But there was also sunlight from the nearby window, a seed cup to sleep in, a water cup to bathe in and seeds to pull from a stick and hide.

Only there was no one to answer his "yank-yank" call except well-meaning people.

One day, near the end of his hospitalization, though he still had a limp, the bird cage was taken outside. The gust of air and the loud rustling noises alarmed him, but then he recognized the wind and aspen leaves.

Just as he settled down, real terror visited.

"What's a tasty morsel like you doing out here all alone?" asked a soft but wicked voice. A blue jay stuck his long bill into the cage.

"You'll never get me!" taunted Hootie from the far side, fluffed to his full 4½-inch height. "You're too fat to fit between the bars!"

One evening three days later his foster parents returned him to the wider world, bringing him to a grove of big cottonwoods and elms – kinds of trees Hootie had never seen before but had heard his parents describe.

"Cheyenne, I must be in Cheyenne!" he thought. Then he heard "yank yank," the call of another red-breasted nuthatch.

"It's party time pardner!" it said.

It was true. In the fall no one argued over mates or nesting territories.

"We're hanging with some Wilson's warblers just in from Canada. They'll only be here for a day or two before heading for Mexico. Aren't you glad we don't have to migrate that far? We can just stay and catch bugs sleeping in tree bark and fill in our empty spots with seed."

"My folks want to stay in the mountains this winter. They think the pine and spruce cone crop will be good. Me, I prefer the easy life. Ah, so many seeds, so little time. You hardly have to hit the same feeder twice in one winter!"

And so, dear reader, should you notice a red-breasted nuthatch at your birdfeeder this winter that favors one leg, give him greetings from all his friends: the bicyclists, the vet, the bird rehabilitator, his foster parents and me.

Author's note: This story is based on an actual rescue this summer and information from Birds of North America Online, www.bna.birds. cornell.edu/BNA.

# 204 Letters from a moose hunt

Wednesday, November 2, 2005, Outdoors, page C1

## Writer Barb Gorges documents the pursuit of Snowy Range moose

*A series of letters to my sister, Beth*

-----

Mountain Home, Wyoming
Friday evening,
September 30

Dear Beth,

Mark's finally winning the lottery for a moose license is probably a once-in-a-lifetime event so I thought I'd document it like Elinore Pruitt Stewart did in "Letters from an Elk Hunt by a Woman Homesteader."

If you remember, she was writing to her former boss in Denver about the trip by wagon in 1914 from Burntfork in southwest Wyoming to somewhere north of Pinedale. Elinore and her husband, older children and friends spent several weeks just getting to elk camp and another two weeks hunting.

We drive two hours from Cheyenne, but only for weekends. Instead of tents, we stay at Wiggams' cabin up here between Foxpark and Mountain Home.

It's humble, but it has heat, lights, a stove, a refrigerator and an indoor composting toilet.

We went out twice on scouting trips earlier this month. Mark's moose area, Snowy Range, overlaps areas for his deer and elk licenses. We had beautiful fall weather, but didn't see any big game, just tracks and scat. At least it's easier hiking in preferred moose habitat, along the creeks, than stepping over downed trees in deep timber where Mark hunts elk.

We left today as soon as Jeffrey got home from school. It's hard to believe at age 16 this is his fifth hunting season.

Tomorrow is also opening day for deer for both Jeffrey and Mark and since the weather is supposed to be warm, Mark wants to concentrate on finding deer. As thick-bodied as moose are, temperatures in the 70s could mean the meat wouldn't cool down fast enough to prevent spoiling.

Mike Wiggam will be in soon. He has a deer tag too. I don't have any tags this year. Ever since that antelope I'm happy enough to let Jeffrey use my rifle while I help spot game and carry it out. What do you and Brian have for tags this year? When do seasons open in Arizona?

I often wonder what Dad would think of his daughters taking up hunting and fishing – things he never did. But remember how much he liked climbing mountains!

Love, Barb

-----

Mountain Home, Wyoming
Saturday afternoon, Oct. 1

Dear Beth,

Boy, are we going to be sore tonight from all this walking! We got up at 5 a.m., had the standard Gorges hunting breakfast of instant oatmeal and V8 juice and were in the field by the beginning of shooting time, half an hour before sunrise.

There were lots of stars, but they faded fast and the tops of the aspens quickly lit up in neon yellow.

Mark and Mike have an amazing sense of direction. They rarely follow any of the numerous roads and after two hours or so, we always break out of the trees right at the vehicles. Of course, Mike's been hunting around here all his life and Mark's been hunting with him the last 15 years.

This morning we heard several shots but saw only squirrels. Back at our vehicles by 8:30 a.m., we ate the first half of our lunches.

On our second foray we were luckier. I was bringing up the rear when we crossed over a beaver dam on a small creek. There's always some ankle-breaker hole waiting under the long grass so I take my time. I thought everyone was waiting for me. Instead, they were watching a bull moose playing peek-a-boo among the tree trunks.

Mark's license is for a cow moose, but one without a calf. He followed the bull and discovered the cow and calf in the willows not more than 20 yards from where I'd floundered across. Mark says though it looks like a family unit, the bull is usually not the calf's sire and is only following the cow because it is rutting season.

This was also the excursion I noticed something dark along the game trail decorated with what looked like red seed beads. They were seeds from rose-hips. It was bear scat. Good thing in south-central Wyoming we only have to worry about black bears, but still....

We ate the second half of our lunch on a sunny ridge within sight of the sparkling dome of the observatory on top of Jelm Mountain, along a narrow track that was a regular highway for pickups and ATVs. Everyone drove sedately and nodded greetings. One couple stopped to chat. The woman was the only female hunter I've seen so far.

It seems like hunting outfits come in two types, either an old beater like Mike's (and at 9 years old, our Explorer is heading that way) or a brand new, $40,000 extended cab 4x4 diesel. However, maybe in deference to high gas prices, most of those giant trucks were carrying three or more hunters.

I've noticed too, some hunters dress like Cabela's catalog models. We, on the other hand, wear old jeans and flannel shirts with orange vests, though Mark likes his loose wool pants.

Remember those matching flannel shirts with the geese on them I got for us when we went to the Becoming an Outdoor Woman weekend in Raton, New Mexico, one red, one blue? Mine is wearing out, perfect for the dirty end of hunting.

We spread out to walk the ridge after lunch. Mike got to the end first and glassed the opposite mountain, finding a cow moose on an open hillside, followed by a bull. They are a wonderful chocolate color. But the funny thing is their legs are whitish, looking like they're wearing Mom's white nurse's uniform stockings over brown hair.

We decided to get a closer look, following a game trail that took a near vertical dive. All of us were thinking how the heck would we get a moose back up this mountain?

Jeffrey, Mike and I waited at the bottom. Mark crossed the creek and headed up toward the cow moose on the other side. Another bull appeared. Then there was a lot of bawling which Mark determined came from twin calves hidden in the trees.

We're back at the cabin now, taking our mid-afternoon nap before having an early dinner, the traditional pan of lasagna Mark makes and freezes at home. Then we'll head out again until shooting time ends.

Love, Barb

-----

Cheyenne, Wyoming
Sunday evening, October 2

Dear Beth,

No more moose yesterday or today. An hour before sunset last night Mark got a nice buck deer, a 3 by 3, and we all worked hard getting it and our gear uphill, thinking what it would be like to pack out a moose that weighs three or four times more.

On the way, Mark gashed the top of his bare head on a lodgepole branch and spilled more blood than the deer. We realized we had three first aid kits – back in the vehicles – so Mike and I sacrificed our personal wads of toilet paper and Mark stuffed them under his knit cap.

This morning Jeffrey got a small buck near the road between Foxpark and Lake Owen. I was the one who spotted it. Looked sort of like a stump with mule ears. Mike didn't get anything and can't come out again this season, but we always share since he is so generous with the cabin. Mark will be busy cutting and wrapping deer meat in the evenings this week.

Love, Barb

-----

Mountain Home, Wyoming
Friday evening, October 7

Dear Beth,

We stopped in Laramie to drop off some deer steaks with our starving college student. Bryan has too much schoolwork to come with us.

We drove up to the cabin later than last week. The sky was that rich turquoise, the moon just a new crescent and the mountains black. Saw only one deer with a death wish standing on the side of the road.

This weekend is cool enough to take a moose. It may snow by Sunday or Monday.

Love, Barb

-----

Pelton Creek, Medicine Bow National Forest, Wyoming
Saturday, 10:30 p.m., October 8

Dear Beth,

We have moose! After a whole day of hiking around and not seeing anything but tracks, just as I'm about to pull onto the Pelton Creek Road and turn towards the highway and the cabin, Mark says, "Let's go down the creek instead."

There was still the half hour of official shooting time after sunset.

Then Mark says, "There she is! Stop! Stop!"

As I pulled over he wondered if it would be too hard to haul her across the creek, but Jeffrey and I said, "Go for it!"

We waited while Mark went down the bank and made sure she didn't have a calf hiding. Then came the shot. I don't know how Mark manages to be so accurate since he never practices. Maybe it's because he's had the same rifle for 32 years.

I was official flashlight and leg holder for over three hours and finally had to come back to the Explorer and put on some more layers. The work is about as slow as that ranch buffalo you helped us with in Montana years ago, even though we have a Wyoming knife this time.

Mark suggested I take a nap while I'm here so I'll be the one in shape to make the 20-mile drive back to the cabin. I can see two little stars of flashlight bobbing in the distance below. The stars above are covered with clouds.

Moose have such expressive faces compared to antelope and deer, or maybe I've seen too many Rocky and Bullwinkle cartoons. But if I'm going to be a meat eater, I need to be brave. Anyway, in Wyoming animals are the most efficient way to harvest and process the vegetation that grows here.

Both Elinore and her friend, Mrs. O'Shaughnessy, shot elk on their hunt. But they let the men deal with the rest of the process.

It's a little spooky out here. The wind keeps sounding like someone is walking back up the bank.

Love, Barb

-----

Cheyenne, Wyoming
Monday morning, October 10

Dear Beth,

We got half the moose back to the cabin late Saturday night – or rather, Sunday morning. It was nearly 2 a.m. when we rolled into bed. Then, later that morning it took another two to three hours to get the rest of it. Everything would have taken substantially longer without Jeffrey's tireless teenage energy.

Unbelievably, Mark and Jeffrey got the moose and our gear to fit in the Explorer and we drove home very carefully in snow mixed with rain.

We had our first bites of moose meat for dinner last night. It's not tough or gamey-tasting despite the fact that moose eat mostly willow.

I'm glad it doesn't have a strong taste, the way antelope meat can sometimes be like a mouthful of sagebrush. We won't have to disguise it in sausage and spaghetti sauce.

We'll have over 250 pounds by the time the butcher is finished with the quarters and Mark cuts up the rest.

Mark took the hide to a taxidermist. It's been promised for about June. Mark took the head to Game and Fish so they could get the lymph nodes tested for chronic wasting disease.

I'm a little disappointed that we won't have to hike around the other three weekends of moose season. I love abandoning town commitments for an off-trail ramble and was having fun taking nature shots to enjoy on my computer desktop this winter.

Jeffrey still wants to see about filling his elk tag. I don't know what we'll do for freezer space if he gets one – maybe we'll have to throw a potlatch. I think you'll be getting some moose jerky from us anyway.

It'll be five years before Mark is allowed to try his luck drawing for another moose. Too bad we can't make the meat last at least that long without freezer burn. I suppose now Jeffrey will want to get his own moose. Hope he'll take Mark and me along!

Love, Barb

# 205 Snowy Range moose

Wednesday, November 2, 2005, Outdoors, page C1

The moose that live in the area roughly west of Laramie and east of Saratoga are part of the Rocky Mountain subspecies known as the Shiras moose. Adults average 600 to 800 pounds and stand 5 to 6 feet at the shoulder.

In summer, 60 to 90 percent of their diet is willow shrubs, said Eric Wald, a University of Wyoming graduate student. They supplement it with grass, sedges, wildflowers and aspen leaves. In winter if willows run short, they may resort to other shrubs, aspen bark and subalpine fir.

Generally, moose are most active foraging at dawn and dusk. They like to stay cool so on a hot day they go into thick timber or deep willows to ruminate.

A moose's four-chambered stomach digests woody material so well their winter droppings are like sawdust pellets.

Moose tracks show two toes a little over 5 inches long followed by two nickel-sized indentations made by the dew claws.

Wald estimates the Snowy Range moose population at 150 to 200, but because moose are loners, normal big game herd survey techniques are not accurate.

Wald has radio-collared eight moose to find out more about their seasonal migration patterns and what corridors they use. One young bull this fall used a road and collided with a vehicle.

Moose were transients in the Snowy Range area before 1978 when Colorado transplanted a population into nearby North Park. The young quickly dispersed over the border into Wyoming. In the 1990s the population took off and by 2000, the Wyoming Game and Fish Department set up Moose Hunt Area 38.

This year the area's quota was 10 licenses for any moose (hunters usually read that as bulls) and 10 licenses for antlerless (cows). Neither category includes cows with calves. The quota was allocated between resident and non-resident preference point and random drawings.

Al Langston, Game and Fish biologist, said odds ranged from 0.21 percent for the 966 residents who applied for the two bull licenses drawn randomly, to 100 percent for the non-resident who was the only one to apply for the preference point draw for the two non-resident antlerless licenses (the other license then went in the resident drawing).

This year a non-resident moose tag cost $1,201 and a resident tag $91. If your name wasn't drawn, the license fee was refunded and you were awarded a preference point.

If you draw a moose tag, you are not eligible to apply again for five years. But once you have your license, getting your moose is almost a sure bet. Wald said last year's success rate in Area 38 was nearly a hundred percent, except for the hunter who had a shot at a bull moose but waited for a bigger one.

Moose license applications are taken in January and February. There is a total of 42 other, mostly smaller, moose hunting areas in Wyoming in the western quarter of the state, the Big Horns and the Jeffrey City area. Contact Game and Fish, 777-4600, for more info.

# 206 Time's here for annual bird counting

Wednesday, November 9, 2005, Outdoors, page C2

The juncos are in – let the counting begin!

I heard the first twitters in my backyard at the end of October, two weeks after the first snowfall – a little late, but so is the killing frost. My uncovered geraniums were still alive as of Nov. 1 – our new backyard record.

Juncos always migrate down from the mountains in time for the beginning of the Cornell Lab of Ornithology's Project FeederWatch season which starts this Saturday.

If you enjoy feeding birds in your backyard and pausing to watch them, you can contribute to this citizen science effort.

No field guide is needed. For the $15 participation fee you get a full-color poster of common backyard birds. Whether or not you choose to participate, Cornell has a terrific Web site to help you identify more birds, www. birds.cornell.edu/programs/AllAboutBirds.

No binoculars are really necessary either for Project FeederWatch, especially if your feeders are as close to a window as ours.

Scientific background is not required. Participants get full instructions with a resource guide to bird feeding, plus counting protocol is very simple. How many birds of one species can you see the first time you look? If you see a larger number the next time, cross out your first number on the tally sheet and write in the new number.

Limited time shouldn't be a constraint either. The Project FeederWatch season runs through early April. You choose a reporting schedule of two consecutive days once every week or once every two weeks. I chose weekends so the rest of the family can participate. But if we're away sometimes, that's OK. We aren't allowed to make up for missing dates. Thousands of other birdwatchers iron out the data blips.

Though I have friends that will spend eight hours watching the birds on their deck, at our house we count less than an hour over the two days. I leave paper and pencil in front of the window and as we pass by we check to see if there are more birds than previously recorded.

A computer isn't necessary. You can ask for paper data sheets and mail them all in at the end of the season. But it's much more fun to enter in the information online and be able to compare this year's data with your previous years.

Whether or not you sign up, you can look at North American species maps online to see if the lack of goldfinches in your yard is part of a trend.

It takes about five minutes to enter each session's checklist. Besides bird numbers, I fill in the amount of snow cover and if and for how long it rained or snowed. I get the high and low temperatures from the newspaper.

So, for just a bit of time and $15, not only do you get a subscription to the quarterly newsletter Birdscope, but you can help scientists learn more about the birds you love to watch.

Register until Feb. 28 in one of three ways. Online, go to www.birds.cornell.edu/pfw. By phone, call 1-800-843-2473. Or mail your check to Project FeederWatch, Cornell Lab of Ornithology, PO Box 11, Ithaca, NY 14851-0011.

We're only two weeks away from another tradition – the Thanksgiving Bird Count. It takes a mere hour on Nov. 24.

Imagine a cylinder 15 feet in diameter around the feeding station in your backyard or wherever you are visiting.

Count the numbers of each species within the area. Your final tally will be the largest number seen at one time. Otherwise, without marking individual birds, it's impossible to know how many repeat.

Should you take part in this fun bit of birding, send the results to me by e-mail or voicemail (see contact info below). Also include your name, mailing address and phone number, the address where you counted, what time your hour started, the weather and temperature during your count and a brief description of the yard's vegetation, feeders, birdbaths, etc.

Our local data is compiled with the rest of the west and the report is mailed in early spring.

It's not too soon for a heads up on the great-granddaddy of them all. This year is the 106th Christmas Bird Count.

The Cheyenne Christmas Bird Count is sponsored by Cheyenne-High Plains Audubon Society and will be held Saturday, December 17.

You can be a Christmas Bird Count feeder watcher. Keep track of how many birds you see and how long you watched and call in the results.

The count runs dawn to dusk, but it isn't necessary for the field observers to stay out all day. The main group finishes the crucial early morning counting in a couple hours and takes a hot chocolate break before driving other routes. All levels of birding expertise (and cooking expertise for the evening potluck) are welcome.

Send me your mailing address if you are interested in this or the Guernsey-Fort Laramie Christmas Bird Count on Dec. 29 and I'll mail you the details.

If neither date works for you, check the CBC Web site, www.audubon.org/bird/cbc, for the count closest to your holiday destination.

Guess I better put in a supply of black oil sunflower seed or I won't have any birds to count!

# 207 The best medium for bird notes

Wednesday, December 7, 2005, Outdoors, page C2

Wendy Morgan's bird art appeals to me. Over the years, every time I would come across it at wildlife refuge and nature center gift shops, I bought it, even before I knew who Wendy was.

Luckily, note cards are very affordable art. But of course I have mailed all the cards to friends and family and only have those sent to me by other Morgan aficionados.

When I finally read the fine print on the back of a card, I was surprised to find that Crane Creek Graphics, the company producing the cards, is based in Wilson, Wyoming, within the golden rectangle Rand McNally has drawn around the super-scenic northwest corner of our state.

I figured Crane Creek Graphics was just another case of someone locating a business to get the scenic address in the wildlife viewing capital of the west.

Not so, I discovered when I called recently.

Wendy Morgan is Crane Creek Graphics. She was born in Jackson and left to earn a degree in anthropology that included a few art classes and a minor in biology and then traveled before coming home and starting the business.

Wendy's distinctive style, airbrush watercolor with stencils, fills the canvas of a note card with one or more birds of a single species. The background is sometimes a single color, with maybe a shoreline indicated. Props, if any, are a twig, a few leaves, berries or blooms.

The birds themselves are beautifully reduced to the simplest, most distinctive field marks.

In 1979 Wendy's first designs started with birds which she silk screened on cards for sale at fairs. They were popular and in 1980 she started the business named after Crane Creek Ranch with a loan from her mom and a bit of advice from the Small Business Administration.

She was soon working with a printer in Salt Lake City who printed on post-consumer recycled paper. Her notecard menagerie expanded to include a few mammals.

"I pick critters I want people to know about," she told me. On the back of each card is information about the featured species, including conservation status.

Wendy began going to trade shows and Crane Creek Graphics sales expanded all over the country. Now with 3,000 wholesale customers and promotion by a fleet of sales reps, Wendy travels less often.

In the 1990s, she added Wyoming artist Jocelyn Slack. Her watercolors feature North American fish, frogs and ocean life in a style distinct from Wendy's but share the simplicity.

Six other area artists are also carried by Crane Creek Graphics. Marty Anderson illustrates butterflies and other insects, and Henry Evans specializes in grasses and garden and wildflowers on white that look like colorful pressed specimens. Foott Prints is a line of cards with wildlife photos, September

Vhay focuses on wildlife and horses, Greta Gretzinger specializes in fish and Corrina Johnson illustrates a variety of wildlife.

The notecards are available singly, in boxes of the same image or assortments, or matted or framed, but they aren't the only products. The artwork is also available on the covers of blank books and address books, coasters, tile boxes, holiday cards, magnets and three different 2006 calendars.

Morgan's images offered for sale are not a static collection. One of the new ones this year, No. 239, features a pair of cardinals on a leafy background.

Only 92 of her images are currently available, so it looks like I'm now out of luck trying to start the definitive Wendy Morgan collection beginning with No. 1, though she said she might branch out into limited edition prints.

Blank cards are harder to sell than greeting cards, Wendy said. People are hesitant to express their own thoughts and have urged her to include the basic messages. So far, only holiday cards are inscribed with a simple generic greeting.

Card sales have dropped a bit, probably due to the internet, but Wendy said running the company is still more than a full-time job for her. January through May is time for new product development. July through November is the busiest time, shipping product all over the country. Even though she has five employees, time for

artwork is minimal.

She watches birds too, but more with an eye to the way they look and behave rather than trying to increase her life list. She also has a ranch to take care of.

One outlet in Cheyenne carries Crane Creek Graphics products – the Wyoming State Museum Store, 2301 Central Ave., open Monday through Friday, 9 a.m. – 4:30 p.m.

The complete line of artwork can be seen in the Crane Creek Graphics online catalog, www.cranecreek-graphics.com, or call 800-742-7263 to order.

I always make sure to have Crane Creek Graphics note cards on hand. In this age of e-mail a hand-written note is a special occasion and deserves a special card.

But maybe I shouldn't mail all of them. I'm rather fond of those early sandhill cranes. Luckily, I can get them singly or along with an assortment of egrets, herons and avocets in the "Longlegs" collection.

And then what about all those colorful butterflies, fish, frogs and flowers in the other collections?

I guess I'll have to look for more birthdays, weddings, new babies and occasions to write thank you notes.

---

## Info

The complete line of artwork can be seen in the Crane Creek Graphics catalog.

Write Crane Creek Graphics Box 367, Wilson, WY 83014 call 307-733-3696, FAX 307-739-0744 or visit cranecreek@wyoming.com

# 2006

## 208 Of birds and beetles

Wednesday, January 4, 2006, Outdoors, page C1

**Woodpeckers find silver lining in devastation done by insects**

Are they the feathered equivalent of American Red Cross volunteers? Or are they just looking to build a home a short commute from where they make a good living?

Whichever best describes them, American three-toed woodpeckers and black-backed woodpeckers show up in forests that are under siege from bark and wood-boring beetles.

Not only do the birds make homes in dead trees, they make meals out of the beetles and their larvae, filling an important niche in the ecosystem there.

Other woodpeckers in Wyoming, including flickers, sapsuckers and downy and hairy woodpeckers, are fairly common, but the three-toeds and black-backeds are quiet and hard to detect.

The three-toed woodpecker can be found in most mountain ranges in the Cowboy State, but the black-backed sticks to the northwest corner of the state and the Black Hills in the northeast.

Outside the state, the range of both species covers forests in much of Canada and interior Alaska plus the Rockies in the northern United States.

Recently, three-toed woodpeckers in North America were split into a species separate from those ranging from Scandinavia to northern Japan.

The birds are relatively unknown despite pages of references for studies of three toed and black-backed woodpeckers listed in "Birds of North America."

Many of the three-toed studies were done in Scandinavia; it isn't known if the findings hold true in North America.

### A relationship with beetles

The three-toed and black-backed woodpeckers both have three toes that help them rise to the challenge of finding wood-boring and bark beetles.

Other woodpeckers have four toes, two facing front and two facing back.

These three-toed specialists use a different stance and grip on a tree trunk that gives them greater force in drilling than all but the much larger pileated and ivory-billed woodpeckers.

Other general woodpecker-evolved adaptations are: stiff tails used as props; long, sticky tongues; and skulls with built-in shock absorbers to withstand all the hammering.

While they might make huge holes to find beetles deep inside the tree, the woodpeckers also can scale or peel off the bark to find beetle larvae.

The birds put their nests in dead trees, called snags. This takes them a few weeks, but a pair will do it once a year.

Old nest holes then are snapped up by bluebirds, chickadees and other cavity nesters that don't have the beaks for the job.

In an average forest, there are always a few trees in decline that provide beetles for a few woodpeckers. But then Mother Nature provides a bonanza every so often when wildfire strikes.

Possibly both birds and beetles can smell the smoke. Beetles are on the scene of the fire within a couple of weeks, and black-backed and three-toed woodpeckers are right behind them.

Steve Kozlowski, a U.S. Forest Service wildlife biologist in Laramie, said three-toeds are normally at such a low density that you are lucky to find one on any given day. But after the Gramm fire near Foxpark in 2003, 14 were seen in one day.

### Trees in distress

Even in healthy forests, beetles congregate where a single tree is succumbing to disease, lightning strike or windthrow, said Jeff Witcosky, a regional entomologist with the U.S. Forest Service in Golden, Colorado.

Old-growth coniferous forests are more vulnerable to beetle kills because young trees have better defenses against beetles.

One way that beetles might find distressed trees, Witcosky said, is through hydrocarbons known as terpenes that are given off by injured tree tissue. The beetles may be able to sense the compounds with their antennae, he said.

In lodgepole pine forests stressed by drought, mountain pine beetles might randomly land and chew bark before deciding whether to attack or try another tree.

Although there is about one beetle for each species of coniferous forest tree, the different kinds of beetles can be lumped into two main groups.

The bark beetles, favorites of three-toed woodpeckers, lay their eggs just under the bark, where they hatch as rice-grained sized larvae that chew little tunnels in the tree's cambium, or growing layer.

These traceries in the layer that would become the newest tree ring eventually reach all the way around and prevent sap from moving between roots and needles, killing the tree. When the bark falls off, the tunnels look like shallow etchings in the wood.

The other general category includes all the wood-boring beetles. The larvae of the beetles, favorites of black-backed woodpeckers, eat deeper into the wood. Growing as long as 1½ inches, when they reach the adult stage they chew their way out, fly off, mate and deposit eggs under the bark of another tree victim.

These beetle life cycles can last from one to three years, depending on the kind of beetle. All that time, they are at the mercy of chisel-billed birds that like beetle larvae more than any other insect flesh.

Surrounded by acre upon acre of lodgepole, ponderosa pine or spruce, how does a woodpecker search efficiently for hidden food?

Doug Faulkner, University of Wyoming bird biologist, said it would make a good master's thesis project to find out.

Do the birds key in on other signs of a tree's distress? Can they smell the same terpenes that attract beetles?

Jane Dorn, co-author of "Wyoming Birds," said she has heard larvae chewing when she's close enough to an infected tree. Woodpeckers presumably have better hearing than people.

### Populations fueled by fire

Arvind Panjabi, a bird biologist with Rocky Mountain Bird Observatory, spent several years studying the effects of the 85,000-acre Jasper fire that burned in the Black Hills National Forest in South Dakota and Wyoming in 2000.

In the areas of ponderosa pine that burned the hottest, black-backed woodpeckers increased tenfold and peaked the second year after the fire.

Because the Hills are isolated, Panjabi guesses the boost was due to an increase in reproduction rather than the unlikely scenario of birds flying in over hundreds of miles of grasslands.

Because the woodpeckers at a burned area are eating well and are at maximum good health, they lay larger clutches of eggs. But the boom lasts only three to five years.

No one is sure what happens to the increased numbers of woodpeckers if another part of the forest isn't burning by then.

The bark beetles eaten by three-toeds seem to prefer singed spruce trees. But within a few years these burns lose their appeal and the beetles also have to leave for blacker pastures.

Panjabi found that in the Black Hills, numbers of Lewis's and red-headed woodpeckers in the burn area continue to slowly increase even as the three-toed and black-backed are decreasing.

Sometimes, black-backeds and three-toeds on a new burn have competition in the form of salvage logging of burned trees. They must be harvested within six months to make good lumber, but without burned trees there are no beetles and no birds.

To share the bounty, wildlife biologists recommend leaving groups of snags standing. Single snags spread out won't support these two species of woodpeckers. Neither will clear cuts.

It is important to be accommodating since without regular woodpecker numbers, normal beetle populations might grow out of hand, forest ecologists say.

Wyoming forests are currently experiencing beetle epidemics too large for woodpeckers or people to control. Beetles are not only attacking sick trees, but apparently healthy ones as well.

The only sure control, Witcosky said, will be a long enough episode of extreme cold which will kill the larvae. Winters have been too warm lately.

# 209 Resolving to do better this year

Wednesday, January 4, 2006, Outdoors, page C2

### With the right choices, we can have our energy and our sage grouse, too

It seems New Year's resolutions come in two categories. The first shows restraint – I won't eat all the chocolate chips before they get in the cookies; I won't waste money on junk, and I won't watch junk TV.

The other category shows motivation – I will get more exercise; I will work harder and save more and I will study that CD of bird songs so I'm prepared for spring.

If we don't make resolutions, whatever time of year, then we are like willow twigs tossed in the creek. Depending on luck, we could be battered by boulders or become rooted in a fertile location where willow twigs can sprout or just go with the flow until we decay into nothing.

Personally, I'd rather be like a fish, determining for myself whether to go up or downstream. Or better yet, as a bird I could find new streams.

Not all our resolutions are of the personal improvement kind. We can be altruistic and resolve to make that charitable contribution or teach that class for free or adopt the homeless dog because we were brought up to be responsible people.

One of my favorite charitable ways to spend time and money is bird conservation work. I think it is necessary because some of the changes in nature threaten the existence of birds.

But what is nature?

What is natural?

If ivory-billed woodpeckers are near extinction because people sawed down their trees, isn't that natural? People are natural, aren't they? If you aren't a plant, mineral, fungus or bacteria, you must be an animal, same as birds and other wildlife.

By this reasoning, whatever people decide to do, it has to be natural. However, civilization has put legal boundaries on us, even extending protection to non-human life.

For instance, when people let loose European starlings in New York in 1890, they probably didn't know starlings would crowd out bluebirds and other native cavity nesters. They thought they were adding variety. But biologists know better today and the introduction of exotic species into the wild is no longer allowed.

But why should we protect our native species? The world is not a static place. If man progresses from burning buffalo chips to burning natural gas, and sage grouse are displaced by the intense drilling in their habitat, so what?

Whether they adapt or whether they move or die, isn't it natural?

Surveys so far show they are not adapting to the disruption, nor are they leaving the only kind of sagebrush habitat that works for them. To wait for the total demise of sage grouse is to act like the powerless twig in the creek.

But I believe people can creatively resolve what looks like an either or situation, such as either energy or grouse.

People can make a resolution to show restraint in not mining an energy source just because it's there.

We can be motivated to find better, less disruptive energy alternatives.

We can act responsibly for sage grouse well-being, even if we can't see a direct benefit to ourselves.

Environmental philosophers tell me that even if I never see a sage grouse or ivory-billed woodpecker myself, just knowing they continue to exist improves the quality of my life.

Conservation biologists

tell me that every species is one of the building blocks of life and if we remove too many, it will eventually affect my health and well-being.

We also have to be careful what we wish and work for.

When a native species expands its range, how do the species already living there respond?

Picking the proper conservation resolution is tricky, but not a lot different from making personal improvement resolutions.

What do I eat when I eat less?

What investments do I make when I save more?

If I'm not careful, my resolutions can have bad results.

I think we – at least some decision makers – are still stuck on the question of whether a few animals living above mineral reserves are as important as the nation's energy needs. The debate is continually framed by the old idea that you can't have your cake and eat it too.

But I think you can have both, as long as you choose a cake that is a

renewable resource.

We can have our energy and our sage grouse too. We just need resolve to act with restraint and motivation.

I wish I were the engineer that could put solar voltaic shingles on the market next week that are hail proof, windproof and affordable.

Just imagine what the world would be like if we didn't have to drill or dig for or argue over energy, or have to scrub away pollution. Clean air!

We would all benefit directly – even the people who make smokestack filters have to breathe.

What are my New Year's resolutions, besides the usual eat less and walk more – preferably while birdwatching?

Well, after this discussion, maybe the best Mark and I can do is to install new energy efficient windows while we wait for the household-sized solar furnace to be invented.

In this case, what's good for the grouse can also be good for the goose and gander.

# 210 Bird flu and you, so far, so good

Wednesday, February 1, 2006, Outdoors, page C2

Bird flu is everywhere I look. The virus keeps popping up in the news and has even become the punch line of jokes.

Bird flu is also, literally, everywhere. Wild birds can carry many subtypes of Influenza A without getting sick, but some can be deadly for poultry like chickens, turkeys and ducks. Wild ducks mingling with domestic ducks are thought

to have sparked the ongoing bird flu epidemic in Asia and eastern Europe.

Not all bird flu viruses are highly pathogenic, but the subtype making the news, H5N1, is deadly to poultry. Ninety to 100 percent of infected birds die within 48 hours. During the Asian outbreaks in 2003 and 2004, flocks were also killed by farmers and officials to try to control the spread of bird flu.

It has not yet spread to birds in the United States.

In 1997, the first cases of humans infected by poultry surfaced in Hong Kong. The risk to people from bird flu is normally low, but of 140 reported cases of people with H5N1 since the beginning of 2004, about half have died. So far, no people have contracted it in North America.

As of Jan. 7, human cases are being reported in

Cambodia, China, Indonesia, Thailand, Vietnam, and Turkey.

But it's a good idea to practice poultry hygiene, especially because of all the other avian-transmitted diseases. So don't breath near a sneezing duck, don't wipe a chicken's nose and don't touch your face with your hands after cleaning the chicken coop. The U.S. Centers for Disease Control

2006

205

recommends poultry workers wear protective suits and treat all birds as if they are infected.

Most of us don't come in contact with live chickens, a sad commentary on the disconnect between consumer and food source, but it is possible to contract bird flu from inadequately cooked poultry from countries presently dealing with outbreaks. You'd have to travel to those countries to eat it since the U.S. has embargoed their unprocessed poultry products.

Bird flu vaccines are in the works, but meanwhile people should get regular flu shots, say experts at the CDC. Should bird flu come to the U.S., you have a better chance of survival if you are in good health. Plus, there's a slight chance that if human flu and bird flu come together in the same host, the dreaded evolution may happen – bird flu transmissible from human to human. It has apparently happened only once so far, between a mother and a closely-held child.

We birdwatchers are also concerned with the ramifications of bird flu. Two dozen Asian wild bird species have been reported to have died from H5N1. However, there are no cases yet of the virus being transmitted from wild bird to human.

What does worry us is how migrating birds will spread H5N1. So far, birds from "infected" countries have not spread it along migration routes through Taiwan, the Philippines and Australia.

We backyard birders also want to know if it's safe to continue feeding birds. It is, as long as we follow the precautions we've always had to promote the health of birds and birdwatchers.

Keep bird feeding areas and feeders clean. Disinfect them every few weeks with a mild bleach solution and rinse well.

If you notice a sick bird, stop feeding. A sick bird acts lethargic, has feathers out of place, is fluffed up more than the other similar birds or

might have crusts around its eyes. Clean up spilled seed and debris, then disinfect and put away your feeders for a week to encourage healthy birds to stay away.

Take precautions for your own health, remembering that there are other diseases carried by birds, including West Nile Virus and salmonella. Never handle birds, dead or alive – or anything full of bird droppings – without disposable gloves or plastic bags over your hands.

Be careful not to breathe the dust when sweeping up old seed hulls, and keep your hands away from your face until you can wash them well.

Now that I've assured you that you can safely eat chicken and feed birds in our country, that's not to say that new wrinkles in bird flu won't develop while this edition of the Outdoors section is going to press.

Since sound bites can be maddeningly uninformative for people with above average interest in a topic, let me recommend the CDC

website, www.cdc.gov/flu/avian. I found it to be informative, clearly written and frequently updated with advice, especially for poultry workers, travelers and people caring for bird flu patients.

For people who work with wild birds or hunt birds or mammals, information from the National Wildlife Health Center's Wildlife Health Bulletin #05-03 is extremely useful. Find it at www.nwhc. usgs.gov/research/WHB/WHB_05_03com.

The media and the experts frequently look back to the 1918 global flu pandemic to try to forecast what will happen when this strain of bird flu evolves the potential to transfer from human to human.

Hopefully, the disadvantages of our modern global mobility will be offset by the advantages of modern science, medicine and communications. Meanwhile, do something to protect yourself. Promote your personal health. Take a walk. Go birding.

# 211 Consider it your war paint

Wednesday, March 1, 2006, Outdoors, page C1

## Outdoors people should be extra guarded in the battle against cancer

At the end of last August, a zit appeared to have formed on my upper lip.

As a survivor of teenage acne, I thought, "No big deal."

But several weeks later, I was getting tired of it bleeding every time I blew my nose. Eventually, it healed, and all that was left was a permanent, flesh-colored bump.

In November, when another family member had a

dermatologist appointment, I decided to make one too.

"It's a basal cell carcinoma," proclaimed dermatologist Sandra Surbrugg after she took one look at the spot.

Surbrugg, who owns the Cheyenne Skin Clinic, said it was the result of fair skin being inadequately protected from too much sun over too many years.

**An unwelcome diagnosis**

Typically, skin cancers

appear to be weirdly shaped moles that are off-color or changing in size. But the bleeding was a tip-off for mine.

Surbrugg sent the biopsy sample to a lab for confirmation, and two weeks later she deftly removed the growth and surrounding tissue before neatly stitching me up.

The surgery should be between 95 and 99 percent effective. But I have a 40-percent chance of growing

another basal cell carcinoma. Considering a few horrible sunburns I got as a child and my years working outdoors, my chances may be higher.

Surbrugg performs six to eight surgeries a day.

"Basal cell is most common, about 1,000 cases a year just in our office, plus 300 to 400 squamous cell and 50 melanoma cases," she said.

Skin cancer accounts for close to 30 percent of her

206        CHEYENNE BIRD BANTER

practice, and while she says the high numbers are because there's more cancer now, people are also more aware.

Wyomingites are particularly prone to skin cancer, she says.

"We live at 6,000 feet," Surbrugg explained. "There's a 5-percent increase in ultraviolet) for every 1,000 feet so we get 30 percent more (than people at sea level). Then we have many, many sunny days. And we're an outdoorsy bunch."

Ultraviolet rays are the destructive light waves in sunshine.

The three bands of UV light are UVA, UVB, and UVC. UVC are absorbed by the upper atmosphere and do not reach the Earth's surface; UVA and UVB rays do. It's these rays that burn the skin and the eyes with too much exposure.

Those with fair complexions face the highest risk of developing skin cancer.

At Christmas I showed off my nearly healed incision. My mother and sister both informed me they had been there, done that. Both have blue eyes and were blonde as children.

Surbrugg said it's a great advantage to have darker skin. It's the people with Celtic heritage – red hair, light eyes and freckles – that are most at risk. The best research on melanoma is from Australia, she said, where the government takes an active role in promoting skin cancer prevention.

She said natives of the British Isles, immigrating from such a cloudy climate to Australia, have had no natural defenses against a landscape overflowing with sunshine.

**Slip, slap, slop**

Slip on a shirt, slap on a hat and slop on sunscreen. That is a popular mantra for those who want to avoid future incisions.

When you protect your skin with clothing, any fabric will offer some sun protection. But Surbrugg recommends the tight weave supplex nylon clothing developed by dermatologists.

Nylon clothing is anathema to those of us spoiled by the breathability of natural fibers, but the designs I have seen are well vented. Surbrugg said she finds the clothing cool and lightweight and, in the long run, cheaper than applying high SPF sunscreen all the time.

Companies like Sun Precautions carry long-sleeved shirts, pants and a selection of gloves for covering hands, another key location for skin cancers.

The company even carries masks to protect the lower parts of the face that may not be completely protected by the shade of a hat brim.

Birdwatcher and retired pediatrician Jim Hecker is never without a hat. His first brush with skin cancer happened in his 20s while in medical school.

The suggested hat brim depth for sun protection is four to five inches all the way around, a size he finds is sometimes incompatible with Wyoming's wind.

Hecker's alternative on windy days is the ball cap style with a cloth drape around the neck. It catches less air and gives a little Lawrence of Arabia cachet.

And don't forget dark, UV-filtering sunglasses to protect your eyes.

**Improved sunscreen**

My first experience with sunscreens 25 years ago was that they felt like a layer of war paint on my skin. They smelled and, after a few sweaty hours, burned my eyes.

But as the wife of a freckled, red-haired man, I have adopted his habit of applying sunscreen whenever we head out to bird, fish, hunt or hike.

I think where I fell down on the job is not reapplying with the recommended frequency of every two hours. Also, drinking out of water bottles probably takes sunscreen off my lip prematurely.

Newer sunscreens seem to be absorbed better by the skin. Now I use an unscented product designed for babies or one that is sweat proof. Either of these, plus washing my face when I come inside, keeps my eyes happy. Hecker said he has supersensitive skin, so his favorite is a cream called Solbar-zinc, SPF 38.

For days at the office, Surbrugg recommends one application in the morning of a minimum of SPF 15. More regular skin products contain this level of protection.

What exactly does an SPF or sun protection factor number indicate? It means how many times longer you can stay out in the sun before burning. So if you would get sunburned in 10 minutes, SPF 15 gives you 150 minutes, or 2½ hours.

Surbrugg put it another way, SPF 15 blocks 93 percent of UV rays; SPF 30 blocks 95 percent; and SPF 50, 97 percent.

**Seek shade at peak hours**

Pamphlets from doctors' offices recommend staying out of the sun from 10 a.m. - 4 p.m. to avoid the most UV damage. Luckily, outdoor activities in early morning, late afternoon and early evening are appealing at the height of summer heat or for the best light for photography.

These can also be the most productive hours for fishing, hunting and birdwatching, though they are limiting for long hikes or winter recreation.

Summer is not the only sunburn season in Wyoming. Sun reflected off snow can be as potent as any day at the beach.

When Surbrugg goes skiing, in addition to wearing a helmet and goggles, she wears a mask on the lower part of her face. She said she shrugs off the odd looks. Her sun protection probably also decreases her chance of frostbite and windburn.

"The message I try to give my patients is: You don't need to be a hermit," Surbrugg said. "I ski, bike, and jog. It's just using common sense. I have many, many hats. Despite the hat hair they give me, I wear them."

## Types of skin cancers

Precancer or actinic keratosis is the name for a small crusty, scaly bump of any color that most frequently forms on skin exposed to sun or tanning machines. A precancer can develop into the more serious forms of skin cancer.

Basal cell carcinoma and the more dangerous squamous cell carcinoma are most common. Left untreated they can cause major damage. Squamous cell has a greater chance of spreading and becoming life-threatening.

Melanoma is the deadliest form of skin cancer. Left untreated it will spread to vital organs.

For more extensive information on what to look for and treatments available, ask a dermatologist for a brochure or visit The Skin Cancer Foundation at www.skincancer.org.

## Preventing skin cancer

The Skin Cancer Foundation recommends these sun-safety habits:

--Avoid unnecessary sun exposure, especially during peak hours of 10 a.m. – 4 p.m.

--Seek the shade.

--Cover up with clothing, including a broad-brimmed hat, long pants, a long-sleeved shirt and UV-blocking sunglasses.

--Wear a broad-spectrum sunscreen with a sun protection factor of 15 or higher.

--Avoid tanning parlors and artificial tanning devices.

--Examine your skin from head to toe every month.

--Have a professional skin examination annually.

## UV sun protective clothing sources

Specialty companies make clothing with fabrics rated for the Ultraviolet Protection Factor and catalogs are available from dermatologists. Sierra Trading Post carries some sun protective clothing made by Sportif, Simms, Mountain Hardware and Ex Officio. Gart Sports carries Columbia sun protective shirts, pants and hats and will carry more in the spring.

-- www.coolibar.com
-- www.solartex.com
-- www.sungrubbies.com
-- www.sunprecautions.com
-- www.uvsunware.com
-- www.exofficio.com

# 212 Research station is, should be for the birds

Wednesday, March 1, 2006, Outdoors, page C2

From a bird's eye view, Cheyenne is an island of trees in a sea of grass.

During migration it makes an ideal place for a layover. More than 150 species of birds were identified in one day mid-May by members of the Cheyenne - High Plains Audubon Society.

We stick to places with public access such as Lions Park and get permission to visit the U.S. Department of Agriculture's High Plains Grasslands Research Station west of town, where 170 bird species have been documented over time by Bob and Jane Dorn.

The station began as a federal horticultural research facility in 1928. Earlier, the city of Cheyenne had acquired a 649-acre section from the state for the Round Top waterworks. It obtained another two sections through trades and sales with local landowners including Zelda Ketcham and Minnie Cox-Brown. It then leased just over 2,000 acres to the federal government for 199 years.

Many of the experimental trees and shrubs there still survive and will become the basis for a future arboretum sponsored by the Cheyenne Botanic Gardens. However, a large part of the station is grassland. This suits the station's mission as of 1974, which is to study range-land management.

Last November, I learned the city planned to annex their station property with an ordinance stating:

"That the City of Cheyenne is dedicated to keeping all of this land in the public domain as land of community wide significance. Should the USDA see no further use on all or part of the land, the City of Cheyenne will preserve this land for parks and recreation department uses. Cheyenne's urban areas are growing and steadily increasing in population over time. Cheyenne's growth will place further pressure on existing park facilities and require the addition of new areas for parks, recreation and open space for the community of Cheyenne."

From the bird's eye view, this is wonderful.

But the person who gave me the information, station neighbor Bill Cox, was not happy.

Neither was Laramie County Commissioner Jeff Ketcham when I spoke with him a couple weeks ago after the ordinance passed. By the way, Zelda Ketcham was his great-grandmother.

The city, to its credit, took the least intrusive annexation option. The typical annexation requires banning septic systems, firearms discharge and livestock. It's good the city took the Wyoming Statute 15-1-407 option instead, since there are few toilets at the station, the deer population needs to be controlled to protect the tree and shrub specimens, and experiments call for cattle grazing.

This option also doesn't give the city the usual developmental oversight within one mile of the city boundary, nor does it allow landowners to use this island of city as a basis for more annexations.

So why are the neighbors and the county, in separate lawsuits, trying to get the annexation appealed?

They don't see an ordinance as a very permanent thing. Ketcham knows how little it would take to change it, or even to have the city decide to sell the property. The lease with the USDA is renewed annually.

CHEYENNE BIRD BANTER

Perhaps the station's neighbors are watching the county's development of the former Archer agricultural experimental facility east of town. Plans call for a business park along I-80, the relocated fairgrounds, a motocross track and shooting range. Only a third to a quarter of the 800 acres is slated for open space.

When I spoke with Mayor Jack Spiker and Councilman Tom Seagraves who sponsored the ordinance, both assured me of the pureness of their intentions.

Seagraves said annexation would give the city's property protection by city police and city fire departments and would deal with a water issue brought about when the new water treatment facility left Round Top high and dry.

From a bird's eye view, the landscape around the research station has become a checkerboard of new residential development. Ironically, the owners of these new homes want the freedom of the county's less restrictive regulations, but many opt for exclusivity provided by very restrictive covenants that may limit numbers and kinds of livestock and outbuildings.

Studies show here on the high plains this rural development changes the kinds of birds seen from grassland species to urban species.

What kills the native small mammals and ground-nesting birds that have adapted to life without trees are people allowing dogs and cats to prey on them, people mowing during nesting season (June through mid-July) and people planting trees from which avian predators can more successfully search for prey.

Right now the island of city property known as the Grasslands Research Station is a great stopover for migrating birds, and a refuge for our native breeding grassland birds.

I want to believe that annexing the research station really is the farsighted action claimed in the ordinance's language. One hundred years from now citizens will applaud the preservation of open space in what might become the new focus of the city. Maybe it will be known as the Central Park of the West.

From a bird's eye view, something more permanent than a mere city ordinance is needed. This special landscape needs some kind of binding city-county agreement that allows for the continuation of the research station's lease for the next 121 years and a plan for if they decide some year not to renew it.

Conservation easements allow private landowners to preserve land in perpetuity.

Does the station's neighbors' attorney, Gay Woodhouse, know of a legal tool available to cities?

If we are serious about protection – and no one I've talked to disagrees with preserving this open space - let's find a way to show we mean it.

# 213 Better binoculars make for better birders

Wednesday, March 29, 2006, Outdoors, page C1

It is a fresh spring day at Lions Park. Each twig glistens in a fresh, new coat of frost against a crystal blue sky.

Each birdwatcher holds a fresh, new pair of binoculars courtesy of Brunton Optics. More than half the group on this Audubon field trip is new to birding.

Jim Danzenbaker, naturalist manager for Brunton, is our guide. He's a crack birder who leads trips to distant destinations and banters about birds with experts for Brunton at bird festivals and bird stores nationwide.

Today, he is teaching Binoculars 101. The first order of business is to get us harnessed in the new strap systems designed to protect our necks and shoulders from pain.

Next, we adjust for the difference between our left and right eyes if opticians haven't already done so with our prescriptive eyewear.

Jim shows Kim how to lower the eye cups if she's going to keep her sunglasses on. Mary wonders why she's still seeing black around what she's looking at, and Jim shows her how to adjust the distance between the barrels to match the distance between her eyes.

Founding chapter member John and I, both old hands at birdwatching, finally get instruction on the right way to clean lenses.

First, brush away grit with a special soft brush or blow really hard. Then moisten the lens and polish it with a lens cloth. Jim advocates licking the lens to get it wet enough.

Now we are ready to find a bird to focus on.

Jim hears pine siskins in the treetops, but it is the house sparrows which cooperate. Everyone gets a good look – again and again – at the male's distinctive black bib.

Jim tells us because of the house sparrow's propensity to search for scraps at fast food places, a friend has rechristened them "Burger Kinglets."

Deb needs a brisker pace to warm up, so we decide to make a full circuit of Sloans Lake. We hit the jackpot on the far side.

Most of the lake has a skin of ice except for this open water close to shore filled with waterfowl. Jim sorts through the mallards and Canada geese and finds us a pair of northern shovelers, several gadwalls and five redheads.

To my amazement, all of the birding newbies can follow along. Sunlight hits the redheads right and Mary is impressed with the beautiful chestnut coloring. She has the help of the $1,500 binocs with the many special lens coatings.

While we examine the ducks, we notice a sparrow in the reeds. It is the clearest view I've ever had of a song sparrow. I'm impressed. Maybe it is the binoculars.

Luckily for me, the pair I'm impressed with is under $500. I also like the slim roof prism style. My short fingers more easily reach and roll the focus wheel even with mittens on.

On the way home, Sharon says she's impressed with her morning's experience. Maybe she will look into signing up for the beginning bird class that Audubon is offering at Laramie County Community College at the end of April.

# 214 Spring break trip should be warm

Wednesday, March 29, 2006, Outdoors, page C2

## Arizona mountains still chilly in March

Spring break is magic, taking you from snow shovel, cold wind and icy roads south to warm up.

On your return, hopefully spring weather has caught up with home and you effortlessly segue into tulips, green grass and warbler migration.

My spring break didn't quite follow this plan, except for flying to Phoenix. I marveled over the palm trees people have planted everywhere and reveled in the delightful, 65-degree breeze while I waited for the shuttle.

It was St. Patrick's Day, and I was headed to Odegaard, Arizona, in the White Mountains, to the home of my sister Beth and brother-in-law Brian.

From Phoenix at elevation 1,072 feet, the van from the White Mountain Passenger Line climbed upward for three hours.

I wonder, in its years of operation since 1937, if the drivers ever tire of the progression from blooming bougainvillea sprawled along the interstate, to scrubby hills sprouting saguaro cactus, to the pinyon-juniper woodland, to the ponderosa pine forest.

Odegaard is 450 miles south of Cheyenne, so in ecological terms, spring there should be nearly five days ahead. But it is 500 feet higher than us at an elevation of 6,600 feet, so you then subtract five days which makes it even with us.

The weekend before, Beth and Brian measured 29 inches of snowfall after an otherwise dry winter. But it melts fast late in winter so much farther south, and evidence of a traffic-stopping storm was hard to find.

I always get a kick out of the name of one area town, Snowflake. I heard during this trip that its name is actually the combination of the names of the founding families, the Snows and the Flakes.

On my first day, we joined members of the White Mountain Audubon Society on their field trip to Petrified Forest National Park which is celebrating its centennial.

We found no snow at its treeless lower elevations except in protected gullies but the wind was equal to one of Cheyenne's brisker days. The most numerous birds were horned larks flittering by. They weren't too shy, and we were able to get nice looks at the feathered "horns" of some of them.

The common raven is common there and each time one flew over, I had to examine it to make sure it wasn't a hawk. Several raptors were identified: red-tailed hawk, American kestrel, northern harrier and merlin.

In the rabbit brush down by Agate Bridge – a huge petrified tree trunk spanning an eroding wash – we found a white-crowned sparrow.

So far, nothing I wouldn't see on a southeast Wyoming field trip.

But then we walked along a riparian area and things got more interesting. The first birds we saw were petroglyphs, but what species? Who knows what the artists were thinking 700 years ago.

Then ahead of us, birds began to scamper across the road. Two larger birds in the brush were identified as curve-billed thrashers. Several people identified scaled quail. Now those are species I can't find at home!

Back at Beth and Brian's, we found more birds than we'd seen all morning. Pine siskins and juncos swarmed the feeders and ground underneath.

Every time I looked at the gnarly old juniper closest to the window, I saw something new. The house is in a forested subdivision so pygmy nuthatches felt very comfortable working over the suet block – five at a time. In my backyard I might see an occasional red-breasted nuthatch.

Whereas I might find an American goldfinch, downy woodpecker or blue jay in my yard, in this yard I saw lesser goldfinches, a hairy woodpecker and both Steller's and pinyon jays.

Whereas I might hope to see a mountain bluebird west of town this time of year, Brian said the western bluebirds I saw in Arizona, with their robin-like red breasts, are there all year round.

Whereas I might find a migrating spotted towhee once in my yard in May, Brian thinks the one I saw in their yard nests in a bush along the fence.

The rest of the feeder list included one each: red-naped sapsucker, house finch, white-crowned sparrow, northern flicker (red-shafted), mountain chickadee and white-breasted nuthatch plus several robins (also year rounders) and a few starlings. Except for the sapsucker, I might find any of these in my northern, manmade, backyard forest.

But then the local tree squirrel came along, and I remembered I wasn't anywhere near Cheyenne. Our squirrels do not have fluffy white tails and ear tassels like the Abert squirrels.

Beth and I took a hike nearby the next day and found mourning doves, a killdeer and a turkey vulture, signs of spring for Wyoming. In Arizona, however, only the "TV" is a sign since the others are found year round.

We also drove over to Show Low Lake and found Arizona's winter ducks: buffleheads, gadwalls, goldeneyes and a ruddy duck – signs that spring might be delayed. After all, it was snowing.

Of course here at home the day before I got back, despite

CHEYENNE BIRD BANTER

being the Spring Equinox, there was enough fresh

snowfall to require plowing. Winter weather isn't

over till the fat Wilson's warbler sings.

# 215 Pup retrieves backyard of wonders

Wednesday, April 26, 2006, Outdoors, page C2

We had no idea so much of our backyard was edible. Euell Gibbons, author of "Stalking the Wild Asparagus," would be proud of us. Grass, leaves, sticks, weeds – of course I'm talking about edible from the perspective of a 3-month-old puppy.

Our yard is only about 100 feet wide by 50 feet deep, but it is amazing how many microhabitats it has as examined by the nose of a puppy.

First, there are all the dead leaves under the shrubs. Sally dragged out some really disgusting looking black specimens which turned out to be dried mushrooms. Maybe I should take her to France to look for truffles.

Then, there are the ants which she licks up. I never knew golden retrievers had anteaters in their lineage, but I guess she provides a natural way to control their population.

The pile of composting leaves by the garage is proving irresistible. Sally flings herself on them, like any child, tunneling to dig out choice bits of decaying roughage.

Wind-blown leaves scuttling across the patio also attract her notice. One such leaf took a sudden turn and as it blew closer, it became a little brown mouse intent on making it to its home in the corner of the raised bed only

10 feet from where we sat.

I'm out a lot now supervising Sally, and I've discovered backyard bird life is more interesting from outside than from a window.

The robins are not perturbed by us at all. One fearlessly marched on the bird bath – even though a panting puppy watched from a pounce away. The squirrels, however, seem to be avoiding the yard. This means a respite for the trees they've been gnawing, especially this last dog-less year.

Also, without the squirrels to antagonize them, a pair of blue jays has decided to build a nest in the spruce tree. They don't hesitate to come down into the yard for whatever bits they need while we watch.

Our cats are avid window birders and Sally seems to have the same avian interests. She noticed a flock of noisy gulls sailing over the house, though she hasn't tuned in yet to the whistling wing beats of the occasional mallard.

Walking the backyard with Sally half a dozen times a day lets me observe the changing pattern of sunlight and shade and the chronology of snowmelt. Each day, the ratio of green to brown in the lawn improves, tulips emerge a little more and leaves grow.

The dawn chorus steadily increases. I can pinpoint the

day the mourning doves added their calls, April 9. The singing of grackles, house finches, starlings and house sparrows increases in frenzy daily. On the other hand, the juncos seemed to have departed the yard for the mountains April 10, but I won't know if it's for good until we get the next snowstorm.

This year, the Eurasian collared-doves are even more abundant in the neighborhood. There must be some nesting nearby since I can hear their croaky cooing almost every time I'm out.

Will our usual spring visitors be as tolerant of a puppy and still show up in mid-May?

I'm on the lookout for the spotted towhee, green-tailed towhee, white-crowned sparrow, western tanager, rose-breasted grosbeak, lazuli bunting and indigo bunting. I hope to find the yellow birds too, goldfinches and different kinds of warblers.

Our dog before the last one was also a golden retriever which Mark took bird hunting. We'll see if Mark wants to get back into it and if Sally proves adept.

Someone asked me if we could hunt birds and still be members in good standing with the National Audubon Society. Of course. Audubon is not an animal rights group. There is no more honorable way than hunting, short of

raising it yourself, to put meat on the table – Vice President Dick Cheney's canned hunts not included.

If all bird species had always been given as much thought, management, study and funding as game birds, Audubon could become a mere birdwatching club. But interest in non-game birds continues to increase so someday there may be parity.

Sally, to her discomfort, has discovered not everything in the yard is safely edible. We were surprised to see juniper and chokecherry on the Cheyenne Pet Clinic list of plants toxic to pets. So we've put the sheep fencing back up.

But even while accompanying a sick dog, there is something beautiful about the backyard in spring at 4:30 a.m. That particular morning, there was only the slightest breeze, and it was warm enough to stand barefoot. The full moon was setting behind the neighbor's trees, and the Big Dipper was visible overhead. Sleepy robins could already be heard.

With Sally's taste for wood products, I'm thinking maybe I can train her to carry my bird field guide.

She can walk next to me as we explore the world beyond the gate.

# 216 Woodpecker visits might bring home improvements

Wednesday, May 24, 2006, Outdoors, page C2

After receiving half a dozen phone calls this spring from people inquiring what to do about woodpecker damage to their wood-sided houses, I decided to investigate and find the best recommendations available.

The primary culprit in our area is the northern flicker. You can identify it by its size which is about three inches longer than a robin. It is named for its red wing linings that flash as it flies. It has a white rump patch, black necklace, polka-dotted breast and a bill that looks about as long as its head is deep.

It's the bill that is used by the males to drum on reverberating surfaces to proclaim territory ownership and attract mates. Both males and females hammer at wood and even synthetic stucco to excavate a nest cavity.

Wood siding has a nice hollow resonance that reminds the birds of a rotten tree, their natural alternative.

Woodpeckers also drill for insects, but in the case of flickers, they're much more likely to be seen on the ground probing for ants, their favorite food. Cheyenne's dry climate is unlikely to produce siding infested with insects.

The first suggestion for avoiding woodpecker damage is to live in a brick house.

The next suggestion is to scare the flickers away. When they started drumming on his house, Cheyenne resident Chuck Seniawski went to that spot, but from the inside, and hammered the wall with his fist. He said it has worked the last two years.

When I searched for "woodpecker damage" on the internet, I found a short but comprehensive article put out by Colorado State University Cooperative Extension, www.ext.colostate.edu. Click on Natural Resources and look for woodpeckers. The authors recommend taking immediate action, "because woodpeckers are not easily driven from their territories or pecking sites once they are established."

I also found www.bird-controlsupplies.com which offers a complete assortment of deterrents, some of which you could make yourself. Starting at $8 there is the windsock approach, with big eyes printed on bright yellow to evoke a scary predator. Its mylar streamers flash in the wind.

Then there's the bucket of wood filler with a chemical deterrent added. Also, for $190 you can get an electronic device that intermittently gives off the sounds of a dying flicker. For $225 you can invest in a 25-foot square net to fence the birds off. You'd only have to keep it up during the courtship and nesting season – beginning about March, based on calls I got, through June.

Then there's the accommodation suggestion. Put up a flicker nest box over the hole the pair has been excavating in your house. The benefit is you get free insect and ant extermination services all summer.

One of the best sources for information on bird houses is the Cornell Lab of Ornithology's monitoring project, The Birdhouse Network, www. birds.cornell.edu/birdhouse.

Whether you build or buy, Cornell says to keep in mind these recommended features for all bird houses:

--Untreated wood at least ¾-inch thick, pine or cedar.

--Extended, sloped roof for protection from predators and weather; it can be fit into a slot on the back board of the nest box so water doesn't run down into the box.

--Rough or grooved interior walls so young can climb out to fledge.

--Recessed floor with drainage holes.

--Ventilation holes at top of sides.

--Easy access for monitoring and cleaning, usually a side panel that swings out but latches closed.

--Galvanized screws or nails.

--No outside perches to aid predators.

--Predator guard if mounted on a pole.

--Hole diameter sized for species, 2.5 inches for flickers.

The latest feature in flicker houses you can buy from places like Wild Birds Unlimited stores is the entrance hole surrounded by slate. It looks like an eighth to a quarter-inch thick square of slate has been drilled with the same 2.5-inch hole as the entrance and then mounted to prevent squirrels from chewing and enlarging the hole and taking over the nest box for themselves.

Cornell has plans for a flicker nest box, though it shows a hinged roof rather than the preferred access through the side. All you need is a 2-inch by 8-inch by 10-foot board cut into these lengths: 32-inch back, 24-inch sides (2), 24-inch front (center of entrance hole drilled 19 inches from the bottom), 4.25-inch floor and 10.75-inch roof or longer. You might want to look at the drawing before assembly.

The key to a successful flicker house is to fill it with wood chips. It helps the flickers think they're excavating a hole in a rotten tree.

Mount the nest box on a pole or the side of your house 6 to 30 feet high with the entrance facing southeast. It would be interesting to know if most damage occurs to that side of people's houses.

Once the birds decide to move in, you can monitor the nest for The Birdhouse Network and viola – the bird problem becomes a home enhancement for both you and the flickers.

# 217 Use a light touch with lawn chemistry

Wednesday, June 21, 2006, Outdoors, page C2

A pair of mallards shows up every spring in the ditch that runs below our neighborhood. I don't know if they try to nest because I never want to be intrusive enough to find out.

On a recent early morning walk, the puppy and I spotted the ducks swimming while above them a black cat, a tabby cat and a red fox crouched along the grassy edge.

But what may be even more hazardous to duck survival is the water in the ditch.

Because it is meant to collect storm-water runoff for a neighborhood of several hundred houses, the ditch also collects any excess pesticides and fertilizers applied to those hundreds of lawns.

Nationwide, homeowners overapply lawn pesticides to the tune of 78 million pounds annually. This is up 50 percent over the last 20 years. If the neighbors apply them to the same degree, then pesticides in the ditch water are a given whether the intended pests were animal, plant or fungus.

So Sally, a water-loving pup, is not allowed to jump in the ditch.

Sally is also fond of rolling on any weed-free, deep green lawns. I drag her off as quickly as I can, remembering what Catherine Wissner said the other day.

Catherine is the horticulturist for the Laramie County Cooperative Extension Service. She explained that lawn pesticides contain neurotoxins that kill pest insects, grubs and other small animals by harming their nervous systems.

She said most people don't find this news alarming until she mentions that the toxins put children and pets at risk. They are susceptible because of their small body size and tendency to play on the ground and put things in their mouths.

In a study of 110 preschool children in a Seattle neighborhood, traces of garden chemicals were found in 99 percent of them. I found this statistic online at the National Audubon Society's Audubon At Home page

A link there said an estimated seven million wild birds are killed "due to the aesthetic use of pesticides by homeowners," meaning the use of pesticides to make our yards look nice.

The Audubon At Home program, in cooperation with the U.S. Natural Resources Conservation Service, is promoting the Healthy Yards initiative. This program promotes the idea that our yards can provide wildlife habitat and can make a difference in the health of wildlife, especially since 2.1 million acres are converted to yards every year.

In an Audubon At Home pamphlet on lawn pesticides, New York State Attorney General Eliot Spitzer is quoted saying, "Pesticides pose health risks, even when used and applied in full compliance with manufacturers' recommendations and legal requirements."

The footnotes and references on this subject are extensive, so let's skip to answering the obvious question: What can a wildlife-loving homeowner do instead of using "weed and feed" and other chemically based products?

First, remember some of the critters that live in your lawn are beneficial, meaning they eat the destructive ones or help to convert organic matter into food for your grass. So it is important to figure out whether you even have a pest problem.

Even if you do have pests, how many are there? Can the birds keep their numbers in check? I enjoy watching flickers and grackles aerating my lawn while searching for grubs.

Can you manually control the pest? My husband Mark patrols our lawn for dandelions, though if we let them go to seed, we might attract more goldfinches.

If we had real pests, we'd confer with Catherine. There are many non-toxic controls available.

Another option is to reduce turf and replace it with native plants. They require less water, less care and attract more wildlife. Ask at Wyoming Game and Fish Department offices for a free copy of "Wyoming Wildscape, How to Design, Plant, and Maintain Landscaping to Benefit People and Wildlife."

From a state with a similar environment, "Colorado Wildscapes, Bringing Conservation Home," is available for about $15 through bookstores (West Cliffe Publishers) or at the Boulder County Audubon Society, www.boulderaudubon.org/wildscaping.

Audubon At Home lists local resources, including our own Cheyenne Botanic Gardens. Visit them in person or online at www.botanic.org.

Mark and I have been lucky not to have any serious lawn pest problems. Or maybe it's not luck but years of using healthy alternatives.

Mark uses a national brand of organic fertilizer he buys at a local garden supply store. He cuts the grass at the highest setting on the lawnmower to shade the roots, and we use the clippings for garden mulch. We water moderately but deeply on the city's summer watering schedule.

While our lawn is not artificial-turf green – except where Sally adds her own fertilizer, it is green enough that she enjoys rolling everywhere. And we don't worry about poisons.

Pesticides shouldn't be used like vitamin tonics. If there aren't any pests, they are a waste of money, a source of water pollution and a threat to everyone's health.

As Sally and I walk the neighborhood and observe the little warning signs planted in our neighbors' yards after visits from their chemically-based lawn care companies, I'm hoping someone with entrepreneurial spirit will fill the niche for alternative lawn care.

It could be, should be, the next big thing.

2006

213

# 218 Shorebird experts reveal their secrets

Wednesday, July 19, 2006, Outdoors, page C2

Shorebirds made a big showing on the Cheyenne High Plains Audubon Big Day Bird Count back in May. Seventeen species were recorded by the end of the day.

Now in July, we are in the midst of shorebird migration again as they straggle back from their northern breeding grounds.

Back in May we had half a dozen crack birders with us who could glance through a spotting scope and proclaim obscure names, but now I'm on my own.

I know a few common, unique-looking shorebirds like the killdeer and avocet, but the rest just seem to blend into a mob of brown birds with long legs. I don't see them often enough to practice identification.

How do the experts do it? Three of them, Michael O'Brien, Richard Crossley and Kevin Karlson, reveal their secrets in a book released this spring by Houghton Mifflin, "The Shorebird Guide." Their technique is based on "jizz" as they pronounce the acronym for "general impression of size and shape."

To help readers get a feel for jizz, they've included multiple photos of each of the 50 shorebird species that can be seen in North America, plus the few that might blow in.

A typical field guide will give you a perfectly lit profile of one individual per species, but here are photos of flocks as seen in the orangy glow of sunrise or sunset, from a distance or with other species, giving an idea of relative size. In some photos, the birds may be molting, or their feathers show wear, or maybe they are this year's young.

After 300 pages of photos, there are 160 pages of text and small range maps. Here's where you get the skinny on population health, migration patterns and South American wintering grounds.

The best way to learn birds is to hang out with people who know them. This book is like that and I think with study, I might come closer to distinguishing the 30 species of shorebirds that pass through here once or twice a year.

A second book released by Houghton Mifflin this spring was "Pete Dunne's Essential Field Guide Companion." There are no photos and no range maps. Instead, Dunne gives you a sense of a species' "gizz," as he spells the nickname for not only general impression of size and shape, but also typical behaviors and activities.

Dunne and his wife traveled all over North America to refresh their first impressions, coming up with nicknames for each species. Black-crowned night-heron is "Waterfront Thug"; American robin is "Lawn Plover" and house sparrow is "Horatio Alger in Feathers (an American Success Story)."

These nicknames only work if you've seen the night-herons at Holiday Park hunched at the water's edge waiting to mug a fish, or if you know how plovers run along the shoreline, stopping suddenly to pluck invertebrates, or if you know that a few house sparrows were brought to North America from England, and now they number in the millions and are seen everywhere.

One of the highlights of Cheyenne's spring count was the golden-winged warbler. Dunne nicknames it "Chickadee-bibbed Warbler." The bird is an eastern warbler few of us had seen before, and the nickname does describe the unique and easy to see field mark.

Pertinent to the other vagrant eastern warblers we often see here during migration is Dunne's "Vagrancy Index." The golden-winged warbler rates a 3, "an established, widespread pattern of vagrancy. Ignore the range descriptions. This bird could be sighted almost anywhere."

One disappointment is that although Dunne gives information on some species' wintering grounds, he doesn't for this warbler. But perhaps science doesn't have the answer yet. A lot of the information Dunne gleaned from the great 17,000-page opus, "The Birds of North America." He nicely translates the scientific terminology for the reader.

Dunne gives a lot of gizz characteristics, but I need a vision in my mind's eye to apply them to, such as the photos in the Shorebird Guide or other field guides. However, this book is represented as only a field guide companion – only 700 pages' worth.

But it's the rainy season here, so it's a perfect time for pulling out the paperback edition that came out this spring of Richard Rhodes' biography, "John James Audubon, The Making of an American." The story is pretty amazing and I read every page.

Rhodes draws different conclusions than other Audubon biographers. He said Audubon was not a bad businessman because his first business failed, but rather, was a victim, considering 90 percent of businesses also failed in the financial panic of 1819. Later, Audubon, a consummate salesman, convinced people to spend thousands (in 1800s dollars) on subscriptions to the four volumes of "Birds of America."

The biography is one part love story, one part starving artist's tale and one part frontier saga. But it will also help you understand why, over 100 years ago, bird watchers concerned with the conservation of birds decided to name their fledgling organizations after Audubon.

# 219 A summer of mountain birding

Wednesday, August 16, 2006, Outdoors, page C2

Everyone in Wyoming is lucky to live less than two hours from an outcrop of granite, evergreens and mountain birds.

While I frequently visit the Sherman Mountains in the Laramie Range just west of Cheyenne, summer is the time to travel farther.

The Black Hills were the location for Cheyenne – High Plains Audubon Society's annual campout at the end of June. The 25 bird watchers were from seven Wyoming communities and Colorado and included members of three of Wyoming's Audubon chapters.

Our mission was to survey several different timber sale units between Moskee and the Wyoming – South Dakota state line and generate a species list for each.

My group first walked an area that looked as though it might have been thinned 20 years ago. We heard a variety of singing birds and thanks to the expert ears of Chris Michelson of Casper, we were able to put names to them such as western wood pewee, plumbeous vireo, warbling vireo, hairy woodpecker, red crossbill, Townsend's solitaire, mountain bluebird, black-capped chickadee, Mac Gillivray's warbler and red-naped sapsucker.

Our second unit was old growth – or at least any stumps left from logging were no longer visible. It was definitely quieter. The species composition was different, and we picked up our first ruby-crowned kinglets.

Our anecdotal evidence, gathered with a low amount of scientific protocol, supports the theory that there is more bird species diversity in a disturbed forest. On the other hand, the disturbed forest is missing species that prefer old growth.

Several birds that caused excitement were dickcissels, a grassland species we found sitting on fence posts just outside the forest boundary north of Sundance.

Dickcissels flock unpredictably, especially here on the western edge of their range. About the only thing you can say for sure is that their overall population numbers keep dropping as grasslands endure disruption. But conditions were right this spring – they like moisture – and I was finally able to add them to my life list.

Returning to the Hills is to return to my memories of the summer I was on a Forest Service boundary survey crew. If only I had been paying more attention to birds then!

One of my favorite spots was the meadow below Warren Peaks where, having camped alone, I woke the morning of my 22nd birthday to a bright mist hanging over a hillside massed with blue and yellow flowers.

This year, I found I had missed the peak for arrow-leaved balsamroot and lupine. The meadow was still thick and green, but sported a new power line right through the middle.

A power line also decorated the landscape for part of our hike in Yellowstone National Park in July. A teaching commitment took me to Powell, and my husband Mark thought it would make a good excuse for a trip to Yellowstone afterward.

Our hike took us up above Mammoth Hot Springs, starting on the Swan Flats along Glenn Creek, up and over Snow Pass and back by way of the Hoodoos. The hike was nearly seven miles total.

We weren't hiking in an actual Bear Management Area, but I took the precaution of wearing a bear bell on my boot. Grizzly bears are not on my list of mega fauna I want to see – at least not while on foot.

The avifauna, however, was abundant. Every time I stopped and the bell fell silent, the forest was filled with bird song: Vireos, chickadees, robins, juncos, chipping sparrows, a red-tailed hawk and even red-winged blackbirds at the pond at the pass.

None of these were birds we couldn't see in our local forest, but it was special to know we were in Yellowstone, the Mecca for wildlife observation. However, other than birds, we saw only a mule deer, ground squirrels and five other hikers.

On our evening drives, along with great vistas, we saw lots of elk and occasional buffalo.

We got a chuckle when we realized another roadside distraction was a lone pronghorn, an animal easily seen outside Cheyenne. It calmly stood while people illegally approached.

Jeffrey, our son the antelope hunter, wondered aloud why animals in Yellowstone don't run from tourists. Simple. They've never been shot at with anything more than a camera.

One thing Yellowstone has that our local mountain range doesn't is a bird book devoted exclusively to it. "Birds of Yellowstone" by park ornithologist Terry McEneaney is a must if you hope to get the most out of your trip bird-wise. It highlights the characteristic birds of Yellowstone, provides a list of good birding places, and its checklist is the best I've ever seen for indicating what time of year and in what habitat and abundance a bird may be found.

Under "Sandhill Crane," Terry writes that the park sandhills winter in the Rio Grande valley of New Mexico and Mexico. They come back and build their nests in late April and early May. Nests are mounds of vegetation in shallow, wet meadows. There are usually two eggs and the young fledge in late August to mid-September.

A pair of cranes we saw one evening had built a nest on the island in Floating Island Lake.

They were both there, nattering to each other quietly as twilight deepened.

Too far away and too dim to photograph, they will have to remain as part of an image from our travels I can't stick on the page of a photo album.

# 220 Small, flat memories of East Coast visit

Wednesday, September 13, 2006, Outdoors, page C2

Our family vacation last month had way too much on its agenda – visiting family, colleges, historical attractions and quilt museums.

It would have been easy to look up birding hot spots and experts in the Adirondacks and New England, but why look them up if there wasn't going to be enough time to make use of them?

We did, however, take binoculars and a field guide, and we assembled a nice assortment of birds we can't see at home and familiar birds in habitats we don't have around here.

I think of them as my bird souvenirs with two valuable characteristics: they took up only as much room as paper and pencil, and they were of no interest to airport security.

Here are a few of my mementos, in chronological order:

--Cedar waxwing, High Peaks Trail, Adirondack Mountains, Newcomb, New York

A whole flock was working over cedar trees, white pine and birch. We had a hot and humid hike with an abundance of raspberries to pick and toads and frogs to avoid stepping on.

Adirondack State Park encompasses 6 million acres, although about half is privately owned. The park has long attracted vacationers, including Vice President Teddy Roosevelt who was climbing Mount Marcy (highest New York peak at 5,344 feet) when he learned of President McKinley's death.

--Common loon, Oliver Pond, High Peaks Trail

This quintessential bird of the northern woods is also found in the million acres of designated wilderness in the Adirondacks.

--Olive-sided Flycatcher, High Peaks Trail

When all you have to go on is the memorable "quick-three-beers," song, it is hard to look up a bird's identity in a field guide if you don't remember who sings it. It can be heard in our western coniferous forest, too.

--Great blue heron, Minerva Lake, Adirondacks

My brother-in-law and sister-in-law have these nifty one-person, fiberglass canoes that weigh only 11 and 13 pounds. Wielding a kayak-style paddle, I was able to quietly observe the heron fishing.

--American black duck, maybe, Minerva Lake and other water bodies

Listed on the same field guide page as mallards, the black duck, this eastern species, looks like mallard females to my uneducated eye.

--Slate-colored junco, Blue Ledge, Hudson River, Adirondacks

This race of the dark-eyed junco is the only one that occurs in the east. Cheyenne gets it, plus four others.

--Osprey, near Fort Ticonderoga, on the shores of Lake Champlain

All along our trip we saw man-made nesting platforms erected near power lines, and birds using them.

--Double-crested cormorant, Harvard Bridge over the Charles River between Cambridge and Boston

Where there are fishing birds, there are fish, with water clean enough to keep them alive.

--Wild turkey and poults, National Monument to the Forefathers, Plymouth, Massachusetts

It's serendipity that at this granite pillar commemorating the Pilgrims, the bird species most associated with them walks out of the brush and into view.

--American Goldfinch, Marble House, Newport, Rhode Island

For a house that features a drawing room with walls painted in gold leaf, why wouldn't they have gold birds on the grounds?

--Northern Flicker, Saint Patrick's Cemetery on the Mystic River, Connecticut

The yellow-shafted variety of northern flicker swooped through the trees. Cheyenne's are red-shafted. A pair of kingfishers was working the river, as was a great egret.

--Mute swan pair and two cygnets, tidal river, coast of Connecticut

The last time I wrote about mute swans being an invasive species in North America, I was reprimanded by someone who believes they exist in this continent's paleontological record. But if you ask a birder or biologist on the east coast, these birds are an increasing menace.

--Gulls on the beach, Rocky Neck State Park, Long Island Sound, Connecticut

There were big gulls with red spots on their bills – herring gulls; smaller gulls with black bill spots – ringed-bills like we have here; a big black-headed gull – likely a laughing gull; and tons of people not worried about skin cancer.

--Great horned owl, Yale University, New Haven, Connecticut

Among the campus's Gothic crenellations and cornices, I spotted the familiar shape sitting on a cross-piece of a tall spire, except it had weathered copper for feathers.

--Rock pigeons, Columbia University, New York City

Although Columbia is protecting its buildings and students from pigeon droppings by installing mesh over potential nest sites, the fountain in the square was attracting a multi-colored mob.

Did you know "Columbidae" is the family name for pigeons and doves?

Also, we saw the lions guarding the New York City Public Library patiently enduring pigeon toes tickling their metal hides.

--Northern cardinals, The Bronx, New York City

My brother-in-law's backyard is crammed with vegetation under a towering sweetgum tree. Our last evening, sitting on the deck, we enjoyed the antics of several cardinals chasing each other from yard to yard. Who knows how long it will be before we see a cardinal in Cheyenne?

--Sandhill crane, I-25 just north of Denver International Airport

In the twilight, we saw the unmistakable silhouette

fly over the highway. Mark and I looked at each

other in surprise – fall migration already!

Where did the summer go?

# 221 The last to go

Wednesday, October 11, 2006, Outdoors, page C2

## Warblers put on a late season show

As the leaves turned yellow-green in late summer, you may have noticed them shaking without benefit of a breeze.

Did you see small, greenish-yellow warblers picking through the foliage for insects?

Since 1993, Cheyenne-High Plains Audubon Society has documented 29 species of warblers on its Big Day counts held mid-May, at the peak of spring migration.

We haven't given the same scrutiny to fall migration since warblers trickle through Cheyenne beginning late August and on into October. In the spring, the timing of their presence is more concentrated.

Most of the reports I've received this fall are for easily recognized warblers: Townsend's with its mask, Wilson's with its black cap and beady black eyes and yellow-rumpeds with their yellow rump in contrast to blue-gray back and wings.

The yellow-rumped warbler stands out in many ways. First, it's just about the most common wood warbler species, which is probably why it has a well-known nickname.

Can't you just hear the ornithologist tracking the quick-flitting unknown bird deep in the bushes and finally exclaiming, "It's just another butter butt!"

The yellow-rumped comes in two forms that were previously two separate species.

One, the myrtle warbler, has a white throat and is considered the eastern form. The other, Audubon's warbler, has a yellow throat. It breeds in the Rocky Mountains and winters in the southwestern United States and Mexico.

We see both forms in Cheyenne, so it is always worthwhile to scrutinize this common bird.

The yellow-rumped has odd habits for a warbler. Last month, friends and I hiked up to Emerald Lake in Rocky Mountain National Park, elevation 10,000 feet. Although there was fresh snow on the peaks and aspen in full color lower down, we found yellow-rumpeds busy catching flying insects, and in a most unwarbler way.

They perched on a picturesque dead tree at the lake's edge, then flew out over the water after their prey and circled back to their perches. This activity is called hawking, and flycatchers are the group of birds that use it most often. According to my books, it's a recognized feeding behavior for yellow-rumpeds too, but not for most warbler species.

What also sets the yellow-rumped apart is its wide range of gastronomic preferences. Other warblers have to head south when it is too cold to find live insects, but the yellow-rumped starts picking berries. That's how the myrtle got its name – it likes to eat wax myrtle berries.

Apparently, yellow-rumpeds have a digestive system that can deal with the berries' waxy coating. I don't think around here we have any myrtle, or bayberry, its other favorite food.

But both Audubon's and Myrtle forms stick around Cheyenne quite late, eating other kinds of berries and seeds. Robert and Jane Dorn list records as late as the first week in December.

Other warbler species' latest dates are in mid-October.

Pete Dunne's Essential Field Guide Companion calls yellow-rumpeds "The Swarm Warbler." I have seen this phenomenon myself, in Lions Park.

Warblers migrate at night. By morning they are ready to come to earth and refuel.

As I walked the dog one spring morning, between the new community house and the lake, yellow-rumped warblers tumbled across the path at my feet like wind-blown leaves.

While they may swarm during migration, yellow-rumpeds prefer to spread out for breeding in the coniferous forests of the mountains and the north. Little is known about this part of their lives compared to that of other, more gregarious songbirds.

David Flaspohler, author of the extensive account in Birds of North America, made a lot of observations while completing his

dissertation on metapopulation dynamics and reproductive ecology of northern forest songbirds in the upper Great Lakes.

It was already observed that the female is usually the sole nest builder, though the male may sing and keep her company while she works.

Flaspohler was able to watch eight nests in northern Wisconsin in 1996 and documented that incubation is almost entirely done by the female.

"Male often sings in vicinity of nest during incubation," he wrote.

When it's time to feed the young, the male helps, in between bouts of singing. In other species, parent birds are very quiet near the nest because they don't want to attract predators.

Many other sections of the account, however, state "No information." It looks like aspects of butter butt life history could provide many more topics for theses and dissertations.

For instance, the last time yellow-rumpeds were tested for the effects of spruce budworm pesticides was 1987, in only one place and for only one kind of pesticide.

Or, why was the yellow-rumped the most abundant warbler found in collisions with towers in Florida, but rarely in Pennsylvania?

Yellow-rumped populations are said to be stable or increasing, but standard avian demographic data is lacking.

Another species, the greater sage-grouse, is suffering from rampant oil and gas development in Wyoming today and is finally attracting lots of research funds.

Had more research been done earlier, wildlife biologists may have been able to make better recommendations sooner to stave off the disastrous situation we have now. Then again, sound biological recommendations need to fall on willing White House ears to have any effect.

Meanwhile, enjoy warbler watching.

Consider posting your bird observations on eBird.com. Every little bit helps us figure out the puzzle that is life on Earth.

# 222 Grouse losing ground fast

Wednesday, November 8, 2006, Outdoors, page C2

Sage grouse have the misfortune of living directly in the path of natural gas development. In Wyoming, we are talking about the bird known by the formal common name, Greater Sage-Grouse.

While some species benefit from human activity – think Norway rat, pigeon, starling, coyote, cockroach – the sage grouse is not one of them.

Matt Holloran spoke at the Cheyenne High Plains Audubon Society meeting last month about his studies on the natural gas fields near Pinedale as a University of Wyoming doctoral candidate. He's now a consultant with Wyoming Wildlife Consultants.

While much of his talk was still couched in the language used to successfully defend his dissertation, accompanied by graphs and charts, Matt's findings are clear enough. No sage grouse, much less anyone else, wants to live next to a drill rig or producing gas well, whatever the season.

First, there are the dancing grounds, or leks, where males perform on spring mornings, and females come to observe and decide on a mate. Leks have to be open areas with good visibility, but a little thing like a flyover by a predator like a golden eagle is enough to cancel the show for the rest of the morning.

Matt calculated the success of each lek by the number of males that continued to attend.

During five years of study, it was obvious that the negative effect of drill rigs and producing wells on grouse increased as structures got closer to leks. It also increased where the wells were positioned closer together.

Matt even determined that leks downwind from a rig were affected more negatively than leks upwind. Noise was the factor.

Also, the closer a road was to a lek and the busier it was, the greater the negative effect. One has to travel more than 5 kilometers away from drilling or wells before leks appear to be unaffected.

Mated female grouse leave the lek to find perfect nesting habitat. They like to lay their eggs under sagebrush where tall grasses also help screen them from predators. They react poorly to drilling, and fewer chicks survive.

Later, when females from disturbed areas share summer habitat with females from undisturbed areas, the former are more likely to die. One of the reasons may be that, having learned to ignore human activity, they are no longer paying close attention to predators.

When leks lose male sage grouse, do the birds leave a disturbed area and move to a new area, or do they just stop reproducing the next generation?

There aren't a lot of places for sage grouse to go.

As a range management student 25 years ago, I learned all the techniques for killing sagebrush to encourage more grass for cattle to graze. Fire is especially effective, retarding sagebrush growth for over 200 years in some cases. Sheep, however, are browsers, and since they nibble shrubs, sagebrush isn't managed the same way for them.

It would be great if we could provide habitat for sage grouse somewhere away from the gas fields. Energy companies are used to thinking in terms of mitigation.

But no one knows yet exactly how to build and connect all the habitats needed for sage grouse, nor does anyone know exactly how other factors, such as West Nile Virus, predators and drought work together to affect numbers of grouse. Their populations have been declining since the 1960s.

Matt said when peregrine falcon numbers were dropping, all it took was a ban on DDT and the population rebounded. In comparison, sage grouse are a puzzle.

He does, however, have several suggestions. One is to increase the distance between gas field activity and leks and nests, as stipulated by the Bureau of Land Management, based on his findings.

Keeping well density to less than one per 699 acres, which is a little larger than one square mile, can be done with directional drilling. Multiple wells could share the same site and access road, pumping gas from pockets up to a mile away in all directions, leaving more land surface to sage grouse.

Something as simple as garbage control on the well sites would quit attracting ravens to the area. They eat sage grouse eggs.

Another key to sage grouse survival that Matt recommends is that "intact sagebrush-dominated habitats be protected and managed for suitable understory conditions."

If you would like to read any of the 223 pages of Matt's dissertation, "Greater Sage-Grouse Population Response to Natural Gas Field Development in Western Wyoming," email him at matth@wyowildlife.com and ask for the PDF version. It also includes a summary of other relevant sage grouse studies.

Natural gas is not a renewable resource. I don't understand the federal government encouraging drilling everything as fast as possible. The resource will just run out faster.

Slower development would give an area of exhausted wells a chance to be reclaimed for sage grouse before new areas are disturbed. And, sage grouse are not the only ones to suffer from high-speed development.

Consider our small western Wyoming towns.

Procrastination is a hallmark of being human, so pessimist or realist that I am, I don't think alternative energy will be given the brain power it needs to find the

most inspired solutions until it is absolutely necessary.

Let's hope it's not too late for the Greater Sage-Grouse by then.

# 223 Get creative with gifts for birders

Wednesday, December 6, 2006, Outdoors, page C2

The last four digits of several of these phone numbers are 2473 because they spell BIRD.

Gift resources for bird watchers

--**American Bird Conservancy:** membership, research, advocacy, publications, gear. www.abcbirds.org, 540-253-5780.

--**American Birding Association:** membership, publications, books, optics, gear, travel. www.Americanbirding.org, 800-850-2473.

--**Birder's World:** magazine, www.birdersworld.com, 800-533-6644.

--**Bird Watcher's Digest:** magazine, bird info, bird festival listings and gear for sale. www.BirdWatchersDigest.com, 800-879-2473

--**Cornell Lab of Ornithology:** membership, bird info, Citizen Science projects. www.birds.cornell.edu, 800-843-2473

--**International Migratory Bird Day:** education, online store. www.birdday.org, 866-334-3330.

--**National Audubon Society:** membership, magazine, research, advocacy, directory of chapters. www.audubon.org (Cheyenne, 634-0463)

--**North American Birds Online:** Internet data base. www.bna. birds.cornell.edu/BNA, 800-843-2473

--**Thayer Birding:** software. www.thayerbirding.com, 800-865-2473.

--**The Nature Conservancy:** membership, publications, gear. www.nature.org, 800-628-6860.

A good gift is useful, educational or edible, if not homemade. If someone on your gift list truly cares about birds, they don't want energy and resources harvested from sensitive bird habitats wasted on making junk.

Here's my list, sorted somewhat by a recipient's degree of interest in wild birds.

First, for anyone, armchair bird watcher to ornithologist, Houghton Mifflin has three new illustrated books.

"Letters from Eden, A Year at Home, in the Woods," by Julie Zickefoose ($26) includes her beautiful watercolor sketches. A frequent contributor to Bird Watcher's Digest, her bird and nature observations are often made in the company of her young children on their 80-acre farm in Ohio or from the 40-foot tower atop her house.

Zickefoose's tower may have been her husband's idea. He is Bill Thompson III, editor of Bird Watcher's Digest and editor of "All Things Reconsidered, My Birding Adventures," by Roger Tory Peterson ($30).

Peterson was the originator of the modern field guide. From 1984 until his death in 1996, he wrote a regular column for the Digest. Peterson had the gift of writing about birds, bird places and bird people so anyone could enjoy his choice of topics. Anyone can enjoy this photo-illustrated book.

The third book, "The Songs of Wild Birds," ($20) is a treat for eyes and ears. Author Lang Elliott chose his favorite stories about 50 bird species from his years of recording their songs. Each short essay faces a full-page

photo portrait of the bird. The accompanying CD has their songs and more commentary. My favorite is the puffin recording.

The field guide is the essential tool for someone moving up from armchair status. National Geographic's fifth edition of its "Field Guide to the Birds of North America," ($24) came out this fall.

New are the thumb tabs for major bird groups, like old dictionaries have for each letter. It has more birds and more pages plus the bird names and range maps are updated.

Binoculars are the second most essential tool. If you are shopping for someone who hasn't any or has a pair more than 20 years old, you can't go wrong with 7 x 35 or 8 x 42 in one of the under $100 brands at sporting goods

stores. You can also find an x-back-style harness ($20) there, an improvement over the regular strap.

Past the introductory level, a gift certificate would be better because fitting binoculars is as individual as each person's eyes.

Spotting scopes don't need fitting. However, if you find a good, low-end model, don't settle for a low-end tripod because it won't last in the field.

For extensive information on optics, see the Bird Watcher's Digest website.

Bird feeders, bird seed, bird houses and bird baths are great gifts if the recipient or you are able to clean and maintain them. To match them with the local birds at the recipient's house, call the local Audubon chapter, or check the Cornell Lab of Ornithology's Project

2006

219

FeederWatch and Birdhouse Network sites.

You can make a gift of a Lab membership ($35), which includes several publications, and of course, there's Audubon and its magazine ($20 introductory offer). Bird Watcher's Digest ($20 per year), mentioned above, makes learning about birds fun, as does Birder's World magazine ($25 per year).

For someone who wants to discuss identifying obscure sparrows and other topics of interest to listers, they might be ready for membership in the American Birding Association ($40). The ABA also has a great catalog available to everyone online. It's filled with optics, gear and every bird book and field guide available in English for the most obscure places in the world.

The ABA tempts members with mailings for trips to exotic birding hotspots, as well as its annual meetings held in different parts of the country. Also check Bird Watcher's Digest for nationwide bird festival listings.

One subscription valuable to an academic type who doesn't already have access, is the Birds of North America Online ($40). Every species has as many as 50 pages of information and hundreds of references to studies.

For the computer literate, Thayer Birding Software's "Guide to Birds of North America," version 3.5 ($75), includes photos, songs, videos, life histories, quizzes and identification search functions.

After the useful and educational, there's the edible. Look for organically grown products because they don't poison bird habitat. The ABA sells bird-friendly, shade-grown coffee and organic chocolate through its website.

Coffee and other items are available also at the International Migratory Bird Day website, and sales support migratory bird awareness and education.

If the person on your list is truly committed to the welfare of wild birds and wildlife in general, skip the trinkets such as the plush bird toys that sing, and don't add to their collection of birdy t-shirts.

Look for products that are good for the environment. These are items that are energy efficient, rechargeable, refillable, fixable, recyclable, made from recycled or organic materials, or are locally grown or manufactured.

Or make a donation in your giftee's name to an organization like Audubon, the American Bird Conservancy or The Nature Conservancy, which work to protect bird habitat.

Presents along these lines would be great gifts for your friends and family members and for birds and other wildlife, any time of year.

# 2007

## 224 Out = healthy

Wednesday, January 24, 2007, Outdoors, page C2

**Author finds direct correlation between outdoor activities and our mental and physical health**

How are you doing with your New Year's resolutions? Mine are the same as last year, but I hope to be more successful.

One was to go outside more often. This year I also want to see more birds and improve my birding skills.

The outdoors, with the addition of sunscreen, should be good for me. Studies show gazing at the natural contours of land and vegetation improves both our mental and physical health.

I've recently been reading "Last Child in the Woods, Saving Our Children from Nature Deficit Disorder," by Richard Louv.

His main premise is that it's a problem that today's child is no longer allowed to, or even interested in, hanging out in the natural environment.

For parents, letting kids spend free time in front of a TV or computer screen is preferable to the outdoors where they might meet dangerous strangers, even if it is a statistically rare event. And the woods may no longer be down at the end of the street or even within biking distance.

Supervised outdoor experiences are better than none, but they're not the same as dinking around on one's own or building a tree house or fort with friends. Louv sees these experiences as developmentally necessary.

I mostly grew up in an inner suburb of Milwaukee, Wisconsin, sometimes playing pioneer in the undeveloped end of the cemetery. But at my grandfather's home in Illinois, there was a stream at the bottom of his hill just right for imaginative play.

One year we lived in Oak Park, Illinois, with a city park across the street. We neighborhood kids spent hours there. The only time the woods were literally at the end of our street was the six months or so we lived in Paducah, Kentucky.

My own children played outside until we got a computer. Even though we limited its use, they turned instead to reading. Outdoor activities became confined to scouting and family outings.

Louv cites many studies that correlate lack of outdoor exposure with ill health in children, especially obesity. But the biggest loser may be the environment if children don't make connections with it.

Why worry about clean air and water if you can filter everything before it comes into the house? Why worry about wildlife if you can clone them on the computer screen or see them on high-definition TV?

Louv gives examples of school programs and housing developments that safely allow kids to explore nature on their own, but they sound impossibly utopian. How do you convince today's young couples to choose those kinds of alternatives if they themselves are already part of the indoor generation?

Louv was amazed by a group of university microbiology students he met who couldn't identify local fauna and flora. He says opportunities for microbiology study have flourished while the study of mega animals and plants, or natural history, has declined.

Bird watching is a bright spot, continuing to grow by leaps and bounds. Recently, 23 students signed up for the local Audubon chapter's winter birding class. They wanted to know everything the instructor knew, not just identification. But most were from the outdoor generation.

Although one might never get further than backyard birding, it can lead to concern for the wider world since for instance, the Swainson's hawk soaring over Cheyenne in the summer eats grasshoppers in Argentina in winter.

However, an interest in birds doesn't always lead to ecological enlightenment and advocacy as I would hope.

The other book I've been reading is "To See Every Bird on Earth, A Father, a Son and a Lifelong Obsession," by Dan Koeppel. It finally explained for me the phenomenon of listing.

Listers keep records of the first time they identify a bird species for their life list. They may also keep a list by state or county, and a yard list, and a year list, and a country list, if they travel as much as Koeppel's father did.

Richard Koeppel became addicted to adding birds to

his life list. Unfortunately, he didn't channel his interest into the field of ornithology, or pass it on to his sons, but followed his parents' wishes to study medicine.

As an itinerant doctor, he had the money and the time to book passage on birding tours and became one of the world's top ten listers competing for the longest life list.

Koeppel saw 7,200 of 9,600 species found worldwide before health problems made him decide to quit a few years ago.

A mild obsession can give life more depth, but to take it to the depth of the "Big Listers" is unhealthy, I think.

Sure, they were outside, but it doesn't seem like they saw the forest for the birds.

No chance I'll get overly obsessed with birding now if I haven't already in 25 years of Audubon field trips.

I have Richard Louv to thank for making my resolution a health prescription since he touts the benefits of green landscapes for all ages.

I hope "green" is just a synonym for "natural." I'd hate to wait for my local landscape to turn green. This time of year, the prairie is mostly 29 shades of brown with white accents, but it works for me.

# 225 This spring watch for 'TVs' soaring above the area

Wednesday, February 21, 2007, Outdoors, page C2

Do you know where Wyoming turkey vultures spend the winter? It could be Venezuela.

Keith Bildstein, director of the Acopian Center for Conservation Learning at Hawk Mountain Sanctuary in Pennsylvania, is working on a migration study in which turkey vultures wintering in northwestern Venezuela have been tagged. He predicts bird watchers in western North America will see them this spring and report back to him.

Little is known about the migration of the "TV," as birders refer to it, and even though it is a species with a stable population that is increasing northward, it's better to do your research in advance of problems, said Bildstein, when I talked to him recently. Also, sometimes the new information will translate to less fortunate species.

Bildstein studied raptors for his dissertation, so it is quite natural to find him at Hawk Mountain, the famous place where so many hawks pass on migration. What bothered him was that observers would refer to "just another turkey vulture." He thought they deserved more respect than that.

Turkey vultures in eastern North America don't migrate much except to get out of the cold – it is hard to chip meat off frozen carcasses.

Bildstein said satellite studies show TVs travel independently and individual birds may not travel the same route each year. They stop along the way and share roosts and food with the local vultures. Maybe they pick up pointers on great Florida real estate.

Meanwhile, when birds of the western subspecies head south, they travel down through Mexico, Central America and into Columbia and Venezuela, possibly heading as far as Argentina. On the wintering grounds they raise the population of vultures to four times that of the year-round resident subspecies. The residents, being smaller birds, are crowded into marginal habitat, Bildstein said.

Two of the places the "gringo" vultures like to hang out are the zoos in Barquisimeto and Maracay. Last winter, zoo folks told Bildstein about a tagged TV they found that turned out to be part of a study in Saskatchewan, Canada.

Bildstein and Adrian Naveda, a biologist from Maracay, put their heads together and designed the northward migration study, counting on the help of the legions of birdwatchers in North America.

It was easy to gather the visiting vultures. Zoo management cleared out one of the aviaries, stocked it with dead chickens from the market, waited for the vultures to walk in, closed the door and tah-dah, 100 vultures ready for tagging. However, grabbing birds with 67-inch wingspans probably wasn't easy.

Bildstein has had one report of a tagged bird so far. It was found shot 45 miles north of the release site. The rest of the birds should be migrating along the coast. In early morning, the warm ocean creates small thermals near shore, giving the TVs an early start. Then, it's a matter of riding one thermal after another over land all day long, day after day. As many as 2,000,000 turkey vultures have passed by an observation point in one season.

Here in Cheyenne, we may see TVs as early as March and definitely will by April. They have favorite roosting spots in the Avenues and are often seen circling over the cemeteries. Recently, they have been noticed far into the summer, however, the nearest nest is probably near Guernsey, according to Doug Faulkner, who is working on the definitive book about Wyoming birds for the University of Wyoming.

OK, this is where you come in. Let's review TV i.d. The most common large birds flying over Cheyenne in the spring, summer and fall are the turkey vulture and the Swainson's hawk, which, incidentally, spends the winter in Argentina and shares the vultures' migration route.

As they soar overhead, look at the underwing patterns. The leading (front) edge of the Swainson's is light and the trailing edge is dark. Turkey vultures have the reverse: dark on the leading edge and light, actually silvery, on the trailing edge. Seen up close, they have red-skinned, featherless heads.

If you see one of the marked birds, it will have either a red tag with white numbers or a blue tag with black numbers wrapped over the leading edge of the wing, visible from top or bottom.

If you see one of these birds, you need to make note of the date, specific location, color and number of the tag, which wing it is attached (the bird's right or left) and the circumstances of the sighting, whether the bird was alone or in a group of vultures, flying, perched, feeding or roosting. Dead birds should also be reported.

Report sightings to: Keith Bildstein, Hawk Mountain Sanctuary Acopian Center for Conservation Learning, 410 Summer Valley Road, Orwigsburg, PA 17961; Bildstein@hawkmtn.org; 570-943-3411, ext. 108. All reports will be recognized and individuals reporting tagged birds will get summary information about the study.

If you would like to print your own copy of the "Wanted" poster, go to www.hawkmountain.org.

The February issue of Smithsonian magazine tells of the demise of millions of vultures in India in just 10 years due to ingestion of a new livestock antibiotic while feeding on dead cattle. It has led to a terrific increase in wild dogs and, in turn, human cases of rabies. Valuable time was lost puzzling it out and there is no guarantee vulture populations will ever recover.

Though turkey vultures are the widest ranging of the vulture species, from Canada to Tierra del Fuego, and probably the most numerous vulture species in this hemisphere, everything that can be learned about them helps keep them that way.

To learn more to marvel over about turkey vultures such as their terrific sense of smell, the way the young protect themselves and the sounds they make, go online to All About Birds, www.birds.cornell.edu/AllAboutBirds/BirdGuide/Turkey_Vulture.

# 226 Visiting cedar waxwings learn lesson

Wednesday, March 21, 2007, Outdoors, page C2

Driving under the influence is a tragedy waiting to happen. Flying inebriated can also have fatal results as was illustrated in our backyard.

It was Feb. 23, and I was assembling my lunch when I heard two thwunks, one right after the other on different windows facing the backyard. Birds, no doubt.

As I brought my sandwich to the table, I realized one of the cats was looking intently out the window to the patio below. I followed his stare and saw a cedar waxwing lying on the concrete.

In the seconds it took me to run out the back door, it died, the milky-white nictating membranes pulled over its eyes. As I picked it up, the head lolled. Broken neck.

I'd been hearing the flock for the past few days, a faint, wispy, background sound, not to be heard on a regular basis. Waxwings don't migrate predictably so much as they go where the berries are. And the neighbors' juniper bushes are full. Over the winter, berries can sometimes ferment, leading to intoxication and poor flying judgment.

I wonder if drunk birds also smack into natural obstacles?

Windows confuse birds, reflecting landscapes like mirrors. However, half of each of our windows is screened and I hope that breaks the image or allows a bird to bounce off the flexible material rather than hard glass.

Mark and I leave the shiny side of the windows dirty, tape junk CDs shiny side out, and add those static cling window stickers. This time there was even a staring cat.

Waxwings would be easy to carve. Their pale brown plumage is so fine you don't notice individual feathers, except for the wing tips with their little drops of red "wax," a carotenoid pigment substance.

They have a black mask, a crest on the back of the head and a band of bright yellow on the tip of their tail, as if someone had dipped them in paint.

I couldn't bring myself to bag and toss in the garbage this gorgeous bird so I left it to naturally decay where the dog wouldn't get it.

I forgot all about the second thwunk until I let the dog out but called her back before she noticed the second bird.

It was another cedar waxwing, huddled on the cement, but breathing. Just stunned.

I set it on the back wall to give it a better vantage point to fly from when it recovered. It didn't have the red drips on its wing tips and later I read that birds hatched last year won't have them yet by this spring. One lucky teenager.

It occurred to me to run back in the house and grab my camera. I got one bad photo and while lining up for a second, the bird flew over my head. Good.

Being a sunny, 50-degree day, I decided to eat my neglected sandwich on the back steps, in the company of two dozen waxwings in the tree overhead. They sat still and quiet, blending in with the bare gray-brown branches, except for their pale yellow bellies.

Were they having a moment of silence for their fallen comrade? Or had they decided to sober up before flying again? I couldn't ask them and find out.

When son Jeffrey came in from school he announced there was a dead cedar waxwing on the front step. I checked the wing tips. Another youngster. It left a few tiny feathers stuck to the front window.

It is now March 13 as I write this, three weeks since I first heard the flock. It is still here. We have found no other accident scenes. Evidently, the waxwings have learned to avoid the windows and it seems they learn from each other since we would have noticed if all 30-40 of them had had to bump into our windows to learn the lesson.

Every time I take the dog out, I listen for the waxwings. They are either across the street in the junipers picking berries or resting in our trees. Jeffrey can't wait for them to move on so it will be worth his while to wash the car he parks in the driveway under a tree.

This is the longest cedar waxwings have ever stayed in the neighborhood and I know they will leave at some point. And, just like their larger and even more nomadic counterparts, the Bohemian waxwings, their return will be as unpredictable.

If waxwings were as common as house sparrows and starlings, would I enjoy them as much? Perhaps I should take a closer look at those abundant species. But then, they hardly ever crash into windows.

Spring migration has been going on for a couple months already, imperceptibly if you didn't know to watch the waterfowl. In the last week I've seen robins, heard mourning doves and heard reports of bluebirds, and so the rush begins.

Most migrating birds won't be drunk, but they don't stay long enough to learn the local obstacle course, so I guess we will leave cleaning windows until later. I should hang more deterrents. And then it will be baby bird season, and they fly into everything. And then it's migration season again.

Well, if it looks like we never wash our windows, you'll know why.

Besides, who has time for chores when there are birds coming in?

# 227 Duck diversity

Wednesday, April 18, 2007, Outdoors, page C2

## How to separate those mallards from shovelers

Spring is a wonderful time to learn to identify ducks because all the drakes (males) are in full breeding plumage. No matter how old they are or where you see them, they look just like their pictures in the field guides – which you can't say about many other bird groups.

Plus, ducks are large and easy to see, especially here on Sloans Lake in Lions Park. A spotting scope is handy to have, but the lake is small enough you can see important field marks with binoculars.

First, let's dismiss all the geese, the large gray birds with the black necks and heads that are here all year round. Be sure to impress your friends that this species is properly named "Canada Goose," not "Canadian Goose."

Next, let's sort out all the brown ducks. These are the females of all duck species. Without inspecting their body shape and particular wing feathers, the best way to identify them is to see with which males they swim.

Then, let's review the field marks of the "Mallard," the most abundant, most recognizable local duck here year round. Many no longer migrate because they've figured out misguided people will feed them.

The mallard drake has the bright, iridescent green head (except when molting in the fall), bright orange legs, bright yellow bill and a tail that curls. There are many other features that could be described, but these field marks distinguish the mallard from other ducks we see locally.

Also, they are the only ones, along with the Canada geese, that will approach you for a handout.

Another duck with a bright green head, the "Northern Shoveler," has an elongated body shape ending in a black bill that looks like a shovel or spatula. Its breast is bright white, and its sides are chestnut brown.

Two species of ducks have plain red heads, as in the bright brown chestnut color of red hair.

One, aptly named "Redhead," has a nice rounded head like the mallard's, with its bill jutting out at about a 90-degree angle to its forehead. The other, the "Canvasback," has a sloping forehead continuous with the slope of its bill, what the field guides like to call a "ski slope."

A third red-headed duck, the "Green-winged Teal," has a section of green feathers that can be seen when it extends its wing, but it also has a green, wing-shaped marking that encircles its eye and extends to the back of its neck.

The "American Wigeon" has the same green wing shape over its eye, but it has a spectacularly wide white stripe from the top of its bill to the back of its head.

A white stripe up either side of its neck distinguishes the "Northern Pintail," an otherwise grayish bird. The stripes are easier to see than the long, pointy tail.

A white crescent on each side of its face, between eye and bill, sets the "Bluewinged Teal" apart, since many other ducks also show a blue speculum (section) on their wings.

If a duck is all chestnut red, and shaped like a mallard, it is the "Cinnamon Teal." If it has a tail that sticks up stiffly, and it sports bright white cheeks and a bright blue bill, it's a "Ruddy Duck."

There are several black and white duck species, two of which we see in winter, "Common Goldeneye," and "Bufflehead." But by spring, you are most likely to see the "Lesser Scaup." Its head and tail ends are black, and its middle is white. The "Ringnecked Duck" looks just like it but with good optics, the black tip on the blue and white bill is noticeable.

The "Gadwall" has the distinction of being the plainest duck, just a sort of fine, tweedy gray, but it is the only duck solidly black under the tail.

It is possible to see mergansers on Sloans Lake. All three species, "Common Merganser" being most likely, have long thin bills for catching fish. They have small heads with feathers that sprout like a bad hair day.

Not all birds that can swim

are ducks. The "American Coot" doesn't have webbed feet. It has a compact all-black body like a rubber ducky and a distinctive bright white bill.

The grebes are not ducks either. The "Western Grebe," a larger gray bird with a long, white-fronted neck, will be back soon, entertaining us with its water-dancing mating rituals.

The "Pied-billed Grebe," small, short-necked and brown with a black and white bill, is not as noticeable.

Horned and eared grebes are more likely on larger reservoirs.

And then there are the big, dark brown "Double-crested Cormorants" that float low, as if waterlogged, or fly overhead looking like sticks with wings.

There are a couple unidentifiable ducks at Sloans Lake that are the offspring of domestic white ducks mating with mallards. You'll know that's what they are because they have the mallard shape, if not size and colors,

and they hang out with the mallards.

Also, there are all those other swimming birds that make surprise visits: pelicans, swans and unusual gulls and terns.

I meant to make this a simple guide to the most common ducks to be seen at Sloans Lake this spring, and already I've mentioned 16 and referred to a dozen other water birds.

At least it will make it easier for you to decide what birds to study in your

field guide.

If you want more help, sign up for the beginning bird class at Laramie County Community College. It includes two, 2-hour, Thursday evening classroom sessions and two, 2-hour, Saturday morning local field trips May 3-12.

I know it's hard to believe that so many kinds of ducks can show up on a lake in the middle of town, surrounded by people walking dogs. But that's the magic of migration. All you have to do is look.

# 228 They look helpless, but they probably don't need rescuing

Wednesday, May 9, 2007, Outdoors, page C1

## Everyone wants to be a springtime hero, but is that tiny bird on the ground in dire straits?

Nothing is as appealing as rescuing a helpless baby bird fluttering on the ground. Everyone wants to be a springtime hero. Becoming one appears to be as easy as scooping up the tiny bird, but is that the right thing to do?

Laura Conn, veterinary technician, knows first-hand how many times a year nestlings are rescued because they all seem to end up at the Cheyenne Pet Clinic where she works.

"People find them on the ground or don't see Mom for a while, or they want to remove a nest from the threat of outdoor cats," Conn said.

Last year it was 39 common grackles, a dozen robins and well over 50 house sparrows besides an assortment of other bird species.

"It can be a couple nests a day. Sometimes because of construction, workers will bring them in," said Conn.

The clinic's standard

advice is that unless birds are in immediate danger, it is safer to return them to the nest. "But everyone wants to bring them in," said Conn.

One myth is that a baby bird alone on the ground has been abandoned. However, if it has feathers already, it may have been pushed out of the nest by its parents who are probably nearby, keeping an eye on the youngster as it makes the transition to independence.

This would be especially true of birds that hatch precocial young, the young of meadowlarks, killdeer and other ground nesting birds. The chicks hatch with feathers and can practically run as soon as they depart the shell.

Altricial young are those helpless, naked nestlings like robins and sparrows that need a few weeks for feathers to grow in.

If the nestling is found completely or

semi-featherless, the best thing to do is put it back in the nest. If the nest has been destroyed, fashion one from a basket or bucket.

A second myth is that once a human has touched a baby bird, the parents will abandon it. Not true, said Conn.

As for marauding cats, Conn said young birds probably have a better chance of surviving under the protection of an angry parent bird than if they are brought into the clinic.

The rate of survival of young birds transferred to the clinic is one in three.

At the clinic the bird is assessed for damages. Falling may produce injuries, making the bird impossible to rehabilitate. Injuries from cats are seldom seen since there's usually nothing left of the baby bird after a feline encounter, Conn explained.

If the nestling is in good shape, it is popped into the

incubator. Then it's time to mix up special mash, either meant for young poultry, or special mixes for wild birds.

A rescued baby bird needs feeding every two hours, at least until 10 p.m., when the last clinic employee goes home. All the employees pitch in at the height of nestling season, even the front desk, said Conn.

The Cheyenne Pet Clinic is the only local facility with the necessary federal permit for handling wild birds. It is not legal to tend wild birds without a permit.

Robert Farr, the clinic's founding veterinarian, said he would be interested in hearing from anyone with previous experience who would like to help by taking orphans home. Volunteers can work under the clinic's permit and the clinic will provide the food.

Conn has been caring for baby birds since she started

at the clinic as a volunteer 19 years ago. While injured large wild birds also come in, such as the great horned owl which was recently recovering from tangling with a barbed wire fence, most are sent on to the veterinary hospital in Fort Collins. But small birds are cared for at the clinic until their release.

Ducklings are also brought in occasionally, but said Conn, "We try to find someone to take them quick – they don't do well here."

Some birds that come in are very prone to stress and succumb quickly while others seem hardier, said Conn.

Some summers it seems like all the young survive and other summers they don't. The older the nestling is, the better its chances of surviving. Some birds just seem to be tougher, like robins.

Depending on how old the bird was when it came in, it can take two or three weeks before it is ready for release.

First, it has to be able to eat on its own. "It's hard to train them to eat," said Conn. "We can't do it as well as their mothers. And they have to be able to fly well on their own, too."

Typically, the birds are taken to the park where there are plenty of trees, since most of the rescued young are tree-nesting species. Sometimes employees will release the birds in their own backyards where they can leave food out, but the young birds don't stay around long.

Rescuing a helpless young bird is a noble act, but knowing when a bird needs rescuing is even nobler.

## How to best help wild birds

### Dazed adult bird on ground
Most likely it has run into a window. Carefully set it on a branch where a cat can't reach it while it recovers. If birds often hit your window, consider applying a shiny decal to the outside of the glass, or hang netting or something shiny in front of it during the spring and early summer.

### Injured adult or young bird
The Cheyenne Pet Clinic routinely provides assessment and first aid. Call 635-4121. For hawks and owls and other large species, please consult the staff on how best to transport the bird to avoid further injury to it or injury to you.

### Feathered young bird on ground
Most often, the parent birds are waiting for you to go away so they can feed their almost independent youngster. So, go away! However, if the neighbor's cat is crouched nearby, see if you can get the youngster to perch on a tree branch.

### Featherless young bird on ground
Try to return it to its nest. Retired wildlife biologist Art Anderson said that if the young are about the same age or size, just about any nest will do.

### Damaged nest
If the nest has broken or can't be set back up, make one from a basket or bucket filled with dry leaves and grass. Attach it to the original location, if it is safe, or nearby. Place the remains of the old nest and the young birds in the container and the parent birds will find them.

### Ground nesting birds
During prime nesting season, May through mid-July, refrain from mowing the prairie or allowing dogs and cats off leash. Planting trees also adversely affects the survival of ground nesting birds such as killdeer and meadowlarks. Predators – hawks and eagles – will use the trees for perches while they scan for small bird prey.

### Habitat improvements
Tree-nesting birds benefit from the planting of more shrubs and trees for food and cover. Cover is the vegetation into which they can disappear to avoid predators or bad weather. Think about adding a water source too. And eliminate pesticides. Check the National Audubon Society's "Audubon at Home" website at www. audubon.org/bird/at_home.

### Keep cats indoors
Cats don't need to be allowed to run free, killing small birds and animals, in order to have a full and happy life. Just ask any contented kitty lying on a cushy pillow in a sunny window. Plus, indoor cats have longer and healthier lives. For help in turning your mini-tiger into a real house cat, visit the American Bird Conservancy site at www.abcbirds.org/cats or get information from the Cheyenne Pet Clinic.

### Wild bird rehabilitator permit
The first requirement is 100 hours of experience. Check other qualifications at www.fws.gov. Look under Permits, then Applications, then "MBTA," short for Migratory Bird Treaty Act. Or call the U.S. Fish and Wildlife Service's Migratory Bird Permit Office, 303-236-8171.

# 229

Wednesday, May 9, 2007, Outdoors, page C3

# Some good guides for days exploring in great outdoors

"Kaufman Field Guide to Insects of North America" by Eric R. Eaton and Kenn Kaufman, 2007, Houghton Mifflin, 392 pages, flexible cover, $18.95

The ideas Kenn Kaufman brought to his bird field guide have been applied to this new book to great advantage, especially for someone beginning to study insects.

Four pages at the beginning show photographic examples of every group of insects. Each is color coded to correspond with pages featuring species in that group.

Every entry has a full color photo and commentary written by entomologist Eric R. Eaton whose prose is lively, yet succinct.

Kaufman indicates the actual size of insects without numbers. All insects illustrated on one page are in proportion to each other. Whatever insect is featured in the upper right-hand corner, next to it is a gray silhouette of that insect life-sized. On page 35 it took a second to realize the tiny gray smudge was the actual size of a human flea. In another

226 CHEYENNE BIRD BANTER

case the silhouette of a lubber grasshopper is much larger, and scarier, than the photo.

One disappointment is that this field guide cannot picture all of the 90,000 known insect species in North America, but it has 2,350 photos. You can narrow your search down to a family, perhaps identifying an "Ebony Boghaunter" or "Alabama Shadowdragon."

The 15-page introduction covers finding insects, their life history and anatomy, identification and classification, conservation, activities with insects and importantly, how to keep healthy and safe while insect watching.

"The Songs of Insects" by Lang Elliot and Wil Hershberger, 2007, Houghton Mifflin, 227 pages plus CD, softcover, $19.95.

Last year Lang Elliot came out with "The Songs of Wild Birds." This new book features insects that sing, 77 species of crickets, katydids, grasshoppers and cicadas. While the emphasis is on eastern species, small maps show that 17 range as far as Wyoming.

Each species gets at least two portraits, one on white background and one full page in its habitat. They are all quite wonderful to look at, in a book. In fact, you can order note cards with photos of six of them. Applied to insects, the meaning of the word "song" is stretched a bit, especially if you consider the "Slightly Musical Conehead" found in southeastern states.

But when you listen to number 11 on the included CD, the "Snowy Tree Cricket," it brings back memories of late summer evenings.

There is a lot of information about these insects, including how to collect and maintain your own orchestra. You can also find more at www.songsofinsects.com.

"The Singing Life of Birds, the Art and Science of Listening to Birdsong" by Donald Kroodsma, 2005, Houghton Mifflin, 482 pages plus CD, softcover, $16.95.

Now out in softcover edition, Kroodsma's book is a detailed study of birdsong even the casual birder can afford.

Kroodsma gives an account of how he came to be interested in birdsong, how it is recorded, how songs can be compared through transcription into sonograms, and what singing means in the life of a bird. The CD of birdsong recordings is as enthralling as any story Kroodsma tells in the book. Together, they were awarded the John Burroughs 2006 Medal Award.

"Why Don't Woodpeckers Get Headaches? And Other Bird Questions You Know You Want to Ask" by Mike O'Connor, 2007, Beacon Press, 212 pages, softcover, $9.95.

Most of Beacon Press's catalog is heavy reading. This is the only book with a cartoon on its cover: Little chickadees hold their wings over their ears as a pileated woodpecker drills a hole in a tree.

Author Mike O'Connor dispenses all of his bird advice with a solid dash of humor. He writes answers to readers' bird questions for the Cape Codder, his local weekly newspaper.

"Dear Bird Folks: I want to get a new birdbath for my wife. Do you have any suggestions? –Mel"

"A question for you Mel, how big is your wife? She might be more comfortable in a hot tub."

O'Connor then proceeds to cover the topic of birdbaths with good, honest information, such as, "Animals love to knock over birdbaths and because of this, birdbaths tend to break. You may want to just buy a top and simply place the top on the ground. Birds are used to drinking on the ground (from puddles, ponds, etc.) and they probably rather come to a bath that's low. Placing a bath on a pedestal is more for the esthetic benefit than for the bird's benefit. There is nothing wrong with using a pedestal, just remember to buy a few dozen extra tops."

Having answered scores of bird questions myself, I can admire O'Connor's thoroughness and realistic approach. Most of the advice is suitable for Cheyenne birdwatchers. However, don't get excited about purple martins. We don't have them here. Yet.

And finally, O'Connor reminds Mel to keep his new birdbath clean, "If that is too much work, you could always hire a pool boy to do it. I'm sure your wife wouldn't mind."

# 230 Professor sheds light on crossbills

## Research points to more than one species of red crossbill in North America

Crossbills are species of birds that eat seeds from the cones of spruce, pine and other evergreens. Over time they've evolved mandibles that cross at the tips. These odd beaks can be slipped between the scales of cones to pry them apart far enough for the crossbills to reach seeds

with their sticky tongues.

Both the white-winged and red crossbills can do this. However, the red crossbill was the star of the program given by Craig Benkman last month for the Cheyenne Audubon chapter.

Craig is a professor at the University of Wyoming, in the Zoology and Physiology Department, and is well recognized internationally for his crossbill studies. He has determined that there may be as many as nine types of red crossbills that are candidates for designation as separate species. Red crossbills are found in North

America and Europe.

To be considered a species means the members of it either don't breed outside the group or crossbreeding produces no successful young.

Types of red crossbills are different in size and in the calls they make. They also have different tastes in cones.

While Craig can show you really cool 3-D graphics of his statistics at www.uwyo.edu/benkman, let's cut to the chase here.

Over the last 5,000 to 7,000 years, red crossbills have evolved their seed extraction methods to such an efficiency that the bills of different populations match seeds of particular conifer tree species. For instance, the red crossbills that work over lodgepole pine have different sized bills than those that specialize in spruces.

Watch a crossbill husk a seed (see the video on Craig's web page), and you will see it uses the same side of its bill each time. It has a sharp ridge in the upper mandible that acts as a wedge or hammer and in the lower mandible there is a groove that is perfectly sized for the preferred seed.

Using its tongue to hold the seed in the groove, the crossbill bites down and breaks the seed husk and then removes it. It's a very efficient system if the bird and the seed are perfectly matched.

Serendipitously, Craig discovered that instead of "collecting" birds, he could use dental impression material on live birds to measure their palates, the inside of their mandibles. He can statistically prove that crossbills with a particular seed preference have a particular groove measurement.

While a crossbill with a pinecone bill can get seeds out of a spruce cone and husk them, it isn't as fast and it can't eat as much per minute as the right type crossbill. In this case, time is food is life.

Craig can show that those birds forced to make do with the wrong seed are less likely to survive and less likely to produce young, even if they are in the same forest as the type of crossbill that matches the cones available.

Crossbill breeding is very dependent on having enough seeds. They breed any time of year, except perhaps late fall, so if a spruce-beaked bird can't find a good spruce cone crop, it won't breed until it does.

Craig has studied the crossbills of the South Hills range in Idaho extensively. The lodgepole pinecone crop is quite dependable, and the type of crossbill found there has been able to settle down over several thousand years and evolve to match its food source closely.

While individuals of other types of crossbills can be found in the South Hills, in among this resident type, Craig has found that they are unlikely to breed or crossbreed.

If mating should occur between crossbill types, the young might be inefficient at eating seed from all cone types and be doomed to never getting enough seed to be in condition to breed.

The evolution of crossbill beaks continues because trees continue to evolve tougher cones. The trees with the toughest cones will have seeds that survive to grow and produce more seed and trees.

On the other hand, crossbills with the best beaks for the toughest cones will produce more young. Craig calls this the co-evolutionary arms race. However, in many locations squirrels throw a monkey wrench in the works, which is another story.

If red crossbills really are nine separate species, birdwatchers will have to differentiate them by calls. That seems to be how the birds do it themselves.

As a flock, they travel nomadically across the northern forests. They land on a tree and begin sampling cones, calling out their results, so to speak, and attracting others of their type.

Within a few minutes a flock can decide a tree doesn't have enough seeds per cone to make it worth their expenditure of energy. Craig has been able to determine that the flock is more efficient at this than a lone bird. Remember, time is food is life, so it is better to be part of the flock.

And once in the flock of one type of crossbill, individual birds are unlikely to meet individuals of another type.

Craig and his graduate students continue to research answers to red crossbill questions.

The next question is how much data will it take to prove to the American Ornithologist's Union, the arbiter of North American bird species designation, that there is more than one species of red crossbill on this continent?

A note in the June 2007 issue of the national magazine, Birder's World, mentioned Craig's study and that the South Hills type is currently under review.

Stay tuned!

# 231 'Prairie Ghost'

Wednesday, June 13, 2007, Outdoors, page C1

## The 'Where's Waldo' of the wilderness

**The mountain plover has its disappearing act perfected – so much so that some people were convinced it was an endangered species.**

If you stare really hard at the rocky soil, you may see a ghost of a bird, the "prairie ghost," but only if it moves.

If you are good at it, you can distinguish its white belly from a pale-colored rock. But if it turns its light brown back to you, it is indistinguishable from the surrounding tilled earth.

The nickname for the mountain plover is apt. Its disappearing act may be partly responsible for people thinking there were so few of them that the species would be a good candidate for listing as threatened or endangered.

On a damp morning in late May, just 50 miles east of Cheyenne and a few miles north of Bushnell, Nebraska, mountain plovers were present, right in the middle of alternating, mile-long strips of winter wheat, millet and fallow ground.

Not only were they present, but the plovers were nesting on the stony ridges of the fallow strips. A nest is harder to find than the birds though, because the eggs are on bare ground between the stones and they don't move. It's like playing "Where's Waldo?"

A mountain plover nest is a mere scrape made by the male with his feet. He makes several. The female lays three eggs in one and three

eggs in another and then each parent incubates a nest. The parents will flick small pebbles at the eggs occasionally, but that's as far as nest building goes here.

Larry Snyder is good at seeing ghosts. His first encounter was about six years ago. While working one of his own fields, an odd-looking killdeer, one without the usual double neck band markings, flew up in front of him. He was able to find its nest and avoid driving over it.

A short time later, on a fishing trip with his daughters, he bumped into Chris Carnine of the Rocky Mountain Bird Observatory. She was setting up the Nebraska Prairie Partners program, which was to include mountain plover nest surveys.

Chris identified the mystery bird. Its nest in Larry's field became the first documented mountain plover nest in the NPP program and also for Nebraska Game and Parks.

Chris found that Larry had a good eye for plovers, and he was hired to find more.

He still farms, but weekdays he works for RMBO.

In the spring he rides his neighbors' fields searching for nests.

He also does burrowing owl and raptor surveys.

The first farmer to sign up for the plover program was Larry's friend and neighbor, Bernie Culek.

As the third generation of his family on his farm south of Kimball, Nebraska, Bernie

is always looking for better ways to make farming pay.

In 1992, when he came back to the farm, he changed it to a certified organic operation producing wheat, millet and several other grains.

Funding of the NPP program from a Nebraska Environmental Trust grant and Nebraska Game and Parks makes each mountain plover nest on his place worth $100.

He allows RMBO to find and mark nests and then he plows around them. He feels he should take some responsibility for wildlife.

Signing up for the program is a risk some of Bernie's farming neighbors have not been willing to take, he said. The reluctant think the federal government might get too interested in plovers found on their land, even though the petition to list them was rejected in 2003 because there were more plovers than originally thought.

Bart Bly, currently in charge of the NPP program, said the long-term goal is to turn the program over to the landowners. They found a fifth of the nests last year.

But, said Bernie, for farmers like him, spring is very busy, and it is unlikely that spending hours to find a nest would be a good use of his time.

It can take two days – the longest interval Larry and summer field technician Cameron Shelton have had between nests this spring.

However, the morning of my visit we found two nests.

Larry put out an invitation for volunteers a couple months ago. I thought it would be a good chance to see another mountain plover, my first being last summer on a field trip with Larry and the folks from the Wildcat Audubon Society of Scottsbluff, Nebraska. Five other volunteers have been or will be out this spring.

The catch was learning how to drive a four-wheeler. It rates right up there with snowmobiles in obnoxiousness in my book. But it's a tool, a modern-day mule.

We rode half the length of the mile-long fallow strip at 6 miles per hour, three abreast, about 30 feet apart from each other, Larry, me, then Cameron. Then we rode back and out again, eventually sweeping the whole width of the strip.

I was watching the ground for rocks and hills instead of birds when a plover flew across in front of me, like a deer in the headlights.

Larry said "she" seemed to have shot out from under his front tire. It is impossible to tell the sex of a mountain plover sitting on a nest, but Larry and Cameron refer to them as "she" anyway. The only time in the field one can be certain of gender, Larry said, is when birds are copulating or the male is performing a courtship display or scraping a nest site.

Larry carefully examined the ground to make sure he hadn't run over the nest and wasn't going to step on it.

At a short distance he found three pale olive eggs with black splotches, each about an inch and a half long. He marked the nest with florescent orange stakes set 40 feet out in four directions.

Meanwhile, Cameron brought over a plastic jar of water for a float test. He examined each egg closely for any signs of pipping, where the hatchling might have picked a hole in the shell. If there was a hole, the float test could drown the chick. The test determines the age of the egg – the higher it floats in the water, the closer it is to hatching.

Incubation takes about 30 days, but the parents aren't tied to the nests. If it isn't too cold, they let solar energy work for them. But if it gets too hot, they stand, casting a shadow over the eggs, even holding out their wings sometimes, Larry said.

So temperature plays a big part in how successful nest hunting is on any given day. On a cold day or a hot day, where the adult flies up from is likely to be the nest. Otherwise, they might be out anywhere, stalking beetles, grasshoppers, crickets and ants – the extent of their food diversity.

After Larry took a location reading and filled out a nesting record form, he explained that we couldn't walk back from the nest to our four-wheelers the way we came. We must continue past the nest and circle back so that predators finding our scent later will also circle away.

A second plover flew, but when a little investigation didn't get a nest, we backed off and waited for the bird to return.

Larry is a patient person. He just hunkered down with his binoculars and waited. He said some birds have an attitude. While some are straightforward, others fly off over the hill and then sneak back.

Even though he sent Cameron around to the other side of the strip, neither of them could re-find the bird until Larry changed location. And then I saw a pale rock move. It was the bird again and the nest could be found. This one had only two eggs.

Larry said 13-lined ground squirrels are the most common nest predators, along with snakes.

Overall, Fritz Knopf, author of the Birds of North America Online account for mountain plover and the one who originated the "field clearing" idea in Colorado, told me the survival rate is very good, greater than 50 percent, sometimes even 90 percent, compared to maybe 25 percent for another ground nester, the mallard.

Last year the RMBO crews found 87 nests. This year, they are already up to 54. Larry hopes they break 100.

With his eye for prairie ghosts and the help of Cameron and the other two-man crew, they probably will.

## A bird of contradictions

The mountain plover is a bird of the prairies. The naming mistake can be attributed to John James Audubon. The species was first collected by John Kirk Townsend along the Sweetwater River in Wyoming in 1832, said retired plover researcher Fritz Knopf.

Townsend shipped the specimen back to Audubon who thought that Townsend's description of the bird's location near the Continental Divide must mean it was found among mountain peaks. But the divide in Wyoming often runs through desert and wide-open prairie.

Also, even though classified as a shorebird, it doesn't spend time at the shore.

Historically, mountain plover breeding habitat is the short-grass prairie of the Great Plains, from Montana to New Mexico, but today populations can be found on tilled fields.

Even on the prairie, the mountain plover prefers disturbed ground, such as burns or areas overgrazed by cattle or trimmed by prairie dogs.

Thus, what may be considered good ranching practice in Wyoming, which Fritz considers the major breeding landscape for plovers, may not be compatible with the plover's bare ground nesting requirements.

Researchers are also looking into the effects of pesticides on mountain plovers, not only on their breeding grounds, but in California's Imperial Valley where most of them winter.

Fritz said mountain plover populations were decimated by an outbreak of plague in prairie dogs in the late 1800s, but they prospered during the Dust Bowl of the 1930s. Bare ground to a mountain plover means no predator ambushes. The hordes of grasshoppers must have been like manna from heaven.

## Online

Rocky Mountain Bird Observatory and the Prairie Partners program: www.rmbo.org

Mountain plover information, including sound clip: www.birds.cornell.edu/ AllAboutBirds/BirdGuide/ Mountain_Plover

Birds of North America Online (an annual $40 subscription fee allows access to over 700 very detailed species accounts): http://bna. birds.cornell.edu/BNA.

# 232 Plenty of bird action despite low count

Wednesday, June 13, 2007, Outdoors, page C2

## Day of memories from morning in Lions Park to evening near Terry Bison Ranch

The Cheyenne – High Plains Audubon Society Big Day bird count this year on May 19, at 108 species, was not our lowest count, but it was a long way from our record count of 169 species in 1993, or the 137 species average over the last 15 years.

Was it too late with too many leaves on the trees already? Was it lack of a cold front to cause a fallout of migrants? Compared to 1993, were counters lacking in numbers and identification expertise, or in the wrong locations? Has there been a change in bird populations or a change in their migration route?

Those of us birding the days before and after the count didn't notice any day that would have been better.

Despite the low count, there was a lot to see. Nothing beats starting out the morning at Lions Park with a peregrine falcon rushing overhead.

Did you know peregrines are found all over the world, with the exception of the steppes of east and central Asia, the Amazon, the Sahara and Antarctica? Their name means "wanderer." Peregrines, barring pesticides, are very successful because they will eat a variety of animals, including ducks.

In a corner of Sloan's Lake was a mallard hen with her very young brood of ducklings. They moved about her erratically, like tiny bumper cars wound too tightly.

Beyond the west end, a pair of Cooper's hawks was getting intimately acquainted at the top of a pine tree.

The black-crowned night herons were already on nests at Holiday Park. But a pair of ducks, redheads, was not shy at all. They swam over as if they'd learned the bad habit of begging from the mallards.

Wilson's phalaropes were still at the Wyoming Hereford Ranch. The small shorebirds migrate from South America and nest in Wyoming and the northwest.

Females may mate with more than one male, leaving behind eggs in several different nests to be incubated by the males. So it's the dads that are less colorful because they need to avoid detection. And if predators get too close, they do the broken wing act.

I'd seen the fuzzy head of a great horned owlet up in a nest of sticks weeks before, but the day of the count, we saw a pale youngster on a pine branch. The darker parent was on a nearby stump. A study in South Dakota showed owlets remain with parents most of the summer, becoming completely independent by October.

Two male robins defending territories jumped at each other, feet first, fanning their wings like roosters in a cock fight. Elsewhere, we found a robin sitting on a nest stuck to the side of a skinny tree, its beady eye on us as we passed.

A belted kingfisher raced across a reservoir to evade a red-winged blackbird. The kingfisher splashed into the water just before collision with the far bank. The red-wing pulled up safely and returned to its nest. The kingfisher perched on a concrete structure to dry off.

At F.E. Warren Air Force Base, the suave airman checking identification asked for what reason we wished to access the base. He did a double take when we said bird watching.

The base lakes were worth the admittance inconvenience. A male wood duck swam in view of our scopes. I read that they are the duck North American hunters harvest the most often, after mallards, but they are rare here.

We had seen American white pelicans at Holliday Park and Wyoming Hereford Ranch Reservoir #1, but at the base three came very close. One had a classic example of the protrusion on its upper bill that both males and females acquire during the mating season. "A highly fibrous epidermal plate" is the way scientists describe it.

The second pelican had a mere bump and the third had none at all, making us think it might be immature.

Apparently, it takes three years to become an adult.

The third pelican also was the only one with black markings as if it were wearing a sparse toupee. One bird book said black or gray markings on the crown are individualistic and not a sign of age or sex.

By 4:30 p.m. I was finished, but Mark wanted to check one more place after dinner.

Off I-25 at the Terry Bison Ranch Road interchange, we scoped a peaceful reservoir in fading light.

The odd shaped bird Mark fixed on turned obligingly to make its best field marks visible. It was a red-necked grebe. It winters off the northern coasts of the Pacific and Atlantic and breeds on small lakes in western Canada and Alaska. It's a rare migrant for Cheyenne.

The warbler turnout at seven species was nowhere near the high of 17 seen in 1993, 2001 and 2002. However, counts this spring in Riverton and Jackson listed only three.

A female bay-breasted warbler, an eastern species, made our thirtieth kind of warbler observed in Cheyenne (of 54 in North America) on the Big Day Count over the last 15 years.

We can't wait to see what next year brings.

2007

# 233 The early birder gets the bird in hand

Wednesday, July 18, 2007, Outdoors, page C2

**Taking part in a banding provides one with a rare and incredible experience**

A songbird in the hand is such an incredible thing, yet insignificant, a bit of tiny feathered flutter against your fingers – or a painful pinch of a beak.

Bird banding allows one to hold what normally can only be held from a distance with optics. It is a gift to hold these 1-ounce wonders and contemplate just how far they might have flown to nest in Wyoming, possibly from as far as South America.

The morning of July 1, friends and I volunteered to help with the MAPS banding station west of Laramie.

MAPS, the Monitoring Avian Productivity and Survivorship Program of the Institute for Bird Populations, has over 500 banding stations across the country. While there is an internship program that provides some of the labor, for 80 percent of the stations trained volunteers are essential.

At this station, even the site has been volunteered, by landowners Fred and Stephanie Lindzey. Ten 12-meter long mist nets have been set up in the same locations deep in the riparian zone along the Little Laramie River each year beginning in 2000.

The nets are unfurled six or seven times during the breeding season, late May through early August, from sunrise to noon each banding day.

This is also mosquito breeding season, and they were whining as I followed volunteer Larry Keffer across a soggy hay meadow and into thick willows and cottonwoods to check nets.

Larry is a retired welder from Casper and got into banding three years ago there, up at the Audubon Center at Garden Creek.

The center is currently managed by his nephew, Ken, who, along with Larry's wife, also was working with us. Larry never thought he could do more than record data since his hands are crippled with arthritis, but he found he had plenty of dexterity with the fine crochet hook used to extract the birds from the nets.

Mist nets are made from very fine black threads. Set up in thick vegetation, the nets are invisible, and birds fly into them. Feisty species like the black-capped chickadee can manage to wrap several threads around their necks and feet in the 30-40 minutes between net runs.

Retrieving birds from a mist net is not for the squeamish. Kim Check, the Audubon Wyoming Community Naturalist now responsible for this MAPS site under the guidance of a master bander, took on a chickadee so deeply embedded, she was afraid it would die from stress.

At one point in the 15-minute ordeal, she asked me to lift one of the tiny threads hung up on the bird's emerging pin feathers. That's when I realized I hadn't brought my reading glasses. For bird outings I usually think binoculars.

Since the chickadee's life was more important than having to repair a hole in the net, the minute it quit fighting and seemed to go into a stupor, Kim didn't hesitate to start snipping. She decided not to take it back to the processing table which would stress it even more.

However, since this bird already had a band, we copied the numbers before Kim let it go. The chickadee flew into the bushes and disappeared with only one backward glance.

Most of the morning's other species – yellow warbler, house wren, veery, red-naped sapsucker, gray catbird, various sparrows – were much easier to extract.

Back at the picnic table, the birds, transported in white cloth bags, were identified by species, sex and age.

It is surprisingly difficult to identify some birds in the hand. Behavior and song are so much help in field identification.

Aging songbirds is particularly difficult. Two fat handbooks, "Identification Guide to North American Birds, Part 1 and 2," by Peter Pyle, help. But when you get right down to the nitty gritty, sometimes it just says, "more study is needed to tell the difference," Kim quoted.

The females, and the males of some species, have a brood patch on their belly, a big, bare patch of skin so that they can more directly transfer their body heat to incubating eggs. The condition of the brood patch skin, especially any sign of feather re-growth, tells how long ago the eggs hatched.

When a female cowbird was checked, she had no brood patch. But then we realized she wouldn't since cowbirds are parasitic nesters—dropping their eggs in other birds' nests for them to incubate, hatch and feed.

All kinds of information are recorded and eventually get into the hands of scientists.

Data for this MAPS station has been entered for 2000-2003 and some is published at www.birdpop. org. It shows 79 species have been banded or observed, of which 33 have been classified as breeding.

Another page shows that, as of 2003 (I think they must need data entry volunteers also) there have been 11 other MAPS sites in Wyoming.

You can wade through scientific discourse and learn the value of the data collected and how it is used to inform management decisions to combat what seemed to be a drastic drop in songbird populations around 1989 when the program began.

If you prefer to wade through wet meadows, call the Audubon Wyoming office in Laramie, 307-745-4848, to sign up to help at one of the remaining banding sessions.

It is an incredible experience.

# Rosy-finch survey provides bird's eye view

Wednesday, August 8, 2007, Outdoors, page C2

A bird's eye view is not for the faint of heart – or the faint of leg or lung. I found myself seated at the edge of a precipice at Schoolhouse Rock, about 11,500 feet, in the middle of July. We had just hiked up the Medicine Bow Peak trail in the Snowy Range.

While not actually dangling my feet over, I was close enough to the edge to touch it and appreciate Lake Marie 1,000 feet straight below.

To my left was the Diamond, another 500 feet higher.

No guard rails, no ropes and thankfully, no wind.

Five of us staunch members of the Audubon Society were looking for brown-capped rosy-finches (not to be confused with the other two species, black and gray-crowned) for the fourth year of a citizen science nesting survey.

You may remember me discussing it here the summer of 2004, when a University of Wyoming graduate student was seriously injured while using technical climbing equipment to reach one of the known nests hidden in a crack on the side of a cliff.

The three rosy-finch species breed at the highest altitude of any species in North America north of Mexico.

The brown-capped is the southernmost breeding of the three, found mostly in Colorado. The nests in the Snowy Range, in southeastern Wyoming, mark the northern limits of its known breeding range.

If you want to add all three rosy-finch species to your life list at one time, visit the top of Sandia Crest outside of Albuquerque, New Mexico, between November and March. You can take the tram up instead of navigating the road that rises 5,000 feet in elevation. Check www.rosyfinch.com for more information.

While white-crowned sparrows accompanied us on this hike up, they seemed to be birds of terra firma. It was the juncos that were willing to explore bits of vegetation clinging to rocky cliffs. The violet-green swallows shot out over the edge into empty space, following flying insects.

One junco, exploring the face of our rocky observation point, flitted right over me as if my shoulder was a geologic continuation. Its wings nearly brushed my ear. We spooked each other, I think.

While our survey party did hear the distinctive monotone call of the brown-capped rosy-finch, we were never able to pinpoint one of the milk-chocolate brown birds with raspberry tints, much less watch one slip into a crack.

The rosy-finch nest is inserted in rocky and inaccessible places, under large rocks in rock slides and moraines and on the walls of caves, abandoned mines and railroad tunnels, as well as cliff faces protected by overhangs. The nest itself is a cup shaped of woven grasses.

What was disturbing was that there were very few

snowfields left in the vicinity. In summer, rosy-finches feed on the frozen insects exposed on melting snowfields and seeds surfacing along their margins.

There was lots of chittering noise from the swallows, making it tough to listen for finches. And then there was a human voice relaying information from across the lake, something about a broken leg.

With our binoculars and spotting scope (if your hiking party includes someone with younger legs and lungs like our son Bryan accompanying us this day, handicap them with equipment – they usually enjoy showing off their superior fitness), we found the harbinger of bad news heading for the Mirror Lake Picnic Area, just a glacial moraine beyond the far side of Lake Marie.

We watched vehicles of various agencies gather. Ant-sized rescuers wearing bright red and yellow hardhats scaled the boulders of the talus slopes and began climbing a smooth-faced peak.

I heard later that an experienced climber had fallen only 12 feet but landed hard on a rock ledge, breaking his leg. At the time we were pretty sure it wasn't another rosy-finch survey member since none of the 20 of us divided into the four parties had had enough time to get that high up. I wonder if the victim distracted himself while waiting for rescue by watching birds, some of which may have been rosy-finches.

Summer is so short at

10,000 feet. That's why rosy-finches can't wait for the snow to completely melt.

Back on June 23, though Lake Marie was ice-free, Lookout Lake, two lakes up and also part of the rosy-finch survey, was still mostly ice-encrusted and the trail along it snow-drifted. Bright yellow glacier lilies bloomed profusely wherever the snow had just melted.

By July 18, three days before the rosy-finch survey, the same hillsides the length of Lookout Lake were covered in columbine and there was no snow left to melt. The columbine were pale looking and probably past their prime, but other flowers were brilliant rose, yellow, blue, purple or white.

Looking closely at the emerald green carpet of vegetation, I could find the seed heads of previous blooms.

Every bird on the ground we saw seemed to be a white-crowned sparrow busy harvesting seeds. Every flock in a tree seemed to be mostly pine siskins. Also working the conifer crop were jays and pine grosbeaks.

I hope to make another pilgrimage to the high country yet this season, barring nasty weather. It isn't too soon to see snow falling up there.

I'm like a junco visiting the mountains in the summer, though I'm ingesting mountain scenery instead of mountain seeds. We're both storing up for spending the long winter back in town.

# 235 A refuge for birds in Alaska

Wednesday, September 5, 2007, Outdoors, page C2

## An old dairy farm now provides nearly 3,000 acres of safety

Heading north in August runs counter to bird migration. However, Mark and I enjoyed our trip to Fairbanks, Alaska, 125 miles south of the Arctic Circle.

The city reminded me of Gillette, Wyoming, home of hardworking pickup trucks, but with a sprinkling of white Princess Cruises tour buses and spruce trees in undeveloped areas instead of sagebrush.

Fairbanks has the patina of culture and academia thanks to the University of Alaska, but it is also the supply point for folks heading into the bush. There is a pond at the airport with a flotilla of float planes at anchor.

The city population is about 30,000, three-fifths of Cheyenne's. The Fairbanks borough has 87,000 people spread over 7,631 square miles, compared to here in Laramie County which has 81,000 people in 2,688 square miles.

With all that space available, there was a lot of local support for the state to establish Creamer's Field Migratory Waterfowl Refuge on the edge of Fairbanks, expanded now to nearly 3,000 acres.

Creamer's Field was a dairy farm from 1915 until the 1960s. Now it is known for the annual Sandhill Crane Festival at the end of August, which celebrates the peak of crane and Canada goose migration. We visited too early to take part.

Mark and I, and Mark's brother Peter, joining us from Sitka, Alaska, opted for the free, daily, naturalist-led hike across a corner of the old pasture and into the forest. It was a quick way to learn the basics of the local ecology.

Interior Alaska is part of the boreal forest that stretches across Canada and into the northern Midwest and New England. Our guide, an Alaska Fish and Game biologist, said there are only six tree species in this part of the boreal forest: balsam poplar, aspen, paper birch, white spruce, black spruce and tamarack. The numerous willows and alders don't count.

Permafrost is a building nightmare if you don't keep it cold. In the forest it naturally creates six-sided potholes and tipsy-looking trees. The refuge has to rebuild the board walk through the boggy parts of the nature trail every few years.

It was too late in the morning for much bird activity in the woods, but in the field was a pond full of ducks that got buzzed by a peregrine falcon.

An offshoot of Creamer's Field is the Alaska Bird Observatory. Among other projects, it has a permanent banding station, banding every suitable day April until October.

The mist nets are along a nature trail and identified with permanent signs. The station is a semi-permanent tent with steel arches and a plywood floor. There are three processing tables and a crew of employees, interns, college students and volunteers, including children under the tutelage of mentors.

While we visited Aug. 14, an American tree sparrow was the first of its species to be banded in the fall migration already underway. However, it is a migratory bird for Fairbanks rather than a winter bird as it is for us in Cheyenne.

On the board listed as the cool birds of the week were many warblers we see in Cheyenne during migration including yellow-rumped, orange-crowned, Wilson's and Townsend's.

The two coolest to be banded the week we were there were slightly off-course Arctic warblers. In 16 years, these were only the twentieth and twenty-first the ABO had banded. We're unlikely to see them in Wyoming as they nest in the Arctic and winter in the Philippines!

However, Wyoming does share two famous naturalists and conservationists with Alaska, Mardy and Olaus Murie. They were mentioned more than once in various museums. Several gift shops carried Mardy's book, "Two in the Far North," which chronicles her early life in Fairbanks and the couple's research in the Arctic beginning in the 1920s. They later made their home in Moose, Wyoming.

My list of bird sightings includes two willow ptarmigan. They were on the side of the road on the tundra at Denali National Park and Preserve. We also saw five grizzlies, numerous caribou and Dall sheep.

Even though we couldn't leave the shuttle bus between official stops, our driver didn't hesitate to make pauses for wildlife photography out the windows.

In Fairbanks, over at Pioneer Park, we saw something unexpected. Instead of begging mallards swimming in the pond at the re-creation of a gold mining sluice, we saw American wigeon. They haven't learned to beg yet, but they weren't too shy either.

Our last evening in Fairbanks, I stood out in the driveway of the bed and breakfast about 10 p.m. to admire the blue sky above, and to watch a large V of Canada geese winging south.

The commentator on the Riverboat Discovery cruise had said that by Halloween Fairbanks would be dark early and the Chena River would be frozen solid enough to become a landing strip for small planes equipped with skis.

Because Fairbanks' high latitude produces extremely long days of summer sunshine, 45-pound cabbages, incredible flower gardens and lots of food for insectivorous birds, I think migration is a great strategy.

However, the chickadees, ravens and gray jays tough out the minus 45-degree winter temperatures, just like the remarkable people we met who choose to live in interior Alaska.

# 236 Going by the book doesn't prove bird's existence

Wednesday, October 3, 2007, Outdoors, page C2

If a bird flies through the forest and there is no one to see it, does it exist?

Conversely, if the annual conference of the American Ornithologists' Union is held in your state, will birds be found never before seen there?

Yes and yes. Several of the 500 attendees of the conference held in Laramie in early August observed what may be the first two records of lesser black-backed gull for Wyoming, if accepted by the Wyoming Rare Bird Records Committee.

One of the gulls was hanging out at Lake Hattie on the Laramie Plains and the other at North Gap Lake high in the Snowy Range.

Wyoming does not have a huge number of resident ornithologists or expert birders to cover our vast plains and mountain ranges so one has to wonder how many lesser black-backeds have visited previously.

The lesser black-backed is essentially a European species, but gulls are likely to travel long distances scouting new territory. North American birders started seeing this species in the winter along the Atlantic coast in the early 1970s.

Field guide range maps indicate at least a single record up to a few sightings every year for states in the eastern half of the U.S., but with a heavy concentration along the Front Range of Colorado.

This makes me smile. Several years ago, I went on a late fall field trip led by Tony Leukering and Doug Faulkner to look for gulls at the reservoirs around Fort Collins. We saw several rarities.

Are these guys gull magnets? Or do they have more knowledge of how to recognize species that aren't expected?

So I asked Doug about the new gull in Wyoming. He wrote back:

"You should look at Sibley's Lesser Black-backed range map. That one is pretty accurate, although as with most publications, it was already out-of-date before it hit the printers.

"LBBG is annual in winter in Colorado in small numbers (about 8-12 per winter; I often see 6 or more). In fact, it is regular enough that the Colorado Bird Records Committee no longer requests documentation.

"Colorado's first record is from 1976. It wasn't until the 1990s, though, that the species really took off and started to occur annually, then in the early 2000s in relatively high numbers for an inland state.

"The Wyoming Bird Records Committee is reviewing documentation of one at Casper from winter 2004. If accepted, that would be the first state record.

"LBBG has been slowly expanding, geographically, westward as evidenced not only by Colorado's records, but also those from other states. More interestingly, the species has broken out of its "rut" of only occurring in winter inland. It is now being found more often in summer (Wyoming's two birds this year, plus several for Colorado and Nebraska in recent years), as well as earlier in the fall and later in the spring."

How many observations does it take before the field guide maps are altered? Last winter a lesser goldfinch, easily distinguished from our usual American goldfinch, was seen at a Cheyenne feeder almost daily, for months. The Sibley Guide to Birds shows a couple green dots meaning that there were already a few records for Wyoming.

But then came this summer. We had one visit our feeder. And so did people from Green River, Casper, Buford and Newcastle who posted their observations on Wyobirds, the e-list for learning about birds in Wyoming (http://HOME.EASE.LSOFT.COM/ archives). Doug posted a report of small flocks around Guernsey when he birded the area.

Is this the beginning of a trend, an expansion of the lesser goldfinch range north or a one-time phenomenon? Time will tell.

Wyoming is woefully short of qualified observers, though not short of people interested in watching birds. I take a lot of bird identification questions over the phone from people who want to know more about the birds in their yards.

On the other hand, I've also taken calls from visiting birders, who, having looked at the Atlas of Birds, Mammals, Amphibians and Reptiles of Wyoming edited by the Wyoming Game and Fish Department (http://gf.state.wy.us/wildlife/nongame), are positive that they have seen a first Wyoming record for a species they are very familiar with back home.

Are these visitors making a familiar species out of one of our similar local species, or have we locals not recognized an unusual species because we aren't familiar with it?

The Wyoming Bird Records Committee judges the credibility of all rare bird records for the state. A few folks looking to bag state records have been deeply disappointed at the slow speed of our committee, but it is staffed by volunteer experts with full-time jobs, and they do the best they can.

Us average birdwatchers are as important as the expert in documenting changes in the ranges of species. So how do we make our observations useful?

Study birds. Participate in data collection efforts like Project FeederWatch and eBird. Learn when and how to file a rare bird form. To request one, call the Lander Game and Fish regional office, 307-332-2688. And keep looking.

After all, as birdwatchers are fond of pointing out, the birds don't read the books.

# 237 In search of great black-backed gulls

Wednesday, October 31, 2007, Outdoors, page C2

## Massachusetts trip nets first look at a variety of birds not commonly found in Wyoming

I had a target bird for our trip to northeastern Massachusetts early in October. No, it wasn't a clay pigeon, but it was a bird I wanted to see.

Houghton Mifflin sent me a review copy of the Peterson Reference Guides series volume "Gulls of the Americas" by Steve Howell and John Dunn.

It has 500 large pages – it's not meant to be a field guide – of which multiple photos of the 36 gull species of this hemisphere in all their various plumages fill 300 pages.

Whenever I had to wait on the phone at my desk the last few months, I would peruse the photos, accounts and range maps. Knowing a trip to the North Shore (north of Boston) was coming up, I checked for a new gull species I might see.

The great black-backed gull caught my attention because it might be easy to identify. It's the largest gull in the world, according to Howell and Dunn. And of two black-backed gull species on the east coast, it's the one with pink legs.

The upshot of reading Jonathan Livingston Seagull is that I noticed gulls, tried to identify gulls and gave up gulls by age 16. So it was much to my surprise that another literary connection gave me my first look at a great black-back.

We went to Salem for the day and walked over to the House of Seven Gables that inspired Nathaniel Hawthorne's novel. Tickets were $13 each for a 30-minute tour. We balked and followed other tourists down a side street for a great view over the fence. We were also at the edge of the harbor and when I turned around, there were a couple of great black-backed gulls. Eureka!

The next day, out on the half-mile long breakwater in Gloucester Harbor, at Massachusetts Audubon's Eastern Point Sanctuary, great blackbacks were lined up on the slabs of granite, dozens of them imperturbable as people passed within five feet.

I overheard one woman tell another what pests they were, pooping on her boat cover.

Oh no, my target bird turned out to be a trash bird – too common and pesky to be appreciated by the locals.

Of the 33 photos of great black-backed gulls in the gull guide, I wonder if the photographers were able to stand as close to their subjects as we were. Even the few herring gulls were patient, their yellow eyes glinting at me.

It was quite another experience to have a chickadee with a glint in its eye hover five inches from Mark's nose later that day when we found our way to Ipswich River Audubon Sanctuary, a few miles southwest of Ipswich.

It was way too muggy and hot for the middle of an October afternoon – and autumn so far had been too warm for the trees to have changed color much.

We expected the birds to be quiet, but as soon as we stepped from field to forest, we were surrounded by a flock, including the one that inspected Mark.

How odd. Was it some kind of enchanted chickadee? And wasn't that other inquisitive bird some kind of titmouse? Why couldn't I find one like it in my field guide?

The next day, while waiting for our son to finish classes at Worcester Polytechnic Institute, we explored Mass Audubon's (as the state organization is known informally) Broadmeadow Brook sanctuary near Worcester.

The woman at the desk explained the Ipswich River sanctuary was known for its tame chickadees that will eat out of your hand.

Oh no, if only we'd known the local custom! Now there's a whole flock that thinks birders from Wyoming are stingy.

Broadmeadow Brook turned out to have invisible birds. For 15 minutes we listened to a flock making loud chipping calls high in the trees, but we never caught sight of one single bird.

However, the bird checklist mentioned tufted titmouse. I looked in my field guide again – but no tufted titmouse listed.

Duh! At home, in my hurry to pack and get to the airport, I'd picked up my favorite field guide, Sibley's western guide, which is not adequate back east.

For a week we explored the northeast quarter of Massachusetts, an area approximately the size of our Laramie County. And still, we didn't see everything or spend enough time anywhere.

We avoided Boston and most museums and attractions this trip in favor of natural areas, accidentally finding Harvard Forest, where signs invited hikers to follow a trail through the hemlocks.

We visited four Mass Audubon sanctuaries, including the tiny one on Marblehead Neck surrounded by houses. There are 40 more, for a total of 32,000 acres. There are also a number of state parks and forests.

Away from the cities, along the small, winding roads, there is the occasional yellow sign cautioning "Thickly Settled" when approaching some tiny hamlet.

The area is thick with trees as well. I have to say that though I enjoy trees as much as anyone, I was happy to return to the Great Plains where they can be individually contemplated – and birds are neither enchanted nor invisible.

And here at home, if the great black-backed gull chooses to make its first Wyoming visit, having already been as far as Colorado a few times, there aren't a lot of yacht covers to ruin.

# 238 How to find birds in strange places

Wednesday, December 5, 2007, Outdoors, page C2

Travel is a good way to add bird species to your life list. Conversely, birding is a great way to enrich your travels – even if it's as simple as watching brown pelicans at sunset across the street from your nephew's apartment in late October, and that street happens to run alongside San Francisco Bay (pre-oil spill).

Serendipity is nice, but birders like to improve their chances. Mark took a look at the map and noticed another park on the bay, Coyote Hills, managed by the East Bay Regional Park District. Water is always a good place to look for birds.

With Mark's brother Mike showing us the way, we found the park and a bird list in the visitor center, but the list didn't have any indication of species seasonality or abundance.

One of the rangers put us on alert for golden eagles. I suppose the white-tailed kites were too common a raptor for him to be excited about, but they made our day – and our life lists.

Another way to find birds in an unfamiliar locale is by recommendation from someone who has already been to the area. Taking the auto tour through the Sacramento National Wildlife Refuge with our longtime friends Pam and Dave was a great way to spend the day together.

The refuge has the perfect bird checklist. After each bird name is a space divvied up by month and if a species appears on the refuge during a particular month of the year, there is a horizontal line. If it shows up in great abundance, it is a very thick line. It was easy to see that the thousands of ducks, geese and coots we saw were going to be spending the winter at the refuge.

Following another outdoor pursuit usually produces bird sightings. We chose to hike where one of the nephews is a ranger, the Sunol Regional Wilderness, also managed by the park district.

A hot, weekday morning left us pretty much alone with the cows as we trudged the water department's access road. But it was inspiring to be in the middle of 6,800 acres of hill country that wasn't decorated with houses or other buildings, something the City of Cheyenne should keep in mind as it looks to "develop" its own ranch.

Our favorite observation was the seven acorn woodpeckers disappearing one after the other into the top of a dead snag. There was such a ruckus of squeaks before they popped out through side openings.

If you were traveling an area without the benefit of friends and family, you could look for a local to give advice. I am that local half a dozen times a year because I allow my phone number and email address to be published in the American Birding Association directory.

While some members indicate that they charge for giving tours, sometimes I have time to invite a birding friend along to meet the visitors at Lions Park, one of our local hotspots.

It's always surprising what visitors get excited about. Two women from California were entranced by the only bird we could find on a windy day, a yellow-headed blackbird. The man and wife from Texas this spring were excited to see migrating birds in their breeding plumage – birds they otherwise only see in their dull-colored winter feathers.

People visiting the Cheyenne – High Plains Audubon Society website, http://org. lonetree.com/audubon, make inquiries. A man calling from Britain asked if he and his friend could join us for the spring sharp-tailed grouse field trip the next week. Certainly. And so they did, and after ticking off that bird on their list of most wanted species, they drove off to find another.

Occasionally, travelers find the Wyobirds elist, http://home.ease.Lsoft.com/ archives, and post a request for information on finding a particular species or birding a particular area. If I can be of help, I reply and invariably, I get a report later about what birds they saw and how much they enjoyed the trip.

Many states have a similar list. If you search "Wyoming birds" however, you may get references to the University of Wyoming coach's notorious recent hand signal.

If you are too shy – or in a hurry – here's another way to find birding destinations. Go to www.eBird.org. It's a great place to store your personal birding records for free. It's also a great resource for planning a trip. Observer information is not available to anyone looking at the data, but the lists of hotspots and the corresponding bird lists are almost as good as meeting one of the locals.

While the list of 1,400 names of hotspots in California is probably not meaningful to visitors, a new map feature does allow you to see where they are. At this time, you can access the maps by pretending you are going to enter data and choosing the option to select a pre-existing hotspot. Those areas that are parks or sanctuaries probably have more information on the internet about hours and access.

Since eBird is relatively new, the checklists generated may be a little sketchy, but at least you can see where the local birders like to go.

Many birds are great travelers themselves, some migrating thousands of miles. I wonder how they pass on information about good places for wintering, eating and breeding. With their bird's-eye view of the world, perhaps it's easier for them to recognize a bird-friendly spot than it is for us.

2007

# 2008

**239** Wednesday, January 2, 2008, Outdoors, page C2
## Birds stay warm, despite cold
**Diving into snow drifts just one strategy used to deal with winter**

Wool, fleece and down clothes; insulated rubber boots; vigorous shoveling and skiing; hot chocolate and soup; lap cat and household thermostat; quilts and down comforters; and maybe a trip to Albuquerque - that's how I survive winter.

Birds have similar strategies. The chief one is migration, whether down from the mountains or down to South America. It's about balancing food-as-fuel availability with air temperature. The colder the air, the more food birds must find to turn into calories to burn.

Insect eaters depart, except for the gleaners, like the brown creepers which are willing to eat frozen bug bodies found in bark crevices.

Seed, berry and bud eaters stay behind, as well as predatory species. Water birds stay as long as ice doesn't prevent them from getting to the pond weeds and animals.

Migration is a matter of following the food. Snowy owls are perfectly happy wintering in the Arctic unless there aren't enough lemmings to go around. Then they head south.

About 50 species out of a total of 325 on the Cheyenne bird checklist are observed regularly on the Christmas Bird Count. How do they survive?

First, there's fat. Songbirds will put on enough on average to weather three days of storm. More fat than that and they would be too overloaded to evade predators. Sea ducks pack on as much as 10 days' worth. That's how long they can go without being able to eat before they run out of fuel and die.

Then there's shivering. When the pectoral (breast) muscles contract and relax quickly, they produce heat.

That heat warms air trapped by feathers which work better as insulation than mammal hair. Most birds have some down feathers and birds of cold climates have especially good down. Think of eider down, which comes from eider ducks. But the more archaic species, ostrich, emu and kiwi, have none.

A cold bird's other feathers will lift away from its skin and trap more warm air, making the bird a completely different shape in cold than in hot weather. Others, like the common redpoll, have more feathers in winter than summer. Some north country birds have feathers on their legs, like the rough-legged hawk.

Otherwise, birds keep their bare parts warm by burying their beaks under their wings or hunkering down, fluffing feathers over their feet. But you've seen the silly Canada geese at Holliday Park, the ones refusing to migrate because people in the past fed them. They walk on the ice. How do they stand it?

Cold adapted birds have a sophisticated heat exchange system. Warm, arterial blood, traveling from the heart to the feet, passes cooled venal blood returning from the feet and warms it before it is pumped back through the heart. Basically, birds have cold feet so we don't have to worry about them getting stuck on cold metal perches on bird feeders.

Sometimes a sleeping bird's core temperature will drop 20 degrees from 108 degrees (chickadee body temperature) for a fuel savings akin to turning down our thermostats at night. However, if we humans let our body temperatures drop from 98.6 to 78.6 degrees, we'd die. Somehow birds can pull it off and wake up for another day of foraging.

A few birds even manage to enter a deeper state of torpor for months. A study in the 1930s of a poor-will over 85 days showed its body temperature fluctuating with air temperatures in the 60s.

Birds search out warm microhabitats. Grouse dive into snow drifts and stay as toasty as any Boy Scout in a snow cave – unless the surface gets iced over and then they are toast – the birds, that is, not the boys.

Starlings standing around chimneys are smart – until the fumes knock them out and they topple over.

The higher a perching type of bird is in the flock's pecking order, the closer to the trunk of an evergreen tree or the middle of a thicket it can roost and the more protected it will be from winter storms. So the more dominant a bird is, the more likely it is to survive. Besides, dominant birds eat better too and may have more fat going into a storm.

And then there's the "dog pile" effect. A researcher in

Maryland cut the top of a hollow wooden fence post where eastern bluebirds had nested so he could remove it from time to time over the winter. Once he found 13 bluebirds packed inside, and another time found more, but two of those were dead from suffocation.

Hawks and owls prefer winter solitude since they are in competition with every other raptor for the limited supply of prey animals.

But for some species, especially the songbirds, sociability is the key to survival. Secretive during the summer nesting season, in winter small birds – chickadees, nuthatches, sparrows – form what is referred to as "mixed species flocks." It means more eyes watching for predators and potential food sources.

I think sociability is a good way for people to survive the winter, too. On January 5, members and friends of Cheyenne - High Plains Audubon Society will be gathering for the annual Christmas Bird Count and the tally party afterwards.

If you are interested in helping, either counting birds in your backyard or counting with the group, contact Greg Johnson, 634-1056, or gjohnson@west-inc.com, or check the newsletter on the chapter website, http://org.lonetree.com/audubon.

The more of us there are, the warmer we'll be!

# 240 Doves continue territory expansion, including here

Wednesday, February 6, 2008, Outdoors, page C2

I don't know about your neighborhood, but mine has a gang of doves loafing around on the street corners and they all sport the distinctive gang insignia: black marks tattooed on their necks. They also wear their tails squared off.

I've witnessed a gathering of as many as 28 of these large, pale gray birds raiding my neighbor's juniper hedge for berries.

I just hope the berries don't ferment, causing gang members to fly drunk. It's bad enough that they defecate in my driveway after every berrying spree!

The Eurasian collared-dove (the American Ornithologists' Union code is "EUCD") has been taking over neighborhoods for centuries. It is thought to have started as a native species in India, Sri-Lanka and present-day Myanmar (formerly Burma). In the 1600s it expanded to Turkey and the Balkans.

Next, EUCD flew through Europe: Yugoslavia, 1912; Hungary, 1930; Germany, 1945; Norway, 1954; Britain, 1955; and Portugal, 1974.

Invasion of northern China and Korea is thought to have come through India. Japan was invaded via China in the 18th or 19th century.

In the mid-1970s, a breeder brought EUCD to the Bahamas where a few escaped and 50 were released.

They were seen in Florida in the late 70s and verified there in 1986, quickly followed by sightings in Georgia and Arkansas. The invasion of the U.S. continued: Alabama, 1991; Texas, 1995; South Dakota, 1996; Iowa and Montana, 1997; Minnesota and Wyoming 1998; and Oregon, 1999.

By the time the species account was published in 2002 for Birds of North America Online, EUCD had also been documented in Colorado, Illinois, Kansas, Nebraska and Oklahoma. There is evidence that some birds were intentionally released in California, Missouri and Texas.

The most up-to-date map available on www.ebird.org shows New Hampshire, Vermont and Maine are the only states where birders have yet to report EUCD to that data base.

Some birds introduced to North America, such as house sparrows, thrive at the expense of native species or, in the instance of European starlings, at the expense of agriculture. In Pakistan, EUCD is considered an agricultural pest. Other released species, such as the ringed-turtle dove (now to be known officially as the African Collared-dove), fail to thrive.

EUCD appears to be prospering and enjoying our winters. There are no studies yet showing impacts on mourning doves returning in the spring looking for similar nesting habitat.

EUCD likes nesting in trees, preferably in urban areas. They hang out at bird feeders and at agricultural operations where spilled grain is available. And they will eat berries, as they do in my neighborhood. They roost on utility lines and in trees and other high places. They have a distinctively unmusical coo.

The federal government has classified EUCD as an unprotected species, just like house sparrows and starlings. In 2006, the Wyoming Game and Fish Department announced that EUCD can be hunted any season, anywhere, any method. Of course, with their fondness for urban landscapes, finding EUCD where the discharge of firearms is permitted could be a challenge.

And then pity the poor hunter in Nebraska. Jeff Obrecht, of the Wyoming Game and Fish Department, told me Nebraska's regulations refer only to "doves" and so any EUCD taken has to be counted towards a hunter's bag limit. I'm sure Nebraska never figured on anything but mourning doves when its rules were written.

Even though the first three EUCD in Wyoming were documented outside Cheyenne at the Wyoming Herford Ranch between May 16 and Oct. 9, 1998, someone in the Cody area stole a march on us, submitting the first breeding record in 2001. Since then, a second breeding record has been submitted for the Sheridan area for 2005.

Cheyenne birders, we must unite! Burns, Pine Bluffs, Albin, Carpenter, Meridan – please join us. The glory of the 28th Latilong is at stake! I'm sure EUCD is procreating in our latilong, defined

2008

239

by one degree of latitude and one degree of longitude, but we need to document it. We need evidence. Even though spring is a couple months away, we don't have research to tell us how early EUCD will breed in Wyoming. Here's your chance.

In winter EUCD is comfortable flocking. A pair from the previous year may still be bonded. But later, the doves get territorial. The male will give an advertising coo and then from his high perch he will fly up at a steep angle with a lot of wing-clapping (just like pigeons) before descending in a spiral, tail spread. The account in Birds of North America Online says he then gives the "excitement" call.

The males bring the females twigs, stems, roots, grasses and urban litter and in one to three days the female has a nest built in a tree, or sometimes, on a building. Two eggs are laid, one after the other, and the parents take turns incubating for about two weeks. The young need around two-and-a-half weeks to fledge but aren't fully independent for another two or three weeks.

In friendly climates, EUCD may start nesting as early as February and produce up to six broods. No one has documented what happens in Cheyenne.

If you observe Eurasian collared-doves sitting on a nest or feeding young, let me know and I'll help you fill out the official paperwork. Who knows, those avian invaders could make you famous, at least in Latilong 28.

# 241 On tail of secretive goshawk

Wednesday, March 5, 2008, Outdoors, page C2

## Twice it's been denied endangered listing because too little is known

The northern goshawk is not a bird on my life list despite my having lived all of my life within its North American range, roughly north of Interstate 80 or in the Rockies and west.

However, I have never lived within its habitat, the forest. Years of forest recreation and birding with experts has never led me to a glimpse of the gray-with-white-bellied hawk, even though it is a large bird: 2 pounds, 21 inches long with 41-inch wingspan.

The Sibley Guide to Birds describes the goshawk and the other two accipiter species, Cooper's and sharp-shinned hawks, as difficult to identify – though I'm pretty sure it is the sharp-shinned terrorizing my feeder birds. Accipiter species have the short, rounded wings and long tails that allow them to navigate in the trees. Other hawk species prefer the wide open spaces.

The goshawk is listed as a sensitive species in six of the eight U.S. Forest Service regions, including Region 2 which includes most of Wyoming.

It is a "management indicator species" which means it is potentially sensitive to habitat changes, especially since it requires large trees for nesting.

Twice the goshawk has been denied listing as a threatened or endangered species because it was ruled that there was not enough information to support either status.

How does one study a bird a field guide describes as secretive? How does one find a set of needle-like talons within a forest of needle-leaved trees?

Jeff Beck and two colleagues came up with a system a couple years ago based on a suggested national protocol. He was the local Audubon chapter's guest speaker last month and is assistant professor of Wildlife Restoration Ecology in the Department of Renewable Resources at the University of Wyoming.

He said traditionally biologists monitored any known goshawk nests for activity each spring. But over time this method shows a downward trend as, I would presume, even with a thriving population, preference for a particular pine tree nesting site might wane as the health of the tree declines over time.

Beck's team's job was to design a more statistically satisfying sampling method for the forest bioregion in Wyoming and Colorado.

First they studied 58 known goshawk territories and statistically analyzed them to see what they had in common. Was it slope, the steepness of the mountain side? Or aspect, the direction the slope faced? Or elevation? The predicting factor turned out to be vegetation.

Nesting territories were 4.6 times more likely to be in lodgepole pine than spruce/fir forest.

This may not be news to falconers who prize goshawks and are also looking for nests – from which they are licensed to pluck young to train for their sport.

Next, knowing from research that a goshawk nesting territory averages 688 hectares, or 1,700 acres (and may contain several nests per pair over the years), Beck and his team laid a grid of 1,700-acre sampling units on Forest Service land and chose 51 units for a pilot monitoring program.

Finally, field biologists hit the ground twice, once during the time studies showed would coincide with nesting and once during fledging.

In each 1,700-acre study unit, the field biologists played a tape of goshawk calls to elicit a reaction from one if it was present. But because the auditory range was only 150 meters (492 feet), there were a lot of acoustical sampling points to cover in each unit.

It's one thing to draw dots on a map, but it's another to get to them, to overcome thick timber, steep slopes, swamps, bad weather and bears. Also, goshawks deal with intruders by slamming into them, so everyone had to wear hardhats. Next time you see field biology statistics, imagine all the sweat and stories that go into them.

What Beck and his

colleagues found was a 65 to 75 percent probability of detecting goshawks. Also, 33 percent of the samples had goshawks. And goshawks were 6.5 times more likely to be seen in their primary pine habitat than their secondary spruce/fir habitat.

Using statistics, the team was able to determine how many sampling units would be required to detect a 20 percent change in goshawk population. And they were able to determine that sampling sites closer to Forest Service offices produced nearly the same results as sampling remote sites and so monitoring costs could be reduced.

A long-term monitoring program like this could show if the goshawk population rises and falls in a natural cycle. It could show the effects of timber harvesting or recreation pressure, including that of falconers.

But with the pine beetle infestation launching itself throughout Colorado and Wyoming, it will undoubtedly document the goshawks' reaction to a much larger natural cycle in which, at this moment, their preferred nesting trees are dying.

Will it be easier for goshawks to find prey (rabbits, grouse and squirrels) when the trees are mere skeletons? If so, will that advantage be offset by the ease with which their main adversary, the great horned owl, devours their young?

Goshawks are also found in Britain, Scandinavia, northern Russia and Siberia and south to the Mediterranean region, Asia Minor, Iran, the Himalayas, eastern China and Japan. So, we can look for answers from biologists there and share our findings with them.

Meanwhile, I'll take a closer look at piles of sticks in large pine trees. Perhaps I should wear a hardhat, so I don't risk my head while adding to my life list.

# 242 Authors explore our fine feathered friends

Wednesday, April 2, 2008, Outdoors, page C3

Frequently, this winter I filled frigid weekends and long dark evenings reading four books about birds. And since we can still expect a few blizzards between now and June, I thought you might want to look for and read one of them yourself.

No matter your taste in literature, one will suit you. The first is a "how to," the second a "what to do," the third is historical/travel and the fourth, spiritual.

Finding Your Wings: a Workbook for Beginning Bird Watchers by Burton Guttman, Houghton Mifflin, available March 2008, softcover, 75 color photos, 224 pp, $14.95.

This addition to the Peterson Field Guides series is not a field guide. It really is a workbook in which you are expected to write and draw. Drawing a rudimentary bird is a way to note distinctive features of an unknown bird to help you identify it later with a field guide.

Most of the workbook exercises require having either the "Peterson Field Guide to the Birds of Eastern and Central North America," 5th edition, or "The Peterson Field Guide to Western Birds," 3rd edition (1990).

An example is Exercise 5-18. "Baltimore and Bullock's Orioles [E317 (eastern guide, p. 317) or W313 (western guide, p. 313)] are very similar and for a time were considered a single species. How do the wings of the males differ?" The answer is at the back of the book.

The three other kinds of activities are field exercises, such as studying crows in flight, quizzes and games.

Guttman, longtime teacher of birding workshops, wanted to write a book that will help people get to know and love nature so they'll protect it.

He says beginners need to work on three goals at once: learn how to see as a birder sees, learn about the categories of birds, and learn as many of the easily identified common birds as possible.

My birding "sight" needs restoration after a long winter so I think I will work through the exercises myself.

Silence of the Songbirds by Bridget Stutchbury, Walker & Co., 2007, hardcover, 256 pp, $24.95.

A review copy of this book arrived in my mail last fall and it took me months to get past the ominous title and read it.

Stutchbury, a professor at York University, holds a Canada Research Chair in Ecology and Conservation Biology and divides her time between homes in Ontario and Pennsylvania.

Her book is a chapter-by-chapter description of songbird perils: deforestation, forest fragmentation, shade-grown versus sun-grown coffee, pesticides, lights, windows, cats and cowbirds.

Adding her personal experiences highlighted by her animated prose style, Stutchbury explains exactly how each hazard affects birds. The facts are much more interesting than what the popular press has time for, and the book is much more cohesive than a collection of journal articles.

Unlike other science writers, Stutchbury's sentences do not need diagramming in order to extract their meaning. The citations for the underlying scientific studies are quietly listed in the back of the book, along with an index.

In the epilogue, Stutchbury reminds us how important birds are to people as pollinators, insect eaters, scavengers and nutrient recyclers.

Most importantly, helping readers avoid a feeling of hopelessness, she gives us a "to do" list: buy shade-grown coffee; buy organic if the produce is from Latin America where so many songbirds overwinter; and buy organic or try to avoid crops that are the greatest pesticide risk to birds: alfalfa, blueberries, celery, corn, cotton, cranberries, potatoes and wheat.

Also, buy wood and paper certified by the Forest Stewardship Council; buy toilet paper, paper towels and tissues made from recycled paper to protect the northern forests where so many songbirds nest; turn off lights at night in city buildings during migration; and keep your cat indoors.

Be brave, buy the book and read it. Through the York Foundation, Stutchbury is donating proceeds to support research on migratory birds. Or don't buy the book and borrow my copy. Then with the money you save, make a donation to a bird conservation organization. Or spend it on organic cotton handkerchiefs and shopping bags.

Falcon Fever: A Falconer in the Twenty-first Century by Tim Gallagher, Houghton Mifflin, available May 2008, paperback, 336 pp, $25.

The first thing you'll recognize is that author Tim Gallagher is the one who recently wrote "The Grail Bird," about his experience finding the ivory-billed woodpecker. However, you'll get little insight into that venture here, even though the book begins in the autobiographical mode.

In the mid-20th century in California, a 12-year-old Gallagher could read about falconry, roam the woods searching for hawk nests and meet adult falconers who generously offer to mentor him. In his teenage years, reminiscent of Kenn Kaufman's "Kingbird Highway," he might escape home and drive a rattletrap with a friend to a national falconry convention a thousand miles away.

Despite this idyllic life (not counting a truly tough home situation), Gallagher longs to be a contemporary of Frederick II, 13th century Emperor of the Holy Roman Empire known for the quintessential book on falconry still consulted today. Frederick was once accused of letting hunting with his hawks interfere with attending to a crucial bit of warfare.

While the beginning of the book is autobiographical, the latter part is travelogue, in which Gallagher spends a year visiting other falconers and makes a pilgrimage to Frederick's Italian castles.

One chapter of interest to Wyoming folks documents Gallagher's visit to falconer and filmmaker Steve Chindgren's hunting lodge near Eden to witness hawking sage grouse.

Chindgren's name may sound familiar since his sage grouse movie (minus falcons) was rescheduled for a free showing at the March Cheyenne Audubon meeting.

Falconry is a very different way to enjoy birds. You'll know much more about it by the end of the book – its centuries of history as well as its modern-day incarnation.

Sightings: Extraordinary Encounters with Ordinary Birds by Sam Keen, illustrated by Mary Woodin, Chronicle Books, 2007, hardcover, 114 pp, $14.95.

Perhaps this book could be classified as a spiritual autobiography in essay form, in which Keen's encounters with birds are the prompts for musings on the various elemental philosophical questions.

Keen is a former professor of philosophy and religion and now a lecturer, seminar leader and consultant. He is also a storyteller, evoking his childhood among staunch Presbyterians, as well as an historian. Consider this partly tongue-in-cheek sampling from the last essay.

"Careful observation has convinced me that birders, far from being just quaint old ladies in sensible shoes and nerdy zoology students, are involved in something strange, archaic, and clandestine – something more like a pagan religion than a hobby....I suspect that the growing number of enthusiastic birders are converts to an ancient cult of bird worship...."

And then Keen explains that bird worship goes back to the Phoenicians, Persians, Greeks and Egyptians. "There is speculation that prior to 100,000 BCE (Before the Christian Era) a culture devoted exclusively to birds existed in America."

Things haven't changed much. The miracle of spring migration is still celebrated. The more science explains it, the more awe inspiring it is.

# 243 How to get energy and save our sage grouse

Wednesday, April 2, 2008, Outdoors, page C2

## Difficult task lies ahead to keep both resources valuable in Cowboy State

Is geology destiny? Geology is rocks. A particular weathered rock makes a particular kind of soil which, with water, grows particular vegetation. Particular vegetation feeds and shelters particular animals.

Thus, a geologic formation rich in oil and gas can be associated with certain wildlife species.

Using overlays last month at the Cheyenne – High Plains Audubon Society meeting, Alison Lyon-Holloran, conservation program manager for Audubon Wyoming, showed Wyoming's oil, gas and coalbed methane fields almost perfectly align with greater sage-grouse habitat.

The sagebrush ecosystem, on which the grouse is entirely dependent, stretches across Wyoming in a wide swath from the northeast to the southwest, avoiding the mountains in the northwest and the grasslands of the southeast.

If you have not driven across the state, it may be hard to believe that so many acres of sagebrush exist, from the ankle-high species on the dry hills to the small forests along riparian (stream) corridors.

It's hard to believe sage-grouse are so dependent on sage, from hiding their nests in a straggly old stand to grazing on the buds while keeping an eye out for predatory golden eagles.

It's hard to believe a

242                    CHEYENNE BIRD BANTER

chicken-like 6-pound male or 3-pound female is so shy and easily distracted that the U.S. Bureau of Land Management's drilling stipulations provide, on average, a 2-mile buffer zone around a lek during breeding season.

Those leks are collections of as many as 50-150 males each spreading spikey tail feathers, popping white-feathered neck sacs and defending small territories. The females stroll through, looking for the best genetic material, which, Alison said, may be the same one or two males for all of them.

Someone in the audience asked how sage-grouse are doing. Fine, Alison said, away from the energy development areas. Two wet years have really made a difference in what was a general decline during drought years. However, despite the moisture, they are not doing well in energy areas. It's too crowded and noisy.

Several energy companies have committed millions of dollars to provide offsite mitigation for wildlife and other land users who have lost the use of lands now in oil and gas production.

It would be nice to think that people could enhance sagebrush habitat away from all the wells to produce more

grouse, but Alison, who studied sage-grouse for her master's thesis and has been immersed in the research and issues for the last 10 years, said there are no studies showing how to produce scraggly 100-year-old sagebrush stands.

The millions of dollars in mitigation money cannot be used to study why some sagebrush is not attractive to sage-grouse and what can be done to improve it.

It is conceivable, said Alison, that the few remaining healthy sage-grouse leks in Wyoming could be compromised, forcing the birds to be listed as either threatened or endangered – something neither energy companies nor environmentalists want to see happen.

If sage-grouse become threatened or endangered, it would mean more development restrictions for energy companies and much more work for the environmental community.

Of Wyoming's total 62 million acres, the federal government owns, and BLM manages, 41 million acres of minerals below the surface (and 18 million acres of the surface).

So far, 14 million acres of federal minerals have been leased for oil and gas. Don't forget state and private

oil and gas leasing because 45 percent of Wyoming's total oil and 37 percent of its natural gas production comes from them. See BLM's 2007 annual report at www.blm.gov/wy.

In the old days, environmental groups would be preparing lawsuits. Instead, Alison and Audubon Wyoming executive director Brian Rutledge came up with the Greater Sage-grouse Species Survival Plan. They have hired Kevin Doherty, who studied sage-grouse for his PhD, to give the issue the necessary rigorous, scientific statistical scrutiny.

The National Audubon Society has taken notice also and has made sagebrush one of its top conservation concerns.

Key players from federal and state government have been working with energy and environmental groups to figure out how, during fluid mineral development, we can have our energy and our grouse, too, here in the state with the most grouse habitat of any in the country. And there are other sagebrush species that will benefit.

The highlight of Alison's presentation was the Steve Chindgren film, "It's Just Sagebrush," a half hour un-narrated look at wildlife in the sage over a year's time.

It was filmed mostly between Farson and Pinedale.

If you haven't yet traveled a two-track, sagebrush tickling the belly of your pickup, pungent sage smell (not the garden variety) wafting through your open window along with a fine wind of dust as you bump over badger holes and glimpse heavy-bodied sage-grouse taking flight like lumbering World War II bombers, you should see the film.

And then you'll be interested in Alison and Brian's plans to begin an e-list to keep you up to date on this issue, letting you know how and when you can be an effective voice for the well-being of an ecosystem.

Contact Alison at 307-745-4848 or aholloran@audubon.org or visit the Audubon Wyoming office at 358 N 5th Street, Unit A., Laramie, WY 82072. Go to www.wy.audubon.org and click on Birds & Science for more information.

So, is geology destiny? Yes, I think so. While geology (and climate) makes some states suitable for farming, geology has made Wyoming rich in fossil fuels and sagebrush. We just have to choose how to keep both resources valuable.

# 244
Wednesday, April 30, 2008, Outdoors, page C2

## 12 practical ways you can help keep birds safe

All winter our relationship to wild birds is confined to observation and, perhaps, feeding them. But now with migration and breeding seasons intersecting with an increase in human outdoor

activity, we need to think about bird safety.

**1. Litter** – The cigarette stubbed out in the driveway disappears, but probably blew onto the neighbor's lawn where, if it isn't picked up, it

will, like other loose trash, break down and its unnatural components will pollute soil and water. Before that is able to happen, litter could end up in the digestive system of curious babies, puppies and

other animals. And remember all those photos of birds hampered by fishing line and other plastic debris.

**2. Windows** – If you are dreading the annual cleaning chores, skip your windows and tell people dirty ones are not as dangerous for birds. If you do wash your windows and find that one is particularly prone to getting messed up by birds thumping into it, you need to put some stickers on the outside. Cornell Lab of Ornithology, Student Conservation Association and Wyoming Public Radio send me those nice static cling type stickers every year so I can advertise my affiliations at the same time.

**3. Cats** – Nasty winter weather made it easy to keep your cat indoors. Just continue to keep it in and buy a harness and leash for little excursions, or build an outdoor pen with a screened roof. If you put a bird feeder outside a window, your indoor cat will be very happy. Just make sure the window screen is strong enough to withstand your cat's aborted bird attacks. If you don't have a cat and are tired of the neighbor's eating the birds that come to your feeder, borrow a cat trap from the animal shelter or get a dog to scare it off.

**4. Feeders** – Cold winters are marvelous for keeping bacteria in check around feeders. Don't quit feeding now in warm weather when migrating birds will make feeder watching even more interesting. But be sure to clean your feeders and feeding areas with a mild

bleach solution every few weeks. If you see any lethargic house finches, perhaps with warty growths around their eyes, quit feeding for at least a week so the healthy birds don't come in and get infected.

**5. Water** – If you provide a bird bath, make sure it has sloping sides or a sloping rock in the middle so birds can wade in. Brush the scum out every day when you refill it. Think about disinfecting it periodically. If you have tanks for watering livestock, make sure they have bird ramps to avoid drownings.

**6. Pesticides** – If toxic chemicals are sprayed on your lawn, you can keep small children and pets off for the necessary period of time, but birds can't read those cute little signs. Plus, pesticides wash into ground and surface water used by people and wildlife. Instead, try non-toxic lawn and garden care. Talk to Catherine Wissner and the Master Gardeners at the Laramie County Cooperative Extension Service, 633-4383, or check out Audubon at Home, www.audubon.org/bird/at_home/IPM_Alternatives.html.

**7. Mowing** – So you bought the house with five acres of prairie, and a riding mower, and you can't wait to get out there. Please relax, take a hike or go fishing instead, and let the ground

nesting birds, including the meadowlarks everyone enjoys, get the next generation started. Give them till at least mid-July.

**8. Dogs** – During the crucial season for ground nesting birds, late April to mid-July, keep dogs on a leash so they don't raid nests.

**9. Nest Boxes** – A birdhouse that is meant to be safely used by birds will have certain crucial features. The opening will be sized precisely for the intended cavity-nesting species: house wren, mountain bluebird, tree swallow, flicker, etc. There's no perch sticking out below, where starlings can stand while reaching in to raid the nest. Some kind of latch allows the nest box to be opened for cleaning. The box is the right dimensions, has proper ventilation, is not painted a dark color and is situated at the right height. Check the library for a book with particulars, or go to www.birds.cornell.edu/birdhouse/resources.

**10. Baby Birds** – Short of a catastrophe killing their parents, baby birds seldom need our help. It is best to leave them alone. If you watch long enough, you'll probably see parents bringing food to the grounded fledgling until it gets up the gumption to fly. You can try setting featherless nestlings back in their nest or in a small bucket

with twigs and grass hung somewhere safe near where you found them (but not if they are a ground-nesting species). Trying to feed baby birds yourself is usually not successful and deprives other wildlife species that depend on baby birds for their own food supply.

**11. Shrubs and Trees** – Cheyenne is in the midst of the grasslands, and if we are to promote the welfare of the beleaguered grassland bird species which have lost habitat due to plowing and development, we shouldn't promote planting trees and shrubs away from creeks and lakes. But right around our homes natural shade and windbreaks conserve energy, shelter migrating birds and attract birds we wouldn't see otherwise out here on the plains. Choose native fruit and seed producing vegetation.

**12. Energy** – There is no energy source yet that doesn't have some negative impact on wildlife. Remember, stuff you buy takes energy to produce so recycle and reuse, of course. And if you reduce the size of the house you need to heat and maintain and reduce the amount of stuff you buy that always seems to take additional energy and maintenance, guess what? You'll save money and have more time to enjoy life and watch birds!

244  CHEYENNE BIRD BANTER

# 245 Birding naked

## It's not nearly as fun as it sounds

Wednesday, May 28, 2008, Outdoors, page C2

Birding naked is all the rage in southeastern Arizona. That's what Gloria Lawrence of Casper told me May 17 while she and five other Casper birders helped local Audubon members with the Cheyenne Big Day Bird Count.

Gloria was not enthusiastic about the birding naked field trips. Actually, going au natural is what most people do when they look at birds without optics. Birding without binoculars and scopes means using your naked eye – eye glasses and contacts excepted.

The morning of the count I put on my binoculars. With the elastic strap harness so many of us birders use, it really is like getting dressed.

And getting dressed the last two weeks has been complicated by having my right hand in a splint. After 23 years, I opted to have pregnancy-induced carpel tunnel syndrome repairs.

You know what? Binoculars, in addition to being made for two eyes, are made for two hands. While my left hand is pretty adept at many things, it couldn't hold the binocs up and focus them too. I'd like to find a skinnier and lighter pair anyway.

When there was a blackpoll warbler directly overhead in a willow at Lions Park, I was able to lay the binocs on my upturned face and use the focus barrel more easily. But every time the bird flitted, I had to lift them off and locate the bird again.

My usual technique of staring at the bird and then putting the binocs between me and the bird didn't seem to be working.

At least the western tanager was close enough to be identifiable naked. Bright yellow body, black wings and red-orange head, there's no mistaking it for anything else. But I soon gave up on any birds that had to be differentiated by spots and streaks. I just couldn't get my binocs focused on them fast enough.

Finally, I found my useful niche and pointed out bird blurs to other people in the group, "I saw a flash of orange head up into that tree." "Oh yes," someone replied, "a Bullock's oriole."

I said, "A sparrow went into those reeds." "Ah, a Savannah sparrow," they said. Of course, I've never been able to tell that species apart from any of the other obscure sparrows anyway, so no loss.

A boss I had years ago was blind in one eye and bought himself a monocular. Imagine – you could afford twice the optical quality if you were buying only half a pair of binocs.

I looked forward to our stop at Wyoming Hereford Ranch Reservoir #1 where we'd set up spotting scopes to look for water birds. Once they are set up, it takes only one hand to use them though often the focus knob is on the right side.

For years Mark and I didn't own a scope and pretty much bypassed checking out reservoirs on our own. The birds are always on the far side, and it takes more imagination than I have to make blurs into birds. But the cheap scope (now selling for $300) we finally bought really made a difference. I can see the field marks and appreciate the variety of species.

There are still many birds to be enjoyed even if you are birding naked. As long as your ears are working, a spring morning is full of different birds, starting with robins at 3 a.m.

Many birds are large and unique. I can tell a turkey vulture (leading edge of the underside of wing is black, trailing edge is silver) from a Swainson's hawk (trailing edge is dark, leading edge is light) from a red-tailed hawk (tail is "red", reddish brown) even at 75 mph with sunglasses on providing the birds aren't soaring too high.

Then there was the redwinged blackbird strolling toward me on the walk around Sloans Lake the morning of the count. I'd fallen behind the rest of the group and decided I might as well enjoy a bird close up if I could.

We stopped about two feet from each other, our eyes locking as if we were characters in a Harlequin romance. His black feathers were glossy and his fire-engine red epaulets were puffed out. His black eyes glinted in the sunshine. I hated to break our rapport, but I knew there was a female in stripy plumage who would appreciate him even more.

By evening Mark and I hit the lakes at F.E. Warren Air Force Base. I was tired of not really seeing birds and having to depend on everyone else to identify the blurs. It was such a good birding day otherwise – little wind, warm, sunny, trees hardly leafed out, the crabapples at their peak and the appearance of a mourning warbler that almost everyone else saw.

And then I saw them, a whole raft of sleeping pelicans. The lake was small enough that they were close enough to enjoy.

All together I'd say birding naked is just about as frustrating as birding without clothes would be uncomfortable – imagine sunburn, bugs, and thorns. What I didn't miss was my usual role as note taker for the group. But I missed having a close look at all my favorite little migrating feather balls and learning to identify new ones.

Remind me to avoid scheduling surgery on my left hand too close to the Christmas Bird Count. It would be kind of cold for birding naked.

## Cheyenne Big Day Count May 17, 2008

123 species overall
L – Lions Park Wyoming Important Bird Area, 48 species
W – Wyoming Hereford Ranch Wyoming Important Bird Area, 81 species
R – High Plains Grasslands Research Station, 28 species
B – F.E. Warren Air Force Base, 34 species
O – Other observations, 39 species

| Species | L | W | R | B | O |
|---|---|---|---|---|---|
| Canada Goose | L | W | | B | O |
| Gadwall | | W | | B | |
| American Wigeon | | W | | | |
| Mallard | L | W | R | B | |
| Blue-winged Teal | L | W | R | B | O |
| Cinnamon Teal | | W | | | O |
| Northern Shoveler | | W | | | O |
| Northern Pintail | | W | | | |
| Green-winged Teal | | W | | | O |
| Redhead | | W | | B | O |
| Ring-necled Duck | | W | | | |
| Greater Scaup | | W | | | |
| Lesser Scaup | L | W | | B | |
| Bufflehead | | W | | | |
| Ruddy Duck | | W | | B | |
| Pied-billed Grebe | | | R | B | |
| Eared Grebe | | W | | B | |
| Western Grebe | L | W | | | O |
| American White Pelican | | | | B | O |
| Double-crested Cormorant | L | W | | | O |
| Great Blue Heron | | | | | O |
| Black-crowned Night Heron | L | | | | O |
| White-faced Ibis | | W | | | O |
| Turkey Vulture | L | | | | |
| Cooper's Hawk | L | | | | |
| Broad-winged Hawk | | | | | O |
| Swainson's Hawk | | W | R | B | |
| Red-tailed Hawk | | W | R | | |
| American Kestrel | | W | | | |
| Prairie Falcon | | W | | | |
| Sora | L | W | | | |
| American Coot | | W | | | O |
| Killdeer | | W | R | | |
| American Avocet | | W | | | |
| Lesser Yellowlegs | | W | | | |
| Willet | | W | | | O |
| Spotted Sandpiper | L | W | | B | |
| Marbled Godwit | | | | | O |
| Baird's Sandpiper | | | | | O |
| Stilt Sandpiper | | W | | | |
| Wilson's Snipe | | W | | | |

| Species | L | W | R | B | O |
|---|---|---|---|---|---|
| Wilson's Phalarope | | W | | | |
| Red-necked Phalarope | | W | | | |
| Ring-billed Gull | | | | B | O |
| Black Tern | | | | | O |
| Rock Pigeon | L | | | B | |
| Eurasian Collared-Dove | L | W | R | B | O |
| Mourning Dove | L | W | R | B | O |
| Great Horned Owl | | W | R | | |
| Chimney Swift | L | | | | |
| Belted Kingfisher | | W | | | |
| Downy Woodpecker | L | | R | | |
| Hairy Woodpecker | | | R | | |
| Northern Flicker | L | | | | |
| Olive-sided Flycatcher | | | R | | |
| Least Flycatcher | | W | | | |
| Dusky Flycatcher | L | | | | |
| Say's Phoebe | | W | | | |
| Cassin's Kingbird | | | R | | |
| Western Kingbird | | W | | | |
| Eastern Kingbird | | W | | | |
| Plumbeous Vireo | L | W | | | |
| Blue Jay | | | | | O |
| Black-billed Magpie | | | R | | |
| American Crow | L | | | B | O |
| Horned Lark | | W | | | |
| Tree Swallow | | W | | | |
| Violet-green swallow | | W | | | |
| N. Rough-winged Swallow | | W | R | | |
| Bank Swallow | | W | R | | |
| Cliff Swallow | | W | | | |
| Barn Swallow | | W | | | |
| Black-capped Chickadee | | | R | | |
| Mountain Chickadee | | | | | O |
| Red-breasted Nuthatch | | | R | | |
| House Wren | L | W | | | |
| Ruby-crowned Kinglet | L | W | | | |
| Blue-gray Gnatcatcher | | W | R | B | |
| Eastern Bluebird | | | R | | |
| Townsend's Solitaire | | | R | | |
| Veery | | W | | | |
| Swainson's Thrush | L | W | | B | |

| Species | L | W | R | B | O |
|---|---|---|---|---|---|
| American Robin | L | W | | B | O |
| Gray Catbird | L | | | | |
| Northern Mockingbird | L | | | | |
| Brown Thrasher | | | R | | |
| European Starling | L | W | | B | O |
| Orange-crowned Warbler | L | W | | B | |
| Virginia's Warbler | | W | | | |
| Yellow Warbler | L | W | | B | O |
| Black-throated Blue Warbler | L | | | | |
| Yellow-rumped Warbler | L | W | R | | |
| Blackpoll Warbler | L | W | R | B | |
| American Redstart | | W | | | |
| Ovenbird | | W | | | |
| Northern Waterthrush | L | | | | |
| Mourning Warbler | | W | | | |
| Common Yellowthroat | L | W | | | O |
| Wilson's Warbler | | W | | | |
| Western Tanager | L | W | | | |
| Spotted Towhee | | W | | | |
| Chipping Sparrow | L | W | | B | O |
| Clay-colored Sparrow | | | R | | |
| Field Sparrow | | | | B | |
| Lark Sparrow | L | | R | | O |
| Lark Bunting | | W | | | |
| Savannah Sparrow | | | R | | O |
| Song Sparrow | L | W | | B | |
| Lincoln's Sparrow | L | W | | | |
| White-throated Sparrow | | W | | | |
| White-crowned Sparrow | L | W | | | |
| Lazuli Bunting | | | | | O |
| Red-winged Blackbird | L | W | R | B | |
| Western Meadowlark | | W | | B | |
| Yellow-headed Blackbird | L | | | B | O |
| Common Grackle | L | W | | B | O |
| Great-tailed Grackle | | | | | O |
| Brown-headed Cowbird | | W | | B | |
| Bullock's Oriole | L | W | | | |
| House Finch | L | | | B | O |
| Pine Siskin | L | | | | O |
| American Goldfinch | L | W | | | O |
| House Sparrow | L | | | B | O |

## 246

Tuesday, July 1, 2008, Outdoors, page D2

# Nesting season a time for activity

Nesting season means a variety of building activities.

Nesting season means the dawn chorus of birdsong quiets down as the business of assuring the continuation of species gets underway.

Great horned owls get an early start. By mid-May a stick nest in a cottonwood west of town sported two owlets that with all their fluff seemed as large as their parents. But, apparently, owlets in the same nest don't hatch at the same time. Jana Ginter told me this spring she rescued two after a windstorm blew down their nest near Carpenter.

The rehabilitators at the Rocky Mountain Raptor Center in Fort Collins, Colorado, suspected, based on the size difference between the owlets, that there must have been a third that hatched in between. Detective work by Jana showed the third owlet was rescued and taken to rehabbers in Nebraska.

The big stick messes in the cottonwoods, along the shores of Lake Minnehaha at Holliday Park, belong to a colony of black-crowned night herons. It's hard to see the large birds on the nests once the trees leaf out, so many park users have no idea what's going on overhead.

Gulls are fond of nesting in colonies too, but on the ground, preferably on an inaccessible island.

May 31: Mark and I were in Casper for the 8th annual Wyoming Audubon Chapters campout and visited Soda Lake as part of a group of guests of Murie Audubon Society. The reservoir is not publicly accessible, and the birds seemed less skittish there than other reservoirs.

Double-crested cormorants and California gulls were shoulder to shoulder on a bit of sand, many quietly sitting on nests which were merely scraped-together mounds of natural debris. Then someone with a spotting scope called out, "Look, chicks!"

Sure enough, around the standing adult gulls were little gray fluff balls, hardly different from chicken chicks. Even though gulls quickly attain adult size, this species will go through five plumage variations in the first four years before getting complete adult coloration.

Walking along the North Platte River at Edness Kimball Wilkins State Park earlier in the day we were surrounded by noisy yellow warblers (the species is also named "Yellow Warbler") high up in the cottonwoods. Every distinctive call was easy to match up with another bright, daffodil-yellow bird. We were surprised when one flitted closer, to eye-level branches of a small tree. Then we spotted the cup-shaped nest in a junction of branches.

Robins build similar, but bigger, cup-shaped nests since they are nearly twice the length of a warbler, but they don't always build in trees. Every year, I get calls about them nesting on front porch light fixtures and dive-bombing homeowners.

We've had robins nest in the bushes under our window where we could look out and watch the nestlings develop. This year, when the male kept zooming past the window with his beak full of nesting material, the trajectory didn't seem right. I realized finally he was building on an exposed beam under the roof overhang, free from disturbance by squirrels, cats and human onlookers.

Canada geese will nest on manmade platforms, but so far, I've never seen them use the upended concrete culverts at North Crow Reservoir. Unlike a Christo landscape art project, these spoil an otherwise beautiful setting and I hope the state parks people will remove them. Most years I've found evidence of geese nesting on the far shore and observed at least two families of goslings per year.

As for other nesting strategies, I didn't stumble over any killdeer or meadowlark ground nests out on the prairie this spring, but I did see tree swallows flitting in and out of nest boxes probably meant for mountain bluebirds.

If you have been watching a nest this season, consider sharing your information through Cornell Lab of Ornithology's NestWatch citizen science program. Unlike Project FeederWatch, participation in this project is free. It is funded by the National Science Foundation and developed in collaboration with the Smithsonian Migratory Bird Center.

Go to www.nestwatch. org and learn how to monitor nesting birds so that the data collected can be of scientific importance. Even if you don't join, you can explore the data. Or learn best nest box construction and maintenance practices and other things at www. nestinginfo.org.

June 16: My robin fledglings left the safety of the nest. The one that insisted on sitting in the middle of the lawn was soon discovered by crows and could not be saved by half a dozen angry adult robins and my belated approach. Crows have to eat too, I guess.

But I glimpsed two other, smarter, fledglings deep in the bushes. They've already learned two things: it's a bird-eat-bird world out there and, you can't go home again – at least until you build your own nest.

# 247 Drive the Big Horn loop

Monday, July 14, 2008, ToDo, page C1

### Back-road drive has magnificent views, dino tracks, sacred Indian landmarks and more.

As fascinating as the cities and towns along the eastern slope of the Big Horn Mountains are, and worth the 300-mile drive from Cheyenne, let me recommend a back-road excursion with no entrance fees that my husband, Mark, and I recently took.

Ten miles north of Sheridan, take U.S. Highway 14, switchback by switchback, 40 miles to Burgess Junction.

On the way, Sand Turn Overlook has a magnificent view of the plains and is the last chance for cell phone reception for a while. It's where the Bighorn National Forest fire crew stationed at Burgess Junction, including our son, go to make calls.

Two miles before the junction of U.S. 14 and Alternate 14 is the Burgess Junction Visitor Center. Get your bearings there and a forest map.

Three nearby resorts offer restaurants and rooms, or drive a mile of gravel to the North Tongue campground and picnic area.

Or, maybe you stopped a few miles earlier at Sibley Lake to wet your fishing line.

In early July there was still snow on the peaks, meadows stuffed with wildflowers, moose on the prowl and deer fawns making their first appearance. Watch out for black bears.

Continue on U.S. 14 down Shell Canyon early in the morning when shadows emphasize craggy canyon walls. The rest area at Shell Falls is closed this summer for reconstruction, so substitute one of the campgrounds or picnic areas. There's also a 10-mile hiking trail above the highway.

You are out of evergreens and the forest by the time you reach the dozen buildings making up Shell.

Four miles later, look for the sign for the Bureau of Land Management's Red Gulch Dinosaur Tracksite. It's worth the 5-mile detour on gravel on the Red Gulch/ Alkali Back Country Byway.

Amazingly, you are allowed to walk the mudstone in soft-soled shoes. Ask the volunteer ranger for help finding the two kinds of petrified tracks.

Back on U.S. 14, drive through Greybull and take U.S. 310 north. It's a chance to experience the Bighorn Basin's solitude.

See how many vehicles you can (or can't) count in the next 34 miles.

At Lovell, visit the Bighorn Canyon National Recreation Area Visitor Center, which also has information on BLM's Pryor Mountain Wild Horse Range nearby.

Lovell to Burgess Junction via U.S. Alternate 14, about 60 miles, cuts across the end of Bighorn Lake.

About 10 miles later, when you spot the cliff-hanging highway ahead, turn off for BLM's Five Springs Falls Campground, accessible via the old highway, a narrow road engineered in the 1930s.

The potholes are large, but picnic tables are in deep shade. Hike the short trail to see the falls.

Back on the modern highway, back in the national forest, it's all up hill, including a 1.5-mile hike to the peak experience of the day: Medicine Wheel National Historic Landmark.

White rocks were laid out like a wagon wheel 300-800 years ago and no one in the 81 tribes that consider the site sacred can explain their meaning, but the 360-degree view is inspiration itself.

# 248 A splendid book for curious kids

Saturday, July 26, 2008, Cheyenne Frontier Days special section, page CFD10

### Actually, "The Young Birder's Guide" is great for new birders of any age

Parents of curious children understand this dilemma: Adult books on a subject are too boring or too much, while non-fiction for children often seems not to have evolved beyond picture books.

Enter two exceptional parents, the editor of Birdwatcher's Digest, Bill Thompson III, and his artist/author wife Julie Zickefoose, whose work appears in the magazine and many other places.

Bill and Julie have put together a splendid bird guide for children within the standard field guide format.

I would recommend it for beginning bird watchers of any age, not just the suggested 8 to 12-year-olds.

When you are ready for other field guides, pass on your copy of "The Young Birder's Guide" to the grandchildren or the neighbor children.

I especially like the preliminary chapters that explain bird watching and bird conservation.

"WOW!" sidebars highlight each species with cool info: Did you know that black-capped chickadees actually grow extra brain cells to help them remember where they've stashed seeds?

About 40 of the 200 species in the guide are unlikely to be seen here in Wyoming (which has about 400 species total), but don't wait for the western edition-- kids grow up too fast.

"The Young Birder's Guide to Birds of Eastern North America" (Peterson Filed Guides series) by Bill Thompson III, illustrations by Julie Zickefoose, Houghton Mifflin, 2008, flexible cover, 300 color photos, 200 black and white drawings, 200 maps, 256 pp., $14.95.

Also: Young Birder's Guide Companion, download for iPod or CD version, Mighty Jams, LLC, songs and more photos for 160 birds, $14.95.

# 249 Should a landfill go here?

Monday, August 11, 2008, ToDo, page C1

## Public can chime in on plans for Belvoir Ranch

**The city-owned ranch, which has teepee rings and other historic features, could soon be home to wind turbines and a garbage dump.**

Where wildlife and pre-historic people once discovered an easy travel route, it isn't surprising that everyone from stagecoach drivers to fiber optic companies have followed.

And that route passes through the Belvoir Ranch, bought by the City of Cheyenne in 2003.

It begins five miles west of Cheyenne and stretches for 15 miles farther west, with Interstate Highway 80 as its northern boundary.

While some residents see the 18,000-acre purchase as a boondoggle, others see it as acquiring water rights and sites for a landfill, wind turbine farm and recreation. It is also a chance to preserve a microcosm of western cultural history.

Chuck Lanham of the

Cheyenne Historic Preservation Board, the guide for a recent ranch tour, pointed out tepee rings at least 140 years old and other archeological features that will be studied.

Arapaho tribal elders have visited recently, sharing their knowledge of the land. Eastern Shoshone, Northern Cheyenne and Lakota tribes also have ties.

Ruts across the rolling, shortgrass prairie show the route of the Denver to Fort Laramie stage line. Other ruts are thought to be Camp Carlin supply wagon tracks to frontier forts. There are vestiges, too, of the old Lincoln Highway, precursor to U.S. Highway 30 and Interstate 80.

When the Union Pacific Railroad came, it built water tanks and "columns" to fill its steam-powered engines. Today, the ranch is crisscrossed by three sets of rails.

Eventually, the early homesteads became part of the huge Warren Livestock Company holdings. F.E. Warren called the main ranch house his "cabin," complete with tennis courts, pool and professional horse racing track. Remains are barely visible today.

Because of the 1962 Cuban Missile Crisis, Atlas missiles were installed on what soon became known as the Belvoir Ranch. The above-ground launching facilities were deactivated in 1965, but the concrete structures can be seen south of I-80 at exit 348.

Currently, the 1,800-acre Big Hole area is under a conservation easement with The Nature Conservancy. Local ranchers hold grazing and haying leases on the rest. The fees they pay cover most of the ranch's operating expenses.

A utility corridor provides easement for a power transmission line, four pipelines and two fiber optic lines. The Borie oil field continues operations, but mineral rights are not owned by the city.

A contract with Wyoming Game and Fish Department through their Hunter Management Area program allows some hunter access, the only legal public access to the Belvoir Ranch at present.

### Tour the Belvoir Ranch and let your voice be heard

Public comments on the plans for the Belvoir Ranch will be taken at the City Planning Commission meeting Aug. 18 at 6 p.m. in the City Council Chambers, 2101 O'Neil Ave.

To view the photo-filled plan, go to www.BelvoirRanch.org or call 637-6200 to find a copy.

If you click on "Build" and then on the map, you will see the proposed locations of the landfill, 5,000-acre wind farm, link golf course, equestrian facilities, trails and other non-motorized recreation ideas.

To get a spot on a Belvoir Ranch tour or arrange for a group visit, call 637-6283.

# 250 A fort with stories to tell

Sunday, August 17, 2008, ToDo, page D4

**Fort Robinson is the place where Crazy Horse died, an Olympic team trained, more than 2,000 dogs were trained and prisoners of war stayed.**

Once nicknamed the "country club of the Army," historic Fort Robinson is still a magnet for recreation today as a state park located in the Pine Ridge country of Nebraska's panhandle.

It is just west of Crawford, less than 200 miles from Cheyenne.

In 1874, Camp Robinson, named after a victim of the Indian hostilities, was established to protect the Red Cloud Indian Agency which distributed goods to the Sioux, Cheyenne and Arapaho tribes.

The log cabin where famous Chief Crazy Horse was killed in 1877 still stands.

Historical museums on the grounds document how a year later Camp Robinson was upgraded to permanent status and renamed Fort Robinson.

But Indian troubles didn't end until the Wounded Knee Massacre in 1890. Beginning in the 1880s, it was home to the 9th and 10th U.S. Cavalry, regiments made up of all black soldiers.

Its reputation as a country club started when Fort Robinson became a Quartermaster Remount Depot in 1919, where horses and mules purchased for the Army were trained and bred.

The Fort's social life revolved around polo, fox hunts, trail rides and steeple chases. In the 1930s, the U.S. Olympic equestrian team trained here.

During World War II, Fort Robinson housed as many as 2,000 dogs being trained for combat. It also had a German prisoner of war camp.

Today, lodging is available April 1 through Nov. 30 in many of the historic living quarters. Motel-type rooms are available in the Lodge, formerly the Enlisted Men's Barracks, circa 1909, along with a restaurant famous for its buffalo dishes from the resident herd.

The "Adobes," officer quarters circa 1887, are set up for housekeeping. The "Bricks," the 1909 officers' quarters, similar to the large homes at F.E. Warren Air Force Base here in Cheyenne, can host groups up to 20 people.

Comanche Hall, the 1909 Bachelor Officers' Quarters, holds up to 60 people and has multiple bedrooms, bathrooms and a fully outfitted kitchen, ideal for family reunions.

RV hookups and primitive camping sites, as well as horse stalls, are available. Call 303-665-2900 or check www.ngpc.state.ne.us.

The printed list of

2008

activities making use of the park's 22,000 acres includes: swimming pool, trail rides, jeep rides, horse-drawn tour, stagecoach, pony rides, mountain bike rental, hayrack breakfast, hayrack steak cookout, chuck wagon cookout, rodeo events, historic building tours, fishing, kayak and tube rides as well as summer theater at the playhouse.

If all that activity is too much for you, drive the six miles of gravel road beyond the buildings, along Soldier Creek, and hike into the Nebraska National Forest, perhaps into the 9,000 acres of the Soldier Creek Wilderness Area for some fishing.

# 251 What birds are in your backyard this summer?

Newspaper copy not found, 2008

Mark and I have been herding cats this summer. Our two are indoor cats, for their own and the birds' sakes. For the last three years in spring and summer they have received nightly visits from loose cats in the neighborhood. Even with the windows shut, the visits get them all upset, yowling and fighting each other when otherwise they would be snuggled up.

One morning back in May I found one of the cats, and the walls, spattered in blood. So now we round them up every night and confine them in a bathroom without windows. We get a good night's sleep and they shed no more blood.

As a reformed loose cat owner myself, I doubt the owners of the roaming felines realize how much they inconvenience their neighbors, or realize how many birds are killed by their pets.

Someone else's Cheyenne backyard hosted a different night creature recently. The friend of a friend called to say she'd been hearing an odd noise in the evenings. Because it seemed high up in the trees she thought it might be a bird and I would know what it was. She tried to imitate it and it sounded to me like a rush of air, like an open-mouthed exhalation.

I was stumped. "Sure it wasn't an owl hoot?"

"Sure."

"Some kind of mammal?"

"It seemed to be above the trees."

"An insect?"

"Too large a sound."

"Any weird mechanical noises from the neighbors?"

"Too far away."

"The only other night bird I can think of is the nighthawk."

"Don't know that one."

So she hung up to go look up nighthawks in her field guide. A few minutes later she called back.

"Yes, I think that's it! The book says, 'In aerial display, male dives steeply, then pulls up with a rushing or booming sound.'"

The Sibley Guide to Bird Life and Behavior describes it as "a loud, humming, whooshing, hoooov, produced as air rushes through the primary feathers of the diving male. This display may be directed at a female, another male, a young bird or an intruder."

Our best summer bird species in the yard has been the lesser goldfinch, considered a rare migrant in our area. Since June we've seen a pair every day at our thistle feeder.

At first the lessers were joined by American goldfinches and it was easy to compare them. The male lesser's head is almost completely black versus the American's yellow with a black forehead. The lesser has an olive-green back instead of yellow. The lesser is half an inch shorter—which is a lot when looking at 5 versus 5.5 inches. And they sound different.

This last week a second lesser male is sharing the feeder at the same time. The competitive breeding season must be over. Several other folks in the Cheyenne area and west of town have also had lessers this summer. I wonder if any will spend the winter as one did a couple years ago.

Aug. 3 was my first hummingbird in the yard this summer. Cheyenne residents begin to notice them mid-July on their fall migration. This one was probably our most common, a broad-tailed. The light was too dim at 6 a.m. to identify it as it sank its bill into a red geranium.

But there are other hummers out there. Visiting an Audubon member's home up by Buford mid-July we saw the rufous (color of a light red golden retriever) and a calliope with the stripy looking feathers, its gorget, around its neck. They are North America's tiniest breeding bird, only 3 inches long, and are considered an uncommon summer resident in our area.

Black-chinned hummingbirds have been working a friend's feeders west of town. That might just be a first record for that latilong. They are considered rare summer residents in Wyoming.

What you see in your own backyard depends on what bird species you can recognize, I think. We were at a cookout over in the Sun Valley neighborhood at the end of July when I looked up and noticed a medium-sized, yellow-breasted bird in a tree. Maybe a western flycatcher? No, it had a bright red head. The only name that fits that description is western tanager. Wouldn't they still be in the mountains mid-summer?

Then, as the light dimmed and we were saying good-by, a blue and white bird flashed over the backyard, one with a distinctive crest on its head, but also a chubby body: belted kingfisher! That surprised both us and our hosts.

As for our greater neighborhood, Wyoming being just one long Main Street, I'd like to thank the Governor for his interest in, and support of, the pro-sage grouse faction. I think it will make a difference. Our "neighborhood" birds need our protection, whether from excessive oil and gas drilling or loose cats.

# 252

Thursday, August 28, 2008, ToDo, page D4

## Book review: "Flights Against the Sunset: Stories that Reunited a Mother and Son"

### Bird-lover memoirs are tear-jerkers

**Author Kenn Kaufman hid his love for birding from his family until he found out his mother was dying from cancer**

Kenn Kaufman documented his teen years as an extreme birder in "Kingbird Highway" and has settled in as author of birding field guides and magazine articles.

This time, he is writing a memoir, relating his birding adventures to entertain his dying mother.

We learn in this small book that he did his best to keep his birding persona from his family, as young people do when they are establishing their separate identities. He didn't share much of his birding knowledge and adventures with his family.

For years Kaufman gave short shrift to his mother's tentative claims to have heard a chickadee in their nearly treeless, suburban Kansas neighborhood, but a lot can change in 30 years. You might need a hankie for some of the narrative between the essays.

Many of the 19 essays are adapted from Kaufman's column in Birdwatcher's Digest magazine. A couple, like "The Birder Who Came in from the Cold" might strike you as tall tales.

My favorite essay goes a long way towards explaining boys I knew in junior high, though their obsession was engineering rather than birds.

Imagine being a 13-year-old girl and meeting a boy from your class after school who can't talk coherently about the latest TV episodes because he won't tell you there's no TV at his house. He admires your hair and says the color reminds him of a buff-breasted sandpiper. Yikes, he used a b-word!

And then, just when first base might be in view, he isn't paying attention at all. He won't mention he's trying to identify the singer of a buzzy warbler song that he knows would make him the envy of the local Audubon members.

Other essays take the reader to Venezuela, Peru, the Amazon, Kenya, Mexico, across our country and into Kaufman's own neighborhood.

The book is great to read aloud, just as Kaufman must have for his mother.

---

### At a glance

"Flights Against the Sunset: Stories That Reunited a Mother and Son" by Kenn Kaufman, Houghton Mifflin, 2008, hardcover, 225 pp., $24.

# 253

Monday, September 1, 2008, ToDo, page D4

## Ayres Natural Bridge offers cool respite to travelers

Pioneers trying to escape the rat race of the Oregon Trail and its wide-open spaces probably found this little side trip up La Prele Creek to be as refreshing as today's travelers on Interstate 25 do.

Perhaps you've noticed the sign for Ayres Natural Bridge as you approach Exit 151, about 140 miles north of Cheyenne, between Douglas and Casper. The Ayres Natural Bridge Park is well worth the ten-mile round trip detour. Along the way, you'll cross the Oregon Trail and Overland Stage Route.

Early travelers had to clamber over rocks and thrash their way through bushes to drop down into the canyon where the creek flows under a natural bridge of rock 90 feet long and 30 feet high.

In 1870, Ferdinand V. Hayden and photographer William H. Jackson, on an expedition sponsored by the U.S. Geological Survey, visited, according to Mae Urbanek, in her book, "Wyoming Place Names."

The Ayres family donated the land, 150 acres, to Converse County in 1921. Today the red sandstone canyon walls and large cottonwoods shade green lawns and picnic tables of the well-kept park. La Prele Creek, named for the French word referring to a plant growing along the water's edge, scouring rush or horsetail, flows at the foot of cliffs decorated with swallow nests before turning and rippling under the arch of rock.

Admission is free. Pets are not allowed, however. They must be kept in vehicles, if they can't be left at home.

Five tenting spots and five spots for small recreational vehicles are available free on a first come, first served basis. There are no hookups. The park is available for family reunions, weddings and other gatherings. Reservations are required. Call the caretaker, 307-358-3532.

# 254

Wednesday, September 17, 2008, ToDo, page D4

## "A Guide to the Birds of East Africa" is a charming novel of a birding contest and love

When Houghton Mifflin sent me a copy of "A Guide to the Birds of East Africa," I was miffed.

How could I review a field guide for a part of the world I know nothing about?

But it's actually a charming novel in which Everyman, Mr. Malik, can compete with Money, Harry Khan, and maybe win True Love, Rose Mbikwa, through a birding competition confined to one week in Kenya and its 1,000 species.

The winner gets the right to ask Rose, the leader of the Tuesday morning bird walks sponsored by the East

African Ornithological Society, to the Nairobi Hunt Ball.

Modern Kenya is a tricky place to live, as well as watch birds. There are plenty of geographical as well as political obstacles to overcome: mountains, storms, corrupt officials, highway men, muggers, soldiers and slavers.

It is interesting to see how the peace-loving, semi-retired Malik, happiest in his own Nairobi backyard, adds to his bird list, as compared to Khan and his methods. Khan, who is visiting from America where he is now a successful businessman, spares no expense and takes

with him everywhere two visiting birding experts from Australia.

It doesn't help that Khan bullied Malik all through school and even now, white-haired, he wants to steal the woman of Malik's dreams.

At least even in a country where one doesn't dare report stolen vehicles, a good man like Malik can find unexpected resources.

As you begin to read "Guide to the Birds of East Africa," Alexander McCall Smith's "The No. 1 Ladies' Detective Agency" series set in Botswana comes to mind, but author Nicholas Drayson

is not hampered with solving murders.

On the surface, both Malik's character and Drayson's prose appear simple. However, by the time I finished reading, I realized I'd soaked up a lot more understanding of Kenya's cultures, history and problems than I expected – not to mention a beginning knowledge of the birds of East Africa.

"Guide to the Birds of East Africa" by Nicholas Drayson, Houghton Mifflin, available Sept. 18, 2008, hardcover, 202 pages, $22.

# 255

Saturday, September 20, 2008, ToDo, page D4

## A must-do hike for fossil fanatics

### ...and it's not too far from Cheyenne! Agate Fossil Beds National Monument

If anyone in your family is a fossil fanatic, surprise them with a day trip to Agate Fossil Beds National Monument, less than a 140-mile trip northeast of Cheyenne, in Nebraska's panhandle.

On a hot summer's day, I thought I'd take the hike out to the two small hills where paleontologists from the Carnegie Museum, Yale University, the American Museum of Natural History and other institutions dug for fossils in the early 20th century.

It didn't look far and so

I left my water bottle in the car. A cool breeze came off the Niobrara River bottom as I crossed over, but I was soon parched as I followed the concrete trail through the thick prairie grass into the rolling hills.

Later, I found out it was 97 degrees, the loop trail was 2.7 miles long and lack of water is what killed the Miocene-era mammals whose fossilized remains have been dug up 19 million years later.

The drought back then caused Stenomylus (small,

gazelle-like camel), Menoceras (three-toed, pony-sized rhino) and other animals to stick to the remaining waterholes, starving to death when they'd eaten up all the surrounding plant life.

Agate Springs Ranch owners Kate and James Cook found a petrified leg bone in the 1880s and by 1892, the first researchers began working the fossil beds. Discoveries continue to be made.

The visitor center has interpretive videos and displays, including the dig site

diorama. It also houses James Cook's collection of gifts from Oglala and Cheyenne tribal members.

Educational programs for groups are available with advance notice.

Another interpretive trail visits the site of large, fossilized, corkscrew-shaped burrows of a prehistoric beaver.

A shady picnic area is available, but camping is not. The closest camping is at Fort Robinson State Park, 50 miles north, or in the Scottsbluff area, 45 miles south.

### Agate Fossil Beds National Monument

**Hours:** 8 a.m. – 4 p.m. from Labor Day through Memorial Day; 8 a.m. – 6 p.m. during the summer months. It's open year-round, except for Thanksgiving, Christmas and New Year's.
**Cost:** $3 per person or $5 per car.
**Directions:** From Cheyenne, drive around 75 miles north on

U.S. Highway 85 to Torrington. Head east 28 miles to Mitchell, Nebraska. This is your last chance for gas, food and lodging before turning north on Nebraska Highway 29 for the last 37 miles to the monument.
**More info:** www.nps.gov/agfo or 308-668-2211.

CHEYENNE BIRD BANTER

# 256 New Peterson's field guide has it all together

Tuesday, October 7, 2008, Outdoors, Commentary, page D1

This year, publisher Houghton Mifflin celebrated the 100th anniversary of Roger Tory Peterson's birth with a brand-new edition of his flagship birding field guide. Without Peterson, who died in 1996, the updating took the skills of six people.

In 1934, Peterson was the first to publish a field guide for identifying birds. Before that there were books about birds, but the descriptions were so thorough, from beak to toe, it was hard to find what set one species apart from another.

Peterson was inspired by an Ernest Thompson Seton story in which a character was able to identify ducks at a distance using their easy-to-see markings. As a trained artist, Peterson could make simple illustrations of birds showing their color pattern – just what the modern birder needed who wanted to bird with binoculars rather than a shotgun.

The Peterson name has since expanded to hundreds of natural history titles including field guides for identifying everything from shells to stars.

If you grew up using one of the famous blue or green-covered "Peterson's," you'll soon realize this isn't the same old, same old. This one is titled "Peterson Field Guide to the Birds of North America." That's right. The eastern and western guides have been combined.

It's about the same thickness, but each page, 6 inches by 9 inches, has 23 more square inches than previous editions – hardly a book to slip in a pocket anymore.

I've always had a problem using Peterson's. The second western edition (1961) had all the plates of bird illustrations in the middle of the book and all the range maps at the end, unlike the Golden Guide of the same era where all information for a species was on the same page.

The third western edition (1990) finally put the pictures and descriptions together but left the maps at the back of the book.

When I opened this new Peterson's, I was looking not only for improvements, but innovations over other recently published field guides.

I looked up "Mountain Chickadee." First is the family description which characterizes the size and behavior of chickadees and titmice and lists what they eat. It mentions the family's worldwide range, which few other North American field guides do. Also, the family name appears on the bottom of each page, color coded for speedier referral.

In this larger format all six chickadee species are on one page, with their trademark (yes, actually trademarked) arrows pointing at the important field marks that distinguish each from the other. All the portraits seem more detailed than the last edition's but that's because they've been digitally enhanced. The birds are bigger, which is good for those of us fighting the need for reading glasses.

Finally, there are thumbnail range maps right next to each bird's description so those cryptic, written range notes, oversimplified and heavily abbreviated, are gone. A picture is worth a thousand words, after all. The larger maps, with notes, are still in the back.

Also new, in the heading for each species, is an abundance rating. How common is the mountain chickadee within its range and preferred habitat? "Fairly common."

Continuing the third western edition's tradition is the habitat description – good for people out in the field working i.d. the other way around and searching for a species. The voice descriptions are more detailed, harking back to the first western edition. And when appropriate, names of similar species are listed.

One of the recent improvements in field guides is a one-page index of bird names. In this guide it is inside the front cover rather than in the back which makes it much easier for right-handers to balance the guide while consulting the index.

Inside the back cover are all of Peterson's original silhouettes for shorebirds, roadside birds, etc.

And here's what no other major bird field guide has so far: 35 video podcasts. Even before you buy this book, you can go to www.petersonfieldguides.com and download these to your iPod or watch them on your computer.

Other sources have excellent free online bird information in a species-by-species format. These videos have family overviews that complement the field guide. I recommend also watching the videos with birding tips and Peterson's biography. Bill Schmoker, a bird photographer from Colorado who spoke at a Cheyenne Audubon meeting once, is listed in the credits for many of the videos.

Field guides just keep getting bigger and better. For one, we know more about birds than we did when Peterson published the first edition of his eastern guide. And also, technology keeps improving, whether printing books or providing digital information.

But I still like the feel of the green, cloth-covered 1941 western edition. It is slim, but with the dense feel that promises so much. I like its creamy, uncoated pages of text and simple, schematic bird drawings. Maybe someone could put a nice green cloth cover on one of those new-fangled, handheld electronic "bird finder" contraptions.

The measure of success of the new Peterson's bird guide will be not how many copies are sold, but how many battered copies lie next to binoculars in years to come.

## On the bookshelf

"Peterson Field Guide to Birds of North America" by Roger Tory Peterson, 2008, Houghton Mifflin, 544 pages, $26.

# 257 Book review

Thursday, October 16, 2008, ToDo, page C4

## Wild horses won't let you put this book down

No matter how long you live here, if you weren't born here, you'll never be a native. But you aren't a tourist either.

Modern wild horses have the same problem. The current herds didn't evolve in North America, although there were horses on our continent until the Ice Age.

Author Deanne Stillman elucidates the wild horse's other identity problem: Not being native wildlife, it can't be managed like big game, and because of North American cultural aversion to horsemeat, treating it like cattle has been outlawed.

Stillman was inspired to write this book because of the senseless massacre of wild horses in Nevada in 1998 by three good-old boys having fun. When she looked into the wild horse back story, she discovered a true saga.

Stillman takes us through horse paleontology and the reintroduction of horses to North America by the Spanish conquistadors in detail.

She writes in a comfortable style about American western historical events from the point of view of the wild horse: the Battle of the Little Big Horn (or Greasy Grass), Wild Bill Hickock and the era of Wild West shows, cow ponies, the life and times of Steamboat – the famous bucking bronco finally ridden at Cheyenne Frontier Days in 1908, and even about Hollywood westerns and TV shows.

When she takes up the story of Wild Horse Annie, the woman who finally got some respect for wild horses built into law and government management, she finally arrives at the purpose of her book: to discuss current wild horse problems. Basically, there are more wild horses than forage allotted to feed them.

My only quibble with Stillman is she seems to use the terms "wild horse" and "mustang" interchangeably when a mustang should be defined as a horse of nearly pure Spanish descent and a wild horse is, well, one that is fending for itself. A wild horse isn't necessarily a mustang and a mustang isn't necessarily a wild horse.

It's easy to see by Stillman's other word choices that this book is not an objective discussion of wild horse history or wild horse politics, as both continue to be divisive subjects. She is a wild horse advocate. However, her extensive bibliography is valuable for further study.

This is a book to read if, like me, you've seen wild horses in Wyoming and you want to know more about this aspect of your adopted state, and the American West.

"Mustang: The Saga of the Wild Horse in the American West" by Deanne Stillman, c. 2008, published by Houghton Mifflin, 348 pages, hardcover, $25.

# 257a Hawaii: Hilo-side

Tuesday, November 18, 2008, ToDo, page C1

## Local family experiences "other side" of Hawaii's Big Island

## The sand is black instead of white. And you can't see the sunset...but the sunrise is amazing.

Hawaii: Visions of honeymooners on a white sand beach at sunset at a swanky resort, right? Yes, if you are recalling a travel poster for Waikiki Beach.

If you know my husband, Mark, and me, you know we had a completely different, but wonderful experience during our nine-day stay in late October on Hawaii, the Big Island.

First, we aren't honeymooners. Jeffrey, our younger son on fall college break, joined us in visiting our older son Bryan, who moved to Hilo for a job last January, hired by a University of Wyoming alum.

Bryan's girlfriend was able to join us often. She's a third-year marine biology major at the University of Hawaii-Hilo so we had a personal guide at beaches and tide pools who hailed from Greeley, Colorado.

Second, the beaches Hilo-side are black, made from eroding lava. The more popular white beaches Kona-side, the west side of the island, are tiny bits of bleached coral.

Kona gets the sunsets. Hilo gets magnificent sunrises, but you have to be up at 6 a.m. to see them.

Hilo isn't as popular or sunny as Kailua-Kona, and since we were visiting before the height of the season, which is late November through March, we found a great place for $65 per night.

Since no other guests were staying at the Na'ali'i Plantation Bed & Breakfast, we had our own complete household.

Every morning we had a hot breakfast with fresh, ripe papaya and apple bananas served on the lanai (porch) overlooking the forest. Birds twittered everywhere. On clear days we could see Hilo Bay, a 15-minute drive away.

There are still anthurium blooming under rows of tree ferns, since at one time the plantation was a commercial operation. Our hostess, Annie Maguire (cousin to Cheyenne resident Helen Hart), has filled her two-acre garden with local specialties such as papaya, guava, avocado, coffee, ginger and orchids.

We didn't spend time sunbathing on beaches as both Mark and I have had skin cancer scares. But we did pack fleece jackets and mittens for a trip up Mauna Kea, the White Mountain, which gets snow in winter.

Vacationers prefer Kona-side where popular resorts are in a sunny, 10-inch annual precipitation zone. Hilo is the wettest place in the U.S., rated at 140 inches, but it is in a drought right now. For us, from the 15-inch precipitation zone, the rain is a novelty we can enjoy since it comes warm, without lightning and wind, usually, or hail and snow. The locals mostly wear "rubbah slippahs" (flip-flops) and carry umbrellas.

We ate out almost every lunch and dinner at whatever establishment was recommended or handy or still open, including Ken's House of Pancakes, a 24-hour family restaurant beloved by locals. We tried Korean, Thai, natural food and roadside cafes, even a hotdog stand at the beach and the Friday night seafood buffet at the Hilo Hawaiian Hotel. It was at the Mongolian BBQ where we heard live Hawaiian music.

Hawaii is a mid-ocean crossroads with no ethnic majority. However, all street signs are in Hawaiian which has only 12 letters from the English alphabet: a, e, h, i, k, l, m, n, o, p, u, and w. It's hard to differentiate names. Our son lives at the corner of Haile and Halai.

We didn't spend much time in museums or shopping or on traditional beach activities except snorkeling. With advice from Bryan's friends and our guidebook, "Hawaii: The Big Island Revealed," we didn't need to sign up for tours.

Instead, we hiked to the top of 13,796-foot Mauna Kea and hiked 900 feet down into the Waipi'o Valley. One day we hiked through a lava tube and another day across a lava field. One day, the endangered Hawaiian goose, the nene, nearly tripped over us, and another day we nearly tripped over a napping green sea turtle, another endangered species.

There were myriad flowers blooming in gardens and along roadsides and so many sights unlike home. We filled our cameras and our memories. At the Hilo airport departure lounge, two hula dancers wished us farewell.

Mark Twain, after his 1866 visit, always hoped to return but never could. Isabella Bird, a visitor in 1873, wrote fondly of her magnificent adventures in "Six Months in the Sandwich Islands," which remains in print. More than 130 years later, the islands still are spellbinding, no matter what color of sand you step upon.

## Mauna Kea

Bryan's job as software engineer with the Joint Astronomy Centre allowed him to give us the inside tour of the James Clerk Maxwell Telescope and United Kingdom Infrared Telescope, two of the group dotting the top of the 13,796-foot mountain, just 43 miles from Hilo. The University of Hawaii offers tours of its telescope.

The visitor center at 9,000 feet elevation is worth stopping at for information as well as a chance to acclimate. Tour companies will take you from Kona-side up the steep gravel road for the sunset and provide warm coats. In the evenings, the visitor center has regular telescopes set up.

## Hawaiian Volcanoes National Park

The park entrance is only 28 miles from Hilo, but at an elevation of 4,000 feet. The quickest way to check the status of the eruption and lava flows is to go to links at www.wizardpub.com.

Bryan's friends took us on the Bird Park trail loop. It wasn't very birdy in October, so it became a botany tour instead.

The next day we drove out to the Hilini Pali ("pali" means cliff) Overlook and encountered no one else except the endangered nene. At the Thurston Lava Tube the parking lot was full so we pulled in at the Kilauea Iki Overlook and walked back along part of the Kilauea Iki trail. The crater views were great, and the tree ferns were magnificent.

Some roads in the park have been closed due to geologic cracks or lava flowing over them. Sometimes "vog," sulfurous fumes from the eruption that can be deadly if concentrated, will temporarily close a section of the park.

## Waipi'o Valley

Only 50 miles north of Hilo, steep-sided and flat bottomed, opening onto the ocean, the valley was well-populated by farmers until the 1946 tsunami wiped them out. A few inhabitants remain, but the only way out is by sea or a steep trail requiring four-wheel-drive.

Tour company vans will take you in, but it is only a strenuous mile hike each way, with the advantage of being able to stop and enjoy the scenery.... frequently.

The valley is sacred to Hawaiians and so a representative of the Waipi'o Circle Association will meet you at the top and give you a brochure on the history, proper etiquette and safety rules for enjoying your visit.

## Hawaii Tropical Botanical Garden

The northeast side of the Big Island is fringed with gulches growing impenetrable vegetation. Imagine one which has been landscaped and planted with amazing blooming tropical plants from all over the world, with paved paths leading from the visitor center down to Onomea Bay.

The Hawaiian Tropical Botanical Garden, 8 miles from Hilo, charges $15 per person but it is worth it for the two or more hours it will take to enjoy every vista.

Even if you aren't interested in reading every identification tag, the juxtaposition of color, texture, waterfalls and ocean views is blissful to the eyes.

Having seen tropical plants only in conservatories, I kept expecting glimpses of glass between distant branches, but the only glint was from the water.

## Hilo

For a former sugar capitol, population 40,000, Hilo is interesting in its own right. It's also conveniently close to beaches and a few tide pools you'll want to explore, besides having three excellent museums to visit on rainy days: the Pacific Tsunami Museum, 'Imiloa Astronomy Center and the Lyman Museum, which features Hawaiian history.

The extensive Farmers' Market is held Wednesdays and Saturdays under awnings and has beautiful flowers cheap, and interesting produce you won't see back home any time.

*Hawaii cutlines*

1. Hilo Bay - On the Big Island of Hawaii an old Hilo neighborhood, with tin-roofed bungalows, overlooks Hilo Bay which is protected by a breakwater. Gardens are filled with blooming tropical plants.

2. Hilo-side sunrise - Located on the east side of the Big Island, Hilo, Hawaii, has magnificent sunrises. This one is embellished with silhouettes of Alexandra palms.

3. Anthurium plantation - Anthuriums of many varieties were raised commercially in the shade of tree ferns on the island of Hawaii. This planting still exists at the Na'ali'i Plantation Bed and Breakfast near Hilo. Modern commercial growers use shade cloth.

4. Mauna Kea visitors - Near the top of 13,796-foot Mauna Kea, visitors gather near two of the group of world-renowned telescopes to watch the sun set over the island of Hawaii. Tour operators provide warm coats. The name of the old volcano means "White Mountain" because it is snow-covered during the winter months.

5. Mauna Kea – United Kingdom Infrared Telescope at sunset - Near the top of Mauna Kea, the United Kingdom Infrared Telescope (UKIRT) opens its dome at sunset. A dozen other telescopes are located atop the old volcano on the island of Hawaii because the view of the sky is less affected by air turbulence at 13,000 feet and vehicle access is usually good, though snow can make it a problem.

6. Hawaii Volcanoes National Park, Pu'u O'o steam - The island of Hawaii continues to grow, thanks to Kilauea Volcano, which began its most recent eruption in 1983. Its activity continued to put up a cloud of poisonous, sulfurous "vog" from the Pu'u O'o vent as of October 2008, seen from Hawaii Volcanoes National Park.

7. Hawaii Volcanoes National Park, 'ohi'a lehua blossom - 'O'hi'a is the dominant tree native to the Hawaiian uplands. At Hawaii Volcanoes National Park, they may be 100 feet tall. The blossoms have their own name, lehua, and are especially important sources of nectar and insects for forest birds found only on the Hawaiian Islands.

8. Hawaii Tropical Botanical Garden path - One of the Island of Hawaii's "gulches" is filled with tropical plants from around the world at Hawaii Tropical Botanical Garden. Family members Bryan, Jeffrey and Mark Gorges are dwarfed by leafy fronds.

9. Hawaii Tropical Botanical Garden orchids - Orchids are everywhere in the Hawaiian Islands, growing on tree trunks even. These are part of a large display at Hawaii Tropical Botanical Garden near Hilo.

10. Waipi'o Valley beach - The feathery branches of ironwood trees frame the beach at Waipi'o Valley, the Valley of Kings, on the windward side of the island of Hawaii. The steep walls of the valley are called pali, or cliffs. When it's wet, 1000-foot waterfalls may gush from them.

11. Waipi'o Valley from above - A mile-long, black sand beach fronts Waipi'o Valley on the windward side of the island of Hawaii. It's a steep hike down to discover the peacefulness Hawaiian kings returned to again and again.

## 258 Visiting Hawaii helps bird lover add birds and endangered species to life lists

Tuesday, December 9, 2008, Outdoors, page B5

Zebra doves calling and Japanese white-eyes twittering as they flit through the hibiscus – who would have thought a year ago I'd be waking up on the Big Island?

When our older son, Bryan, accepted a job in Hilo, Hawaii, nearly a year ago, Mark and I knew it would be an opportunity to travel.

With copies of "Hawaii's Birds" from the Hawaii Audubon Society and "Hawaii: The Big Island Revealed" guidebook, plus what Bryan had learned, we were able to add birds to our life lists, including endangered species.

The Hawaiian Islands are a mid-ocean crossroads. Volcanoes built them 70 million years ago. A small amount of flora and fauna was able to make it by water or wind and establish itself.

However, no terrestrial mammals, amphibians or reptiles made it, so surviving species evolved without the ability to evade those kinds of predators. Also, adaptive radiation occurred. For instance, from one species of honeycreeper, several species evolved, each particularly suited for one of the various habitats within the islands.

Far flying bird species still accidentally find Hawaii, but none seem to establish breeding populations now.

About 1,600 years ago, the Polynesians found the Hawaiian Islands and colonized them, bringing pigs, chickens, dogs and rats as well food crops like taro. It appears as many as 35 bird species and subspecies became extinct as a result.

Following Captain Cook's visit in 1778 was a flood of introduced alien species with unfortunate results for the natives. Even today, illegally released parrots may increase and spread, threatening commercial crops.

256        CHEYENNE BIRD BANTER

Still, the Audubon field guide lists 71 species and subspecies that are endemic, meaning they are found nowhere else in the world. Of those, 30 are on the U.S. Fish and Wildlife Endangered Species List.

At Hawaii Volcanoes National Park, efforts are being made to restore the landscape by fencing out wild pigs and controlling mongoose and feral cat populations. We saw one endangered species there, the nene, the Hawaiian goose.

Following a trail at the park in a deserted walk-in campground, we looked for a picnic table in the shade. No such thing in a lava field with scrubby vegetation. But we spotted the goose grazing on berries. We stopped and watched as its route brought it closer. Then we walked away.

After lunch the goose was alongside the return path, and we had to step around it. No wonder there were so many signs warning against feeding them and caution signs on the road picturing them. They are too trusting for their own good.

These days there's only one hawk species on Hawaii, the 'Io. As Mark and I took a break in our labors hiking up out of the Waipi'o Valley, we saw two soaring on the thermals. This was fitting since the valley has always been considered the Hawaiian kings' favorite place and the 'Io is a symbol of royalty in Hawaiian legends.

At Kaloko-Honokohau National Historical Park we plodded along the beach, stepped around a napping, endangered green sea turtle and arrived at a birding hotspot, 'Aimakapa Fishpond, to find several Hawaiian coots, also endangered. They look like ours except the hard white material of their beaks extends up over their foreheads.

We saw familiar birds: wild turkeys introduced from North America in 1815 and now at home in the grasslands on the slopes of Mauna Kea, house sparrows introduced via New Zealand (1871), cardinals from eastern North America (1929) and cattle egrets from Florida (1959).

We even saw a black-crowned night-heron, a bird that nests in Cheyenne parks. Apparently, this species made it to Hawaii on its own some time ago.

The 'Apapane, a small red bird which works ohi'a blossoms for insects and nectar is one endemic forest bird that is fairly common. I identified one only because the field guide had a photo of what it looks like when it is perched high overhead.

We needed to start on our world bird lists after finding an Erckel's francolin (Africa 1957), junglefowl (original Polynesian settlers), zebra doves (Asia), mynas (India) and yellow-billed cardinals and saffron finches (South America).

We weren't visiting at the right location or time of year for the marine birds. The visiting shorebirds were difficult to i.d., as usual. I'm sure of only one, the Pacific golden plover. It breeds in Alaska, with many wintering in Hawaii and others migrating south another 2,500 miles.

Some say the golden plover was the bird that intrigued the ancient Polynesians who lived on islands south of Hawaii. It made them wonder where the birds went when they left for the summer, if it was to land farther north. It caused them to load up their double-hulled canoes to find out.

How was the golden plover to know so many of us would follow?

# 2009

Tuesday, January 13, 2009, Outdoors, Commentary, page B5

## 259 It's possible that wolves in Yellowstone are having a positive effect on songbirds

Back in the early 1990s, the National Audubon Society lobbied for the reintroduction of wolves in the Yellowstone ecosystem.

Why, some folks wondered, would an organization with a name equated with bird conservation be interested in wolves?

An Audubon member myself, I agreed with the ecologists who were saying it was important to have all the pieces of an ecosystem, from top dog predator on down to burying beetle and I lobbied for wolves on ecological principles.

There were a couple people who thought wolves shouldn't be reintroduced because, based on a few anecdotes, wolves might already be present.

If there were wolves in Yellowstone immediately before reintroduction, and not just casual stragglers or hybrids and captives dumped by people, they were nearly invisible, awfully quiet, well behaved and unproductive.

Today, commercial enterprises will take you on a wolf tour (www.wolftracker.com). Reintroduced wolves multiplied so quickly, they also became a noticeable nuisance to livestock operators.

Wolves are apparently having an effect on the Yellowstone ecosystem that their predator stand-ins, the coyotes, were not able to achieve between early 20th century wolf eradication programs and wolf reintroduction in 1995. The willows are increasing, which means increasing numbers of songbirds.

Doug Smith, leader of the Yellowstone Wolf Project for Yellowstone National Park, in a reply to my email query, was quick to point out that studies are still ongoing and that some people believe there is more than the wolf at work in the growth of willow shrubs. Papers are in the process of being written and Smith said, "As far as wolf impacts on songbirds, we are on our way to establishing the link that goes through willow and elk."

Willow grows in riparian zones, the areas along creeks and rivers. A healthy riparian zone, with lots of vegetation, absorbs rainfall and snowmelt like a sponge and releases it slowly into creeks. In an unhealthy situation, with little vegetation present, water runs off quickly, eroding the surface, depositing sediment in the stream where it suffocates fish eggs and the invertebrates that feed fish.

In a healthy riparian zone, vegetation slows the runoff water. Slow water can't carry as much sediment and organic matter, and so it drops it on the plants adjacent to the stream, rather than in the stream. Riparian plants, such as willow, thrive on and grow through the sediment deposits, eventually providing more and more vegetation.

If the lack of vegetation in a riparian zone is from overgrazing by wildlife or livestock, managers can reverse the trend by either fencing the animals out for a period of time or reducing the number of animals grazing and/or the time and amount they graze.

National park managers have restrictions that prevent them from removing elk which have kept the willows trimmed too well while the wolves were out of the picture.

However, the willows seem to be recovering and expanding. One theory is that climate change is providing a longer growing season. Another theory is that the Yellowstone fires of 1988 provided a huge increase in forbs (non-woody plants including wildflowers) which elk like better than shrubs, and they grazed the willows less.

A third theory is that elk no longer get to graze willows at their leisure since wolves are constantly nipping at their heels and running them off.

Range management scientists have spent years conducting studies of the effects of various grazing schemes and could probably make some predictions, but every ecosystem has its quirks.

Whether the wolves are totally or partly responsible for the regeneration of Yellowstone willows, we can reasonably predict healthier riparian zones.

From my birdwatcher's perspective, this means more and greater diversity of songbirds which are attracted to the insects associated with the willows and the shelter provided. Smith listed willow flycatcher, yellow warbler, common yellowthroat,

Lincoln's sparrow and song sparrow in particular.

Improved riparian habitat means improved fisheries. It also means ephemeral and intermittent streams will flow a little longer each year.

The increase in vegetation can support more critters (even livestock outside the park). Wyoming's riparian zones are important to something like 70 percent of wildlife species.

Bureaucracy and politics will continue to plague the Yellowstone wolves, but if studies show wolves have helped repair an important part of their ecosystem, reintroduction has been worthwhile.

## 260 Wednesday, January 28, 2009, ToDo, page D4
# Hawaii travel guidebook won't disappoint

"Hawaii The Big Island Revealed" by Andrew Doughty, 5th edition, 2008, Wizard Publications, Inc., 307 pages, $16.95.

I found the third edition of this guide at the Laramie County Public Library. My husband, Mark, and I found it so entertaining and packed with so much information that we ordered the latest edition to plan our trip to visit our son in Hawaii.

There's also a lot of free information and updates available at the publisher's website, www.wizardpub.com. If you aren't going to the island of Hawaii, Wizard Publications also has guides for the islands of Oahu, Maui and Kauai.

I was a little worried that our vacation experience wouldn't match up to the beautiful photography in the guidebook, but it did.

I also was worried that "ground-truthing" the guidebook would show it full of over-simplifications and generous ratings, but it wasn't. In one case, it had more accurately drawn hiking trails than a government publication.

The author (and whomever he refers to as "we") has investigated every trail, resort, restaurant and attraction anonymously and takes no advertisements or freebies.

He is enthusiastic about good experiences and good return for money spent. He points the way to less crowded, as well as spectacular beaches.

This guidebook is well-regarded by locals I talked to, and we used it extensively to plan day trips from our base in Hilo in October.

The top five destinations we visited on Hawaii, in no particular order, were:

1. Mauna Kea – home to world famous telescopes.

2. Hawaii Volcanoes National Park – see flowing lava if you're lucky.

3. Waipi'o Valley –a remote valley beloved by Hawaiian kings.

4. Hawaiian Tropical Botanical Garden – was spectacular even though the height of flowering was in June.

5. Kaloko-Honokohau National Historical Park – features ancient Hawaiian structures.

My top five things to do on the Big Island next time are:

1. Find flowing lava at Hawaii Volcanoes National Park.

2. See Hawaii Tropical Botanical Garden in June.

3. Go whale-watching during the December – March season.

4. Take a submarine tour of the reef in Kailua Bay.

5. Check out the Lyman Museum in Hilo.

6. Check out more beaches—there are no entrance fees!

7. Hike into more birding hotspots in spring.

8. Hike into Waipi'o Valley again and continue into the more remote Waimanu Valley.

9. Oops, that's more than five, isn't it? What about a tour of a Kona coffee farm? Or the macadamia nut factory? Or the famous Parker Ranch? Or taking out a kayak?

Luckily, "Hawaii: The Big Island Revealed" has advice on finding the cheapest plane fare for our next visit.

## 261 Friday, March 20, 2009, ToDo, page D4
# On the Wyoming range the mountain bluebird is a harbinger of spring

The colorful birds move back to their perches and fence posts in late February and early March

A true Wyoming sign of spring is the migration of mountain bluebirds. Forget robins – there aren't too many to be found out on the open range.

By late February or early March, we see mountain bluebirds take up their perches again on fence posts and other high spots out in the country – the open country – even though winter weather isn't quite over.

The flash of that incredible shade of blue, like a piece of sky, as a male mountain bluebird flies out from his perch is enough to endear the species to you forever.

The female is gray with just a trace of blue. If you look hard, you'll see she isn't very far away from the male.

Ironically, mountain bluebirds prosper amidst farming and grazing or where forests have been cut as long as there are fruits and seeds to forage for in the winter and insects in the summer. They've also been aided by people willing to put up nest boxes for them.

If you have rural property, consider starting your own bluebird trail, a series of nest boxes. But not any old bird house will do. The openings must be 1 and 9/16 inches in diameter to keep out house sparrows.

The boxes must be deep enough to keep marauding raccoons (yes, we have

2009

259

raccoons around Cheyenne), cats and birds from reaching inside and grabbing or pecking the nestlings. Raw wood is better than wood finishes. And it is necessary to install an easy way to open the box for cleaning.

No need to reinvent the wheel. Lots of advice is available through the North American Bluebird Society. While the two red-breasted species, the eastern bluebird and the western bluebird, are more recognizable, the society also addresses our solid blue mountain species.

The website, www.nabluebirdsociety.org, has lots of information on monitoring nest boxes, keeping them safe from predators, and enticing bluebirds to use them. You can also call 1-812-988-1876 between noon and 3 p.m. EST, Monday - Friday.

The website has plans for several kinds of nest boxes. The Peterson Bluebird House is the style you see on the fence posts out at Curt Gowdy State Park, 30 miles west of Cheyenne. The park is an excellent place to catch a glimpse of mountain bluebirds. Watch for flocks of sparrow-sized birds that flash blue as they turn in the sun.

# 262 Friday, March 27, 2009, ToDo, page D4
# Book review: "Prairie Spring: A Journey into the Heart of a Season"

## East Coast naturalist records a memorable spring on our prairie

"Prairie Spring: A Journey into the Heart of a Season" by Pete Dunne, published by Houghton Mifflin, hardcover, 288 pages, $24. Publication date: March 2009

It's the rare nature book that relates to us out here on the prairie. Even rarer is the nationally recognized author who leaves the east coast to write it.

It is always interesting to see familiar places through the eyes of a newcomer, especially one who has both an easy-to-read style and who has done his research, not only on prairie places but on spring itself. Author Pete Dunne even took time to interview locals while his wife Linda photographed their adventures.

In his first of a projected four-season, four-volume series, he has a way of personalizing history and ecology in storyteller mode that keeps you reading, even though you know the outcome of this plot.

The plot isn't only about the advance of spring, but how European farming traditions, weather cycles, economic recessions and other human actions changed or will change the grasslands.

As director of New Jersey's Cape May Bird Observatory, a famous spring migration Mecca itself, Dunne is able to rhapsodize about our spring, too.

Able to look for the mysteries and miracles of spring anywhere, why would Dunne choose the grasslands? Perhaps it is to bring attention to the struggling grassland ecosystem.

Dunne chose particular locations to examine at particular times during the spring of 2007. You remember it as the spring all those white evening primroses carpeted the pastures along I-25 between here and Fort Collins.

He visited Pawnee National Grasslands in Colorado, just 25 miles southeast of Cheyenne, several times during the season, beginning on Ground Hog Day, the real beginning of spring. He also visited the sandhill crane migration in Kearney, Nebraska, Comanche National Grasslands in southeast Colorado, Milnesand Preserve in northern New Mexico and Custer State Park in South Dakota.

Hmm, he never mentioned Wyoming. Spring on our prairie is just as remarkable as the places spotlighted in this book, but it will be our gladly-kept big secret. O.K.?

# 263 Sunday, April 5, 2009, Outdoors, page E2
# Wyoming has 48 places that are important to birds

Spring means a birder's thoughts turn to migration and those hotspots where birds will be thickest.

Some spring hotspots are on a national list of Important Bird Areas. Two of those IBAs are right here in Cheyenne. No entrance fees required.

BirdLife International has identified places important to birds on every continent, in 100 countries and territories. Places like Fiji, Romania and Peru. They work with local agencies to help implement conservation and education plans.

In 1995, the National Audubon Society became the sponsoring organization for the IBA program in the U.S. While some places are rated as globally important, such as Yellowstone National Park, others are recognized as important at the national and state level.

Audubon Wyoming has recognized 48 places as important to birds in our state so far. Coordinator Alison Lyon-Holloran is still taking nominations. Contact her in Laramie at 307-745-4848 or at aholloran@audubon.org. Check www.AudubonWyoming.org to see which sites are already in the program.

What makes a place important to birds? It is important to birds during migration, and/or breeding and/or wintering seasons for one or more species, meaning birds can find food, shelter and water and whatever else they require during a particular season. Alison also requires approval from the landowner before reviewing the nomination.

Nominating Lions Park was a no-brainer for my local chapter, Cheyenne – High Plains Audubon Society (CHPAS). We have people traveling to Cheyenne from a 200-mile radius because the park's trees, shrubs and lake attract so many species during spring migration. At the mid-May peak one year, I counted 60 species in two hours.

The nomination stalled at first when the ornithologists on the technical committee countered that we only saw a lot of birds at the park because a lot of people birded there. Yes, but we could probably find the same diversity and abundance of songbirds, if not the waterfowl, in all of the old neighborhoods. We couldn't very well walk through everyone's yard.

CHPAS continues to monitor the park's bird life through seasonal surveys and evaluates the impacts of new park developments.

The Wyoming Hereford Ranch has had a long and friendly relationship with the local birding community. Anna Marie and Sloan Hales welcome inspection by binocular, as long as no one disturbs the livestock, hops the fences, or intrudes on the residents of the ranch.

Again, I'd venture to say that other properties along cottonwood-filled creeks in southeastern Wyoming might have similar abundance and diversity. The difference is the Hales.

Not only have they welcomed birders, but they were thrilled to be part of the nomination process. They've worked with the Laramie County Conservation District to improve wildlife habitat and in cooperation with Audubon Wyoming to install signs this spring educating visitors about why their ranch is an IBA. The Hales have also created a little nature trail.

The ranch, as an oasis of wildness on the edge of Cheyenne, will only become more and more important a refuge as high-density housing and commercial enterprises continue to move into their neighborhood. Who knew over 100 years ago, when the ranch was established, people would want to build houses in cow pastures 10 miles from the State Capitol building?

IBA designation doesn't bind any landowner to any course of action. But it does make people aware that their actions will have an impact on birds. It makes us stop and think about beings besides ourselves and we get back to the original question: Does a bird have any value if you aren't a birdwatcher?

Sometimes it has an obvious usefulness, such as keeping pests under control. If nothing else, birds are a part of nature, and contact with nature is being scientifically proven to improve our mental health.

With the onset of spring, many of us are looking for an excuse to get outside. Here in Cheyenne we don't have to travel to Important Bird Areas, even our local ones, to see special birds. We just need to keep our eyes and ears open in our own backyards.

# 264 Figure out those chirps with Birdsong CD, book

Tuesday, May 26, 2009, ToDo, page D4

"Birdsong by the Seasons, A Year of Listening to Birds" by Donald Kroodsma, c. 2009, Houghton Mifflin Harcourt, 366 pages, 2 CDs, hardcover, $28.00.

Four years ago, Donald Kroodsma wrote the book that documented his life's work and won him wide acclaim, "The Singing Life of Birds."

It was a big book, describing his passion for recording and studying birdsong plus how to record and read sonograms yourself and what it all means.

So what can Kroodsma do for an encore? Tell birdsong stories by the season. Although there's still an index, two appendices, notes, a bibliography, and two CDs this time, this writing is more like a series of 24 short stories.

For instance, at the beginning of January, Kroodsma, like a detective, goes under cover with recording devices in the center of a winter roost of hundreds of robins in western Massachusetts. He listens to every "piik" and "tut," weaving together meanings, drawing conclusions, trying to stay awake and warm on his stakeout.

If a picture is worth a thousand words, then a recording is worth at least as many. After you read the robin thriller, immerse yourself in the robin tracks, imagining a dark night, hearing mysterious footsteps, hoot of a predatory great horned owl, rustle of wings of departing robins at dawn and follow along, if you want to, with the recording notes.

Or listen to the CDs first, checking on the subheadings in Appendix 1 when you can't figure out what you are hearing. I've never heard alligators growl, have you?

Kroodsma doesn't always wait in one place for the seasons to pass. He often runs out to meet them, parabolic microphone in hand: the Everglades, the Platte River, Corkscrew Swamp, Nicaragua, Costa Rica, Virginia, Pawnee National Grasslands (just over the state line southeast of Cheyenne), and various locations back home in Massachusetts.

This book is definitely not your typical linear reading experience. You, your kids – and your pets – will find many ways to enjoy it.

2009

261

# 265   Why can't we encourage the country to "glow locally"?

Sunday, May 31, 2009, Outdoors, page E3

As the daughter of an engineer who installed turbines in power plants, I admire the sleek industrial-sized blades of a wind farm.

But I am also your local "bird lady." I have the usual concerns about birds flying into those blades. At least current designs are less lethal than when the Altamont Pass array was first installed in California.

I am also concerned about visual pollution. Cheyenne will soon be surrounded by wind farms, like giant picket fences, many producing energy for people outside Wyoming. Also, what about those industrial-sized solar arrays spreading over the "wastelands" of the West? I don't think there are any natural landscapes we can afford to waste by covering them with solar panels.

Why do the power experts continue to use the point source model, useful for coal, nuclear and hydro plants, for sustainable power sources that occur, to some degree, everywhere? Shouldn't the Department of Homeland Security be encouraging dispersed power production?

I think it is time to plan how we can "glow (as in turn on the lights) locally." Locally grown energy has advantages similar to locally grown food.

Why put solar arrays over our deserts (which are actually full of life) when we already have asphalt deserts that are nearly barren? Instead, imagine parking in the shade of a solar array while shopping at WalMart. Imagine small windmills strung along existing utility corridors; their blades whipping so fast, birds see and avoid the blurry disks.

Point source energy is sent long distances but the energy lost to transmission isn't as much as I thought. Sadrul Ula, University of Wyoming professor of electrical and computer engineering, specializing in power, told me perhaps 10 percent would be lost over 800 miles, depending on the size of the lines.

He pointed out energy storage, guaranteeing 24-hour availability of electricity from intermittent power sources like wind and solar, is still problematic. But I don't think that should keep

us from using them wherever we can, tying them into the grid for now.

In nature, prosperous animal and plant species use diversity and redundancy strategies.

The sage grouse in Wyoming is in trouble because it requires a certain type of sagebrush habitat that is disappearing. Conversely, crows are becoming more abundant here because they are smart enough to find new foods in new places, such as French fries in fast food parking lots. Developing diverse energy sources is our human equivalent.

The most prolific and widespread plants, such as dandelions, produce multiple, redundant seeds so some can be eaten by goldfinches and others can blow about and pioneer new land.

The energy equivalent is many small wind turbines and solar arrays. Siting small devices would be less problematic. If solar and wind collectors are spread everywhere, the sun will always be shining and the wind blowing on them somewhere.

I have complete

confidence engineers will put their minds to improving sustainable energy production. They are already working on the "smart," more efficient, power grid.

We don't have to wait for a shortage of fossil fuels to drive innovation. We consumers can drive innovation ourselves. Witness the increased demand, production and improvement of fuel-efficient cars in the last five years.

I think we can look forward to solar technology that ekes out electricity from gloom (haven't you gotten sunburned on a cloudy day?) and wind turbines that respond to the slightest breeze. Then all parts of the country can produce sustainable energy.

Bird lady that I am, I would like to see sustainable energy production incorporated into the already built environment. I'd rather not see Wyoming continue its role as energy colony, giving up resources, again, to the detriment of its valuable, one-of-a-kind, wild landscapes.

# 266   Q & A with author Rachel Dickinson

Wednesday, June 24, 2009, ToDo, page D4

## Book follows life of man who hunts with falcons

"Falconer on the Edge, A Man, His Birds, and the Vanishing Landscape of the American West" by Rachel Dickinson, 2009, Houghton Mifflin Harcourt, 220 pages, hardcover, $24.

Rachel Dickinson

examines the life of hardcore falconer Steve Chindgren in her new book, "Falconer on the Edge." For several months of each fall, Chindgren lives in a cabin in southwest Wyoming, near Eden, to hunt sage grouse with his

falcons every day. Dickinson researched the book by making a number of trips to Eden. She recently responded to questions by email.

**Question:** When you decided to learn about falconry from somebody besides your

husband, Tim Gallagher (author of "The Grail Bird" and "Falcon Fever"), were you thinking as a freelance writer or were you more concerned about getting a handle on your husband's obsession?

**Dickinson:** I really

262          CHEYENNE BIRD BANTER

wanted to understand what was going on and did think it would make a great book, but I guess rather than stress the relationship by learning about the sport from Tim, I decided to go further afield and find the most hardcore of the hardcore falconers. And all roads led toward Steve Chindgren.

**Q:** How would you handle it if Tim, like Steve, decided he was going to fly birds in the middle of nowhere for several months every year?

**Dickinson:** As Steve's wife, Julie, told me, she knew that this was the way Steve was when she married him. So whenever there was a fork in the road or a choice to be made, falconry always won. If Tim decided that this was what he really wanted to do with his life, I would say, "See you later and have a nice life." I think Tim's obsession is a bit more under control.

**Q:** Did you develop a taste for sage grouse? No catch and release for falconry, huh?

**Dickinson:** We always ate what the falcons caught. Sage grouse has a strong sage taste but if you take the breast meat and marinate it and then throw on the grill for a few minutes – not too long because the meat is so lean – it's darn yummy. What we didn't eat, the falcons ate.

**Q:** Energy development in Wyoming is a subplot in your book, the menace lurking in the background, taking sage grouse before Steve's birds can. How did that come about?

**Dickinson:** I spent a couple of years going to Wyoming for a week or so at a time and I saw enormous change just over those two years. I knew early on that the changing landscape due to energy exploration would be a strong sub-plot in the book – it had to be because it affected everything that Steve loved.

**Q:** The local Audubon chapter showed Steve's sage grouse film which supports sage grouse conservation. Does his obsession with falcons carry over to sage grouse?

**Dickinson:** Steve knows more about the natural history of the sage grouse than probably most wildlife biologists working out in the field. When he's not flying his birds, he's driving around checking on the grouse – looking for leks, looking for wintering grounds, looking for evidence of bird strikes on fences – because if you don't really understand the prey species, you can't really understand how to be a falconer. He's as hardcore about the sage grouse as he is about his falconry because it's a part of his falconry experience.

**Q:** What does Steve think he'll do if they are listed as threatened or endangered?

**Dickinson:** If the sage grouse gets listed, Steve says he's going to fly his birds on jack rabbits – that will require a real paradigm shift for him since he's hunted sage grouse for so long. He loves his spot in Wyoming and is determined to keep it as a falconry lodge, so he's got to do something. I know he's just hoping and praying it doesn't come to that.

**Q:** Did you have trouble adapting to the open spaces around Eden?

**Dickinson:** I come from the northeast where everything is pinched in with hills and gorges and lakes and streams, so it took me a little while to get used to all that space. Once that happened, there was no looking back. What a fabulous place. I miss it and hope to get back there, maybe this fall for the annual grouse dinner at Steve's cabin.

# 267 I-80's Fort Steele isn't just the rest area

Sunday, June 28, 2009, ToDo, page D4

For frequent I-80 travelers, "Fort Steele" is the name of the rest area between Walcott and Rawlins. Be sure to stop and use the facilities first, since the real fort is mostly building foundations or skeletons.

Since there are no picnicking or camping accommodations and no large boat access for the North Platte River, you and the wildlife will likely be the only visitors.

You must walk from the parking area down the sidewalk to the river and under the railroad bridge to get to the site.

Fort Steele was established in 1868 to protect a section of the Transcontinental Railroad which was finished in 1869. It was named for Major General Frederick Steele, a Civil War hero, shortly after his death.

The troops also helped out with civilian law enforcement at nearby mining camps.

Abandoned by the military in 1886, Fort Steele reached its zenith when the Lincoln Highway passed through, beginning in the 1920s. But then the highway moved in 1939 and the site was abandoned again.

Don't forget to walk into the Bridge Tender's House for more information about the area's historic economy.

### Fort Fred Steele State Historic Site

**Directions:** I-80 Exit 228, north, then north on first road east of the exit.

**Open:** May 1 – Nov. 15, every day, 9 a.m. – 7 p.m.
**Admission:** Free.

**Phone:** 307-320-3013.
**Website:** http://Wyoparks.state.wy.us.

**Attractions:** Self-guided tours, river habitat.
**Time:** Allow at least 1 hour.

2009

263

# 268 Alaska's Kenai Peninsula

Sunday, July 5, 2009, ToDo, page D1

## Wildlife watchers investigate Anchorage, Seward and Homer, Alaska

It is impossible to escape wildlife in Alaska. It is everywhere: bridge railings formed as a series of jumping salmon, business names, likenesses on every imaginable item and items made from fur, feathers, bones and tusks.

The good news is that you can also see the animals in person.

In mid-June, my husband, Mark, and I met his brother, Peter, in Anchorage for a taste of the Kenai Peninsula during the 50th anniversary of Alaska's statehood.

If you imagine the outline of the southern border of the state, locate the Kenai Peninsula at the apex (Anchorage just above it) between the Aleutian Islands trailing to the southwest and the southeast coast bordering Canada.

Besides trying to eat as much fresh seafood as possible, our unstated goal was to see as much wildlife as possible. Here are some of our highlights.

### Anchorage

We started with a couple days in Anchorage, finding a moose grazing next to a scenic overlook and waterfowl nesting half a foot from a busy bike and pedestrian path, both within city limits.

The Tony Knowles Coastal Trail, 13 miles long, borders the city on the northwest along Knik Arm. You can rent a bike and enjoy the long, colorful evenings.

Sunset was 11:30 p.m. and sunrise at 4:30 a.m. while we were there, though Anchorage considers itself to have 24 hours of "functional" daylight at the summer solstice.

Chugach State Park covers the mountains at the eastern city limits. We chose the Flattop Mountain trail for a close look at the tundra. You need an early start because by noon on a summer weekday, the parking lot is full of both residents and tourists.

The Alaska Native Heritage Center, admission $25 per adult (discounts available, small children free), is over-priced unless you spend the whole day investigating the work of onsite Native craftspeople, walking to replicas of Native life for five distinct regions and catching all the performances of Native singing, dancing and game playing.

The Kenai Peninsula is a small percentage of Alaska, but don't let that deceive you when figuring driving distances. To access it from Anchorage, one must drive east about an hour on a twisty two-lane highway along the shore of Turnagain Arm. Try to avoid driving when the Anchorites are using it for a weekend escape route.

It's a total of 127 miles from Anchorage to Seward. Be sure to schedule enough time for all the scenic turnouts and especially the Begich-Boggs Visitor Center at Portage Glacier. It was

well-worth the U.S. Forest Service's minimal admission fee to get an entertaining, hands-on education on local natural history.

### Seward

In Seward we took a 6-hour boat trip with Kenai Fjords Tours to see a bit of Kenai Fjords National Park. Our captain took us through the brash ice to the foot of Holgate Glacier. On the way she pointed out three kinds of whales and other marine life. Mid-June is the season for seeing newborn sea lion pups and nesting kittiwakes and puffins on rocky islands.

Back on shore, we saw the sea animals again (minus whales) up close at the Alaska SeaLife Center. It has a camera working the sea lion rookery and images can be viewed live on the internet or on a local Seward TV station.

We took a hike up to look at Exit Glacier, part of the national park, and marveled at how far it has retreated since 1815, and even in the last 10 years. Because the Harding Ice Field, of which it is a part, is in the way, it is 174 circuitous miles to drive to Homer.

### Homer

Don't miss the Alaska Islands and Ocean Visitor Center operated by the Alaska Maritime National Wildlife Refuge and Kachemak Bay Research Reserve, not just because admission is free.

There are lots of jokes

about Homer Spit, 5 miles of sand sticking out into Kachemak Bay, but that's where you'll find the water taxi, tour companies, seafood restaurants, Land's End Resort, gift shops, fishing charters, and kayak rentals. If you are worried about tsunamis, don't camp here.

We stayed at a charming bed and breakfast, "A Rosy Overlook," halfway up the bluff above town, with a wonderful garden and view of the bay. The hosts, Rosie and Erless Burgess, long-time Alaska residents, entertained us with local stories in the evenings.

We took a day-long boat trip across the bay to Seldovia, which is not accessible by highway. It was settled by the Russians in 1870 as a fishing village. The captain, again a woman, gave us historical background and great looks at nesting gull colonies.

It was 225 miles back to Anchorage, but again, it required a full day at 45 to 55, seldom 65, miles per hour and several stops, especially to see the line of at least 100 fishermen standing in the Russian River, perfectly spaced two rod lengths apart, trying to snag migrating sockeye salmon.

We saw lots of wildlife, ate lots of fresh salmon and halibut, and need to schedule another trip to see everything we missed.

## Planning your own Alaskan adventure

Remember, Alaska is two hours behind Cheyenne. When it is 11 a.m. here, it is 9 a.m. there.

**General**
--Alaska Geographic (nonprofit, trip planning help), www.alaskageographic.org, 1-866-AK-PARKS.
--Bed and Breakfast Association of Alaska, www.alaskabba.com.
-- Kenai Peninsula Tourism Marketing Council, www.kenaipeninsula.org, 1-800-535-3624.

**Anchorage**
--Anchorage Convention and Visitors Bureau, www.goanchorage.net, 907-276-4118.
--Tony Knowles Coastal Trail, http://dnr.alaska.gov/parks/aktrails/ats/anc/knowlsct.htm, 907-786-8500.
--Alaska Native Heritage Center, www.alaskanative.net, 1-800-315-6608.
--Chugach State Park, www.dnr.state.ak.us/parks, 907-269-8400.
Portage Glacier
--Begich-Boggs Visitor Center, Chugach National Forest, http://www.fs.fed.us, 907-743-9500.

**Seward**
--Kenai Fjords National Park, http://www.nps.gov/kefj, 907-224-7500.
--Kenai Fjords Tours, www.kenaifjords.com, 907-276-6249.
--Alaska SeaLife Center, www.alaskasealife.org, 888-378-2525.

**Homer**
--Rainbow Tours, www.rainbowtours.net, 907-235-7272.
--Alaska Islands and Ocean Visitor Center, free, http://IslandsandOcean.org, 907-235-6961.

# 269 Birders: If you want to see variety, visit Alaska

Sunday, July 5, 2009, Outdoors, page E3

I have a habit of unexpectedly running into old friends in my travels, miles and/or years from where we first met. A recent trip to south-central Alaska was no exception, though it was birds rather than people.

Also, it is funny how the local birdwatchers where I travel will speak reverently of a bird I think is common, but shrug at a bird I've always wanted to add to my life list. A species' "high value" apparently depends on a bird's lack of abundance in a particular area.

I didn't check, but probably the bird hotline in Anchorage doesn't get excited about multiple Arctic terns flitting over the city's Westchester Lagoons, like over-sized white barn swallows. But they wintered in Antarctica and made a 12,000-mile journey back. Anchorites do brag about that on interpretive signs.

My husband, Mark, and I, and his brother, Peter, who lives in Sitka, Alaska, walked up to a group of obviously dedicated birders with an array of scopes pointing to a lagoon's island where gulls were nesting (this was mid-June). Turns out they were on a multi-week Alaskan bird tour. One gentleman was excited about the mew gulls he had just learned to identify and carefully pointed out their chicks to me.

Mallards were in short supply at the lagoons and not looking for handouts. Two pairs were nesting six feet from the parking lot and completely immovable.

We saw the same reaction from the pair of red-necked grebes nesting two feet off the bike path, except in this case, it gave us a very good look at a bird that rarely visits Wyoming.

Up in the Glen Alps in Chugach State Park, on the outskirts of Anchorage, Peter pointed out the one-note song of the varied thrush, a bird finding itself in Wyoming only accidentally. Another loud singer turned out to be an orange-crowned warbler, a regular, but quiet, spring visitor back home.

On another forest hike, I recognized the sweet double notes of a hermit thrush, a species I'd recently heard on a recording. I stopped to watch it sing while at the same time wondering where the bears were.

At the bed and breakfast in Homer, decorated with a birdhouse theme, the hosts had a bird feeder out and I was finally able to put a name to the large sparrow I'd been catching glimpses of, the fox sparrow. The ranges of its four subspecies seem to bypass Cheyenne.

There was also a pair of familiar red-breasted nuthatches nesting in the yard (ours go to the mountains in the summer) and black-capped chickadees. I was hoping to find boreal and chestnut-backed chickadees on this trip, but they are hard to spot in all the foliage.

Down at the waterfront in Seward, the line of ducks bobbing in the waves and diving in synchrony were harlequin ducks. To see them in Wyoming, you have to go up by Yellowstone National Park.

The highlight of the trip for me was the seabirds. Last time we were out on Alaskan waters, I was too sick to appreciate anything. This time we took two trips, and the sea was nearly perfect, or maybe it was the ginger tablets, or staying outside, which worked.

It was nesting season and our boat captains took us right up to cliffs covered with nests of black-legged kittiwakes (Seward to Kenai Fjords National Park) and glaucous-winged gulls and common murres (Homer to Seldovia).

We even saw puffins, both tufted and horned. To draw them as cartoon characters is to draw them accurately. We watched one at the Alaska SeaLife Center in Seward and it flies underwater as well as it does the air.

One tour boat captain's assessment was that finding bald eagles on forested hillsides was like looking for golf balls, and we saw them everywhere, but they still seemed majestic to me.

If you love birds and are visiting the Kenai Peninsula, be sure to visit the free Islands and Ocean Visitor Center at Homer, operated by the

Alaska Maritime National Wildlife Refuge and Kachemak Bay Research Reserve.

They do a wonderful job of recounting the history of Alaskan seabird conservation (including Wyomingite Olaus Murie's seabird surveys in the 1930s) and the research being done in Kachemak Bay and along the Aleutian Islands. It might take a few seasickness-prevention patches, but I think the islands are where I'd like to go next.

Meanwhile, out behind the center was a lone sandhill crane along the Beluga Slough trail and whimbrels on the beach.

Did I mention the black-billed magpies we saw everywhere? What a surprise. But it's always nice to see familiar faces far from home.

# 270

Monday, July 6, 2009, ToDo, page D4

# Quilt Wyoming 2009 draws many show entries and vendors from three states

Have an appreciation for quilts? View the Quilt Wyoming 2009 quilt show for free Friday and Saturday [July 10 and 11] at Laramie County Community College, in the Centennial Room of the Center for Conferences and Institutes on the southwest side of campus.

You can even meet the quilters at a special reception Saturday from 4:30-5:30 p.m.

Have an addiction to quilting, quilting notions, patterns and fabrics? Sixteen vendors from around Wyoming, Colorado and South Dakota will be selling their wares at the same time and location as the quilt show. They will also be open Thursday, July 9, from 6:30-8 p.m.

This year's opportunity quilt was designed by Molly Wilhelm of Wheatland and features architectural details of the Cheyenne Depot, supporting the theme of the event, "Quilting at the Crossroads."

Entries for the annual quilt challenge with the same theme will be exhibited and judged in three categories: "High Road (professional)," "Low Road (everyone else)" and "Back Road ('the dream was there, the sewing machine was threaded, but life got in the way' of finishing the entry)." Challenge kits will be available for next year when the convention will be held in Rock Springs.

Quilt Wyoming is the annual conference of the Wyoming State Quilt Guild, formed 12 years ago. This year, the Southeast Region is hosting more than 200 registered participants who are staying in the dorms, eating in the cafeteria and taking classes from 13 teachers from Wyoming, Colorado and Nebraska in a variety of techniques and patterns.

The featured national teachers are Rachel D.K. Clark, from California, known for her embellished quilted clothing, and Jan Krentz, also from California, famous for her Lone Star and Hunter Star quilts and patterns.

Late registrations are being accepted for classes that still have openings. Check the website, www.wsqg.org, for more information or call Vicky (Laramie), 307-742-9153, or Nancy (Cheyenne) 630-1743.

# 271

Tuesday, July 7, 2009, ToDo, page D4

# Expedition Island commemorates historic Green River adventures

Starting out as a stage station on the Overland Trail, the town of Green River became a division point on the Transcontinental Railroad and the jump off for many expeditions down the Green River.

William Ashley descended the Green in 1825, during the fur trapping era, but in 1869, just after the railroad was completed, John Wesley Powell wanted his trip to be a scientific survey. After him came attempts at navigating the river in a paddlewheeler, speed records, women steering their own craft and commercial float trips before the river was dammed at Flaming Gorge.

The historical interpretive signs along the path around Expedition Island are worth reading, but after a long, probably hot drive, you and your family will appreciate the free splash park more. Afterwards, you can retire to the shade with something cold from the concession stand.

At the entrance to the footbridge on the other side of the parking lot, check out the map of local pathways that connect several local parks and natural areas.

## Expedition Island Park

**Directions:** I-80 Exit 91, south on Uinta Dr. (State Hwy. 530), right on 2nd St.
**Open:** Year round, 7 a.m.-10 p.m. Splash park open 10 a.m. 8 p.m. Memorial through Labor Day.
**Admission:** Free.
**Address:** 475 S. Second East
**Phone:** 307-872-6151
**Website:** http://www.cityofgreenriver.org
**Attractions:** splash park, changing rooms, playground, concessions, picnicking, shade, historical interpretive signs around the edge of the island. White-water kayaking venue expected to reopen summer of 2010.
**Time:** Allow at least 1 hour.

# 272 Another I-80 attraction

Thursday, August 6, 2009, ToDo, page D4

## See wild horses on loop drive or in Rock Springs

Whatever your beliefs are about wild horses, prized native species or feral cow ponies, you should also drive this loop tour for the wildflowers and vistas. Early morning and late afternoon are the best times to find horses and give some feeling of depth to views of distant ranges as you drive the long crest of White Mountain.

The White Mountain Wild Horse Herd Management Area covers 392,000 acres of checkerboard lands. Square miles alternate between private ownership and U.S. Bureau of Land Management land, but there are no fences and few other manmade structures, although a wind farm is planned.

You can't miss evidence of horses along the road. Stud horses build up piles of droppings to mark their territory and locating them on the road gives the stud piles better visibility.

If the horses stay off in the distance, or the weather or your vehicle make the road unsuitable to drive, check out the Wild Horse Viewing Area in Rock Springs. The corrals can have 500 horses after a roundup, with many available through BLM's adoption program.

*Cutline:*

Wild horses roam on 392,000 acres of the White Mountain Wild Horse Herd Management Area. A scenic drive takes you through its heart. Photo courtesy of the Wyoming Bureau of Land Management State Office.

### Pilot Butte Wild Horse Scenic Loop Tour

**Directions:** I-80 Exit 104 north on State Hwy 191 for 14 miles from Rock Springs, then left on Co. Rd. 4-14 for 2.5 miles, left on Co. Rd. 4-53 for 21.5 miles to Green River, coming out east of I-80 Exit 89. Return to Rock Springs on I-80.
**Open:** Year round, weather permitting. High clearance vehicle preferred.
**Admission:** Free.
**Wild horse corral viewing area:** I-80 Exit 104, north on Elk St., right on Lionkol Road and 1.2 miles to corral overlook. Free, open year round.
**Phone:** Rock Springs Bureau of Land Management office, 307-352-0256
**Website:** www.blm.gov/wy
**Attractions:** Wild horses. Bring your binoculars. Also views of Rock Springs and Green River from the top of White Mountain, and 8 interpretive signs.
**Time:** Allow at least 1 hour just for driving.

# 273 Laramie prison sheds decay

Sunday, August 16, 2009, ToDo, page D4

## The Territorial Prison has a cleaner, brighter look and has plenty of exhibits that are worth checking out.

If you haven't been to the prison in the last few years, you are in for a treat. The native sandstone building has been restored. Clean and bright and more like a gallery – a rogue's gallery – it is hung with larger-than-life portraits and stories about notorious inmates, including Butch Cassidy.

The prison's setting, with a view of mountain ranges, enticed 25 percent of the inmates to escape during its first three years, before the stockade was built.

The prison also features exhibits about women inmates, the wardens and the prison's relationship with the local community. It opened in 1872, four years after Wyoming became a territory, and closed in 1903.

There's barely any sign of the livestock the University of Wyoming housed there for most of the 20th century.

Across the prison yard is another newly renovated building, the prison broom factory. Today its products are sold in the gift shop.

The Territorial Park includes other historic buildings and exhibits, plus access to Laramie's greenway. Check for more information about the Horse Barn Theater summer productions and the ghost tours held in October.

*Cutline:*

Native sandstone was durable enough for a prison for Wyoming Territory, so it endures today. Newly renovated, the Wyoming Territorial Prison is one of several historic buildings at the site on the outskirts of Laramie. Barb Gorges

### Wyoming Territorial Prison State Historic Site

**Directions:** I-80 Exit 311, then east on Snowy Range Road less than 1 mile.
**Open:** May 1 – Oct. 31, every day, 9 a.m. – 6 p.m.
**Admission:** $5/adult, $2.50/12-17 years old. Free for 11 and under and State Parks pass holders.
**Address:** 975 Snowy Range Road, Laramie.
**Phone:** 307-745-6161.
**Website:** http://wyoparks. state.wy.us/index.asp.
**Attractions:** Self-guided tours, guided tours, living history, special events, gift shop.
**Time:** Allow 1 – 3 hours.

2009

267

# 274
Tuesday, August 18, 2009, ToDo, page D4

## Another I-80 Roadside Attraction:

### No. 6: Point of Rocks Stage Station State Historic Site

Overland Trail relay station is a precursor to the truck stop.

The rocky cliffs rise high above the convenience store at the Point of Rocks exit, located on the north side of I-80. Stop there for gas, food, water and restrooms since the original stagecoach stop has been out of business for over 100 years.

When you're refreshed, cross under the Interstate and explore the precursor to the truck stop.

Imagine the hustle and bustle in the years before the railroad arrived. Around the barn, now only a sandstone foundation, new teams are being hitched to stagecoaches, and many passengers and supplies are transferred to wagons for the trip north to the gold mining districts.

Ben Holladay bought the overland mail delivery contract, but in 1862, the U.S. government asked him to find a safer alternative to the Oregon Trail across Wyoming. Even after the arrival of the Transcontinental Railroad in 1868, the Overland Trail continued to be used, even as late as 1900.

It continued west, on to Fort Bridger, Salt Lake City and California.

*Cutline:*

Recent restoration of the Point of Rocks Stage Station makes it easier to visualize pre-railroad days. Ruts to the left of the building mark the route of the Overland Trail. Barb Gorges

### Point of Rocks Stage Station State Historic Site

**Directions:** I-80 Exit 130, south, then west on frontage road about ¼ mile, then south over railroad tracks.

**Open:** Year round.
**Admission:** Free.
**Address:** Point of Rocks
**Phone:** 307-332-3688

**Website:** http://wyoparks.state.wy.us
**Attractions:** self-guided tour. No visitor amenities.

**Time:** Allow 1 hour.

# 275
Saturday, August 22, 2009, ToDo, page D4

## Another I-80 Roadside Attraction:

### No. 7: Flaming Gorge National Recreation Area at the Confluence

Visit a quieter stretch of the popular Flaming Gorge Reservoir

Flaming Gorge Dam was built in the 1960s on the Green River about 30 miles south of the Utah – Wyoming border. The reservoir it created stretches into Wyoming and is designated a National Recreation Area.

If you want to avoid the marinas and all the people, check out the road to the confluence of the Green and Black's Fork rivers, preferably arriving in time for sunset when the red sandstone cliffs "flame."

The cliffs are larger closer to the dam, but here, especially in the middle of the week, you are more likely to encounter solitude and wildlife.

The turnoff from the highway is labeled "Lost Dog." We weren't sure if the name pertains to the road, the area, or a notice tacked along the roadside perpetuated as a standard highway sign.

Even if you don't fish, the road is worth the drive. A rattlesnake crossed in front of us, curious antelope stood before us, and sage grouse flew up beside us.

Down on the water, swallows were feeding on a swarm of non-biting insects. Gulls and terns winged along, above floating western grebes and gadwall.

*Cutline:*

The glow of the setting sun on the sandstone cliffs illustrates the origins of Flaming Gorge's name. At the confluence of the Green and Black's Fork rivers, the cliffs aren't so high, but neither is the noise level. Barb Gorges

### Flaming Gorge National Recreation Area at the Confluence

**Directions:** I-80 Exit 91, south on State Hwy 530 through Green River, about 8 miles. Turn left, beyond the overlook turnout, at the Lost Dog sign, and drive 9 miles on rough gravel to water.

**Open:** Year round, weather permitting. High clearance vehicle preferred.
**Admission:** $5/day, day use or National Parks and Federal Recreational Lands Pass.

**Phone:** Ashley National Forest, Vernal, Utah: 435-784-3445.
**Website:** http://www.fs.fed.us/r4/ashley/recreation/flaming_gorge

**Attractions:** Fishing, boating, camping, picnicking (BYOB—bring your own blanket), restroom.
**Time:** Allow 1 hour to drive round trip.

# 276 Young birds stage a summer drama series

Sunday, August 23, 2009, Outdoors, page E2

Late summer is late to be hatching ducklings. Mama Mallard either lost her first brood or having raised them, thought she could slip in another before fall.

I saw her swimming with her very young family in a corner of Lake Minnehaha at Holliday Park early on a mid-August morning. Two ducklings suddenly propelled themselves forward, as if they were trying to catch flying insects. There was a flurry of peeps and I realized Mama was having a hissy fit, launching herself at a specific spot of water and then another and another while all five ducklings suddenly crowded together near shore.

That's when I saw the snakelike head of a young double-crested cormorant emerge. I doubt its underwater intentions were merely to tickle the ducklings' toes. I didn't stay long enough to see if getting beat up by a mad mom discouraged it or if it was finally successful in grabbing a snack. Sometimes the final act of the play is just too sad to stay and watch.

A tragedy played out at the park earlier, but I missed the coup de grace. The four big white domestic geese are prominent park citizens. Back in July, I counted four

goslings in their midst, still in their fluffy grayish down. Considering the number of dogs being walked every day, I thought it brave of them to stand around on the mainland rather than on the island. But any dog with a brain would recognize their malevolent gaze and the damage their wedge-shaped bills can do to intruders.

So, I was surprised the day I counted only two goslings. And no, the white birds sleeping down by the water's edge are domestic ducks, not geese. You can tell because they have, umm, duck-shaped bills.

This brings up the latest chapter in an ongoing daytime drama. Since white domestic ducks are descendants of mallards and the mallards at Holliday Park act domesticated, not bothering to migrate anymore since they learned people will feed them illegally, it wasn't surprising that proximity would breed, well, hybrids.

Like a litter of stray kittens, no two of these hybrids seem to look the same. There's been more than one beginning birder who has flipped through their new field guide in frustration. Some hybrids come out as giants with perfect mallard

coloring. Some are almost black. Currently, one has a big white spot on the back of its head that gives it a daffy look as it stands around with the other young ducks.

Not all gawky teenagers want to hang out with the others. For several weeks, I could depend on seeing one of the young black-crowned night-herons patrolling the lawn on the east side of the park, like a gargantuan sparrow. It didn't budge an inch as long as I didn't approach it directly and stayed on the sidewalk, even though I had a 125-pound dog walking beside me.

I don't know what our unflinching hero was finding in the grass, but four or five of its cohorts, also born in the colony of nests overhead and wearing identical stripy brown plumage, could be found stooped over at the edge of the island waiting to stab fish. Where are their parents? Sleeping days, I expect, now that they don't have to feed their young every minute.

It isn't unusual to see small birds harassing large birds. But a couple days ago, the David and Goliath story was a bit different. In front of Henderson Elementary School, I saw a Eurasian

collared-dove flying after a Swainson's hawk. The hawk landed on the top of a utility pole and the dove on the attached streetlight, less than two feet away.

As I passed by, I could see the hawk was ignoring the live bird in favor of the meat under its talons. I can't imagine what would drive the dove to want to watch a hawk eat. Was it feeding on another dove or some other animal? I couldn't tell.

Thousands of coming-of-age stories are playing out in our backyards in late summer. Every time you see a speckled-breasted young robin on its own, somewhere there's a parent getting down to the business of finding enough to eat to prepare itself for migration. Or notice the grackle being pursued by a youngster thinking it is entitled to a few more free meals.

Just this morning I heard, but didn't see, two red-breasted nuthatches. Back from their summer in the mountains already? Were these the birds which visited my feeders all last winter, or their kids?

So many stories, so many conjectures, so little time.

# 277
Friday, August 28, 2009, ToDo, page D4

## Another I-80 Roadside Attraction:

### Overland Trail along Bitter Creek

Follow a few miles of the Overland Trail sans pavement.

Bitter Creek, much of the time only a trickle of undrinkable alkali water, is responsible for providing the canyon followed by railroad, Interstate and the historic Overland Trail, between the old Bitter Creek stage stop and its confluence with the Green River at Green River.

Where the wide gravel road from the Interstate crosses the railroad, there was a livestock loading facility 30 years ago, complete with corrals, old boxcars and loading chutes. Today, all that is left is a tipsy metal structure and a concrete skeleton.

As you follow the railroad and creek to the north and west, to Point of Rocks, you will be following the Overland Trail just as the stagecoaches did – without pavement.

From 1862-68, it was the official alternative to the Oregon Trail, which was plagued by Indian attacks. It branched off of the Oregon Trail at Julesburg, Colorado. Coming from Laramie, as you drove across the flank of Elk Mountain on I-80, you followed another section.

Even after the advent of the Transcontinental Railroad in 1868, stagecoaches followed the Overland Trail as late as 1900.

*Cutline:*

The Overland Trail doesn't get as much mention as the Oregon Trail, but present-day highways and railroads follow it. Not much remains at the location of the Bitter Creek stage station, named for the undrinkable alkali water in Bitter Creek. Barb Gorges

### Overland Trail along Bitter Creek

**Directions:** I-80 Exit 142, south on Bitter Creek South Road for 6 miles to railroad tracks, then 18 miles, generally west, then north along tracks and creek, to Exit 130 at Point of Rocks. Keep tracks close on your right until then.

**Open:** Road may be impassable during inclement weather. High clearance vehicle preferred.

**Admission:** Free.
**Attractions:** Historic trail, operating oil and gas field.
**Time:** Allow 1 hour.

# 278
Monday, August 31, 2009, ToDo, page D4

## Another I-80 Roadside Attraction:

### If scary is your thing, go visit former state pen

This is the prison to take your kids to as a deterrent to a life of crime.

I took a tour the year after the old state penitentiary closed in 1981, and it was hard to believe people had lived in it so recently, it was so run down. A few years later, it was used as a movie set.

But it was brand new in 1901. The massive sandstone structure looks medieval, complete with turrets.

If you decide to pay for the hour-long tour, you'll see the cell blocks and the Death House and hear the stories of escapes, hangings and riots.

There is no charge to tour the museum area at the entrance. It features inmate stories and their handiwork plus a gift shop.

One room is dedicated to the Wyoming Peace Officers Museum, and in another a video plays interviews with employees at the current prison.

At the back of the museum, a sign advertising a nature trail points to a door leading outside. Lucky for you, escaping the dreary surroundings is that easy.

*Cutline:*

The former Wyoming State Penitentiary building faces a residential street in Rawlins. Take a tour and get the inside story on executions, escapes and prison riots. Barb Gorges

### Wyoming Frontier Prison Museum

**Directions:** I-80 Exit 211, east on Spruce, north on 7th to Walnut.
**Open:** Memorial through Labor Days, every day, 8 a.m. – 6 p.m.

**Admission:** One-hour tours every hour on the half hour until 4:30 p.m., $7/adult, $6/ senior or child, $30/family.
**Address:** 500 W. Walnut St., Rawlins.

**Phone:** 307-324-4422.
**Website:** http:// wyomingtourism.org/ overview/Wyoming-Frontier-Prison-Museum/4748.
**Attractions:** Tours for a fee, free exhibits, gift shop, nature trail. Special off-season tours and hours available.
**Time:** Allow 1-2 hours.

# 279
Wednesday, September 2, 2009, ToDo, page D4

## Gilbert's "Flyaway" captivates reader

"Flyaway: How a Wild Bird Rehabber Sought Adventure and Found Her Wings" by Suzie Gilbert, illustrations by Laura Westlake, c. 2009, HarperCollins, $25.99, hardcover, 340 pages.

Author Suzie Gilbert is the kind person who never said no for five years when called about another injured bird.

She found rehabilitating birds at her home in upstate New York to be her life's work, but it could be overwhelming, even when approached with the same effervescent spirit that permeates her writing.

I enjoyed the book a lot and found myself wanting to read funny parts aloud to anyone who would listen and tearing up over the sad parts.

Suzie fielded some questions for the WTE by e-mail.

**What kind of bird-related work are you pursuing currently? Have you rehabbed any birds since George the crow? If so, did you find a way to specialize?** Post-George, I decided to specialize in raptors (hawks, owls, eagles) and crows – raptors because they don't require the constant care of songbirds, crows because they're so cool and I promised George I would. Supposedly I am now closed for a couple of months, in order to promote my book and to deal with two teenagers who are home and need to be driven everywhere. But as you might have gathered from my book, I have a hard time saying no, so I have one red-tailed hawk and three fledgling crows. Oh … and two grackles, but don't tell anyone.

**I love Laura Westlake's bird illustrations. Is she someone you recommended to the publisher for the book?** I recommended Laura – she is a great friend, a fellow rehabber, and an incredible artist. They took one look at the two samples she sent and ordered twenty drawings! I sent Laura photos – so the illustrations you see really are the birds I'm talking about in the stories. You can see her work at www.hadleylicensing.com.

**You mentioned 95 percent of the injuries suffered by wildlife are the direct result of human activity. Short of removing all vehicles, buildings and other structures, what one change in behavior would you most like to see people make that would help protect wildlife?** Keep your cats inside! Cats are safer and healthier inside. They are pets – they are not part of nature. Anyone who insists on letting their cat outside should be sentenced to work for a bird rehabilitator for one day, just to see what these pets do to the birds they catch. I guarantee they'd never let a cat outside again.

**Have you met anyone else that tried combining raising children with rehabbing wildlife? You talked about how complicated it would be to have volunteers, but it seems like your kids were a big help. How do they see those five years now?** Oh yes, there are a few of us crazy rehabber/parents out there! My kids were a great help, and I think they have fond memories. We have a wonderful time re-telling stories about those years – the time a Canada goose ended up chasing our dog around the kitchen, the time one of their friends opened the bathroom door and came face to face with a great horned owl. They will still help me if I need them, though neither seems to want to follow my footsteps. Skye is still into photography; two months ago, Mac traded his guitar for a paintball gun, which I'm hoping is temporary.

**You are quick enough to chase down an injured vulture, gentle enough to bandage a songbird and smart enough to research how to** care for different species. **What other qualities does a wildlife rehabber need?** Well, thank you! A rehabber also needs to be resourceful, flexible, and have an iron stomach, as well as a hardy sense of humor.

**Many chapters read as short stories. Do you have a storytelling tradition in your family? Were you keeping a journal of rehabbing adventures in addition to the required reports?** My storytelling ability was honed in my home town of Oyster Bay, New York, where being a bore at dinner was considered a cardinal sin. I kept (and still keep) the reports required for each bird which comes in, but every bird is so fascinating and has such unique quirks that each of my reports is several pages long. When I started, I had no idea how handy this journal-keeping would be!

**For several years Elizabeth T. Vulture, your alias, wrote a regular column for your local newspaper chain, giving everyone a piece of her mind on various environmental issues. Any chance she'll make a comeback?** I'd love to get Elizabeth back to work again … I just need to figure out how to expand a day from 24 to 36 hours!

# 280
Monday, September 7, 2009, ToDo, page D1

## All Aboard!

### Get the lowdown on train travel: It definitely is better than driving

When our son asked if he could take the family clunker back to school, we said OK.

When he asked me if I'd like to drive the 2,000 miles with him, I said sure. And how would I return?

By plane was the expected choice, but I decided to try the train instead.

How was it, you ask?

It was fine, mostly. How

2009

271

does it compare to driving or flying?

### Time

We spent three days driving from Cheyenne to Worcester, Massachusetts, averaging 660 miles per day.

I was able to get on the train in Worcester. One day and 21 hours later, including 6 hours for two scheduled layovers, my husband picked me up in Denver.

The plane takes between six and seven hours for a direct Boston to Denver flight, plus the two additional hours for security clearance. Add an hour or more for the shuttle to the Boston airport.

Let's not forget the bus. Greyhound can drive me from Worcester to Cheyenne in two days and five hours.

### Cost

Driving at $2.65 and 20-22 miles per gallon cost us $250. Add in the costs of food and lodging for three days and two nights.

My Amtrak train ticket, Worcester to Denver, coach class, cost $178. Many kinds of discounts are available. As the train fills, seat prices rise to as much as $349. Shorter trips cost less.

A "roomette" in the sleeper car, which includes meals on the train, would have been an additional $231 for the night between Worcester and Chicago. The roomettes sleep two but are priced by the room, so it would be half the cost per person if you share with a traveling companion. The bunks convert to your seats during the day, giving you privacy you don't get in coach class.

Reasonably priced food was available on the train

and at stations where I had layovers.

Airline ticket prices are notoriously fickle. At the moment, figure $200 to $400 or more round trip plus a meal or two. But figure another $40-$50 at each end, each way, for the shuttles to and from the airports.

Greyhound, Worcester to Cheyenne, runs as little as $94 (21 day advance booking) to $209 (refundable ticket).

### Carbon footprint comparison

For one person to travel 2,000 miles by:

--Ford Explorer, 1.25 tons
--Plane, 0.58 tons
--Bus, 0.38 tons
--Train, 0.06 tons

The calculator I used didn't say how calculations were made but the numbers are proportionate.

### Connectivity

Take I-25 to Exit 213 along the south side of Coors Field for a few blocks to Union Station at 1701 Wynkoop. Parking on either side is available at $17/day. Call the station, 303-825-2583, to find out about other parking options.

Or catch our local Greyhound-affiliated bus (1-800-231-2222). Depart Cheyenne on the 3 a.m. bus and arrive at their Denver station, 1055 19th St., at 6:30 a.m. and catch the 8:05 a.m. westbound (San Francisco) train. Or depart Cheyenne at 3:25 p.m. for the eastbound (Chicago) train leaving at 8:10 p.m. But you have to walk eight blocks to Union Station.

You can use your favorite method to get to Denver International Airport

and then take an RTD bus (www.rtd-denver.com) to the station. SuperShuttle (1-800-258-3826), which takes Cheyennites to DIA for $40, has a Denver affiliate (303-370-1300) that can take you from DIA to the station door for $19.

The larger Amtrak stations have car rental offices and access to commuter rail lines. Amtrak itself frequently offers its own motorcoach option where it doesn't have rail service.

### Comfort

With your car you have the ability to stop any time or anywhere or take a detour.

You can also take detours on the train. Schedule your trip in multiple legs and rent a car for a few days at any of your stops.

Since you don't have to drive the train, obviously, or help navigate or keep the engineer awake or worry about snowstorms, you can nap, take a walk to the lounge car, read a book, listen to your iPod or watch movies on your laptop. One gentleman I saw was pursuing leatherwork and beading in the lounge car. And you can make cell phone calls at any time you can get a signal.

Train seats are so spacious that they didn't fit me ergonomically. Next time I'll bring a sturdy carry-on I can put my feet up on and one of those neck pillows.

Check on a particular train's accommodations if you have a disability.

### Eating

The lounge or café car offers sandwiches and snacks. The dining car on long-distance routes offers real white

table clothes and silverware, flowers and five or six dinner entrees in the $12-$22 range.

Dinner seating is by appointment, and you share a table with other passengers. I met a woman returning from a job interview, a man in public relations who'd visited his mother and an aristocratic woman who takes train trips frequently for fun.

Breakfasts, such as "Classic Railroad French Toast," run $6-$9, including beverages. Over orange juice I met a retired art teacher from St. Louis and a grandfather going to his granddaughter's first birthday party. Other passengers included students, young families and a nurse and patient.

On my 4-hour layover in Chicago, I left the station (gasp!) and walked across the Chicago River to a normal restaurant for lunch.

### Sleeping

My first night in coach, the train was crowded. My seatmate went out for all the smoke breaks at stops across New York and Ohio. I'm a non-smoker.

The second night I got two seats to myself, flipped up the leg rests, curled up and got considerably more sleep.

Both nights, the crew passed out little pillows to everyone and turned down the lights.

### Security

There is no mandatory security check of you or your baggage, so you need only arrive more than 30 minutes in advance if you have baggage to check. They just want to see your picture I.D. Normal sharp objects and liquid amounts are allowed. Weapons and hazardous items are not. Check with Amtrak for more information.

Trains are nearly as safe as airplanes, which are many times safer per passenger mile than cars.

It's nice the train crew circulates regularly—and makes sure you get off at your destination.

**Baggage**

Amtrak is generous. It allows two carry-ons under 50 pounds, no larger than 28 x 22 x 14 inches (large suitcase), plus personal items like purses, laptops and coats.

You can check up to three pieces of baggage under 50 pounds each no larger than 36 x 36 x 36 inches at no

extra charge.

Don't lose your baggage claim check. You must present it at your destination.

**Destinations**

Amtrak won't be convenient for every trip. But a train ride is more than transportation. It's sightseeing, too, including historic train stations. Amtrak even offers

vacation packages.

The part of the California Zephyr route (each route has its own name) west of Denver is so scenic, Amtrak schedules it for daylight hours. They include a dome car and a couple of park rangers.

Hmm. Denver to Grand Junction is only $45 each way....

### More info on Amtrak

Amtrak.com
Or call 1-800-USA-RAIL (1-800-872-7245) and ask to receive the

timetable which includes all train travel information found on the Web site except for ticket prices.

## 281 Another I-80 attraction
Friday, September 11, 2009, ToDo, page D4

**Fish, float, find wildlife along the North Platte**

Presumably named for having had to dig out space for a road next to the North Platte River, there is no reason to recommend this site, Dugway, unless you are fishing and/or boating and love sun. Campsites have tables, but there is no shade. Consequently, it is very quiet, especially during the week.

The www.recreation.gov website recommends putting in your canoe or kayak upstream (south) because of a control crest installed below the dugway. Put in at the Interstate bridge or Fort Steele.

Look up the Bureau of Land Management's map of "Public Fishing Opportunities—South and West Central Wyoming." Cheyenne folks can pick up a copy at BLM's

information desk, 5353 Yellowstone Road, during business hours Monday through Friday.

It shows rainbow trout, brown trout, cutthroat trout and walleye can be caught here.

If your boat is too big for the river, continue another 30 miles up the road to Seminoe Reservoir, at Seminoe State Park.

The fishing map indicates Dugway is good for wildlife watching. We saw a pair of common mergansers swim by.

*Cutline:*

There's good fishing and floating at this access to the North Platte River.
Barb Gorges

### Dugway, Bureau of Land Management Recreation Site

**Directions:** I-80 Exit 219, north on Seminoe Road (County Road 351), about 8 miles.
**Open:** Year round, weather permitting.

**Admission:** Free.
**Phone:** Rawlins BLM field office, 307-328-4200.
**Website:** http:// www.recreation.gov/

recFacilitySearch.do.
**Attractions:** Boating (canoeing and kayaking), camping, fishing, picnicking, wildlife watching.

**Time:** Allow at least 1 hour – or more if the fish are biting.

## 282 Local birder makes it to "Bird" Mecca
Sunday, October 18, 2009, Outdoors, page E3

"Mecca - a place that is an important center for a particular activity or that is visited by a great many people." Encarta Dictionary

"The Cornell Lab of Ornithology uses the best science and technology – and inspires the widest range of people and organizations – to

solve critical problems facing wildlife. Our mission: to interpret and conserve the Earth's biological diversity through research, education, and citizen science focused on birds." www.birds.cornell.edu

Any active birdwatcher, or anyone who has been

reading my columns the last 10 years, has heard of the CLO, especially when I'm trying to recruit participants for Project FeederWatch (see accompanying box) or the Great Backyard Bird Count or the Christmas Bird Count or eBird, the free bird sighting archive.

The CLO's address is quaint: 159 Sapsucker Woods Road, Ithaca, NY. I never thought I'd get a chance to visit.

In August, though, I drove with my younger son, Jeffrey, back to school. There are many ways to get to Massachusetts, and I found the one

2009

273

that led through Ithaca. It wasn't a hard sell to schedule a stop since his good friend Eric is a student at Ithaca College, just across town from Cornell University.

Central New York State is marked with 11 long, skinny, very deep, north-south oriented natural lakes, the Finger Lakes, set in wooded hills. Ithaca is at the end of 40-mile-long Cayuga. It's wine country – a vacation destination even if you aren't a bird watcher.

Since Jeffrey and I were racing the calendar, we allowed ourselves only a morning in Ithaca, and most of that at Sapsucker Woods.

The woods are 225 acres just outside Ithaca, protected by the Lab while the surroundings are farmed and built on. The Lab has done some building, too.

The I.P. Johnson Center for Birds and Biodiversity is no down-home affair. It is a modern office building where 200 people work: staff, faculty, grad students and visiting scientists. And where 100,000 people visit per year, says their website.

Luckily, the Lab understands its role as Bird Mecca and has provided a visitor center, complete with an in-house Wild Birds Unlimited store, an auditorium, gallery, multi-media presentation and a hands-on sound laboratory.

And there's a two-story bank of windows facing an incredible bird feeding station with pond and woods beyond. There are even spotting scopes set up. I took note of the eastern species, various woodpeckers and sapsuckers, black-capped chickadees, etc., but after so many days in the car, I was ready to hit the trails.

Jeffrey, Eric and I have been on many field trips in our Cub Scout days. They both appreciate the outdoors, even when we discovered how mosquitoey and humid it was. Eric was the one who noticed the submerged bullfrogs in the pond. Once you learned how to see one, you could see the others.

It just wasn't much of a bird day – everything we could hear was hidden up in the leafy canopy. No wonder the CLO is so big into bird song recordings – there's more to hear than to see in their country.

With a little more time and planning, we might have attended an educational program or hired someone listed in the American Birding Association directory to help us navigate the unfamiliar avifauna.

There were a lot of cars in the second, more remote parking lot (it probably makes for a nice walk in the woods on the way into the office each morning). I didn't think the Citizen Science programs I mentioned earlier needed that many employees, so I did a little research.

Much of the CLO's $16 million budget activity comes from research: Bird Population Studies, the Bio-acoustics Research Program (58 staff around the world) and the Evolutionary Biology Program.

The Macaulay Library (21 staff) archives wild sounds. Sometimes they are featured on special segments on National Public Radio news. You can listen to thousands of snippets online for free at www.macaulaylibrary.org.

If you can't get to Sap-sucker Woods, the next best thing is to go to www.birds.cornell.edu. The website is a gateway to an incredible amount of information. Even if you intend to only travel as far as your own backyard, check out the link to the Lab's website, www.AllAboutBirds.org, and get a taste of the bird-watcher's Mecca.

As for me, I'm going to have to go back to see the natural features that result in the visitor's bureau slogan, "Ithaca is Gorges."

## Project FeederWatch

**Project FeederWatch begins its new season Nov. 14.** There are certain simple counting conventions to follow, but once you sign up, you'll get a kit with simple instructions and even bird i.d. help.

**During the five months of the season,** you'll establish a weekly or bi-weekly observation schedule, but if you miss, it's OK. Your observations can be entered online and then you can view and print your own data as well as see what's going on elsewhere. If you prefer, you can opt for the paper version and turn your data in at the end of the season.

**There are three options for signing up** (and it is never too late to sign up) and paying the $15 fee--$12 if you are already a Cornell Lab of Ornithology member.

**Option A:** Sign up online: www.birds.cornell.edu/pfw.

**Option B:** Call toll-free, 877-741-3077, Monday-Thursday, 8 a.m. to 5 p.m. EST, and Fridays 8 a.m. to 4 p.m. EST (8 a.m. here is 10 a.m. there).

**Option C:** Send your check, name, mailing address, phone number, email address if you have one, and whether you will be recording your data on paper or online to: Project FeederWatch, Cornell Lab of Ornithology, P.O. Box 11, Ithaca, NY 14851-0011.

Any questions? Call me, Barb Gorges, 634-0463. I'm signed up for my eighth, or maybe it's my ninth, season.

# 283 To feed or not to feed?

Sunday, November 29, 2009, Outdoors, page E2

## Local birdwatcher battles with whether her bird feeder is a good idea.

To feed or not to feed, that is the question this time of year.

On one side are the purists who say bird feeders are an unnatural source of food for birds. They blame the invasion of the East Coast by a western bird species, the house finch, on feeding. They'll point to avian diseases transmitted when unnaturally high numbers of birds congregate in the same location day after day.

The purists will mention birds die when they fly into windows near feeders or when they are attacked by loose cats. They argue that some birds may decide not to migrate if they have a ready food source. That is true for the Canada geese in Holliday Park and Lions Park.

But let's keep this discussion centered on the songbirds fond of sunflower seeds.

The purists are right: A bird in its native habitat does not need supplemental feed to survive the winter. If its preferred seed crop had poor production or becomes covered in snow, it will fly. Grosbeaks, redpolls, waxwings, crossbills and siskins are all noted for travelling when they need food, sometimes hundreds of miles from their expected wintering grounds.

Yes, the backyard feeding station can be hazardous to small birds, but probably not any more so than natural predators and hazards.

So why feed birds? Do it for your own enjoyment. Do it for the cheerful chatter, the bright colors, the bustle and hustle. If watching fish swim in a bowl relieves stress, as I've heard, then watching birds out the window not only relieves stress but is life affirming. It is for me.

Wildlife is elusive enough that most people have little contact with it unless they hunt or fish or have spotting scopes or long lenses on their cameras. Without some other kind of personal relationship, how can we expect the general population to begin to buy into any kind of conservation ethic? Most wild animals are too dangerous to approach or feed. Chickadees seldom are.

Is it important to have a conservation ethic? Yes. What makes wildlife and land healthy makes people healthy. If you want the footnotes and scientific references, read one of Michael Pollan's recent books on food.

Meanwhile, let's talk about ethical bird feeding. For more information see the American Bird Conservancy, www. abcbirds.org.

Grow diversity in your yard by providing native flowers, shrubs and trees for shelter and habitat, and even food. Reduce or eliminate pesticide use. Pesticides are toxic to birds and can kill insects beneficial to them. Even seed eating birds feed their young insects.

Provide water and keep it

clean and fresh. If you don't offer food, water will attract birds. Get one of those little heaters meant for bird baths or dog water dishes or use a portable pan or plastic dog food dish you can bring inside to thaw the ice.

Feed the good stuff, black oil sunflower seed – and thistle seed if you can afford it. Forget the mixes with red and white milo which tend to attract the non-native house sparrows and Eurasian collared-doves. They compete well enough with our native birds already.

You don't need to run a soup kitchen. A couple of feeders are enough. If the birds empty them in the morning, then wait until mid-afternoon or the next morning to refill them. We don't want to upset the natural balance too much.

Keep feeders clean. At our house, we no longer use the feeders with the little saucers at the bottom—those get really gross. Our feeders are hung over the concrete patio so we can sweep up the debris regularly. If the weather gets warm, it is important to wash the feeders weekly before organisms can grow. If you notice sick birds, stop feeding for a week and clean everything.

Keep feeders within three feet of the window, so birds will be aiming for the perches instead of the glass or at least won't hit the glass so hard. Leave the

window screen on so birds will bounce off or put decals on the outside of the window to break up the reflection. If feeding birds is for your enjoyment, there's no point in putting the feeders where you can't see them easily.

Keep your cat indoors— they look better if they haven't lost the tips of their ears to frostbite. If it isn't your cat lying in wait under the feeder, then send the dog out for a while to clear the area.

If you won't keep your cat indoors, make sure it doesn't have a place to hide in ambush within 25 feet of your feeder. And if you can't do that, don't feed birds.

If you have trouble with deer horning in, either don't feed the birds or put the feeders where troublesome wildlife can't reach them.

If sharp-shinned hawks start picking off seed-eaters at your feeder, congratulate yourself on attracting the next level in the food chain. Life in the wild is about death as well.

Get a field guide from the bookstore or the library and find out what birds are visiting. Take a close look at the LBJs and LGBs (little brown jobs and little gray birds) and you might be surprised how many kinds you've attracted. Many are just here for the winter, so enjoy them while you can.

# 2010

## 285 Meditation on pine beetles
Sunday, January 31, 2010, Outdoors, page E3

### Is there life after tree death?

For anyone who doesn't regularly recreate in or travel through the forests of south-central Wyoming and north central Colorado, the photos of pine beetle damage shown at January's Cheyenne-High Plains Audubon Society meeting might have been a shock. Especially the photos of grown trees blown down like straws, and campgrounds denuded by the removal of hazardous trees.

Many of the 75 people in the audience, however, judging by their questions and comments, have mountain property and are in the midst of the battlefield.

The largest mountain landowner, the U.S. Forest Service, was represented by the evening's speaker, Steve Carrey, director of renewable resources for the Medicine Bow-Routt National Forest. One irate audience member demanded to know why the Forest Service hadn't headed this epidemic off when it started.

The simple explanation is that pine beetles are always with us but were at a high point in their cycle when drought was weakening trees of an age beetles prefer, and warm winters didn't freeze any beetles dead following the initial outbreak in 1996 west of Denver. It created, as Steve said, a perfect storm. Lack of funding hasn't helped either.

Even with limitless funds, one cannot spray every pine tree in the forest or change the climate quickly.

One can only clean up the mess, clearing dead trees before they fall across roads, trails, power lines and campgrounds and before they begin to burn.

No one seems to want the dead trees – the price of timber is still too low to reopen more than one of the local sawmills. Some are being turned into pellets for pellet stoves, and there is talk of building a plant that uses wood to generate electricity.

Lodgepole pine is the main tree being killed. The stands we are used to seeing on the Medicine Bow are 80 to 150 years old, the re-growth after initial logging. Most of us in the audience will not be around to see this second re-growth reach maturity.

In fact, many people in the audience looked old enough to have been recreating on the forest more than 50 years (30 for me) and may not be around in 10-15 years when the trees are finished falling over and are no longer hazardous except as fuel in wildfires. Even then, a stroll off the trail will entail climbing over the deadfall.

Downed trees may, happily, slow illegal off-road driving.

It tears at my heart to see ponderosa pines turning red between Cheyenne and Laramie, along my favorite Pole Mountain trails, knowing that soon it will be unsafe to roam there – for a while. But I don't feel the same about the mountain sides of lodgepole monoculture over west of Laramie and have never yearned for a cabin in that dense forest.

Having, on quests for elk, tramped through the endless monotony of tree trunks as far as the eye can see, with no underbrush, no bird song, only squirrel chatter and the occasional break for a birdy, spruce-lined creek and beaver pond or rocky outcropping with a view of soaring hawks, I'm ready for a change.

Having driven endless miles of roads lined with future telephone poles right down to the shoulder, wondering when a deer will spring out to meet my bumper, I'll appreciate the change.

A connoisseur of cloud formations and sunsets, I look forward to vistas opening up. Let's just hope that all the mountain cabins and structures that come into view are picturesque.

There may be a lack of shade, but the forecast is that the sunny slopes will produce lots of grass and shrubs, even aspen, before the pines shade them out again, not unlike a clearcut or burn.

Just exactly which wildlife species will disappear, and which will appreciate the change, is ripe for research by a generation or two of grad students.

Doesn't this remind you of the 1988 Yellowstone fires?

The difference is that the Medicine Bow isn't quite done with the epidemic. For a few more years, each year's generation of beetles will fly to new trees mid-summer, where they'll burrow under the bark and lay eggs that hatch into larva that eat the trees' cambium layer, girdling and killing the trees over the winter, with the red needles showing the following summer as the next generation of adults flies off.

The year the new beetles can't find any live trees to bore into and lay eggs will be the year their population plummets.

If you want to see how our forest will soon look, visit central Colorado. Visit the website www.fs.fed.us/r2/bark-beetle .

Losing the forest we know and love is like losing our old dog, the one whose body language we know so well, he doesn't even have to ask to be let outside. The new forest will be as dynamic as a puppy, full of surprises and excitement, anxious to grow.

## 286 Backyard bird count needs you

Monday, February 8, 2010, ToDo, page D4

The 13th annual Great Backyard Bird Count is Feb. 12-15 and you are invited, no matter your age, location or bird watching experience.

Participation is free, but the data you collect has high value for the scientists who study it, even if you are a beginning birdwatcher.

Folks in Cheyenne have been taking part since 1998, the first year of the count. House finch, house sparrow and dark-eyed junco show up most often, but the Eurasian collared-dove, first recorded in 2002, is moving up in the top ten list.

To take part, count the birds you see in one location for at least 15 minutes, or longer if you like, on one of the count days. Your final tally for any species can only be as many individuals as you were able to see at one time – it is impossible otherwise to know if you are counting a bird twice. You can make separate counts on the other days and in additional locations.

Then, go to www.birdcount.org to record your observations as a checklist for that particular location and date, including information about where you were, what time, for how long, and in what kind of weather. If you are unsure about a bird's identification, just mark that you are not reporting all the birds you saw.

For help with bird identification, you can generate a list on the website of Cheyenne's winter birds. Each bird name is linked to an online field guide page.

There's also a video, kids' page with games, educators' page, photo contest gallery, plus previous year's results searchable by species, region, year and location.

Last year, Wyoming observers counted 85 species in 36 communities and four national parks and forests, a small but important part of the 11.5 million birds counted continent-wide.

With your help, we'll find out where the birds are this winter (and maybe send in more checklists than the birders in Casper!). If you need help, call Barb, 634-0463, or email bgorges4@msn.com.

## 287 Rite of Spring: Watch sandhill cranes along the Platte River

Monday, February 15, 2010, ToDo, page D4

One of the world's great spectacles of spring is 600,000 sandhill cranes stopping over on the Platte River during migration. And it takes place only a five-hour drive east of Cheyenne.

The first four-foot tall, elegant gray birds appear just before the beginning of March, winging in from Mexico, Texas and New Mexico to spend three or four weeks on the Platte. Numbers peak late March, and the last of them straggle out mid-April, continuing on their way to breeding grounds in Canada, Alaska and Siberia.

The cranes, with a few whooping cranes mixed in, roost on river sandbars at night and forage in the surrounding grain fields by day, between Lexington and Grand Island, Nebraska.

Crane watching, according to the University of Nebraska-Lincoln, adds $10.3 million to the local economy. A beautifully designed website, www.Nebraskaflyway.com, tells you everything you need to know about cranes, crane watching, where to stay and what else to see and do. For a print version, call 1-877-855-2951.

Plenty of cranes can be seen in the fields during the day and from viewing decks 1.5 miles south of Interstate 80 at Exit 285 and 2 miles south at Exit 305. Remember crane watching etiquette: don't approach them, startle them or trespass on private land. Pull to the side of the road and use your vehicle as a viewing blind.

But consider treating yourself to viewing them from a riverside blind at sunrise or sunset at either Rowe Sanctuary, run by Audubon, at Gibbon, ($25 per person), or the Nebraska Nature and Visitor Center at Alda, just west of Grand Island ($30 per person, 308-382-1820, www.nebraskanature.org).

Kent Skaggs, at Rowe, said as of Feb. 11, weekend slots are pretty full, except for Easter weekend, but there are plenty of weekday slots available. Call 308-468-5282 during business hours (Central Standard Time) to make a reservation.

Rowe is hosting the 40th Rivers and Wildlife Celebration Mar. 18-21, celebrating the sandhill crane. A few workshops and field trips are already full. Guest speakers

Prize finalist Scott Weidensaul. Call Rowe for more include natural history book author and Pulitzer information or see www. RoweSanctuary.org.

# 288 Wyoming Quilt Project's quilts now searchable online

Wednesday, February 24, 2010, ToDo, page D4

Quilts documented by Wyoming Quilt Project, Inc., are now on view through the Quilt Index website, www. quiltindex.org.

WQP was organized in 1994 by a group of quilters in Laramie to record information about old quilts before they disappeared.

"We wrote to 16 other state documentation projects and from that, compiled a Wyoming documentation form," said founding member and one of the current directors, Anne Olsen.

The form asks for the quilt's size, pattern, fabrics, techniques and information about the quiltmaker. Photos of each quilt are taken.

Donations from Cheyenne Heritage Quilters, other guilds, people whose quilts are documented and WQP members help cover the costs.

WQP members held documentation days in many Wyoming communities until 2004. Now all documentation is done in Laramie.

The information for 2,800 Wyoming quilts is being entered into the Quilt Index by volunteers and a student who is paid through a grant provided by the Quilt Index.

The Quilt Index is run jointly by the Alliance for American Quilts and Michigan State University. It has 50,000 quilts in its free, publicly accessible, searchable database – many from museum collections.

To look at just the Wyoming quilts, on the Search page use the drop-down menu titled "Contributor/Institution" and select "Wyoming Quilt Project." So far, each quilt is represented by at least a photo and pattern name. It will be several years before each quilt's full record is added.

The emphasis has been on finding pre-1940s quilts. The oldest is one made in England, dated 1808. Another founding member and current director, Wendye Ware, said a large number of the quilts are from the 1920s and 1930s.

A search of that era's popular patterns shows WQP has documented 120 versions of Double Wedding Ring, 76 of Grandmother's Flower Garden and 27 of Sunbonnet Sue.

The goal is to document another 200 quilts for a nice round number total of 3,000.

Anyone residing in Wyoming who would like to have a pre-World War II quilt documented can contact Anne Olsen, 307-742-9144, to make arrangements.

# 289 Listening for birds doesn't get easier with age

Sunday, March 21, 2010, Outdoors, page E3

A couple months ago, our field trip reverberated with a bit of icy snow crunching underfoot, a tiny breeze rattling dry leaves and the murmur of bird watchers. When we stood perfectly still, straining our ears for a sound of bird life in the trees along Crow Creek, I finally heard the faintest whisper from a brown creeper, found the bird and pointed it out.

The creeper cooperated and everyone had a good look as it flitted to the base of a tree, spiraled up the trunk looking for dead and slumbering insects in the bark, and started over on the next tree.

Creepers have a very distinct but faint call as they work and I wanted everyone to hear it, but almost no one else could, even though the 5-inch-long bird was close enough to see without binoculars – if you could pick out the bark-colored feathers from the bark.

I realized finally that nearly everyone on this field trip was older than me and perhaps they couldn't physically hear the creeper.

A fact of aging is losing the ability to hear high-frequency sounds. It isn't uncommon for an older birder to think that the population of kinglets in his favorite birding spot has decreased over time, only to discover it was his decreased hearing that diminished the number of the tiny, high-pitched voices he could hear.

Binoculars and spotting scopes are expected paraphernalia for birdwatchers. But if the birds aren't out in the open, or if you don't catch their movements flitting in the branches 50 feet overhead, hearing is the only way you'll know which direction to point your binoculars.

Acute hearing partly explains the extraordinary abilities of hotshot young birders, especially if they have been too busy birding to ruin their hearing with loud music.

Unfortunately, many people pick up birding in mid-life. Optics make up for failing sight, and field guides and all kinds of handheld devices make up for failing memory.

For failing hearing there are a few choices. I don't have any experience with any of these and if you do, tell me more about them.

The first step is to visit an audiologist and make sure the dearth of high-frequency birdsong can't be attributed to a dearth of birds. Depending on the type of hearing loss, there are kinds of hearing aids that will help birdwatchers.

There are lots of ads in birding magazines for binoculars, but not for aids to hearing. One I came across is the Songfinder (see at www. nselec.com). It is a little case on your belt connected to

CHEYENNE BIRD BANTER

slim headphones. It picks up sound and translates it to your ears at a lower frequency. You choose from three settings how much lower.

Songfinder presupposes the user can hear the frequency of the human voice well. At $750, it is cheaper than the audiologist's special hearing aids mentioned by blogging birdwatchers. The drawback would be, if you've been birding for years, to suddenly hear a brown creeper sing alto, tenor, or even bass, relatively speaking.

The old-fashioned option is to get a parabolic microphone. You've seen photos of the scientist holding a big dish with a microphone in the middle connected to earphones. You can also connect it to a recorder so you can tell people you are recording birdsong, and they will think you are a science nerd instead of hard of hearing.

The Cornell Lab of Ornithology is offering an eight-day course in wildlife sound recording in June out in California, but I found only one small advertisement for this kind of listening equipment in my birding magazines, www.stithrecording.com.

The problems with fancy microphones are you have to lug them around, and they only amplify sound. You'll get a lot of other amplified noise. But you'll be in the same boat as the rest of us trying to pick out birdsong over traffic noise and wind.

There are bird sounds to be heard in winter. Some other soft, high frequency calls are Bohemian and cedar waxwings communicating as they search for another berry-full tree, the golden-crowned and ruby-crowned kinglets scampering high in evergreen treetops, and horned larks, blowing to and fro over country roads and fields.

When birds start thinking spring, they, mostly the males, bring out their songs for some practice. They need to be in top form if they are going to keep other males from invading their territory and also attract the attention of the best females.

Luckily for us, the birds still sing even after a spring snowstorm. Also lucky for us is when Wyoming's state bird, the western meadowlark, projects its loud arias from roadside fences, it will be hard to miss no matter how old our ears get.

# 290 Nebraska spring festival is for the birds

Sunday, April 18, 2010, Outdoors, page E3

*Note: It's Ft. Kearny, but the city is spelled Kearney.*

If a late winter-early spring trip to Belize, Mexico, the Everglades, southern Arizona or Hawaii wasn't on Mark's and my calendar this year, I thought, why not central Nebraska instead?

At an elevation nearly 4,000 feet lower than Cheyenne, spring would be farther along.

We packed our snow gear anyway and headed for Kearney for the first weekend of spring and the 40th Annual Rivers and Wildlife Celebration.

I have always thought this was a weekend to avoid when planning a trip to view the spring migration of sandhill cranes. But having become increasingly intrigued with the idea of attending a birding festival, we paid the registration fee and signed up for one of the pre-conference, daylong field trips.

We didn't sign up for the crane-viewing blinds. That just seemed futile with the number of people coming for the conference. Plus, we've done it before.

It was 70 degrees Thursday afternoon when we arrived in Kearney at the conference hotel. On the way out to Rowe Audubon Sanctuary, 12 miles further east, we stopped to admire a field full of cranes as thick as cows in a feed lot and quietly dozing or picking up the odd bit of food. They seemed to be anticipating their evening performance.

Rowe's educational displays provide the background to appreciate the Platte River, its history and the unique phenomenon of 600,000 sandhill cranes stopping over on their way to northern breeding grounds.

Forty years ago, the cranes could barely find the scoured river sandbars they need to roost on at night to avoid predators. The controlled flow of the Platte didn't give it the flooding needed to keep it clean.

Ron Klataske, working for Audubon, inspired the troops during those early years, and the original spring meetings were rallies for river protection. At lunch, Ron, now director of Audubon of Kansas, reviewed the progress made.

A lobbying workshop featuring a panel of Nebraska lawmakers was scheduled Saturday afternoon. But after a morning learning about sandhill crane behavior and the state of whooping crane research, Mark and I opted for a walk out to the river on the Fort Kearny State Recreation Area's Hike and Bike Trail.

An old railroad bridge spans a perfect treeless, crane-roosting stretch of the river, but we were too early for the evening performance of incoming cranes.

Instead, we'd paid to attend the banquet. At our table we met folks from New York and Nebraska, a few of the 150 people from 22 states registered for the weekend.

I was looking forward to the after-dinner speaker and Pulitzer Prize finalist Scott Weidensaul. I've enjoyed several of his two dozen books about birds and natural history.

It turns out he is good at speaking, too, with great photos. His theme was from his book, "Return to Wild America," in which he retraces Roger Tory Peterson's 1953 trip across North America and notes the changes.

The family of the previous evening's speaker, Nebraska photographer Michael Forsberg, was around all weekend selling his incredible photographs and his new book, "Great Plains: America's Lingering Wild." He gave

2010

279

us a look behind the scenes of the professional wildlife photographer's life. Not only do you need to know your camera, you need to know your wildlife, more than a few landowners, and how to set up a camera trap or figure out other ways not to disrupt your subjects' lives while shooting them.

Chris Wood came out all the way from the Cornell Lab of Ornithology in Ithaca, New York, to encourage

us at breakfast Saturday to record our bird observations in eBird. More about that in a future column.

But how was the birding, you ask. Fine.

For $25 apiece, we rode a 20-passenger shuttle bus all day Friday with huge windows and Kent Skaggs from Rowe Sanctuary at the wheel. He knows every road and bird. Hefty sack lunches were provided, plus plenty of interesting passengers, as

well as enough potty stops at small towns. It was cold and snowy and downright raw when we clambered out for stops to explore the Rainwater Basin Wetland Management District south of Kearney, but other stops required only cozy armchair birding from the bus.

The highlights included greater prairie chicken, Lapland longspur, eastern meadowlark and a rare glaucous gull. The other birds were

all species we see regularly around Cheyenne, except for the flock of eastern bluebirds we saw Saturday afternoon – a great way to mark the first day of spring.

We'll see what famous name in birdwatching or conservation is invited next year and maybe even risk registration roulette and sign up for a sunrise or sunset in the crane viewing blinds, too. Everyone needs a little inspiration after a long winter.

# 291 Sunday, May 16, 2010, Outdoors, page E3
## "eBirding" our backyards gives science important knowledge

What year was it when I saw seven western tanagers in our yard at one time? How often do lazuli and indigo buntings visit? I have a few notes scribbled on old calendars stored in the basement, but otherwise, 20 years' worth of backyard spring migration sightings are just fond memories.

A few months ago, I received an email from Brian Sullivan, eBird project leader, gently extolling the virtues of using eBird, the free online avian data system from Cornell Lab of Ornithology and the National Audubon Society, to track my sightings and share them with scientists and birdwatchers. I've submitted a few in the past but didn't get into the habit. However, this spring, I think eBird is finally becoming part of my routine.

I'm still jotting cryptic notes on scrap paper, but I'm taking them to my computer and entering information on the eBird website, www. eBird.org, before I forget. When I check "My eBird"

I can see how many species I've observed so far in 2010, and how many I have total. I can look and see if I was the first one in Wyoming to report a species this year.

For the serious birder, eBird offers that kind of competition. It allows uploading records from other avian record keeping systems and downloading of personal records from eBird and viewing data in different ways. You can be alerted to sightings of birds seen in your area you haven't got on your life list yet, or you can use the database to find the best place and time to see target species.

For the rest of us, especially beginning birdwatchers, a look at the list of local, public birding hotspots and their respective checklists is invaluable.

It is also easy to mark a personal birdwatching location and then have eBird generate a list of potential species.

If you accidentally type in "300" for the number of

peregrine falcons you saw, you'll get a polite question. Or perhaps your sighting is unusual for the time of year. If you say you are sure it isn't a mistake, eBird might ask for documentation. If you can't provide enough, your observation can stay in your personal data but won't be shared with the public – birders or scientists.

You can always go back and make corrections to your entries.

All the cool, free tools eBird offers are inducements to get us to share our bird sightings. Our data is most useful if we take a little extra effort to record time spent observing, distance traveled, or size of area birded and estimate the numbers of birds of each species seen. The hardest part is to notice all the birds where you are, including those annoying background species like starlings and house sparrows.

For instance, when I walk the dog around Holliday Park, my focus is looking for what is unusual. On different

days in April, the lake hosted a white pelican, half a dozen cormorants, hooded mergansers, a pair of wood ducks and a pair of redheads. The 60-70 Canada geese are just background, not to mention the starlings and pigeons, but eBird prefers I submit a checklist of all the birds I can identify.

The use of eBird data is free to ornithologists, conservation biologists, educators, land managers and anyone who likes to play with raw numbers.

Doug Faulkner cites eBird as a reference in his new book, "Birds of Wyoming."

But don't worry, no nosy scientist is going to knock on your door. No contact information for observers shows on the website. There are several ways to remain nearly anonymous.

But in looking through Wyoming data identified by observer, no one here has chosen "Anonymous" or a fake-sounding name. Many folks on vacation submit Wyoming sightings, too.

280   CHEYENNE BIRD BANTER

Because eBird only started in 2002, there are a lot of gaps, though historic data can be added. Bird life at Wyoming Hereford Ranch is fairly well documented for spring, but apparently local interest dies off in winter. For all the birds I've seen over the years in Lions Park, the checklist for it as a birding hotspot has few species.

Just how many people are taking part in eBird as it gets ready to go global? Here in Wyoming this year so far, 41 observers have observed 152 species. Since 2002, 6,948 checklists (a checklist is a list of birds observed for a particular time and location) have been submitted for Wyoming. Natrona County (Casper) has the greatest number of checklists, 1,561. Our county, Laramie County, with similar population, is in 8th place, having only submitted 275. Now you know why Brian Sullivan emailed me and other Wyoming birdwatchers.

In the rest of the country, urban areas like Los Angeles County (23,000 checklists) have a lot of eBirders, as does a birding travel destination getting a lot of scientific research like the Aleutians Borough in Alaska, 39,000 checklists submitted. If you want to get your name in the records, there are a few counties in Alabama and other Southeastern states for which no checklists have ever been submitted.

Birdwatching is a satisfying hobby for many of us, and eBird allows us to contribute to serious science. Go to www.eBird.org and look around and register for free.

I look forward to seeing more balloon markers on the map, showing more Wyoming birders are "eBirding."

# 292 Sunday, May 30, 2010, ToDo, page D1
# When not to rescue wildlife

## The scenario: You're hiking in the woods and discover a fawn lying under a shrub, no mother in sight. Does it need your help?

It happened to our family once. We'd walked off the trail to admire wildflowers and practically tripped over a deer fawn lying partially obscured under a shrub, no mother in sight.

The fawn's instinct was to sit tight. Ours was to beat a hasty retreat, but apparently not all people have the same reaction.

Every spring, deer and antelope fawns are needlessly "rescued" by well-meaning people who want the Wyoming Game and Fish Department to take care of them. They don't realize the young are normally, and often, left alone while the mother is feeding out of sight.

"They may seem abandoned, but chances are that handling will cause them to be abandoned, or human scent might attract predators," said Reg Rothwell of Wyoming Game and Fish Department's Biological Services Division. "Game and Fish has no facilities to care for abandoned wildlife."

Rothwell said that the WGFD once again has an employee designated to answer questions brought about by the seasonal increase in interactions between people and wildlife.

Is a nest of rabbits or birds abandoned? It takes hours of patient, non-disruptive observation to determine if parents are not returning.

What about the nest that blew out of the tree? Put the remains and young in a container back in the tree and the adults will be happy to continue caring for the nestlings. If you need additional nesting materials, use shredded newspaper or paper towel, not green plant material.

What about young birds on the ground, completely feathered but unable to fly? Put them up on a branch and keep them safe from dogs and cats. It will only take a couple more days for them to learn to fly, said Rothwell.

If you have mallards nesting on your lawn, be patient. They'll leave when the eggs hatch.

If there are no trees around, chances are you've discovered a grassland bird from a ground nest. Your only option is to put your dog on a leash and leave the bird in the care of its parents, members of a species that has been distracting would-be native predators for eons by faking a broken wing.

Even if a wild baby actually needs rescuing, you cannot legally take it home and take care of it yourself unless you are working directly under a licensed wildlife rehabilitator.

What about injured wildlife? If it is a natural injury, caused by the natural environment or other wildlife, keep in mind that the misfortune of one animal is fortune, or dinner, for another, and you needn't do anything that changes the natural order.

But often the injury is human caused. While prevention is best (cats indoors, decals on windows, careful driving, little ramps out of window wells and stock tanks), we would not be human if we didn't want to help.

The first rule is to protect yourself during the rescue. Rothwell said rescuers are always at risk of being bitten or contracting diseases. Western grebes, he said, go for your eyes with their sharp beaks. Even the smallest songbird will nip or could carry interesting parasites.

For large birds and mammals, it is best to call the experts for help. They will know the best way to safely transport the animal and where to take it.

2010

281

## Want to help?

The Cheyenne Pet Clinic has recruited 15 volunteer "Nestling Nursemaids" so far this season. Veterinarian Dr. Robert Farr is licensed to rehabilitate wildlife and his staff is very knowledgeable. Call them at 635-4121.

For various wildlife dilemmas, including nuisance wildlife, call the Wyoming Game and Fish Department, 777-4600, Monday through Friday, 8 a.m. – 5 p.m. For after-hours emergencies, city police, the sheriffs' department and the Wyoming Highway Patrol can reach WGFD officials. In the Cheyenne area, WGFD Warden Mark Nelson can also take calls at 638-8354.

For more preventive advice, look online at WGFD's website, http://gf.state.wy.us, and click on Wildlife. Under the heading Habitat Home Page, click on Extension Bulletins.

For more about when to rescue wildlife, an excellent tutorial is available from the Champaign (Illinois) County Humane Society. Go to http://cuhumane.org. Click on Resources, then CCHS Library, and then choose Wildlife.

# 293 Balloon Fiesta dazzles amateur photographers

Saturday, June 20, 2010, Journey, page E1

Ever been inside a Kodak moment? That's where I was last October, at the 38th Albuquerque International Balloon Fiesta, looking for the quintessential photo op of bright colors on a cloudless blue sky.

Albuquerque, a 550-mile drive from Cheyenne, is a drive I've done often over the last 30 years, but this was my first Balloon Fiesta.

Hot air balloonists are attracted by the "Albuquerque Box," a phenomenon in which winds in the valley blow in opposite directions at different elevations, making it possible to return close to the starting point.

The event is held for nine days over the first two weekends in October. This year's is Oct. 2-10. Each day, Balloon Fiesta Park opens at 4:30 a.m. At 5:45 a.m., the Dawn Patrol, a dozen balloons launching in the dark, go up to "prove the wind."

At least one other event is held at sunrise each day and additional events at dusk on the weekends, weather permitting.

Be sure to plan to be in town more than one day and attend the first good day. Last year, the final Mass Ascension had to be cancelled due to wind over 12 mph – which we Cheyenne folks consider a faint breeze.

In the dark of the day we went, we saw a row of nylon envelopes stretched out on the grass being inflated for balloon rides. Visitors were milling around, keeping out of the way of crews hauling on ropes as the balloons began to stand up.

A Special Shapes Rodeo, one of two held the second weekend, was inflating on the other side of the field. Various cartoon animals were becoming airborne one limb at a time: squirrels, bees, pigs, penguins, along with buildings, trains, trees and other objects. Each time one of the more than 80 balloons floated off, the surrounding crowd cheered.

Then the field was quickly cleared by polite "zebras," volunteers wearing black and white (some in zebra costumes), and the Albuquerque Police Department's horse patrol mounted on Percherons.

From a launch site a mile away, the regular-shaped balloons began to appear, aiming for the field where they could take part in two competitions. One is to drop a flagged marker on a target, and the other is to grab for a prize envelope atop a pole. The crowd got a close look at many of the 600 balloons registered for the fiesta, 82 from 17 other countries, as they sailed down the field.

By about 9 a.m., the last balloon straggled by, and it was time to peruse the vendors and grab a breakfast burrito.

Except for Monday through Wednesday, each evening at 5:45 p.m. there is a Balloon Glow or Special Shapes Glowdeo, when tethered balloons are inflated and the burners fired up, creating a stain-glass effect.

When the balloons aren't flying, there is the Albuquerque International Balloon Museum, a fun hands-on experience any time of year, plus all the other local attractions: other museums, Old Town, trails along the Rio Grande River (where you might see balloons "splash and dash") and the tram to the top of Sandia Crest.

Although there are package ticket deals available, and shuttle options, our family drove to the park, crept along in a line of cars for about 30 minutes, paid the $10 parking fee and the $6 per person entrance fee.

Free "Survival Guide" brochures are handed out, but the $6, 108-page, full-color program is a worthwhile investment. It gives a lot of hot air balloon and gas balloon (there's a special competition for them the first weekend) information and a photo directory of all the registered balloons.

A folding chair might be a good idea. And warm clothes. Make sure you work out how to regroup because any party over one person in size will want to go in different directions, looking for the best balloon and the best light.

Everyone has a camera in hand. Fiesta officials say it is the most photographed event in the world. Luckily, extra batteries, memory cards and disposable cameras are available everywhere at the park. Each opportunity to photograph a balloon is as ephemeral – and as colorful – as a butterfly.

**More info**

--Albuquerque International Balloon Fiesta, 4401 Alameda Blvd. NE, Albuquerque, NM 87113, toll-free 1-888-422-7277, balloons@balloonfiesta.com, www.balloonfiesta.com.

--Anderson-Abruzzo International Balloon Museum (on the park grounds), 9201 Balloon Museum Dr. NE, Albuquerque, NM 87113, 505-880-0500, www.balloonmuseum.com.

# 294 "Birds of Wyoming" is a must have treasure

Wednesday, July 7, 2010, ToDo, page D4

Birds of Wyoming by Doug Faulkner, c. 2010 by Roberts and Company Publishers, Greenwood Village, Colorado, 404 pages, 8.25 x 10.25 inches, full color, $45.

The book, "Birds of Wyoming" by Doug Faulkner is here. You can find a copy at local and national booksellers.

The birdwatching community, state and national, has been waiting for this book ever since the University of Wyoming announced hiring Faulkner, a professional wildlife biologist and super birder, for the project enabled by a generous donation from Robert Berry.

This is not a field guide. Although it has color photos of our state's 244 resident species, it won't give you tips on identifying them. There are another 184 species, migrants and other regular visitors, with no photos.

Nor is it Oliver Scott's "A Birder's Guide to Wyoming" which gives directions to birding hotspots, but as you browse the new book, you'll see some place names pop up again and again.

This book most resembles the Wyoming Game and Fish Department's bird atlas and Jane Dorn and Robert Dorn's "Wyoming Birds" but with much more discussion and information.

Each account will give you an idea of where, when and with what abundance a species occurs in Wyoming, and how widespread it is in the world.

I found myself referring to the accounts in "Birds of Wyoming" often this spring as each migrating species made an appearance. I was able to find out if they breed in Wyoming and if so, in what habitat, and found out just how uncommon it is to see a rose-breasted grosbeak in my backyard.

If you are new to birding in Wyoming, this book gives you much of that intimate knowledge of its avian life without having to be, or hang out with, an old timer.

In the first chapter, Jane Dorn introduces the history of Wyoming ornithology, beginning with a French-Canadian fur trader's notes in 1805. Other chapters describe Wyoming bird conservation and management challenges. Robert Dorn neatly lays out the landforms of Wyoming and associated plants and birds. Unfortunately, unlike other scientific publications, no credentials are given for the eight authors of the chapters.

I hope the next edition comes with a more conventional map inside the covers – one with major landforms, cities, towns and public birding spots named on the map rather than numbered, with an accompanying alphabetical index with reference grid locations.

While Doug is listed as the author, he is quick to acknowledge the numerous people, including photographers, who contributed to the project. However, for many species he writes that more information is needed.

We need to get out and bird more and put our observations into a public database like www.eBird.org, instead of in a shoebox, before the next edition comes out in five or 10 years.

This is a big book, but if you want to learn about the birds of Wyoming, you'll want your own copy.

# 295 Wyoming Roadside Attraction

Saturday, July 10, 2010, ToDo, page D4

**Take a free soak in mineral hot springs**
**The State of Wyoming will even rent you a swimsuit and towel for $1 each**

Bet you didn't know you could rent a swimsuit from the State of Wyoming for only $1, and a towel, too, for another $1. And 20 minutes of soaking in either the freshly renovated indoor or outdoor mineral hot springs pools at the State Bath House in Hot Springs State Park is free year round.

You can bring your own suit and towel, but if you weren't planning to spend the day playing on the slides at one of the two commercial hot pools in the park, or one of the Thermopolis motels offering naturally heated pools, it's nice to know you can still sample the therapeutic waters. The Bath House attendant is very good at estimating your suit size.

Free soaking was a provision requested by the tribes signing the treaty in 1897 that gave the land to Wyoming. As many as 200 people take advantage on a summer day.

The shady park grounds are a good place for a picnic. Also check out the terraces of mineral deposits behind the Bath House and try out the swinging bridge over the Big Horn River.

*Cutline:*

If you are passing through

2010

283

Thermopolis, stop by for a free, 20-minute soak in the mineral hot springs at the State Bath House.

Barb Gorges

## State Bath House, Hot Springs State Park

**Directions:** From State Hwy 789/U.S. Hwy 20, look for state park signs at Park Street directing you east across the Big Horn River. The State Bath House is one block north, on Tepee Street.
**Distance from Cheyenne:** about 300 miles.
**Open:** Monday – Saturday 8 a.m. – 5:30 p.m., Sunday noon – 5:30 p.m. Open on summer holidays, closed for winter holidays.
**Admission:** Free.

**Address:** 538 N. Park St., Thermopolis (park headquarters).
**Phone:** 307-864-2176
**Website:** http://wyoparks.state.wy.us
**Attractions:** mineral hot springs, picnicking, boat docks, Volksmarch trail.
**Time:** 30 minutes or more.

# 296 Spotting nests isn't easy, but here's an idea of where to look

Sunday, July 25, 2010, Outdoors, page E3

Back in the dark ages of bird appreciation, people collected wild bird eggs and nests to display them. It's illegal now unless you have a permit. But what I've always wondered is how they found the nests.

Or how about the researchers that measure eggs and nests? How do they find enough to make statistically valid statements?

And then I started counting the number of nests I've seen this season – without really trying.

First, there were two different great horned owl nests. You can't miss the big bulky affairs, though you could miss the bit of feathered head sticking up above the rim. Don't get any closer or you could get a talon in the face.

Then there was the face-off between a robin and a squirrel on the roof of the house next door. Robins have nested between the houses before, but it wasn't until early June that I saw the bulky cup balanced on the rafter up under the roof overhang and got the hairy eyeball from one of the parents.

The island in the middle of the lake at Holliday Park was so crowded this spring that some of the 60-70 Canada geese were perching in the trees and on the picnic shelter, trying to figure out how to balance eggs on a branch or roof ridge. By the end of May, abandoned eggs littered the island and three families hatched two, seven and 17 goslings respectively.

One pair of mallards hatched seven ducklings, but after about a week there were only five. Another pair hatched eight and lost only one the first week. Ducklings are bite-sized compared to goslings.

Interestingly, it took the 15 white domestic geese until July to hatch four goslings.

The nests of the black-crowned night-herons became invisible as soon as the park's trees leafed out. The occasional squawk during the nesting season didn't begin to match the 49 birds counted earlier.

What about all the birds chirping around town? House sparrow nests are easy to spot – look for messy stick piles stuck into the letters of three-dimensional signs or anywhere they can squeeze themselves.

But what about interesting songbirds? Keep your eyes open. Memorial Day weekend, Mark and I birded along Crow Creek just as the trees were finally getting fully leafed out. We were staring into a clump of willows, trying to identify small songsters. Familiar, noisy birds were flying in – a yellow warbler, a robin and a western kingbird. They distracted me, and as I let my gaze follow them, I discovered their three nests, all in one spacious tree. What a treat.

A month later in similar habitat – mosquito infested – I saw an oriole's nest freshly woven and heard household murmurings from a nest plastered under a deck by a pair of Say's phoebes.

I might have to start a life nest list.

Nests aren't always the quintessential robin's cup of mud and twigs. Besides the oriole's woven sack hanging from a tree branch, burrowing owls and belted kingfishers use burrows, many shorebirds barely scrape out a depression on the ground, flickers peck out holes in trees (and house siding), loons float their nests, herons build tree top rookeries and peregrines nest on cliffs and building facades.

Nests are used only temporarily. As soon as the young fledge, which may be before they even learn to fly, the nest is abandoned. Some species may fix it up and start a second clutch right away. Big sturdy hawk nests may be used again the next year – by owls. Songbird nests disappear, broken down by wind and weather. Other birds and animals may steal the building materials for their own nests.

The nest is not a permanent home. Home is where a bird can find food, water and shelter. And for migratory birds, is home where they spend the winter, or where they spend the short few months reproducing?

If you want to know more about how and where your favorite birds nest, go to www.AllAboutBirds.org.

284     CHEYENNE BIRD BANTER

# 297 Summer reading list for birders

Monday, July 26, 2010, ToDo, page CFD9

1. Peterson Field Guide to Birds of Western North America, 4th edition, by Roger Tory Peterson, Houghton Mifflin Harcourt, 2010.

Two years ago, the "Peterson Field Guide to Birds of North America" was updated and combined Roger Tory Peterson's eastern and western guides into one volume for the first time. This spring, the information was published in separate guides again.

The publishers must have decided it was easy enough to cater to both birders who like the entire continent in one book and birders who like the regional field guides which are divided by the 100th Meridian, vertically bisecting the Dakotas, Nebraska, Kansas and Texas, and cutting off the Oklahoma panhandle.

Unfortunately, in the western edition, the individual species range maps cut off half the continent, so you can't get a feel for continent-wide distribution when a species has one.

Cheyenne is frequently visited by eastern warblers during spring migration, and while the western guide has their pictures and descriptions, no range maps are provided to give you an idea how far away their normal range is.

If you live out here in the middle of the continent and you want a Peterson guide to birds, famous for its trademarked field identification system and Roger Tory Peterson's classic illustrations, go for the big one, "Peterson Field Guide to Birds of North America," only $6 more than this new $20 western guide.

You'll get a more complete view of our birds and be able to use it wherever you travel in North America – and get more muscles carrying it.

2. Molt in North American Birds, by Steve N. G. Howell, Houghton Mifflin Harcourt, 2010.

Part of the Peterson Reference Guide Series, this book addresses molt, the process of birds growing new feathers. It's a confusing topic but necessary for identifying birds beyond their characteristic breeding plumage.

When do birds grow new feathers, pushing out the old worn ones? Do all birds have different winter and summer plumages? Can they fly when they are molting wing feathers? What causes a molt cycle to begin? When is the best time to molt?

All birds molt, but not the same way or as often, which is why there is now a 267-page book to explain it.

What's even more confusing is that there are different systems used to talk about molt.

Howell has written 67 pages explaining the different classification systems as well as bird molt strategies. Once you've digested those pages with the help of Howell's clear writing style, move on to the bird families such as the gulls, champion molt artists.

Even if you aren't particularly interested in molt, this book is jam-packed with bird photos, almost all taken by Howell himself in the last five years. He leads birdwatching tours for WINGS, Inc., is affiliated with the Point Reyes Bird Observatory and lives in California.

3. Bayshore Summer, Finding Eden in a Most Unlikely Place, by Pete Dunne, Houghton Mifflin Harcourt, 2010.

Following his book, "Prairie Spring," a three-month tour of the Great Plains, Pete Dunne, director of the famous Cape May Bird Observatory in New Jersey, has decided to stay home for this installment in his seasonal series.

The Bayshore is southern New Jersey, where summers are marshy, hot, humid and swarming with insects.

Dunne provides a fascinating trip through an area mostly unfamiliar, even to folks going to the Jersey Shore.

He explores the intricate relationship between the 400-year-old human adaptations to nature, and nature's adaptation to man when he tries his hand at harvesting salt hay or goes out with the watermen to pull crab pots.

The heart of the red knot problem (knots are shorebirds) gets Dunne and his photographer wife, Linda, immersed in tidal flat mud. Later, he catalogs the many kinds of insect and arachnid species locally available. He is a wall flower on a party boat searching for weakfish. He expounds on the Jersey tomato and why the state's nickname is "the Garden State." And he spends a night with a state game warden on a stakeout for a habitual deer poacher.

Dunne makes you feel all the summer sweat and all the itches, so maybe you'll want to save this small book for next winter or your vacation in cool mountains. Despite the discomforts of his climate descriptions, it makes me want to visit the Bayshore myself, but maybe before Memorial Day or after Labor Day.

2010

# 298

Sunday, August 1, 2010, Cheyenne Frontier Days, page CFD12

## Wyoming Roadside Attraction: Lake Marie, Snowy Range Scenic Byway

### Breathtaking views abound at 10,000 feet at Lake Marie

The Snowy Range rises out of the Medicine Bow Range. Along the juncture, a series of lakes collect snowmelt.

Lake Marie, named by an early government surveyor for his wife, May Bellamy, who later became the first woman elected to the state legislature, is the most photogenic and accessible. Small parking lots on each end are connected by a flat, paved walk.

On the west end are the restrooms and the trailhead for climbing Medicine Bow Peak, elevation 12,013 feet. Following the trail a little way will give you some great views, but hiking the peak demands preparation, physical fitness and a very early start.

On the other side of the highway is a nice sample of the trail system alongside a mountain stream.

From the smaller parking lot on the east end, you can find the trail up to the Mirror Lake Picnic Area for different views of Lake Marie. Mirror Lake is also accessible by vehicle from the next turnoff east. The trailhead there leads to views of more alpine lakes.

If you hike, leaving your dog at home is easier than following the leash regulation – and safer. Visit early in the day so you aren't caught by thunderstorms, and remember, you'll be out of breath just standing along the highway at 10,000 feet.

But the wildflowers are breathtaking, too.

*Cutline:*

A view of Lake Marie on July 4 from the east shows some of the snowdrifts blocking area trails. Barb Gorges

### Lake Marie, Snowy Range Scenic Byway

**Directions:** From I-80 Exit 311 at Laramie, drive about 35 miles on Wyo. State Hwy. 130 west through Centennial.
**Distance from Cheyenne:** about 90 miles.
**Open:** June – September, whenever the road is snow-free.
**Admission:** Free.
**Address:** Laramie Ranger District, Medicine Bow National Forest, 2468 Jackson St., Laramie.

**Phone:** 307-745-2300.
**Website:** www.fs.fed.us/r2/mbr
**Attractions:** Scenery, hiking, fishing with Wyoming fishing license, wildlife viewing, picnicking at adjacent Mirror Lake Picnic Area.
**Time:** 20 minutes to 2 hours.

# 299

Sunday, August 8, 2010, Outdoors, page E3

## New order, new names and new species of birds dictated by AOU

The American Ornithologists' Union has come out with the 10th supplement to the seventh edition (1998) of the Checklist of North American Birds, which means you can start penciling in changes in your current field guide or buy a new edition next year.

It can also mean that, like one subscriber to the Wyobirds e-list, you may be able to add two species to your life list without even looking out the window.

In the AOU's first days in 1883, birds were classified by appearance and habit. With study, a fine distinction could be made between similar birds that together never produced fertile young – separate species – and similar birds that were variations within a species.

Similar species were grouped into a genus and similar genera were grouped into a family. It all made sense to birdwatchers in the field.

Then, as we became more globally aware, we tried to align the common and scientific names of birds with their counterparts overseas. Thus, our "sparrow hawk" became the "American Kestrel" some years ago.

Now DNA testing has come into common use, and ornithologists are making discoveries and adjustments regularly to reflect the evolutionary relationship of species to each other.

The latest changes are documented in The Auk, the AOU's journal, and they are meaningless to the casual birder who may have, like me, not learned the scientific bird names in Latin.

However, the American Birding Association has done somewhat of a translation that shows a lot of the changes are a shuffling of species between different genera and a shuffling of the order of species and genera. For instance, green-tailed and spotted towhees will still be in the genus "Pipilo," but other towhees will be in "Melozone."

The AOU goes through cycles of splitting and lumping species. This time the whip-poor-will has been split. This isn't a big deal for us in Wyoming since we don't get them here (we have poorwills), but if you saw one in the southwest and one in the eastern U.S., you can now amend your life list and have "Eastern Whip-poor-will" and "Mexican Whip-poor-will" instead.

The winter wren got both global and continental splits. Now there will be the Eurasian Wren and in North America, there will be the Pacific Wren and the Winter Wren.

Luckily, "Birds of Wyoming," by Doug Faulkner, refers to the now former subspecies by nearly the same name as the Pacific wren, so we won't be too confused. In Peterson's field guides, it is noted that west of the Rockies winter wrens sound different than in the east, which is part of the AOU's justification for the split as well as DNA differences.

This latest catalog of changes is all of the AOU's decisions only between January 1, 2009, and March 31, 2010. I'm sure more will continue to come.

What is the point to being picky about bird names? For ornithologists, it's scientifically precise labels in English and Latin. For the ABA listers, it's an accurate count of species on their life list.

But for us backyard birdwatchers, it's being able to communicate with each other, and with the scientists who want our observations for citizen science projects.

So my advice is to make it your priority to keep track of common names and the species getting split and lumped. Learning genera and families is secondary.

After a while, when you talk to someone new about birds, you'll be able to tell how old their field guide is by what common bird names they use. That means in addition to learning the new names, you can't forget any of the old names!

# 300 Bird IDs can be tricky, so a photo is always welcomed

Sunday, August 15, 2010, Outdoors, page E3

Spring and early summer are when I get the most bird calls, questions about woodpecker damage, inconvenient robins' nests, but mostly bird identification.

Unless they can email me a defining photo, I usually give callers a few possibilities to look up and let them decide for themselves.

For instance, in spring Cheyenne regularly gets six species with dark or black heads, backs and wings, and orange breasts, the most obvious being American robin, which we compare everything to.

The others are orchard oriole, Bullock's oriole, black-headed grosbeak, spotted towhee and the American redstart.

In early May, a friend mentioned having a flock of painted redstarts at her house. Was she misnaming American redstarts? She insisted on painted redstart.

At home I looked both up. They are both small (American is 5.25 inches and the painted is 5.75 inches) black-headed birds with red markings. The American has a white belly and red patches on its black wings and tail. The painted has a red belly and white patches on its black wings and tail. I saw it once in 1996 in southeastern Arizona.

There are no documented records for painteds in Wyoming as of 2008. Sibley's shows them in Arizona and New Mexico, in oak and pine canyons, with records of sightings in north-central Colorado.

There are two possible scenarios here. One is familiarity breeds complacency on my friend's part. She may have spent some time in the Southwest where she identified painted redstarts. When a similar bird showed up in her yard in Cheyenne, she assumed it was a species she knew and loved seeing previously. Who needs to look closely and look it up in the field guide again?

Me. I've been known to look through binoculars to enjoy common birds 15 feet outside my window, but I wouldn't expect everyone does that, so a general impression of small bird flashing black, white and red could remain misidentified, causing no harm until the observer talks about it to someone with too many field guides, like me.

The second scenario is familiarity breeding complacency on my part. Although I see maybe one American redstart every other spring, I page past the entry every time I look up other warblers in my field guide. The Cheyenne bird checklist (compiled by more knowledgeable people than me) says they are uncommon migrants. They normally hang out around riparian (stream) areas, so it would be unlikely for them to be on high prairie where my friend lives.

There is of course, a third scenario. The bird in question is not a redstart at all.

The future scenario I'd like is my friend gets a close look at and takes a photo of her visiting birds, double-checks her field guide, and based on her previous familiarity, is quite convinced she sees painted redstarts – and based on the species range map in her field guide, she realizes it is a rare species for Wyoming.

Next, she convinces the Wyoming Bird Records Committee she saw painted redstarts. It's a challenge. Observer credibility is as essential as good digital photos.

How does she get credibility? She becomes an active part of the birding community. By joining other birders on field trips, they will get a feel for her birding ability, and her ability to say, "Gosh, I guess that cerulean warbler was something else," which is what one of Wyoming's best birders said last month after some additional study.

There are advantages to birding with others. If everyone can see the same rare bird at the same time, they can confirm the identification. The records committee likes those kinds of reports, especially if a detailed description of the bird's look and behavior is submitted, along with justification for not identifying it as a similar species.

The field guide is sort of a birder's Bible – but with one main difference: the

2010

birds don't read it. They have wings and travel intentionally, looking for new habitat, or unintentionally, caught by wind. The range maps are just a measure of likelihood.

Birdwatching as a hobby shares something with gambling and fishing. We go out hoping for the next big thing, the next rare bird, even while we enjoy all the other birds we see.

So, next time painted redstarts show up, take a photo and then give me a call, and I'll be right out.

# 301 Wyoming Roadside Attraction: Miner's Cabin Trail

Sunday, August 15, 2010, ToDo, page D4

## Miner's Cabin Trail is worthy stop on Snowy Range Scenic Byway

The Snowy Range Scenic Byway between Centennial and Saratoga has many turnouts and small parking lots with views that give flatlanders plenty to photograph, but don't neglect this one and the Miner's Cabin interpretive trail.

Be forewarned there are no restroom facilities, so be sure to stop first at Libby Flats to the east or Lake Marie to the west.

If hiking at 10,000 feet is too much for you to even contemplate, this is still a worthwhile stop as the entire Snowy Range is visible, and all the peaks are identified on an interpretive sign in the small parking lot.

The trail, barely three-quarters of a mile, may be encumbered with a few snow drifts as late as July. With plenty of signs interpreting the area's mining history and natural history to stop and read, you'll never get too out of breath.

Besides the miner's cabin, look for the picturesque remains of the Red Mask Mine built in the 1920s in hopes of making money in copper, gold and silver.

*Cutline:*

The Miner's Cabin Trail circles remnants of mining history along the Snowy Range Scenic Byway between Centennial and Saratoga.

### Miner's Cabin Trail on Snowy Range Scenic Byway

**Directions:** From I-80 Exit 311 at Laramie, take Wyo. State Hwy. 130 west about 35 miles.
**Distance from Cheyenne:** about 90 miles.
**Open:** June – September, or whenever the road is snow-free.
**Admission:** Free.
**Address:** Laramie Ranger District, Medicine Bow National Forest,

2468 Jackson St., Laramie.
**Phone:** 307-745-2300.
**Website:** www.fs.fed.us/r2/mbr
**Attractions:** Scenery, hiking, wildlife viewing, restrooms available at other turnouts.
**Time:** 20 minutes to 1 hour.

# 302 What the "Bird of the Week" has taught me

Sunday, October 10, 2010, Outdoors, page E2

In the summer of 2008, I committed to writing "Bird of the Week" for the Wyoming Tribune-Eagle's ToDo section for two years.

My idea was to help readers learn about birds more often than this column, known informally as "Bird Banter," with its publication dependent on the WTE's available space.

Two other developments inspired this idea. The first was the redesign of the WTE which introduced "sky boxes" at the top of pages that feature paragraph-long bits of information accompanied by attention-grabbing headlines and photos.

The second was Pete Arnold's bird photography which he shares via email. I asked him if he would be interested in sharing his photos via newspaper.

Next, I examined Pete's list of bird photos (I really don't know how bird photographers get shots of such "flighty" animals!), identified 104 species WTE readers might see easily in the Cheyenne area and then assigned them to a week in which they might actually be seen here. Naturally, there was a dearth of species for winter and an abundance of species for summer.

How to sum up such interesting creatures in few words is a challenge I faced about 10 years ago when I wrote an educational CD for the National Audubon Society and the Wyoming Game and Fish Department called "Wyoming Birds." It was easy to write for children who might never have noticed birds before.

However, "Bird of the Week" readers would span bird appreciators, owning no binoculars or field guides, and local bird experts.

I decided on a mix of generalization – a glimpse of the bird in its Cheyenne area habitat – plus some unsuspected trivia I was betting more experienced birders might not know or had at least never mentioned to me.

My reference was Birds of North America Online, www.bna.birds.cornell. edu, available for the annual subscription fee of $42 per year. Accounts include video, photo and sound files as well as updates and search capability. Glad I didn't have $2,000 to buy the original 18,000-page print version when it first came out.

So how long does it take to write 60-80 words? About five minutes. But first it takes one to two hours to read the BNA species account. And it took an infinite amount of time to edit each bird's paragraph – I make changes every time I read through my own writing.

Eventually, I learned to give preference to the interesting factoids appropriate to the season the bird was featured. If it was spring and I was writing about a warbler that only visits Cheyenne during migration, I might focus on its interesting migration facts rather than its nesting or wintering habits.

Sometimes the reading in BNA is pretty tough sledding, unraveling sentences that are little more than diagrams of technical terms. Bird of the Week was a lot like a two-year home-study course in ornithology. So now I have a much broader understanding of how different birds solve problems of survival of the individual and perpetuation of the species.

There are another 220 more obscure species on the Cheyenne bird checklist. However, the problem with continuing the series would be whether Pete has photos or if readers would be disappointed if a species is not as easily seen as the first group.

Reader feed-back has been positive. Once I heard from a reader who noticed her first green-tailed towhee the same day it was featured. I consider that "mission accomplished."

Thanks to the cooperation of Pete and the WTE staff, more people are more aware of what's around them.

I wonder what else readers would like to know – and what I would learn by researching it.

# 303 Holliday Park summer bird counts total 43 species

Sunday, November 7, 2010, Outdoors, page E2

Between the last week in April and end of September, I counted 43 species of birds at Cheyenne's Holliday Park while walking a friend's dog about three times a week.

I recorded my observations at the free website, www.eBird.org, so now I can look back and tell you that there were 60-70 geese at the end of April, a high of 180 mid-July, including a crop of about 25 goslings, and in September there were around 130. I hope some more will migrate and not eat all the grass in the park.

Yes, there were the other usual urbanites: mallard, European starling, house sparrow and pigeon, but there were surprises.

From late April until the first week in August, I could count on up to a dozen double-crested cormorants each visit and maybe around five American white pelicans, both species trolling the lake for fish.

I caught glimpses of wood ducks and turkey vultures during spring and fall migration. Warblers passed through too, but identification is difficult because I don't take binoculars. It takes two hands to walk a 125-pound dog known to sometimes walk me when she sees other dogs, squirrels or her park friends.

Holliday Park has had a colony of black-crowned night-herons for more than 20 years. May 29, before the cottonwoods completely leafed out, Mark, my professional wildlife biologist husband, counted 49 adults working on their treetop nests. Two days later, they were invisible except for the occasional adult collecting another branch for repairs. But they were noisy.

And then August 2 I saw 20 brown-striped youngsters strung out along the edges of the lake, perched on rocks and branches, waiting to pounce on passing fish. This was not all the young that hatched. Park caretakers told me earlier they picked up a lot of dead young herons under the trees after a windstorm.

I noticed the chimney swifts sporadically June through August. They really do look like flying cigars compared to swallows. I wonder if I was too busy counting ducks to look up and see them more often.

There was a lull in the number of crows I counted mid-July. That could have been while they were protecting their eggs and nestlings by not drawing attention to themselves.

I think the flickers successfully fledged their young. I was used to hearing or seeing around two per visit, and then Aug. 24 there was a family of four together on the ground.

Common grackles weren't seen after Sept. 4 and red-winged blackbirds after Sept. 10, reducing the park's noise level.

The verdant lawns of the park look like perfect robin habitat. I was surprised not to see more hunting for worms. Perhaps regularly applied pesticides have reduced the food that lawn-loving species can find, or there is too much human and dog activity.

Of species not already mentioned, the spring and fall migrants on my Holliday Park list were: northern shoveler, redhead, lesser scaup, spotted sandpiper, western wood-pewee, yellow warbler, yellow-rumped warbler, common yellowthroat, Wilson's warbler, chipping sparrow and Brewer's sparrow.

The species that flew over only once were osprey, sharp-shinned hawk, Swainson's hawk, Franklin's gull, ring-billed gull, common nighthawk, belted kingfisher, downy woodpecker and black-capped chickadee.

The summer regulars were Eurasian collared dove, mourning dove, blue jay, cliff swallow, and American goldfinch.

Of the winter birds that should be returning from the mountains, I first heard the red-breasted nuthatch in late July – perhaps they nested in town instead – and the first junco flashed by me Sept. 27.

Who knows what else will drop by during the rest of fall migration?

2010

289

## 304 eBird: How to use a scientific database as a vacation planner

Sunday, November 14, 2010, Outdoors, page E2

Because most of our long-distance travels are to see family, Mark and I haven't had time to sign up for bird tours. But we have figured out how to add birding to our family visits by using eBird.

eBird is a free, publicly accessible, scientific database and isn't designed as a vacation planner, but here is how we've used it to find birding hotspots for our future trip to eastern Massachusetts.

Step 1. Go to www.eBird. org. Sign-in is free.

Step 2. Of the tabs along the top of the page click on "Submit Observations."

Step 3. Click on "Find it on the Map."

Step 4. This next page, about location, says Step 1, but that's for the process of submitting an observation. We don't know the counties in Massachusetts, so we only filled in the "State/Province" box with "Massachusetts" and made sure the "Country" was United States. Put in your destination and click on "Continue" at the bottom of the page.

Step 5. A Google Map of Massachusetts popped up for us. If you aren't familiar with Google maps, note that you can click on "Map," "Satellite," "Hybrid" or "Terrain" buttons. The last one will give you an illustration of the topography and a road map.

Step 6. Zoom in on your destination by clicking on the "+" sign. To slide the map to one side, click anywhere on the map, hold the mouse's left button down and drag the hand icon across the screen. If you have a slow internet connection, it might take a bit for the map to catch up with each operation.

Step 7. As you zoom in, you'll see the red balloons with cross marks change to groups of unmarked red balloons, each representing a birding hotspot location. Click on a balloon to find out its name, which will pop up in the box above the map. As long as you don't click "Continue," it won't matter where you click the map. Write down the names of likely looking hotspots.

Step 8. At the top of the page, click on "View and Explore Data."

Step 9. Click on "Bar Charts."

Step 10. On the "Choose a Location" page that comes up, Mark and I scrolled down to Massachusetts and clicked on it, and then in the list on the right, we selected "Hotspots." At the bottom of the page click "Continue."

Step 11. From the list of Hotspots, choose the one (or more) you want to use to generate a bird list. Mark and I were interested in Parker River National Wildlife Refuge, so we clicked all 16 of the refuge's hotspots. Click "Continue" at the bottom of the page.

Step 12. Review your bar graph. Ours showed us how abundant 342 species are at every quarter-month interval of the year.

Step 13. If there is a species you'd really like to see, click on its name. I have never identified an American Black Duck, so when I clicked on it, a Google map of the refuge popped up with red and yellow balloons. This time, the red balloons mean observations at those locations are more than 30 days old. Yellow balloons indicate less than 30 days.

Step 14. Click on one of the balloons. You'll see on what dates your chosen species was observed at that location, plus how many individual birds were seen each time, and by which observer.

Step 15. Go see the birds! We looked up Parker River on the internet and learned all about visiting.

So who are all these people sharing their observations with the public and the scientific community? They are any birdwatcher or birder who takes the time to enter their data online at www. eBird.org. You can contribute your information too. Click on the tab at the top of the page, "About eBird," and learn how to enter your backyard and vacation bird sightings.

# 2011

## 305 Birds in fiction need facts, too

Sunday, January 2, 2011, Outdoors, page E3

**Local author C.J. Box may write fiction, but his Wyoming-based books should still reference wildlife that is actually found in Wyoming.**

Cheyenne's national best-selling crime novelist either needs to do more scenery research or needs to make a rare bird report. In "Below Zero," C.J. Box's hero, Joe Pickett, is hiking into the Hole in the Wall, Butch Cassidy's famous hideout, and the description includes bluebirds and cardinals.

Cardinals are rare in Wyoming, and no observations have been documented for Johnson County where the Hole in the Wall is located.

If Box were from back east, cardinal country, I would say he added a splash of color to the scenery using a species he was familiar with. But Box is a Wyoming native, so he needs to contact the Wyoming Game and Fish Department's non-game bird biologist and report the cardinal if he saw it while researching the novel. Apparently, Pickett, the fictional game warden, was too busy looking for trip wires to realize what he'd seen.

Can authors mention any but the most common birds in fiction or can they give unusual birds enough context so non-birdwatcher readers will understand their significance? In "Freedom," a literary novel by Jonathan Franzen which I just finished reading, cerulean warblers are a bit more than scenery – they and their predicament, diminishing habitat, are both metaphor and plot device. And a main character is identified as a birdwatcher to explain his anti-social tendencies.

Realistic fiction has to be more believable than real life, it seems. Would it be believable that on the November Audubon field trip in the middle of Cheyenne we saw a Cooper's hawk knock a mallard drake on its back, less than 50 feet from where we stood at the railing at the edge of Sloan's Lake in Lions Park? They don't usually go for ducks.

Of course, if it were fiction, we'd only include the anecdote to illustrate character or to move the plot along. I was all for leaving the duck, seemingly close to death, for the hawk. Pat wanted to tip it over onto its feet and Art tipped it. Fifteen minutes later, it was walking with an occasional stumble. I suppose the three of us were characterized, even though we are not fictional: I think hawks deserve to eat, Pat is a retired nurse and Art has handled a lot of gamebirds.

Some of the best storytelling I've read recently was about real birds and real birdwatchers: "The Big Year, a Tale of Man, Nature, and Fowl Obsession," by Mark Obmascik, the featured author at this year's Laramie County Library Foundation's Booklover's Bash.

In 1998, three men independently decide to break the record for the number of bird species seen by one person in one year in North America. It takes them half a year to realize they are competing against each other.

No Hollywood scriptwriter could come up with such craziness as these men risking their lives in storms on Attu, the farthest west point of Alaska, where lost Asian bird species blow onto our continent.

In fact, the story is so crazy that Hollywood bought it and made it into a movie to be released early in 2011, starring actors you've heard of: Owen Wilson, Jack Black and Steve Martin.

What I like about Obmascik's writing is how deftly he explains the birding world without bogging down the story.

He knows what needs to be explained to non-birders because he was one, yet he understands birders, too, having recently become one.

For non-birding authors of fiction, adding avian color to realistic fictional scenery is simple enough. Check a recently published field guide, and then call the local Audubon chapter for confirmation on exactly what time of year and on what kind of bush your chosen bird species might be seen. Millions of birdwatchers, who tend to be well-read folks, will appreciate your effort.

# 306 Patchwork birding benefits birds

Sunday, February 13, 2011, Outdoors, page E3

Patchwork. The word draws my eye the way "quilt" does because both describe my indoor hobby the way "bird" describes my outdoor hobby.

But why was Ted Floyd, editor of "Birding," the American Birding Association magazine, making an obscure reference to patchwork in a recent issue? I emailed him, and he sent a link to a blog post he'd written about it and how it relates to green, environmentally friendly, birding, http://blog.aba.org/2010/10/green-birding.html.

Patchwork birding refers to birding in your own patch – your yard or a local park where you go often, versus jumping in the car or on a jet to see a rare bird.

Ted is concerned that birding has evolved into the hobby of the affluent who indulge in expensive travel and equipment, as has quilting, I would add, leaving huge carbon footprints right across

great bird habitat. Of course, extreme birders wouldn't know about most rarities if local birders weren't regularly examining their local patches.

Just the week before reading Ted's patch reference, I finished reading "Life List" by Olivia Gentile, a biography of Phoebe Snetsinger. Phoebe was the woman determined to see as many of the world's bird species as possible.

She started birdwatching in 1965 but became obsessive about it after being diagnosed with terminal melanoma in 1981. Aided by an inheritance from her father, she went on multiple foreign bird tours every year. She valiantly endured bad weather, bad trails, and bad men, finally dying in a vehicular accident in 1999 in Madagascar, leaving a worldwide record of nearly 8,400 bird species, the most anyone had seen at that time.

We can charitably say

Phoebe was birding before carbon footprints were in our vocabulary, and that extreme birding kept her sane and kept professional bird guides and tour operators employed. I hope someone has transferred her carefully kept note cards to eBird, the digital archive where scientists can make use of our personal birding observations.

Soon after Ted's reply, I got an email from the Cornell Lab of Ornithology describing a new eBird feature: patch and yard birding record keeping set up to allow for friendly competition within one's county. It will also give ornithologists more intensive information about birds. I imagine Ted knew all about this when he wrote his blog post – the world of professionals in birding is very small.

So now there is a name for the kind of birding most of us do. Most of us who begin to keep notes on the birds in our own backyards are

already patchwork birding. I highly recommend www.eBird.org as a record keeping alternative to notebooks and scraps of paper.

Ted thinks patchwork birding is the responsible, green way to bird – no great amounts of fuel are wasted in long distance travel.

It's amazing how many species of birds pass through my favorite patches: 50 in my backyard and a different 50 in Holliday Park here in Cheyenne since April 2010, when I began recording sightings on eBird. That's not a lot of species among obsessed birders. However, frequently birding those areas helped me know exactly where to find an American kestrel for the Cheyenne Christmas Bird Count.

I've been thinking about how to control the size of my patchwork quilt making carbon footprint. Maybe I should spend less time quilting and more time walking around town watching birds.

# 307 Bird feeder quarantine was good for the birds, hard on the observer

Sunday, February 27, 2011, Outdoors, page E3

There he was, the lone house finch on the tube feeder, contentedly pulling black oil sunflower seeds out and munching them thoughtfully, left behind when the rest of the flock scattered.

I took a closer look and just as I feared, he showed outward signs that he was not a well bird, despite his glowing red head and chest. Eye disease. One eye was

encircled in rings of crusty featherless wrinkles.

Sick birds conserve their strength. They don't fly off with the flock for every little perceived threat. When they get really sick, I've seen them huddle on my windowsill.

There isn't anything anyone can do for them, but I can protect the rest of the birds by taking down my feeders which will get the

house finches to disperse and be less likely to pass diseases to each other or to other finch species.

This winter, I've had quite a regular crew showing up every day: two Eurasian collared-doves, two mountain chickadees, two red-breasted nuthatches, a downy woodpecker, 10 or so house finches and a few juncos with occasional appearances by pine

siskins and goldfinches. I really hated to disappoint them.

After I took the feeders down, dumped out the birdbath and swept all the seed debris off the patio, I watched later as the gang sat on the powerline while one or two individuals would sally forth and fly a circle around the last known location of the sunflower seed feeder. Then they left.

In a week, after Mark scrubbed the bird poop off the railing and patio and washed out the feeders with a mild bleach solution, he refilled the feeders, and nearly all the previous birds began to reappear within a day. The chickadees took five days.

I don't know where the sick house finch contracted his disease, but I do know that we had gotten behind in cleaning up our feeders and the area around them.

Feeding birds is something we do because we enjoy watching birds up close. The birds usually don't depend on feeding – they have plenty of naturally occurring seeds to forage, but they sure enjoy the convenience.

Songbirds can get a pox that affects their eyes. But there is also house finch eye disease, mycoplasmal conjunctivitis, caused by a bacterium common to domestic turkeys and chickens, according to the Cornell Lab of Ornithology. It was first noticed in house finches in 1993 on the east coast, where house finches were introduced 50 years earlier.

The Lab has been studying the spread of the disease for the last 15 years using observations provided by birdwatching citizen scientists. After a major outbreak on the east coast a few years ago, the disease is no longer quite as prevalent. Some sick individual birds actually survive but apparently do not become immune to the disease.

The eastern house finches seem to be more susceptible, and one reason might be that most of them are thought to be inbred descendants of a small group that was introduced in the east in the 1940s from western North America where they are native. Inbreeding can cause susceptibility to disease.

The good news is, this is one avian disease that does not pose a health risk to people, except that quarantining our feeders and losing "our" birds for a week felt like a mental hardship. But then again, perhaps I accomplished more in that time because I wasn't distracted by the comings and goings I like to watch out the window above my laptop screen.

Some people watch fish swim in a bowl to reduce stress. Watching birds outside my window works for me. I'm glad the gang came back.

# 308 Book Review

Monday, March 14, 2011, ToDo, page D4

## New field guide is so much more than its title implies

"Identifying and Feeding Birds," new in the Peterson Field Guides series, has a misleading title.

It is about so much more.

Author Bill Thompson III, also editor of Bird Watcher's Digest, covers the four basics of what birds need: food, water, shelter and a place to nest. The chapter covering bird feeders and different kinds of bird food is what you would expect. He also talks about bird-friendly plants for your yard, the birdbaths rated most popular by birds and how to build a birdhouse and situate it properly.

Thompson writes in a breezy, fun to read style and includes his personal backyard bird experience, but he's not afraid to point out it's not enough to just hang the right feeder.

Birds don't usually depend on us to supplement their wild food.

"Knowing (as we do now) that we feed birds so that we can enjoy them up close, we also need to understand that we owe it to our avian friends to feed them responsibly," Thompson writes. "By this I do not mean simply that we feed them the proper foods in the correct feeders….Rather, I mean that we need to make sure our backyards – feeders and all – are safe for birds."

Thompson covers safety issues such as cats, lawn pesticides and moldy seed.

Half the book is devoted to accounts, photos and range maps for 125 common backyard birds in North America. This means novice birders in Cheyenne need to check the maps before putting out food meant to entice a species we don't normally see. Go to http://org.lonetree.com/audubon/cheyennechecklist06.pdf to get a better idea of birds in our area.

But otherwise, as someone frequently asked to give advice on attracting backyard birds, I highly recommend this book and Thompson's catchphrase: "Feed more birds, have more fun."

Identifying and Feeding Birds (Peterson Field Guides) by Bill Thompson III, c. 2010, Houghton Mifflin Harcourt, paperback, 256 pages, $14.95.

# 309 Are roadrunners enroute to our residential neighborhoods?

Sunday, March 20, 2011, Outdoors, page E3

Out for a walk with the dog in early February, I noticed when her attention was galvanized by wildlife on the lawn across the street.

The motionless animal was the same color and size as a cottontail, but not the same shape. It was a roadrunner, a long-legged, long-tailed, slender bird with a shaggy crest. We were in my mother's neighborhood in Albuquerque, and she had been telling us about the roadrunners that frequent her yard and nested last year in the neighbor's pine tree.

Roadrunners are the classic symbol of the desert

2011

293

Southwest, but Mom's neighborhood is a dead ringer for any of Cheyenne's, except that the houses are flat-roofed and adobe-like, and the shrubs and big trees, both coniferous and deciduous, are different species.

So, when did the epitome of wide-open desert move to town?

I looked up "Greater Roadrunner" in my favorite compendium of avian research, The Birds of North America Online, http://bna.birds.cornell.edu/bna/, and found they have been expanding their range since early in the 20th century, north to southern Colorado and Kansas, and east to Arkansas and Louisiana. They often move into atypical habitat on the fringes. But since they also like desert riparian areas – shrubs and trees along creeks – they must find long established residential neighborhoods similar.

If roadrunners keep progressing northward, we might see them here. Cold doesn't seem to be a problem. Albuquerque had subzero temperatures and this bird survived to walk across the street in front of us a week later.

**Does Cheyenne have what a roadrunner wants?**

--**Roads.** Check. They really do prefer to travel roads, trails and dry creek beds, running as fast as 18 mph. They seldom fly, usually just taking short flaps to get to a perch or nest.

--**Food.** Check. Some fruit and seeds in winter, but mostly snakes, lizards, spiders, scorpions, insects, birds, rodents, bats – well, we might be a bit short on lizards and scorpions, and I might have a problem with their penchant for hanging around bird feeders and picking off songbirds, but we already

provide prey for sharp-shinned hawks. Roadrunners can also nab hummingbirds which might explain why Mom hasn't had as many at her feeders recently.

--**Shelter.** Check. In cold weather, roadrunners roost in dense shrubs rather than migrate. Cheyenne has lots of spruce, pine and juniper that could fit the bill. Persistent snow cover over a large area could be a problem, but not often.

--**Dust and sun.** Check. Roadrunners take dust baths, not water baths. They also like to sit with their backs to the sun, wings lifted a bit, for hours at a time.

--**Water.** Check. They only need a drink if their food hasn't been juicy enough. Their physiology has evolved to recycle water in their digestive tract, a handy thing for a desert dweller.

--**Nest sites.** Check. Just need dense bushy shrubs or

trees. As we might expect from their cartoon acting experience, not only can the adults draw a predator's attention away from the nest by faking a broken wing, they can also fake a broken leg. I wonder what Cheyenne's feral and loose cats would make of that?

--**Subscription to Equality State values.** Check. Both parents build the nest, incubate the eggs and feed the nestlings, although the female makes a whining call if the male gets distracted in his search for twigs for nest building.

It might be a while yet before roadrunners run this far north. In the next 20 years, if you hear their distinctive coo calls – they are in the cuckoo family – don't be surprised. After all, Native American tradition identifies roadrunners as symbols of courage, strength and endurance.

# 310 A year in search of butterflies

Monday, April 4, 2011, ToDo, page D1

## Butterfly "Big Year" captures heart of one man's passion

Competitive birders will attempt a Big Year, but in 2008, Bob Pyle was the first to see how many butterfly species can be counted by one person in one year in the U.S., and he reports his results in "Mariposa Road."

Bob is a writer, naturalist and lepidopterist who has authored several butterfly field guides and who, since the 1960s, has cultivated a shrewd knowledge of butterflies, their favorite plants, and people who know where to find both, and when.

He traveled on a

shoestring, often camping along the roadside in his 1982 Honda Civic hatchback, affectionately named Powdermilk.

Bob has affectionate names for his favorite butterfly nets, too, Marsha and Akito, and has an endless supply of affection for every butterfly and butterfly lover he's ever met.

Every sentence sparkles with optimism like a Florida purplewing bouncing across a swampy hammock. Every foray into the field holds hope for rarities and beauty, even

if experience would point out the chiggers and thorns. There is always a refreshing mug of beer on tap afterward, or dinner with friends.

Bob did make it to Wyoming for a couple pages, mostly reminiscing about Karolis Bagdonas and his Flying Circus, a band of students that "careened around the Rockies doing butterfly counts and sampling little-known habitats, subsisting on Hamm's and trout...." I remember hearing about them more than 30 years ago.

Bob's goal was to see 500

of the 800 known species in the U.S., including Hawaii. He made it to 478 species certified by three experts. He found 30 of the 40 "holy grails," hard-to-find species he'd hoped for. And almost as a footnote in his appendix, he mentions 600 donors to his Butterfly-a-thon raised $46,000 for the Xerces Society which protects wildlife through the conservation of invertebrates and their habitat.

At more than 500 pages, this could be heavy reading if you aren't already a butterfly fan. But if you like a good road trip and have a butterfly field guide handy to supplement the color photos on the end papers, I think you'll enjoy the read.

Mariposa Road, The First Butterfly Big Year, by Robert Michael Pyle, c. 2010, Houghton Mifflin Harcourt, hardcover, 558 pages, $27.

# 311
Thursday, May 19, 2011, ToDo, page D4

## Wyoming Roadside Attraction: The Wyoming Dinosaur Center, Thermopolis

### See the Supersaurus

Supersaurus. Barnum and Bailey would never believe it. But you can see it at The Wyoming Dinosaur Center in Thermopolis, all 106 feet of it. "Jimbo," as it is nicknamed, was excavated by the center.

Castes of more dinosaurs, their actual fossilized bones and fossils of pre-dinosaur life forms are professionally presented in the museum.

Through windows, visitors can watch lab assistants painstakingly pick away at the rock matrix that surrounds other fossils that come from the center's nearby dig sites.

The center is on the Warm Springs Ranch, where 60 dig sites have been identified in the Morrison Formation of the late Jurassic, the time of the well-known species Stegosaurus, Allosaurus, and Aptosaurus, previously known as Brontosaurus.

Visitors can tour several of the dig sites, and in summer, the center offers "Dig-for-a-Day," a chance to help scientists make more finds. Kids' camps are also offered.

The gift shop's selection of educational toys, books and fossils rivals those of major museums and is also available online.

*Cutlines*

"Stan," a 35-foot T-Rex, and a Triceratops (the Wyoming state dinosaur) were excavated near Thermopolis and fill the Wyoming Dinosaur Center in Thermopolis. Courtesy of The Wyoming Dinosaur Center.

### The Wyoming Dinosaur Center, Thermopolis

**Directions:** In Thermopolis, at the junction of US Hwy 20/Wyo. Hwy 789 and Wyo. Hwy 120, turn east on Broadway Street and cross the Big Horn River. Follow it as it becomes East Broadway and turn south (right) on C Avenue which becomes Carter Ranch Road within a few blocks. Or, from Hot Springs State Park, the center is just a few blocks south of the park boundary.
**Distance from Cheyenne:** about 300 miles.
**Open:** 7 days a week, 8 a.m. – 6 p.m. summer (May 15-Sept. 15), 10 a.m. – 5 p.m. winter.

**Admission: Museum:** Adults $10; children, seniors and veterans: $5.50; children under 3: free. Dig site tours run $12.50 to $8.75 and museum/dig site packages are available for individuals and families. Dig-for-a-Day is $150 for adults and $80 for children.
**Address:** 110 Carter Ranch Road, Thermopolis, WY 82443
**Phone:** 800-455-3466
**Website:** www.wyodino.org
**Attractions:** Fossils and castes of dinosaurs and other prehistoric animals, tours of dig sites, dig participation, gift shop.
**Time:** Museum: 1 hour, Dig tour: 1 hour, Digging: a day.

# 312
Sunday, May 22, 2011, Outside, page E3

## How to raise a birder: take a child outside

There are three attributes most really good birders share: terrific eyesight, terrific hearing and a mind like a sponge. These attributes describe most children, too, unless they ruin their eyesight with too much screen time or their hearing with loud music or fill their minds with rules for arcane video games.

Can children become really good birders? Yes. Years ago, our Audubon chapter received a call from a mother wondering if her junior-high-aged son, Jason, could come with us on a field trip. He had birded regularly with folks in California before the family moved to Cheyenne. So, we said sure. If he'd been birding with adults before, he knew what he was getting into spending a day with us.

Jason turned out to be a very personable young man and his young eyes and ears helped us find species we might have missed otherwise. Plus, he'd studied up on the birds in our area. Thanks to Jason, I saw my first green-tailed towhee.

Sad to say, he didn't grow up to become an ornithologist. Last we heard, he was at Harvard and on his way to becoming a pediatrician, but even pediatricians have hobbies, as illustrated by famous Wyoming birder and pediatrician Oliver Scott. I wouldn't doubt Jason is still adding to his life list.

How does a child become a birder? Famous birders usually point to a "spark" bird that sparked their interest as a child. For Roger Tory Peterson, famous for inventing the modern field guide, it was a blue jay he saw as a grade schooler.

RTP was an independent-minded boy who spent days roaming the local woods on his own and looking things up at the library. Eventually, he discovered other people interested in birds, finding that accompanying a birder better than him out

2011

295

in the field is faster than reading a book for improving birding skills.

The American Birding Association sponsors two summer birding camps for young people ages 13-18, and they don't lack for applicants though it seems it would be much more difficult now for children to catch the spark, with parents less likely to let their children roam and more likely to over-schedule them for after school activities.

How does someone who may not know much about birds encourage a child to develop an interest in them?

First, children have to see that birds exist almost everywhere. The more time they spend outside, the more birds there are to notice. Wondering what kind of birds they are leads to looking them up in a field guide. Later, binoculars become important for seeing details.

A field guide is a most wonderous thing. Years ago, our local Audubon chapter gave out plaques to winners we chose at the school district's elementary science fair. But for the same cost, we started to and still do, give the winners "adult" field guides which are as full of colorful pictures as any children's book.

Not every child with an interest in birds is going to grow up to be an ornithologist, just as every high school violinist isn't going to go on to play with the New York Philharmonic. But that's OK. They can still have a lifelong love of birds (or music) and appreciation for people who make birds (or music) a career.

Beginning birding skills, such as discrimination between species, noticing details and researching bird information, are all transferrable to other aspects of life.

So now, as the birds increase in number for the summer breeding season, take a child outdoors and see what you can see. You never know what it will lead to.

# 313 Wyoming Roadside Attraction: Nici Self Museum

Wednesday, May 25, 2011, ToDo, page D4

## Centennial museum specializes in mining, ranching and timber history

Museums in small western towns have collections of predictable items, but at the Nici Self Museum in Centennial, 28 miles west of Laramie, there are a few unusual mementoes, such as the hand-cranked sock knitting machine.

Some artifacts are too large for the main museum building, such as the 1914 fire engine on loan from the Laramie Plains Museum or the tipi-shaped sawdust burner, which is as large as a building.

Housed in the former depot for the now defunct Laramie, Hahn's Peak and Pacific Railroad, the museum specializes in local history back to 1875 when the new local gold mine was named for the nation's upcoming centennial.

When the mine closed, the nearby town by the same name continued, supported by prospectors, ranchers and timber men. Now it has become a gateway for people wanting to picnic, hike, fish, hunt, camp, photograph, ski and snowmobile in Medicine Bow National Forest.

Centennial has a full calendar of weekend events. Check http://www. centenniallibrary.net/calendar.html.

*Cutline:*

Though the Laramie, Hahn's Peak and Pacific Railroad and its tracks are long gone, the depot at Centennial remains, housing a collection of artifacts illuminating a history of prospecting, mining, ranching and timber harvesting. Barb Gorges/Special to the WTE

### Niki Self Museum, Centennial

**Directions:** Take I-80 Exit 311 at Laramie, then Wyo. State Hwy. 130 west 28 miles.
**Distance from Cheyenne:** about 75 miles.
**Open:** Thursday through Monday, noon to 4 p.m., Memorial Day through Labor Day. In September, Saturdays and Sundays, 1-4 p.m.
**Admission:** Donations appreciated.

**Address:** 2734 Hwy. 130, Centennial
**Phone:** 307-742-7763 off-season, 307-745-3108 during museum hours.
**Attractions:** Objects and buildings illuminate the area's past in ranching, lumbering, mining and railroading.
**Time:** 30 minutes to 1 hour.

# 314 Pelicans at Holliday Park: Why do they stop here?

Sunday, June 5, 2011, Outdoors, page E3

The "American White Pelican," with its fantastic orange bill, seems more akin to a unicorn than any bird. My first sighting, above a river on the dry plains of eastern Montana, seemed out of place. But to see a flock at Holliday Park, in the middle of Cheyenne, seems miraculous.

Last year I tracked my sightings of pelicans in the park on Lake Minnehaha between April and July. On the three or four mornings a week I was there, I could almost always see a few.

I didn't observe any of the courtship antics like the kind the geese go through. The pelicans were either clumped together on the north side of the island, like a forgotten snowbank, ignoring the ruckus of nesting geese a few feet away, or they were out elegantly sailing on the water,

296        CHEYENNE BIRD BANTER

dipping their bills in and catching fish, or just loafing. A few times I saw them sail into thermals overhead, wing flapping hardly a necessity.

Why do we see pelicans at Holliday Park? In April, some may be migrating. They winter along the Gulf of Mexico and head for the western states, Colorado, Wyoming, Montana, the Dakotas and western Canada, to breed. Pelicans on the other side of the Continental Divide spend winter in coastal California.

They look for big lakes in isolated places that have islands for breeding safely, such as the Great Salt Lake. In Wyoming there are three colonies: Bamforth National Wildlife Refuge, Pathfinder Reservoir and Yellowstone Lake. Any disturbance will cause them to abandon their nests.

Each breeding pair looks for a sub colony where everyone else is at about the same stage in the breeding cycle. The nest is whatever a sitting pelican can pull up around itself with its bill. A parent keeps the two eggs warm underfoot and every day or two, for 30 days, swaps places with its mate and goes out as far as 30 miles to forage.

Three weeks after hatching, the chicks are all out of their nests and huddle at night in a group called a crech. Ten weeks after hatching, they are in the air, exploring everywhere until it's time to head south. By September, most pelicans have left Wyoming.

It takes three years before pelicans are old enough to breed so these immature birds spend their summers hanging out. Apparently, Holliday Park, with its shallow water, is perfect despite the burgeoning flock of domestic and Canada geese and all the people walking dogs.

I sometimes wonder what the uninformed observer thinks pelicans are. Mutant swans or domestic white geese? At 16 pounds, they are much bigger than most geese and a little smaller than trumpeter swans. But at 108 inches, or nine feet, their wingspan is much longer than a swan's.

The brown pelican found on the coasts dives for fish, but the white pelican merely dips its pouched bill while swimming. These are the pelicans that are famous for synchronicity, forming a line and forcing small fish into the shallows of rivers, lakes or marshes where they can be scooped up. They also fish at night, by feel, rather than by sight.

Down around the Mississippi Delta, they are the bane of aquaculture, feeding at fish farms, but out here, they eat the rough fish and are generally not in competition for sport fish.

In the early 20th century, their populations decreased, but in the 1960s they began to make a comeback. So, your chances of seeing a pelican, compared to a unicorn, are pretty good, especially if you stop by Holliday Park on summer mornings.

# 315 Wyoming Roadside Attraction: South Pass City State Historic Site

Thursday, June 9, 2011, ToDo, page D1

## Wyoming's second town to be incorporated survives as historic site

Gold fever produces towns in the most unlikely places.

But when it cools, it's hard for those towns to make ends meet.

South Pass City was the second town to be incorporated in Wyoming – Cheyenne was the first.

But the fates of the two couldn't have been more different. The second is now a professional ghost town.

South Pass City was founded in 1868 with the discovery of the Cariso Lode and the opening of the subsequent Carissa Mine.

With the bust in 1872, it was forced to diversify with ranching, timbering and market hunting. By 1949, it became a ghost town and was run as a private tourist attraction for 20 years.

Today, the State of Wyoming maintains the two dozen buildings, including cabins, stores, saloons and hotel, looking as fresh as their previous owners would wish.

Pick up the newspaper at the visitor information area/ dance hall that describes everything. Ask questions there or at the general store where you can pick up Wyoming souvenirs.

The Carissa Mine's head frame and trestle reconstruction were completed in 2008.

Tours of the mill house are given on Saturdays and Sundays, June through August, by reservation only. Call 307-332-3684 or talk to the Dance Hall attendant.

*Cutline:*

In South Pass City, the front of the Carissa Saloon receives a new coat of paint, even though it has been out of business since the 1940's. Barb Gorges

## South Pass City State Historic Site

**Directions:** From Lander, take U.S. Hwy. 287 south 9 miles. Turn off on State Hwy. 28 and follow it about 20 miles to the turn off.
**Distance from Cheyenne:** about 290 miles.
**Open:** Buildings are open May 15 – September 30, 9 a.m. – 6 p.m. daily. Grounds are open year round, sunrise to sunset.
**Admission:** $2 per vehicle for state residents.

**Address:** 125 South Pass Main, South Pass City.
**Phone:** 307-332-3684
**Website:** http://wyoparks.state.wy.us
**Attractions:** Historic buildings and furnishings, gift shop, staff on site to answer questions.
**Time:** 1 – 2 hours.

2011

# 316 Planning for serendipity makes for a satisfying birding trip

Sunday, June 19, 2011, Outdoors, page E3

Mark and I are learning how to add birding opportunities to obligatory travel. A graduation, in the middle of spring migration, has to be the best excuse for a birding adventure.

Our younger son, Jeffrey, graduated mid-May from Worcester Polytechnic Institute in Massachusetts (mechanical engineering) so Mark and I studied the maps and www.eBird.org in advance to determine where and what birds would be present.

Our first destination was Cape Cod, more for its national icon status than the birds. We stayed at a B & B in Yarmouth Port, in a house built in 1730. A chance encounter with the director of the Edward Gorey Museum next door led us down a path to a whole network of woodland trails and ponds.

Luckily, this deciduous forest was less than half leafed out. We were delighted to be able to see our first eastern birds of this trip – northern cardinal, tufted titmouse and black-capped chickadee – birds we don't see at home.

At Cape Cod National Seashore, it was a bit past prime seabird migration. Things didn't get birdy until we got to Plum Island, managed mostly by Parker River National Wildlife Refuge and located off the northeast shore of Massachusetts, by Newburyport.

We walked softly along the boardwalk on the Hellcat Interpretive Trail, looking for any signs of life and then, in one magic spot, warblers began to appear. Other birders told us the weather was driving migration inland, but we were satisfied with the total of 13 warbler species we found on our trip, half of which would be considered rare in southeast Wyoming.

Quite by chance, our visit coincided with the regular Massachusetts Audubon (always abbreviated in print and speech as "Mass Audubon") Wednesday morning field trip leaving from Audubon's Joppa Flats Education Center, across from the refuge's headquarters. Yes, it cost $15 per person, but we rode in Mass Audubon vans and the leaders showed us northern gannets and Caspian terns – birds we would have missed.

Hearing we would be in Boston our last day, everyone told us to go to Mount Auburn Cemetery, which is actually in Cambridge. Three of our relatives were happy to join us.

Mount Auburn was developed in 1831 at the beginning of a movement to make cemeteries a place for the living to enjoy and not just rows of tombstones. The 175 acres of hills and dells had a reputation for birds before it was landscaped as an arboretum and cemetery.

Fine rain plagued our visit on and off, but the blooming dogwoods, azaleas and rhododendrons were spectacular, contrasting nicely with elaborate tombs and the Gothic chapel and tower.

Scattered along the paths and charmingly named lanes were other birders, some more outgoing than others. One woman went out of her way to see if she could refind a special bird for us. Instead, we were all treated to a memorable performance of a wood thrush singing from a bare branch. It sounded better live than all the recordings I've heard.

Then we were down to our last afternoon and my sister's request to really experience the Atlantic. Mark remembered an earlier visit we made to Marblehead. It is less than an hour from Boston and was very quiet before the summer season. I was happy to see once more common eiders, those sea ducks with fanciful, lime green bills.

After five trips to Massachusetts in four years, this will be the last one for a while. We're expected to show up in a couple years for our son's next graduation, in Seattle. More new birding opportunities!

# 317 Boysen State Park

June 28, 2011, ToDo, page D1

## Make Boysen Reservoir a short stop for shade or a long layover for fishing

Like so many of the large reservoirs in Wyoming, the country around Boysen Reservoir is rather bleak and treeless except in the dozen Boysen State Park campgrounds/picnic areas where trees have been planted. But the exposed geology is interesting.

The biggest trees are in the campgrounds below the dam, growing in the silt left at the site of the original dam and hydro-electric plant built by Asmus Boysen in 1908 to service the growing gold and copper mining industry.

The big draw here is boating and fishing on a 15-mile-long body of water. You'll have to bring your own boat as the marina no longer offers rentals. You might check with the boat dealer in Shoshoni.

The main game fish include walleye, sauger, perch, crappie, ling and rainbow, cutthroat and brown trout.

State fishing licenses are available at the marina.

*Cutline:*

A hydro-electric dam was built at this site by Asmus Boysen in 1908. When it silted in and flooded railroad tracks, it was dismantled and the current dam was built further up the Wind River in the 1950s, providing a reservoir popular for recreation.  Barb Gorges/Courtesy

## Boysen State Park

**Directions:** U.S. Hwy. 20 between Thermopolis and Shoshoni.
**Distance from Cheyenne:** about 280 miles.
**Open:** Year round, except drinking water and restrooms are not available Oct. 1 through April 30.
**Admission:** For residents, the daily use fee is $4 per vehicle per day, and the camping fee, which includes the daily use fee, is $10 per night.

**Address:** 15 Ash, Boysen Route, Shoshoni
**Phone:** 307-876-2796
**Website:** http://wyoparks.state.wy.us
**Attractions:** Campsites are available by reservation as well as first come first served. Fishing, boating and picnicking are popular activities.
**Time:** 20 minutes or a whole weekend.

# 318

Sunday, August 7, 2011, Outdoors, page E3

## Killer kitchen window adds to national bird death toll

It became an almost daily occurrence this past May: a soft thump on the glass as another visitor to our backyard and bird feeders hit the kitchen window.

I tried putting up big stripes of blue painter's tape and that helped a little. Mainly, we needed to remember to check the yard for dazed birds before letting our bird dog out. Luckily, the couple we saw her take seemed to be the only window fatalities – that we knew about.

Migration seemed to be over after the first week in June. With just the regulars now, including the house finch with the white head and her friends, there have been no more collisions.

It's the visiting birds, 25 other species in our yard, mostly during spring migration and not so much in the fall, which get confused by a window that reflects the trees. At least with the feeders within only a few feet of the window, they didn't have

a lot of momentum when they hit. But it wasn't always feeder birds, and it wasn't always that window. In fact, it is windows everywhere.

Audubon Minnesota, a state office of the National Audubon Society, came out with a booklet available online for free, Bird-Safe Building Guidelines, that explains the problem and solutions at the architectural level. Go to www.mn.audubon.org, where there is a link to the 40-page pdf or put the booklet title in the website's search window.

People love buildings with lots of natural light. If they are well-insulated, windows can save energy on lighting. But those glass-sided high-rises, and even residential windows, are calculated to kill hundreds of millions of birds per year in the U.S. Spring and fall migration are the most problematic times of year.

Anything that breaks up the glass expanse, like my painter's tape (engineered to

peel off easily), or screening, helps. It has to be applied to the outside surface, though. Waiting to clean the dust off the outside of your windows until after migration might help, too, as can drawing drapes and shades to prevent birds from seeing straight through your house to a window on the other side.

Scaring the birds away with items hanging in front of the window, things that move in the breeze, like flagging, or that move and shine, like old CDs, might do the trick.

At the architectural level, Bird-Safe Building Guidelines discusses ideas for new buildings and retrofitted buildings: netting, fritted glass, films applied to glass, etched glass designs, glass sloping to reflect the ground, taking into account proximity to habitat and feeding sites, and special glass making use of birds' ability to see ultraviolet patterns we can't. Problems occur mostly at the

ground level and first few stories of buildings.

Then there is the problem of lit-up buildings attracting night-flying migrants, especially during bad weather, and all the ill effects of light pollution in general. Turning off interior building lights at night saves money and birds, and so does making sure outdoor lights are not needlessly lighting the sky.

Our killer kitchen window, a six-foot wide replacement with sliding halves, currently is only half screened. The track is still there for the full screen, so we should order one and put it in place from early April to mid-June. All the glass will be less reflective then, and birds still colliding may bounce off the screen.

But next spring, before we let Sally out, we'll check the area below the window for dazed birds. No need to increase the dog's "life list."

# 319

Sunday, August 14, 2011, ToDo, page D4

## Try the Wind River Canyon, by road or river

It's disorienting to have to have a river flow north, which the Wind River does. And after it goes through

the Wind River Canyon, it becomes the Big Horn River, which flows through Thermopolis and eventually flows

into the Yellowstone River in Montana. What's even more disconcerting is that as you drive up the river, you could

swear you were dropping in elevation.

The canyon cuts through the Owl Creek Mountains,

2011

299

through 10 different geologic formations. Look for the names on signs as the rock walls change color. The canyon stretches 15 miles long and 2,400 feet deep.

The safest way to stop and enjoy the scenery along the way is to drive south from Thermopolis. That way all the pullouts along the river will be on your right side, and you won't have to cut across oncoming traffic to access them.

The more exciting way to enjoy the canyon would be to make a reservation for a raft trip.

At the south end of the canyon, both the highway and the railroad tracks on the other side of the river go through several short tunnels.

*Cutline:*

Ten geologic formations are on display in the Wind River Canyon as U.S. Hwy 20 squeezes between the rock walls and the river.
Barb Gorges

### Wind River Canyon Scenic Byway

**Directions:** The 15 miles of U.S. Hwy. 20 south of Thermopolis.
**Open:** Year round except when weather causes highway closures.
**Admission:** None. The highway is the responsibility of the Wyoming Department of Transportation and is within the Wind River Indian Reservation.

**Attractions:** Geologic formations and tunnels. The reservation recently granted permission to the Wind River Canyon Whitewater company to guide raft trips between Memorial Day and Labor Day. Call 888-246-9343 for information.
**Time:** 20 minutes if no stops are made.

# 320 Rendezvous site attracts living history

Saturday, August 27, 2011, ToDo, page D4

### Re-enactors bring the era of the mountain man back to life.

The era of the mountain man fur trapper was brief, but it lives on at the actual site of the 1838 rendezvous site.

On the south edge of the present-day city of Riverton, on the banks of the Wind River, near the confluence with the Little Wind River (also known as the Popo Agie), you can find markers about famous characters along primitive paths.

Squint a bit and you can imagine hardy men wearing buckskins trading furs and sharing jugs of whiskey and stories at their annual summer gathering.

If you are very lucky, instead of ghosts, you will see real buckskins worn by re-enactors if you arrive in time for the annual event put on by the 1838 Rendezvous Association. The activities are usually scheduled in late June, everything from hatchet throwing to hide scraping. Visitors need not wear authentic garb.

*Cutline:*

Use your imagination to see the 1838 rendezvous of mountain men trading furs and stories, or visit June 29-July 3 this year when the rendezvous is reenacted.
Barb Gorges

### 1838 Rendezvous Historic Site

**Directions:** From downtown Riverton, follow Wyo. State Hwy. 789 (South Federal Blvd.) south and turn east on Monroe Avenue, past the gravel pit ponds, following signs.
**Open:** year-round
**Admission:** none
**Address:** east end of E. Monroe Avenue, Riverton, Wyo.

**Phone:** 1838 Rendezvous Association, John Boesch, 307-856-7306
**Website:** www.1838rendezvous.com
**Attractions:** Location of an actual mountain man rendezvous site with historic markers.
**Time:** 15 minutes – 2 hours.

# 321 Improve your bird and butterfly eye

Tuesday, September 6, 2011, ToDo, page D4

### New bird and butterfly book can improve your yard and a new field guide will improve your bird ID skills.

Bill Thompson always writes with the casual birdwatcher in mind, the person who appreciates birds but is always saying, "Someday, I want to know more." And then he provides the hook.

This time, he has concentrated on hummingbirds while his co-author, Connie Toops, brings us butterflies. First, there is everything you wanted to know about hummers, a few myths dispelled (no, they don't hitch rides on geese during migration) and the basics of putting up a hummingbird feeder (4 parts water to 1 part white table sugar – no substitutes, additions or changes, please, and keep it clean).

But Bill offers not only a field guide to 15 North American hummer species, he has a chapter on plants hummingbirds like and planting ideas to maximize views of them feeding on flower nectar.

Connie does the same things for butterflies in the second half of the book, including how to photograph them.

Is this book worthwhile

300                    CHEYENNE BIRD BANTER

for folks in southeastern Wyoming? Yes, even if we have mostly broad-tailed hummingbirds during the summer in higher country, rufous hummingbirds during migration in July. And keep in mind, only 17 butterflies profiled are expected here in summer. The plant information is sorted by parts of the country, and as you might expect, all the recommended species would be colorful additions to your yard.

If you already garden, you'll find what a little tweaking might do to improve your chances of observing butterflies and hummingbirds. And if you don't garden, you might be inspired to begin with a container of colorful blooms.

-----

Kenn Kaufman has totally rewritten his 20-year-old guide to advanced birding. It's not only because he has

better ideas for identifying difficult species, but he understands better the kinds of mistakes birdwatchers make – though perhaps the most important lesson for an ardent birder to learn from this book is that sometimes a bird cannot be identified.

Anyone who has mastered identification of the common and colorful birds soon finds that there are groups that are difficult: gulls, sparrows, hawks, shorebirds, flycatchers, swallows and seabirds.

And what about hybrids? Subspecies? Birds with white feathers where they shouldn't be? Birds that are molting and missing feathers?

While there are chapters for each of the difficult bird groups and a reader might be tempted to jump right to his nemesis species, the first seven chapters of general information are worth studying, especially the list

of 14 Principles of Field Identification and the 14 Common Pitfalls.

Study is key to improving bird identification skills, and not just studying books, but going outside and finding more birds, Kaufman points out frequently.

One frustration with this book is that there are more than a few photos of unidentified birds. The caption might say the four gulls pictured could be identified as five or six different species if compared to various field guide illustrations, but there is no key in the back to check to see if you can identify them correctly. And sometimes Kaufman's sentences get long and twisty because some bird i.d. problems are as much about the exceptions as they are about the rules.

A lot of information is packed into this book and it can be overwhelming.

But Principle 14 says you don't have to take on all of the challenges. You can note merely "Gull sp. (species)."

Kaufman says, "As long as you're not causing serious disturbance to the birds, their habitat, or other people, there is no 'wrong way' to go birding."

Amen.

---

### 2 books worth reading

Hummingbirds and Butterflies (a Peterson Field Guides Backyard Bird Guide), by Bill Thompson III and Connie Toops, c. 2011 by Bird Watcher's Digest, Houghton Mifflin Harcourt Publishing, 288 pages, softcover, $14.95.

Kaufman Field Guide to Advanced Birding, Understanding What You See and Hear, by Kenn Kaufman, c. 2011, Houghton Mifflin Harcourt Publishing, 448 pages, flexible cover, $21.

---

## 322 Crows come home to roost, bringing entertainment and fertilizer

Sunday, September 18, 2011, Outdoors, page E3

This summer, our street became the unwilling host to loud and raucous juveniles coming by at dusk and leaving messes in our driveways. They were young crows, coming home to roost in our neighborhood's big trees.

Since they settled down shortly after sunset, I could live with the noise, but they were back at it at 5 a.m., as good as an alarm clock through our open windows.

Mid-spring this year, I noticed crows flying with sticks. I think they built a nest in a neighbor's spruce tree. It wasn't too surprising in early July to hear

inexpert cawing.

One evening three young crows put on a performance for us. Up on the cross arms of the utility pole they restlessly moved about, one having to flutter its wings every time it wanted to step over an obstacle. Another one, on the high wire, kept flapping crazily to keep its balance. All three kept up a conversation with inflections of babies learning their mother tongue.

With binoculars it was easy to see the young crows' feathers weren't fully grown out, but otherwise, their plain awkwardness gave them away. For a week or two, I

could find four of them in our trees or somewhere on our street, walking together, inspecting lawns.

Walking out one day while an adult was with them, I was suddenly the focus of a tirade of abusive adult language. What was the backstory? One should not daydream while walking the dog.

Last summer, my bird dog quietly picked up a young crow out of tall grass on the side of the road while I wasn't looking. It probably had been hit by a car. A study of banded nestlings in Illinois shows that the mortality rate within a crow's first year

is 57 percent.

The dog proudly carried the limp bird home and was all smiles all the way, even though the parent birds followed us, yelling. Was this current, irate parent one of the birds that followed me and the dog home last year?

At a recent meeting of the Cheyenne – High Plains Audubon Society, we'd watched a documentary, "A Murder of Crows," about researchers proving that crows recognize faces and will convey to their young who to watch out for. I guess I am a marked woman, except it wasn't me that killed the young crow last year. I only provided it with a decent burial so it wasn't left on the side of the road for the foxes to find.

Crows don't migrate seasonally in the middle latitudes of North America, so I'm wondering if the family now roosting in our neighborhood will continue to noisily welcome the day an hour before sunrise, or if they will make friends with the flock I noticed last winter at the VA hospital. I sure hope they aren't so hospitable that they bring home friends, especially ones coming down from Canada for the winter.

Twenty years ago, crows were hard to find in Cheyenne. And then, like so many other immigrants to our fair city, they've decided it's a good place to raise a family.

We've planted trees on the prairie and made it inviting for crows, just as we've planted oil and gas machinery in sage grouse habitat, giving ravens, close cousins to crows, hunting perches and greater success preying on sage grouse eggs. But that's a story for another day.

I hope our crows only provide entertainment and a little extra fertilizer for our neighborhood.

# 323 Deliberate littering leaves local citizens wondering

Sunday, November 6, 2011, Outdoors, page E2

We enjoyed our summer Sunday mornings. They were golden. The low angle of the early light filtered through the leaves along the creek and a few birds sang.

The air was so still, we could hear every note, yet mosquitoes were absent. It is a perfect time for bird watching along a country road.

From week to week, we noticed changes. One week it was vociferous western kingbird families everywhere, another week it was tiny grasshopper sparrows on the barbed wire.

Once, a middle-aged pickup passed us and we waved to the old guy driving. He towed a light trailer with mesh sides, carrying a little mound of dried plant material and a garden hose.

I briefly wondered where a man would be going at 7 a.m., coming from town, hauling a trailer. Twenty minutes later we found out.

The hose and the dried weeds, plus some soiled paper towels, were on the side of the road. It was a well-watered spot where the weed seeds will easily sprout.

I am sure it is the same weeds and hose, because years of distinguishing birds and their various shades of color has sharpened my eye. No, the hose didn't fall out accidently. It was poorly tucked into the roadside vegetation – just the way you'd expect a slob to try to hide something quickly.

Public lands near towns are always victims of hit and run littering. My sister, who works for the U.S. Forest Service, tells me tales of contractors dumping building debris – and there is some along this local road – as well as an assortment of tires and furniture.

But here the land is privately owned. Perhaps this dumping is a personal statement to the owners, but more likely, the slob doesn't know to whom the land belongs.

I wonder if he's the one who, just this year, left evergreen branches, fireworks trash, a chair, a love seat, a mattress and box spring, an old computer and porn magazines.

Perhaps the litterer doesn't know that the Cheyenne compost facility takes any vegetation for free. It is open 10 a.m. to 5 p.m., closed on Tuesdays and Sundays (from March through October, it's open every day but Tuesday.).

Anyone in the county can dump anything for free at the Cheyenne waste transfer facility, 200 N. College, on the middle Saturday of May. If you can't wait, they'll take your old tires the rest of the year for $1.55 apiece, more if they are still on their rims.

Useable furniture and other things should be donated to charities, but whatever items are beyond repair only cost $10 per pickup load to take them to the transfer station – if you live in the city – more if you live outside city limits. For a small fee, the sanitation department will pick up things that don't fit in your regular bin.

Call the sanitation department at 637-6440 for particular costs and hours.

The city does its best to make it easy to dispose of trash properly. The recycling businesses even make it profitable, in some cases.

But teaching personal responsibility is the best way to control deliberate litter. In a state like Wyoming, where its citizens worship private property rights, you'd think there would be zero tolerance for making a mess of someone else's land, if only to preserve your own.

However, my husband Mark and I do have something in common with the litterer: We like having an excuse to enjoy a Sunday morning in the country. If that old guy reforms, he'll need a new reason for a drive. You see, we have this extra pair of binoculars....

# 324 Plan to refresh Lake Minnehaha would benefit park visitors, including birds

Sunday, December 11, 2011, Outdoors, page E2

What stinks at Holliday Park in the summer?

The waters of Lake Minnehaha, 6.5 surface acres in the middle of the park, are stagnant. There isn't enough movement of water and so it provides a perfect habitat for blue-green algae. In hot weather, it dies and produces the putrid smell.

This particular algal species can at times be toxic, killing dogs that drink it or sickening people coming in contact with it. It spreads on the water surface and blocks sunlight that would otherwise encourage growth of healthy organisms. Storm water runoff brings in more gunk and debris.

Teresa Moore, planning manager for the Cheyenne Parks and Recreation Department, invited me to read the recently compiled report from Ayres Associates proposing how to clarify the water.

**1.** Deepen the lake, from 3 feet to 8 or 9, with gradual slopes where there are now eroded banks. The island would not be rebuilt.

**2.** Instead of aerators, which have been tried before, install a SolarBee. The 300 already installed nationwide show they are effective in circulating water which creates surface turbulence that keeps blue-green algae from growing. It would be in the middle of the lake and powered by attached solar cells.

**3.** At the storm water inlets, put in SNOUTS, ingenious technology that collects gunk in an underground vault before it can go into the lake. Vacuuming the vault once a year would be easier than the maintenance department's current methods.

**4.** Develop wetlands-- cattails and rushes--by the inlets to catch remaining sediment so it doesn't fill in the lake over time.

**5.** Route "reuse" (treated waste) water through the lake. Cheyenne has plumbed itself to use it for irrigating other parks, cemeteries and athletic fields. The water would constantly flow through new inlets and out through a new automatic outlet (the current one has to be adjusted by hand), helping prevent blue-green algae growth. Reuse water would also irrigate Holliday Park.

From my birdwatching observations at the park, blue-green algae doesn't affect the Canada geese. By the middle of last June, I was counting 200 of them, including 40 goslings.

The adults were molting and unable to fly. By fall, wing feathers grown back in again, daily numbers (between 8-9 a.m.) were running 100-150. They spend time in the water, but mostly they graze the grass.

There are also a few dozen mallards and domestic ducks, three dozen white or gray domestic geese and occasional wild visitors: wood ducks, redheads and shovelers.

Removing the island would make me sad, but it would take with it the major location for goose nesting. By all standards – especially the standards of people trying not to step in goose poop – there are too many geese.

By clean water standards, there is too much nitrogen in the water, some of it from goose poop. Removing the island hatchery could encourage wild geese to disburse and nest elsewhere. The island is not used as a refuge from potential predators. When the geese feel threatened, they, and their goslings, head for the water, not the island.

The black-crowned night-herons used to nest in the island's trees, but when the big trees disappeared, they moved to the big cottonwoods to the north. Pelicans sometimes rest on the island in spring and summer, but I've seen them enjoy island-free lakes on F.E. Warren Air Force Base and they like a thick stand of cattails just as well.

What attracts the non-water birds are the trees. If willows are added to the shoreline, as suggested in the plan, over time, they will make up for the loss of the scrubby foliage on the island.

All of the improvements would clarify the water, allowing other organisms to grow, including the food chain that leads to fish. We might see more of the fish-eating bird species that we see at Lions Park, like the grebes.

Altogether, the proposed improvements would have a positive impact on birds – and other park users.

So, when can the digging begin? As soon as $1.5 million can be found. The city does not have a budget line for construction in the parks, though there are many repair and improvement projects needed.

Park damage just doesn't get the same respect a pothole does.

Because the Holliday Park project involves water and engineering, Moore said there are some funding options. She's an expert when it comes to writing grants, but if you have ideas, contacts or appropriate funding sources, be sure to contact her at 638-4375, or tmoore@cheyennecity.org. If you have comments on other park topics, call 637-6429.

# 325 A hawk ate my songbird!

Sunday, December 25, 2011, Outdoors, page E3

## Bird feeder or bird feedlot, it's all a part of the food chain

Coming home from errands recently, I let the dog in and glanced out the window. What was that on the grass in the backyard? It was a sharp-shinned hawk sitting on its prey – a sparrow, perhaps. How exciting!

Mark and I have fed wild birds for years, and though we've seen plenty of sharp-shinneds patrol our yard, this was the first time one of us saw one be successful.

Most people think of small songbirds when they think about bird feeding, so watching hawks feed on feeder birds can come as an unwelcome surprise.

I've had callers who ask me how to protect "their" birds from hawks. They aren't always happy to hear me explain how wonderful it is that they are witnessing the next step in the food chain.

For a small hawk like the sharp-shinned, which has

the aerodynamics to navigate the urban forest easily, our birdfeeders must seem like feedlots. But when the feeders/feedlots are right outside our windows and we welcome the same cheerful chickadees day after day, I think we forget their role in the food chain.

Mark and I are on a first-name basis with several farmers and ranchers who raise our meat, if not with the actual animals, plus we hunt and fish, commiserating with predator species. But even if we were vegetarian, we would be wrong to transfer that ideology to wild, meat-eating animals. Carnivorous, omnivorous and carrion-eating animals need animal protein to stay healthy.

Most of the little songbirds, including our seed-eating feeder visitors, prey on insects and spiders when

they are feeding their young and need lots of protein. No humans complain.

Conversely, some of the birds that we would consider meat eaters occasionally pick up the odd seeds or berries.

But, looking through my copy of Kenn Kaufman's "Lives of North American Birds," I found plenty of birds that eat only non-plant material: some of the grebes, all of the seabirds, pelicans, herons, cormorants, egrets, osprey, hawks, falcons, eagles, some shorebirds, many gulls, all of the terns, owls, nighthawks, swifts, most of the swallows and wrens, the dipper, both shrike species and some of the warblers – and warblers are the quintessential songbird!

Granted, warblers are eating insects and although insects are animals, their deaths don't seem to bother many people.

The day after I wrote the rough draft of this column, the dog and I, leaving for a walk around the neighborhood, witnessed a sharp-shinned hawk doing acrobatics a few feet over the driveway, fighting to hang on to a starling. There are so many of those invasive starlings that this seemed like a good thing, except that our feeders remained unvisited for the next six hours due to hawk fright. Oh well.

We, who feed birds, do so for our own enjoyment, to bring wild birds in close to us. I think if we are very lucky, we feed a hawk or two.

Since all of us feeding birds don't put out the same seed, I wonder if the hawks notice what their prey species have been eating. I can hear it now. "Ah, I just enjoyed a Gorges free-range, sunflower seed-fed sparrow!"

# 326 Arctic Autumn: third volume of seasonal quartet is chilly and chilling

Print copy not found, 2011

Arctic Autumn, a Journey to Season's Edge, by Pete Dunne, photos by Linda Dunne, c. 2011, Houghton

"Arctic Autumn" is the third volume in Pete Dunne's seasonal quartet, so far including "Prairie Spring" and "Bayshore Summer."

No simple drive through places known for autumn color for Pete. Instead, he has picked the Arctic: Alaska and northern Canada, where life

responds as early as the summer solstice to shortening day length, which is where his book begins, because by the fall equinox, an Inuit guide told Dunne, "All birds gone in September."

As good naturalists can't help but do, Pete and his wife Linda, who provides the photos, show how the tundra ecosystem operates and how life adapts, including the native humans, the Inuits.

But these days, one can't travel the Arctic without noticing that climate change, that 13-letter dirty word, is making inroads.

On a polar bear photography tour out of Churchill, Manitoba, that uses a structure on treads to move across the polar ice, Dunne reflects on the disappearance of that ice which will leave the bears on shore without the sea ice they need to fish for seals. He

attempts to explain to a bear peering at him from below how these changes might fit into the larger time frame.

Read this book for the well-written overview of the Arctic ecosystem, as well as the poetic prose from a man who delights in the details.

As you read, keep in mind our winter birds in Cheyenne, some of which, like the American tree sparrow, come to us from the tundra.

# 2012

## 327 Robins take up year-round residence

Sunday, February 5, 2012, Outdoors, page E3

Spoiler alert: I'm about to disclose to you that one of the time-honored symbols of spring never entirely left last fall.

I'm talking about robins. I grew up in Wisconsin where the robin is the state bird and the prime grade-school example of avian seasonal migration. Imagine my surprise years later when I found my first wintering robin on a zero-degree day in December in southeastern Montana.

Wyoming has robins in winter, too, as does every one of the lower 48 states, with the greatest density in the southern states – where we imagine robins should be in winter.

Range maps in bird field guides plainly show robins all across the lower 48, year round, with the exception of parts of the Gulf Coast, Florida and the Southwest being winter-only. Conversely, Canadians and Alaskans should see robins only during the spring/summer breeding season.

Do robins breeding in Wyoming migrate? After reading the species accounts in "Birds of Wyoming" by Doug Faulkner and "The Birds of North America Online," I found no one has a definitive answer. Doug's assessment: "Movements of American robins in fall are highly complex and poorly understood in Wyoming."

During September and October, we see large flocks of robins, but these may be northern robins passing through. We don't know if some of the northern robins spend the winter here, thinking it's balmier than Canada, or if they gather up some of our local robins and take them along to Florida.

It seems robins are fickle about where they spend their winters. Berries and other fruits are acceptable substitutes for their favorite warm-season food, earthworms, and so they will only stick around where there is fruit, and only while it lasts.

This winter, my neighbors' junipers have a good crop of berries, and just about every January afternoon I saw one or two robins over there snacking. In rural areas of the west, wintering robins are most likely to find food along rivers and creeks full of fruit-bearing shrubs or up in the junipers. The more fruit, the more robins.

So, why do we consider the robin a sign of spring? I think most people aren't outside enough in winter, in the right place – near the fruit – to see the few robins around.

When spring comes, robins flocking during their migration peak in April are much more noticeable. People are spending more time outside then, or they might have the window open and hear the robins beginning to sing to establish territories and attract mates.

I'd like to suggest a different bird, and one just as noticeable, as a better sign of spring in Wyoming. We need a sign of hope since winter weather spans as many as eight or nine months and February, the shortest month, drags on forever, especially this year being Leap Year.

Mountain bluebirds could work, except they fly past town. They cross our southern state border as early as the beginning of February, with migration picking up in March. The bright blue males are easiest to see. I see them west of town usually, flashing around fence posts as we go out for one last ice fishing trip to North Crow Reservoir or an early hike dodging snow drifts at Curt Gowdy State Park.

Interestingly, mountain bluebirds and robins are in the same family, the thrushes. Like robins, bluebirds concentrate on animals (invertebrates) for food during the breeding season and fruits in the winter.

If you check your field guide range map, you'll see that there are mountain bluebirds wintering just south of Wyoming. With predicted climate changes, we could easily end up with bluebirds all winter too. Well, geez, that would leave the warblers as the only reliable, easy to see, true sign of spring. But they don't show up until mid-April and May. That's just too long a wait. I'll stick with looking for bluebirds.

# 328 Snowy owls' visit a sight to behold

Sunday, March 25, 2012, Outdoors, page E3

Thanks to Hedwig, millions of children may recognize a snowy owl.

She was Harry Potter's companion in the books and movies. How many of those children, some now adult, have caught a glimpse of these nearly snow-white owls during this winter's invasion?

Snowy owls in North America usually leave their summer breeding grounds by the Arctic Ocean and head south. Adult females travel the shortest distance before establishing winter territories, with males and juvenile females continuing on. Juvenile males travel farthest, especially if prey becomes scarce.

Cold is not a problem. All snowies stay warm with feathers covering their toes and nearly engulfing the tips of their beaks.

There are migrant snowies to be seen every winter, according to Christmas Bird Count data, in southcentral Canada, Montana, the Dakotas and New England. They will fan out a little further 50 percent of the time, to the Pacific Northwest and the upper Midwest and east through eastern Canada. They make it as far south as Wyoming and the central Great Plains 30 percent of winters. Experts admit they don't know entirely what drives this species' nomadic migration.

This year has been note-worthy for the number of sightings in 31 states as far south as Texas. It could be that birdwatchers are better connected than ever before, thanks to the internet. Observations aren't just scribbled in someone's notebook – they are shared and mapped.

But there are also a lot of birds, sometimes in groups. The thought is that the lemmings were particularly fertile last summer and provided enough prey that a bumper crop of young snowy owls fledged. Each pair can raise up to 12 young, compared to the two or three chicks our resident, similarly sized, great horned owls raise per year.

It's most likely that it was juvenile snowy owls that people observed and photographed this winter. They have the brown barring – horizontal stripes – across their bodies, though adult females have some also. The adult males are pure white.

In the February issue of Prairie Fire, an alternative Nebraskan newspaper, Paul Johnsgard reports that in the previous 35 years, a total of 21 snowy owls had been brought into the Raptor Recovery Nebraska facilities in Lincoln. However, between December and mid-January, 10 more snowy owls were picked up. Nine were emaciated and didn't survive.

It makes me wonder how well the snowies compete with local hawks. Do they prey on the same species of rodents?

Or maybe the problem is their hunting techniques. Unless a landscape is totally snow-covered, snowies really stand out. I saw my first one in a spruce tree by Old Main on the University of Wyoming campus winter of 1980-81, but more typically people see them in the grasslands, perched on a rise, a fence post or a utility pole.

They sit motionless, waiting for prey, or walk about looking rather than flying. Unlike most owls, they are diurnal – active in the daytime. Their hearing is sharp. They can locate potential prey under the snow. They also take ducks and shorebirds.

Snowy owls are circumpolar, meaning they breed in the polar region from Alaska through Canada, Greenland, Scandinavia, Russia and Siberia. "The Birds of North America Online" species account reports their conservation status – whether populations are increasing or decreasing – is unknown. Apparently, no one has spent enough time in the polar region to find out, though one scientist who went found snowy owls can seriously wound humans too close to their nests.

If you missed the invasion, probably ending at the end of March, take a look at a three-minute YouTube video at Cornell Lab of Ornithology's channel, www.youtube.com/user/LabofOrnithology for close-ups.

Visit http://www.allaboutbirds.org, another free CLO resource, to read about snowy owl natural history and hear one calling.

At CLO's free Great Backyard Bird Count site, http://www.birdsource.org/gbbc/, you can "Explore the Data" and see where snowy owls were seen Feb. 17-21.

And finally, if you are willing to register, at no cost, you can access www.eBird.org and its huge database to look at maps and compare sightings from year to year.

Most of what is known about sightings in Wyoming this winter, including one near Burns, appeared in posts to the Wyobirds e-list. Sign up for free at HOME.EASE.LSOFT.COM/archives. Click on Wyobirds. It's a great way to keep in touch with birders across the state.

# 329 It's quite clear – birds losing war on the windows

Sunday, April 15, 2012, Outdoors, page E3

Wyoming is a tourist destination. We love statistics on how many visitors come from how many other states and countries. We also try to keep visitors safe, reminding them to stay hydrated at our high, dry elevation, to stay away from dangerous wildlife and to avoid summer lightning storms.

Spring migration is like the beginning of tourist season for birds. On May 19, members of Cheyenne – High Plains Audubon Society and friends will again hit the local birding hotspots, hopefully at the peak of migration, to see how many different species of birds can be counted.

Some years we hit the shorebird migration just right, and others it's the flycatchers. But every year we hope for a warbler year. We scour the tree branches for those smaller-than-sparrow-sized, color-coded birds which are scouring the same branches for insects to devour.

Over the previous 18 years, we have had 31 of North America's 50 warbler species visit. Only four have made it every year: yellow warbler, yellow-rumped warbler, common yellowthroat and Wilson's warbler – probably because they are part of the 12 warblers breeding in Wyoming, and because they are abundant species.

Others we've seen only once because they breed in eastern North America and, for some reason, they take the scenic route through Cheyenne. They include golden-winged warbler,

black-throated blue warbler, worm-eating warbler, prothonotary warbler, and six others.

Almost all of these were observed at one of Cheyenne's two Wyoming Important Bird Areas, Lions Park and Wyoming Hereford Ranch. But there is reason to believe that all of Cheyenne, wherever there are trees and shrubs hosting insects, is hosting common and rare warblers, if only people look.

Casual observation of Mark's and my yard has turned up nine species including regular appearances of Wilson's and yellow-rumpeds, sometimes a MacGillivray's and once, a chestnut-sided warbler.

Between mid-April and mid-June, who knows how many warblers pass through our yard? Maybe our retriever knows. Last year I caught her eating at least two after they were injured flying into our window.

While I'm fine with continuing to keep our remaining cat indoors year round (the other passed on last month at nearly 14 years old),

we need the dog on squirrel offense duty. But even if we didn't, there would still be injured birds.

Short of plywood over this one deadly window, how can we keep birds from hitting the glass? I tried a small sticker in the middle, but over the winter we collected a wreath all around it of lovely imprints of Eurasian collared-dove wings and tails, outlined in feather dust on the glass where they tried to avoid the sticker.

Hanging dangly, shiny objects in front of the window probably wouldn't work with the caliber of breezes we get – the objects would end up stuck in the gutter or perhaps banging on – and breaking – the window.

The American Bird Conservancy has come out with a new product this spring, BirdTape, which we are going to try. It sticks to the outside of the glass, breaks up the reflective surface that fools the birds, and is translucent – like frosted glass.

The strips of ¾-inch-wide BirdTape can be applied vertically four inches apart,

or horizontally two inches apart. Studies show that our backyard birds will try to zoom between obstacles spaced any greater distance. It obviously takes less tape to do the vertical arrangement.

The tape also comes in rolls three inches wide. These can be cut into squares placed in a pattern leaving spaces between them four inches horizontally and two inches vertically.

I didn't do the math to see which tape size's pattern is more economical. Your choice might have more to do with whether you prefer bars or floating squares.

Currently, BirdTape is available through www. ABCBirdTape.org or call 1-888-247-3624.

I'm not sure I like the idea of anything impeding my view of our backyard, but with up to a billion birds hitting home windows each year, according to ABC, I want to give this product a try. It's the least I can do to protect avian tourists on their annual spring and fall visits to Wyoming.

# 330 Peregrines back with a little help from friends

Sunday, May 13, 2012, Outdoors, page E3

Peregrine falcons were listed as endangered in the U.S. two years before I opened my first bird field guide in 1972.

The guide, "The Birds of North America," published by Golden Press in 1966, did not allude to the peregrine's diminishing population. It only said it was "a rare local falcon."

However, in the era of an awakening environmental

consciousness, we all heard about the peregrine, a very handsome poster child for the drive to ban DDT, one of the pesticides responsible for poisoning birds of prey and causing their eggshells to be too thin for un-hatched young to survive.

One doesn't expect to meet an endangered species in the wild, especially when ornithologists had declared it extirpated in the eastern

U.S. by 1970 and in trouble in other parts of the world (peregrines are found everywhere except the Sahara, the Amazon and Antarctica). But I had another encounter with a peregrine last month, just outside Cheyenne.

My six peregrine observations, all since 2003, have been around Cheyenne, at either Wyoming Hereford Ranch or Lions Park. All but one was in spring.

I remember the first sightings, on Audubon field trips, for which I was relying on more experienced birders for identification. Once, at WHR Reservoir No. 1, we saw a peregrine in one of those legendary dives – once clocked by a scientist at 200 mph.

It slammed into an unsuspecting duck standing on a sandbar. The peregrine's former common name was "duck hawk" – ducks being a favorite among the many kinds of birds they eat.

Last month, my husband Mark and I saw a bird sitting in a cottonwood below the same reservoir, watching us. It had all the peregrine field marks, including the dark cheek patches, which must have been the inspiration for those cheek pieces for first-century Roman centurions' helmets.

Peregrines have been favorites of falconers for 3,000 years. While the young can be taken from wild nests and raised by humans, they are also bred in captivity. In 1970, the founder of The Peregrine Fund, Tom Cade, began breeding them in earnest, as did Bill Burnham of Fort Collins, future president of TPF, beginning in 1974.

By 1984, TPF had opened the World Center for Birds of Prey in Boise, Idaho. By 1997, 4,000 peregrines had been bred and released into the wild. By 1999, the peregrine was off the Endangered Species list. The fund continues to work to conserve raptor species around the world.

It isn't quite the same as the old days for the peregrines. Someone thought of also introducing – or hacking – them into cities that have plentiful pigeon prey and tall buildings that would imitate their cliff-face nesting habitat. Urbanites could be seeing peregrines much more often than we do.

While peregrines went missing in the eastern U.S., what happened to them in Wyoming? I asked Bob Dorn, co-author with his wife, Jane Dorn, of the book, "Wyoming Birds." From his research, he was able to give me a list of more than a dozen observation dates going back to 1929.

In 1939, Bob said O. C. McCreary categorized the peregrine as "a rather rare summer resident," usually indicating that they are breeding, and "an uncommon migrant," meaning not quite so rare during migration. As Bob put it, "When you're at the top of the food chain, you are in scarce numbers." (Somehow, that isn't true of humans.)

The Wyoming Game and Fish Department's species account states that by 1970, Wyoming had no viable breeding population. They formed a partnership with TPF and over 15 years, 1980-1995, introduced 384 captive-bred peregrines. It was successful. There were 90 breeding pairs recorded in 2009, the most recent information available.

Today, breeding peregrines tend to be found in the northwest part of the state. Down here in the southeast, we have the potential to see migrants from April through May.

The most recently published field guide I have, "Peterson Field Guide to Birds of North America" (2009), does mention the peregrine was endangered – small concession to the idea that the hobby of bird identification can no longer be divorced from bird conservation.

The new Peterson range map shows there is still a big empty area in the middle of the country where the Golden guide had indicated

wintering peregrines nearly 50 years before. But it also shows summer range, presumably breeding range, where the Golden guide did not.

Unfortunately, many threatened or endangered birds are not as charismatic as the peregrine. Experience with captive breeding may be nonexistent and the reason for a species' plummeting population may not be as simple as a particular pesticide. The commonality, however, is that human experiments with new technology often produce unexpected, bad consequences for some birds, while accidently promoting the unwanted reproduction of others – think starlings.

Meanwhile, birders continue to collect and share observations, causing range maps to continually be redrawn. Mark's and my single peregrine sighting on April 8 becomes part of the larger story.

Keep your eyes open, too.

# 331 Bird Count yields results labeled as "a crazy spring"

Sunday, June 10, 2012, Outdoors, page E3

A successful Big Day Bird Count is all about knowing where to find the birds.

When it rains buckets during the first hour of the count, you have to know where the birdwatchers go too. They went to Starbucks.

I don't know how long Cheyenne – High Plains Audubon Society has been conducting the annual Cheyenne Big Day, probably 30 or 40 years. It's a bit of an anomaly, since an official

American Birding Association Big Day is something an individual birder or small team conducts to see how many species they can count in 24 hours in a specified area. All the team members have to see all the species.

Our count compiler, Greg Johnson, did his own Big Day for Laramie County on May 20, the day after the chapter's count, and saw 93 species.

The chapter count,

however, starts out as a large group enterprise, but by afternoon we are birding in smaller parties. Everyone's results are pooled for a final tally. This year, it was 104 species, not much more than Greg's, considering we had over 20 people scouring the city and countryside. It's the lowest number of species I can remember in the last 20 years.

I also don't remember getting soaked on a count and

having to track down half the participants to a downtown coffee shop where they were warming up and drying off. I don't think we missed many birds during the rain – they were tucked away as well.

It stopped raining and we had an incredible fallout of goldfinches by the time we reached the Wyoming Hereford Ranch. The dandelions were finished, but a flock of 50-plus bright yellow males gobbling the seeds made one of the lawn areas look as though it were blooming again.

We hit another snafu at the High Plains Grasslands Research Station. For the first time, our permit to walk the back road didn't come through in time so we may have missed a few species – last year there were 10 we saw there and nowhere else.

And then there was the spring itself, warmer than usual, with trees leafing out and plants blooming as many as three weeks ahead of schedule. Was that the reason we had only 104 species compared to a more normal 130-140? Had some migrating species already come and gone?

Greg Johnson is puzzled.

"My take is that this is a completely crazy spring – much lower numbers of both species and individuals. I have been birding nearly every day since May 8," Johnson said in an email. "I have seen six species of warblers all spring. To put that into perspective, I remember seeing 11 species of warblers once over a lunch hour at the Wyoming Hereford Ranch.

He noted that Cobirds, an elist for Colorado birders, showed that a banding station at Barr Lake had about 40 percent fewer birds banded than a year ago.

"So it seems to be a region-wide phenomenon," Johnson said.

This year we missed 40 species we saw last year – but saw 20 we didn't see a year ago. We're always looking for the birds that "don't read the maps," such as those rarely seen eastern warblers. This year, common migrants apparently weren't even reading the calendar.

## 2012 Cheyenne Big Day Bird Count Results

The abundance ratings, A - abundant, C - common, U - uncommon, R - rare, indicate how likely experts think a species will be seen in Cheyenne in its preferred habitat in spring.)

Canada Goose, A
Gadwall, C
American Wigeon, C
Mallard, A
Blue-winged Teal, C
Cinnamon Teal, C
Northern Shoveler, C
Northern Pintail, C
Green-winged Teal, C
Canvasback, U
Redhead, A
Common Merganser, C
Ruddy Duck, U
Ring-necked Pheasant, R
Pied-billed Grebe, C
Eared Grebe, C
Western Grebe, C
American White Pelican, U
Double-crested Cormorant, C
Great Blue Heron, C
Black-crowned Night Heron, C
Turkey Vulture, C
Northern Harrier, C
Swainson's Hawk, C

Red-tailed Hawk, C
American Coot, A
Killdeer, A
American Avocet, C
Greater Yellowlegs, C
Lesser Yellowlegs, C
Willet, U
Spotted Sandpiper, C
Whimbrel, R
Least Sandpiper, U
Baird's Sandpiper, C
Long-billed Dowitcher, U
Wilson's Snipe, C
Wilson's Phalarope, A
Red-necked Phalarope, R
Ring-billed Gull, C
Rock Pigeon, A
Eurasian Collared-Dove, A
Mourning Dove, A
Great Horned Owl, C
Belted Kingfisher, U
Downy Woodpecker, U
Northern Flicker, C
Olive-sided Flycatcher, U

Western Wood-Pewee, C
Least Flycatcher, U
Western Kingbird, A
Eastern Kingbird, C
Loggerhead Shrike, U
Blue Jay, C
Black-billed Magpie, C
American Crow, A
Common Raven, U
Horned Lark, C
Tree Swallow, U
Violet-green Swallow, U
N. Rough-winged Swallow, U
Bank Swallow, U
Cliff Swallow, A
Barn Swallow, A
Red-breasted Nuthatch, C
White-breasted Nuthatch, R
House Wren, C
Ruby-crowned Kinglet, C
Mountain Bluebird, C
Veery, U
Swainson's Thrush, A
Hermit Thrush, U
American Robin, A
Gray Catbird, U
Brown Thrasher, U
European Starling, A

Yellow Warbler, A
Yellow-rumped Warbler, A
American Redstart, U
Northern Waterthrush, U
Common Yellowthroat, C
Wilson's Warbler, C
Western Tanager, C
Chipping Sparrow, C
Clay-colored Sparrow, U
Vesper Sparrow, C
Lark Sparrow, C
Lark Bunting, C
Song Sparrow, U
White-crowned Sparrow, C
McCown's Longspur, U
Chestnut-collared Longspur, R
Red-winged Blackbird, A
Western Meadowlark, A
Yellow-headed Blackbird, C
Common Grackle, A
Great-tailed Grackle, R
Brown-headed Cowbird, C
Orchard Oriole, U
Bullock's Oriole, C
House Finch, A
Pine Siskin, U
American Goldfinch, C
House Sparrow, A

# 332 New guidebooks take guesswork out of birding

Sunday, June 17, 2012, Outdoors, page E3

What the Robin Knows: How Birds Reveal the Secrets of the Natural World, by Jon Young, c. 2012, Houghton Mifflin Harcourt, hardcover, 241 pages, $22.

Includes science and audio editing by Dan Gardoqui and corresponding audio clips at www.hmh.com/whattherobinknows.

Author Jon Young takes what native trackers, other human mentors, and birds have taught him and passes it on to others through workshops and his website, www.BirdLanguage.com. Now he's reaching a wider audience with this book.

There's a slight New Age ring to it – after all, he's moved from his boyhood home in New Jersey where he roamed the woods, to live in California.

In a way, this is also a self-help book. Young contends if you learn to pay attention in nature, specifically, distinguishing different bird calls and songs to understand "what the robin knows" about what is going on around you outdoors, it will make a difference to you spiritually.

But even if you only want to learn to puzzle out wildlife secrets, you'll find reading Young's advice is time well spent.

Life Everlasting, The Animal Way of Death, Bernd Heinrich, c. 2012, Houghton Mifflin Harcourt, 256 pages, $25.

We all know about the food chain, but we don't often want to think about how animals are recycled to become sustenance for future generations.

It is the request from a friend for a natural burial on author and scientist Bernd Heinrich's land in Maine that causes him to examine the strategies of nature's undertakers.

Heinrich's in-depth look at how beetles, whales, ravens,

vultures, salmon, among others, gracefully take part in the cycle of life contrasts sharply with what he shows us about human cultural practices that either use a huge amount of energy or poison extensive amounts of land with formaldehyde.

Heinrich is not only a scientist, but also a storyteller and a philosopher. If you enjoy this book, be sure to look up his others, including "Mind of the Raven" and "Why We Run."

The Young Birder's Guide to Birds of North America, by Bill Thompson III, c. 2012, Houghton Mifflin Harcourt, 364 pages, softcover, $15.95.

As predictable as the spring migration of birds is the spring publication of a new bird field guide,

especially one in the Peterson Field Guides series.

This one is for children old enough to read and is written by Bill Thompson III with help from his kids (whose mother, Julie Zickefoose, is one of the book's illustrators) and Mrs. Huck's fifth grade class.

If you buy this field guide for a child you know and hope to turn into a birdwatcher, even if you aren't one, go ahead and read the introduction. You may find you want to buy a copy for yourself.

Only the book-wormiest kids will actually read the introductory chapter. The rest will go straight to the photos and the "Wow!" factoids – the surprising tidbits about each bird.

A generous 300 species of the 800 North American birds are included (200 of those can be expected in Cheyenne), but the volume's dimensions still remain child-sized.

My only quibble is with the range maps. You have to infer that during migration a species might be seen anywhere between the winter and summer ranges.

If this book sounds vaguely familiar, it is because Thompson and Zickefoose came out with "The Young Birder's Guide to the Birds of Eastern North America" in the same format in 2008, which would be a better option for your grandchildren living east of the Mississippi, unless they are coming out to visit.

# 333 Gardener reports from backyard: Bird life, death, and allegiance

Sunday, July 22, 2012, Outdo ors, page E3

What I know about the birds in our backyard I've mostly learned from watching from the window.

It's different being out there with them. This spring and summer, I've spent more time than usual out in the yard, working on my new vegetable garden.

We've always had robins, but now I've learned they recognize that a person digging is not only non-threatening, it can also be a source of earthworms. The male of our local pair waited just a few feet behind me, not even flinching when I turned to look at him.

A couple weeks later, our robins brought their speckle-breasted youngsters to

show them how to wash up in the birdbath, how to find earthworms and how to pick the ripening Nanking cherries from our hedge.

This year we have a huge cherry crop, but it seems that only the local robin family is picking. It would be nice to think they are defending their territory and our cherries from other hungry wildlife. Whatever the reason, we are harvesting plenty since a flock hasn't come in to eat them all in one day.

I've been trying to listen to "what the robin knows" ever since reading the book by that title. That means when I heard a robin squawking without ceasing very early one morning, I went out

to investigate, finding a long-haired black cat – a potential nest robber – waiting under the bushes.

In return for food, water and cat eviction, our robins are not only defending the cherries, they perch on the garden fence posts, adding fertilizer and planting cherry pits.

We saw blue jays often in May and by early June, I observed one fly into the vegetable garden with something white in its bill. It hopped up to one of the tomatoes and carefully inserted the white object next to it, under the leaf mulch. It was a fecal sac, collected from one of its nestlings, removed to keep the nest clean, and buried

so predators wouldn't track down the nest. But it was also tomato fertilizer.

But I don't think the blue jays were successful.

One evening, while working outside, Mark and I heard a plaintive blue jay call high overhead in one of the big green ash trees. And then, several times, we could see a blue jay attack something in a clump of leaves, creating a ruckus.

After about the third attack, four crows left the clump one at a time. I'm pretty sure the blue jay got the raw end of the deal, losing nestlings. I try not to have favorites in the bird world – perhaps the crows were teaching their offspring how to feed themselves.

And then there was the incessant twittering one weekend. It's a familiar background sound, but I finally connected it to downy woodpeckers.

Hour after hour, as we worked out in the yard, I could hear this calling between three or four downies. It must have been the young fledging. A month before, we'd seen a lot of one pair picking over the bark of our tree limbs for bugs, the male announcing his presence by hammering on some metallic part of the utility pole.

We sent the house finches and Eurasian collared-doves packing when we took down the sunflower seed feeder at the end of May. The doves had already produced one brood. You can tell the youngsters even though they are the same size as their parents – their black neck markings are a bit indistinct and they gaze around at the world wide-eyed.

But the thistle feeder is still up, and we can count at least two pairs of goldfinches visiting multiple times a day. They nest later than other songbirds, waiting until the source of food for their nestlings, seed, especially wild thistle seed, is available.

Did you know that goldfinches are the only songbirds that feed mashed seed to their nestlings? Other seed eaters switch to insects – more protein for building bones. It makes one wonder how goldfinches manage without that source of protein?

On June 30, I saw my first hummingbird of the year in our neighborhood – two weeks early. It was time to do the hummers a favor and hang their feeder – our tubular-type red garden flowers weren't blooming yet.

And the flickers came and worked on the ant invasion on the front lawn.

Isn't it nice when we and the birds can help each other? It's what it's like to be part of a healthy ecosystem.

# 334 Colorado black swift wintering grounds are found in Brazil

Sunday, August 26, 2012, Outdoors, page E3

Imagine that in 2009 there was still one bird species whose wintering location was still unknown. And imagine that for that same bird species, few of its nesting colonies had even been found until the late 1990s.

Let me introduce the black swift, the North American subspecies (not that the southern subspecies is better known).

At 7.5 inches long, the black swift is longer than our local chimney swift by 2 inches and its wingspan is an 18-inch curve. Swifts are perpetual bug-eating flying machines that might be mistaken for swallows but look more like flying cigars with wings.

The first black swift was documented in 1857 on Puget Sound in Washington State, and the first nests in 1901 in California sea caves where ocean spray kept them moist. By 1919, intrepid egg collectors found their nests behind mountain waterfalls.

In the 1950s, Owen A. Knorr made the black swift his master's thesis at Colorado University in Boulder, making a concerted effort to look for nests in Colorado by learning mountain climbing skills and developing a system for predicting which waterfalls would be nest locations. He found 25 colonies, each with a handful of mossy nests stuck to tiny rocky ledges, each one holding one nestling.

In 1997, Kim Potter was one of two biologists beginning a new search for swifts. A year later, Rich Levad got hooked on looking for them and joined her in organizing surveys through the Rocky Mountain Bird Observatory, infecting others with swift enthusiasm along the way.

I met Levad and Potter in 2005 when Wyoming Audubon members helped them find flammulated owls in Wyoming's Sierra Madre range. Already one year into a diagnosis of Lou Gehring's disease, Levad was soldiering on impressively.

When he had to cut back on field work, Levad started writing "The Coolest Bird: A Natural History of the Black Swift and Those Who Have Pursued It," still making edits the day before his death in 2008. You can find the 152-page, free edition provided by the American Birding Association online at www.aba.org/thecoolestbird.pdf.

It's a great read about an exciting bird and many memorable characters – check out the scathing exchange between Knorr and a dignitary in Arizona who believed a bird species only existed if he could hold the collected specimen – the dead body, in other words.

I spoke recently with one of Levad's protégés, Jason Beason, director of special projects at the Rocky Mountain Bird Observatory and lead author of an article about a black swift breakthrough published in the Wilson Journal of Ornithology this past March about finally discovering the black swift's wintering grounds.

Every August, black swift adults leave each morning to collect food and later at twilight they slip back to feed the young. This is when researchers hope to see them.

Levad learned that training field observers increased their abilities to find swifts, upping known Colorado colonies from 27, including Knorr's found in the 1950s, to 86, but it wasn't until mist netting was tried in a couple of narrow canyons that it became apparent how many swifts were eluding detection.

Banding the captured swifts and recapturing many of them the following years showed how loyal they are to nest sites.

Beason, Potter, and another of the paper's authors, Carolyn Gunn, wanted to strap recorders on the birds to find out where they go in winter, but most equipment is designed to attach to a bird's leg and swifts hardly have a leg. They never walk. If they land at all, they cling to vertical surfaces. It's thought that for some swift species, non-breeders stay aloft for a year or two.

Enter the British Antarctic Survey, which had developed a micro geolocator that works off day length to determine location and archives the data every 10 minutes for a year. One was strapped on the back of each of four black swifts about to leave Colorado in September 2009.

Beason and his team were able to recapture three of the four swifts in the fall of 2010 and download and process the data. If you want the technical description and don't subscribe to the Wilson Journal, email Beason, jason. beason@rmbo.org, for the digital manuscript.

Beyond doubt, at least these black swifts, from two colonies in Colorado, winter in the Amazon basin of western Brazil. Next summer, Beason plans to outfit a few swifts from Idaho to see if they winter there, too.

There are also a few other documented black swift colonies in the West, including Montana and Utah, and of course, the gazillion in Colorado, but none in Wyoming, probably "just because nobody's gotten out and looked up there," Jason told me.

So I asked him how we could help, thinking of that flammulated owl survey, but also realizing that few of those same people are capable of climbing up to waterfalls off the beaten track, much less hiking out in the dark after the swifts come home.

Beason said to let him know of any small grants he could apply for. It wouldn't take much, maybe $1,000, to add a stop next summer on his way to Idaho, to check out where Knorr thought he once saw a black swift flying at Grand Teton National Park. Grants, schmantz. I have a better idea: crowd sourcing, or the Tinkerbelle solution. If all of us made a small contribution, we might add a breeding bird species to the Wyoming records.

To support next summer's survey, please send contributions by the end of January 2013 to: The Richard Levad Memorial Fund (earmarked for Wyoming Black Swift Research), Rocky Mountain Bird Observatory, P. O. Box 1232, Brighton, CO 80601-1232.

If you contribute online at www.rmbo.org, click on the "Chip in" button on the home page and then, in the first step's drop-down menu, choose the "Other" option. Or call Rachel, 303-659-4348, ext. 17, during business hours.

# 335 Mother laid tinder for my "spark bird"

Sunday, September 23, 2012, Outdoors, page E3

Birders talk about their spark bird: the one that hooked them on birdwatching, boosting their awareness of birds from background noise to center stage whenever they walk outdoors or even look out the window.

But I think that spark has to land on some tinder. In my case, that tinder was laid by my mother.

Both my parents enjoyed the outdoors. I remember many soggy camping trips in the 1960s flavored by the wet canvas smell of a great uncle's hand-me-down, umbrella-style hunting tent, so old it featured a wooden center pole.

Dad always talked about his teenage escapade driving "out West" from Illinois with friends to climb Longs Peak, while Mom referred to childhood visits to the family dairy farm near Madison, Wisconsin.

We didn't hunt or fish, but we did enjoy watching wildlife when it crossed our path – except for the black bear that raided our campsite in the Smoky Mountains. However, this alone doesn't account for my sister and me getting degrees in natural resource management and consequently marrying wildlife biologists.

There was Girl Scouts. Mom signed us up for Brownies and became a troop leader. I stayed with it through high school, mostly for the summer camps and weekends in the woods.

There was also the spring of my junior year when Mom came home from a trip with a souvenir for me, "The Golden Guide to the Birds of North America." I flipped through it thinking, "Nice, but none of these birds, besides the house sparrow and robin, are in our neighborhood."

It was a month later while biking along the Menomonee River Parkway, which runs through my hometown of Wauwatosa, Wisconsin, that two brightly colored birds caught my eye, an indigo bunting and a rose-breasted grosbeak, my spark birds. Then I identified a Canada warbler, bright yellow, perched on a branch hanging over the back door. Having a field guide made it easy to start my bird life list.

Mom was also the one to clip a little notice about volunteering that summer with the Student Conservation Association (www.thesca. org). I was selected for a crew to rehabilitate backcountry trails and campsites at Rocky Mountain National Park. Most of the 15 students, from all over the country, knew they wanted a career in conservation or biology. One was even a "bird nerd," an entirely new species of male teenager, in my experience.

From then on, the outdoors, the natural resources, or the environment, as we began calling it in the 1970s, became a big part of my life.

Mom never became a birder, other than to enjoy the hummingbirds in her yard in Albuquerque, where she'd moved to more than 30 years ago, or to try to follow our pointing fingers on trips. She had other interests, though, that demanded the same kind of research and attention to detail: knitting, needlepoint, gardening, doll collecting and repair, and her career as a registered nurse.

CHEYENNE BIRD BANTER

My sister and I spent most of this summer taking turns helping Mom through the aftermath of a severe stroke. The only bright spots for me were the birds in her backyard and, at sunrise, the dawn chorus along the nearby boulevard.

I've almost always avoided Albuquerque in summer because I thought it'd be too hot. It is. But I had also missed Mom's summer birds – the family of kingbirds, the curved-billed thrasher, a flock of bushtits and the incessant cheeriness of lesser goldfinches calling to each other every day, unseen in the treetops.

Mom lived until the end of August, slipping away before I could return from another furlough home. I never said those last things I wanted her to know because during weeks of rehab it didn't seem appropriate – we all thought there was more time.

But if you are a birdwatcher who can remember a spark bird, search your memory for those who laid the tinder and tell them thanks. Don't wait for their last minute.

# 336 Celebrity field guide author visits Cheyenne

Sunday, October 28, 2012, Outdoors, page E3

In September, Ted Floyd was a guest of the Cheyenne – High Plains Audubon Society. He is the editor of the American Birding Association's magazine, the author of the Smithsonian Field Guide to the Birds of North America, and a really sharp birder.

On the field trip along Crow Creek, he was able to identify a first-of-the-year female chestnut-sided warbler (rare in Cheyenne), in a treetop. It didn't look very distinctive, but behavior and voice helped him identify it.

Ted has a rather humble attitude towards his birding and literary talents, as evidenced by the following interview.

**Question.** What was your first field guide when you started birding in 7th grade?

**Answer.** The fourth edition of Roger Tory Peterson's "Eastern Birds." It was brand-new at the time.

**Q.** How many bird field guides do you have in your collection?

**A.** Hundreds. Literally.

**Q.** When did you start dreaming about writing your own bird field guide?

**A.** I've been thinking about writing a field guide almost from the very beginning. When I was in the eighth grade, I created my own "checklist sequence" – I thought it was better than Peterson's.

**Q.** How did you get to be the author of the Smithsonian Field Guide to the Birds of North America?

**A.** Honestly, I'm not entirely sure. The folks at Scott & Nix contacted me; then we had a long series of informal chats; and, eventually, we all agreed that we'd do the field guide. My name is on the front cover, but it's been a collaborative effort.

**Q.** How did Charles Nix and George Scott think this field guide could be different and better than all the North American guides currently available?

**A.** The Smithsonian Guide is holistic. It encourages birders to employ a "whole-bird" approach to bird identification: Look at the bird, listen to the bird, pay attention to molt and behavior, take note of ecology and the environmental context – and do that all at once. It's a very natural approach for beginners. For more experienced birders, who can be very rigid and compartmentalized about bird identification, some amount of reprogramming may be required.

**Q.** How long was it after signing the contract before the books were on store shelves? How much of that time did it take you to actually write the field guide?

**A.** The guys at Scott & Nix are slavedrivers, and I mean that in a good way. They were excellent at keeping the project on schedule. I would say the project took about a year of organization, and then it took me a year to write the book.

**Q.** When you were writing the 28-page introduction to birdwatching and the species accounts, what kind of birdwatcher did you have in mind?

**A.** Anybody who's interested in nature and open-minded about new ways to engage the natural world.

**Q.** In your research for this guide, what was the most surprising thing you learned about a bird you thought you knew?

**A.** As I listened to recordings of duck vocalizations, I was enthralled by how beautiful they are. I have come to believe that the Redhead has one of the most arrestingly beautiful songs of any North American bird species – right up there with the Hermit Thrush or Winter Wren. I wonder how many birders even know that Redheads say anything at all!

**Q.** If Paul Lehman is the go-to guy for range maps for North American field guides, including this, plus National Geographic's and the Sibley Guide, was your personal knowledge of bird ranges added to any of the maps?

**A.** Yes, but it's not as if I "overruled" Paul Lehman on anything. Rather, the folks involved with map production (including me) had conversations about range limits for certain species. We also had conversations about the best color scheme to use.

**Q.** On pages where more than one species account appears, they are often laid out side by side, but when they are laid out one over the other, it is easy to miss the lower one when rapidly flipping pages to identify a bird from a large group such as warblers or shore birds. Am I the only one who has problems with that?

**A.** You need to slow down when you read, Barb...No, seriously, layout is a huge issue with this or any field guide. I can't begin to convey to you how much time all of us labored over where to place the species. In the linear format of a book--you go from page 1 to page 2 to page 3 – it's impossible to present the multi-dimensional

2012

313

problem of comparing species. I think we got it right in most instances.

I'm grateful for comments like yours. That's because a second edition is in the offing, and we'll be tweaking the formatting in places. If you or anybody has suggestions, please tell me about it (tfloyd@aba.org). You'll make a difference.

**Q.** What other kinds of changes will you be making in the second edition?

**A.** New taxonomy, a few new names even, and some changes to range maps. We'll also correct the single typo from the first edition.... On a substantive note, look for some new photos. There will be more photos showing distinctive geographic variation and more photos showing cool bird behaviors.

# 337 Goose population success is messy problem for parks

Sunday, November 25, 2012, Outdoors, page E2

As your resident bird lady, it's time for me to bring science to the issue of too many geese in Cheyenne parks.

The domestic geese that Teddie Spier mentioned in her letter to the editor Nov. 6 are not a problem. The city can round them up any time, which they did this summer at Holliday Park, leaving behind four whites and a gray.

Mallards are common park ducks, but here they are a fraction of park waterfowl. Mid-winter, the large flock of ducks on the open water at Lions Park is made up of species eating aquatic invertebrates, not mallards begging for handouts.

It's the wild geese, properly known as Canada geese. (If one of them hails from Canada, you could refer to it as a Canadian Canada goose.)

Over the last three years, I have been counting the birds at Holliday Park around 8 a.m., 10 days per month on average, recording the results at www.eBird.org. In the spring of 2010, Canada geese were numbering 60-100 per day. This spring they were running over 200. You can access my data for free by setting up your own login and password at the website.

Cheyenne geese move between the parks, golf courses, F.E. Warren Air Force Base and rural fields, so to get the big picture, look at the Cheyenne Christmas Bird Count, which strives to count geese all over town at the same time. The data is available for free at: http://netapp.audubon.org/cbcobservation/Historical/SpeciesData.aspx.

I found eight Canadas were recorded in 1974, then none until 50 in 1983. In the 1990s, numbers jumped into the hundreds, and by 2000, to over 2,000. Last year's count was 1,332, probably not a sign of a downward trend but instead some geese may have been in fields outside the count circle.

The increase in geese, and geese that aren't migrating, is nationwide over the last 50 years. Hunting (2 million harvested in 2002) hasn't held back the Canadas. Plus, the birds in most parks, including ours, are safe by law. No one can hunt within city limits.

So yes, there is more goose poop in our parks than before. Because it is recycled grass, I don't find it as objectionable as dog droppings.

At Holliday Park, goose nesting was confined to the island, but this year there were three pairs nesting off-island – three ganders hissing at park visitors trying to walk the sidewalks – for four weeks of incubation. I worry geese beaks are about

eye-level with small children.

Because Canada geese and other migratory waterfowl are protected by international treaty and congressional acts, the city can't touch them without permission from the U.S. Fish and Wildlife Service. Before the city could try addling eggs to slow population growth, FWS asked the city to have a ban on feeding birds in the parks.

According to Birds of North America Online, www.bna.birds.cornell.edu, which is summaries of scientific bird studies available to the public for an annual subscription fee, Canada geese eat grass-type plants almost exclusively, adding berries and seeds in the winter, though they've learned to find waste grain in farm fields.

People objecting to the city's feeding ban, saying it's bad for the geese especially in winter, need to keep several things in mind:

--People often feed the geese junk food rather than dried whole corn, which is what farmers have determined works for domestic geese.

--Handouts represent very little of the total diet of the 1,500-plus Cheyenne geese – most of which are too busy grazing far from the parking lots.

--According to research, urban geese have adapted to a year-round diet of grass.

--Our geese often fly to nearby fields for grain.

The urban Canada goose is looking for lawns next to ponds, say the studies referenced by BNAO. I don't foresee the city draining lakes and paving parks since people like grass and water as much as the geese do.

Where it is imperative to keep geese away, such as airport runways, harassment by dogs has some effect. But don't try this yourself since it's illegal to harass a federally protected species. It probably isn't realistic to fence Holliday Park and turn it into a dog park, either.

I haven't seen much evidence of predation except for the cormorants eyeing goslings. What we need is a way to harvest Canada geese within the city without using firearms, and to be practical, feed them to the hungry. Wild geese are very nutritious, especially when park employees work hard to grow the grass they eat.

Instead, we have to wait and see if a feeding ban and egg addling will limit the goose population. If not, we'll be stepping around more droppings and territorial ganders.

# 338 Nationally known birders have nothing on birds, the true celebs

Sunday, December 16, 2012, Outdoors, page E3

Studying birds in your field guide will help you identify them when you finally see them in the field. But don't neglect to study the author photo on the back cover – you never know when you'll have a chance to identify them as well.

Over Thanksgiving, Mark and I attended a family wedding in Philadelphia. One of my new shirttail relations, John, is a birder and came with us to Cape May for a day.

The southern tip of New Jersey has long been recognized for its numerous and varied migrating birds and has lots of public access for birding. We stopped at the Cape May Bird Observatory hawk-watching platform first, figuring that official migration observers might still be around and help us Westerners and John, an Irishman living in England, identify local birds.

The first ID I made was that the observer on deck was Pete Dunne, author of several books I've reviewed for this paper, including "Pete Dunne's Essential Field Guide Companion" and the first three of his seasonal quartet beginning with "Prairie Spring." He's also a co-author of "Hawks in Flight," along with other entertaining books and articles for birding magazines.

Pete's day jobs are director of CMBO and chief communications officer for New Jersey Audubon Society. But they let him out of the office to count birds. He can distinguish a turkey vulture from a black vulture, even when they are mere specks overhead. Without his help, we would not have identified the red-shouldered hawk or determined that the Cooper's hawk in flight was not a sharp-shinned.

If you go to bird festivals, you too, will meet nationally recognized birders. Nearly 30 years ago, I shook hands with Roger Tory Peterson at a National Audubon convention. A few years later at another one, I correctly identified, from a distance, without using binoculars, the man in the middle of a flock of middle-aged women was author Kenn Kaufman.

But we met Pete at home, in his own habitat, with no printed agenda or groupies to indicate his status.

The next day we visited the John James Audubon Center at Mill Grove, in Audubon, Pennsylvania, where one of the world's most famous bird artists lived when he first came to the U.S. in 1803 at age 18.

Hiking the trails around the house, now a museum, through the woods and fields overlooking Perkiomen Creek, I was able to add a singing Carolina wren to my life list, imagining Audubon first hearing it here as well.

A week later, the wedding couple, my uncle and my new aunt – both knowledgeable birders – were able to re-find Pete on the same deck at Cape May.

Pete remembered the three of us from the week before, and conversely, said the sighting of Wyoming birders was a lot rarer event than meeting John, even though he came further. There are just over half a million Wyomingites, after all, compared to 62 million Brits. Many of us from the Cowboy State, by nature of our choice of residence, and especially those attracted to birding, prefer travelling to remote places rather than congested coasts.

We were too late to meet superstar storm Sandy, luckily. Cape May was untouched because Sandy landed some miles up the coast at Atlantic City where she devastated Brigantine National Wildlife Refuge and the barrier islands.

We still noticed Sandy's damage in the suburban-rural area north of Philadelphia, mostly toppled pines and hardwoods, including one that impaled the second story of a house.

What happens to birds in severe storms? My aunt forwarded a New York Times article by Natalie Angier, published Nov. 12, reporting how perching birds have toes that automatically lock around a branch when they bend their legs. So they are as safe as the branch they sit on.

Angier reported birds feel the changes in air pressure from an approaching storm and those migrating may steer around it or correct course afterwards. Some get a boost, having been documented as flying into a storm at 7 mph and coming out the other side at 90.

No matter how many well-known birders I may meet, the birds are the true celebrities. The migrants propelling themselves over dangerous distances, as well as the ordinary house finch weathering another winter, are to be celebrated.

So, if I attend the National Audubon convention in Stevenson, Washington, in July, I will be sure to identify the keynote speakers, but the local birds will be the real stars.

# 2013

**339** Sunday, January 20, 2013, Outdoors, page E2
## Winter is good time to spot unusual birds

Last winter, snowy owls irrupted. Meaning, there were sightings all across the northern tier of the lower 48 states. Apparently, more owls fledged than usual, and there weren't enough small rodents to go around in their Arctic winter territories so, they headed for more productive habitat.

This year in Cheyenne, it's the seedeaters that are irrupting, or at least coming down from the mountains.

My first inkling was the Steller's jays I saw at a friend's, up on the north edge of Cheyenne, enjoying the pine-juniper windbreak and the birdfeeders. They are dark blue with black heads, unlike the usual blue and white blue jay. Five made an appearance for the Cheyenne Christmas Bird Count on Dec. 22, as they have eight out of the last 38 years.

Named for Georg Steller, the first to find this bird and describe it for science while serving as the naturalist travelling with Vitus Bering in 1740-42 to what became Alaska, Steller's jay is found in western mountains down to Central America.

Both its usual plant (seeds, nuts, fruits) and animal (small vertebrates) foods must be in short supply in nearby mountains. Even if the Birds of North America lists cookies and other picnic provisions as preferred food, don't be tempted. Give them black-oil sunflower seed.

Making its first-ever appearance on the Cheyenne bird count was the pygmy nuthatch. A flock was noted several weeks before on the west side of town, and we were able to re-find it. The five individuals were mixed in with white-breasted and red-breasted nuthatches and mountain chickadees, all in the same pine tree.

The pygmy nuthatch is another mountain species, but it seldom comes down. It needs dead or partially dead trees with cavities, not just for nesting, but also to stay warm. Studies show families, even whole flocks, will pile into a cavity when it's cold. The birds at the bottom are the warmest, but the entire space will be several degrees warmer than it is outside.

There must be empty food caches and a dire lack of frozen insects to pick out of the bark of mountain pine trees for pygmies to leave their known hollow trees for an urban area where we keep dead wood to a minimum.

I was thrilled to see evening grosbeaks on the Guernsey-Ft. Laramie Christmas Bird Count on Dec. 29. They were at a feeder between Guernsey and Hartville, looking like over-sized goldfinches. Another mountain species, they expanded their range east from the Rockies in the mid-1800s. It's thought that the planting of box elder (their favorite seeds) and ornamental fruit-bearing trees, and the invasion of spruce budworms, led them on.

Today, they are not quite so common back east – the reason they were brought to attention as the 2012 American Birding Association Bird of the Year. They are well known for their irruptive behavior, usually every other winter. We've had a handful of them on each of nine Cheyenne Christmas Bird Counts over the last 38 years.

The range map for common redpolls in Douglas Faulkner's "Birds of Wyoming," shows that every winter they will show up in the northeast corner of Wyoming. They breed in the Arctic. This year, they are all over the state, according to multiple reports on the Wyobirds elist, with 24 present for the Cheyenne Christmas Bird Count.

Redpolls, too, seem to show up at bird feeders on alternate winters. If you are familiar with house finches at your feeder, scan them closely for redpolls, slightly smaller, streaky brown birds with a small red spot on the forehead and sometimes a wash of pale pink on the breast.

On their home turf, redpolls eat the very small spruce and birch seeds. At your feeder, small seeds like white millet would be a good replacement.

In his report of the Cheyenne Christmas Bird Count, compiler Greg Johnson said the rarest sighting was a red-bellied woodpecker, a species of the eastern U.S., particularly the southeast. The first ever recorded sighting of one in Wyoming was in Cheyenne in 1992. Then there were two other sightings of single birds in eastern Wyoming in 1993 and 2002, followed by three sightings at the Wyoming Hereford Ranch outside Cheyenne in 2002, 2006 and 2008.

Irruptive is not the explanation for the appearance of this woodpecker in Cheyenne. Lost is more like it, though lost seems to be coming a regular habit. Officially, the term is "vagrant." This individual may have been caught in some weather in October and was lucky enough to find Mike Schilling's feeding station, where it has been since.

There is only one way to see species uncommon for Cheyenne, and that is to look. A well-stocked feeder helps, but the best way is to get outside and keep your eyes open. And when you see some unusual bird, tell someone.

# 340   Game and Fish needs our help

Sunday, February 10, 2013, Outdoors, page E3

It's easy to support the work the Wyoming Game and Fish Department does. Just buy a hunting license. Or show up at the Capitol during the state legislative session to testify on the merits of license fee increases.

The department gets 80 percent of its funding from license fees, but it is the legislature that approves any fee changes every six or seven years. By this session, the fees approved in 2007 had 20 percent less buying power, thanks to inflation, but the legislation did not pass.

Yet, there are more expenses. There are more people coming to work here who need wildlife education. And there's more baseline data collection and monitoring work to be done in the face of more energy development.

Surprisingly, at the committee meeting Feb. 1 to hear testimony on the second bill proposing increasing hunting fees, there was a lobbyist for a minor sportsmen's group opposed. His board members begrudge having to pay more to hunt, even when it is apparent that the cost of everyday agency work gets more expensive.

In my testimony, I mentioned that my husband and I hunt and fish, we enjoy nongame wildlife, and we made an investment to support the Game and Fish by buying lifetime fishing licenses for our family. Later, the lobbyist told me we bird-watchers ought to be paying something, too.

He is right. There are more people in Wyoming enjoying looking at nongame wildlife, including birds in their back-yards, than are hunting it. We are indirectly benefiting from the 6 percent of hunting license fees spent on nongame species work.

However, grants and legislative funding cover most of the $9.5 million (14.5 percent of the total Game and Fish budget) spent on nongame species: programs to prevent aquatic invasive species invading; programs to prevent "sensitive species" from requiring listing as threatened or endangered; programs for wolves and sage-grouse; and work on brucellosis and chronic wasting disease.

There is also one biologist who tracks all the bird species not hunted.

How can a non-hunter support Game and Fish?

First, we need better terminology. Rather than "non-hunter," say "wild-life watcher." Rather than "nongame," I like "watchable wildlife," a term the department already uses, even if it does seem to include the huntable megafauna.

Some states sell special vehicle license plates to support wildlife. That was suggested here a few years ago, but apparently, the University of Wyoming is going to be the only entity with the sacred right to raise funds that way.

Some states have a check-off on their income tax forms to give people an easy option to contribute a few dollars, but it will be decades before any Wyoming legislator wants to prematurely end her career by suggesting instituting state income tax.

Colorado uses the majority of its lottery income to support its wildlife programs. Wyoming considers legislation to join one of the national lotteries every year. If it ever passes, could funds be earmarked for wildlife?

In other places, a special license allows a person access to special state land. With so much federal land available for recreation, that probably wouldn't work in Wyoming, either.

The federal government once proposed a minor tax on outdoor gear that would be shared with states, but the gear companies nixed that.

Game and Fish does have a nice selection of items available in their gift shop here in Cheyenne and online, but seriously, who needs another mug or T-shirt if you already belong to one wild-life organization or another?

What we really need is a voluntary wildlife watching license: Something on the order of $25 per family, with the option of contributing more and being listed in the back of Wyoming Wildlife magazine, as supporters of other organizations are in their publications.

Besides being listed in the magazine, one's support could be shown with a small sticker on the car window, maybe pasted right next to the annual state parks entrance pass. We could also charge visitors non-resident fees if they also wanted a wildlife watching license.

And then, as sometimes happens, maybe third parties would offer perks for license holders – perhaps a discount from local purveyors of outdoor gear. Or maybe each year, license holders would be put in a drawing for a pair of super-duper binoculars or a spotting scope.

But really, for some of us, just knowing we are contributing to the well-being of all the wildlife in Wyoming – and there are a lot more kinds of critters out there than the ones sportsmen hunt – would be worth it.

If you have any other ideas, please contact Wyoming Game and Fish Department Deputy Director John Emmerich, 777-4501.

2013

# 341 Florida in February is full of fab birds

Sunday, March 17, 2013, Outdoors, page E2

Florida is where people see their favorite Disney characters come to life. Florida is where I finally saw 14 species of birds that I've paged past in my field guide for 40 years.

I wasn't sure what to expect last month when Mark and I went down to visit his relatives. Images of beaches with high-rise hotels from TV alternated with those of cattle ranches, alligator swamps and orange groves.

I was familiar with historical accounts of the first wardens hired by the precursor of the National Audubon Society to protect the wading birds of Florida being hunted for their plumes used to adorn ladies' hats. In 1903, President Theodore Roosevelt set aside Pelican Island as the first unit of what became the National Wildlife Refuge system.

Those early efforts have led to preserving numerous other Floridian acres, and so I was able to see those plumy wading birds 110 years later.

Our first bird outing was Caladesi Island State Park, on the Gulf about 1.5 miles offshore from Dunedin. You have to take a ferry since this is a barrier island without a bridge.

We were welcomed by the park manager, Bill Gruber, who back in 1999 was the Wyoming Tribune Eagle Outdoors editor and invited me to write this monthly bird column.

The day we were there, staff and volunteers were setting up poles to rope off part of the award-winning, white sand beach where four kinds of plovers would soon be nesting. I also met my first gopher tortoise. Florida lists it as a threatened species.

The island has a long human history, including a family who homesteaded from 1890 through the 1930s. The daughter, Myrtle Scharrer Betz, banded birds and had an island list of 158 species. Remarkably, Caladesi managed to avoid development, the fate of so many other Florida barrier islands. Betz's book, "Yesteryear I Lived in Paradise," explains how and why.

On a whim, I bought a copy of the DeLorme Atlas and Gazetteer for Florida. The amount of human development becomes quite plain – all those little blue lines in perfect grids are drainage canals. Florida's peninsula would otherwise mostly be a freshwater swamp surrounded by saltwater ocean, which is exactly what makes it so interesting for birding.

At Loxahatchee National Wildlife Refuge, on the north end of the swampy Everglades, we nearly managed a complete roster of all the large wading birds we saw on the whole trip: great blue heron, great egret, snowy egret, little blue heron, tricolored heron, cattle egret, green heron, black-crowned night-heron, yellow-crowned night-heron, glossy ibis, white ibis (also seen on lawns everywhere), roseate spoonbill and wood stork. The reddish egret eluded us.

On our last day, at the Storm Water Treatment Area 1 East, adjacent to Loxahatchee, among many coots and common gallinules (known before their name change last year as common moorhens), we spied something that looked like a purplish-blue, over-sized gallinule. It was a purple swamphen! It is an exotic from Eurasia, but the week before, the powers that be decided it was countable for anyone pursuing an official life list.

We also picked up a copy of "A Birder's Guide to Florida," by Bill Pranty, published by the American Birding Association. It lists almost every good birding spot in Florida, explains Florida's landscape history and its many habitats, and provides information about all the birds in the state, showing where and when they can be found.

Florida is, for me, upside down. We saw many familiar Wyoming summer birds there in February. Unlike Wyoming, summer is when Florida has its fewest bird species. The problem is that in winter, our familiar birds may not be in the breeding plumage we northerners are used to seeing them in.

Besides Caladesi Island, the ABA guide missed another fabulous birding spot, maybe because it was too new. We heard about it from another birder we met on a trail.

Palm Beach County acquired 100 acres of farmland in Boynton Beach for a third of its value from a couple who stipulated it must be made into wetlands. Construction of Green Cay Wetlands began in 2003. Besides catering to native wildlife and human visitors, the park is a water reclamation facility.

It has 1.5 miles of boardwalk so you don't have to worry about alligator encounters, and the wildlife, including the alligators we saw, doesn't have to worry about encounters with people. No birds flinched as we walked by, and so the photo ops for my little point and shoot were fabulous.

I imagine we'll go again next year – there are 300-plus other birdy parks and preserves to visit, and more of Florida's 481 native bird species to find.

# 342 Wyoming Birding Bonanza strikes again

Sunday, April 7, 2013, Outdoors, page E2

Are you ready for the second annual Wyoming Birding Bonanza? Polish your binoculars because you can be a winner.

The competition was dreamed up last year by James Maley and Matt Carling, both from the University of Wyoming's Department of Zoology and Physiology. James is collections manager of the Museum of Vertebrates and Matt is an assistant professor.

Their goal is to increase the number of bird observations for Wyoming during spring migration that are recorded in the eBird.org database and to get birders into the habit of submitting information. The data is used by scientists.

Last year, the contest ran from mid-April to mid-June, but this year, it is being pared back to May 1-31, concentrating on the peak weeks.

And again, thanks to sponsors like last year's, Cheyenne - High Plains, Laramie and Meadowlark Audubon societies, as well as UW's Biodiversity Institute, Audubon Wyoming and eBird, there are prizes.

Registered contestants who enter at least 15 checklists will receive a WBB T-shirt. A checklist is a list of bird species and number of individuals of each, seen in a particular location during a period of time. James promises this year's T-shirt will be a work of art. Everyone who turns in at least 10 checklists will be entered in a grand prize drawing.

Also, for each Wyoming county, the participant reporting the most species will win a prize. Last year, I was the Laramie County winner and received the latest edition of the National Geographic field guide. This year, our county is up for grabs since I'm going to be out less often.

For better odds, try birding Big Horn, Converse and Sublette counties, where no checklists were turned in last year, James said.

"April, May, and June of 2012 are now the top three months of all time for number of checklists statewide," he said. There were 1,282 turned in, compared to 424 for the same months in 2010. A total of 266 species was observed in 2012.

I know I paid closer attention to the birds around me because of the competition. I found a summer tanager in our backyard May 11, considered rare for Wyoming.

James passed on a list of other rare bird sightings from 2012:

--1 Glossy Ibis at Meeboer Lake (west of Laramie) on April 17

--1 Lesser Black-backed Gull also at Meeboer Lake on April 17

--1 Black-and-white Warbler at Holliday Park on April 21

--1 Juniper Titmouse at Guernsey State Park on April 22

--1 Long-tailed Jaeger at Hutton Lake NWR on May 3

--1 Northern Cardinal in Laramie on May 4

--5 Short-billed Dowitchers at Hutton on May 5

--1 Snowy Owl at Keyhole State Park on May 15

--1 Blackpoll Warbler at Hereford on May 15

--1 Cattle Egret in Rock River on May 17

--1 White-eyed Vireo near Lander on May 28.

So, are you ready to earn that WBB T-shirt? You can do it by simply counting the birds in your backyard for a few minutes at least 15 different times. Here's what you need to do.

First, sign up at www. eBird.org, if you haven't already. It's free. Click on the "About eBird" link, and then the "eBird Quick Start Guide," the first link on that page.

When setting up your observation locations, select a hotspot marker if there is one at one of your locations already, such as Wyoming Hereford Ranch or Lions Park. Otherwise, on the map your personal marker may be hidden underneath the hotspot's. You can view your data for a hotspot alone or collated with everyone else's. If you have questions about eBird, call me.

Next, sign up for the Wyoming Birding Bonanza at http://www.uwyo.edu/biodiversity/vertebrate-museum/birding-bonanza/. It's also free.

Here are the rules.

**Counting**:

--Participants will count only full species as defined by the current American Birding Association checklist.

--Birds identified to a taxonomic level above species may be counted if no other member of the taxonomic level is on the checklist. For example, duck sp. can be counted if no other ducks are seen.

--Birds counted must be alive and unrestrained. Sick and injured birds are countable. Nests and eggs do not count.

--Electronic devices are allowed but see ABA's Code of Ethics for guidelines.

**Time:** We will extract final eBird data for the Bonanza on 30 June 2013.

**Area:** Anywhere in Wyoming.

**Conduct:**

--Participants must only count birds unquestionably identified. If in doubt, leave it out.

--Know and abide by the rules.

--Share information with other birders – they'll thank you.

Good birding to all!

2013

319

# 343 Early birds yield clues

Sunday, May 5, 2013, Outdoors, page E3 and Southeast Wyoming Extra

First birds of the season can tell us about climate change

The great joy of springtime, if you are a birdwatcher of any sort, is seeing your first robin, first bluebird or first mourning dove of the year.

There is sometimes a friendly bit of competition where serious birders gather – to see who is first to report their "FOYs," first of the year observations, especially of more obscure migratory species, say "Greater Yellowlegs," a long-legged shorebird.

Those of us in southeastern Wyoming have the advantage over the birders in the rest of the state posting on the Wyobirds elist, as many spring migrants often funnel up against Colorado's Front Range and across Cheyenne before spreading out over the rest of Wyoming.

At eBird.org, where ordinary folks file their bird observations for free, for their own record-keeping and for use by scientists, FOY data is constantly updated and can be found in the Explore Data section under "Arrivals and Departures."

A check of the Wyoming statistics shows that Del Nelson got the jump on all of us this year by birding January 1 near Crowheart, Wyoming, and reporting 28 species

– mostly birds we expect to see mid-winter.

However, Del's list included a single western meadowlark. On occasion, individuals of migratory species like that miss the bus south in the fall and sometimes find a perfect pocket of habitat that allows them to survive the winter. Insectivorous birds like meadowlarks usually prefer live insects, not the foods of wintering birds: frozen bugs picked out by flickers, seeds preferred by finches, or warm-blooded creatures preyed on by hawks.

A few days later, Del reported a mountain bluebird, another bird uncommon in winter, which I always thought of as a sign of impending spring. Even robins aren't reliable – one was listed for Wyoming January 2.

Studying the "Birds of Wyoming" compendium by Doug Faulkner, I found that many migratory bird species often have a few individuals observed in Wyoming during Faulkner's designated winter months of December through February. If you don't count those species, the first true spring migrant (no over-wintering records so far), the 97th species listed by eBird for 2013 in Wyoming, is the group of sandhill cranes seen February 12 in Riverton – by Del Nelson, the birder

who must be spending more time afield than anyone else in the state.

Here in Laramie County, our signs of spring, our FOYs, were observed more seasonally:

Killdeer – March 5
Robin – March 6
Mountain bluebird – March 7
Meadowlark – March 16
Turkey vulture – March 29
Mourning dove – March 30
American avocet and Swainson's hawk – April 7.

Granted, in a state like Wyoming with a sparse population of birdwatchers, it is quite possible the first flock of anything to flit over the county line goes unnoticed. Sometimes we are hiding at home during snowstorms.

Brian Kimberling, a columnist writing for the New York Times, recently posed the idea that FOYs might help us track climate change the way tracking plants has. There is a website, www.BudBurst.org, asking citizen scientists (you and me) to track when particular perennial species bloom. Changes are already noticeable when historic records of eccentric gardeners and naturalists are examined, showing blooming times advancing as much as a week over a few decades.

I'm not sure the migrations

of birds are as useful as the bloom times of plants. After all, plants sit in one place and accumulate degrees of heat necessary to bloom, while birds will push the envelope in their quest for food, sometimes losing their gamble when, after a pleasant spring weather spell, disaster hits. Many dead birds were reported after our April 15-17 snowstorm.

The opposite of FOY, what I think of as LOS, "Last of the Season," might be more accurate a measure. Sometime in April, just when I think I've seen the last of the juncos until fall, another bout of cold blows in and they reappear briefly, pushed back into town from their summer homes in nearby mountains.

But there are whole groups of birds, mostly the insectivorous species, which have never been reported in winter in Wyoming: hummingbirds, vireos, most of the shorebirds, flycatchers, swallows, swifts, terns, dickcissels, bobolinks, and most warbler species.

When those birds begin to show up earlier and earlier, establishing a trend over the years, it will be one way we'll know that global climate trends apply to us, too, right here in the Magic City.

# 344 Spring migration surprise delights birdwatchers

Sunday, June 9, 2013, Outdoors, page E2

The highlight, and maybe rarest bird, was pretty flashy.

A male scarlet tanager

was spotted May 18 during the Cheyenne - High Plains Audubon Society's Big Day

Bird Count.

If you've never seen one back East, think bright red

bird with black wings. More than 20 people were able to see it at Wyoming Hereford

Ranch as it perched on a low branch on the front of a bush - unlike other rare birds which would rather skulk behind.

Every year in mid-May, members and friends of the chapter hope to encapsulate Cheyenne's spring migration in one day, but with more people discussing the birds they see on the Wyobirds elist before and after the chosen date, it's apparent this year's migration was not so tidy. For instance, white-crowned sparrows had nearly all moved on when the black-headed grosbeaks and some of the unusual warblers moved in.

We counted more species this year than last, 111, perhaps because it was a nicer day. Last year's 104, the lowest in the 21 years for which I can find records, were counted in rain and cold. But we still are nowhere near the 140-150 species average

from the 1990s.

In all, 25 species recorded last year didn't show up this year, including nine kinds of shorebirds. A look at the local reservoirs on May 18 showed that it was the fault of the previous month's weather. All that snow melted and filled them, leaving no bare, sandy ground along the shoreline for the shorebirds.

Another group missing was the flycatchers, except for our most abundant species – eastern and western kingbirds and Say's phoebe. The ones not seen, including the Western wood pewee, which should be somewhat common around Cheyenne, were running late.

But we more than made up for missing species with 32 that weren't seen last year.

The warbler family was in fine fettle. Last year, we had only six species, this year 11 (on the accompanying list they begin with

northern waterthrush, going on through yellow-breasted chat).

In the 21 years I've been keeping track, we've had as few as six warbler species (2005 and 2012) and as many as 17 (2001 and 2002), averaging 11, with an overall total of 31. Many of these species "belong" back East.

In some ways, the Cheyenne Big Day is a game we play: How many species can we see in one day? The American Birding Association has rules for Big Days. If you have a team, all team members must see the same birds. Team Sapsucker, top birders from Cornell Lab of Ornithology, broke the North American Big Day record this spring with 294 species, by birding in Texas during the right 24 hours.

If we could see all in one day the species recorded on our Cheyenne spring counts over the last 21 years,

we'd have 263.

But here in Cheyenne, we are more egalitarian. We invite anyone willing to bird with us, or bird on their own, to contribute to our total, so our results might be influenced by the number of birders and their level of expertise as much as by the weather. This year we started at 6 a.m. with about 15 people and were helped out by a dozen more from Casper and Laramie who found birds we missed and in areas the main group didn't get to.

Our results from as far back as 1994 have now been entered into www.eBird. org, where they will be of use to scientists, so that our Big Day isn't just a game or an excuse to indulge in our favorite hobby. But it would seem more like work if it wasn't full of surprises like the scarlet tanager.

## 2013 Cheyenne Big Day Bird Count Results

The abundance ratings indicate how likely experts think a species will be seen in Cheyenne in its preferred habitat in spring.
A – abundant  C – common  U – uncommon  R – rare  V - vagrant

| | | | |
|---|---|---|---|
| Canada Goose, A | Black-crowned Night Heron, C | American Kestrel, C | Swainson's Thrush, A |
| Gadwall, C | White-faced Ibis, U | Say's Phoebe, C | Hermit Thrush, U |
| American Wigeon, C | Turkey Vulture, C | Western Kingbird, A | American Robin, A |
| Mallard, A | Cooper's Hawk, U | Eastern Kingbird, C | Gray Catbird, U |
| Blue-winged Teal, C | Swainson's Hawk, C | Plumbeous Vireo, U | Northern Mockingbird, U |
| Cinnamon Teal, C | Red-tailed Hawk, C | Warbling Vireo, C | European Starling, A |
| Northern Shoveler, C | Ferruginous Hawk, U | Red-eyed Vireo, U | Cedar Waxwing, U |
| Northern Pintail, C | Sora, U | Steller's Jay, R | Chestnut-collared Longspur, R |
| Green-winged Teal, C | American Coot, A | Blue Jay, C | Northern Waterthrush, U |
| Redhead, A | Killdeer, A | Black-billed Magpie, C | Orange-crowned Warbler, C |
| Ring-necked Duck, U | American Avocet, C | American Crow, A | MacGillivray's Warbler, U |
| Lesser Scaup, A | Spotted Sandpiper, C | Common Raven, U | Common Yellowthroat, C |
| Bufflehead, U | Wilson's Phalarope, A | Horned Lark, C | Yellow Warbler, A |
| Common Merganser, C | Common Tern, R | N. Rough-winged Swallow, U | Black-throated Blue Warbler, R |
| Ruddy Duck, U | Rock Pigeon, A | Bank Swallow, U | Yellow-rumped Warbler, A |
| Pied-billed Grebe, C | Eurasian Collared-Dove, A | Cliff Swallow, A | Blackpoll Warbler, U |
| Eared Grebe, C | Mourning Dove, A | Barn Swallow, A | Black-and-White Warbler, R |
| Western Grebe, C | Great Horned Owl, C | Black-capped Chickadee, R | Wilson's Warbler, C |
| Double-crested Cormorant, C | Belted Kingfisher, U | Mountain Chickadee, C | Yellow-breasted Chat, U |
| American White Pelican, U | Downy Woodpecker, U | Red-breasted Nuthatch, C | Spotted Towhee, C |
| Great Blue Heron, C | Northern Flicker, C | White-breasted Nuthatch, R | Chipping Sparrow, C |
| | | House Wren, C | Clay-colored Sparrow, U |
| | | Blue-gray Gnatcatcher, U | Vesper Sparrow, C |
| | | Ruby-crowned Kinglet, C | Lark Sparrow, C |

| | | | |
|---|---|---|---|
| Lincoln's Sparrow, U | Lazuli Bunting, U | Great-tailed Grackle, R | Lesser Goldfinch, R |
| Song Sparrow, U | Red-winged Blackbird, A | Brown-headed Cowbird, C | American Goldfinch, C |
| White-crowned Sparrow, C | Western Meadowlark, A | Orchard Oriole, U | House Sparrow, A |
| Scarlet Tanager, V | Yellow-headed Blackbird, C | Bullock's Oriole, C | |
| Western Tanager, C | Brewer's Blackbird, C | House Finch, A | |
| Black-headed Grosbeak, U | Common Grackle, A | Pine Siskin, U | |

# 345 Birder learns to look more closely

Sunday, July 7, 2013, Outdoors, page E2

Maybe it's because I've been spending more time outside, or maybe I'm becoming a better observer, but I've seen more bird nests and young birds this breeding season than in years past.

It helps to know how different birds nest so you can recognize them. Knowing that belted kingfishers nest in burrows in stream banks helped me make sense of watching one disappear into rocky dirt, especially when another came to take its place. Surprisingly, the cliff with the nest burrow was 100 feet above and a hundred yards from our local stream.

For nearly half of June, Mark and I were on the road to and from our younger son's graduation (master's, mechanical engineering, from that other UW, the one in Seattle), and we checked out local birds whenever time allowed.

Our first stop was the Bear River Migratory Bird Refuge, a U.S. Fish and Wildlife Service unit on the edge of the Great Salt Lake outside Brigham City, Utah. The 10-mile, one-way loop drive takes visitors along acres of shallow water that in mid-June has hundreds of wading birds and their nests, including white-faced ibis, black-necked stilts and American avocets. I was lucky enough to see one fuzzy avocet chick close up. Their nests are just vegetated high spots above the human-managed water level.

When we arrived at the beginning of the loop, we saw an open-sided shelter and thought it might be a shady place in the treeless landscape to eat our supper sandwiches – except cliff swallows had daubed half a hundred nests in the roof rafters. Old nests and droppings littered the benches below. Each gourd-shaped nest was built from probably a few hundred swallow-sized mouthfuls of mud from the nearby canal. We watched from inside the car while the cloud of swallows swooped after the cloud of horseflies. Later, down the road, when strong wind came up, we found young cliff swallows clinging to the gravel road surface for dear life.

In Seattle, at the north end of the Washington Park Arboretum, there is a board walk out over the edge of Lake Washington. We were enchanted by wood ducks, mom and seven ducklings, feeding around the lily pads.

It was clear the family was attracted to us, following us as we travelled the walk, but, strangely, not in the begging-for-handouts way. After taking many pictures of them, we finally looked up to see the bald eagles and osprey overhead. Oh. We were being used as human shields. That's OK.

Bushtits are common little brown birds in Seattle, but not common for us Wyomingites, so we stopped to get a better look at one, trying to follow it through the thick vegetation. It kept disappearing. Then we realized it was disappearing with its mouth full and reappearing with it empty. Sure enough, we soon spied the nest, a pendulous ball of plant material tucked in among the leaves. I learned later that spider web holds it together.

Bullock's orioles also have elongated, but woven, hanging nests. One we saw was made mainly of fishing line. Discarded fishing line is dangerous for many kinds of wildlife, including the birds using that nest – they could get toes stuck in it and become trapped.

In Wenatchee, Washington, mitigation for a hydroelectric dam on the Columbia River has resulted in lovely parks and a natural area. Mark and I enjoyed the number of bird species from a variety of habitats: river, wetlands, grasslands and huge cottonwoods. I was thinking about how intelligent the land managers were to leave dead trees standing and what a nice hole that was in that big dead tree, and there it was, a motionless young American kestrel eyeing us – proof that dead trees are productive.

Also in Wenatchee, in Ohme Gardens, a manmade coniferous forest on a dry bluff overlooking the city, we observed an Oregon junco feeding a fledgling on the ground. But wait, the fledgling was decidedly larger than the junco. It was a cowbird, hatched from an egg left by its own parents in the junco's nest for the junco to raise, something we've read about but never seen before.

Now, in July, the first flush of nesting is over here in Cheyenne, but the goldfinches are just beginning – they wait for the thistles to go to seed. Some birds, like robins, will attempt a second brood. If you missed the early days of the mallard ducklings and Canada goose goslings at Lions Park at the end of June, it will get harder and harder to tell them from adults. Just watch for that gawky teenage look.

# 346 The generosity of other birders improves travel experience

Sunday, August 18, 2013, Outdoors, page E2

Mark and I have yet to hire a birding guide or join an organized bird trip in our travels, but someday we may have to if we travel somewhere with unfamiliar birds.

Meanwhile, we find birding hotspots by using www.eBird.org. But then there is also the generosity of local birders.

Mid-July, Mark and I were walking the trails at the Sitka National Historical Park in Sitka, Alaska, overwhelmed by the sounds of small birds high in the rainforest canopy. We knew the singing came from thrushes, but even though we'd gone online the night before to listen to the three possibilities, varied (definitely different), hermit and Swainson's, our audio memory wasn't very good.

This was supposed to be a vacation, so I was trying to just let it go. I can't reliably sort out Swainson's and hermits by sight in my own backyard during migration, where they never sing, much less these invisible birds.

But enter Lucy. When you are wearing binoculars, it is not considered rude to walk up to a total stranger also wearing binoculars and ask, "What are you seeing?"

Some birders, I have heard, are curmudgeons but not this woman. We chatted more than 5 minutes before she invited us to come with her to an opening in the bushes along the shore where she'd seen four species of gulls the day before.

Lucy also explained an easy way to identify the singing thrushes: the Swainson's fluting song spirals up and the hermit's makes a little rise before spiraling down.

We were surprised any birds were still singing in mid-July, still advertising for mates and establishing territories. I later read that the Alaska Natives call the thrushes "salmonberry birds," because they sing at the time of year salmonberries are ripe.

They were right about that. Sitka's brambles were full of these raspberry-type berries, either a deep gold or a deep red when ripe. Mark's brother, Peter, who lives in Sitka, pointed them out. They are kind of seedy, so the best way to eat them is to pop one in your mouth, smash it with your tongue to get the juice and then swallow it whole. It was hard to concentrate on birding with so many berries to pick. Luckily, no bears were competing with us.

Lucy still had to get to work that morning, so we bid adieu.

The next day, we stopped by the Fishermen's Eye Gallery where she works and gave her an update on the birds we'd seen, including nine young bald eagles checking out the first returning salmon of the season and glaucous-winged, mew and Bonaparte's gulls.

We were very lucky to meet Lucy, a local, who has led bird tours in the past.

In Juneau, it was a fellow traveler who alerted us to birds. We took the tram up Mount Roberts, hoping to see ptarmigan on the trail at the top. Instead, we got an unsolicited heads-up on a family of sooty grouse (formerly spruce grouse) from a man wearing all black – and binoculars.

At Mendenhall Glacier, we struck up a conversation with another man with binoculars. He said we should talk to his buddy, a "real birder," an energetic, white-haired man who spent half our conversation promoting birding his own neighborhood and encouraged us to call him if we ever travel to Point Reyes National Seashore in California.

We watched for seabirds from the Alaska Marine Highway – the ferry – from Sitka to Skagway and all the way back to Bellingham, Washington, and from all our stops in between, but we had better luck identifying whales. The fast-winging black dots remained inscrutable. They might be worth a guided trip.

One of the crew members did point out the flock of pink, plastic flamingos perched in a tree on Highwater Island, near Sitka.

These migrate from China, he said, by way of the U.S. Coast Guard's training center.

# 347 Encourage birding as a lifelong addiction

Sunday, September 15, 2013, Outdoors, page E3

Ask a simple question of a man pulling weeds in a public garden in Juneau, Alaska. It is always possible you will discover you both know the same people.

Alaskans, like Wyomingites, are always interested in where visitors are from.

They, like us, often are from somewhere else themselves.

This summer, when I told Merrill Jensen, manager of the Jensen-Olson Arboretum (the co-founder is no relation of his), that I was from Cheyenne, he said he graduated from Cheyenne's East High.

We both graduated in 1974. But I graduated from an altogether different East High, 1,000 miles east of Cheyenne. So, the only person I could think of that might have graduated with him is actually one of Merrill's old buddies – and the husband of the friend I walk with every morning.

As my husband, Mark, called our attention to a nearby pair of harlequin ducks in the bay just yards from the edge of the gardens, Merrill remarked that he is also an avid birder.

2013

323

So that precipitated discovery of another mutual acquaintance, May Hanesworth.

May was the "Bird Lady" of Cheyenne when Mark and I moved here. We went to the 1989-90 Christmas Bird Count tally party held at her elegant apartment, and the next thing I knew, I'd been recruited to type bird lists from Christmas and spring counts for submission to the newspaper, which I still do.

May, born in 1900, was of the generation that believed real ladies didn't type. But she had elegant hand-writing. And she must have been an elegant music teacher in the Cheyenne school district.

Merrill remembers going to Audubon meetings in her living room in the 1960s. His parents discovered he had led his first-grade classmates on a "bird field trip" around his elementary school playground, so they indulged his interest in birds by tracking down local Audubon folks.

In a recent email, Merrill remembered those early days:

"(May) gave me a lot of encouragement and was able to persuade my parents to install a bird feeder/bath in the back yard. It was one of my kid duties to keep it filled in the winter. I remember we didn't have much diversity coming through to the feeder; lots of house sparrows, house finches and juncos.

"I don't remember going on any actual birding trips with May, just going to her home in the winter for meetings and watching birds out her window. I was the youngest member of the group by a long shot!" Merrill wrote.

By the time I met May, she was entering her 90s. Though she no longer went out birding, she continued to compile the bird count lists, calling all her local contacts. She was the go-to person for bird questions, remembering where to find the regular species and the particulars of the rare bird sightings.

May was in her late 90s before she was willing to become "Bird Compiler Emeritus," finally passing on in 1999 at age 99. But her influence lives on for Merrill.

"As I went through junior high and high school, there were too many other demands on my time, and I didn't go to any more meetings past probably 1968. I have continued to be an avid birder and take my binoculars everywhere.

"As to my further Auduboning, I participated in the Christmas Bird Counts while I was at Washington State with the head of the zoology department and in the Boise area with staff from Deer Flat National Wildlife Refuge.

"Here in Alaska, I've led several bird walks, do the CBC and I've just rotated off of the Juneau Audubon Society's board where I served for 4 1/2 years. Even though I'm the resident plant geek, birds still play a large part of my outdoor experiences and will continue as long as I'm able to."

Here it is, about 45 years after Merrill last sat in May's living room, and her example of the birding volunteer spirit lives on.

But let's not forget those parents who recognized, indulged and enabled their son's life-long birding addiction.

Do you know children who notice birds? Indulge them today. It will add a layer of richness to their lives, wherever they go.

# 348 Fall migration kicks up kites, but not the kind found on strings

Sunday, October 6, 2013, Outdoors, page E2

Recently, we've discovered some new kids on the block – well, over the block to be precise.

Three Mississippi kites were soaring, fluttering, soaring, fluttering high up above our house on September 2. This is a type of hawk rarely recorded in Wyoming.

I'm not sure Mark and I would have run outside with binoculars as the crow-like birds soared overhead if our friend, Chuck Seniawski, failed to mention seeing them on his way home just a day earlier, a mile away.

About three years ago, Wayne McNicholas told me about seeing one of these kites hanging around College and South Greeley Highway. He's seen them elsewhere and was familiar with them.

The Birds of North America Online describes the Mississippi kite as a "sleek, acrobatic, crow-sized raptor," so I wonder how many others we might have just shrugged off as odd-behaving crows, especially since they don't look pale gray with the sky's light behind them.

The BNA account describes the kite's range as central and southern Great Plains, but I also checked www.eBird.org.

In Greeley, Colorado, in July 2012, seven birds were documented with photos, and in August 2013, two birds. Our only previous local sighting recorded was James Maley's in August 2012, at the Cheyenne Airport Golf Course.

A number of other sightings in the last 10 years have been along the North and South Platte rivers in western Nebraska and northeastern Colorado. It's no accident kites are seen along rivers – big trees are favorite nesting habitat.

As the kites expand their range, nesting in colonies in groves as well as in urban areas, more people have learned they eat insects on the wing, as well as small animals such as mice and snakes. That's good. But kites also defend their nests when people get too close, and that has been a problem in some urban areas.

We don't know where exactly in South America they go in the winter, but the number of birdwatchers there is increasing.

The other excitement was my identifying a great crested flycatcher, an eastern species, in our backyard September 17. That day, after a week of rain, the backyard was suddenly full of birds.

There were several kinds of nearly identical gray flycatchers I'm not comfortable identifying. However, around 11 a.m., I noticed something larger than a warbler but not as hefty as a robin.

The bird had a bright yellow belly and a rufous (birder-talk for the color of an Irish setter) tail. The breast was a plain gray and the back was darker. I paged through my field guide but then had to get on an Audubon conference call.

Mid-way through the call, the bird came back, a mere

15 feet from my window. When I looked through the field guide again, I came up with a perfect match. Two days later, I spotted it again, briefly – and Mark missed it again.

Like the Mississippi kite, great crested flycatchers are seen regularly in western Nebraska and northeastern Colorado but hardly in Wyoming. Doug Faulkner's "Birds of Wyoming," published in 2010, lists no reports for spring, a few for summer (including one in Cheyenne in 1967) and four reports for fall, but all in the northern part of our state.

Doug's summary of great

crested flycatcher distribution is, "They are most likely to occur at wooded migrant traps and along river systems characterized by a mature cottonwood overstory."

That's a good description of the Wyoming Hereford Ranch, on the edge of Cheyenne, where Ted Floyd also saw a great crested flycatcher a couple weeks earlier, and why the ranch was designated an Important Bird Area in the state. Thousands of migrating birds appreciate the big old trees along the creek as a place to rest and refuel on insects.

I don't have Crow Creek running through my yard,

but I am in a 50-year-old neighborhood where the first homeowners planted many trees, though not always in the right place. Lately, some have had to be removed.

There are many neighborhoods like ours that if more people were paying attention during spring and fall migration, we could prove my contention that Cheyenne is one big migrant trap.

So, keep planting trees and shrubs (but not over sewer lines or under utility lines or too close to buildings and fences), and keep your eyes open for the next rare bird.

# 349 Curiosity, generosity rewarded by UW's Biodiversity Institute

Sunday, November 10, 2013, Outdoors, page E2

It's wonderful when friends are recognized for a lifetime of work they enjoy.

Last month, the Biodiversity Institute recognized Chris Madson of Cheyenne, and Jane and Robert Dorn, formerly of Cheyenne, now residing near Lingle.

The Biodiversity Institute, established in 2012, is a division of the University of Wyoming's Haub School of Environment and Natural Resources. It "seeks to promote research, education, and outreach concerning the study of living organisms in Wyoming and beyond (www.wyomingbiodiversity.org)." This was the first year for what will be biannual awards.

Chris's award for "Contributions to Wyoming Biodiversity Conservation" highlights his 30 years as editor of Wyoming Wildlife, the magazine published by

the Wyoming Game and Fish Department. The week before the awards ceremony, he retired.

Each issue has been a compilation of the work of the best nature and outdoor photographers and writers who were attracted to the prize-winning magazine. Judith Hosafros, longtime assistant editor, should also be credited for her attention to graphic details and proofreading that made it easy to read all these years.

Most subscribers turned to page 4 first, to read Chris's monthly elucidation of issues or hosannas to nature, and then they looked for any articles he authored.

Getting in touch with Chris for what might have been a minute could turn into a conversation exploring a topic in nearly any field – not surprising for a man with

degrees in biology, English, anthropology and wildlife.

Chris's dad was also a writer and conservationist in Chris's native state of Iowa. He remembers his dad interpreting the scenery on long car trips. When I spoke to two of Chris and Kathy's three daughters at the awards, Erin and Ceara, they both mentioned long drives as favorite times with their dad.

Chris made Wyoming Wildlife much more inclusive than the typical hook and bullet publication – for instance, the October issue had three major non-game bird articles. Illuminating the conservation ethic was always uppermost for Chris, and that's why he was nominated for this biodiversity award.

The Dorns received the Contributions to Biodiversity Science Award. Both Bob

and Jane trained as scientists: Bob with a doctorate in botany, and Jane with a masters in zoology. They met in 1969 at UW, he coming from Minnesota and she from Rawlins. They have been a productive partnership ever since.

When Bob first started his studies at UW that year, he realized there was no single good plant guide for Wyoming, and he set out to correct that, publishing "Vascular Plants of Wyoming" in 1977. It's essentially a key he made for identifying hundreds of plants, based on his and many others' research, and Jane has provided scientific illustrations for it. The third edition, still with a humble, plain brown paper cover, is available through UW's Rocky Mountain Herbarium. It's considered the bible by anyone working in botany in Wyoming.

Bob has had his own biological consulting business, working on clearances and inventories for threatened and endangered species, reclamation evaluations and wetland determinations. But he has continued to have scientific papers published and other books. Many of his contracts called for inspecting remote areas, and at this point, out of the 448 units he divided the state into back in 1969, he has botanically surveyed 445.

Jane is no slouch, botanically. Growing up, she spent a lot of time on her grandparents' ranch, and her parents impressed on her that everything has a name. I'm not sure it is possible to divide Bob and Jane's joint interests in botany and birds, but when researching in the nation's great scientific libraries, Jane tends to find the birds.

Having met them through the local Audubon chapter, Bob and Jane became my mentors when I first started writing this bird column in 1999. They put their research into two editions of their book, "Wyoming Birds." Doug Faulkner continually credits them throughout his 2010 book, "Birds of Wyoming." Jane wrote the chapter for him on the history of Wyoming ornithology, and Bob wrote the chapter on landforms and vegetation.

While both books often save me from having to make phone calls, the Dorns' book also has 70 pages of Wyoming birding hotspots and directions on how to get to them.

What Jane, Bob and Chris have in common is not only intelligence and education, but insatiable curiosity that has and will keep them going long after any official retirement; the afternoon before the awards ceremony on campus, I found Bob doing research in the herbarium.

And they also share a huge spirit of generosity, making all of us, maybe unknowingly for many people, beneficiaries of their scientific and conservation passions.

# 350 Project FeederWatch needs you

Sunday, December 15, 2013, Outdoors, page E2

OK, listen up, people. I want YOU for Project FeederWatch.

While I can't draft you like Uncle Sam, I would still like to recruit you.

Project FeederWatch is one of the citizen science programs of the Cornell Lab of Ornithology. This is the 27th season backyard birdwatchers in North America have contributed data about the birds that visit their feeders during the winter. The information is becoming increasingly important to scientists, yet it is so easy to submit, even a child can do it – and children are welcome.

It takes only a glance at the participant map to see that the Great Plains region is vastly under-observed. Even in a populated place like Cheyenne, the last few years there has been only one red dot – me, and possibly someone else too close by to show up as a separate dot. A few years back several dots showed up across the city.

I'd hate for the scientists to consider my backyard typical, or to have them completely drop our area in studies because of insufficient data, so that's why I'm inviting you to join me. Besides, it's fun, and it doesn't have to take much time. Also, like me, you can learn a lot about the birds in your backyard.

Here's what to do:

Visit the Project FeederWatch website, www. feederwatch.org.

Go to the "About" tab for an introduction and a step-by-step explanation of how to participate. Under the "Learn" tab, you can find out about feeding and identifying our local birds. The "Community" tab is where you'll find tips and photos from other participants and the FeederWatch cam.

At the "Explore" tab, you'll find a bibliography of studies that used PFW data and nifty animated maps.

Next, click on the "Home" tab and then the Join Now button. Yes, it costs $15 ($12 if you are already a CLO member), but it's a contribution to bird conservation. You have the option of paying over the phone, 1-800-843-2473, 8 a.m.-5 p.m. ET (6 a.m. – 3 p.m. Mountain Time).

All new participants get a handbook, a calendar and a full-color poster of common feeder birds in the mail. You may send your data online or mail in tally sheets at the end of the season.

Once you receive your identification number, you can log in and set up your count site by describing it: number of trees and shrubs, bird feeders and birdbaths, and so forth. Sprinkling black-oil sunflower seeds on the ground where they can be seen from a window is perfectly acceptable.

Scientific protocol requires selecting your count days in advance. Each set of two consecutive days must be at least five days apart. Mark and I have chosen Saturday and Sunday each week.

It's OK if you miss some of those count days. Project FeederWatch officials don't expect you to stay home for the whole season, which is early November through early April. You can sign up after the season has started.

It's also not necessary to sit by the window continuously. Mark and I leave pencil and paper on the table in front of the window, and whenever we are in the vicinity, we check and see if there are any new species for the current count days or more individuals of any species than previously recorded.

The other bit of protocol is that you only count the birds you can see at any given time. You can't add the 15 house sparrows you saw in the afternoon to the 10 you saw in the morning. You can only record the largest number you saw at one time.

Record the high and low daylight temperatures over the two days. We use the weather reports published the next day in this paper, figuring the coldest temperatures

326                    CHEYENNE BIRD BANTER

are pretty close to dawn.

What do I have to show for 14 years of submitting data? With the newly redesigned website, I can see very colorful graphs for each of the 25 species I've observed. I know that 11 of those seasons we've had goldfinches, and that 2004 was the first winter we had any Eurasian collared-doves – and only twice.

But mostly, by participating, I find satisfaction in knowing that "my birds" are contributing to scientific knowledge.

While the current season has already begun, it isn't too late for you to share that satisfied feeling, or even provide it for someone else as a gift.

# 2014

## 351 Owls are among us
Sunday, January 5, 2014, Outdoors, page E2

Here's how to tell if the elusive bird is lurking in your Cheyenne neighborhood

In late November, Mark and I became aware that a flock of crows, also known as a murder of crows, was convening just before sunset in a neighbor's big spruce tree.

They were very loud, very raucous, as if they were a lynch mob yelling for noose justice.

Our double-paned windows are somewhat of a sound barrier, but when we let the dog out, we were bombarded with enough noise to overwhelm a backyard cookout.

Was there an owl roosting in the spruce? It's a big tree, probably planted when the neighborhood was new 50-60 years ago, so you can't easily see inside, even when standing beneath it.

Or had the crows decided to establish a roost in our neighborhood? That was an unbearable thought.

Thanksgiving morning, while I was out sweeping up sunflower seed hulls from under our bird feeder and throwing the ball for the dog, the crows sounded even more agitated – gathered in a spruce even closer to our house. "There must be an owl within those thickly-needled branches," I thought. "And he isn't getting any sleep after a night of hunting."

The next morning, just before sunrise, I lifted the window shade and saw a lump on the bare branch of our big green ash tree. Yep, a great horned owl. I told the dog she would have to wait a few minutes before she could go out.

The owl was perched about a foot away from a small squirrel nest made of dry leaves stuffed into a vortex of small branches. Leaving the kitchen lights off, I pulled out my binoculars and there was just enough light to see which way the owl was facing. It wasn't surprising that it was facing the squirrel nest, bobbing its head up and down in a circular way, to get a better fix on a squirrel probably trying desperately not to be heard breathing.

There's a bigger nest, or drey, on the other side of the alley. Ours looks like it is barely big enough for one squirrel, much less the three scampering around our yard every day, teasing the dog.

I was surprised that the owl didn't just poke a taloned foot or sharp beak into that pile of leaves. But great horned owls prefer to feed in openings where they can perch and then wing after prey they hear or see, and pounce, pinning it to the ground. Eventually, this owl spread its wings and flew off.

No more mobbing crows here, however, owls have come up in recent conversations with two women I know, one living east of town and one on the northwest edge of Cheyenne. Both women were pretty sure their local owls were knocking off rabbits, the great horned's favorite food. And both women seemed fine with that, noting that there seemed to be bunny abundance this year.

I've talked to my share of folks who complain when an avian predator grabs a meal, especially if the prey is a cute songbird or furry animal. So, in addition to getting reports on owl activity, it was gratifying to hear people appreciate owls, even for their feeding habits.

If you are connected to any sources of birding news, you know that this winter there is another irruption of snowy owls, but in the Northeast and upper Midwest rather than the Great Plains, as it was two years ago. Another shortage of lemmings in the Arctic, forcing them south, I guess.

Snowy owls like to be out in the open, being birds of the tundra, even if it's the middle of the day, making them relatively easy to pick out when there isn't too much snow acting as camouflage.

So how many great horned owls are among us, shrouded in a cloak of nocturnal invisibility or daytime coniferous cover? What about the smaller, less common owls of southeastern Wyoming: eastern screech-owl, long-eared owl, short-eared owl?

Is there a great horned owl in your neighborhood? Look for the signs: angry crows, the odd rabbit leg on the sidewalk, a large bird flashing through the beam of your headlights, and even the chunky silhouette, the size of Harry Potter's snowy owl, in a tree or on a fencepost at dawn or dusk.

Don't begrudge your dog's request to be let out on a winter's evening or just before dawn. Follow and take a look around.

# 352 The great migration

Sunday, February 9, 2014, Journey, page E1

## Head east this spring to meet the famous sandhill cranes

One of the great annual events of the natural world, especially for North America, happens just down the road from Cheyenne every spring. Yet it isn't as well-known, much less well-attended, by Wyomingites as it is by people from all over the country, even the world.

I'm talking about the spring migration of sandhill cranes.

Yes, there are millions of migrating birds, but most don't stand nearly 4 feet tall in flocks of thousands, out in the open, making such a racket that they can't be missed.

More than 500,000 birds, representing 80 percent of the entire sandhill population, come in for a landing along a stretch of the Platte River, between wintering in New Mexico and Texas and breeding in Canada and Alaska.

The peak time for Nebraska is the month of March into the first week in April, about when I get my annual spring urge to travel.

Driving Interstate 80 five hours east (and don't forget to account for the lost hour entering the Central time zone), to an elevation 4,000 feet lower, is to meet spring a couple weeks early. Central Nebraska has a Midwestern flavor with birds to match, so it's even more like getting out of Dodge for a vacation.

When Mark and I first went to see the cranes, our boys were younger than 12, too young to be allowed in the blinds at the Rowe Audubon Sanctuary. Can you imagine how quickly the cranes would leave if small children staged a temper tantrum, echoing through the plywood construction? So, we left them with a friend in Kearney for a few hours. We've been back a couple times since.

I love the openness. The only trees are in the river valley. But those trees are exactly what the cranes don't want.

The Rowe Audubon Sanctuary, since its establishment in 1974, has worked diligently to remove trees from its stretch of the river, leaving unvegetated sandbars for the cranes to roost on at night, with no place for predators to skulk unseen. Damming the river upstream has eliminated spring floods that would normally clear the channels regularly.

The blinds at Rowe, near Gibbon, 20 minutes from Kearney, and at The Crane Trust Nature and Visitors Center further east, near Grand Island, allow people to view cranes at sunrise and sunset.

While the cranes (even the occasional whooping crane) are scattered in the local fields and wetlands feeding on corn and invertebrates all day, great for photo ops, it's the blinds that allow you to see the concentration of birds where they roost for the night.

If you want to get closer, sign up to stay overnight in the special photographers' blinds – no heat or light allowed – and pay $200-$300 for the privilege.

It is a privilege to watch these magnificent birds from the blinds, but it may not seem like it if you don't bring your warmest boots and layers of clothing. That's the downside of being further east – the cold is damp.

Once you enter the blind, at 5 p.m. (6 p.m. after daylight saving time starts March 9), you aren't allowed to leave for two to three hours, until it's dark enough to sneak away. Alternatively, if you enter at 5 a.m., 6 a.m. DST, you must wait until after the birds have left before you can leave. The blinds do have adjacent chemical toilets now, but the guides discourage their use.

Not only do you want to wear dark clothes to keep from spooking the birds, but regular flashlights are not allowed, and bright LCD screens are frowned on.

And for heaven's sake, leave your flash at home, and make sure you deactivate the flash on your point-and-shoot or smart phone. If your flash triggers a mass bird departure, everyone in the blind, up to 31 other people, will hate you, because there won't be a second chance to see sandhills that morning or evening.

Blaine McCartney, a photographer at the WTE, recommends a 400mm lens to get close enough to the birds, along with a monopod. Though everyone gets their own little window, there isn't really room for tripods.

Judy Myer, a Cheyenne photographer, went on a shoot with the Fort Collins camera club last year. The club members used the Rowe blinds one morning and the Crane Trust blinds in the evening.

"The evening viewing was dark, but we could hear them," she said. "Is one place better than the other? I can't really answer that except to say I wouldn't do (those blinds) again in the evening."

Instead, she said, she would head to the bridge at the trust, where, for $15, you can watch the cranes fly overhead in the evening to their roosts.

But that just goes to show everyone's experience can be different. I'm not familiar with The Crane Trust blinds. We've had pretty good luck at Rowe, and it's closer.

The Trust exists because of the settlement in 1978 from a lawsuit contending that the Grey Rocks Dam, built on the North Platte in Wyoming, had a negative impact on whooping cranes and other wildlife in Nebraska downstream on the Platte. Like Rowe, they do a lot of work to clear vegetation from the river channels and offer educational opportunities.

Yes, it's half a day's drive each way. Yes, it can be cold.

But no nature film can take the place of being surrounded by a crowd of birds continuing a ritual that's tens of thousands, maybe millions of years old, that's partly instinctual and partly learned from their parents.

Their calling fills your ears with a roar you never forget.

2014

**To visit or make blind reservations**

**Rowe Audubon Sanctuary**

Located at 44450 Elm Island Road, near Gibbon, Neb. Visit www.rowe.audubon.org for details and rules. Call 308-468-5282 weekdays 9 a.m. to 5 p.m. Central time. Reservations are available March 1-April 6. Cost is $25 per person and must be paid in advance. Reservations are refundable up to seven days in advance, with a 5 percent charge.

The Iain Nicolson Audubon Center at Rowe Sanctuary, 44450 Elm Island Road., Gibbon, Neb., is free. From Feb. 15 to April 15, it's open daily 8 a.m.-5:30 p.m.

**Crane Trust Nature and Visitors Center**

The visitor center, at 6611 Whooping Crane Drive, Wood River, Neb., is free and open March 1-April 7, 8 a.m.-6 p.m. daily. Normally, it is open Monday-Saturday, 9 a.m.-5 p.m. Reservations are $25 a person and are available March 1-30. Another option is to view cranes from their bridge ($15) as they fly overhead in the evening to their roosts. Visit www.cranetrust.org or call 308-382-1820.

## Crane festivals

Festivals are held all along the cranes' Central Flyway migration route, and on their breeding and wintering grounds.

The biggest (with the most cranes) is Audubon's Nebraska Crane Festival (formerly Rivers and Wildlife Celebration), scheduled March 20-23 in Kearney, which includes speakers, kid activities, field trips, vendors, etc. See www.nebraskacranefestival.org.

**Other Central Flyway crane festivals:**

Whooping Crane Festival, Port Aransas, Texas, Feb. 20-23.

Monte Vista Crane Festival, Monte Vista, Colo., March 7-9.

Crane Watch Festival, Kearney, Neb., (includes Audubon's Nebraska Crane Festival), March 21-30.

Tanana Valley Sandhill Crane Festival, Fairbanks, Alaska, August 22-25.

Yampa Valley Crane Festival, Hayden, Colo., September.

Festival of the Cranes, Bosque del Apache National Wildlife Refuge, New Mexico, November.

**Other crane festivals:**

Othello Sandhill Crane Festival, Othello, Wash., March 28-30.

Sandhill Crane and Art Festival, Calhoun Co., Michigan, Oct. 11-12.

Sandhill Crane Festival, Lodi, Calif., November

Tennessee Crane Festival, Birchwood, Tenn., mid-January 2015.

# 353

Sunday, February 16, 2014, Outdoors, page E2

# Let's rethink mega windfarm on behalf of birds, efficiency

David Yarnold is not happy.

The president of the National Audubon Society writes in the January/February issue of Audubon magazine that our country's wind farms kill 573,000 birds a year, including 83,000 raptors.

The Migratory Bird Treaty Act and the Bald and Golden Eagle Protection Act should be protecting most of those birds, he says.

It's illegal to kill them without a permit, but the Interior Department has only enforced the law once, he says. Apparently, wind farms don't have permits for all the birds killed.

A new federal rule allows wind companies to get 30-year permits. But to Yarnold, that represents too many birds, with no incentives to cut the number of deaths.

Wind energy is a great idea. It's been used for centuries to propel boats, grind grain, pump water.

A structure for catching the wind can be erected wherever the power is needed, though a backup system is essential for windless days. People are working on more efficient battery systems.

But leave it to American ingenuity to take a simple idea and enlarge it, making it industrial-sized, much like family farming morphed into industrial agriculture.

Wind energy is clean, producing no pollution except whatever manufacturing the components entails and maintenance requires. We need cleaner energy sources like wind since the traditional fuel-burning, power-producing businesses are reluctant to make their energy production cleaner. Never mind the climate change debate – we all have to breathe.

But wind energy has an Achilles heel. Developers want to site numerous turbines in the windiest places, which also attract birds. Collisions with the blades, the towers and the transmission lines kill birds and bats. Wind farms have mazes of roads running over habitat, forcing out wildlife.

Audubon suggests targeting development for areas that are already disturbed or developed, avoiding areas known to be dense with birds, such as the Prairie Pothole region, the Texas Gulf Coast, and the northeast's raptor migration bottlenecks.

If you don't care about birds, I suppose you wouldn't see any of this as a problem. But you should care. To sum up Basic Ecology 101, every living thing, including you, is connected to every other living thing. It's hard to predict how a loss of birds may affect you. It could be as simple as insect populations getting out of control and decimating crops.

330      CHEYENNE BIRD BANTER

But there are other reasons to rethink the concept of the mega wind farm.

I am a fan of dispersed power production, placing it among the structures where we live and work. For instance, solar panels over every roof, providing extra roof insulation and hail protection. Solar panels over parking lots would keep cars and asphalt cool. Small windmills could be placed along every highway where power lines are already strung. What if we were to place constellations of pinwheels on the outer walls of a skyscraper to produce power for that building?

The advantage of disbursed power production is we don't lose the power consumed by transporting it over long distances. Plus, any power outages would affect fewer people at a time.

OK, so every location in the country isn't terribly windy, but as a descendent, and mother, of engineers, I think we can engineer our way to more efficient turbines. It's happening already.

Last month, a story in this paper mentioned in passing that Ogin Inc. has invented a wind turbine with cowling, or shrouding as they are calling it. I went online to www.oginenergy.com to see what it was about.

Compare the old-style propeller-driven plane with the more efficient, more powerful jet engine enclosed by cowling. This new wind turbine design is the same thing. According to their information, "energy output is increased up to three times per unit of swept area."

Ogin turbines are smaller, at 200 feet versus the current 500-foot-tall turbines, so they can fit into already developed landscapes more easily. Because they are shorter and the tips of the blades are outlined by the shrouding, it is believed fewer birds would be killed.

Testing of this new design will be happening at the infamous Altamont Pass in California, where some years ago, biologists helped engineers change turbine tower designs from open lattice work into the smooth cylinders we know today – taking away perches for raptors which were otherwise unwittingly launching themselves into the blades.

There are vertical axis wind turbines, identical to the one in Cheyenne at the Children's Village, which, at only 30 feet tall, have far less impact visually and environmentally.

Vertical turbines would even be a good replacement in wind farms, says California Institute of Technology professor John Dabiri. Placing them close together improves their efficiency by a factor of 10, using a much smaller footprint per kilowatt of production than current, giant horizontal axis turbines we see. [http://www.caltech.edu/content/caltechs-unique-wind-projects-move-forward]

Ever since we first felt the wind pushing at our backs, we have been refining ways for it to aid us. The challenge is to make our design choices work for other species as well.

# 354 Sunday, March 23, 2014, Outdoors, page E2
## The bird migration picture gets animation

This year, Feb. 19 marked the first sighting of mountain bluebirds in Cheyenne. This is early, but not unusual. I can't help thinking the next batch of February snowstorms drove them back south again.

Recently, I discovered I can watch animated maps of bird migration. These maps on www.eBird.org take the data of 60 selected bird species and show their journeys across the country, week by week. (Click on "About," then look for "Occurrence Maps" under "News and Features.")

The data in these maps come from bird sightings that citizen scientists – you and me – have submitted over the years.

In a field guide, the mountain bluebird's yearly movements are difficult to depict on a static range map that accompanies their descriptions.

They are settled mostly over the Rocky Mountain West, avoiding the Pacific Coast. But in winter, they leave the northern Rockies and mountains and leak out over the Great Plains (eastern Colorado, Kansas, Texas panhandle), in addition to the interior of California and Oregon, the southwest and Mexico. Some even winter in southern Idaho.

To watch the animation of our sightings of them is fascinating. As you watch, you'll see a spectrum of fire colors flicker across the map.

These colors indicate their rate of occurrence, which is the probability of detection.

Areas with slight possibility of occurrence are a cold, ashy gray. As the possibility increases, the color warms to orange, finally heating up through yellow to white-hot—where the species is thickest.

As the animation cycles from week to week, the "flames" flicker across the land. As someone who was asked to drop statistics before the professor was forced to flunk me, this visualization of numbers, statistical modeling, is magic.

Watching the screen is like watching flocks of mountain bluebirds roaming the prairie. And I notice that even in

January, there is a faint haze of orange in Wyoming. It means someone was outside, or at least looking out the window, noticing bluebirds in the depths of winter.

Other species are completely absent from the U.S. for six months.

In mid-April, the western tanager explodes over the Mexican border in a hurry to find the best breeding locations. Then it spreads out into little islands – islands of preferred breeding habitat scattered over western mountain ranges. Then it seems to drift slowly south beginning in mid-July as young birds explore. It is entirely gone – from the U.S. at least – by October.

Our state bird, the western

2014

331

meadowlark, a short-range migratory species, apparently overwinters in low numbers in southern Wyoming. I'm glad we picked a bird that doesn't completely abandon us.

For each species with a map, there are notes that describe what is going on, and admission that sometimes the numbers have biases. One of those biases is detectability.

In the spring, birds, mostly the males, are often singing during migration. But by the time they head south, they can be rather quiet. The note for grasshopper sparrow, a small, drab, brown bird says, "it appears that the species just disappears when in fact thousands are passing southward…. Ideally, future versions of these maps will be able to incorporate species-specific detectability variables and will start measuring abundance, not just occurrence."

Another bias is caused by birders themselves, and their propensity to flock to where the most birds are. In the discussion of the blackpoll warbler map, the note says, "…there are biases in how birders sample the landscape. For this reason, we have tried to promote the use of random counts so that widespread habitats (with less rare bird potential) are sampled in a proportion that more closely resembles their percentage 'on the ground.'"

Good luck with that!

Where would you prefer to spend a spring morning birdwatching? Along Crow Creek, among the cottonwoods where interesting warblers are known to show up? Or out on some treeless, nameless, numbered gravel road in the hinterlands of Laramie County? It could be worth a look, though.

Don't forget to take your notebook and pencil (or your eBird reporting app) with you everywhere this spring and submit all your bird observations to www.eBird.org.

You may be helping to re-write – and re-visualize – what we know about bird migration.

# 355 Trying out Texas birding trail rewards Wyoming birders

Sunday, April 20, 2014, Outdoors, page E2

The Texas Gulf Coast during spring migration is legendary among birders, especially if weather conditions cause a "fallout" of tired migrants that have just crossed the Gulf of Mexico.

We didn't find birds dripping from branches on our first trip to Texas, since it bridged March and April, a bit ahead of the peak. We missed the 37 species of warblers, but some will arrive in Cheyenne next month.

However, my husband, Mark, and I did add several life-list birds.

When our younger son Jeffrey, and his fiancé, Madeleine, moved to Houston last fall, we started researching the Great Texas Coastal Birding Trail, part of the Great Texas Wildlife Trails.

The idea of a birding trail was born on the Texas Gulf Coast, where 450 species of birds might be seen. First established in 1994, the concept has since been applied to many other places. On the Gulf Coast, it is made up of many loop routes connecting 308 public places to see birds.

"Finding Birds on the Great Texas Coastal Birding Trail" (Eubanks, Behrstock and Davidson, c. 2008) is however, not the best guidebook, though it does work well in tandem with "A Birder's Guide to the Texas Coast" (Cooksey and Weeks, c. 2006, published by the American Birding Association). This book has a bar chart listing all the bird species that shows what section of the coast and what months to find them and other tips.

Mark went online and used www.eBird.org to find more recent information. Under "Explore Data" are two new and very useful functions, "Explore a Location" and "Explore Hotspots."

Our son's neighborhood in Houston, the Heights, is full of old bungalows and trees and the pleasant but unrelenting sound of mockingbirds and white-winged doves.

At the Houston Arboretum and Nature Center, we found forest birds: cardinals, tufted titmice and Carolina chickadees. However, the 155 acres are flanked by Interstate 610, making it impossible to hear any birds on the west edge.

The next day, all four of us headed for Brazos Bend State Park, 5,000 acres of bottomland hardwood forest, river, ponds and grasslands only an hour southwest of downtown.

As soon as we parked, we discovered an amazing sight, a dead tree full of roseate spoonbills, large pink wading birds. This was also where both Mark and I added "Black-bellied Whistling Duck" to our life lists.

This duck is primitive - it's found at the beginning of the evolution-based taxonomic order of the North American birds. When it flies, its long legs stick out behind, reminding one of a cormorant. And yes, its voice is whistle-like.

It acts like a wading bird, chumming around with the white ibises, snowy egrets and spoonbills in shallow water, looking for aquatic plants to eat, but on the other hand, it nests in cavities in trees or in nest boxes, like a wood duck. Its bright orange bill and pink legs add snap to its rich brown-colored body marked by large white wing patches – and a black belly.

A red-shouldered hawk at the park was another life bird for us, and a glimpse of a pileated woodpecker was a first also. Altogether, we saw 27 bird species at the park – and several alligators.

The next two days, while the kids had to go to work and school, Mark and I headed for the coast. Luckily, Matagorda Island was not in our plans, and we left cleanup of the recent oil spill to the experts.

Near Freeport are several notable stops, including the tiny Quintana Neotropical Bird Sanctuary, which is apparently a good place after a fallout. But it was while driving a farm-to-market road between industrial chemical facilities that Mark found

332                    CHEYENNE BIRD BANTER

another lifer for us, a scissor-tailed flycatcher perched way up on a high-tension power line. With a body the same size as the 8-inch kingbirds that perch on Wyoming fences, its extreme tail makes it twice as long.

At Baytown Nature Center, our life bird was the Neotropic cormorant. It was worth Mark lugging the spotting scope everywhere to have it on hand to see the diagnostic little white feathers on the sides of its head.

The nature center is the result of common sense. Back in the 1980s, it was a high-class neighborhood – doctors, lawyers and oil company executives. But damage from subsidence from extensive oil drilling, severe storm surge flooding and hurricanes led to its abandonment. Local and national officials decided to return it to nature. And after sighting 38 bird species there, I'm glad they did.

Like so many other birders, we hope to return – 300 Great Texas Coastal Birding Trail sites left to visit, 380 more species to find.

# 356 Owl family draws visitors to Lions Park

Sunday, May 11, 2014, Outdoors, page E2

There's been quite the parade of admirers trekking to Lions Park to see the pair of owls that nested there this spring, and their three owlets.

By mid-April, they became widely known within the Cheyenne birding community and among regulars at the park. Generally speaking, they can be found in the trees north of Sloans Lake and the Cheyenne-Kiwanis Community House.

I suspect someone aiming a long-lens camera at the top of a tree will have passersby surreptitiously looking in the same direction to figure out what they are shooting. Or, being Cheyenne-friendly, they'll simply ask. Then they, too, become converted to the owl-watching cult.

The day I went to see them, my husband, Mark, and I could only find one adult and one young, but I'd heard that one of the owlets had been seen on the ground, toddling, like a Furby toy, to another tree – and climbing it. It takes a few weeks before owlets are strong enough to fly much.

The owlets will stay with their parents for the summer, so we hope everyone keeps their dogs leashed while in the park. The neighborhood red foxes present enough of a challenge.

When it comes to breeding, great horned owls get an early start in the year. The male can be heard hooting in February to establish his territory. Chances are, his mate from last year is still around. Other than courtship, they don't roost together during the year. They don't build a nest. Instead, they use a tree cavity or an available nest in a tree made by a hawk, crow, heron or squirrel.

The female is the one who incubates the (typically) two eggs. She'll lay more if food – prey animals – is very abundant. For more than 30 days between February and March, she can successfully incubate through winter conditions, even -27 degrees.

The male keeps her fed. Food found in our park could include rabbits, mice, waterfowl and other birds. This was not a good winter to find ducks, since Sloans Lake stayed completely frozen until mid-March.

Research shows owls occasionally take squirrels, and with the overabundance available in the park, that would be my guess as to what they are eating. If anyone finds owl pellets – the compacted balls of bones that are regurgitate by the owls – we could find out for sure.

At 6 weeks old, and nearly equal to their 22-inch-tall parents, young owls climb out of the nest and take a stroll onto nearby branches. Over the next four weeks, they practice flying short distances and may be found at times roosting on the ground.

The siblings hang out together, but the parents, except for occasionally dropping off food, prefer to roost away from the kids, to avoid hearing their incessant begging that starts up whenever the parents come near.

The owlets start out catching insects and eventually learn to catch mammals and birds by the perch and pounce method. By October, they are ready to fend for themselves.

Typically, young owls are 2 years old before they breed. But it really depends on the amount of prey available. If pickings are slim, many can't find a big enough territory to support a family because there are probably more dominant owls in the area chasing them off. The researchers call the unpaired birds "floaters."

Great horned owls don't migrate seasonally. But the young disperse to find new territory, looking for some place that has an abundance of prey. Studies cited in Birds of North America Online show they moved a mean distance of 46 miles. Otherwise, they would have a long wait before they could take the place of their parents' generation – this species has been documented to live more than 20 years. So, for the young, it's about waiting for those years when rabbit reproduction is up.

Whether the current pair nests in the park again next winter depends on the nest they used still being in good shape – or if a replacement is found. But more importantly, is there still enough food?

Great horned owls across North America, the only continent where they are found, work out answers to these questions every year.

It seems, despite people feeding the park squirrels (even though they shouldn't, and the over-abundant population is chewing up and damaging park trees), the owls are here to bring balance. It's another step in making a manmade landscape more natural.

2014

# 357 Wyoming refuge is a treasure hidden in plain sight

Sunday, June 22, 2014, Outdoors, page E3

In early June, Hutton Lake National Wildlife Refuge birds are busy reproducing. They barely notice birders.

The refuge is southwest of Laramie. It's small by national refuge standards, just under 2,000 acres, and relatively unknown compared to others in Wyoming like Seedskadee or the National Elk Refuge.

Hutton Lake has little to offer people: no visitor center, no restrooms, no picnic tables, no fishing, no hunting, no camping, no off-road vehicles, horses or dogs allowed anywhere, no trees, no dramatic landscape, and no decent road – until recently.

Instead, it caters to wildlife, attracting 29 mammal species, six amphibian and reptile species and 146 kinds of birds, including 60 species that have been known to nest there.

What do avian visitors find at Hutton Lake?

Five small lakes, including namesake Hutton, have a variety of wet habitats – shallow water for puddle ducks and wading birds, deeper water for diving ducks, muddy shores for shorebirds and thick reed beds for nesting. On land, there are greasewood thickets perfect for nesting songbirds like the sage thrashers. The short grass of the surrounding plains, as green as a golf course this spring, will have its share of bird nests on the ground – grassland species do without trees.

The comparatively flat (the Snowy Range glimmers in the distance) and nearly featureless topography of the Laramie River Valley does have a few rocky outcrops and ridges. The astute birder will find eagles and hawks perched on them or soaring overhead.

### Hutton is part of a complex

Ann Timberman is the project manager for the Arapaho National Wildlife Refuge Complex, which includes Hutton Lake and two other small refuges nearby, but which are closed to the public because of endangered species work. There's also Pathfinder near Casper, and Arapaho, the main refuge, is where the complex's headquarters are located, outside Walden, Colo.

In some ways, Ann's job, which she's had for 10 years now, is easy. The National Wildlife Refuge System doesn't have to manage for multiple, and often conflicting, uses like the Bureau of Land Management or the Forest Service. Its mission is to benefit wildlife. Hutton Lake was established in 1932 "to provide resting and breeding habitat." Livestock grazing permits are available only in years when it's been determined it will benefit wildlife.

Ann and I toured Hutton Lake together June 2 on a wonderfully windless day. Bringing along the spotting scope did not make for the most efficient interview – we kept losing our conversational focus while focusing on the differences in field marks for immature bald and golden eagles and other birdwatching matters.

### Improvements welcomed

The tour was to show off improvements made last year, the biggest being the roadwork, tons of gravel filling the deep ruts I remembered from my last visit. The road improvement also extends to the two-track across state land between Sand Creek Road, which is the closest county road, and the boundary of the refuge.

Even a small car with minimal clearance can navigate the single lane road, as we found when we saw one at the new gravel parking lot at the end of the road.

One improvement was unglamorous, but very expensive – replacing the infrastructure that regulates the flow of water from one of the lakes to another.

This summer, an interpretive trail and observation platform will be built at one of the lakes.

There's a birdwatching blind now, too, built last year by an Eagle Scout candidate, with funding for materials provided by Laramie Audubon Society.

I went out again to Hutton five days later with some of the chapter members on a field trip. As much as they appreciate the improved road, they are a little sad to lose vehicle access to some of the roads that are now for pedestrians only. Tim Banks, trip leader, pointed out that some of their older chapter members are not going to be hiking in to regain closer views of the lakes.

Laramie Audubon members are just about the only regular visitors and the only interest group which keeps tabs on the refuge. They worked to have it designated as a Wyoming Important Bird Area.

### Partnerships benefit wildlife

In fact, two bird lovers, Gere and Barbara Kruse, were responsible for the recent improvements. In their memory, their daughter, Babs, brought $42,000 to Bob Budd, executive director of the Wyoming Wildlife and Natural Resource Trust, asking for help finding an appropriate wildlife/public use habitat project in Albany County.

The Trust matched the donation. Laramie Rivers Conservation District's Martin Curry, resource specialist, wrote the grant and oversaw most of the work. Other cooperators were the Wyoming Game and Fish Department and the refuge, as well as its parent agency, the U.S. Fish and Wildlife Service. A total of $111,000 will have been spent when the improvements are finished.

There are drawbacks to having a better road. Back in January, kids started a fire even though fires are not allowed, and it got out of control. Thankfully, the refuge is on local law enforcement's beat, and Albany County firefighters put it out. Ann decided to lock the gate for the winter, allowing only walk-in access.

With only 3.5 staff

CHEYENNE BIRD BANTER

members for the whole Arapaho refuge complex, locals become Ann's eyes and ears at Hutton Lake. There are few birds and few people on the windswept plains in winter. But, for instance, deciding when in spring to open the gate will depend on local birders apprising her of conditions. Visitors can also report suspicious or illegal activities – impossible to hide on the open plains.

For Ann, from a management perspective, making the refuge more accessible is a double-edged sword of sorts, allowing in vandals as well as visitors. But, she said, in the long run, it pays to make friends and develop partnerships. In this case, sharing Hutton Lake with people who appreciate it benefits the wildlife. And that fits the refuge's mission.

### If you go

The refuge is open to driving on established roads as conditions permit and to hiking on roads and trails year-round. Wildlife watching and photography are the recreational activities allowed. Spring, especially April, is a great time for birdwatching.

There is no drinking water and no restroom. Please pack out trash. Hunting, shooting, fishing, fires and camping are not permitted.

### How to get there

From Laramie, drive south on U.S. 287. When the huge cement plant comes up on your right less than two miles south of I-80, aim for the plant's front office using one of the crossroads, but instead of entering the plant, veer left (south) and you will be on Sand Creek Road. After about 8 miles, you will see a brown sign for Hutton Lake pointing to the right. Turn and follow the gravel trail to the refuge entrance, which is marked by a large sign and a small parking area.

More information is available at http://www.fws.gov/refuge/hutton_lake/.

# 358 BioBlitz finds birds, butterflies, bees, bats and more

Sunday, July 20, 2014, Outdoors, page E3

"A BioBlitz is a 24-hour event in which teams of volunteers, scientists, families, students, teachers, and other community members work together to find and identify as many species of plants, animals, microbes, fungi and other organisms as possible." National Geographic Society

Microbes?! No one went looking for microbes during the Wyoming BioBlitz.

It was held last month on the longest day of the year at The Nature Conservancy's Red Canyon Ranch near Lander. And hopefully, no one took home any unwanted microbes.

But we did find lots of other life. More than 70 people participated: putting out pollinator traps, extracting birds from mist nets, bouncing over a mountain meadow after butterflies and bees, dip netting for macro-invertebrates, electrofishing a stream, botanizing up the side of the canyon, searching for reptiles and amphibians, setting small mammal traps, attracting moths to blacklight, and until nearly midnight, netting bats, only to roll out of sleeping bags or beds in town the next morning to count birds before sunlight hit the canyon floor.

It's one thing to have scientists come and present their work in a lecture, as they do, for instance, for Cheyenne – High Plains Audubon Society meetings. It's quite another to find out firsthand how difficult it is to untangle a bird from a mist net in order to study breeding patterns and longevity.

Then there was the chance to perfect my butterfly net technique with Amy Pocewicz of The Nature Conservancy. It's like tennis, but butterflies are more erratic, and the court is littered with shrubby obstacles.

Sometimes fieldwork is monotony. I went with Wyoming Natural Diversity Database's (WYNDD) Ian Abernathy and his group to pick up small mammal traps in the sagebrush, little folding aluminum boxes baited with sweetened oats. Each had a tuft of polyester batting thoughtfully provided so the mouse or vole could bed down comfortably for the night in a place not as warm as its own burrow.

To check the traps, we all had to don disposable face masks and gloves to protect us from possible exposure to hantavirus.

We were led by an indefatigable 4-year-old who enjoyed marching ahead to pluck the pin flag marking the next trap.

No critters were captured in any of the 60 traps in the sagebrush and only one in the 20 traps along the creek. Too much human scent from the group setting traps the night before?

Martin Grenier, Wyoming Game and Fish Department non-game biologist, set a mist net over the creek in the evening, and his group was able to catch four bats of three different species.

The same evening, Lusha Tronstad, invertebrate zoologist with WYNDD, hung two white table cloths on the Learning Center's patio, placing one small blacklight against each, and then turned off the regular lights. Moths and nocturnal wasps flocked in, and extremely small insects were "vacuumed" into a glass bottle for close inspection.

One special moth will have to be identified by an expert in Florida.

Audubon Wyoming, now Audubon Rockies, is the originator of Wyoming's BioBlitz, holding the first one in 2008, and has partnered with various organizations, agencies and companies to hold it in different locations around the state.

Wyoming teachers can receive continuing education credits – it's a lot more fun, one teacher from Bighorn told me, than attending lectures.

This year, the Red Canyon BioBlitz sponsors and partners also included, in

2014

335

addition to those mentioned earlier, the University of Wyoming Biodiversity Institute and the Wyoming Native Plant Society. During a creative interlude, an artist from the Lander Art Center had us harvesting cheatgrass—an invasive plant – and making art out of it.

The very first BioBlitz was held in 1996 at a park in Washington, D.C., where National Park Service naturalist Susan Rudy coined the term from the German word "blitz," meaning lightning, or fast.

Search online for "BioBlitz" and you will find 20 more listed in this country plus Korea, Canada, New Zealand and especially, the United Kingdom. It's a plot to infect people with the awareness and joy of biodiversity.

One of my favorite memories of the weekend, besides all the biota, is camping out on the lawn by the Learning Center and going to bed with the stars in my eyes and waking with birdsong in my ears. The other favorite is meeting old friends and new, all interested in the wonderful biodiversity of our home state.

You too, can come along next year, wherever BioBlitz may be.

## Related websites:

Audubon Rockies, http://rockies.audubon.org
Lander Art Center, www.landerartcenter.com
The Nature Conservancy, http://www.nature.org/wyoming
UW Biodiversity Institute, www.wyomingbiodiversity.org

WyoBio, www.wyobio.org
Wyoming Natural Diversity Database, www.uwyo.edu/wyndd
Wyoming Game and Fish Department, www.wgfd.wyo.gov
Wyoming Native Plant Society, www.wynps.org

# 359 Mind your manners to reduce bird stress

Sunday, August 17, 2014, Outdoors, page E3

I'm sure your parents taught you, as mine did me, that it is impolite to stare.

Does this rule apply, in some way, to birds? After all, the point of birdwatching is to watch them.

Know that whenever you enter a bird's environment, it can bother a bird. For instance, even when you are on the other side of a window, it may react to your presence.

I recently heard an anecdote about a hawk nest so close to a public road that it was well known. Birdwatchers regularly showed up to watch and photograph the chicks as they grew.

What these folks apparently missed was that the parents were agitated. The presence of the birdwatchers bothered them. The situation could easily have caused the parents to abandon the chicks. And even though it apparently didn't, stress on the birds could cause some unintended consequences down the road – just as it does for people.

There is a new field guide that came out this spring, one that offers something a bit different from the rest.

The New Birder's Field Guide to Birds of North America, by Bill Thompson III, is recommended if you are a casual backyard birdwatcher who wants to know more.

It explains the hobby of birdwatching, why it's fun, how to get into it, what to wear to be comfortable, how to adjust binoculars. What follows is a page per species with helpful information for identifying each one.

But one brief chapter bears on this column's subject, titled "Birding Manners."

Some of it pertains to birding with others: keep your voice down, treat others as you'd like to be treated, stay with the group, share the spotting scope, help beginners, pish in moderation.

What is pishing? It's an attempt to get a better look at a bird by getting it to come out of the vegetation by making a noise that sounds like "pish," which happens to sound like a bird alarm call. The birds come out to see what's wrong. Playing recorded bird songs to attract a bird that thinks he's hearing a rival is another method to bring it out of hiding.

As Thompson asks new birders to use these techniques in moderation, he explains, "We owe it to the birds we love so much to respect their privacy."

This is the beginning birdwatcher's version of the American Birding Association's Birding Code of Ethics. The part that pertains to the nesting hawks' situation reads:

"1(b) To avoid stressing birds or exposing them to danger, exercise restraint and caution during observation, photography, sound recording, or filming....

"Keep well back from nests and nesting colonies, roosts, display areas, and important feeding sites. In such sensitive areas, if there is a need for extended observation, photography, filming, or recording, try to use a blind or hide, and take advantage of natural cover."

If the ABA members, vying to see as many bird species as possible, can restrain themselves, I think the rest of us can as well.

Given today's optics and cameras, it might have been quite possible to observe the activity in the hawk nest from a less intimidating distance, since building a blind on the side of a public road probably isn't feasible. Contacting the adjacent landowner for permission to erect a temporary blind might have been a solution.

But on the other hand, if we are observing the hawks for our own enjoyment and not as a part of scientific study, two minutes from inside our car would be quite enough, rather than hour after hour, day after day. Try to make part of your enjoyment of birds knowing that your actions haven't endangered or distressed them.

There is so much interesting bird behavior to watch unobserved by the birds if you walk carefully, and stop and stand still often, being the proverbial fly on the wall. If you don't make noise or make sudden movements, birds in the bushes will continue to flit about feeding. If you sit as still as a rock at the shore, the shorebirds may pass close by.

And should a bird look you in the eye, acknowledge it as you would a person, with a nod.

And then look away, so it can continue with its important business of living.

# 360 6 reasons why you should go to "Bird-day"

Thursday, September 11, 2014, page B4

To mark 40 years since the hatching of the Cheyenne – High Plains Audubon Society, you are invited to the "Bird-day" celebration Sept. 26-28.

While the Wyoming Audubon Society was established in 1950, local Audubon members decided to form their own chapter, achieving their goal in 1974.

Judging by the newspaper clippings preserved in the chapter's scrapbook, this was a coming together of environmental activists who enjoyed bird watching, and serious birdwatchers concerned for the future of birds.

Today, the chapter still offers a mix of educational activities, environmental advocacy, citizen science and birdwatching.

So how do Bird-day festivities apply to you? We've come up with at least six possibilities:

**1 You use eBird.** You can hear more about the citizen science programs offered by the Cornell Lab of Ornithology from its director, John W. Fitzpatrick, at the Sept. 27 banquet.

**2 You are a gardener.** This free presentation, "Be a Habitat Hero," talks about a program and its benefits to all the pollinators: birds, bats, butterflies and bees. Connie Holsinger's talk is Sept. 26, 7 p.m. at Laramie County Community College. It's free and open to the public.

**3 You've always wanted to visit the legendary Wyoming Hereford Ranch.** And it happens to be a designated Wyoming Important Bird Area, as it attracts unusual migrants. To help you find and identify them, Ted Floyd, editor of American Birding Association's magazine is leading a field trip Sept. 27. Meet at 7 a.m. at the parking lot at LCCC's Center for Conferences and Institutes.

**4 You have or know kids in kindergarten through eighth-grade.** Take them to the "Adventures with Audubon" programs led by educators from Audubon Rockies and Cheyenne Botanic Gardens Children's Village on Sept. 27, with sessions at 10 a.m. and from 2-4 p.m. at LCCC.

**5 You simply want to know more about birds.** You'll definitely want to attend the talks on Saturday (see schedule).

**6 You like birthday cake** and want to help John Cornelison, founding president of the Cheyenne – High Plains Audubon Society, celebrate 40 years of bird education and conservation in southeast Wyoming.

## If you go

**What:** Cheyenne – High Plains Audubon Society hosts its 40th Anniversary "Bird-day" celebration

**When:** Sept. 26-28

**Sept. 26** – no registration required
6-9 p.m. reception
7 p.m. – Be a Habitat Hero. Speaker is Connie Holsinger of Audubon Rockies.

**Sept. 27** – registration required
7 a.m. – Guided birding at Wyoming Hereford Ranch (meet at the parking lot at LCCC)
10 a.m. – Advanced Warbler Identification for Beginners and the Rest of Us. The presenter is Ted Floyd from the American Birding Association. He'll also sign copies of his books, "Smithsonian Field Guide to the Birds of North America" and "The American Birding Association Field Guide to the Birds of Colorado."
10 a.m. – noon – Adventures with Audubon: Exploring Wildlife (for those in grades kindergarten through eighth grade)
11 a.m. – Bird and Bat Diversity: Why So Many Species? Presenter is Brian Barber of the University of Wyoming Biodiversity Institute.
Noon (lunch) – Moving Conservation Forward through Partnerships. Audubon Rockies Executive Director Alison Holloran will talk about conservation partnerships.

2 p.m. – Feathers and Talons: A Closer Look at our Local Birds of Prey. Get tips on identifying raptors and understanding their ecology from Jeff Birek of the Rocky Mountain Bird Observatory.
2-4 p.m. – Adventures with Audubon: Exploring Wildlife (for those in grades Kindergarten through eighth grade)
3 p.m. – Counting Wyoming's Birds. Learn about bird surveys and upcoming Citizen Science opportunities in Wyoming from presenter Andrea Orabona, a non-game bird biologist for Wyoming Game and Fish Department.
6 p.m. (banquet) – How Birds Can Save the World. The keynote speaker at the banquet is John W. Fitzpatrick of the Cornell Lab of Ornithology, which is a supporter of citizen science programs such as eBird.

**Sept. 28** – registration required
8-10 a.m. Question and answer with Audubon Rockies staff
Where: Laramie County Community College, Center for Conferences and Institutes, 2200 E. College Drive.
Register: Sept. 17 is the registration deadline.
Print the brochure at http://home.lonetree.com/audubon/ and mail in your registration; register online at www.cvent.com/d/j4qcst, or call Barb Gorges, at 307-634-0463. The cost is $10 registration, plus $14 for lunch and $22 for dinner.

2014

# 361 Following flock of birders to Sterling was fun

Sunday, September 21, 2014, Outdoors, page E3

I wasn't sure what to expect when Mark and I decided to attend the annual meeting of the Colorado Field Ornithologists. The group's name sounds so formal.

Their 51st annual meeting was held over Labor Day weekend in Sterling, Colorado, only two hours southeast of Cheyenne. Like other conventions, it included talks, vendors and a banquet with a keynote speaker, but unlike other conventions, there were dozens of field trips.

I worried I might feel out of place, even if the information on the website, http://cfobirds.org/, assured me that beginning birdwatchers were welcome. CFO is all about the study, conservation – and enjoyment of birds.

Enjoy we did. Our first trip leader, CFO member Larry Modesitt, not only patiently explained field marks for common birds to several trip members who needed help, but he was also able to discuss the finer details of flycatcher fall plumage with one of the other members who surveys birds for a living.

CFO takes birding quite seriously. Each day, 14 field trips left every 10 minutes beginning at 5:20 a.m. One even started at 4 a.m.

Each field trip had a designated leader who contacted all their trip participants at least a week in advance to discuss routes, carpooling, rest and lunch stops, and even how much to reimburse drivers for gas.

Our third trip leader, Nick Komar, also a CFO member, consulted with other people on what had been seen where we were going and then worked hard to help us find those birds.

It was while visiting a designated birdwatching bench along the South Platte River at the Brush State Wildlife Area, that our group saw warblers in clear view, all in one bush, six to eight at a time: Wilson's, orange-crowned and yellow-throat. They were flitting about gleaning bugs, sparkling like yellow ornaments. At the other river overlook nearby, we found a plethora of woodpeckers: a redhead family, several red-bellied woodpeckers, a hairy woodpecker and yellow-shafted flickers.

While our first trip was filled with flycatchers and our third highlighted woodpeckers and warblers, the second, with Mark Peterson, was about shorebirds.

The whole idea for Sterling as a convention site was to catch the shorebird fall migration. Usually, the CFO annual meeting is planned around the spring migration. This was only the third time for a fall gathering. The keynote speaker was John Dunn, co-author of the National Geographic Field Guide to the Birds of North America, and shorebird expert.

But all the lovely summer rains keeping the prairie green around Cheyenne and Sterling kept the reservoirs full. And when they are full, there is no shore, no bare sand or mudflats for shorebirds to pick over. But we got lucky and found muddy shores at a small pond at the Red Lion SWA—and shorebirds.

Shorebirds are right up there on my list of difficult-to-identify species, partly because I don't see them often, and partly because I see them in migration when they aren't very colorful. I thought John Dunn's after dinner talk on shorebird identification might help. But it was definitely over my head.

However, if I keep looking at shorebirds in photos and in the flesh, eventually my identification skills will improve. Luckily, there was no quiz afterwards.

There was, however, a quiz the afternoon before: Jeop-birdie. Categories, among many, included identifying famous ornithologists, poorly photographed birds and types of bird nests. Very entertaining.

Larry Modesitt told me CFO began because Colorado needed an arbiter to sort through claims of unusual bird species seen in the state (Wyoming has a rare bird records committee). Many members also belong to Audubon.

CFO also supports bird research. A number of the papers given Saturday afternoon were partly supported by CFO funding. Many looked at facets of bird life that, once understood, such as the impact of oil and gas drilling noise on nesting birds, might make it easier to make land use decisions.

It isn't easy walking into a group of 200 unknown people, but when they are all dressed like me, in field pants, sun-protection shirts or T-shirts printed with birds, large-brimmed hats and binoculars, it's less intimidating. It's very easy to start a conversation with, "What field trip are you going on tomorrow?"

And after birding together, sharing exciting bird observations, many faces become familiar over the course of the weekend.

Next year, the convention will be in Salida, Colorado, first weekend in June. Now, there's a spot I might pick up some new life birds.

Gosh, did I just sound like a dyed-in-the-wool, serious species-nabbing-twitcher? Yikes!

338    CHEYENNE BIRD BANTER

# 362   Can birds save the world?

Sunday, October 26, 2014, Outdoors, page E2

Last month, the National Audubon Society publicized the result of a seven-year study to determine what would happen to North American birds if the change in climate continues as predicted.

The startling conclusion is that by 2080, nearly half our bird species, 314 (588 were studied), would have a hard time finding the food and habitat they need. They probably would not adapt, since evolution normally needs more than 65 years. So, they could become extinct.

"OK," some people say, "big deal, I've never seen more than three kinds of birds anyway."

That attitude was prevalent in the 1960s when eagles began producing eggs with shells so thin, the weight of the incubating parent crushed them.

"So what?" people said back then, especially if eagles made them and their lambs nervous.

The culprit was discovered to be DDT. And it was discovered to do nasty things to people as well. So, you might say that birds saved the world from DDT (except it continues to be produced to control malaria).

Last month, the Cheyenne – High Plains Audubon Society celebrated its 40th anniversary. John Fitzpatrick, director of the Cornell Lab of Ornithology, was keynote speaker at the banquet: "How Birds Can Save the World."

Fitzpatrick's premise is birds are so many species of canaries in the coal mine. Or, to localize the analogy, so many sage-grouse in the oil patch. We should pay attention to what they are trying to tell us before we hurt ourselves.

The Audubon report makes predictions based on two long-term, continent-wide citizen science efforts: the Christmas Bird Count (begun in 1900) and the Breeding Bird Survey (begun in 1966).

The Cornell Lab of Ornithology itself is well-known for citizen science projects, such as Project FeederWatch and the Great Backyard Bird Count. But the one that has mushroomed into a global phenomenon is eBird (www.eBird.org).

People who enjoy birdwatching have learned over the last 10 years to put just a little extra effort into it by counting birds they see and entering their notes online. Scientists can now see where bird species go and when, as if they have radar running year round. The more people enter observations, the clearer the picture emerges. And population changes are clearer, too.

When bird numbers change, or populations move, it's due to one or more changes in the species' environment. Some can be directly attributed to people, such as building a subdivision over a burrowing owl colony, and some indirectly, like climate change causing nectar-producing flowers to bloom too early for migrating hummingbirds.

Back in the 1970s, saving the environment always seemed to mean doing without, like hippies living off the grid. To some extent, curbing our desire for items built with planned obsolescence, like the latest smartphone, would preserve a little more landscape.

But Fitzpatrick's contention is that we can live smarter, rather than poorer, have our cake and eat it too, have our lifestyle and our birds.

We need creative people. For instance, I read 400,000 acres of California cropland is barren for lack of water this year. Yet power companies are stripping vegetation in the Mohave Desert to build arrays of solar panels. What if farmers rented out those barren fields for temporary solar installations?

There's work being done on solar paving. Imagine a sunny city like Los Angeles being able to power itself from all its lesser used streets, rather than depending on the transmission of electricity across hundreds of miles.

What if we put as much effort as we put into getting man on the moon into finding ways for every part of the country to produce energy in a way that keeps birds happy and us healthy?

I'm not an engineer, and probably neither are you. There is a shortage of them in this country. How can we raise more engineers and research scientists?

Take kids birdwatching. No, this isn't exactly one of Fitzpatrick's fixes. It's mine.

What are your kids doing on Saturday mornings? Watching cartoons and competing in athletics are all well and good. But what birdwatching does for children, and the rest of us, is to make us ask questions about the birds and their behaviors, to research, to communicate with others, and now, to search the eBird database.

When children develop curiosity through birdwatching – or other disciplines – they begin to see themselves in the sciences, in engineering, in technology, in all those "hard" subjects. And we will have the creative minds we need.

Our local Audubon chapter, now age 40, will continue with its traditional field trips (open to accompanied children and recorded for eBird, of course), educational meetings and projects, habitat improvements, and conservation advocacy. But watch for those special opportunities to introduce your children, grandchildren or neighbor children to birds. Because birds can save the world.

# 363 Big Bend hosts surprises for local birders

Sunday, November 30, 2014, Outdoors, page E2

Have you heard the rumor that Texas has mountains?

It does. The ranges I saw weren't the Grand Tetons, and I doubt they are ever snow-capped. But in terms of size, they remind me of many of Wyoming's smaller ranges.

Earlier this month, Mark and I visited Big Bend National Park, which entirely encompasses the famous (especially for birders) Chisos Mountains, where the Colima warbler nests. It breeds only in those mountains and Mexico's Sierra Madre Occidental.

If you look at Texas as your left hand, palm down, fingers pointing south, Big Bend is the end of your thumb. It is above a big bend in the Rio Grande which forms the border with Mexico.

The north park entrance station is 39 miles from the closest town, Marathon (pop. 436), and the northwest entrance is 76 miles from Alpine (pop. 6,000). We were able to reserve a room three months in advance at Chisos Mountains Lodge, in the heart of the park, because we were a tad early for the height of the tourist season. Summer, with temperatures over 110 degrees, is the off season.

It is the only lodging in the park unless you bring your own. It isn't fancy, but it's clean, comfortable and the food is good. We learned that reservations for the lodge for 2016 will open this January.

The lodge is tucked into the Chisos Basin, closed in by peaks, including Emory, which is 7,832 feet high. Centrally located, we were 30 miles from Rio Grande Village to the east, at an elevation of 1,850 feet, with visitor amenities and scenic attractions on the river, and 38 miles in the opposite direction from the other visitor amenities near the river at Castolon. It's a big park.

Like the rest of the Southwest, Big Bend has a monsoon season – heavy rainstorms at the end of summer. It wasn't supposed to be raining in early November. But it did. So I wore my rain suit in the desert because after all that driving, I didn't want to miss a thing.

However, it was so foggy the two days we were there that we never saw the tops of the Chisos Mountains. And we couldn't go down to see the famed Santa Elena Canyon because too much water was flowing over the road, and it was closed.

But we did find birds. These days, it is easy to use eBird.org to find birding hotspots. Mark identified Cottonwood Campground. It was a little intimidating reading all the signs that warned how to stay safe in encounters with javelinas, bears and mountain lions, but the big old cottonwoods were all a-twitter.

It sounded familiar – a flock of yellow-rumped warblers frantically feeding in trees and on the ground during a break in the rain, just like I've seen them behave in Cheyenne during migration.

But we also found uncommon Southwestern species. A vermilion flycatcher – incredibly red – alternately perched on treetops and signs. Nicely perched on a picnic table was a black phoebe, another flycatcher. The flicker-like bird was a golden-fronted woodpecker.

We stopped at nearly every pullout, walked out on many trails and added a few more southwest specialties like cactus wren and pyrrhuloxia (faded version of a cardinal), Inca dove, black-throated sparrow and roadrunner.

And we found familiar birds escaping winter – mockingbird, loggerhead shrike, Wilson's snipe, blue-gray gnatcatcher – although for these species, individual birds may make the park home year round.

There are plenty of trails for the adventurous who have real 4-wheel-drive trucks – not SUVs built on car chassis. I'll bet Big Bend has little trouble with people driving off road due to the multitude of tire-piercing cactus.

And what interesting vegetation is out there in the Chihuahuan Desert: 20-foot-tall century plants and other rosettes of sharp-pointed leaves putting up tall flower stalks, along with tiny flowers tucked beneath spiny neighbors, and higher up, southwest versions of oak, juniper and pine, even Douglas fir.

In addition to the one-volume edition of Sibley's field guide to birds of North America (some Texas birds are in the eastern edition and some in the western), the most valuable publication for visiting birders is the park's bird checklist available at the visitor centers. It's by Mark Flippo, one of the local birding guides. The 28-page booklet lists the more than 450 species found in the park, preferred habitat for each, and how likely you are to find them each season. It also points out the specialties, birds that are easier to find in Big Bend than in the rest of the U.S. and Canada.

The only question I have for Mark is, can we go back for another stay in the Chisos Basin maybe during spring migration 2016?

CHEYENNE BIRD BANTER

# 364 Feral cat policy will fail

Wednesday, December 10, 2014, Opinion, page C9

Last month, the Cheyenne City Council passed an ordinance allowing the Cheyenne Animal Shelter to implement a "trap, neuter, vaccinate and release" program for feral cats in the city.

The shelter staff is tired of euthanizing cats – 84 last month alone, many more in spring months – and sees this as a proactive measure.

The Community Cat Initiative allows "community cat caregivers" to bring in feral cats and pay $30 to sterilize and vaccinate, then release them, their ears tipped so they can easily be recognized as neutered.

Normally, unwanted cats, if not adoptable (and there is a barn cat adoption program for the less sociable), are euthanized.

I object to the TNR program, as it is referred to, for several reasons.

One is, I love cats. Our current feline, an indoor cat, is pushing 16 and is curled up on my shoulder as I write this.

I think more inhumane than euthanizing them is leaving cats outdoors. Feral cats as well as roaming family pets encounter life-threatening dangers: vehicles, predators – including other cats, not to mention inhospitable weather.

Conversely, feral cats untrapped – and unvaccinated – are public, human health concerns.

Why tolerate cats running loose, but not dogs?

It's also inhumane to leave wildlife at the mercy of a non-native predator like the cat. Many of our native birds here on the prairie are ground nesters, easy prey, as are small mammals.

In the U.S., free-roaming domestic cats kill an estimated 1.4 billion to 3.7 billion birds and 6.9 to 20.7 billion mammals each year, according to a U.S. Fish and Wildlife Service study. More recent studies show it could be more.

Nowhere in the literature has "trap, neuter, vaccinate and release" been shown to be successful in controlling feral cat populations.

On paper, the program sounds good, and I wish it worked.

Simply put, if you have a colony of cats and neuter all of them, the colony will die out when the last cat dies. Problem solved in the space of a feral cat's lifetime – probably less than five years.

In real life, no agency practicing "trap, neuter, vaccinate and release" has been able to trap enough cats to substantially lower the population.

Cheyenne's policy, waiting for the public to bring feral cats in, is doomed to fail even more rapidly.

Trapping cats is a bit like herding them. Plus, do the soft-hearted have deep enough pockets?

A staff member at the shelter said they are pursuing grants that would allow for a more aggressive "trap, neuter, vaccinate and release" program.

Meanwhile, we'll have an ever-increasing feral cat population (think about lying awake at night listening to cat fights) until nature finally deals with it – probably an ugly new and deadly disease. Not very humane.

Here are some more humane suggestions.

Hunt for nests of kittens and bring them in to be neutered and adopted at the age they can be socialized and become happy indoor cats. But don't allow them to be released outdoors.

Also, instead of charging people to bring in feral cats for neutering and vaccination, pay them $30. Putting a price on a species sent the passenger pigeon to extinction and nearly did the same for the buffalo.

Next, release adult, neutered feral cats, if they cannot be socialized, in a cattery, a place where they are safe and wildlife is safe from them.

Those options I've mentioned take money. Meanwhile, the problem grows.

I don't think it is fair to ask people charged with sheltering animals to do what really needs to be done from the wildlife and public health standpoint.

The wildlife agencies need to step in, as they have in Hawaii, another place where non-native predators, including feral cats, are decimating the wildlife.

Removing feral cats, euthanizing them, is not a happy proposition. Each one looks just like our own cat.

We need the fortitude to take actions to insure the well-being of cats. Releasing them to fend for themselves is not good for them nor for wildlife.

If you want to read a balanced look at this topic, see this Nebraska Extension Service publication, "Feral Cats and Their Management," http://ianrpubs.unl.edu/live/ec1781/build/ec1781.pdf.

# 365 Risking nice Wyoming weather, grebes, loons get caught

Sunday, December 21, 2014, Outdoors, page E2

You probably recognize that sinking feeling I had the morning of Nov. 10 when we cleared Denver traffic and a solid wall of cloud was suddenly visible 50-60 miles away, between us and home.

The rain, predicted for afternoon, started around 9 a.m. at Fort Collins, Colorado. In a few miles it turned to flakes. The road surface quickly iced as we climbed in elevation.

Northbound traffic slowed to a crawl, but only because there was a traffic jam of emergency vehicles gathered near the Colorado-Wyoming state line, where vehicles slid off the interstate earlier. One was lying on its roof.

No matter how slowly, we were happy to still be creeping toward home.

Some birds, however, were not as lucky with the weather.

One would think that migratory birds would be tuned into changes, but even they can be caught unawares.

If you remember that week, along with the snow, the temperatures dropped into negative numbers at night. My husband was contemplating an early start to the ice fishing season.

On Nov. 14, I had a bird call – people wondering how to help a Western grebe found at the plant west of town. They took it

to the Cheyenne Pet Clinic, which is licensed to handle wild birds.

The bird had to be euthanized because a wing joint was broken and couldn't be repaired. Veterinarian Christopher Church said two other grebes were rescued and brought in that week, and staff were able to release them at the Wyoming Hereford Ranch. Later that day, a Wyobirds report came in about a loon stuck in a small bit of open water, unable to take off. Someone in the Riverton area reported eared grebes, I think it was, also getting stuck.

Grebes and loons have bodies evolved for swimming underwater, not walking. Their legs are at the back of their bodies, like an outboard motor, and not under their center of gravity like a normal bird.

Because their feet do all the work when underwater, their wings are small. And their bones are not light and hollow like a songbird's, but dense, making it easier to dive. Flying is difficult for them.

Ornithologist Joel Carl Welty calculated a loon's wing-load, square centimeters of wing area to bird weight in grams, as 0.6. On the other hand, a black-capped chickadee is

6.1--comparatively buoyant. A Leach's petrel, an ocean-going bird, finds flying extremely easy at 9.5.

So, these heavy loons and grebes, hardly ever trying to move around on land, can only take to the air by flapping while pattering their feet against the water surface, Loons need as much as 650 feet for takeoff, according to Arthur Cleveland Bent, another ornithologist.

You can see where this is going. The grebe tucks its head under its wing one night and the next morning looks around at new ice hemming it in. "Oh crud."

The loon in the Wyobirds report kept busy diving for fish, and even tried walking a bit on the ice, but the next day, when the little bit of open water had frozen over, and the same observer went back, she saw no trace of the loon – not a feather or drop of blood, despite the bald eagles hanging around. It's quite unlikely that it flew. Perhaps the ice was finally thick enough for someone to walk out and rescue it, or for a predator to carry it off.

Ducks, which are better-balanced, need little space to take off, but have managed to become trapped in ice also.

What happened to the grebe with the broken wing?

Grebe species migrate at night. Apparently, they can get disoriented in snowstorms or fog or get confused by the sight of a wet parking lot shining in reflected lights and hit it hard, thinking it's water.

The common loon migrates through Wyoming, as do three species of grebes we see most often: pied-billed, eared and Western. Doug Faulkner, in "Birds of Wyoming," describes their fall migration patterns, always mentioning that a few individual birds don't leave until the reservoirs freeze up.

For these risk takers, the later they stay, the fewer birds they have to share the food source with. Some years, the bet pays off, and they are better fed when they arrive on the wintering grounds, reaping the benefits, such as better reproduction. But then again, maybe they don't make it.

After surviving this latest, unexpectedly dicey road trip, our weather forecasting being not much better than the lingerers', I'm wondering if we should have taken a lesson from all the smart loons and grebes that headed out by October for their ice-free wintering grounds.

# 2015

## 366   Archiving bird columns shows changes
*Sunday, January 11, 2015, Outdoors, page E3*

I'm afraid to mention this, lest the editor of the Wyoming Tribune Eagle think I've been doing this too long, but next month is the beginning of my 17th year writing this bird column.

It started because Bill Gruber, the Outdoors editor in 1999, asked me if I'd be interested.

I protested that there were people in town more knowledgeable – and there still are. But I had the time. And I could always research and ask the experts.

Besides Bill, I've worked with these other editors: Ty Stockton, Cara Eastwood Baldwin, Shauna Stephenson, Kevin Wingert and now Jodi Rogstad. All have been kind in their editing, catching style and grammatical errors.

A year ago, I had this great idea to archive all of my past columns as blog posts. I'd taken an online course in blogging as part of my teaching recertification, and I was intrigued. For one thing, I could add a widget that allows me to search all my past posts. So I could find out how many times I'd written about say, the Christmas Bird Count (about a dozen times).

I decided to make it a publicly accessible blog, www.CheyenneBirdBanter.wordpress.com. So far, I have 86 followers from all over the world without actively publicizing it.

Because bird topics are seasonal, and because there might be followers, strict chronological order wouldn't be best. So I used chronological order within each month, starting with February. The first post was the column I wrote that month, in 1999, followed by the one from February 2000, and so on.

Then I realized that these old columns could be outdated. So each one is accompanied not only by the date it originally was published, but by a short update on the topic.

There are some things that just don't change in the bird world, but technology has. I can now find an incredible amount of information online, and I can ask experts questions without having to call them long distance or mail a letter to them.

The most dramatic change in the bird world has been the advent of eBird, of course. The first column mentioning it was in 2003. It seems like every six months they come up with a new way for all of us citizen scientists to explore the eBird database – and more easily contribute to it. Amazing scientific studies are generated by it too.

The birds themselves continue to change. Mostly, it's population numbers and distribution.

For instance, there are more crows in Cheyenne today. There are way too many more Eurasian collared-doves now than there were in 1999, a year after the first one in Wyoming was identified in Cheyenne.

Do we have fewer numbers of any species? Evening grosbeaks don't seem to be visiting anymore. But a few years ago, lesser goldfinches started becoming regular, if still uncommon, visitors.

There is never a lack of topics to explore in the bird world. Feedback shows that many WTE readers are willing to come along on these sometimes intellectual excursions with me.

Hearing from readers is what makes writing these columns better than merely writing in a diary or notebook.

Information from readers has driven me to investigate topics, especially when there are several calls about the same thing. What to do about flickers drilling holes in wood siding is a column I've forwarded often since writing it.

Interestingly, for a while if you googled my name, the column that seemed to come up most often – because a friend in Colorado reposts my columns to his blog – is the one I wrote about the University of Wyoming graduate student studying hummingbird metabolism. In fact, it has been included in some online science anthology I can't access without buying a subscription.

There are now more than 300 Bird Banter columns posted. It has been fun looking back at them, seeing how, between the lines, they reflect my family's life. And I'm happy to have become the community bird lady, a responsibility which I appreciate.

More conventionally, I can be classified as a science writer. Actually, that isn't too far off from my course of study in college – and what one of my professors thought I should be.

Well, thanks, WTE editors and readers, for this monthly privilege. What's up at your bird feeders these days?

# 367 "Habitat Heroes" wanted to grow native plants

Sunday, February 8, 2015, Outdoors, page E2

Sometimes, wildlife issues seem to be out of the hands of ordinary people, people like those of us who are not wildlife biologists, land managers or politicians. Often, it seems futile to write a letter or email stating my opinion.

Connie Holsinger has devised a way for us to do something for wildlife right in our own backyards - literally.

Connie is the founder of the Habitat Hero program, which shows people in the Rocky Mountain area how to turn all or part of their yards, no matter what size, even a container or an apartment balcony, into wildlife habitat for birds, bees, butterflies and, may I add, even bats.

A popular term for this is "wildscaping." Add to that the term "waterwise" and Connie immediately grabs the attention of everyone who pays an increasing amount for watering their lawns as well as those who recently read the articles in the paper about Laramie County's finite water supply.

Connie is a native of Maine, in a zone that enjoys 50 inches of precipitation each year, compared to Cheyenne's 10-15 inches. When she moved to Massachusetts, she discovered birds, as well as the fact she can plant what would attract them to her yard. She volunteered with Massachusetts Audubon's Ipswich River Wildlife Sanctuary on habitat improvements.

Next, at her home on Sanibel Island, Florida, she discovered if she ripped out all the invasive vegetation and planted natives, her once-quiet yard was suddenly full of birds.

Relocating to the Front Range of Colorado in 1998, she learned what semi-arid means, especially when a major drought was just getting started. And she also learned that some native plants like the semi-arid life – after she killed her plantings of native penstemons two years in a row because she was rotting their roots with too much water.

It's no surprise that a smart woman like Connie then put "waterwise" with "wildscaping," a natural fit here in the arid West.

Also, the decline in the numbers of bees and butterflies documented in recent years makes even more important the idea of converting conventional urban/suburban landscapes into nectar and pollen havens, in addition to providing seeds and berries and cover for birds. Not to mention that native plants can take less work and water (read money) than a lawn.

With funding from the Terra Foundation, her private foundation that supports projects restoring the Colorado River Basin, Connie launched the "Be a Habitat Hero" campaign in 2013.

Anyone who would like to pursue the designation of "Habitat Hero" can apply through the website, www.HabHero.org, in September to see if their yard measures up. Last fall, 28 people, including Laramie County master gardener Michelle Bohanan, earned the designation.

While most of Cheyenne's homeowners and renters have mastered the basics of lawn care and keeping shrubs and trees alive, and many have a flair for flowers and vegetables, wildscaping requires a little change in horticultural practices, and a little change in mindset.

Explaining exactly how to transform all or part of a conventional yard or commercial landscape into a wildscape will be the topic of a Habitat Hero workshop scheduled March 28 at Laramie County Community College, 9 a.m. – 4:30 p.m. The $15 registration fee covers lunch, handouts and a tote bag for each participant full of donated items.

The three speakers will be Susan Tweit, plant biologist and author of "Rocky Mountain Garden Survival Guide;" Jane Dorn, co-author of "Growing Native Plants of the Rocky Mountain Area" (a digital version will be given to each participant); and Clint Basset, Cheyenne Board of Public Utilities water conservation specialist.

The major sponsors are Laramie County Master Gardeners, Cheyenne – High Plains Audubon Society, Audubon Rockies (which now administers the Habitat Hero program), Cheyenne Botanic Gardens and the Laramie County Conservation District.

One of the fun parts of the day will be the panel discussion, when the three speakers take a look at selected yards submitted by participants in advance and make recommendations on how to transform them into wildlife destinations.

Registration is available online at www.BrownPaperTickets.org, key words "Habitat Hero Cheyenne." Registration will also be available at the door, provided there are seats left. The workshop is limited to 100 participants.

"Plant it and they will come," Connie has said often.

This approach to landscaping benefits wildlife, but Connie said it speaks to her soul too when she sees the birds, bees and butterflies.

Her biggest aha moment came when she realized, "I can create a habitat in my yard, and take it beyond looking pretty" – making a difference in the world – in her own backyard.

# 368 Florida full of great birds and people

Sunday, March 8, 2015, Outdoors, page E3

Last month, I had a chance to visit Florida's birds a second time.

And I learned what it is like to have a nemesis bird – the reddish egret – that eluded me again despite visiting the right habitat at the right time with 40 people on the lookout.

Mark and I took part in a Reader Rendezvous weekend at Titusville, Florida, sponsored by Bird Watcher's Digest, www.birdwatchersdigest.com, a bi-monthly birding magazine read worldwide and celebrating its 35th year of publication.

Editor Bill Thompson III, son of the founders, was one of the weekend's event team members which included six magazine staff – all birders – and three local experts.

Having only 34 participants meant the birding experts were easily available for questions and to help spot birds. Although it was billed as a weekend for beginners, many of us were experienced, though not so much with Florida birds.

About a year or so ago, I noticed Bird Watcher's Digest was beginning to offer these Reader Rendezvous trips. Among them, one featured their humor columnist on a trip to the famous Sax-Zim Bog in northern Minnesota in winter (you needed humor to enjoy the temperatures), and another with optical experts to try out a variety of binoculars.

I asked Bill how the idea for the Reader Rendezvous weekends came about. He said he has been a speaker and field trip leader at birding festivals for 20 years and was looking for another way to reach readers. He said, "I love to show people birds."

Mark and I met several folks who had been on previous weekends, but they didn't strike me as groupies, though I have to say Bill has amusing takes on the birding life. What is appealing is the event team's interest in every participant, learning our names and asking often if we were enjoying ourselves.

The other participants were pleasant people who enjoyed the intense weekend of birding. And they didn't mind indulging Bill in his requests for group selfies. We even agreed to look silly doing "lifer dances."

The three days (Mark and I opted for the additional Friday trip) wore everyone out, but since all of us had invested time and money to be there, I heard no complaints about meeting the bus at 5:30 a.m. each day. At least we got a break on Sunday – 6:30 a.m. instead.

The Space Coast of Florida (area code 321 – no kidding!) is known for the Kennedy Space Center, and among birders for the Space Coast Birding Festival held mid-January.

While it seemed like the ducks had mostly migrated by the time we arrived Feb. 20, the group still logged 123 species over three days. I documented only 104 because sometimes the group split up. But of those, 13 were life birds for me, bird species I've never seen before.

We had a list of target birds – those that were advertised and those requested by participants.

On Friday, we went in pursuit of the red-cockaded woodpecker, a federally listed endangered species that makes a brief appearance at dawn when leaving its nest hole in a longleaf pine. The March-April 2015 issue of Audubon magazine has an excellent article detailing its life history and population ups and downs.

The half-mile hike in the dark and cold (frost on the grass in Florida!) was worth the minutes we were able to watch the small black and white woodpeckers.

Another target bird we saw in that same piney woods was the Bachman's sparrow, a species of concern that benefits from habitat work done for the red-cockaded woodpecker.

The Florida scrub-jay, a federally listed threatened species, is easy to find. We saw three sitting in treetops. Harder to find are the remnants of its necessary habitat, oak scrub.

While waiting in line at a potty stop, everyone got a long look at another threatened species, the wood stork. Three of the enormous birds scrutinized us from a nearby tree.

Our first look at the crested caracara, a threatened hawk, was fuzzy, but the next day it swooped over our heads. The endangered snail kite, another hawk, required a spotting scope to be identified.

Perhaps this weekend should have been billed as the "Threatened and Endangered Species Tour."

At any given birding festival we might have done as much birding, but in the course of several separate excursions with different people each time. With the Reader Rendezvous format, not only did we become acquainted with new birds, we made new birding friends. We may meet up again on another Reader Rendezvous, or here in Cheyenne since some folks were thinking about heading west.

While the weekend was somewhat of a marathon, the equivalent of three of our all-day Cheyenne Big Day spring bird counts plus two evening programs like our Audubon chapter's monthly meetings, my binocular hand-eye coordination is all warmed up now and I'm ready for spring migration.

# 369 Are you a bird expert?

Sunday, April 12, 2015, Journey, page E1

Raise your hand if you've been reading my bird column for the 16 years I've been writing for the Wyoming Tribune Eagle.

Good for you!

And even if you haven't been reading it that long, here's a quiz to see what you've learned so far. All the birds mentioned are listed on the Cheyenne Bird Checklist posted at the Cheyenne-High Plains Audubon Society website, http://home.lonetree.com/audubon.

**Ready?**

**1.** What "sign of spring" shows up most often on Cheyenne Christmas Bird Counts?

a. Western Meadowlark
b. Red-winged Blackbird
c. Mountain Bluebird
d. American Robin

Answer: (d) The robin has been seen on almost every Cheyenne Christmas Bird Count since our first, 60 years ago, and the red-winged blackbird about half as often. Meadowlarks have been seen six times, and bluebirds never.

**2.** Which large white birds visit Cheyenne in small flocks each spring to go fishing?

a. Snow Goose
b. American White Pelican
c. Tundra Swan
d. Great Egret

Answer: (b) While lone great egrets are seen occasionally, flocks of pelicans show up regularly. Snow geese and swans don't eat fish.

**3.** Which is the only blue bird that would have been seen here in pioneer times?

a. Mountain Bluebird
b. Blue Jay
c. Blue Grosbeak
d. Indigo Bunting

Answer: (a) Pioneers would have seen mountain bluebirds. Farmers planting windbreaks made our high plains friendly to blue jays which were noticeably present by 1939. Indigo buntings were recorded by the 1950s and blue grosbeaks by the 1960s.

**4.** Which black-colored bird seen around Cheyenne never raises its own offspring?

a. Common Grackle
b. Red-winged Blackbird
c. Yellow-headed Blackbird
d. Brown-headed Cowbird

Answer: (d) Brown-headed cowbirds always leave their eggs in nests of other birds. Historically, they needed to be off right away to follow the buffalo.

**5.** Which woodpecker is more often seen pecking Cheyenne lawns instead of trees?

a. Downy Woodpecker
b. Hairy Woodpecker
c. Northern Flicker
d. Red-headed Woodpecker

Answer: (c) I get more calls about strange, polka-dotted birds digging for grubs in people's front lawns, but yes, the flicker is a woodpecker. Just ask anyone whose wood siding has been pecked.

**6.** Which bird only nests on the ground?

a. Western Meadowlark
b. Wood Duck
c. Black-crowned Night-Heron
d. Great Blue Heron

Answer: (a) Wood ducks nest in tree cavities or nest boxes. Great blue herons and black-crowned night-herons almost always nest in colonies in trees. But the meadowlark, like many grassland birds, always nests on the ground. Don't mow your prairie until July, after nesting season.

**7.** Several species can be seen in both southeastern Wyoming and the Middle East. Which one didn't require human help to get here?

a. Caspian Tern
b. Rock Pigeon
c. Ring-necked Pheasant
d. Eurasian Collared-Dove

Answer: (a) The Caspian tern, a rare visitor here, occurs naturally on all continents except Antarctica. Pigeons, pheasants and collared-doves started out as Eurasian (including the Middle East) species.

**8.** Which big hawk likes the Magic City so well it flies from Argentina to nest in our neighborhoods?

a. Red-tailed Hawk
b. American Kestrel
c. Swainson's Hawk
d. Northern Harrier

Answer: (c) The Swainson's hawk winters in Argentina. The other three may get as far south as Panama.

**9.** Besides Wyoming, what other states claim the Western Meadowlark as their state bird?

a. Kansas
b. Nebraska
c. Montana
d. North Dakota
e. Oregon

Answer: (a, b, c, d, e) All claim the western meadowlark.

**10.** How many warbler species have been observed since 1993 on the Cheyenne Big Day Bird Count, at the height of spring migration?

a. 23
b. 27
c. 31
d. 35

Answer: (c or d) There have been 31 warbler species on our Big Days, by my count. However, there are 35 listed on the Cheyenne checklist. Take credit for either answer.

**11.** What sandpiper likes nesting on our prairie, and even on football fields?

a. Killdeer
b. Greater Yellowlegs
c. Lesser Yellowlegs
d. Wilson's Snipe

Answer: (a) Killdeer. Both yellowlegs pass through on their way to nest in Canada. Snipe nest here, but only on the edge of water.

**12.** Three of these species spend only the winter in Cheyenne. Which one leaves instead?

a. American Tree Sparrow
b. Dark-eyed Junco
c. Rough-legged Hawk
d. Lark Bunting

Answer: (d) Lark buntings leave Wyoming to winter in the southwest and Mexico. The others arrive: juncos after nesting in the mountains, tree sparrows from Alaska and Canada, and rough-leggeds from the Arctic.

**How did you do?**

Whatever your score, by taking part in the quiz, you show you are part of the community of inquisitive birdwatchers.

Remember, whatever local wisdom about birds is cited here, your careful observation could turn it on its head. Birds never stop teaching us new things.

# 370 Birds are always around to fascinate the young

Sunday, May 24, 2015, Outdoors, page E2

I've been invited to visit with a group of mothers to share with them a few tips on birdwatching with young children.

It's been more than 20 years since our kids were preschoolers and I'm trying to remember just what my husband, Mark, and I did. We must have done something right: One son recently bought his own binoculars, and the other reports interesting birds from his travels.

I doubt we were very different from other parents who have a serious interest in a field or outdoor pursuit: You just take the kids with you. We took ours birdwatching, fishing, hunting, camping and hiking.

Of course, it isn't necessary to be the child's parent in order to mentor them – just more convenient.

Babies are as mobile as the distance you are willing to carry them – as long as diapers and milk hold out.

When kids start walking, you are suddenly limited to how far they are willing to go. They are a lot more willing if they are comfortable: warm and dry – or cool in the summer, protected from sunburn and bites, not averse to outhouses or bushes, with plenty of food and water, and naptime far in the distance.

How do you make a child interested in birds? Like so many other traits, you model it. No guarantee it will take.

The great thing about fostering an interest in birds is there are always birds around. And they have color, movement and sound. All you have to do is point. You don't have to know much at first.

As kids get older, it isn't hard to introduce them to a field guide full of colorful illustrations, then working together to figure out the names of bird species or going online some place like www.allaboutbirds.org.

Toy binoculars are perfect for imitating adult birdwatchers. By grade school, kids might be able to appreciate what they are seeing through higher quality binoculars.

Don't make up stuff when you don't know the answer. It's OK to say "I don't know." Look it up…or call me.

It's OK if your child doesn't become an ornithological know-it-all by age 7. Perhaps you observe your child is often distracted by rocks, weeds, sticks or worms instead. There's a lot of nature to enjoy out there besides birds.

It doesn't hurt to point out a pretty flower (try not to pick it) or insect or an animal track. Birders often learn something about a bird by observing its whole environment – food sources, nesting materials, perches, predators.

It's OK if your kids find other ways to entertain (or distract) themselves while you bird. Maybe they would like to take along a sketchbook, a camera, or a butterfly net. But sometimes kids just have to dig in the dirt or chase around and be silly.

Every child is different so it is up to you to figure out what will keep yours interested in coming outdoors with you again.

If you want philosophical guidance, look for a new book called "How to Raise a Wild Child, The Art and Science of Falling in Love with Nature," by Scott D. Sampson, known for his appearances on the PBS Kids show "Dinosaur Train."

His worry is if more kids don't get outside and learn to love it, nature will lose her constituency and the Earth will be ravaged until it can no longer support human life.

He wants to see more "hummingbird" parents rather than helicopter parenting, allowing kids to make discoveries. He wants to see school playgrounds filled with natural landscapes and objects, not asphalt and gravel. He wants kids to get dirty.

2015

347

He has a bibliography listing studies proving why spending time in nature is good for kids – and the rest of us.

Sampson's book is a direct descendent of Richard Louv's "Last Child in the Woods" (www.ChildrenandNature.org), but it has "Nature Mentoring Basics" and lists of things you can do at different age levels.

If you need more local ideas, check out WY Outside, http://www.wyoutside.com. This year they are holding the WY Outside Challenge.

My family camped a lot when my sister and I were kids, but I don't remember either of our parents doing anything much, beyond sending us to Girl Scouts, to guide both of us into our love of the outdoors, except supporting us as we began to seek the outdoors on our own.

Not every child delighting in a wild bird is going to become an ornithologist. That's OK. It is their appreciation for birds and the rest of nature we are after, hoping that it will foster good stewardship and a healthy life.

# 371 Changes in spring bird count bring up questions

Sunday, June 14, 2015, Outdoors, page E2

A Virginia's warbler was the celebratory guest at the Cheyenne – High Plains Audubon Society's Big Day Bird Count May 16.

This southwestern bird is a rare migrant in our area. Two other rare migrants were broad-winged hawk, an eastern species, and black tern.

This year 110 species were counted. This is lower than a typical count the last several years – and way lower than the counts in the 1990s, averaging 140-150 species.

It could be the result of a change in the birders participating. For many years, the Murie Audubon Society put on a bird class in Casper every spring and many of the students made an overnight excursion to be here at the crack of dawn for the Big Day. More eyeballs equals more birds seen. This year only one person came down.

However, the Laramie Audubon Society has taken to scheduling a field trip to the Wyoming Hereford Ranch on our Big Day. This year they brought 14 people to augment our 20.

Possibly another change is that back in the 1990s, Bob and Jane Dorn birded the High Plains Grasslands Research Station at 6 a.m. Now we don't get there until nearly lunch time, after birding Lions Park and the ranch. Birds are more active early in the day.

In the world of birdwatching, a big day is a marathon to see how many species an individual or a small team can see in 24 hours. The area birded may be limited. The American Birding Association, for the sake of competition, has rules that describe how many people can be on the team and what percentage of the species counted have to be seen by all team members.

By contrast, Cheyenne's count starts out as one big group and slowly dissolves into individuals by afternoon. Perhaps we should lean more toward the Christmas Bird Count model and have groups of people birding each hot spot simultaneously at dawn.

There's also the possibility that the birds have changed over the years. While Cheyenne residents have planted more trees, inviting more songbird species, areas of prairie we used to check are now developed and thus, no burrowing owls or longspurs found on the day of the count.

Typically, spring migration is a short burst, compared to fall migration, which begins sometime in July with shorebirds and still finds some species straggling south in November and December.

Now we can look at observations for this May in Laramie County at www.eBird.org to see where the peak of migration was. There were 173 species observed for the month. Keep in mind many pass through within a week's time or less:

1st week – 79 species
2nd week – 99 species
3rd week – 145 species
4th week – 128 species.

The third week includes our Big Day, but had 35 more species than we saw on May 16, which was a cold day so perhaps birds were sitting tight and were more visible the rest of that week.

Even in the age of eBird, our Big Day is worth the effort, I think. It's a chance to learn to identify, with the help of the best local birders, species that are here rarely or for a short time, like the Virginia's Warbler.

Simply, it is a great time for birders to flock together and enjoy the magic of migration.

## Cheyenne Big Day Bird Count 2015

| | | | |
|---|---|---|---|
| Canada Goose | Lesser Scaup | Turkey Vulture | Wilson's Snipe |
| Gadwall | Ruddy Duck | Cooper's Hawk | Wilson's Phalarope |
| American Wigeon | Ring-necked Pheasant | Broad-winged Hawk | Franklin's Gull |
| Mallard | Pied-billed Grebe | Swainson's Hawk | Ring-billed Gull |
| Blue-winged Teal | Eared Grebe | Red-tailed Hawk | Black Tern |
| Cinnamon Teal | Western Grebe | American Kestrel | Rock Pigeon |
| Northern Shoveler | Double-crested Cormorant | American Coot | Eurasian Collared-Dove |
| Northern Pintail | American White Pelican | Killdeer | Mourning Dove |
| Redhead | Great Blue Heron | American Avocet | Eastern Screech-Owl |
| Ring-necked Duck | Black-crowned Night-Heron | Spotted Sandpiper | Great Horned Owl |

| | | | |
|---|---|---|---|
| Chimney Swift | Common Raven | European Starling | White-crowned Sparrow |
| Broad-tailed Hummingbird | Horned Lark | Northern Waterthrush | Dark-eyed Junco |
| Belted Kingfisher | Northern Rough-winged Swallow | Orange-crowned Warbler | Western Tanager |
| Downy Woodpecker | | Virginia's Warbler | Black-headed Grosbeak |
| Northern Flicker | Tree Swallow | Common Yellowthroat | Red-winged Blackbird |
| Western Wood-Pewee | Bank Swallow | American Redstart | Western Meadowlark |
| Willow Flycatcher | Barn Swallow | Yellow Warbler | Yellow-headed Blackbird |
| Least Flycatcher | Cliff Swallow | Yellow-rumped Warbler | Brewer's Blackbird |
| Cordilleran Flycatcher | Black-capped Chickadee | Wilson's Warbler | Common Grackle |
| Say's Phoebe | Mountain Chickadee | Green-tailed Towhee | Brown-headed Cowbird |
| Cassin's Kingbird | Red-breasted Nuthatch | Spotted Towhee | Orchard Oriole |
| Western Kingbird | House Wren | Chipping Sparrow | Bullock's Oriole |
| Eastern Kingbird | Blue-gray Gnatcatcher | Clay-colored Sparrow | House Finch |
| Loggerhead Shrike | Ruby-crowned Kinglet | Vesper Sparrow | Red Crossbill |
| Plumbeous Vireo | Swainson's Thrush | Lark Sparrow | Pine Siskin |
| Blue Jay | American Robin | Lark Bunting | American Goldfinch |
| Black-billed Magpie | Gray Catbird | Song Sparrow | House Sparrow |
| American Crow | Brown Thrasher | Lincoln's Sparrow | |

# 372

Sunday, July 26, 2015, Outdoors, page E3

## High-end binoculars, mid-level prices from Wyoming's Maven

There's a new Wyoming company making binoculars.

You may have seen Maven binoculars mentioned in hunting and birding circles last fall when they came on the market. So far, reviews are good. I'll add to that, six months after I bought a pair of my own.

The Maven Outdoor Equipment Company, located in Lander, offers three models, each in two sizes of magnification:

--There's the B1: 8x42, 10x42 ($900);

--B2: 9x45, 11x45 ($1,000);

--B3: 8x30, 10x30 ($500).

I went for the B3 8x30 not only because it is in my price range, but also because it weighs 16 ounces (compared to the larger models at 26 ounces or more). Binoculars classified as compacts usually have a narrower field of view, but not these: you get a 430-foot view at 1,000 yards.

The B3s are a hit with birdwatchers, as well as archers, who want to travel light, said Mike Lilygren, one of the three co-owners.

Ironically, back in January, I was only sort of in the market for new binoculars as I was growing increasingly unhappy with my Brunton binoculars.

Lilygren and the other Maven co-owners, Brendon Weaver and Cade Maestas, used to work for Brunton's optic division before forming their own company. (Brunton is now out of the optics business.)

Maven was mentioned to me by someone at a store that caters to birders – very generous, considering no retail outlet sells this brand. For now, it is only available online, unless you happen to actually be in Lander or are at an outdoor or birding equipment show Maven is attending.

Without the middleman, consumers can pretty much double the quality of optics they can buy for their money.

Suddenly, $900 for the favorite of many birders, 8x42, looks like a bargain compared to the top-of-the-line Leica, Swarovski and Zeiss models that cost more than $2,000.

But do the optics compare? For someone like me who has never paid more than $200, the B3 is a big improvement. I notice the difference in distinguishing details on birds, especially in low light situations. (If you want an extended technical discussion and comparison, check out www.BirdForum.com.)

Ergonomically, the B3 suits my short fingers, and it doesn't take much to change the focus from close to far – two quibbles I've had with other binoculars I've owned.

But the adjustable eye cups do have a tendency to collapse a bit after an hour. I bird without glasses and have the eyecups pulled all the way out. People with glasses leave them all the way down. Lilygren said they've noticed the problem in-house,

but I'm the first customer to mention it. Possibly, most people use them with glasses or sunglasses.

Standard advice has always been not to order binoculars sight unseen, but Maven will mail you a demo that's easy to return. So far, only one person has returned theirs – but not because of dissatisfaction, Lilygren said.

Ordering online, www.mavenbuilt.com, allows for customizing the look of the binoculars beyond standard black and gray. Try camo-print bodies and your choice of various pieces of orange, silver, red or pink trim, plus up to 30 characters of engraving – adding as little as $10 or as much as $250 to the price.

Pink trim? There were many requests, and purple may be coming soon.

Lilygren said 75 percent of online customers chose some customization, but the people buying at shows do not. Overall, half are buying custom.

While the glass is ground in Japan by the famed Kamakura Company, the binoculars are assembled in the U.S., then shipped to Lander where they have to pass inspection by the company owners. Lilygren said he was going through a stack of 25 pairs when I called.

A customized pair can take three weeks to arrive, but a stock pair, like mine, can arrive practically overnight with Wyoming's typical one-day in-state postal delivery.

Besides adding purple, what's next for Maven? Next year, it will offer a 10x56 and 15x56. Not something

birders would tote around. But a spotting scope will also come out.

So, what's with the name "Maven"? The word means "trusted expert" or "one with knowledge based on accumulation of experience." And that is their forte, compared to other outdoor gear companies, said Lilygren. He

and Weaver and Maestas are passionate hunters. The company is based in Lander, the center of the outdoor recreation universe, because that's where they want to live.

And the three have so much faith in their products, they offer a lifetime warranty. They expect Maven binoculars to last a lifetime.

# 373 Hummingbird rescue reveals beauty and mystery

Sunday, August 30, 2015, Outdoors, page E2

Terry Masear has a soft spot for hummingbirds yet has survived the hard realities of rescuing and rehabbing them for 10 years.

Her new book is destined to wet your eyes now and then, as well as open them to the beauty and mystery of hummingbird life. She talks about her work in this interview by email. You can also listen to her NPR interview at http://hereandnow.wbur.org/2015/07/28/hummingbird-rescue-masear.

First, we should note in Cheyenne, hummingbirds typically visit only during migration, nesting at higher elevations.

**Q. How is your 2015 season going? How about your success rate? Any favorite stories from this year?**

A. Southern California hummingbird rehabilitators admit over 500 injured and orphaned birds into rescue centers annually. I release between 70 and 80 percent of my intakes. Due to promotional events for the book, I could not participate in hands-on rehabilitation this year, but I answered 2,000 calls, saved 200 birds over the phone, and sent another 200 to rehab centers in the

Los Angeles area. I helped rescue a pair of Allen's nestlings that got entangled in a bizarre drama between their mother and her frustrated hybrid daughter (named Rosie by webcam viewers) from last year. This fascinating event, along with footage of several webcam nests and fledges this year, can be seen on Bella Hummingbird clips posted on YouTube.

**Q. Have you noticed the drought affecting your hummingbird work?**

A. The drought is leading to more mite-infested nests. But we have been able to save and keep most of these nestlings in their natural environment by having finders dust the nests and chicks with diatomaceous earth, which in no way deters the mothers from continuing to feed their young.

**Q. What makes bird rescue in Hollywood different from other places?**

A. Los Angeles has a larger and more diverse hummingbird population than any city in the world. Females often nest in backyards and near houses, which leads to encounters with humans and makes rescue more necessary. We see seven species

– Allen's, Anna's, black-chinned, rufous, Costa's, broad-tailed, and calliope – in rescue. Rehabbers also believe the Allen's and rufous have hybridized in Southern California as we are noticing extensive rust coloration in many young males.

**Q. What makes hummingbird rescue different from other bird rescue?**

A. Hummingbird babies are extremely high maintenance. They have to be hand fed every 30 minutes for 15 hours a day until they fledge and can be feeder trained. So a lot of bird rescue centers refuse to take them, which is why private rehabbers stay busy.

**Q. What are the biggest hazards for hummingbirds in L.A.?**

A. Tree trimmers and weekend gardeners are by far the greatest threats to young hummingbirds. So we are trying to educate the city and private citizens to refrain from trimming trees in the spring when birds are nesting. Also, a lot of well-meaning finders pick up grounded fledglings and carry them home, which takes the young birds away from their mothers who are still

feeding them. Other dangers to hummingbirds include windows, domestic cats, termite tenting, and weather hazards like heavy wind and torrential rain.

**Q. All things considered, do you think hummingbird feeders are good for hummingbirds?**

A. As long as people keep them clean, sugar feeders benefit hummingbird populations and, along with introduced vegetation, have allowed species like the Anna's and rufous to expand their ranges considerably.

**Q. Which is more difficult, dealing with emotionally distraught callers that have found an injured or abandoned hummingbird, or dealing with the birds?**

A. Of course, serious injuries present challenges for the rehabber and some losses will haunt you. But ask any rehabber on the front lines what the most difficult part of their work is, and they will say dealing with the public. The majority of callers are compassionate and caring, but a certain percentage do not have the wildlife's best interests in mind. Some callers don't want to make any effort and will let helpless nestlings die if rehabbers don't show up immediately. Others insist on keeping young birds as pets. When we explain why they cannot do this, legally or in good conscience, some get abusive. These conversations strain the patience of even the most forgiving rehabber, especially during peak season when the pressure is on.

**Q. Record keeping is required for your permit, but are you also keeping notes that helped you write this book – all the anecdotes about particular birds and their personalities and challenges?**

A. As far as overall intakes and releases, my records are pretty precise, so I referred to those when writing "Fastest Things on Wings." And through these records I can recall certain birds because of their unique histories. Other remarkable characters, like Pepper, Gabriel, Iris, and Black-top, are easy to remember because their stories are so extraordinary.

**Q. A PhD in English doesn't necessarily translate into being able to write a riveting story, as you** have. **What writing experience did you have before writing this book?**

A. I taught research writing at UCLA for years and wrote a textbook for ESL students. Five years ago, ironically, I wrote a nonfiction book about a unique and mysterious experience my husband and I had with our cats. While I was trying to sell that manuscript, editors kept asking about hummingbird rehab, which led to this book.

**Q. Were you out on book tours this spring and summer, and if so, who held** down the fort?

A. I have been doing book signings and interviews all summer, which is why I could not do rehab. But my phone hasn't stopped ringing for six months, so I've been deeply involved in the rescue business. And as exhausting as it is, I miss the powerfully rewarding experience of rehab and can't wait to get back to it.

"Fastest Things on Wings, Rescuing Hummingbirds in Hollywood," by Terry Masear, Houghton Mifflin Harcourt, 2015, 306 pages, indexed, $25 hardcover.

# 374 Many mountain birds mean summer of no regrets

Sunday, September 27, 2015, Outdoors, page E3

This fall I have no summer regrets. I made it to the mountains several times.

For me, the best reason for living in the West is access to mountains – living within commuting distance of timberline.

From Cheyenne, the alpine tundra along Trail Ridge Road in Rocky Mountain National Park is about two hours away, as are the trailheads for the Snowy Range in the Medicine Bow National Forest.

The national park was my introduction to mountains when I was 6 years old. I had traveled from Wisconsin with my parents and grandparents on the occasion of my uncle's graduation from Colorado University. This year, the occasion was his memorial, and I was helping introduce his 6- and 3-year-old grandchildren to mountains.

The Snowies, on the other hand, I found on my own, when the lotto game that is federal seasonal work brought me to Wyoming. I was lucky this summer to visit three times, twice above 10,000 feet.

On Father's Day afternoon it was rather appalling to see the traffic on Rocky Mountain's Trail Ridge Road, amplified by the park's 100th anniversary celebration.

It's a pilgrimage. At every comfort station, one parks and walks a trail out into the landscape. But other than a selfie with magnificent mountains in the background, I'm not sure if many of the pilgrims know what they are seeking as the stiff wind makes them shiver in their tank tops, short shorts and flip flops.

A few weeks later in mid-July, all of us on the Audubon field trip at least knew exactly what miracles to look for as pilgrims hurried past us.

We wanted to see the sparrow-like American pipits. It's hard to pick them out from the litter of rocks and plethora of wildflowers. But soon we recognized their calls and realized they were all around us.

Our other goal was the white-tailed ptarmigan – high-altitude relative of sage-grouse – which turns white in winter and brown in summer. Except that in July, the birds are really just a mottled/speckled brown and white, matching perfectly those lichen-encrusted rocks scattered all around.

We found the location of the previous e-Bird sighting of a ptarmigan and resigned ourselves to examining every rock along the way, knowing that unless the wind ruffled the bird's feathers or it decided to move, we might never see it. But we were joined by a birding tour leader on her day off, as well as two other hikers who were lucky enough to find a hen taking a stroll and who pointed it out.

There are some advantages to crowds.

A week later in the Snowies, Mark and I took one of the trails starting at Brooklyn Lake, expecting many fewer people.

But a file of at least 40 teenagers from a Midwestern church passed us, toting serious backpacking equipment. I like to think that, like the crowds in Rocky Mountain, these people will become supporters for preserving this country's wild lands.

The wonderful wildflower displays made up for a lack of birds on this first Snowies trip, but three weeks later, on an Audubon chapter hike, things were reversed. The wildflowers were waning, but the birds were gathering, and we were the big group, 15 people between the ages of 10 months and 75 years old.

We never hit tree line, only getting as far as the trees growing in isolated islands. Gobs of ruby-crowned kinglets flitted in and out of the branches of Engelmann spruce. Three mountain chickadees carried their conversation to the outer branches where we could see them clearly. Pine grosbeaks, larger versions of our house finches in town, were busy grooming their feathers in plain sight. Young spotted sandpipers, their bodies mere halos of stiff white fuzz

perched on impossibly long legs, scrabbled after their parent, negotiating the rubble at the foot of a snowfield still melting and providing the watery habitat they needed.

Juncos were flashing their white outer-tail feathers everywhere. Soon, we will see them down in town.

Not only did the 10,000-foot elevation offer its usual respite from summer heat,

but puffy clouds, dead ringers for snow clouds, sailed by on cold wind, keeping us in our winter jackets, which we were experienced enough to bring. Birdwatchers just don't hike hard enough to warm up, and this day there were 17 species of birds making the 3-mile round trip take more than four hours.

Back at the parking lot, the fall feeling, stirred by the

wind and the gathering birds, was amplified by realizing the meadow grasses had gone to seed and turned brown – in early August.

Now fall is finally here at lower elevations.

Summer is such a fleeting season at high altitude, but at least this year, I didn't let it pass me by.

# 375 Virginia is for bird lovers, but also makes Wyoming birder think about home

Sunday, November 1, 2015, Outdoors, page E3

The greater sage-grouse could be the expected topic for this column, another in the flurry of opinions published since September about its non-listing as an endangered species.

We should be happy it wasn't listed. A lot of hard work went into compromises and new concepts in cooperation, worked out between industry, government agencies and environmental organizations.

But there are biologists afraid the compromises are not enough to keep sage-grouse populations from continuing to fall. There will be more legal battles ahead, from both industry and environmental groups.

What I really wanted to talk about was exploring the birds of Virginia in mid-October. Our older son and his wife moved there last winter, and Mark and I birded with them recently, attending the Eastern Shore Birding and Wildlife Festival.

Birding back east is often about birding by ear, especially with leaves still on the trees. I soon learned

to appreciate the Carolina chickadee's call – just like our mountain chickadee's – but as if it had drunk too much coffee.

Small birds in autumn don't sing much and their call notes are hard to distinguish. Who knows how many warblers we missed? Identifying them visually is difficult because they have molted out of their more identifiable spring plumage.

Maybe that's why I liked the black-throated blue warblers we saw – males have solid blue backs, white bellies, and black faces and throats year round. But, oh the "warbler-neck" pain from looking up into extremely tall trees of the deciduous forest.

Our daughter-in-law was hired this last summer to help The Nature Conservancy with bird surveys in the Allegheny Highlands in western Virginia, to see if the fire management plan for the rare montane pine barrens will give songbirds like the golden-winged warbler what they need.

The International Union for Conservation of Nature

classifies the golden-winged as "near threatened." It has declined in population 76 percent since 1966, and 95 percent in its historic range in Appalachia for a variety of reasons, according to the Cornell Lab of Ornithology.

We didn't get up to TNC's Warm Springs Mountain Preserve to see this warbler this time, but Mark and I did accumulate a nice list of birds in our numerous walks. To name a few, we walked about the grounds of Monticello, on a bit of the Appalachian Trail in Shenandoah National Park, through woods and meadows at the U.S. National Arboretum, across the swamp at Historic Jamestowne, and in the Eastern Shore of Virginia National Wildlife Refuge.

My favorite birds (maybe because they were easy to see) were the gulls on one of the islands along the 20-mile-long Chesapeake Bay Bridge-Tunnel, which is a combination of bridges and tunnels. Where one of the bridges ends and one of the tunnels begins on a manmade island, there is a scenic view pullout. Gulls

paraded around, posing for cameras and expecting tips – food scraps.

I liked the view of open water. It's like home, where so much is out in the open, though much also goes on below the surface.

I thought about the difference between saving sage-grouse and saving golden-winged warblers.

I thought about how easy it is to count football-sized sage-grouse compared to surveying for 5-inch birds that play hide and seek in leaves that are bigger than they are.

I thought about our large tracts of public and private land in Wyoming that seem to be endless and changeless, compared to the eastern forest constantly under attack by invading plants and subdivisions. Use Google Earth to trace the path of our flight from Washington, D.C., to Atlanta, and trace the endless curlicues of brand-new roads. It would be difficult to insert fire as a management tool outside of a remote, unfragmented place like the Allegheny Highlands.

Acres in Wyoming are

352            CHEYENNE BIRD BANTER

not untrammeled. As a range management student at the University of Wyoming 35 years ago, I learned that sagebrush needed to be eradicated to increase cattle productivity. Land managers, especially in the livestock industry, took action.

And now we see the error: that for a few more pounds of beef we may have jeopardized not only the sagebrush, but the sage-grouse and the whole sagebrush community, above and below the soil surface.

In turn, that could jeopardize future kinds of agricultural and wildlife productivity (such as hunting) that we put dollar values on. Energy developers, whether mineral or alternative, despite reclamation and mitigation claims, have the same problem.

I think we are on the right track though, segregating incompatible land uses, just as we zone areas of a city. It's a matter of figuring out, before it's too late, how to have our sage-grouse, and energy too.

Gee, it's great to be home.

# 376 Feed winter birds for fun

Sunday, December 6, 2015, Outdoors, page E3

Feeding birds in your backyard is a time-honored tradition. It makes a great gateway to building your interest in birds. But there are a few things you should keep in mind if you decide to put up a feeder.

**Birds don't need our food.** They are good at finding natural food. Don't worry if you don't have food out for them every day, although being consistent means you are more likely to see interesting birds.

Bird feeding is really about enjoying the birds, so put your feeders close to windows you look out of often. Be sure to put them close so that birds won't hit your windows at high speed when leaving your feeder.

**Keep your feeding operation affordable.** I've had people complain bird seed is expensive. But it's up to you how much seed to put out and how often. Fill feeders at the time of day you can enjoy watching the birds.

**Never put out more feeders than you can keep clean or clean up after.** Feeders can get gunky and can spread diseases. Every couple weeks, clean them with soap and water, maybe a little bleach, and rinse well. If you see a sick bird, don't put the feeders back up for a week. We usually don't feed in the summer because even more disgusting stuff grows in feeder debris.

Be sure to keep the seed hulls swept up every few days or think about feeding hulled sunflower seeds.

**Don't be cheap.** Rather than the bags of mixed seed, go for the black-oil sunflower seed. Seed mixes often contain filler seed – or at least seed that birds around here won't eat – and you'll just be sweeping it up anyway. Black oil sunflower seed attracts a wide variety of seed-eating birds. Buy the 40-pound sack at the feed store for a better price per pound. If it still seems too expensive, feed only the amount you can afford each day.

**Leave the cats indoors.** There are many reasons cats should live indoors fulltime, including their health and safety, but really, is it fair to invite birds to your yard where a predator lurks? The feeder may be on a pole or hanging above the cat, but certain birds prefer to feed on the spilled seed on the ground.

On the other hand, if a neighbor cat stakes out your yard, you can make sure the area around the feeder has no place for a cat to hide. I've also heard of putting up a 2-foot high wire fence around the feeder, maybe at a radius of about 6 feet. The time it takes the predator to jump the fence gives the birds enough advanced warning to get out of the way.

**Offer variety.** Some birds like tube-style and hopper feeders. Others that prefer feeding on the ground can learn to use a shelf feeder. Consider nyjer thistle, which is expensive, but use a special feeder for it designed with smaller seed ports or ports that are below the perches, something goldfinches and chickadees can handle but others can't. Add a suet or seed cake. It may help draw in woodpeckers and chickadees. Offer peanuts and you may get blue jays – and squirrels.

**Don't clean up your flowerbeds in the fall.** The seed-eating birds attracted to your feeders will enjoy the seed heads. Plus, tree leaves, while providing mulch, may also provide a variety of eggs of insects (many beneficial) that the birds enjoy picking over.

**On a frigid day, have open water in a birdbath.** It is almost more attractive than food. Find some kind of shallow bowl, preferably with sloping sides, which won't break if the water freezes. It should be easy to bring in the house to thaw out. Or get an electric heater designed for birdbaths or dog water dishes.

For more detailed feeding information, go to my archives at www.CheyenneBirdBanter.wordpress.com. Look for "Bird feeding" in the list of topics.

**Study your visitors.** From your feeder-watching window, scan your trees and shrubs and garden beds to see if you can get a glimpse of more than house finches and house sparrows, especially in the spring. Of the 85 species I've seen in or above our yard, I've recorded 27 from November through March, prime feeder season.

**Share your bird sightings** at www.eBird.org, or for $18, this winter you can take part in Project FeederWatch, www.feederwatch.org. It isn't too late to sign up. You get a nifty bird calendar poster and a handbook. Even if you don't participate, the website is full of information about bird feeding and feeder birds.

**Have fun.** However, if you find it isn't fun, take down the feeders. Reduce your stress by going for a walk and enjoy the birds along the way.

# 2016

**377** Sunday, January 3, 2016, Outdoors, page E4
**New camera technology can help birders get perfect shot**

I have often wished the view through my binoculars could become a photograph of that colorful warbler high in the tree, the distant hawk or the swimming phalarope.

Then digiscoping was invented – a digital version of trying to take a photo through a scope. The idea is that you don't need a camera with a big lens if you can use your scope instead.

But who wants to carry around the heavy tripod and scope, plus a camera to attach to it? Not me.

But a couple years ago, the Cheyenne Audubon chapter had members Greg Johnson and Robin Kepple give a talk on their birding trip to Australia. The bird photos were fabulous. What camera was used? Canon PowerShot SX50 HS.

The PowerShot series of cameras is really a collection of point and shoots – I have an early one, but it doesn't zoom like the SX50. They all have lots of manual and partially manual ways to adjust speed, aperture and color. You can do a surprising number of things, including macro and video. Some will even connect to your smart phone to transmit photos.

The SX50 (and now there is an SX60 and rumors of an SX70, not to mention Nikon's cameras in this class) is moderately priced, between $300 and $600. That price might get you another lens for a digital single lens reflex camera, the type the professionals use.

And the SX50 weighs only 1 pound 6 ounces, whereas my Brunton 8 x 42 binocs weigh 4 ounces more. A recent publication of the American Birding Association, "Birder's Guide to Gear," features four men who did a photographic Big Day. They could only count bird species they photographed. All of them carried multiple camera bodies and lenses. Imagine the weight.

Among our birding friends, my husband, Mark, was the first to follow Greg and Robin's lead by buying an SX50 in the fall of 2014. By spring of 2015, there were three or four people carrying these cameras on a local field trip. Even our friend, ABA magazine editor Ted Floyd, has one now. It makes his Facebook posts even more entertaining.

Ted mentioned that young birders seem to be forsaking binoculars for these "compact ultra-zooms" as they've been referred to. They have one big advantage over binocs. If you snap a photo of an unusual bird, you can then show it to your birding companions using the 2.8-inch screen on the back, beginning a good half hour or more's discussion of the finer points of feathers.

And it is really handy to have a photo when you submit your field trip checklist to the eBird database where the experts want proof of the rare bird you saw.

Are Mark and his friends practicing the art of photography? I'm not sure. They all seem to be using the camera on the automatic setting. Their goal is to get the bird. They don't worry about whether the background contrasts nicely.

Often, the camera's automatic setting determination is matched by the location's lighting for a really nice shot. Fixing the framing of the subject can be accomplished back on the computer with cropping. Putting the camera on a tripod would probably improve the number of well-focused shots. Though these cameras come with image stabilization technology, it is sorely tested by flighty birds.

Mark has taken 7,500 photos so far. I asked him if he thought about bringing just the camera on birding trips, since it zooms farther than our scope. It's kind of a pain carrying camera, scope and binocs.

No, he said, the camera lacks a wide field of view. It makes it difficult to re-find that speck of distant movement you saw with your naked eye.

With the steady advancement of technology in my lifetime alone, computers have gone from room-sized to hand-sized. Cameras have gone from the wooden box my grandmother used in 1916 to who-knows-what in the next 10 years.

Equally amazing is how birders take the latest technology and use it for learning more about birds. We've learned so much from eBird, for instance – all those observations from birdwatchers being sent in via internet from all over the world.

And how about geolocators? Attached to birds, they allow scientists to track them during migration.

But let's not forget the thrill of photography itself.

Like artist John James Audubon, you can see the bird you shot today displayed for posterity on your computer screen.

# 378 2 bird books suited for winter reading

Sunday, January 31, 2016, Outdoors, page E3

"Owls of North America and the Caribbean," by Scott Weidensaul. Part of the Peterson Reference Guide Series published by Houghton Mifflin Harcourt, hardcover, 333 pages, $40.

"The Living Bird, 100 Years of Listening to Nature," by The Cornell Lab of Ornithology and Gerrit Vyn, photographer. Published by Mountaineers Books, hardcover, 200 pages, $29.95.

Two bird books of note were released last fall because they would make perfect holiday gifts, and now I've finally read them.

"Owls of North America and the Caribbean" is by Scott Weidensaul, whose previous book about migration, "Living on the Wind," was nominated for a Pulitzer Prize.

This book concentrates on a group of birds that he has spent nearly 20 years researching. Surprisingly, there are 39 owl species to write about. Half occur south of the U.S., in Mexico and the Caribbean. Twelve of the northern owl species are regularly seen in Wyoming.

The accounts of the Caribbean owls are each only three to four pages long since little is known about them. Our familiar, comparatively well-studied owls have 12-15 pages each.

This is not a field guide, though lavishly illustrated with wonderful photos. Instead, each account sums up what is known about a species: length, wingspan, weight, longevity, range map, systematics, taxonomy, etymology (how it got its name), distribution by age and season, description and identification by age, vocalization, habitat and niche, nesting and breeding, behavior and conservation status.

In a reference like Birds of North America Online, this information is reduced to tedious technical shorthand, but Weidensaul makes it readable, injecting his experience and opinion. Of the snowy owl's description, he says, "If you can't identify this owl, you aren't trying."

I'll admit, I haven't read this book cover to cover yet. I looked up the owls I'm most familiar with, learning new information, and now I'm curious about the others.

One drawback: there is a reference map naming the states of Mexico but not the states of the U.S. or the provinces of Canada – or the Caribbean countries.

However, all the information presented makes owls more intriguing than ever.

The second book, "The Living Bird," is a joint project of the Cornell Lab of Ornithology and photographer Gerrit Vyn to mark the 100th anniversary of the lab.

In a 10- by 11-inch format, there is plenty of room for Vyn's art, which forms the heart of the book. Some birds portrayed life-sized practically step off the page. All of the 250-plus photos are available as individual prints through www.gerritvynphoto.com.

It would be easy to ignore the text in this book, except for the name recognition of the contributing authors.

If you missed CLO director John Fitzpatrick's inspiring presentation at Cheyenne – High Plains Audubon Society's 40th anniversary banquet in 2014, you can read it here as the introduction, "How Birds Can Save the World." The age-old human attraction to colorful creatures that fly makes us notice bird reactions to environmental degradation. And in Fitzpatrick's additional essay, in stories of rehab success, we find that when we help birds, we help ourselves.

The essay by one of my favorite authors, Barbara Kingsolver, hit home. She was a child forced to accompany her parents on birding field trips. But despite her best efforts to rebel, birds have come to be important to her, as I hope they have to my children who attended many bird events in their early lives.

Scott Weidensaul also has an essay, more of a golly-gee-whiz list of cool things you might not know about birds (including some I didn't), titled "The Secret Lives of Birds."

The other major essayists are Lyanda Lynn Haupt, a naturalist and author who examines how birds inspire us, and Jared Diamond, an ardent birdwatcher who is famous as the geographer and author who wrote "Guns, Germs and Steel: The Fates of Human Societies." He projects what coming decades will hold for birds, after decades of population declines.

There are also three essays written by Vyn about his exhilarating photographic expeditions and three short profiles include a citizen scientist, a researcher, and an audio recordist. CLO is known for its extensive library of recorded bird songs and other sounds of nature, thus the second part of the book's title, "100 Years of Listening to Nature."

After perusing the photos, I could still barely concentrate on the text. But afterwards I enjoyed the photos again and the photo captions. Written by Sandi Doughton, they add insight. The photos themselves are laid out in a thoughtful, coherent way.

Altogether, this is a book to enjoy, a book to inspire, and maybe it is even a book to cause you to take action.

Read now, before spring migration, when you abandon books for binoculars.

# 379 UW songbird brain studies shed light

Sunday, February 28, 2016, Outdoors, page E2

We are used to thinking about many animals standing in for humans in studies that will benefit us: rats, chimps, rabbits. But should we add songbirds to that list? They apparently work well for studying how we learn to speak.

At the February Cheyenne Audubon meeting, Karagh Murphy, a University of Wyoming doctoral candidate in the Zoology and Physiology Department, explained how Bengalese finches help her study how brains learn.

Learning by example, whether bird or human, takes place in two parts. First the student observes, or in the case of male birds learning to sing so they can defend their territory and attract mates, they listen. Then they attempt imitation, practicing by listening to themselves and getting feedback.

What Karagh wanted to know is if Higher Vocal Center neurons in the birds' brains are active at both stages – hearing and doing. It's just a simple matter of plugging a computer into the right place in a bird's brain.

First though, you have to wrangle your subjects, capturing them in the walk-in-sized aviary, then over time, get them used to having the wispiest of cables attached to the tiny instrument on their heads. Otherwise, they are too stressed to sing.

Karagh recorded the firing pattern of the HVC neurons, producing something like the electrocardiogram that shows heart beats and compared it to the spectrogram, another linear graph of peaks and valleys that visualizes the frequencies of the song she played for the bird to hear, and then the song the bird sang. Both spectrograms matched the peaks and valleys of the HVC neuron pattern, essentially showing the neurons are used for both auditory and motor output – the action of singing.

Recently, something very similar has been found in humans, called mirror neurons.

The second speaker was Jonathan Prather, an associate professor in the department's neuroscience program. While Karagh has been studying males learning to sing, Jonathan has been figuring out what the female Bengalese finches want to hear.

Female birds don't sing. At most, they produce call notes to communicate. But they enjoy listening to males sing, and they judge them by their song to determine which one is the fittest potential mate, which will give them the fittest young.

Jonathan thought there might be a "sexy syllable," some part of the song that would get the females excited, measured by how often the females call in response. He measured their responses as he played back songs he had manipulated.

Or maybe it was tempo, so he manipulated the recording to go faster in some trials, then slower in others. Or maybe the female birds would react differently to songs at different pitches. That would be similar to human women who, studies have shown, are attracted to men with deeper voices (connected to higher testosterone levels).

Apparently, female finches are looking for quantity and complexity, for males who sing in the most physically (neuromuscular-wise) demanding way.

That means sweeping from high to low notes a lot, and really fast. Think how opera stars singing the most demanding repertoire get the biggest applause. A bird that can sing well is well-fed, healthy and of good breeding – perfect father material.

The field of neurobiology is more about figuring out human brains, but when birds are used as models, birdwatchers find it intriguing. The questions from the Audubon audience reflected their familiarity with birds.

Our songbirds in Wyoming are only seasonal singers, so birds from equatorial locations that sing year round are used to make trials more efficient. Would there be a difference?

Are female bird brains different from the male brains? Yes, because learning songs increases one part of the male brain, however, females have other roles that increase the size of other parts of their brain.

If a young bird never hears another bird sing, will it eventually sing? Not really, it will only babble in an unformed way, as human babies do when they start out.

If a young bird hears only the singing of a different species, will it learn that song instead? Yes, although not completely perfectly – there is some genetic influence on bird song.

And what about the mimics? What about birds like starlings and mockingbirds that learn to imitate lots of other birds' songs and even some human vocalizations and mechanical noises? Karagh broke out in a grin. That line of study could keep her busy for her entire career.

# 380 Ecotourists enjoy Texas border birds

Sunday, April 3, 2016, Outdoors, page E2

At the beginning of March, Mark and I indulged in five days of ecotourism in South Texas after visiting our son and his wife in Houston.

We met up with avid birders for another Reader Rendezvous put on by the Bird Watcher's Digest magazine staff. Last year, we met them in Florida.

I'd heard about the fall Rio Grande Valley Birding Festival in Harlingen, Texas, but had no idea how well bird-organized the entire lower Rio Grande Valley is until a woman from the Convention and Visitors Bureau spoke to us.

I was expecting McAllen, Texas – where we stayed – to be a small town in the middle of nowhere, but its population is 140,000 in a metropolitan area of 800,000, with a lot of high-end retail businesses attracting shoppers from Mexico.

Outside of the urban and suburban areas, nearly every acre is farmed. But in the 1940s, two national wildlife refuges were set aside and another in 1979, as well as a number of state parks. This southern-most point of Texas is an intersection of four habitat types and their birds: desert, tropical, coastal and prairie, and it is a funnel for two major migratory flyways.

Local birding guide, Roy Rodriguez, has compiled a list of 528 bird species (we have only 326 for Cheyenne), including 150 accidentals seen rarely – though our group saw two, northern jacana and blue bunting.

Many of Roy's common species that we saw are South Texas specialties like plain chachalaca, green parakeet and green jay. We also saw uncommon Texas specialties including white-tailed hawk, ringed kingfisher and Altamira and Audubon's orioles.

From the rare list, some of the species we saw were ferruginous pygmy-owl, aplomado falcon and red-crowned parrot. Interestingly, several Texas rarities we saw are not rare in Wyoming: cinnamon teal, merlin and cedar waxwing.

Most of the Texas specialties have extensive ranges in Mexico. Thus, a species can be rare in a particular location, or just plain rare like the whooping cranes Mark and I saw further east on the Gulf Coast at Aransas National Wildlife Refuge.

What is rare is the co-operative effort shown by nine entities to establish the World Birding Centers, www.theworldbirdingcenter. com, including four city parks, three state parks, a state wildlife management area and a national wildlife refuge. Another partnership has produced a map of the five-county area which locates and describes those and 76 additional public birding sites. The map is helpful even if you are proficient using www.eBird.org to check for the latest sightings.

Wyoming will be coming out with something similar soon, the Great Wyoming Birding Trail map app.

The concentration of birds in south Texas draws people from around the world. We saw the natural open spaces drawing local families, too. But it's the visitors who spend money which the McAllen convention and visitor's bureau counts. They estimate bird-related business is the third biggest part of their economy, after shopping and "winter Texans."

Roy said birdwatchers contributed $1 million to the economy when a rare black-headed nightingale-thrush spent five months in Pharr, Texas, and $700,000 in just a few weeks while a bare-throated tiger-heron could be seen.

The International Ecotourism Society says ecotourism is "responsible travel to natural areas that conserves the environment, sustains the well-being of the local people, and involves interpretation and education."

We mostly think in terms of ecotourists going to third world countries, but it applies here in the U.S. as well.

"Ecotourism is about uniting conservation, communities, and sustainable travel," continues the description. At each of the seven locations the Reader Rendezvous visited, staff or volunteers gave us historical and conservation background. And each location is managed by conservation principles. I'm not sure about the sustainable travel aspect, though we did travel by van and bus, minimizing fuel and maximizing fun.

Short of staying home, travel will not be sustainable until modes of transportation have clean fuel, and restaurants and hotels are more conservation-minded. But experiencing and building understanding of other places and cultures is worthwhile. At Anzalduas County Park, we stood on the edge of the Rio Grande, looking across at a Mexican park close enough to wave. If a bird flew more than halfway across the river, would we have to document it for eBird as being in Mexico? Is there any place to tally the number of Border Patrol trucks, blimps and helicopters we saw at that park?

Besides a few extra pounds from enjoying the always enormous and delicious portions of Texan and Mexican food, I brought home other souvenirs as well: a list of 154 species, 37 of them life birds for me (at least on eBird), photos, great memories and new birding friendships. Now we're back in time to welcome the avian "winter Texans" to Cheyenne as they migrate north.

# 381 Big Day Bird Count results affected by cold weather

Sunday, May 22, 2016, Outdoors, page E3

The 2016 Cheyenne Big Day Bird Count was held May 14. It was cold (33-43 degrees F), wet and foggy. Conditions kept down the number of birdwatchers participating as well as the number of birds observed.

Thirteen Cheyenne-High Plains Audubon Society members and friends birded as a group at Lions Park, Wyoming Hereford Ranch and the High Plains Grasslands Research Station. Seven others birded on their own and contributed to the total of 107 species observed. Last year's total was 110 species.

Only a few flycatchers, vireos and warblers were seen because few insects, their primary food, were around due to the cold. Few kinds of shorebirds were seen at area reservoirs. High water levels from previous rain and snowfall left few areas of shallow water and exposed sandbars for them.

Although many of the species that migrate through

Cheyenne were seen, the day may not have represented quite the peak of spring migration.

A highlight of the count was a black-and-white warbler at the research station. It is considered an eastern

warbler, rarely seen this far west, although it does appear next in the Black Hills.

The Cheyenne Big Day

ran concurrent with the Global Big Day. For a look at global results, see www.eBird.org/globalbigday.

## Cheyenne Big Day Bird Count, May 14, 2016, 107 species total

Compiled by Greg Johnson.

Canada Goose
Gadwall
American Wigeon
Mallard
Blue-winged Teal
Cinnamon Teal
Northern Shoveler
Northern Pintail
Green-winged Teal
Redhead
Ring-necked Duck
Lesser Scaup
Bufflehead
Common Merganser
Ruddy Duck
Pied-billed Grebe
Eared Grebe
Western Grebe
Clark's Grebe
Double-crested Cormorant
American White Pelican
Great Blue Heron
Black-crowned Night-Heron
White-faced Ibis
Turkey Vulture
Osprey

Cooper's Hawk
Broad-winged Hawk
Swainson's Hawk
Red-tailed Hawk
American Coot
American Avocet
Killdeer
Spotted Sandpiper
Willet
Wilson's Snipe
Wilson's Phalarope
Red-necked Phalarope
Bonaparte's Gull
Franklin's Gull
Ring-billed Gull
Forster's Tern
Rock Pigeon
Eurasian Collared-Dove
Mourning Dove
Belted Kingfisher
Downy Woodpecker
Northern Flicker
American Kestrel
Prairie Falcon
Western Wood-Pewee
Least Flycatcher
Western Kingbird

Eastern Kingbird
Loggerhead Shrike
Blue Jay
Black-billed Magpie
American Crow
Common Raven
Tree Swallow
N. Rough-winged Swallow
Bank Swallow
Cliff Swallow
Barn Swallow
Mountain Chickadee
Red-breasted Nuthatch
House Wren
Blue-gray Gnatcatcher
Ruby-crowned Kinglet
Eastern Bluebird
Mountain Bluebird
Veery
Swainson's Thrush
Hermit Thrush
American Robin
Brown Thrasher
European Starling
Black-and-white Warbler
Orange-crowned Warbler
Yellow Warbler

Blackpoll Warbler
Palm Warbler
Yellow-rumped Warbler
Green-tailed Towhee
Chipping Sparrow
Clay-colored Sparrow
Vesper Sparrow
Lark Sparrow
Lark Bunting
Song Sparrow
Lincoln's Sparrow
White-crowned Sparrow
Western Tanager
Rose-breasted Grosbeak
Black-headed Grosbeak
Red-winged Blackbird
Western Meadowlark
Yellow-headed Blackbird
Common Grackle
Great-tailed Grackle
Brown-headed Cowbird
Bullock's Oriole
House Finch
Pine Siskin
Lesser Goldfinch
American Goldfinch
House Sparrow

# 382 Bird count day gives us big picture

Sunday, May 22, 2016, Opinion, page C9

May Hanesworth was ahead of her time. An active Cheyenne birder as early as the 1940s, she made sure the results of the local spring bird counts were published every year in the Cheyenne paper. She recruited me in the 1990s to type the lists for her. She felt that someday there would be a place for that data, and she was right.

A few years ago, members of the Cheyenne-High Plains Audubon Society collected and uploaded that data to eBird.org, a global database for bird observations. The oldest record we found

was for 1956.

We refer to the count we make at the height of spring migration as the Big Day Bird Count. Elsewhere in the world, competitive birders will, as a small team or solo, do a big day to see how many species they can find in a specified area. But the idea of a group of unlimited size like ours going out and scouring an area is unusual, though closer to what the originator, Lynds Jones, an Oberlin College ornithology professor, had in mind back in 1895.

Now eBird has started a new tradition as of last

year, the Global Big Day. This year, it was scheduled for May 14, the same day as ours. Results show 15,642 people around the world saw 6,227 bird species. For our local count, 20 people looked for birds around Cheyenne, and 107 species were counted (see the results on page E3 of today's edition).

Finding our favorite birds in the company of friends is a good incentive for taking part, but there is the science too. Back in the spring of 1956, May saw 85 species. And when Mark and I started in the 1990s, 150 seemed

to be the norm – perhaps because Cheyenne had more trees by then. However, the last 10 years, the average is lower, 118.

Maybe we aren't as sharp as earlier birders. Or we are missing the peak of migration. Or we have lost prime habitat for migrating birds as the surrounding prairie gets built over and elderly trees are removed in town. Or it's caused by deteriorating habitat in southern wintering grounds or northern breeding grounds.

But imagine where we would be without the

Migratory Bird Treaty.

This year marks the 100th anniversary of the first agreement in 1916, between the U.S. and Great Britain (signing for Canada), followed by other agreements and updates. In summary: "It is illegal to take, possess, import, export, transport, sell, purchase, barter, offer for sale, purchase or barter any migratory bird, or parts, nests or eggs."

Even migrating songbirds, like our Wyoming state bird, the western meadowlark, are protected.

But who would want to hurt a meadowlark?

Look at the Mediterranean flyway. Birdlife International reports 25 million birds of all kinds along that flyway are shot or trapped every year for fun, food and the cage bird trade. Perpetrators think the supply of birds is endless. But we can point to the millions of passenger pigeons in North America prior to the death of the last one in 1914, to show what can happen.

The city of Eliat, Israel, is the funnel between Africa and Europe/Asia on the Mediterranean flyway, and to bring attention to the slaughter, the annual Champions of the Flyway bird race is based there. A Big Day event, this year it attracted 40 teams, Israeli and international, which counted a combined total of 243 species during 24 hours.

This year, funds raised by the teams are going to Greece to support education and enforcement – killing migratory birds is already illegal. Some of the worst-hit areas are in forests above beaches popular with tourists. Attracting birdwatching tourists could pay better than killing and trapping birds, a kind of change that has been beneficial elsewhere.

Many factors affect how many birds we see in Cheyenne on our big day, but we do have control over one aspect: habitat. If you live in the city, plant more trees and shrubs in appropriate places. If you live on acreage, protect the prairie and its ground-nesting grassland birds. And then join us on future Cheyenne Big Day Bird Counts and contribute to the global big picture of birds.

# 383

Sunday, May 29, 2016, Outdoors, page E3

## Following individual birds brings new insights

There's more to birdwatching than counting birds or adding species to your life list. The best part of birdwatching is watching individual birds, observing what they are doing.

Thank goodness it isn't rude to stare at them.

While some species may skulk in the undergrowth, most of our local birds are easily seen, even from our windows.

Every morning, I check the view out the bathroom window, and often there's a Eurasian collared-dove sitting in the tall, solitary tree two yards down. By March, I was seeing collared-dove acrobatics. The males, like this one, like to lift off from their high perches and soar in a downward spiral. I'm not sure what that proves to the females, but one of them has taken up with him.

I saw them getting chummy one day, standing together on the near neighbor's chimney cover. I can imagine their cooing reverberates into the house below. Then they kept taking turns disappearing into the upright junipers where last year they, or another pair, had a nest.

But one day I caught sight of a calico cat climbing the juniper. The branches are just thick enough that I couldn't see if the cat found eggs. Eventually she jumped out onto the neighbor's roof and sauntered across to an easier route down to ground level.

More than a month later, I have not seen the calico here again but have seen a collared-dove disappearing into the juniper once more. I'll have to watch for more activity.

If I were authors Bernd Heinrich or Julie Zickefoose, I would be making notes, complete with date, time and sketches. I would be able to go back and check my notes from last year and see if the birds are on schedule. I might climb up and look for a nest. And I might do a thorough survey of the academic literature to find out if anyone has studied the effects of loose cats on collared-dove populations.

However, most of us have other obligations keeping us from indulging in intense bird study, and we don't sketch very well either.

But Heinrich and Zickefoose do. Heinrich is liable to climb a tree (and he's no spring chicken) or follow a flock of chickadees through the forest near his cabin in Maine. Zickefoose, who has a license to rehab birds at her Ohio home, can legally hold a bluebird in her hand.

Both have new books out this spring which allow us to look over their shoulders as they explore their own backyards.

Heinrich is known for his books exploring many aspects of natural history (my most recent review was of "Life Everlasting"). His new one, "One Wild Bird at a Time: Portraits of Individual Lives," has 17 birds, one chapter at a time, in a loose seasonal arrangement. He has also portrayed each species in watercolor, directly from sketches he's made in the field. This is sometimes as close as his own bedroom where he was able to rig a blind when flickers drilled through his cabin siding and nested between the outer and inner walls.

Though Heinrich is professor emeritus, his writing style is pure, readable storytelling.

Zickefoose's goal in her new book, "Baby Birds: An Artist Looks into the Nest," is also somewhat encyclopedic. From the woodland surrounding her home, she was able to document nestling development for 17 species. Finding a songbird nest, she would remove a nestling every day to quickly sketch it in watercolors, feed it and return it. Her drawings are like full scale time-lapse photography. Don't try this

2016

at home unless you are a licensed bird rehabber.

Although she has handled lots of birds in the course of her work, following individual nestlings gave Zickefoose an insight into how those of different species grow at different rates – ground nesters are the fastest.

Either of these books can serve as inspiration for becoming a more observant birdwatcher, but they are also great storytelling, with the benefit that the stories are true and full of intriguing new information.

If you find a nest this spring, consider documenting it for science. See www.nestwatch.org. The site's information includes lots of related information, including plans for building nest boxes.

"One Wild Bird at a Time:

Portraits of Individual Lives" by Bernd Heinrich, c. 2016, published by Houghton Mifflin Harcourt.

"Baby Birds: An Artist Looks into the Nest" by Julie Zickefoose, c. 2016, Houghton Mifflin Harcourt.

# 384 New bird singing, maybe breeding

Sunday, June 19, 2016, Outdoors, page E3

There is a new bird on our block. It's a loud bird. That's how I know it is here, even though it is tiny – only 4.25 inches long – and prefers to hang out unseen around the tops of mature spruce trees while gleaning insects and spiders.

The ruby-crowned kinglet, despite its name, is not a brightly colored bird. It is mostly an olive-gray-green, with one white wing-bar. Only the male has the red crown patch, and he may show it when singing, but the red feathers really stand up like a clown's fright wig when he's around other male ruby-crowneds.

We get a variety of small migrating songbirds in our Cheyenne yard. In May, we had a lazuli bunting, pine siskin, clay-colored sparrow and even our first ever yellow-breasted chat.

This isn't the first time for a ruby-crowned kinglet in our yard. I recorded one at www.eBird.org on April 25, 2012, and another April 24, 2015. They are usually on their way to the mountains to nest in the coniferous forest of spruce, pine and fir trees.

The difference this year is that beginning May 8, I've been hearing one every day. My hopes are up. Maybe it is

going to nest. My neighborhood has the requisite mature spruce trees.

I talked to Bob Dorn May 27, but he thought that it was still too early to suspect breeding. They might have been waiting out cold spring weather before heading to the mountains. Bob is the co-author of "Wyoming Birds" with his wife, Jane Dorn. Their map for the ruby-crowned kinglet shows an "R" for the Cheyenne area, "Resident" – observed in winter and summer with breeding confirmed.

The Dorns' breeding record is from the cemetery, where they saw kinglet nestlings being fed July 18, 1993. They also suspected breeding was taking place at the High Plains Grasslands Research Station just west of the city June 2, 1989 and June 15, 1990.

For more recent summer observations that could indicate breeding in Cheyenne, I looked at eBird, finding three records between July 3 and July 7 in the last five years, including Lions Park. There were also a couple late June observations at the Wyoming Hereford Ranch and Lions Park in 2014.

I first learned the ruby-crowned kinglet's

distinctive song in Wyoming's mountains. You've probably heard it too. Listen at www.allaboutbirds.org. It has two parts, starting with three hard-to-hear notes, "tee-tee-tee", as ornithologist C.A. Bent explained it in the 1940s, followed by five or six lower "tu or "tur" notes. The second half is the loudest, and sometimes given alone, "tee-da-leet, tee-da-leet, te-da-leet."

Those who have studied the song say it can be heard for more than half a mile. The females sing a version during incubation and when nestlings are young. The males can sing while gleaning insects from trees and while eating them. Neighboring kinglets have distinctive signature second halves of the song, and males can apparently establish their territories well enough by singing that they can avoid physical border skirmishes.

Actual nesting behavior is not well documented because it is hard to find an open cup nest that measures only 4 inches wide by 5 to 6 inches deep when it is camouflaged in moss, feathers, lichens, spider webs, and pieces of bark, twigs and rootlets – and located 40 feet up a spruce tree.

The female kinglet builds the nest in five days, lining it with more feathers, plant down, fine grass, lichens and fur. She may lay as many as eight eggs. The nest stretches as both parents feed the growing nestlings tiny caterpillars, crickets, moths, butterflies and ant pupae.

Ruby-crowned kinglets winter in the Pacific coast states and southern states but breed throughout the Rockies and Black Hills and in a swath from Maine to Alaska. If my neighborhood kinglet stays to breed, it will be one more data point expanding the breeding range further out onto the prairie.

While kinglets are not picky about habitat during migration, for breeding they demand mature spruce-fir or similar forest. Some communities of kinglets decrease in the wake of beetle epidemics, salvage logging and fires. However, the 2016 State of the Birds report, www.2016stateoftheBirds.org, shows them in good shape overall – scoring a 6 on a scale from 4 to 20. High scores would indicate trouble due to small or downward trending population, or threats to the species and its habitats during breeding and non-breeding seasons.

As of June 19, the kinglet is still singing – all day long. If it is nesting on my block this summer, I must thank the residents who planted spruce trees here 50 years ago. What a nice legacy.

We should plant some more.

# 385 Kids explore nature of the Belvoir Ranch

Sunday, July 17, 2016, Outdoors, page E2

I was delighted to recognize my neighbor at the Belvoir Ranch Bioblitz last month. She is going to be a senior at Cheyenne East High in the fall and was there with two friends. All three were planning to spend the weekend looking for birds, mammals, herps (reptiles and amphibians), pollinators, macroinvertebrates and plants to fulfill hours required for their Congressional Award gold medals.

The weekend could have served for all four award areas: volunteer public service (we were all volunteer citizen scientists collecting data), personal development (the staff taught us a lot of new things), physical fitness (hiking up and down Lone Tree Creek in the heat was arduous), and expedition/exploration (many of us, including my neighbor and her friends, camped out and cooked meals despite being only 20 miles from Cheyenne).

Mark and I have attended other bioblitzes around the state, but this was the first one close to Cheyenne. With all the publicity from the four sponsoring groups, Audubon Rockies, The Nature Conservancy, University of Wyoming Biodiversity Institute and the Wyoming Geographic Alliance, a record 100 people attended, plus the staff of 50 from various natural science disciplines.

When I asked my neighbor why she and her friends had come, she said, "We're science nerds." That was exciting to hear.

There were a lot of junior science nerds in attendance with their families. Small children enjoyed wading into the pond along the creek to scoop up dragonfly and damselfly larvae – and even crayfish.

A surprising number of children were up at 6 a.m. Saturday for the bird survey. The highlight was the raven nest in a crevice on the canyon wall, with three young ravens crowding the opening, ready to fledge.

Sunday morning's bird mist netting along the creek was very popular as well. Several birds that had been hard to see with binoculars were suddenly in hand.

Because it wasn't at an official bird banding site, the mist netting was strictly educational and the birds were soon released. Several young children had the opportunity to hold a bird and release it, feeling how light it was, how fast its heart beat and feeling the little whoosh of air as it took flight. What I wouldn't give to know if any of the children grow up to be bird biologists or birdwatchers.

The Belvoir Ranch is owned by the city of Cheyenne and stretches miles to the west between Interstate 80 and the Colorado-Wyoming state line. The city bought it in 2003 and 2005 to protect our upstream aquifer, or groundwater, as well as the surface water.

While limited grazing and hunting continues as it did under private ownership, other parts of the master plan have yet to come to fruition, including construction of a wind farm, landfill, golf course or general recreation development. It is normally closed to the public. However, progress is being made on trails to connect the ranch to Colorado's Soapstone Prairie Natural Area and/or Red Mountain Open Space.

A good landowner takes stock of his property. The city has some idea of what's out there, including archeological sites. But with budgets tightening, there won't be funding to hire consultants for a closer look. But there are a lot of citizen scientists available.

The data from the Bioblitz weekend went into the Wyobio database, www.wyobio. org, a place where data from all over Wyoming can be entered. The bird data also went into eBird.org.

The data began to paint a picture of the Belvoir: 62 species of animals including 50 birds, eight mammals, four herps, plus 13 taxa of macroinvertebrates (not easily identified to species) and 12 taxa of pollinators (bees and other insects), plus many species of plants. All that diversity was from exploring half a mile of one creek within the ranch's total 18,800 acres – about 30 square miles.

The members of the City Council who approved the ranch purchase are to be congratulated on making it public land in addition to protecting our watershed. Sometimes we don't have to wait for the federal and state governments to do the right thing. The essence of Wyoming is its big natural landscapes, and we are lucky to have one on the west edge of Wyoming's largest city.

Let's also congratulate the parents who encouraged their children to examine the critters in the muddy pond and pick up mammal scat (while wearing plastic gloves) on the trails among other activities.

Someday, these kids will grow up to be like my high school neighbor and her friends. Someday they could be the graduate students, professors and land use professionals. No matter what they become, they can always contribute scientific data by being citizen scientists.

2016

# 386 Pondering how much eagles can take in life

Sunday, August 14, 2016, Outdoors, page E3

Just when we thought eagles were safe (bald eagles were taken off the threatened and endangered species list in 2007) we discover that golden eagle numbers are still down. And there are plans to build a massive wind farm in Wyoming which will take the lives of both bald and golden eagles.

I should have written a column about this earlier so you could send your comments to the U.S. Fish and Wildlife Service, but I was sidetracked by spring migration.

However, staff at Audubon Rockies and their counterparts at the Natural Resources Defense Council and the Wilderness Society have written extensive comments backed by science and experience.

The Power Company of Wyoming (PCW) is developing the 1,000-turbine Chokecherry/Sierra Madre Wind Energy Project. It is located on 500 square miles in Carbon County, south-central Wyoming, where there is some of the best wind in the country. It will be the largest onshore wind farm in the U.S.

PCW is working with the Fish and Wildlife Service on improving the locations for turbines and has reduced the projected take to 10-14 golden eagles and 1-2 bald eagles per year. The definition of "take" is death incidental to industry activities.

The projected take numbers also account for the eagles that will live because PCW will retrofit 1,500-3,000 power poles per year for eagle safety. Eagles' large wingspans can cause their electrocution when they perch on poles, and then they touch two electrical hotspots at the same time and cause a completed circuit.

PCW is applying for an eagle take permit for the first half of the development. It is voluntary but good insurance. PCW saw a competitor without a permit get hit with a $1 million fine for killing eagles.

The Fish and Wildlife Service is in the process of updating the eagle take rule. It will probably apply to the second half of PCW's development, the other 500 turbines.

The update would give wind power companies across the country 30-year permits, to cover the expected lifespan of a windfarm, rather than the current five years. However, Fish and Wildlife proposes a review every five years.

Audubon Rockies concedes that PCW needs some assurance that they can operate for a longer length of time that will make the investment worthwhile – they can't get investors if there is a possibility eagle deaths will shut down part of the development after the first five years.

However, there are

concerns. In the proposed rule update, any monitoring done by the company would be considered proprietary and not be required to be available to public scrutiny. Audubon feels more transparency is needed on what is happening with our eagles.

And there needs to be more flexibility to manage the windfarm/eagle interactions as more eagle research is done. We don't know yet how eagles will deal with the Chokecherry/Sierra Madre development. It's not just the spinning windmill blades, the tips of which can travel 150 mph. Eagles also collide with the transmission lines and towers.

Because it takes eagles five years to reach sexual maturity, we know their populations can't quickly bounce back like rabbits.

The site of PCW's Chokecherry/Sierra Madre wind project is gorgeous. The thought of developing it is heartbreaking. But the company has done a lot of work and spent a lot of money studying the wildlife problems. They deserve clear answers from the federal government on what they can and can't do.

Eagles are just one of the items addressed in the draft environmental impact statement for the wind farm. Other wildlife, including bats and songbirds, are affected too.

By the end of the year,

we will find out how Fish and Wildlife will react to public comments, not only on Chokecherry/Sierra Madre but also the proposed update of the eagle take rule.

Does clean energy have to come down to this? Do we have to fill Wyoming's open spaces (they are not empty spaces) with industrial clutter? Why didn't the coal companies spend millions on cleaner power plant emissions research instead of on litigation at every turn?

Why does alternative energy, specifically wind and solar, have to follow the old centralized, mega-production model? I still think disbursed ["distributed" is the frequently used term] power production would be better, safer – less of a target for troublemakers.

In comparison, look at how Mother Nature spreads oxygen-producing plants everywhere. Even where natural or man-made catastrophes have stripped the vegetation, it doesn't take long for another little oxygen-producing factory to take hold.

Plus, wouldn't you like to park in the shade of a solar panel while shopping at the mall? Adding solar panels to our rooftops and choosing energy efficient appliances will not only cut our personal utility bills, but in a way, save eagles in the future.

# 387 Collaboration could keep eagles safe

Sunday, September 4, 2016, Outdoors, page E3

Last month, while researching the wind energy/eagle issue, I learned about new technology that could help eagles survive encounters with wind farms.

IdentiFlight uses stereoscopic cameras to detect and identify eagles in flight far enough out to shut down a turbine, preventing a deadly collision.

The idea that cameras hooked up to a computer can learn to "see" eagles, using machine vision technology, is as remarkable as the collaboration behind it.

It starts with Renewable Energy Systems, started in 1982, and now a global company in the business of designing and installing as well as developing wind energy projects.

I spoke with Tom Hiester, vice president of strategy for RES Americas, whose office is in Broomfield, Colorado.

He said RES is funding the development of IdentiFlight and will own the rights to the technology and sell equipment. Other wind companies concerned with avoiding the fines for killing eagles will be the customers.

RES is working with Boulder Imaging, a Boulder, Colorado, tech company specializing in industrial precision applications.

Initial testing of the IdentiFlight system was done through the U.S. Department of Energy's National Renewable Energy Laboratory. Its testing facility, the National Wind Technology Center, is south of Boulder on 300 acres up against the foothills where the wind can be ferocious. Companies, universities and government agencies come to test their turbines for reliability and performance.

Machine vision requires training the computer. In this case, it needed to see how real eagles fly. A golden eagle and a bald eagle were brought in from the Southeastern Raptor Center where birds of prey are rehabilitated. They also happened to be the mascots for Auburn University, Auburn, Alabama. You can see a video at www.energy.gov/eagles.

Hiester told me they have found that eagles are more susceptible to collisions when hunting. Their heads are down, eyes concentrating on the ground. Machine vision has to identify a moving object as an eagle at 1,000 meters to give the appropriate turbine the 30 seconds needed to shut down.

This summer, IdentiFlight is getting tested by a third party selected by the American Wind Wildlife Institute. AWWI was organized about eight years ago. Half its partners are a who's who of wind energy companies. The other half are national environmental organizations such as Audubon and the National Wildlife Federation, as well as wildlife managers represented by the Association of Fish and Wildlife Agencies and scientists represented by the Union of Concerned Scientists.

One of AWWI's interests is minimizing eagle deaths. They expect to publish and share what they learn. Besides detecting and deterring eagles from wind turbine collisions, they are also looking at lead abatement (lead shot in carcasses left by hunters will poison eagles because eagles often eat dead animals), reducing vehicle strikes (by removing dead animals along roads), and improving the habitat of eagle prey species.

AWWI science advisors include Dale Strickland of Cheyenne. His environmental consulting firm, Western EcoSystems Technology, has studied wind and wildlife interactions across the country for years.

AWWI selected the Peregrine Fund to conduct the testing. The Peregrine Fund, established in Idaho in 1970 to protect and reestablish peregrine falcon populations, also works now with other raptors around the world.

The test site is Duke Energy's Top of the World wind farm outside Casper. In general, Wyoming has more eagles than other states, and some of our topographic features that cause strong wind also concentrate eagles.

For the test, IdentiFlight cameras have been set up on a tower with a 360-degree view. When motion is identified as an eagle, and velocity and proximity figured, human researchers in an observation tower confirm it. In the future, the system would be totally automated, and the identification of an eagle would trigger the shutdown of the turbine in the eagle's path. IdentiFlight can also be used to survey for eagles on prospective wind sites.

Hiester said the number of eagles actually killed by wind turbines is minor. There are more deaths from other causes. But, as more and more wind projects are built, that could change, especially in Wyoming where there is a lot of wind and a lot of eagles.

Most other bird species flying through wind farms don't have the federal protections that eagles do. IdentiFlight won't do much for them unless they fly alongside the eagles. Hiester said that thermal imaging techniques could help identify them and bats.

Hiester has been invited to share the results of this summer's IdentiFlight trials the evening of January 17, 2017, at the meeting of the Cheyenne – High Plains Audubon Society, which is expected to be held at the Laramie County Public Library.

# 388 Cranes are a "gateway bird"

Sunday, October 9, 2016, Outdoors, page E3

I visited the Yampa Valley Crane Festival in Steamboat Springs, Colorado, with my husband, Mark, in early September.

Steamboat is known for world-class skiing, but how does that relate to the greater sandhill crane the festival centers around?

It starts with a couple of skiers. Nancy Merrill, a native of Chicago, and her husband started skiing Steamboat in the late '80s. They became fulltime residents by 2001.

Merrill was already "birdy," as she describes it, by that point. She was even a member of the International Crane Foundation, an organization headquartered in Baraboo, Wisconsin, only three hours from Chicago.

She and her husband wanted to do something for birds in general when they moved to Colorado. They consulted with The Nature Conservancy to see if there was any property TNC would like them to buy and put into a conservation easement. As it turns out, there was a ranch next door to TNC's own Carpenter Ranch property, on the Yampa River.

The previous owner left behind a list of birds seen on the property, but it wasn't until she moved in that Merrill discovered the amount of crane activity, previously unknown, including cranes spending the night in that stretch of the river during migration stop overs, which we observed during the festival.

Cheyenne folks are more familiar with the other subspecies, the lesser sandhill crane, which funnels through central Nebraska in March. It winters in southwestern U.S. and Mexico and breeds in Alaska and Siberia. It averages 41 inches tall.

Greater sandhill cranes, by contrast, stand 46 inches tall, winter in southern New Mexico and breed in the Rockies, including Colorado and on up through western Wyoming to British Columbia. Many come through the Yampa Valley in the fall, fattening up on waste grain in the fields for a few weeks.

In 2012, there was a proposal for a limited crane hunting season in Colorado. Only 14 states, including Wyoming, have seasons. The lack of hunting in 36 states could be due to the cranes' charisma and their almost human characteristics in the way they live in family groups for 10 months after hatching their young. Mates stick together year after year, performing elaborate courtship dances.

Plus, they are slow to reproduce, and we have memories of their dramatic population decline in the early 20th century.

Merrill and her friends from the Steamboat birding club were not going to let hunting happen if they could help it.

Organized as the Colorado Crane Conservation Coalition, they were successful and decided to continue with educating people about the cranes.

Out of the blue, Merrill got a call from George Archibald, founder of the International Crane Foundation, congratulating the CCCC on their work and offering to come and speak, thus instigating the first Yampa Valley Crane Festival in 2012.

Merrill became an ICF board member and consequently has developed contacts resulting in many interesting speakers over the festival's five years thus far. This year included Nyambayar Batbayar, director of the Wildlife Science and Conservation Center of Mongolia and an associate of ICF, and Barry Hartup, ICF veterinarian for whooping cranes.

Festival participants are maybe 40 percent local and 60 percent from out of the valley, from as far away as British Columbia. Merrill said they advertised in Bird Watchers Digest, a national magazine, and through Colorado Public Radio.

It is a small, friendly festival, with a mission to educate. The talks, held at the public library, are all free. A minimal amount charged for taking a shuttle bus at sunrise to see the cranes insures people will show up. [Eighty people thought rising early was worthwhile Friday morning alone.]

This year's activities for children were wildly successful, from learning to call like a crane to a visit from Heather Henson, Jim Henson's daughter, who has designed a wonderful, larger-than-life whooping crane puppet.

There was also a wine and cheese reception at a local gallery featuring crane art and a barbecue put on by the Routt County Cattlewomen. Life-size wooden cut-outs of cranes decorated by local artists were auctioned off.

We opted for the nature hike on Thunderhead Mountain at the Steamboat ski area. Gondola passes good for the whole day had been donated. This was just an example of how the crane festival benefits from a wide variety of supporters providing in-kind services and grants. Steamboat Springs is well-organized for tourism and luckily, crane viewing is best during the shoulder season, between general summer tourism and ski season.

Meanwhile, the CCCC has a new goal. Over the years, grain farming has dropped off, providing less waste grain for cranes. Now farmers and landowners are being encouraged financially to plant for the big birds. It means agriculture, cranes and tourism are supporting each other.

Merrill thinks of the cranes as an ambassador species, gateway to becoming concerned about nature, "The cranes do the work for us, we just harness them," she said.

# 389 Turning citizens into scientists

Sunday, November 13, 2016, Outdoors, page E2

A year after I married my favorite wildlife biologist, he invited me on my first Christmas Bird Count.

It was between minus 25 and minus 13 degrees Fahrenheit that day in southeastern Montana, with snow on the ground. He asked me to take the notes, which meant frequently removing my thick mittens and nearly frostbiting my fingers.

I am happy to report that 33 years later, my husband is the one who takes the notes, and the Christmas Bird Count has become a family tradition, from taking our first son at 8 months old and continuing now with both sons and their wives joining us.

The Christmas Bird Count started in 1900 and is one of the oldest examples of citizen science, sending ordinary people (most are not wildlife biologists) out to collect data for scientific studies.

In 1999, I signed up for the Cornell Lab of Ornithology's Project FeederWatch and have continued each year. Last season, 22,000 people participated. In 2010, I started entering eBird checklists, and now I'm one of 327,000 people taking part since 2002. And there are nearly a dozen other, smaller,

CLO projects.

It is obvious CLO knows how to keep their citizen scientists happy. Part of it is that they have been at it since 1966. Part of it is they know birdwatchers. That's because they are birdwatchers themselves.

How do they keep us happy? I made a list based on my own observations – echoed by an academic paper I read later.

First, I am comfortable collecting the data. The instructions are good. They are similar to something I do already: keeping lists of birds I see. The protocol is just a small addition. For instance, in eBird I need to note when and for how long I birded and at least estimate how many of each species I saw. It makes the data more useful to scientists.

Second, I am not alone. The Christmas Bird Count is definitely a group activity, which makes it easy for novice birders to join us. I especially love the tally party potluck when we gather to share what the different groups have seen that day.

Project FeederWatch is more solitary, but these days, there are social aspects such as sharing photos online. Over President's Day

weekend, when the Great Backyard Bird Count is on, I can see animated maps of data points for each species. On eBird, I can see who has been seeing what at local birding hotspots.

Third, I have access to the data I submitted. Even 33 years later, I can look up my first CBC online and find the list of birds we saw and verify my memories of how cold it was in December 1983.

The eBird website keeps my life list of birds and where I first saw them (OK, I need to rummage around and see if I can verify my pre-2010 species and enter those). It compiles a list of all the birds I've seen in each of my locations over time (89 species from my backyard) and what time of year I've seen them. All my observations are organized and more accessible than if I kept a notebook. And now I can add photos and audio recordings of birds.

A fourth item CLO caters to is the birdwatching community's competitive streak. I can look on eBird and see who has seen the most species in Wyoming or Laramie County during the calendar year, or who has submitted the most checklists. You can choose a particular location,

like your backyard, and compare your species and checklist numbers with other folks in North America, which is instructive and entertaining.

I would take part in the CBC and eBird just because I love an excuse to bird. But the fifth component of a happy citizen scientist is concrete evidence that real scientists are making use of my data. Sometimes multiple years of data are needed, but even reading a little analysis of the current year makes me feel my work was worthwhile and helps me see where my contribution fits in.

What really makes me happy is that I have benefitted from being a citizen scientist. I'm a better birder, a better observer now. I look at things more like a scientist. I appreciate the ebb and flow of nature more.

If you have an interest in birds, I'd be happy to help you sort through your citizen science options. Call or email me, or check my archival website listed below or go to http://www.birds.cornell.edu.

The first Wyoming Citizen Science Conference is being held Dec. 1-3 in Lander. All current and would-be citizen scientists studying birds or any other natural science are welcome. See http://www. wyomingbiodiversity.org.

# 390 Winter raptor marvels, mystery show up in southeast Wyoming

Sunday, December 4, 2016, Outdoors, page E4

Mark and I drove over to join the Laramie Audubon Society on their mid-November raptor field trip on the

Laramie Plains.

It was a beautiful day that makes you forget all the previous white-knuckle

drives over the pass. However, what's good weather for driving isn't always good for finding raptors.

Trip leader Tim Banks checked his intended route the day before and found nary a hawk, falcon, owl or

2016

365

eagle. So instead, we drove across the Laramie Plains on a route his chapter frequently takes for general birding.

The reason for our first stop was a mystery, but then the broken branch stub of a lone cottonwood across the road became a great horned owl. However, a rough-legged hawk and a northern harrier were too distant to enjoy.

Finally, at Hutton Lake, out of the birdless sky, the wind picked up and kicked out a golden eagle, two bald eagles, and a ferruginous hawk.

Three weeks before, on a Cheyenne Audubon field trip at Curt Gowdy, we saw two bald eagles in the canyon. Another day at the park, Mark spotted three checking out his stringer of fish.

Bald eagles are marvelous looking, but I also marvel at their history, from endangered species to birds seen three times in three weeks.

Bald eagles were first federally protected in 1940. Later, they were classified as endangered. Banning the pesticide DDT and educating people not to shoot them allowed their numbers to increase. In 1995, they were reclassified as merely threatened. They were completely delisted in 2007, though they are still protected by the Bald and Golden Eagle Protection Act.

While bald eagles do breed in Wyoming, there are more here in the winter, migrating from farther north. Fish are their favorite food (carrion is second choice), so looking for them around reservoirs and Wyoming's larger rivers is good strategy, especially if there are big cottonwoods for them to roost in.

We all recognize the adult bald eagle, dark brown with white head and white tail, but until they are about 4 or 5 years old, they are dark with splotchy white markings like those of young golden eagles.

Golden eagles never came quite as close to extinction as bald eagles, but they were targeted by stock growers. In 1971, one man confessed to killing many of the 700 found shot or poisoned near Casper.

Golden eagles live in Wyoming's grasslands and shrublands year round. They might choose to nest on cliffs. And they prefer eating rabbits, ground squirrels, prairie dogs and the occasional new lamb if the rancher isn't watching.

If you see a massive raptor flying in Wyoming in the winter, it is probably an eagle. Balds and goldens have wingspans about 80 inches long.

But if it is a smaller dark bird, wingspan only 50-plus inches, with a neater black and white pattern under the wing, it might be a rough-legged hawk.

Every winter they come down from their Arctic breeding grounds, sometimes right into Cheyenne, wherever there's a power pole perch, open land, and mice, voles or shrews. It's a break from eating lemmings all summer.

They were also shot at, but like all migratory birds, they are now protected by the Migratory Bird Treaty Act.

For me, the most fascinating raptor we saw on the Laramie Plains is less common: a ferruginous hawk. Its name refers to the color of rusted iron because its top side is a reddish brown. Its belly is a creamy white, slightly spotted, compared to the streaky rough-legged's. Both have feathers all the way down their legs.

However, some sources say the ferruginous shouldn't have been in Wyoming in November. They are almost all supposed to migrate south in October and return in March.

Some field guides show the Colorado and Wyoming border as the north boundary of their winter range. I think that winter range boundary at the state line may have more to do with the greater number of birders in Colorado in the past who could distinguish between ferruginous and rough-legged. But there are now a dozen Laramie Plains and Cheyenne-area eBird records for ferruginous from November through February within the past three years.

Guess I can no longer assume in winter any large dark hawk that isn't a red-tailed hawk is a rough-legged. It might be a ferruginous.

Meanwhile, we can all brush up on our hawk identification skills at www.AllAboutBirds.org or download the free Merlin Bird ID app for help. It will make winter more interesting.

# 2017

## 391 — Eulogy for an indoor cat
*Sunday, January 1, 2017, Outdoors, page E5*

Today I write a eulogy for Joey, an ordinary orange and white house cat who lived with our family.

I offer the details of her life as an example of the advantages of an indoor cat.

Joey died in the fall at the age of 18 ½ years old. She was my writing companion, sometimes draped over my left shoulder, sometimes over my lap. She exuded enough cat hair to melt down my previous laptop by clogging up the fan.

She was opinionated. She talked about a lot of things, her self-assured gaze drilling into you, assessing you.

Joey and her brother were products of a liaison between an unknown father and a footloose mother belonging to a friend. Our boys, in grade school and junior high then, enjoyed building climbing gyms for the kittens and playing catch and release cat toy games with them.

We took the cats outside occasionally on harness and leash, but Joey's brother soon refused after stepping on a bee and getting stung.

Joey was always the one to look for before opening a door. It wasn't that she wanted to go outside. She just wanted to go to the other side, whether into the basement or into a closet. If she did get out the front door, all we had to do was quietly leave the door open, circle around behind her where she was quivering under a bush, and gently herd her towards the door.

However, one time she escaped without us realizing it right away. It took three days for her to come home and start pounding on the aluminum storm door. We were the only happy people that week after 9/11.

One good reason to keep your cat indoors is so you don't have to worry about them. Of course, you could build them a "catio" – safe enclosed space for them to enjoy the outdoors. The enclosure would also prevent your cat from hunting local wildlife.

Even if it isn't important to you to save billions of animals each year – birds, mammals, reptiles and amphibians – from domestic cats, if you have children, you don't want them in contact with cats that roam outdoors.

Cats are the hosts for toxoplasma gondii, a parasite with eggs that persist in soil. We know it causes serious health problems for pregnant women who come in contact with cat feces. But we now know that large percentages of the global human population are infected, and studies suggest toxoplasma gondii can cause behavioral and personality changes and is associated with disorders including schizophrenia.

Outdoor cats, whether owned or feral, are a bigger and more complicated problem than we ever expected. You'll want to read "Cat Wars: The Devastating Consequences of a Cuddly Killer," by Peter P. Marra and Chris Santella, neither of whom are cat haters.

For an introduction to the book, see the video of Marra's talk last month at www.AllAboutBirds.org. Search with the term "cat wars."

One moment Joey was a tiny kitten, and the next moment an adolescent adventurer, then an unflappable middle-aged cat who would still perform amazing acrobatics to catch miller moths buzzing ceiling lights.

And then she became my elder, content to follow the daily rotation of sunny spots around the house, lounging among the house plants while watching birds at the feeder outside.

I believe Joey and her brother, who died of natural causes a few years ago, had better lives, longer lives, than if they had to roam outside in the hazardous world. I know I've had a better life because they were inside with me.

In Joey's memory, please work to keep cats off the street.

## 392 — Birding by app: New adventures in tech
*Sunday, February 12, 2017, Outdoors, page E4*

Mark and I finally made the jump to smart phones last month. Our children are applauding.

What I was really looking forward to once I was in possession of a smart phone was eBird Mobile. My daughter-in-law, Jessie, was using it when we birded together over the holidays. It means that you can note the birds you see on your phone while you are in the field and then submit them as an eBird checklist.

The second day I had my phone, I went to eBird.org to find out how to download it (in the Help section search for "eBird Mobile"). It's free. If you aren't signed up for eBird already, it will help you do that for free also. Then I prepared for a trial run birding out at F.E. Warren Air Force Base with Mark.

Because we are rather miserly with our monthly data allotment, I chose to use the app offline while in the field. But because I was establishing a new birding location for the mobile version, I did that while I was at home and could use our Wi-Fi.

The preparation for offline means you are downloading an appropriate checklist of birds possible for the area. Otherwise eBird Mobile will give you the world list, 10,414 species, to scroll through.

As we birded, I scrolled through the much shorter list of local possibilities and added the numbers of each species seen as I observed them. At the end of the trip, I hit the submit button.

However, on my next eBird Mobile attempt it was bitterly cold. Recording birds while holding a pencil in a mittened hand works, but it was too cold to risk a bare hand to manipulate the touch screen, though I have since invested in "touch screen" gloves.

The mobile app can't do everything the regular checklist submission process does, like attach photos. But that upgrade may be coming soon. Meanwhile, you can edit your mobile-produced checklists on the eBird website whenever it's convenient.

I've also downloaded the free Merlin Bird ID App, http://merlin.allaboutbirds. org/ and tried it. I told Merlin where I was, what day it was, how big and what color the bird was and where it was (ground, bush, tree, sky) and up popped a photo of the most likely candidate, other possible species, general information and bird song recordings.

Both apps are Cornell Lab of Ornithology projects and are designed to get more people excited about birds. More data collected means more understanding, and more

understanding means better conservation of birds.

The lab has even more up its sleeve. At a recent meeting, staff from far-flung places gathered to discuss making animated migration maps that will allow zooming in on particular locations. Recently, Audubon and CLO announced eBird Mobile is available on the dashboard of select Subaru models. That's an update I wouldn't mind seeing the dealer for.

CLO employs a lot of tech people. Job openings on the eBird website list required technical qualifications. Preferred qualifications include "An interest in birds, nature, biology, science, and/or conservation helpful."

So maybe it doesn't surprise you that our son, Bryan, with a degree from the University of Wyoming in software engineering – and exposed to birdwatching from birth – has become not only a birder, but in October moved to Ithaca, New York, to work for CLO.

He can bird to and from work, walking through the famous Sapsucker Woods. He tells us the winter regulars include many of the same

species we see in Cheyenne. However, he says he sees four kinds of woodpeckers: downy and hairy, which we see, but also red-bellied woodpecker and pileated woodpecker, eastern birds.

Surrounded by serious birdwatchers all week, perhaps on weekends you would be forgiven for picking up a different hobby. But no, on the Martin Luther King holiday, everyone from Bryan's office went up near Seneca Falls and found snowy owls, a gyrfalcon, northern shrike and thousands of snow geese.

The next weekend, Bryan and Jessie went back and found two more snowy owls and three kinds of swans.

The full eBird website can help me predict the height of spring migration in Ithaca, and I hope to time Mark's and my visit accordingly. But we must fit in one last trip to Texas to visit our younger son, Jeffrey, before he and his wife move to Seattle for new jobs.

If your children aren't moving back to Cheyenne, at least let them live in interesting places.

# 393 Bird books worth reading

Sunday, March 12, 2017, Outdoors, page E4

If you are the books you read, here is what I've been this winter.

**"The Genius of Birds" by Jennifer Ackerman**

This was a Christmas present from my daughter-in-law, Madeleine, who teaches cognitive psychology. It's an enthralling overview of the latest studies that show how much smarter birds are

than we thought, sometimes smarter than us in particular ways. They can navigate extreme distances, find home, find food stashed six months earlier, solve puzzles, use tools, sing hundreds of complex songs, remember unique relationships with each flock member, engineer nests, adapt to new foods and situations. They can even

communicate with us.

**"Good Birders Still Don't Wear White, Passionate Birders Share the Joys of Watching Birds," edited by Lisa A. White and Jeffrey A. Gordon**

The previous volume, in 2007, was "Good Birders Don't Wear White, 50 Tips from North America's Top Birders."

One of my favorite essays is by our Colorado friend Ted Floyd, "Go Birding with (Young, Really Young) Children." Having frequently accompanied him and his children, I can say he does a terrific job of making birdwatching appealing.

Many of the essays start out with "Why I Love..." and move on to different aspects

368  CHEYENNE BIRD BANTER

of birding people love (seabirds, drawing birds, my yard, spectrograms, "because it gets me closer to tacos"), followed by tips, should you want to follow their passions.

### "Field Guide to the Birds of California" by Alvaro Jaramillo

This is part of the American Birding Association State Field Guide Series published by Scott & Nix Inc. The series so far also includes Arizona, the Carolinas, Colorado, Florida, Minnesota, New Jersey, New York, Pennsylvania and Texas.

Each author writes his or her own invitation to the beginning birdwatcher or the birder new to their state.

While a few birding hotspots may be mentioned, the real service these books provide is an overview of the state's ecological regions and what kind of habitats to find each species in, not to mention large photos of each. I'll probably still pack my Sibley's, just in case we see a bird rare to California.

### "Peterson Field Guide to the Bird Sounds of Eastern North America" by Nathan Pieplow

While including the usual bird pictures and range maps, this book is about learning to identify birds by sound and corresponding audio files can be found at www.petersonbirdsounds.com.

Bird songs are charted using spectrograms, graphic representations of sound recordings.

You can think of spectrograms as musical notation. They read from left to right. A low black mark indicates a low-pitched frequency. A thin, short line higher up indicates a clear sound with few overtones, higher pitched and short-lived. But most bird sounds are more complex, some filling the spectrogram from top to bottom.

Pieplow explains how to read spectrograms, the basic patterns, the variations, the non-vocal sounds like wing-clapping, and the biology of bird sounds.

Once you can visualize what you are hearing,

Pieplow provides a visual index to bird sounds to help you try to match a bird with what you heard.

Taking a call note I'm familiar with in my neighborhood, the one note the Townsend's solitaire gives from the top of a tree in winter, I find that Pieplow categorizes it as "cheep," higher than a "chirp" and more complex than "peep." It's going to take a while to train our ears to distinguish differences.

### "The Warbler Guide" by Tom Stephenson and Scott White, c. 2015, Princeton University Press and The Warbler Guide App.

Spectrograms are a part of the 500 pages devoted to the 56 species of warblers in the U.S. and Canada.

The yellow warbler, whose song we hear along willow-choked streams in the mountains in summer, gets 10 pages.

Icons show its silhouette (sometimes it can be diagnostic), color impression (as it flies by in a blur), tail pattern

(the usual underside view of a bird above your head), range generalization, habitat (what part of the tree it prefers) and behavioral (hover, creep, sally, walk).

Then there's the spectrogram comparing it to other species and maps show migration routes and timing, both spring and fall. We can see the yellow warbler spends the winter as far south as Peru.

Forty-one photographs show all angles, similar species, and both sexes at various ages.

The companion app, an additional $13, has most of the book's content and lets you rotate to compare 3-D versions of two warblers at a time, filter identification clues and listen to song recordings.

This is a good investment for birding in Cheyenne where we have seen 32 warbler species over the last 20 springs.

# 394 Coast comes through with great birds

Sunday, April 30, 2017, Outdoors, page E2

If I added these bird species to my life list last month, where would you say I'd been?

Surf scoter, pelagic cormorant, western gull, band-tailed pigeon, Anna's hummingbird, Allen's hummingbird, Nuttall's woodpecker, California (formerly western) scrub-jay, California towhee, golden-crowned sparrow.

If you guessed California, you would be right. But it

isn't the birds with "California" in their names that is the best clue. That would be the Nuttall's woodpecker, found entirely in the state and the northern tip of Baja California. We saw ours in the arboretum at the University of California Davis.

Five of the species new to me – the hummingbirds, pigeon, towhee and scrub-jay – were in the backyard of the bed and breakfast we stayed at in Olema, California. The

host fills the feeders every morning at 8:15 a.m. just before serving breakfast and his guests are treated to a flurry that also includes numerous California quail, white-crowned sparrows and, just like home, Eurasian collared-doves.

The pelagic cormorant would tell you that we spent time at the ocean. Despite the "pelagic" part of its name, which should indicate it is found far offshore, this

cormorant is a shore dweller. Mark and I saw it way below us, in the rocks, at the lighthouse at Point Reyes National Seashore.

At Point Reyes Beach North, we encountered signs warning us about the protected nesting area for the federally designated threatened western population of snowy plovers. The area of the beach to be avoided was clearly marked with 4-foot white poles and white rope.

2017

369

Mark and I, and our Sacramento friends, formerly of Casper, dutifully gave it a wide berth.

And then the birds flew up in front of us anyway. We watched as five or six of the little white-faced sand-colored shorebirds fluttered away and settled down again nearby – in human footprint depressions.

The American Birding Association's "Field Guide to the Birds of California" says that the snowy plovers breeding on beaches like to find depressions so they don't cast as much of a shadow, avoiding detection by predators. They like the depressions for nesting too. Makes me think someone should walk once or twice through the official nesting area to make some, but who wants to pay the fine for trespassing? Besides, human activity and loose dogs scare the birds and prevent them from breeding.

Snowy plover was not a lifer for us – our first ones were at Caladesi Island State Park, Florida. There too, their nesting area was delineated and protected, though in Florida, they are only on the state-level threatened species list.

Snowy plovers are more than oceanic beach birds. You might find nesting populations across the southwestern U.S. at shallow lakes with sand or dried mud.

One bird I wanted to see was the wrentit. California, western Oregon and northern Baja California are the only places to see it. At the Sacramento National Wildlife Refuge, I found two cute little birds that seemed to match the field guide. Another visitor noticed them popping in and out of a two-foot-long hanging sack made of bits of vegetation woven together, and a red flag went up in my mind.

Didn't this hanging nest remind me of one I'd seen before in Seattle? Made by bushtits? Well darn, those were bushtits. They are only 4.5 inches long, whereas wrentits are 6.5 inches long, and wrentits build cup-shaped nests instead. If you were to draw a line from Seattle to Houston, bushtits can be found south of it, anywhere brushy and woodsy.

This was our first trip to California as eBirders, recording birds we saw at eBird.org. As usual, it came about as the result of a family commitment, which almost all our traveling does. We might have seen more species had we been on a birding tour, like we've done in Texas and Florida, but I think we did well at 86 species. The birds just seemed to pop out and give us a good look. Or maybe you could say they took a good look at us.

We'll have to make a point of visiting our family and friends in California more often. There are 571 bird species left to see – and half would be life birds.

# 395 Citizen science meets the test of making a difference

Sunday, May 14, 2017, Outdoors, page E4

Birdwatchers have been at the forefront of citizen science for a long time, starting with the Christmas Bird Count in 1900.

Today, the Cornell Lab of Ornithology is leading the way in using technology to expand bird counting around the globe. Meanwhile, other citizen science projects collect information on a variety of phenomena.

But is citizen science really science? This question was asked last December at the first Wyoming Citizen Science Conference.

The way science works is a scientist poses a question in the form of a hypothesis. For instance, do robins lay more eggs at lower elevations than at higher elevations? The scientist and his assistants can go out and find nests and count eggs to get an answer.

However, there are hypotheses that would be more difficult to prove without a reservoir of data that was collected without a research question in mind. For instance, Elizabeth Wommack, curator and collections manager of vertebrates at the University of Wyoming Museum of Vertebrates, studied the variation in the number of white markings on the outer tail feathers of male kestrels. She visited collections of bird specimens at museums all over the country to gather data.

Some kestrels have lots of white spots, some have none. Are the differences caused by geography? Many animal traits are selected for (meaning because of the trait, the animal survives and passes on the trait to more offspring) on a continuum. It could be north to south or dry to wet habitat or some other geographic feature.

Or perhaps it was sexual selection – females preferred spottier male tail feathers. Or did the amount of spotting lead directly to improved survival?

Wommack discovered none of her hypotheses could show statistical significance, information just as important as proving the hypotheses true. But at least Wommack learned something without having to "collect" or kill more kestrels.

Some citizen science projects collect data to test specific hypotheses. However, others, like eBird and iNaturalist, collect data without a hypothesis in mind, akin to putting specimens in museum drawers like those kestrels. The data is just waiting for someone to ask a question.

I know I've gone to eBird with my own questions such as when and where sandhill cranes are seen in Wyoming. Or when the last time was that I reported blue jays in our yard.

To some scientists, data like eBird's, collected by the public, might be suspect. How can they trust lay people to report accurately? At this point though, so many people are reporting the birds they see to eBird that statistical credibility is high. (However, eBird still does not know a lot about birds in Wyoming and we need more of you to report your sightings at http://ebird.org.)

370     CHEYENNE BIRD BANTER

Are scientists using eBird data? They are, and papers are being published. The Cornell Lab of Ornithology itself recently published a study in Biological Conservation, an international journal. Their study tracked requests for raw data from eBird for 22 months, 2012 through 2014.

They found that the data was used in 159 direct conservation actions. That means no waiting years for papers to be published before identifying problems like downturns in population. These actions affected birds through management of habitat, siting of disturbances like power plants, decisions about listing as threatened or endangered for example. CLO also discovered citizens were using the data to discuss development and land use issues in their own neighborhoods.

CLO's eBird data is what is called open access data. No one pays to access it and none of us get paid to contribute it. Our payment is the knowledge that we are helping land and wildlife managers make better decisions. There's a lot "crowd sourced" abundance and distribution numbers can tell them.

Citizen science isn't often couched in terms of staving off extinction. Recently I read "Citizen Scientist, Searching for Heroes and Hope in an Age of Extinction," by Mary Ellen Hannibal, published in 2016. In it, she gave me a new view.

Based in California, Hannibal uses examples of citizen science projects there that have made a difference. She looks back at the early non-scientists like Ed Ricketts and John Steinbeck who sampled the Pacific Coast, leaving a trail of data collection sites that were re-sampled 85 years later. She also looks to Pulitzer Prize-winning biologist E.O. Wilson, who gives citizen science his blessing. At age 87, he continues to share his message that we should leave half the biosphere to nature – for our own good.

Enjoy spring bird migration and share your bird observations. The species you save may be the one to visit you in your own backyard again.

# 396 Thrushes take over Cheyenne Big Day Bird Count

Sunday, June 18, 2017, Outdoors, page E2

The spring bird migration of 2017 is leaving people scratching their heads in puzzlement.

Because of safety issues due to heavy snow the two days before which left large trees hazardous to walk under, the Cheyenne Big Day Bird Count was postponed a week, to May 27.

Cheyenne – High Plains Audubon Society members who organize the count assume that the Saturday closest to the middle of May will be the closest to peak migration. However, while the event was held a week later this year, we counted 113 species compared to last year's 110.

In the preceding weeks, we saw posts from Casper birders about sightings of spring migrants we hadn't seen yet, as if they skipped Cheyenne and continued north.

At our house, we eventually had about one each of our favorite migrants (indigo bunting, black-headed grosbeak, MacGillivray's warbler, Wilson's warbler), but most were after the original Big Day date.

In early May, my husband, Mark, and I visited High Island, Texas, a famous landing spot for migrating songbirds crossing the Gulf. It was empty except for the rookery full of spoonbills, herons, egrets and cormorants. A birder we met had visited during the peak in April and said it was a disappointing migration.

Bill Thompson III, editor and publisher of Bird Watcher's Digest, posted similar thoughts about what he saw from his home in southeastern Ohio. Someone responding from New Hampshire said he saw only three species of warblers in the first 25 days of May when he would typically see a dozen.

Everyone hopes that the low number of migrating birds is due to weather patterns that blew them north without stopping over. We hope it isn't a sign of problems on the wintering grounds, breeding grounds or somewhere in between.

For our Cheyenne Big Day, we have one group that birds the hotspots: Lions Park, Wyoming Hereford Ranch, the High Plains Grasslands Research Station and the adjacent arboretum. This year, between 6 a.m. and 3 p.m., the group varied in size from five to 15. Even the most inexperienced birdwatcher was helpful finding birds.

Because we couldn't change the date of the permit we had to access the research station, we contented ourselves with the road in front of the buildings, and that's where we found two eastern bluebirds, a species showing up here more often in recent years.

The long-eared owl seen by two participants at the Wyoming Hereford Ranch is a species last recorded on the Big Day in 1996.

Besides the group canvassing an area roughly the same as the Christmas Bird Count's 15-mile diameter circle centered on the Capitol, five people birded on their own. And though they sometimes visited places the main group did, it was at different times, counting different birds.

The most numerous species this year was the Swainson's thrush. The quintessential little brown bird, like a junior robin, was everywhere. Two days later, there were none to be seen.

Maybe there is no one-day peak of spring migration. Maybe there never was. But spending any day outdoors in Cheyenne in May, you are bound to see more species of birds than if you don't go out at all.

## 2017 Cheyenne Big Day Bird Count results: 113 species

| | | | |
|---|---|---|---|
| Canada Goose | American Avocet | Warbling Vireo | Yellow-rumped Warbler |
| Wood Duck | Killdeer | Plumbeous Vireo | Wilson's Warbler |
| Gadwall | Spotted Sandpiper | Blue Jay | Chipping Sparrow |
| Mallard | Solitary Sandpiper | Black-billed Magpie | Clay-colored Sparrow |
| Blue-winged Teal | Willet | American Crow | Brewer's Sparrow |
| Cinnamon Teal | Wilson's Snipe | Horned Lark | Lark Sparrow |
| Northern Shoveler | Wilson's Phalarope | Tree Swallow | White-crowned Sparrow |
| Northern Pintail | Ring-billed Gull | N. Rough-winged Swallow | Vesper Sparrow |
| Green-winged Teal | Rock Pigeon | Bank Swallow | Savannah Sparrow |
| Ring-necked Duck | Eurasian Collared-Dove | Cliff Swallow | Song Sparrow |
| Lesser Scaup | Mourning Dove | Barn Swallow | Lincoln's Sparrow |
| Bufflehead | Long-eared Owl | Mountain Chickadee | Green-tailed Towhee |
| Common Merganser | Great Horned Owl | Red-breasted Nuthatch | Western Tanager |
| Ruddy Duck | Common Nighthawk | House Wren | Black-headed Grosbeak |
| Eared Grebe | Broad-tailed Hummingbird | Ruby-crowned Kinglet | Red-winged Blackbird |
| Western Grebe | Belted Kingfisher | Eastern Bluebird | Western Meadowlark |
| Clark's Grebe | Downy Woodpecker | Swainson's Thrush | Yellow-headed Blackbird |
| Double-crested Cormorant | Northern Flicker | Hermit Thrush | Brewer's Blackbird |
| American White Pelican | American Kestrel | American Robin | Common Grackle |
| Great Blue Heron | Olive-sided Flycatcher | Gray Catbird | Great-tailed Grackle |
| Black-crowned Night-Heron | Western Wood-Pewee | Brown Thrasher | Brown-headed Cowbird |
| Turkey Vulture | Willow Flycatcher | European Starling | Orchard Oriole |
| Osprey | Least Flycatcher | Cedar Waxwing | Bullock's Oriole |
| Sharp-Shinned Hawk | Hammond's Flycatcher | McCown's Longspur | House Finch |
| Cooper's Hawk | Cordilleran Flycatcher | Northern Waterthrush | Lesser Goldfinch |
| Broad-winged Hawk | Say's Phoebe | Orange-crowned Warbler | American Goldfinch |
| Swainson's Hawk | Cassin's Kingbird | Common Yellowthroat | House Sparrow |
| Red-tailed Hawk | Western Kingbird | American Redstart | |
| American Coot | Eastern Kingbird | Yellow Warbler | |

Sunday, July 16, 2017, Outdoors, page E5

# 397 Bird by ear to identify the unseen

Here on the western edge of the Great Plains, our trees don't grow so thick that you can't walk all the way around one to see the bird that's singing. But it is still useful to be able to identify birds by sound.

I'm a visually-oriented person, so over time I've learned to identify our local birds well enough to often figure out who they are as they flash by. I can only identify bird voices of the most common or unique sounding species.

At the big box stores in town, in the garden departments, there is almost always an incessant cheeping overhead from invading house sparrows.

If you get up at oh-dark-thirty on a spring or summer morning in town, you are likely to hear the cheerful "cheerio" of a robin.

Putting up a bird feeder may bring in house finches with their different chatter. I especially like hearing the goldfinches around the thistle feeder which sound as if they are small children calling questions to each other.

Birding by ear becomes a more important skill in the mountains where the forest is thicker. The Cheyenne – High Plains Audubon Society's mid-June field trip was to the Vedauwoo recreation area on the Medicine

Bow National Forest. We planned to hike the Turtle Rock trail. Since most of Wyoming's birds are found near water, we focused on the beaver ponds.

Some birds, like the flocks of tree swallows flitting across the water, are never hidden away.

But one warbling bird was. It didn't sound quite like a robin. I went through a mental list of birds that like riparian, or streamside, habitats and casually remarked, "Maybe it's a warbling vireo."

Then I realized I could check the free Merlin app on my phone and play a recording of a warbling vireo.

Amazingly, it matched.

Yellow warblers are almost always somewhere around in the brush around water at upper elevations too, and we could hear one. It has a very loud, unique call. Being bright yellow, it isn't hard to spot singing in the willows.

There are species of birds that resemble each other so closely - the empidonax flycatchers – that it is necessary to hear them sing to tell them apart. On the other hand, there are species that sound so much like each other, it causes the problem people used to have telling me and my mom apart on the phone.

For example, robin and

black-headed grosbeak songs have a clear, babbling quality, but if you listen a lot while the grosbeaks are here during migration, you can tell who is the real robin.

On the Turtle Rock trail, chapter member Don Edington picked out a bird at the tip top of an evergreen, singing away. It was yellow, with black and white wings, like an over-sized goldfinch. Its head had the lightest wash of orangey-red. It was another robin voice impersonator, the western tanager.

Visually, the sparrows are mostly a large brown cloud in my mind. The same can be said for distinguishing, much less remembering, many bird songs. I like birds with easy to remember songs, like the ruby-crowned kinglet, another bird to expect in the forest. It is so tiny, your chances are slim for seeing it on its favorite perches in large spruce trees.

After being inundated by Swainson's thrushes this spring (all completely mute while they inspected our backyard), it was a pleasure to catch the trill of one on the trail. But then I checked it against a recording on Merlin and realized we had the thrush that doesn't trill upwards, but the other, trilling downwards, the hermit thrush.

It does help to study the field guides in advance of seeing a bird species for the first time – just knowing which ones to expect in a certain habitat is helpful. Studying bird songs before venturing into the woods again would be as useful.

I need to crack open that new book by Nathan Pieplow, "Peterson Field Guide to Bird Sounds of Eastern North America," and the corresponding recordings at www. petersonbirdsounds.com.

Except, we'll only find the species we share with eastern North America. We won't find our strictly western bird species until he finishes the western edition. But I could work on his technique for distinguishing songs before I spend too much more time in the woods.

# 398 Bird-finding betters from generation to generation

Sunday, August 20, 2017, Outdoors, page E4

When your interest in birds takes you beyond your backyard, you need a guide beyond your bird identification book. That help can come in many forms – from apps and websites to a trail guidebook or local expert.

Noah Strycker needed a bird-finding guide for the whole world for his record-breaking Big Year in 2015. His book, "Birding without Borders," due out Oct. 10, documents his travels to the seven continents to find 6,042 species, more than half the world total.

In it, he thoughtfully considers many bird-related topics, including how technology made his record possible, specifically www.eBird.org. In addition to being a place where you can share your birding records, its "Explore Data" function helps you find birding hotspots, certain birds and even find out who found them. Strycker credits its enormous global data base with his Big Year success.

Another piece of technology equally important was http://birdingpal.org/, a way to connect with fellow enthusiasts who could show him around their own "backyards." Every species he saw during his Big Year was verified by his various travelling companions.

Back in 1968, there was no global data base to help Peter Alden set the world Big Year record. But he only needed to break just over 2,000 species. He helped pioneer international birding tourism through the trips he ran for Massachusetts Audubon. By 1981, he and British birder John Gooders could write "Finding Birds Around the World." Four pages of the nearly 700 are devoted to our own Yellowstone National Park.

When I bumped into Alden at the Mount Auburn Cemetery in Cambridge, Massachusetts, (a birding hotspot) in 2011, he offered to send me an autographed copy for $5. I accepted. However, until I read Strycker's book, I had no idea how famous a birder he was.

As Strycker explains it, interest in international birding, especially since World War II, has kept growing, right along with improved transportation to and within developing countries, which usually have the highest bird diversity. However, some of his cliff-hanging road descriptions would indicate that perhaps sometimes the birders have exceeded the bounds of safe travel.

For the U.S., the Buteo Books website will show you a multitude of American Birding Association "Birdfinding" titles for many states. Oliver Scott authored "A Birder's Guide to Wyoming" for the association in 1992. Robert and Jane Dorn included bird finding notes in the 1999 edition of their book, "Wyoming Birds." Both books are the result of decades of experience.

A variation on the bird-finding book is "the birding trail." The first was in Texas. The book, "Finding Birds on the Great Texas Coastal Birding Trail," enumerates a collection of routes connecting birding sites and includes information like park entrance fees, what amenities are nearby, and what interesting birds you are likely to see. Now you can find bird and wildlife viewing "trails" on the Texas Parks and Wildlife website. Many states are following their example.

People in Wyoming have talked about putting together a birding trail for some years, but it took a birding enthusiast like Zach Hutchinson, a Casper-based community naturalist for Audubon Rockies, to finally get it off the ground.

The good news is that by waiting this long, there are now software companies that have designed birding trail apps. No one needs to print books that soon

2017

373

need updates.

The other good news is that to make it a free app, Hutchinson found sponsors including the Cheyenne–High Plains Audubon Society, Murie Audubon Society (Casper), Wyoming State Parks, and WY Outside – a group of nonprofits

and government agencies working to encourage youth and families in Wyoming to spend more time outdoors.

Look for "Wyoming Bird Trail" app on either iTunes or Google Play to install it on your smart phone.

Hutchinson has made a good start. The wonderful

thing about the app technology is that not only does it borrow Google Maps so directions don't need to be written, the app information can be easily updated. Users are invited to help.

There is one other way enterprising U.S. birders research birding trips. They

contact the local Audubon chapter, perhaps finding a member, like me, who loves an excuse to get out for another birding trip and who will show them around – and make a recommendation for where to have lunch.

# 399 Kitchen window like a TV peering into lives of birds

Sunday, September 17, 2017, Outdoors, page E4

The view out of our 4-by-6-foot kitchen window is the equivalent of an 85-inch, high-definition television screen.

The daytime programming over the summer has been exceptional this year. Not many murder mysteries, thank goodness, and instead, mostly family dramas.

The robins always seem to get on screen first. Walking flat-footed through our vegetables and flowers, the speckle-breasted young, unlike some human teenagers, kept looking towards the adults for instruction and moral support.

Young birds have this gawky look about them. They have balance issues when they land on the utility line. Or they make a hard landing on a branch. They look around, tilting their heads this way and that. Maybe they are learning to focus.

The first hummingbird of the season showed up July 10, nearly a week earlier than last year. Luckily, their favorite red flower, the Jacob Cline variety of monarda, or beebalm, was blooming two weeks ahead of schedule.

We immediately put the hummingbird feeder up (4 parts water to 1 part white

sugar – don't substitute other sugars – boiled together, no red dye, please, but maybe a red ribbon on the feeder). Within a few days we had a hummingbird showing up regularly at breakfast, lunch and dinner – which is when we watch our window TV.

Sometimes we saw three at a time, often two, though by August 25 sightings dropped off. It is difficult to distinguish between rufous and broad-tailed females and juveniles that come. Kind of like trying to keep track of all the characters in a PBS historical drama.

My favorite series this summer was "Father Knows Best." Beginning July 1, a lesser goldfinch male, and sometimes a second one and females, started joining the American goldfinches at our thistle tube feeder.

The lesser goldfinch is the American goldfinch's counterpart in the southwestern U.S., and they are being seen more regularly in southeast Wyoming. They are smaller. Like the American, they are bright yellow with a black cap and black wings, but they also have a black back, although some have greenish backs.

Every day the lesser males

showed up, pulling thistle seed from the feeder for minutes at a time. Unlike other seed-eating songbirds which feed their young insects, goldfinches feed their young seeds they've chewed to a pulp. After a couple weeks, we began to wonder if one of them had a nest somewhere.

August 4, the lesser fledglings made their TV debut. The three pestered their dad at the same time. My husband, Mark, got a wonderful photo of the male feeding one of the young. However, within five days the show was over, the young having dispersed.

Year round we have Eurasian collared-doves. I've noticed one has a droopy wing, the tip of which nearly drags on the ground. She and her mate are responsible for the only X-rated content shown on our backyard nature TV – that's how I know the droopy-winged bird is female.

One morning outside I noticed a scattering of thin sticks on the grass and looked up. I saw the sketchy (as in a drawing of a few lines) nest on a branch of one of our green ash trees, with the dove sitting on it. Every time I went out, I

would check and there she was, suspended over our heads, listening in on all our conversations, watching us mow and garden.

Then one day I heard a frantic banging around where Mark had stacked the hail guards for our garden. It was a young dove. It had blown out of the nest during the night's rainstorm. The sketchy (as in unreliable) nest had failed.

The presence of the trapped squab, half the size of an adult, would explain the behavior of the mother nearby who had been so agitated that she attracted our dog's attention.

I put the dog in the house and went to extract the young bird. It didn't move as I approached and scooped it up. There is something magical about holding a wild bird, even one belonging to a species that has invaded our neighborhoods, sometimes at the expense of the native mourning dove. So soft, so plump. I set it down inside the fenced-off flower garden. Later, I checked and it was gone.

Within a few days, Droopy-wing and her mate were involved in another X-rated performance. Then

374      CHEYENNE BIRD BANTER

I noticed one of them fly by with a slender stick. Sure enough, two days later she was back on her rehabbed throne, incubating the next generation.

# 400 Project FeederWatch tells us a lot about juncos and our backyard birds

Sunday, October 15, 2017, Outdoors, page E4

Despite snow on the ground and pea soup fog at South Gap Lake in the Snowy Range (11,120 feet elevation), on Sept. 27 I saw a flock of dark-eyed juncos. They like snow. I should see the first ones down in my yard mid-October, when alpine winter conditions get too rough.

Juncos are those little gray birds that come in five sub-species and multiple hybrid colorations in Cheyenne, but they all have white outer tail feathers. They are my sign of the start of the winter bird feeding season – and the Project FeederWatch bird counting season.

Project FeederWatch is a citizen science opportunity for people with bird feeders to count the birds they attract as often as once a week (or less) between November and early April. Begun in Canada in 1976 and in the U.S. in 1987, more than 20,000 people participated last year. The data is used in scientific studies, many of which are summarized on the project's website.

Participation costs $18. You receive a research kit, bird identification poster, the digital version of Living Bird magazine and the year-end report.

If you feed wild birds or are considering it, you must visit the Project FeederWatch site, https://feederwatch.

org/, whether you register for the program or not. It is now beautifully designed and packed with information.

For instance, in the "Learn" section, I find juncos prefer black-oil sunflower seeds – and seven other kinds. I stick with black-oil because it's popular with many species in Cheyenne. I also learn juncos prefer hopper-style feeders, platform feeders or feeding on the ground.

Seventy-one species are listed as potential feeder birds in the Northwest region, which stretches from British Columbia to Wyoming. However, about 17 of those species have yet to be seen in Cheyenne, so click on the "All About Birds" link to check a species' actual range.

The Project FeederWatch website addresses every question I can think of regarding wild bird feeding:
--Grit and water provision
--Feeder cleaning
--Predator avoidance
--Squirrel exclusion
--Window strike reduction
--Sick birds
--Tricky identification, like hairy vs downy woodpecker.

In "Community" section you'll find the results of last season's photo contest, participants' other photos, featured participants, tips, FAQs, the blog, and the FeederWatch cam.

I find the "Explore"

section fascinating. This is where you can investigate the data yourself. The "Map Room" shows where juncos like to winter best.

Based on last season's data, in the far north region of Canada, juncos were number 12 in abundance at feeders. In the southeastern U.S., they were number 13. However, in the southwest, which has a lot of cold high elevations, they were number two, as they were in the northeast region, and number three in the central region, the northern Great Plains. Here in the northwest region, they were number one. We have perfect junco winter conditions, not too cold, not too warm.

However, looking at the top 25 species for Wyoming in the same 2016-2017 season (based on percent of sites visited and the average flock size), juncos came in fifth, after house sparrow, house finch, goldfinch and black-capped chickadee. Other years, especially between the seasons beginning in 2007 and 2013, they have been number one.

I looked at my own Project FeederWatch data to see if I could spot any dark-eyed junco trends.

I get in 18-20 weekly counts per year. In the past 18 years, there were three years when the juncos missed none or only one of the

weeks, in 2001, 2005 and 2008. Those seasons also happened to be the largest average flock sizes, 8.65 to 9.72 birds per flock.

Later, there were three seasons in which juncos came up missing six or seven weeks, 2011, 2013 and 2016. Two of those were the seasons of the smallest average flock sizes, 1.6 to 2.5 birds per flock.

It appears my local junco population was in a downward trend between 2008 and 2016. Let's hope it's a cycle. Or maybe our yard's habitat has changed or there are more hawks or cats scaring the juncos away. Or some weeks it's too warm in town and they go back to the mountains.

One yard does not make a city-wide trend, but we won't know what the trend is unless more people in Cheyenne participate.

How many FeederWatchers are there in Cheyenne? We've had as many as four, back in 1999-2004, but lately there's only been one or two of us. Statewide, Wyoming averages 25 participants per year.

If you sign up, you'll have your own red dot on the map (but your identity won't be publicized). I hope you'll become a FeederWatcher this season.

# 401 Wyoming's greater sage-grouse conservation plan is in jeopardy

Sunday, November 12, 2017, Outdoors, page E4

Wyoming successfully addressed the sage grouse issue through a collaboration of state and local government, sportsmen, conservationists, the oil and gas industry, and agricultural interests.

Over six years, the state was able to draw up a plan to establish protected core areas of habitat. Good habitat is the best protection for this species which has declined 30 percent across the west since 1985.

The plan leaves a large majority of Wyoming open to oil and gas and other development.

In 2015, the U.S. Fish and Wildlife Service said state plans across the west were good enough that it wouldn't start proceedings to list the sage grouse as threatened or endangered.

Here in Wyoming, the Sage-Grouse Implementation Team, headed by Bob Budd, is working hard. The team represents all the previous collaborators.

However, the new federal administration is intent on dismantling anything that happened under the previous president. It tasked new U.S. Department of Interior secretary Ryan Zinke with reviewing all state sage grouse plans to either toss them or amend them.

None of the collaborators on Wyoming's plan are happy with this – including the oil and gas people who desire certainty for their business plans. Wyoming Governor Matt Mead is not happy either.

I went to the Bureau of Land Management's public meeting November 6 in Cheyenne to find out more about the proposed amendments to Wyoming's plan.

I heard these criticisms:

--Switching to using sage grouse population numbers to determine an oil and gas producer's ability to drill and plan for mitigation (more sage grouse, more leniency) would leave companies with a lot of unwanted uncertainty. Sage grouse numbers vary enormously from year to year due to weather and other natural effects.

--Basing conservation plans on sage grouse population numbers rather than habitat would discount the 350-plus other species that depend on the sagebrush ecosystem, including 22 "species of conservation concern."

--Messing around with the plan could cause U.S. Fish and Wildlife to decide the sage grouse warrants listing after all. That would close much more land to oil and gas drilling, as well as coal mining and other mineral extraction.

--The current Republican administration thinks states should have more say in issues like this, and the six years of collaboration Wyoming went through is a perfect example of how it can happen. Ironically, it's the Republicans in Washington who now decree they know what is best for us.

--Wyoming's conservation plan has been in effect for only two years, which is not enough time to gauge success. Instituting major changes now would cost a lot of taxpayer money that could be better spent in the field.

BLM invites us to comment during their scoping process. They want to know if we think they should amend the management plans that were developed by the states to protect sage grouse.

They don't make it easy, says my husband, a retired BLM wildlife biologist.

Go to http://bit.ly/GRSGplanning (case-sensitive). Click on "Documents and Reports." This will give you a list of documents. Only "GRSG Notice of Intent" is available for commenting. "GRSG" is ornithological shorthand using initial letters of the parts of the bird's common name.

After you read the document, click on "Comment on Document." You'll have to fill in the title of the document you are commenting on: "GRSG Notice of Intent." And then you have 60 minutes to finish the procedure or everything you've written disappears. You may want to compose your comments elsewhere and then paste them in.

The deadline for comments is either Nov. 27 or Nov. 30 – there's a discrepancy in BLM's handouts from the public meeting. Go with the earlier date if you can.

To educate yourself before commenting, you can visit the Wyoming State BLM office in Cheyenne, 5353 Yellowstone Road, or contact Erica Husse, 307-775-6318, ehusse@blm.gov, or Emmet Pruss, 307-775-6266, epruss@blm.gov.

But if you are most interested in what is best for sage grouse, it may be easier to jump to the analysis provided by conservation groups like the National Audubon Society, www.audubon.org/sage-grouse. The former Audubon Wyoming executive director Brian Rutledge was instrumental in the Wyoming collaboration and is still involved as NAS's director of the Sagebrush Ecosystem Initiative.

Two other interested groups are Wyoming Wildlife Federation, http://wyomingwildlife.org/, and the Wyoming Outdoor Council, https://wyomingoutdoorcouncil.org/.

All three organizations offer simple digital form letters that can be personalized, and they will send them to BLM. However, BLM says it gives more credence to comments sent via their own online form.

I hope you can take a few minutes to put in a good word for the bird that maybe should be our state mascot.

Next month I'll look at what the Wyoming State Legislature did last session that may also negatively affect sage grouse.

CHEYENNE BIRD BANTER

# 402 Critics of sage-grouse captive breeding doubt it will succeed

Sunday, December 10, 2017, Outdoors, page E5

Over the eons, the greater sage grouse figured out how to prosper in the sagebrush.

It's not an easy life. Some years are too wet and the chicks die. Others are too dry with few leaves, buds, flowers or insects and the chicks starve. Some years there are too many hungry coyotes, badgers and ravens.

Every spring the sage grouse go to the meet-up at the lek, the sage grouse version of a bar. The males puff out their chests vying for the right to take the most females, then love them and leave them to raise the chicks on their own.

Experienced hens look for the best cover for their nests. They teach the young how to find food and avoid predators. In the fall, every sage grouse migrates to winter habitat, 4-18 miles away.

In the past hundred years, obstacles were thrown in the path of sage grouse, including in their Wyoming stronghold where sagebrush habitat can be found across the whole state except in the southeast and northwest corners.

The low-flying birds collide with fences, vehicles, utility lines. The noise from oil and gas operations pushes them away. Sagebrush disappears with development.

Each state is responsible for all wildlife within its borders. But if a species heads for extinction, the U.S. Fish and Wildlife Service steps in. Since 1985, the sage grouse population declined 30 percent across the West. It looked like the species might be listed as either threatened or endangered, curtailing oil and gas drilling and other development.

Last month, I explained how Wyoming conservationists, sportsmen, the oil and gas industry, agricultural interests and state and local government collaborated on a state plan to conserve sage grouse. However, the current federal administration wants all the state plans to be examined to see if sage grouse habitat can be more densely developed.

Wyoming's collaborators strongly disagree with the attempt. Public comments were solicited by the Bureau of Land Management through the end of November and the Forest Service is taking comments through January 5.

Meanwhile, a Wyoming man is hoping to change the dynamics of the sage grouse issue by increasing their population through captive breeding.

Diemer True, of the True Companies (oil and gas drilling, support, pipelines, and seven ranches), and former president of the Wyoming Senate, bought Karl Baer's game bird farm in Powell.

True convinced the Wyoming Legislature to pass legislation during the 2017 session to allow him and Baer to apply for a permit to take up to 250 sage grouse eggs from the wild per year and experiment for five years with captive breeding. The idea is that birds can be released, bring up the numbers and maybe allow higher density of development in protected areas.

But no one has been very successful breeding captive sage grouse. No one has successfully released them to procreate in the wild and, if True is successful, he wants his techniques to be proprietary – he won't share them. He wants to profit from wildlife rather than take the more typical route of supporting academic research.

Gov. Matt Mead signed the captive breeding legislation into law this fall. The Wyoming Game and Fish Commission wrote very specific regulations about it, which you can read at https://wgfd.wyo.gov/Regulations/Regulation-PDFs/REGULATIONS_CH60.

Five permits are allowed, for a total withdrawal of 1,250 eggs per year, but it is doubtful that anyone besides True and Baer will qualify. Consensus among wildlife biologists I spoke to is that True will have trouble finding 250 wild eggs for his permit.

The facility requirements mean True is building new pens separated from the bird farm's other operations. Despite these best management practices, there's still a chance captive-bred birds could infect wild birds when they are released.

The Wyoming Game and Fish Department monitors sage grouse leks every spring to see how successful the previous year's breeding was. Numbers naturally vary widely year to year. The effects of captive breeding on these surveys will be included when setting hunting limits.

No one who knows sage grouse well believes they can be bred in captivity successfully. Young sage grouse learn about survival from their mothers. By contrast, the non-native pheasant captive-bred here is acknowledged to be a "put-and-take" hunting target. It hardly ever survives to breed on its own.

We can only hope that this sage grouse experiment will go well. If captive-bred chicks don't thrive in the wild, there will be some well-fed coyotes, badgers and ravens.

# 2018

## 403 Two Christmas Bird Counts – 80 miles apart – compared

Sunday, January 14, 2018, Outdoors, page E7

I took part in two different Christmas Bird Counts last month.

The Guernsey-Fort Laramie 7.5-mile diameter count circle is centered where U.S. Hwy. 26 crosses the line between Goshen and Platte counties, halfway between the towns. Guernsey's population is 1,100, Fort Laramie's is 230, while the Cheyenne count is centered on the Capitol amidst 60,000 people.

Guernsey is 80 miles north of Cheyenne, but 1,600 feet lower. Cheyenne's few small reservoirs were nearly entirely frozen this year. However, within the other count circle are Guernsey Reservoir, on the North Platte, and part of Grayrocks Reservoir on the Laramie River. There was more open water on the day of that count, Dec. 17, so you'll see more ducks listed compared to Cheyenne's, held Dec. 30.

The cliffs along the North Platte have juniper trees with berries, attracting lots of robins and solitaires. Cheyenne, on the other hand, has lots of residential vegetation and more bird feeders.

All the species in the combined list below have been seen on previous CBCs in Cheyenne, except for the canyon wren.

There were 16 people on the Cheyenne count, about 10 on the other. We take the same routes every year and statistical analysis of time and distance travelled smooths things out for scientists using our data.

Jane Dorn, the compiler for the Guernsey-Fort Laramie count, includes certain subspecies in her reports when possible. Of her 14 northern flickers, one was yellow-shafted (yellow wing-linings) like the flickers in eastern North America.

Jane also sorts out dark-eyed juncos. Of the 33 on her count, eight were slate-colored (the junco of eastern North America), one was white-winged (range centered on the Black Hills) and three were Oregon. The other 21 were either difficult to see or hybrids – the reason there are no longer multiple species of juncos with dark eyes.

Jane had four adult and two immature bald eagles. Those of us coming up from Cheyenne missed a chance to see them when we skipped Greyrocks Reservoir while delaying our trip two hours for black ice on Interstate 25 to melt.

The weather for the Cheyenne count put a damper on the number of songbirds out in the morning when we have the most people participating. Dec. 30 was when everything was thickly covered in fluffy ice crystals. Serious birders shrugged off the 7-degree temperature and were rewarded with beauty. By lunchtime, I was shrugging off layers to keep cool when the day's high reached 56 degrees.

Cheyenne count compiler Greg Johnson noted raptors were well represented this year, with 10 species observed, the rough-legged hawk the most abundant with 13 seen, and the two merlins being the most unusual.

Johnson said, "Three lingering red-winged blackbirds were visiting a feeder at the Wyoming Hereford Ranch. Otherwise, no unexpected or rare species were observed."

### Guernsey – Fort Laramie (Dec. 17, 2017) and Cheyenne (Dec. 30, 2017) Christmas Bird Count Comparison

Numbers are presented as **Guernsey-Fort Laramie/Cheyenne**; "cw" refers to "count week," birds seen during the three days before or after the count day but not on the official count day.

| | G-FL | C | | G-FL | C |
|---|---|---|---|---|---|
| Western Grebe | 6 | 0 | Common Merganser | 285 | 0 |
| Canada Goose | 2877 | 1259 | Killdeer | cw | 0 |
| Cackling Goose | 2 | 0 | Bald Eagle | 6 | 1 |
| Mallard | 67 | 76 | Northern Harrier | cw | 5 |
| Common Goldeneye | 2 | 1 | Red-tailed Hawk | 3 | 6 |
| Green-winged Teal | 45 | 0 | Ferruginous Hawk | 0 | 1 |
| Bufflehead | 1 | 0 | Rough-legged Hawk | 0 | 13 |

Numbers are presented as Guernsey-Fort Laramie/Cheyenne; "cw" refers to "count week," birds seen during the three days before or after the count day but not on the official count day.

| | G-FL | C | | G-FL | C |
|---|---|---|---|---|---|
| Sharp-shinned Hawk | 1 | 1 | Black-capped Chickadee | 31 | 0 |
| Cooper's Hawk | 0 | 1 | Mountain Chickadee | 3 | 3 |
| Golden Eagle, Adult | 1 | 0 | White-breasted Nuthatch | 2 | 1 |
| American Kestrel | 6 | 3 | Red-breasted Nuthatch | 7 | 7 |
| Merlin | 0 | 2 | Pygmy Nuthatch | 0 | 1 |
| Prairie Falcon | 1 | 1 | Brown Creeper | cw | 0 |
| Wild Turkey | 11 | 0 | Canyon Wren | 1 | 0 |
| Ring-billed Gull | 7 | 0 | Townsend's Solitaire | 58 | 6 |
| Rock Pigeon | 333 | 463 | American Robin | 144 | 5 |
| Eurasian Collared-Dove | 159 | 83 | European Starling | 202 | 353 |
| Great Horned Owl | 0 | 1 | Unidentified waxwing | 0 | 35 |
| Eastern Screech Owl | 1 | 0 | Cedar Waxwing | 7 | 0 |
| Belted Kingfisher | 4 | 1 | American Tree Sparrow | 8 | 0 |
| Downy Woodpecker | 7 | 2 | Song Sparrow | 3 | 0 |
| Hairy Woodpecker | 1 | 0 | Dark-eyed Junco | 33 | 30 |
| Northern Flicker | 14 | 5 | Unidentified blackbird | 0 | 7 |
| Northern Shrike | 2 | 0 | Red-winged Blackbird | 0 | 3 |
| Blue Jay | 1 | 4 | House Finch | 27 | 40 |
| Black-billed Magpie | 3 | 46 | Pine Siskin | 16 | 0 |
| American Crow | 11 | 168 | American Goldfinch | 102 | 10 |
| Common Raven | 2 | 32 | House Sparrow | cw | 139 |
| Horned Lark | 12 | 37 | | | |

# 404

Sunday, January 28, 2018, Outdoors, page E5

## Year of the Bird celebrates the Migratory Bird Treaty Act

This is the Year of the Bird.

It's been declared by four august organizations: the National Audubon Society, the National Geographic Society, Cornell Lab of Ornithology and BirdLife International. A hundred other organizations have joined them.

My husband, Mark, and I have been members for years of the first three, and I'm on the email list for the fourth, so I've heard the message four times since the first of the year.

The Year of the Bird celebrates the 100th anniversary of the Migratory Bird Treaty Act that protects birds. You can read the act at https://www.fws.gov/birds/policies-and-regulations/laws-legislations/migratory-bird-treaty-act.php.

The Year of the Bird is also about advocating for birds in a variety of ways.

Today you can go to the National Geographic website, https://www.nationalgeographic.org/projects/year-of-the-bird/, and sign the Year of the Bird pledge. You'll receive monthly instructions for simple actions you can take on behalf of birds. The official Year of the Bird website, www.birdyourworld.org, will also take you to the National Geographic page, and the other sponsors' websites will get you there as well.

If it is too cold for you to appreciate the birds while outside, check out National Geographic's January issue with photos by Joel Sartore. More of his bird photos for National Geographic's Photo Ark project, studio portraits of the world's animals, will be in a book coming out

this spring written by Noah Strycker, "Birds of the Photo Ark." Strycker will be speaking in Cheyenne May 14.

Meanwhile, the National Audubon Society, http://www.audubon.org/yearofthebird, is your portal to these articles so far: How Birds Bind Us, The History and Evolution of the Migratory Bird Treaty Act, The United States of Birding and Audubon's Birds and Climate Change Report. My favorite – Why Do Birds Matter? – quotes dozens of well-known authors and ornithologists.

For their part, BirdLife International, http://www.birdlife.org/worldwide/news/flyway, offers ways to think about birds. When you see your next robin, think about where it's been, what it's flown over. Think about the people in other countries who

may have seen the bird, too. Think about the work being done to protect its migratory flyways.

On the other hand, the Cornell Lab of Ornithology begins the year addressing bird appreciation. At one of their websites, https://www.allaboutbirds.org/6-resolutions-to-help-you-birdyour-world-in-2018/, Hugh Powell recommends getting a decent pair of affordable binoculars after reading this guide on how to shop for them, https://www.allaboutbirds.org/six-steps-to-choosing-a-pair-of-binoculars-youll-love/. Powell also recommends CLO's free Merlin Bird ID app to get to know your local birds better (or see www.AllAboutBirds.org). Then you can keep daily bird lists through CLO's free eBird program, including photos and sound recordings.

2018

379

The site also has an article on drinking bird-friendly, shade-grown coffee.

Or you can play CLO's new Bird Song Hero game to help you learn how to match what you hear with the visual spectrograph, https://academy.allaboutbirds.org.

Finally, Powell suggests "pay it forward" – by taking someone birding and join a bird club or Audubon chapter (locally, I'd recommend my chapter, https://cheyenneaudubon.wordpress.com/).

Earlier threats to birds caused conservationist Aldo Leopold to write in his 1949 book, A Sand County Almanack, "We face the question whether a still higher 'standard of living' is worth its cost in things natural, wild, and free. For us of the minority, the opportunity to see geese is more important than television, and the chance to find a pasqueflower is a right as inalienable as free speech."

I would say that people who appreciate birds are not a minority. And many of us agree with biologist and biodiversity definer Thomas Lovejoy, "If you take care of birds, you take care of most of the environmental problems in the world."

Now go to www.BirdYour-World.org and take the pledge and find out each month what simple action you can take on behalf of birds.

# 405 Migratory Bird Treaty Act is under attack
Thursday, February 1, Opinion, Letters to the Editor, page A9

This year is the 100th anniversary of the Migratory Bird Treaty Act. The U.S., along with co-signers Mexico, Canada, Japan and Russia, agreed to protect birds that cross our borders and theirs.

A hundred years ago, there was a battle between conservationists and industrialists, and the birds won. Industry is now held accountable for "incidental take" – birds killed unintentionally during the course of business. That has included birds hooked by long-line ocean fishing, birds attracted to oily evaporation ponds in oil and gas fields, and birds hit by wind turbines.

These kinds of hazards can add up and make a population-threatening dent. Instead, the MBTA has forced industries to pay fines or come up with ingenious solutions that save a lot of birds.

However, Wyoming Congresswoman Liz Cheney is backing U.S. House Resolution 4239, which would remove the requirement to take responsibility for incidental take. Here we are, 100 years later, fighting the battle again.

If you would like to speak up for the birds, please call Cheney's office, 202-225-2311. The polite person who answers the phone only wants to know your name, address and your opinion, so they know which column to check – anti-bird, or pro-bird and the MBTA.

# 406 Raptors are popular birds; new book celebrates them
Sunday, February 18, 2018, Outdoors, page E4

Raptors were the stars of a late January field trip taken by the Cheyenne – High Plains Audubon Society.

We visited the Rocky Mountain Arsenal National Wildlife Refuge on the outskirts of Denver, only 90 minutes from Cheyenne.

The man at the visitor center desk told us the bald eagles were at Lower Derby Lake. He was right.

Farther down the road, we found a bald eagle on top of a utility pole calmly eating something furry for lunch – either one of the numerous prairie dogs or a rabbit. Several photographers snapped away. No one got out of their cars because we were still in the buffalo pasture where visitors, for their own safety, are not allowed out of their vehicles. But vehicles make good blinds and the eagle seemed unperturbed.

Winter is a good time to look for raptors. They show up well among naked tree branches and on fence posts, though we noticed mature bald eagles look headless if they are silhouetted against a white winter sky – or the snow-whitened peaks of the Colorado Rockies. Our checklist for the Arsenal included rough-legged hawk, red-tailed hawk, and some unidentifiable hawks.

On the way home, we stopped in Fort Collins because a Harris's hawk, rare for the area, was reported hanging around the Colorado Welcome Center at the East Prospect Road exit. The center volunteers told us all about it – and that we were several days late. But they knew where the local bald eagle nests were and were proud of the other hawks that could be seen right outside the window.

Raptors, generally defined as hawks, eagles, falcons and sometimes vultures, sometimes owls, are a popular category of bird. When our Audubon chapter sponsored the Buffalo Bill Center for the West's Raptor Experience last spring, more than 100 people crowded into the biggest meeting room at the library to see live hawks, falcons and owls.

Maybe we are fascinated by raptors because their deadly talons and powerful beaks give us a little shiver of fear. Or maybe it's because they are easy to see, circling the sky or perched out in the open. Even some place as unlikely as the Interstate 25 corridor makes for good hawk-watching. I counted 11 on fence posts and utility poles in the 50 miles between Ft. Collins and Cheyenne on our way home from the field trip.

Since I was driving, I didn't give the birds a long enough look to identify them. But I bet I know who could: Pete Dunne.

Dunne watches hawks at Cape May, New Jersey, during migration. After more than 40 years, most as director of the Cape May Bird Observatory, he can identify raptor species when they are mere specks in the sky, the way motorists can identify law enforcement vehicles coming up from behind. It's not just shape. It's also the way they move.

Dunne is co-author of "Hawks in Flight: A Guide to Identification of Migrant Raptors." Last year he authored a new book with Kevin T. Karlson, "Birds of Prey, Hawks, Eagles, Falcons, and Vultures of North America."

This is not your typical encyclopedia of bird species accounts. Rather, it is Dunne introducing you to his old friends, including anecdotes from their shared past.

You will still find out the wingspan of a bald eagle, 71-89 inches, and learn about the light and dark morphs (differences in appearance) of the rough-legged hawk.

But Dunne also gives you his personal assessment of a species. For instance, he takes exception to the official description of Cooper's hawk (another of our local hawks) in the Birds of North America species accounts as being a bird of woodlands. After years of spending hunting seasons in the woods, he's never seen one there.

Dunne is even apt to recite poetry, such as this from Alfred, Lord Tennyson's "The Eagle":

He clasps the crag with crooked hands;

Close to the sun in lonely lands,

Ring'd with the azure world, he stands.

This is not a raptor identification guide, but since there are photos on nearly every page – an average of 10 per species showing birds in all kinds of behaviors – you can't help but become more familiar with them and more in awe.

At 300 pages, this is not a quick read, but it is perfect preparation for a trip to the Arsenal or for finding out more about the next kestrel you see.

# 407 How well do birds tolerate people?

Sunday, March 11, 2018, Outdoors, page E5

Every soaring bird I saw in early February along 1,300 miles of interstate highway between Nashville, Tennessee, and Fort Lauderdale, Florida, was a black vulture or turkey vulture.

However, near Vero Beach, Florida, where we were visiting Cheyenne snowbirds Karen and Fred Pannell, there was a black bird of a different shape, a magnificent frigatebird, a life bird for both me and my husband, Mark.

But about those vultures. Were they really more abundant along the interstate than away from it? Were they waiting for roadkill? We passed a couple landfill "mountains" that were big vulture magnets, too.

We think wild birds go about their lives oblivious to people, or at least avoiding us. Except for birds coming to feeders. Or ducks at the park looking for handouts.

Or Canada geese that enjoy eating the grass on park lawns and the leftover grain in farmers' fields.

We know that some human activities are detrimental to birds. But how many are beneficial to them? Chimney swifts have experienced both. We took down the old hollow trees they used to build their nests in, and they moved into our chimneys.

The speaker at February's Cheyenne – High Plains Audubon Society meeting, Cameron Nordell, relayed interesting research results on nesting ferruginous hawks and their reactions to people that could answer some of those questions. Nordell, Raptor Fellow at the University of Wyoming Biodiversity Institute, is with the Wyoming Raptor Initiative.

In his previous work in southern Alberta, Canada, Nordell and his colleagues experimented in part to see at

what distance hawks would flush from their nests as researchers approached by vehicle or on foot to check the nests for other aspects of the study.

Southern Alberta is a mix of agriculture, oil and gas and other development. The farmers and homeowners have planted trees on the prairie and the ferruginous hawks have found them to be great for nesting – they are a ground-nesting hawk otherwise. The trees give them better protection from predators.

However, along with people came another species that climbs trees and raids nests: racoons. Barns and other structures have helped increase the population of great horned owls and they too prey on the nestlings.

Ferruginous hawks nesting near the busiest roads were more tolerant than birds that had not seen as much traffic.

Approaching vehicles were tolerated better than approaching people.

Raptors have been shown to hang out by roads, looking for injured prey species. The problem is that they risk getting hit by vehicles, too.

The Wyoming Raptor Initiative (see https://wyomingbiodiversity.org/Initiatives-Programs) wants to understand the state's raptors better, including the road problem. It has two goals:

"To synthesize our scientific understanding of raptors in Wyoming so that the public, scientists, land managers and energy companies will be better informed in developing and implementing future conservation strategies and land mitigation efforts.

"To foster appreciation of raptors in Wyoming and the world through education and outreach efforts."

Nordell and his colleagues will be looking at

previous studies of raptors in Wyoming, gathering more data, talking to all kinds of people to get more information, and then they'll relay what they learn.

What will they discover about Wyoming's ferruginous hawks, for instance? What human activities help them or harm them?

Nordell also studied arctic peregrine falcons near Hudson Bay, where there were few direct human impacts. However, the weather was ferocious. Too much rain, and a young bird, poorly nourished, could succumb to the cold rainwater collecting in the cliff-face nest. Better-fed youngsters had better survival rates.

The next questions: What affects the availability of peregrine prey species and the peregrine parents' ability to bring food back to the nest? Is there any human influence on their success? Are humans linked in any way to that Arctic location getting demonstrably rainier?

What will be discovered about peregrines in Wyoming? I watched one nail a duck on a ranch reservoir just outside Cheyenne once. The human-made lake attracted the peregrine's food target – southeastern Wyoming doesn't have many natural water bodies.

I look forward to answers from the Wyoming Raptor Initiative. I'm sure they will also discover many more questions.

# 408 World-record-setting birder and author to visit Cheyenne – and Wyoming – for the first time

April 2, 2018, WyomingNetworkNews.com

World-record birder Noah Strycker is coming to speak in Cheyenne May 14, sponsored by the Cheyenne – High Plains Audubon Society and the Laramie County Library (7 p.m., 2200 Pioneer Ave., Cottonwood Room, free admission, open to the public).

Strycker is the author of the book "Birding Without Borders, An Obsession, A Quest, and the Biggest Year in the World." His talk, humorous and inspiring, will reflect the subject of his book.

Imagine travelling nonstop for a year, the year you are turning 30, taking only a backpack that qualifies as carry-on luggage. At least in this digital age, the maps Strycker needed and the six-foot stack of bird field guidebooks covering the world could be reduced to fit in his laptop.

Also, it was a year of couch surfing as local birders in many countries offered him places to stay as well as help in locating birds. There were knowledgeable bird nerds everywhere that wanted him to set the world record. First, he used https://eBird.

org to figure out where the birds would be, and then he looked up http://birdingpal. org/ to find the birders.

Strycker planned to see 5,000 species of birds, nearly half the 10,365 identified as of 2015, to break the old record of 4,000-some. But he hit that goal Oct. 26 in the Philippines with the Flame-crowned Flowerpecker and decided to keep going, totaling 6,042 species.

Strycker is looking forward to visiting Wyoming for the first time. The day after his talk, May 15, his goal is to see 100 species of birds in our state. This is not an impossible feat at the height of spring migration.

He'll have help from Wyoming's best-known birders, Jane and Robert Dorn, who wrote the book, "Wyoming Birds."

Robert has already plotted a route for an Audubon field trip that will start in Cheyenne at the Wyoming Hereford Ranch at 6 a.m. and move to Lions Park by 8:30 a.m. Soon after we'll head for Hutton Lake National Wildlife Refuge west of Laramie and some of the Laramie

Plains lakes before heading through Sybille Canyon to Wheatland, to visit Grayrocks and Guernsey reservoirs.

There's no telling what time we'll make the 100-species goal, but we expect to be able to relax and have dinner, maybe in Torrington. Anyone who would like to join us is welcome for all or part of the day. Birding expertise is not required, however, brownbag lunch, water, appropriate clothing, and plenty of stamina is. And bring binoculars. To sign up, send your name and cell phone number to mgorges@juno.com. See also https://cheyenneaudubon. wordpress.com/ for more information.

I don't know if Strycker is going for a new goal of 100 species in every state, but it will be as fun for us to help him as it was for the birders in those 40 other countries. I just hope we don't find ourselves stuck on a muddy road as he was sometimes.

Anyone, serious birder or not, can enjoy Strycker's "Birding Without Borders," either the talk or the book. The book is not a blow-by-blow description of all the

birds he saw, but a selection of the most interesting stories about birds, birders and their habitat told with delightful optimism. But I don't think his only goal was a number. I think it was also international insight. Although he's done ornithological field work on six continents, traveling provides the big picture.

Strycker is associate editor of Birding magazine, published by the American Birding Association. He's written two previous books about his birding experiences, "Among Penguins" and "The Thing with Feathers."

You can find Strycker's "Birding Without Borders" book at Barnes and Noble and online, possibly at the talk. He will be happy to autograph copies.

His latest writing is the text for National Geographic's "The Birds of the Photo Ark." It features 300 of Joel Sartore's exquisite portraits of birds from around the world, part of Sartore's quest to photograph as many of the world's animals as possible. The book came out this spring.

# 409
April 3, 2018, WyomingNetworkNews.com
## Enjoy reading nature writing in three styles: essays, trail guide and guide to field guides

Houghton Mifflin Harcourt has three very different new nature books out this spring: a compilation of nature essays; a cross between trail, travel, nature and history guides; and a guide to using field guides.

A Naturalist at Large, The Best Essays of Bernd Heinrich, 2018, $26, 285 pages.

I am a fan of this man who finds so many questions to ask and then looks for the answers, even if it means climbing a tree and waiting hours to see where the ravens come back from, and spending hours watching a dung beetle make its ball.

You'll recognize Bernd Heinrich's topics of interest if you've read his other books including "Mind of the Raven," "Racing the Antelope" and "Life Everlasting."

The essays in this new collection were published in various magazines, mostly in recent issues of Natural History Magazine. So, the book title also means the older Heinrich gets, the better his writing. I agree. If his subjects appeal to you, soil, plants, trees, insects, bees,

birds, mammals and how living things cope with the universe, you'll enjoy this book.

I especially liked his investigation of the mechanics of how yellow iris instantly pop from bud to bloom.

The Guide to Walden Pond, An Exploration of the History, Nature, Landscape, and Literature of One of America's Most Iconic Places, Robert M. Thorson, 2018, $17, 250 pages, full color.

This book won't mean much if you aren't familiar with Henry David Thoreau, essayist, poet, philosopher, abolitionist, naturalist, tax resister, development critic, surveyor, and historian. Or his two-year experiment begun in 1845 living in a tiny, bedroom-sized house he built himself at Walden Pond, outside Concord, Massachusetts. You may want to first find a copy of his book, "Walden."

Thoreau's fame helped the state set aside 335 acres as the Walden Pond State Reservation (see https://www.walden.org). And he has inspired many conservationists with words such as, "In Wildness is the preservation

of the World."

Robert Thorson sets up his book as a trail guide and while taking a Thoreau-styled amble around the pond, the reader gets a mix of history, natural history, biography and lots of beautiful photography.

Peterson Guide to Bird Identification—in 12 Steps, Steve N.G. Howell and Brian Sullivan, 2018, $18 152 pages, full-color.

This is a small book full of well-illustrated information that should be at the beginning of every bird field guide.

The intended audience is everyone, the authors say, "We include some things that may be challenging for beginning birders, and others that may seem too basic for those more advanced, but this is intentional." And that's why you'll want your own copy to study over and over.

Step 1 – Make sure you are looking at a bird. What kind? Duck, hawk, songbird?

Steps 2, 3, 4 – Where are you geographically, habitat-wise and seasonally? Despite some birds getting spectacularly lost (and

becoming the rarities birders dream of), you can assume a species of bird will show up when and where field guides say it will.

Step 5 and 6 – Is the lighting good enough and the bird close enough to identify?

Step 7 and 8 – Is the bird behaving as its presumed species does? What does it sound like? Getting a handle on birdsong will make you a terrific birder.

Step 9 – Structure – size and shape – makes an easy identifier for birds you already know. Think about those plump robins in your yard. But I would argue it is difficult to use on birds you aren't familiar with.

Step 10 – Finally, plumage! What color feathers?

Step 11 – Be aware of plumage variations.

Step 12 – Take notes – and photos.

Howell and Sullivan's book makes a good introduction or review as we fly into spring migration. And you can fit in reading it between field trips.

# 410
Sunday, May 6, 2018, Outdoors, page E6
## Keep birds safe this time of year

It's that time of year that we need to think about bird safety – migration and nesting season.

The peak of spring migration in Cheyenne is around mid-May. If you have a clean window that reflects sky, trees and other greenery,

you'll get a few avian visitors bumping into it. Consider applying translucent stickers to the outside of the window or search online for American ArBird Conservancy's Bird Tape.

If a bird hits your window, make sure your cat is not

out there picking it up. The bird may only be stunned. If necessary, put the bird somewhere safe where it can fly off when it recovers.

How efficient is your outdoor lighting? In addition to wasting money, excessive light confuses birds that

migrate at night. Cheyenne keeps getting brighter and brighter at night because people, especially businesses with parking lots, install lighting that shines up as well as down. It is also unhealthy for trees and other vegetation, not to mention

2018

383

people trying to get a good night's sleep.

Do you have nest boxes? Get them cleaned out before new families move in. Once the birds move in or you find a nest elsewhere, do you know the proper protocol for observing it?

If not, you might be interested in NestWatch, https://nestwatch.org/, a Cornell Lab of Ornithology citizen science program for reporting nesting success.

Their Nest Monitoring Manual says to avoid checking the nest in the morning when the birds are busy, or at dusk when predators are out. Wait until afternoon. Walk past the nest rather than up to it and back, leaving a scent trail pointing predators straight to the nest. And avoid bird nests when the young are close to fledging (when they have most of their feathers). We don't want them to get agitated and leave the nest prematurely.

Some birds are "flightier" than others. Typically, birds nesting alongside human activity – like the robins that built the nest on top of your porch light – are not going to abandon the nest if you come by. Rather, they will be attacking you. But a hawk in a more remote setting will not tolerate people. Back off and get out your spotting scope or your big camera lenses.

If your presence causes a young songbird to jump out of the nest, you can try putting it back in. NestWatch says to hold your hand or a light piece of fabric over the top of the nest until the young bird calms down so it doesn't jump again. Often though, the parents will take care of young that leave the nest prematurely.

Loose cats and dogs should also be controlled on the prairie between April and July – and mowing avoided. That is because we have ground-nesting birds here on the edge of the Great Plains such as western meadowlark, horned lark and sometimes the ferruginous hawk.

There will always be young birds that run into trouble, either natural or human-aided. Every wild animal eventually ends up being somebody else's dinner. But if you decide to help an injured animal, be sure the animal won't injure you. For instance, black-crowned night-herons will try to stab your eyes. It is also illegal to possess wild animals without a permit so call a licensed wildlife rehabilitator like the Cheyenne Pet Clinic, 307-635-4121, or the Wyoming Game and Fish Department, 307-777-4600.

Avoid treating your landscape with pesticides. The insect pest dying from toxic chemicals you spread could poison the bird that eats it. Instead, think of pest species as bird food. Or at least check with the University of Wyoming Extension office, 307-633-4383, for other ways to protect your lawn and vegetables.

Are you still feeding birds? We take our seed feeders down in the summer because otherwise the heat and moisture make dangerous stuff grow in them if you don't clean them every few days. Most seed-eating birds are looking for insects to feed their young anyway. Keep your birdbaths clean, too.

However, we put up our hummingbird feeder when we see the first fall migrants show up in our yard mid-July, though they prefer my red beebalm and other bright tubular flowers.

Make sure your hummingbird feeder has bright red on it. Don't add red dye to the nectar, though. The only formula that is good for hummingbirds is one part white sugar to four parts water boiled together. Don't substitute any other sweeteners as they will harm the birds. If the nectar in the feeder gets cloudy after a few days, replace it with a fresh batch.

And finally, think about planting for birds. Check out the Habitat Hero information at http://rockies.audubon.org/programs/habitat-hero-education.

Enjoy the bird-full season!

# 411 Bird counting

Sunday, July 1, 2018, Journey, page E1

The Cheyenne – High Plains Audubon Society has been holding an annual Big Day Bird Count at the height of spring migration since at least 1956 (see more at https://cheyennebirdbanter.wordpress.com). But this year we essentially did two counts five days apart.

It started with birder and author Noah Strycker visiting mid-May to give a talk at the library about his 2015 record-breaking global Big Year (6,042 species) and his book, "Birding Without Borders." He had the next day free, May 15, before heading for another speaking engagement. Naturally, we volunteered to take him birding.

He said since he'd never been to Wyoming before, he wanted to see 100 species. I enlisted the help of Bob and Jane Dorn, authors of "Wyoming Birds," and Greg Johnson, also a chapter member, whose global bird life list is just over 3,000 species.

An ambitious route was mapped out, starting at 6 a.m. with a couple of hours at the Wyoming Hereford Ranch, then Lions Park, onto Pole Mountain and over to Hutton Lake National Wildlife Refuge and the other Laramie Plains lakes. This would be followed by a drive down Sybille Canyon over to the state wildlife areas and reservoirs on the North Platte.

Thirty-six people signed up in advance for the field trip. Most couldn't come for the whole day, including the two birders from Jackson, three from Lander, one from Gillette and four from Colorado. By dinnertime, there were only 10 of us left.

After the Laramie Plains Lakes, we'd only made it to Laramie and Strycker had seen 118 species, so we had dinner there and returned to Cheyenne by 8 p.m. The day before he saw a life bird in Colorado on the way up from the airport – the lark bunting, Colorado's state bird. The

384                                                     CHEYENNE BIRD BANTER

day after the field trip, Johnson took him to see another life bird, the sharp-tailed grouse, on the way back.

Somehow the carpooling worked out – ten vehicles at the most. Strycker rode at the front of the caravan with the Dorns and saw birds the rest of us didn't. That's the way it is with road birding. But even on foot at the ranch, 30-some people didn't see all the same birds.

It was a beautiful day – not much wind – and we dodged all the rain showers. Strycker is welcome back anytime.

The following Saturday lived up to its terrible forecast so Johnson rescheduled our regular Cheyenne Big Day Bird Count for the next day, May 20, when it finally warmed up a bit and stopped raining.

Only eight of us showed up at 6:30 a.m. and represented a wide spectrum of birding experience. We searched Lions Park thoroughly, then the Wyoming Hereford Ranch and the High Plains Grasslands Research Station (permit required) – very little driving. I think we had about 80 species by 3 p.m. Four other people were birding the local area as well.

The final Big Day tally was 113. Not bad, considering we stayed within a 15-mile-diameter circle centered on the Capitol – essentially our Christmas Bird Count circle. That's consistent with recent years.

Ted Floyd, the American Birding Association's magazine editor (who birded at the ranch with Strycker, his associate editor) and I have discussed whether a birder will see more birds on their own or with a group.

Floyd birds by ear, so not having a lot of people-noise works for him. For me, I appreciate the greater number of eyeballs a group has – often looking in multiple directions – and the willingness of people to point out what they are seeing. Presumably, a group of 30 birders sees more than a group of eight. However, the larger group may be looking at several interesting birds simultaneously, making it hard to keep up.

But there's nothing much more enjoyable in spring than joining gatherings of birds and birders, or any time of year. Look for Cheyenne Audubon's field trip schedule at https://cheyenneaudubon. wordpress.com/.

## Cheyenne Big Days compared

The 119 birds with an "N" before their name were seen by Noah Strycker in southeastern Wyoming May 15. Additional birds he saw are marked *. The 113 birds with a "B" were counted in the Cheyenne area on the Cheyenne Big Day Bird Count May 20. The combined list has 145 species.

| | | | |
|---|---|---|---|
| N B Canada Goose | N Sharp-shinned Hawk | N B Belted Kingfisher | N B Ruby-crowned Kinglet |
| N B Wood Duck | N B Cooper's Hawk | B Red-headed Woodpecker | N Mountain Bluebird |
| N B Blue-winged Teal | N B Bald Eagle | N B Downy Woodpecker | B Townsend's Solitaire |
| N B Cinnamon Teal | N B Swainson's Hawk | N Hairy Woodpecker | N B Swainson's Thrush |
| N B Northern Shoveler | N B Red-tailed Hawk | B Northern Flicker | B Hermit Thrush |
| N B Gadwall | N Ferruginous Hawk | N B American Kestrel | N B American Robin |
| N American Wigeon | N Sora | N B Western Wood Pewee | N B Gray Catbird |
| N B Mallard | N B American Coot | N Least Flycatcher | B Brown Thrasher |
| B Northern Pintail | N Sandhill Crane | N Dusky Flycatcher | N B Sage Thrasher |
| N Green-winged Teal | N Black-necked Stilt | N B Cordilleran Flycatcher | N B European Starling |
| N Canvasback | N B American Avocet | N B Say's Phoebe | N McCown's Longspur |
| N B Redhead | N B Killdeer | N B Western Kingbird | N* Ovenbird |
| N Ring-necked Duck | N Least Sandpiper | N B Eastern Kingbird | N* Tennessee Warbler |
| N B Lesser Scaup | N Long-billed Dowitcher | B Warbling Vireo | N B Orange-crowned Warbler |
| N B Ruddy Duck | B Wilson's Snipe | N B Blue Jay | B MacGillivray's Warbler |
| N* Sharp-tailed Grouse | N B Wilson's Phalarope | N B Black-billed Magpie | N B Common Yellowthroat |
| N B Pied-billed Grebe | N B Spotted Sandpiper | N B American Crow | N B American Redstart |
| N B Eared Grebe | N Willet | N B Common Raven | N Northern Parula |
| N B Western Grebe | N Lesser Yellowlegs | N B Horned Lark | N B Yellow Warbler |
| B Clark's Grebe | N B Ring-billed Gull | N B Northern Rough- | B Chestnut-sided Warbler |
| N B Double- | N California Gull | winged Swallow | N Blackpoll Warbler |
| crested Cormorant | N B Black Tern | N B Tree Swallow | N B Yellow-rumped Warbler |
| N B American White Pelican | N B Forster's Tern | B Violet-green Swallow | B Wilson's Warbler |
| N B Great Blue Heron | N B Rock Pigeon | N B Bank Swallow | N Grasshopper Sparrow |
| B Great Egret | N B Eurasian Collared-Dove | N B Barn Swallow | N B Chipping Sparrow |
| N B Black- | N* White-winged Dove | N B Cliff Swallow | N B Clay-colored Sparrow |
| crowned Night-Heron | N B Mourning Dove | B Black-capped Chickadee | N B Brewer's Sparrow |
| N B White-faced Ibis | N B Eastern Screech-Owl | N B Mountain Chickadee | N B Lark Sparrow |
| N B Turkey Vulture | N B Great Horned Owl | N B Red-breasted Nuthatch | N B Lark Bunting |
| B Osprey | B Chimney Swift | N B House Wren | N Dark-eyed Junco |
| N B Golden Eagle | B Broad- | N Marsh Wren | N B White-crowned Sparrow |
| N Northern Harrier | tailed Hummingbird | B Blue-gray Gnatcatcher | N B Vesper Sparrow |

| | | | |
|---|---|---|---|
| N B Savannah Sparrow | B Lazuli Bunting | N B Brown-headed Cowbird | N B Pine Siskin |
| N B Song Sparrow | N B Yellow-headed Blackbird | N B Brewer's Blackbird | N B American Goldfinch |
| N Lincoln's Sparrow | N B Western Meadowlark | N B Common Grackle | N B House Sparrow |
| N Green-tailed Towhee | B Orchard Oriole | B Great-tailed Grackle | |
| B Western Tanager | N B Bullock's Oriole | B Evening Grosbeak | |
| N Black-headed Grosbeak | N B Red-winged Blackbird | N B House Finch | |

# 412 Burrowing owls materialize on SE Wyoming grasslands

Sunday, July 29, 2018, Outdoors, page E2

Burrowing owls were like avian unicorns for me until this spring. Mark, my husband, and I searched prairie dog towns in southeastern Wyoming to no avail.

It wasn't always like that. Fifteen years ago, there was a spot on the east edge of Cheyenne guaranteed to produce a sighting for the Cheyenne Audubon Big Day Bird Count. But the area around it got more and more built up.

I did some research through my subscription to Birds of North America, https://birdsna.org, and discovered burrowing owls don't require complete wilderness.

These owls are diurnal – they are active during the day, most active at dawn and dusk. However, when the males have young to feed, they hunt 24/7.

The eggs are laid in old animal burrows, primarily those of prairie dogs. Because prairie dogs live in colonies, the burrowing owls tend to appear in groups, too, though much smaller. Besides nesting burrows, they have roosting burrows for protection from predators. They stockpile prey in both kinds of burrows in anticipation of feeding young. One cache described in a Saskatchewan study had 210 meadow voles and two deer mice.

Western burrowing owls, from southwestern Canada to southwestern U.S., winter in Central and South America. However, there are year-round populations in parts of California, southernmost Arizona and New Mexico and western Texas and on south. But there is also a subspecies of the owl that lives in Florida and the Caribbean year round. They excavate their own burrows.

Burrowing owls breed in the open, treeless grasslands. No one is sure why, but they like to line their nesting burrows with dung from livestock. They, along with their prairie dog neighbors, appreciate how grazing animals keep the grass short. It's easier to see approaching predators.

The owls' biggest natural nest predator is the badger. Both young and adults can scare predators away from their burrows by giving a call that imitates a rattlesnake's rattle.

Short grass means it's easier to catch prey by walking or hopping on the ground as well as flying. Burrowing owls also like being near agricultural fields.

The fields attract their primary prey species: grasshoppers, crickets, moths, beetles, in addition to small mammals like mice, voles and shrews.

You would think these owls are ranchers' and farmers' best friends. However, in the Birds of North America's human impacts list are wind turbines, barbed wire, vehicle collisions, pesticides and shooting. I'm surprised by shooting.

Since western burrowing owls can't be blamed for making the holes in pastures (they only renovate and maintain burrows by kicking out dirt), I can only surmise that varmint hunters have bad eyesight and can't tell an owl from a prairie dog. It could be an easy mistake: Owls are nearly the color and size of prairie dogs and have similar round heads. Except the owls stand on long skinny legs. From a distance the owls look like prairie dogs hovering over the burrow's mound – and then if you watch long enough, they fly.

Burrowing owls have been in sharp decline since the 1960s despite laying 6 to 12 eggs per nest. The Burrowing Owl Conservation Network, http://burrowingowlconservation.org, reports the U.S. Fish and Wildlife Service lists them as "a Bird of Conservation Concern at the national level, in three USFWS regions, and in nine Bird Conservation Regions. At the state level, burrowing owls are listed as endangered in Minnesota, threatened in

Colorado, and as a Species of Concern in Arizona, California, Florida, Montana, Oklahoma, Oregon, Utah, Washington, and Wyoming."

In our state, Grant Frost, Wyoming Game and Fish Department wildlife biologist, said "(burrowing owls) are what we classify as a species of greatest conservation need (SGCN), but mostly due to a lack of information, their status is unknown. That is why these surveys were started three years ago. There are 15 surveys being done throughout the state in potential habitat ... each survey route is done three times each year during set times to occur during each of the three nesting stages – pre-incubation, incubation/hatching, and nestling."

When Grant said he could lead an Audubon field trip to see the owls and other prairie birds, 15 of us jumped at the chance.

As might be predicted from the BNA summary of the literature, the owls were in the middle of an agricultural setting of fields and pastures. We watched them hunt around a flock of sheep and enjoy the view from the tops of fence posts along an irrigation canal.

The first sightings of the morning were distant – hard to see even with a spotting scope. But as we departed

CHEYENNE BIRD BANTER

for home, driving a little farther down the road, two burrowing owls appeared much closer, and we all felt finally that we could say we'd seen them and not just flying brown smudges.

# 413 Condor visits Wyoming. Next one needs to find steel instead of lead

Sunday, August 19, 2018, Outdoors, page E2

Exciting news in the Wyoming birdwatching community: A California condor, North America's largest raptor with 9 ½-foot wingspan, was sighted July 7 west of Laramie perched on Medicine Bow Peak. The reporting birder was Nathan Pieplow. He is the author of the Peterson guide to bird sounds. Maybe he recorded it.

Wing tags printed with a big T2 declared this was a female condor hatched and raised in 2016 at the Portland, Oregon, zoo and released in March at the Vermilion Cliffs National Monument in northern Arizona.

Several people from the Laramie Audubon chapter climbed up to see the condor. Brian Waitkus got excellent photos.

Medicine Bow Peak, elevation 12,014 feet, is a popular destination for hikers who want a challenge including lightning and boulder fields. As many as a dozen hikers were congregating near the condor July 9. The condor didn't mind people but was flushed by three dogs off leash, observed Murie Audubon president Zach Hutchinson.

T2 was one of many condors released into the wild by the Peregrine Fund working to re-establish the population of this officially endangered species. In 1982, there were only 22 birds left. Today there are 500, half flying free in Arizona, Utah, California and Baja Mexico. Some are now breeding in the wild. For more, read "Condors in Canyon Country" by Sophie A. H. Osborn and https://www.peregrinefund.org/.

The distance between the Arizona release site and the peak is only 440 miles as the condor flies, not difficult for a bird that can travel 200 miles per day. T2 was spotted earlier, on June 28, near Roosevelt, Utah.

The closest previous Wyoming condor sighting was 1998, in Utah at Flaming Gorge Reservoir, which spans the Utah-Wyoming line.

T2's visit was brief. A Peregrine Fund researcher following the condor using telemetry later got the signal 30 miles away indicating the bird was not moving. By the time he arrived, the bird was dead. It's been sent to the U.S. Fish and Wildlife Service for autopsy. Foul play was not suspected.

Serendipitously, soon after the first news broke about T2, Chris Parish, director of global conservation for the Peregrine Fund, was about to drop his daughter off in Laramie. He offered to give a talk on condors sponsored by the Laramie Audubon Society and the University of Wyoming Biodiversity Institute.

In his presentation, Chris touched briefly on the history of restoring the condor population.

Condors are tough. They survived the large mammal extinction 10,000 years ago. However, they are slow to reproduce, only one chick every two years. At propagation centers, experts can get a pair to lay an extra egg to put in an incubator.

Condors live 50-60 years by avoiding predators and finding new habitat. A few are still being shot, despite condors being as harmless as turkey vultures, eating only carrion – already dead animals. They fly into powerlines and get hit by vehicles, too.

The biggest problem for condors is poisoning from lead ammunition, Chris said. When a deer is shot, the bullet disintegrates into hundreds of fragments. Often, the fragments are in the gut pile, or offal, that hunters leave in the field. Offal is the condor's main dish.

All those little lead fragments add up and eventually cause lead poisoning. Some of those lead fragments also find their way into game meat people eat. Researchers try to check the blood lead levels of all free-flying condors once a year and treat them, if necessary, before releasing them again.

Our national symbol, the bald eagle, also feeds on carcasses. In 1991, lead shot for waterfowl hunting was banned, but upland animals – and birds like the eagle – are not protected.

A few years ago, the Arizona Game and Fish Department asked hunters on the Kaibab Plateau, where condors are released, to voluntarily use steel ammunition or to remove offal. They offered each participant two free boxes of steel ammunition. Participation is now at 87 percent. A similar program is nearly as successful in Utah. California has banned lead ammunition since 2008, said Chris.

The Peregrine Fund holds shooting trials and gives away steel ammunition for hunters to test. Chris, a lifelong hunter, spouts ballistic statistics with ease.

The bottom line is lead and steel ammunition of comparable quality are nearly the same cost. However, manufacturers need encouragement to offer more variety.

Chris also said, yes, steel ammunition takes a little practice for the hunter to become proficient with it, but practice is required any time a hunter switches to the same caliber ammunition made by a different manufacturer.

Steel bullets aren't silver bullets for all wildlife problems. But maybe Wyoming can join the steel states.

That way we'll make it safer here when more condors show up.

# 414 How to prepare for international birdwatching

Sunday, September 23, 2018, Outdoors, page E2

The back-to-school sales reminded me that I have some studying to do. In a few months, Mark and I are going to Costa Rica on our first international birding trip. We are going with Bird Watcher's Digest with whom we've birded before in Florida and Texas.

Our friend, Chuck Seniawski, has been to Costa Rica five times and recommended, as did BWD, "The Birds of Costa Rica: A Field Guide" by Richard Garrigues and Robert Dean. It shows 903 species in a country 20 percent the size of Wyoming, which has only 445 species. About 200 I've seen before because they migrate up here for the summer or their range includes parts of both North and Central America.

I asked local birder Greg Johnson, veteran of many international birding trips, how he learns the birds before heading to a new destination.

Greg said he starts with the country's field guide. "I start reviewing it almost daily beginning several weeks or even months before the trip. For most trips, the tour company should be able to provide you trip reports from previous trips with the same itinerary. The trip reports should have a list of all birds they saw or heard. I then check those birds with a pencil mark in the book to focus only on those I am likely to see and ignore the rest. For example, if your trip to Costa Rica only includes the highlands and Caribbean slope, you can ignore those birds which only occur on the Pacific slope."

Mario Córdoba of Crescentia Expeditions, trip leader, has provided a list of target bird species based on our travel route including several ecolodges we'll stay at near national parks. No Pacific slope.

Greg's email continued, "If you spend enough time studying the birds you are most likely to see, you'll surprise yourself at how easy it is to ID birds you have never seen before at first sight. There are always some groups that are still hard to ID without help from a guide [bird expert] because differences between species are very subtle. In Costa Rica, these would include woodcreepers, some of the antbirds, elanias, tyrannulets, other flycatchers, etc."

There are recognizable genera in Costa Rica: hummingbird, woodpecker, wren, warbler. But then the others seem straight from Alice in Wonderland: potoo, motmot, puffbird.

Mark and I also went to eBird and looked at the bird lists for the hotspots we will be visiting and filtered them for the month we are visiting. Of 421 species we found, 338 will be unfamiliar birds.

There is an alternative to thumbing through the field guide to study the birds. Our daughter-in-law, Jessie Gorges, with a degree in marine biology from the University of Hawaii, got a job one summer surveying birds across the Great Plains. She had a couple months to learn to recognize a few hundred birds by sight and sound.

Jessie's solution is a free program called ANKI, https://apps.ankiweb.net. She created her own deck of digital flashcards with photos and birdsong recordings. It's like a game, and Jessie is the queen of complicated board and card games. The program prepares a daily quiz based on how much review and repetition it thinks you need.

But of course, even to make bird flashcards like I did 20 years ago for kids for Audubon Wyoming, printable from a CD, I need to find photographs. Finding them online or scanning pages of the field guide can help me study.

I take for granted the decades of familiarity I have with bird species in the U.S. There are groups in which I still can't distinguish individual species well, for instance, flycatchers. But at least I know they are flycatchers. On this trip I'll be leaving behind most of the birds I know.

But Greg assured me, "Once you go on an international birding trip, you'll likely get hooked and won't be able to stop. There are so many great birds that don't occur in the U.S. I'll never forget seeing my first keel-billed toucans in Belize or African penguins in South Africa."

Preparing for this trip will make me appreciate the birds I do know when I meet their tropical cousins. I never thought about our northern rough-winged swallow having a counterpart, the southern rough-winged swallow. We could see both in Costa Rica.

Meanwhile, excuse me while I begin studying in ornithological order: "Great tinamou, little tinamou, great curassow, gray-headed chachalaca, black guan, crested guan, buffy-crowned wood-partridge, least grebe, sunbittern, fasciated tiger-heron, boat-billed heron, green ibis, southern lapwing, northern jacana, white-throated crake, lesser yellow-headed vulture, king vulture, gray-headed kite, tiny hawk...."

CHEYENNE BIRD BANTER

# 415 Can you ID that bird?

Sunday, October 14, 2018, Outdoors, page E2

## Cheyenne bird book coming in late October

I'm very good at procrastinating. How about you? But I've discovered there are some advantages.

From 2008-10, I wrote "Bird of the Week" blurbs for the Wyoming Tribune Eagle to run in those sky boxes at the top of the To Do section pages. But they needed photos.

I asked one of the Wyobirds e-list subscribers from Cheyenne, Pete Arnold. Pete invites people to join his own e-list, where he shares his amazing bird photos. He generously agreed.

Using the checklist of local birds prepared by Jane Dorn and Greg Johnson for the Cheyenne - High Plains Audubon Society, I chose 104 of the most common species and set to work figuring out which weeks to assign them to. Pete perused his photos and was able to match about 90 percent.

We eventually met in person – at Holliday Park. Pete stopped on his way to work one morning to snap waterfowl photos and I was walking a friend's dog and counting birds. We discovered we have several mutual friends.

By the time our two-year project was over, I'd heard about making print-on-demand books, uploading files via internet for a company

to make into a book. I rashly promised Pete I'd make a book of our collaboration. After the paper published BOW, I had all the rest of the rights to the text. And I've had college courses in editing and publishing.

Here's where my procrastination comes in. Over the next six years, my family had three graduations, three weddings, three funerals and two households to disassemble, not to mention my husband Mark retired and wanted to travel more.

Finally, a couple years ago, I gave print-on-demand a trial run through Amazon, designing my small book about quilt care. I realized then the bird book would be beyond my talents and software. I considered learning InDesign but also started looking for a professional.

I discovered, through the social media site LinkedIn, that Tina Worthman designed books in her spare time. We'd started talking when she got the job as director of the Cheyenne Botanic Gardens. No more spare time.

However, Tina recommended Chris Hoffmeister and her company, Western Sky Design. What a great match – she's a birder! I didn't have to worry about her mismatching photo and text. And she could speak to

Pete about image properties and other technicalities.

The book features a 6 x 6-inch image of each bird. Chris asked Pete to provide bigger image sizes, since the small ones he'd used for the paper would be fuzzy. He also had to approve all the cropping into the square format. But the upside of my procrastination is he had more photos to choose from.

There were still a few species Pete didn't have and so we put out a call on Wyobirds. We got help from Elizabeth Boehm, Jan Backstrom and Mark Gorges.

Meanwhile, even though the WTE features editor at the time, Kevin Wingert, had originally edited BOW, I sent my text for each species and all the other parts of the book (introduction, acknowledgements, word from the photographer, bird checklist, resources list) to Jane Dorn, co-author of the book "Wyoming Birds." Another friend, Jeananne Wright, a former technical writer and editor, and non-birder, caught a few ambiguities and pointed out where I'd left non-birders wondering what I meant.

The title of the book was the last step. Instead of naming it Bird of the Week, two years' worth of bird images and written bird impressions/trivia are

organized differently. The title is "Cheyenne Birds by the Month, 104 Species of Southeastern Wyoming's Resident and Visiting Birds."

The book is being printed by local company PBR Printing – print-on-demand is too expensive for multiple copies.

While the book will be available late October at the Wyoming State Museum and other local outlets, our major marketing partner is the Cheyenne Botanic Gardens, a natural fit since it is in the middle of Lions Park, a state Important Bird Area.

The Gardens will have the book available at their gift shop and at two book signings they are hosting: Tuesday, Nov. 20, 11:30 a.m. – 1 p.m. and Sunday, Dec. 9, 1 – 3 p.m., 710 S. Lions Park Dr.

You can get a sneak peak, and Pete's behind the camera stories, at our presentation for Cheyenne Audubon at 7 p.m. Oct. 16, in the Cottonwood Room at the Laramie County Library, 2200 Pioneer Ave.

For more information about the book and updates on where to find it, see Yucca Road Press, https://yuccaroadpress.com/.

It took part of a village to make this book and we are hoping the whole village will enjoy reading it.

2018

# 416 Benefit birds (and yourself) with feeders

Sunday, November 11, 2018, Outdoors, page E2

Your backyard may look empty after the leaves fall, but you can fill it with birds by offering them shelter, water and food.

There is some debate on whether feeding wild birds is good for them. But in moderation – the birds find natural food as well – I think it is a great way to increase appreciation for birds.

A bird feeder is no substitute for providing trees and bushes for birds to perch on or take shelter from weather and predators. Birds can also pick the seeds and fruits – or pick dormant insects out of the bark. Provide evergreen as well as deciduous trees and shrubs plus native perennial wildflowers.

Water is nice to have out. The birds appreciate drinking it and bathing in it. But if you can't scrub out the gunk regularly, it's better not to bother with it. In winter you'll want to skip concrete and ceramic baths in favor of plastic since freezing water might break them. The best winter bird bath we ever had was the lid of a heavy plastic trash can – we could pop the ice out.

Feeding seed-eating birds – house finch, goldfinch, junco, pine siskin – is as easy as scattering seed on the ground. But here are tips to benefit you and the birds more.

**1. Black oil sunflower seed is the one best bird seed for our area.** Seed mixes usually have a lot of seed our birds won't eat and then you must sweep it up before it gets moldy.

**2. Put out only as much seed as you can afford each day (and can clean up after).** If it lasts your local flock only an hour, be sure to put the seed out at a time of day you can enjoy watching the birds. They'll learn your schedule.

**3. Tube-type feeders and hopper feeders keep seed mostly dry.** Clean them regularly so they don't get moldy. Consider hanging them over concrete to make it easier to clean up the seed hulls.

**4. If you don't like sweeping up sunflower seed hulls** or are concerned that the hulls will kill your lawn, consider paying more for hulled sunflower seeds.

**5. Spilled seed under the feeder attracts the ground feeders,** like juncos, those little gray birds. They like elevated platform feeders too.

**6. If you have loose cats in your neighborhood,** consider outlining the spilled-seed area under your feeder with 2-foot-tall wire fencing all the way around. It's enough of an obstacle to make approaching cats jump so the birds will notice the break in their stealthy approach.

**7. Put your feeder close to the window you will watch from.** It's more fun for you, and the birds are less likely to hit the window hard as they come and go. They get used to activity on your side of the glass.

**8. Once you have the regulars showing up,** probably the house finches – striped brown and the males have red heads – and house sparrows – pale gray breasts, chestnut-brown backs, consider putting up a special feeder for the nyjer thistle seed that goldfinches and pine siskins love so much.

**9. Seed cakes are popular with chickadees and nuthatches.** They require a little cage apparatus to hold them.

**10. Suet-type cakes are popular** with downy woodpeckers and flickers.

**11. Squirrels like bird seed too.** You can add a cone-shaped deterrent above or below a feeder so they can't get to it. Or ask your dog to chase the squirrels. If you get more than a couple squirrels, quit feeding birds for a week or so and see if the squirrels won't move somewhere else. The birds will come back.

**12. A sharp-shinned or a Cooper's hawk may be attracted to your feeder,** though they are coming by for a finch or sparrow snack instead of seed. This means that you have successfully attracted animals from the next trophic level and contributed to the web of life.

**13. Take pictures.** Look up the birds and learn more about them through websites like www.allaboutbirds.org.

**14. Take part in citizen science programs like www. eBird.org and Project FeederWatch.** Check my Bird Banter archives for more information, www.Cheyenne-BirdBanter.wordpress.com.

# 417 Try these bird and wildlife books for winter reading and gift giving

Sunday, December 16, 2018, Outdoors, page E2

Several books published this year about birds and other animals I recommend to you as fine winter reading – or gift giving.

The first, **"How to be a Good Creature, A Memoir in Thirteen Animals,"** is a memoir by Sy Montgomery, a naturalist who has written many children's as well as adult books about animals.

Montgomery has been around the world for her research. Some of the animals she met on her travels and the animals she and her husband have shared their New Hampshire home with have taught her important life lessons: dog, emu, hog, tarantula, weasel, octopus.

This might make a good read-aloud with perceptive middle-school and older children.

**"Warblers & Woodpeckers, A Father-Son Big Year**

390       CHEYENNE BIRD BANTER

of Birding" by Sneed B. Collard III was a great read-aloud. For two weeks every evening, I read it to my husband, Mark, while he washed the dishes – a long-standing family tradition.

Like Montgomery, Collard is a naturalist and author, though normally he writes specifically and prolifically for children. He lives in western Montana.

When his son is turning 13, Collard realizes he has limited time to spend with him before his son gets too busy. Birdwatching becomes a common interest, though his son is much more proficient. They decide to do a big year, to count as many bird species as possible, working around Collard's speaking schedule and taking friends up on their invitations to visit.

There are many humorous moments and serious realizations, life birds and nemesis birds, and a little snow and much sunshine. Mark plans to pass the book on to our younger son who ordered it for him for his birthday.

Two Wyoming wildlife biologists, Matthew Kauffman and Bill Rudd, who have spoken at Cheyenne Audubon meetings on the subject, are part of the group that put together "Wild Migrations, Atlas of Wyoming's Ungulates." I ordered a copy sight unseen.

We know that many bird species migrate, but Wyoming is just now getting a handle on and publicizing the migrations of elk, moose, deer, antelope, bighorn sheep, mountain goat and bison, thanks to improved, cheaper tracking technology.

Each two-page spread in this over-sized book is an essay delving into an aspect of ungulates with easy-to-understand maps and graphs. For example, we learn Wyoming's elk feed grounds were first used in the 1930s to keep elk from raiding farmers' haystacks and later to keep elk from infecting cattle with brucellosis.

Then we learn that fed elk don't spend as much time grazing on summer range as unfed elk, missing out on high-quality forage 22 to 30 days a year. Shortening the artificial feeding season in spring might encourage fed elk to migrate sooner, get better forage, and save the Wyoming Game and Fish Department money.

This compendium of research can aid biologists, land managers and landowners in smarter wildlife management. At the same time, it is very readable for the wildlife enthusiast. Don't miss the foreword by novelist Annie Proulx.

Thanks, Houghton Mifflin Harcourt, for sending me a copy of the newly revised **"Peterson Field Guide to Western Reptiles & Amphibians"** by Robert C. Stebbins and Samuel M. McGinnis to review. I now know that what friends and I nearly stepped on while hiking last summer was a prairie rattlesnake, one of 12 kinds of rattlers found in the west.

There are 40-plus Peterson field guides for a variety of nature topics, all stemming from Roger Tory Peterson's 1934 guide to the birds of eastern North America. I visited the Roger Tory Peterson Institute in Jamestown, New York, this fall and saw his original artwork.

The reptile and amphibian guide first came out in 1966, written and illustrated by the late Stebbins. In its fourth edition, his color plates still offer quick comparisons between species. Photos now offer additional details and there are updated range maps and descriptions of species life cycles and habitats. It would be interesting to compare the maps in the 1966 edition with the new edition since so many species, especially amphibians, have lost ground.

I would be doing local photographer Pete Arnold a disservice if I didn't remind you that you can find our book, **"Cheyenne Birds by the Month"** at the Cheyenne Botanic Gardens, Wyoming State Museum, Cheyenne Depot Museum, Riverbend Nursery and PBR Printing. People tell us they are using Pete's photos to identify local birds. I hope the experience encourages them to pick up a full-fledged bird guide someday by Peterson, Floyd, Sibley or Kaufman.

# 2019

**418** Sunday, January 13, 2019, Outdoors, page E2
## Costa Rica's birds awe Wyoming birders

"Rufous motmot, collared aracari, bronze-tailed plumeleteer, bare-throated tiger-heron, yellow-throated toucan, golden-browed chlorophonia, white-collared manakin"—these were some of the names that rolled off our tongues as my husband, Mark, and I spotted birds in Costa Rica on a trip in early November.

I saw two species endemic to Costa Rica found nowhere else (remember, it's only 20 percent the size of Wyoming): the coppery-headed emerald, a hummingbird, and Cabanis's ground-sparrow, on the edge of a new clearing for an apartment building.

We saw 32 regional endemics, often meaning the species is found only in Costa Rica and neighboring Panama. My favorite, the slaty flowerpiercer, cleverly pierces the base of large flowers to extract nectar. Later, hummingbirds come by and get nectar too.

We drove up Cerro de la Muerte (Mountain of Death), to 11,400 feet where all the communications towers are, to find the volcano junco. It's another regional endemic, cousin of the juncos under our feeders in winter. It obligingly hopped around in front of us.

Of the 234 species I saw in seven straight days of birding, 187 were life birds. The others, mostly migrants, I'd seen in North America previously.

The top six bird groups I saw were hummingbirds (27 species), flycatchers (23), warblers (17), tanagers (12), woodpeckers (10) and wrens (9). Mario Cordoba H., our guide, explained Costa Rica has a lot of bird diversity (922 species), but not a lot of any one species – no big flocks.

Mario, a native of Costa Rica, has been in the guiding business more than 20 years. Bird Watcher's Digest contracted with his company, Crescentia Expeditions, to plan and guide the trip. Mario included a variety of habitats and alternated hikes in the forest to see elusive birds like streak-headed woodcreepers with stops for nectar feeder stations where bright-colored birds like the fiery-throated hummingbird were the target of everyone's cameras.

Feeding stations filled with fruit at one ecolodge attracted the turkey-sized, prehistoric-looking great curassow. A frequent feeder visitor everywhere was the blue-gray tanager. It reminded me of our mountain bluebird. I even saw it buzzing around our bus, checking out the sideview mirrors and roof, the way the bluebirds do in spring.

There are many aspects to travelling in Central America beyond birding. For instance, lodging. Our first and last nights, we stayed at two different boutique hotels. Hotel Bougainvillea is the one with 10 acres of bird-filled gardens.

The three ecolodges in between were in rural areas and a little more rustic: Arenal Observatory Lodge, Selva Verde and Paraiso Quetzal. Mario picked these for their proximity to bird diversity. There are more independently owned lodges scattered across the country.

For lunch and dinner, we often had "Typical Plate" – rice, beans, vegetables and meat (chicken, beef, pork). Up in the mountains, trout was an option because people farm trout there.

Some of our travelling companions tired of beans and rice, and tired of the rain – we were maybe a little early anticipating the dry season – but otherwise, we were a congenial group of 12, plus Mario; Dawn Hewitt, Bird Watcher's Digest's managing editor; and Ricardo, our fearless bus driver. He was also great at spotting birds and taking photos through the spotting scope with our smart phones without an adaptor. I'm going to have to learn that art. There were no bird snobs. Everyone wanted to help everyone see birds.

Costa Rica has been a leader in eco-tourism. Its map shows a large percentage of land in national parks and preserves.

Mountain farmers have been encouraged to hang on to their wild avocado trees, providing the favorite food and habitat of the resplendent quetzal. It is the green bird with the nearly 3-foot-long tail feathers revered by the ancient Aztecs and Mayans. In return, the Costa Rica Wildlife Foundation's quetzal project brings birdwatchers out to see them, paying the farmer $5 a head – not a small sum in the local economy.

We saw dangerous animals. In the dim light along the trail at La Selva Biological Station, there was a bright yellow eyelash pit viper arranged on the side of a log. The mantled howler monkeys overhead were watching visitors as much as being watched. Mosquitoes, however, were nearly non-existent. Mark and I wore our permethrin-treated field clothes anyway.

I think how neat it would be if Wyoming, too, had a cadre of trained naturalist guides and ecolodges in the vicinity of more of our interesting wildlife – not just the elk and wolves.

# 419 Careful what you wish for

Sunday, February 10, 2019, Outdoors, page E2

## Wind development on the Belvoir Ranch has its downsides

This month marks the 20th anniversary of my first Bird Banter column for the Wyoming Tribune Eagle. I wrote about cool birds seen on the ponds at the Rawhide coal-powered plant 20 miles south of Cheyenne.

This month's topic is also connected to Rawhide. It's NextEra's 120-turbine Roundhouse Wind Energy Center slated partly for the City of Cheyenne's Belvoir Ranch.

Roundhouse will stretch between I-80 south to the Wyoming border and from a couple miles west of I-25 and on west 12 miles to Harriman Road. The Belvoir is within. It's roughly a 2- to 3-mile-wide frame on the north and west sides. All the power will go to Rawhide and tie into Front Range utilities.

The 2008 Belvoir masterplan designated an area for wind turbines. In the last 10 years, I've learned about wind energy drawbacks. I wish the coal industry had spent millions developing clean air technology instead of fighting clean air regulations.

We know modern wind turbines are tough on birds. Duke Energy has a robotic system that shuts down turbines when raptors approach. Roundhouse needs one – a raptor migration corridor exists along the north-south escarpment along its west edge.

But in Kenn Kaufman's new book, "A Season on the Wind," he discovers that a windfarm far from known migration hot spots still killed at least 40 species of birds. Directly south of the Belvoir, 125 bird species have been documented through eBird at Soapstone Prairie Natural Area and 95 at Red Mountain Open Space. Both are in Colorado, butting against the state line.

Only a few miles to the east, Cheyenne hotspots vary from 198 species at Lions Park to 266 at Wyoming Hereford Ranch, with as many as 150 species overall observed on single days in May. With little public access to the Belvoir since the city bought it in 2003 (I've been there on two tours and the 2016 Bioblitz), only NextEra has significant bird data, from its consultants.

There are migrating bats to consider, plus mule deer who won't stomach areas close to turbines – even if it is their favorite mountain mahogany habitat on the ridges. The Wyoming Game and Fish Department can only suggest mitigation and monitoring measures.

There are human safety and liability issues. The Friends of the Belvoir wants a trailhead on the west edge with trails connecting to Red Mountain and Soapstone. Wind turbines don't bother

them. However, during certain atmospheric conditions, large sheets of ice fly off the blades – "ice throw." Our area, the hail capital, could have those conditions develop nearly any month of the year.

The noise will impact neighbors (and wildlife too) when turbines a mile away interfere with sleep. Disrupted sleep is implicated in many diseases.

Low frequency pulses felt 6 miles away (the distance between the east end of the windfarm and city limits) or more cause dizziness, tinnitus, heart palpitations and pressure sensations in the head and chest. The Belvoir will have bigger turbines than those on Happy Jack Road, reaching 499 feet high, 99 feet higher.

A minor issue is the viewshed. In Colorado, the public and officials worked to place the transmission line from the Belvoir to Rawhide so that it wouldn't impact Soapstone or Red Mountain. What will they think watching Roundhouse blades on the horizon?

Because this wind development is not on federal land, it isn't going through the familiar Environmental Impact Statement process. I'd assume the city has turbine placement control written into the lease.

The first opportunity for

the public to comment at the county level is Feb. 19. And in advance, the public can request to "be a party" when the Wyoming Industrial Siting Council meets to consider NextEra's permit in March.

NextEra held an open house in Cheyenne on Nov. 28. They expect to get their permits and then break ground almost immediately. This speedy schedule is so the windfarm is operational by December 2020, before federal tax incentives end.

It doesn't seem to me that we – Cheyenne residents – have adequate time to consider the drawbacks of new era wind turbines – for people or wildlife. Look at the 2008 Master Plan, http://belvoirranch.org. Is it upheld by spreading wind turbines over the entire 20,000 acres, more than originally planned? People possibly, and wildlife certainly, will be experiencing low frequency noise for 30 years.

At the very least, I'd like to see NextEra move turbines back from the western boundary two miles, for the good of raptors, other birds, mule deer, trail users, and the neighbors living near Harriman Road. The two southernmost sections are already protected with The Nature Conservancy's conservation easement.

What I'd really like to see instead is more solar

development on rooftops and over parking lots in Cheyenne. Or a new style of Wyoming snow fence that turns wind into energy while protecting highways.

# 420 BirdCast improving birding – and bird safety

Sunday, March 17, 2019, Outdoors, page E2

Last year, the folks at Cornell Lab of Ornithology improved and enhanced BirdCast, http://birdcast.info/. You can now get a three-night forecast of bird migration movement for the continental U.S.

This not only helps avid birders figure out where to see lots of birds but helps operators of wind turbines know when to shut down and managers of tall buildings and structures when to shut the lights off (birds are attracted to lights and collide), resulting in the fewest bird deaths.

The forecasts are built on 23 years of data that relate weather trends and other factors to migration timing.

Songbird migration is predominately at night. Ornithologists discovered that radar, used to detect aircraft during World War II and then adapted for tracking weather events in the 1950s, was also detecting clouds of migrating birds.

There is a network of 143 radar stations across the country, including the one by the Cheyenne airport. You can explore the data archive online and download maps for free.

CLO's Adriaan Doktor sent me an animation of the data collected from the Cheyenne station for May 7, 2018,

one of last spring's largest local waves of migration. He is one of the authors of a paper, "Seasonal abundance and survival of North America's migratory avifauna," based on radar information.

At the BirdCast website, you can pull up the animation for the night of May 6-7 and see where the migrating birds were thickest across the country. The brightest white clouds indicate a density of as many as 50,000 birds per kilometer per hour – that's a rate of 80,500 birds passing over a mile-long line per hour. Our flight was not that bright, maybe 16,000 birds crossing a mile-long line per hour. A strong flight often translates into a lot of birds coming to earth in the morning – very good birdwatching conditions. Although if flying conditions are excellent, some birds fly on.

I also looked at the night of May 18-19, 2018, the night before last year's Cheyenne Big Day Bird Count – hardly any activity. The weather was so nasty that Saturday, our bird compiler rescheduled for Sunday, which was not a big improvement. We saw only 113 species.

Twenty-five years ago, the third Saturday of May could yield 130 to 150 species. Part of the difference is the greater number of expert

Audubon birders who helped count back then. Birding expertise seems to go in generational waves.

But we also know that songbird numbers are down. I read in Scott Weidensaul's book, "Living on the Wind," published in 1999, about Sidney Gauthreaux's 1989 talk at a symposium on neotropical migrants. He used radar records to show that the frequency of spring migrant waves across the Gulf of Mexico was down by 50 percent over 30 years. Radar can't count individual birds or identify species, but we know destruction or degradation of breeding and wintering habitat has continued as people develop rural areas.

But I also wonder if, along with plants blooming earlier due to climate change, the peak of spring migration is earlier. A paper by scientists from the University of Helsinki, due to be published in June in the journal "Ecological Indicators," shows that 195 species of birds in Europe and Canada are migrating on average a week earlier than 50 years ago, due to climate change.

Would we have been better off holding last year's Big Day on either of the previous two Saturdays? I looked at the radar animations for the preceding nights in 2018,

and yes, there was a lot more migration activity in our area than on the night before the 19th. Both dates also had better weather.

As much fun as our Big Day is – a large group of birders of all skill levels combing the Cheyenne area for birds from dawn to dusk (and even in the dark) – and as much effort as is put into it, there has never been a guarantee the Saturday we pick will be the height of spring migration.

The good news is that in addition to our Big Day, we have half a dozen diehard local birders out nearly every day from the end of April to the end of May adding spring migration information to the eBird.org database. It's a kind of addiction, rather like fishing, wondering what you'll see if you cast your eyes up into the trees and out across the prairie.

I recommend you explore BirdCast.info (and eBird.org) and sign up to join Cheyenne Audubon members for all or part of this year's Big Day on May 18. See the chapter's website and/or sign up for the free e-newsletter, https://cheyenneaudubon.wordpress.com/newsletters/.

# 421 Four book reviews: Birds and bears

Sunday, April 21, 2019, Outdoors, page E2

"Peterson Reference Guide to Sparrows of North America" by Rick Wright. Copyright 2019, Houghton Mifflin Harcourt.

Birders can be nerdy. This is a book for sparrow nerds and would-be nerds.

There are three main parts to Wright's multi-page treatment of each of 76 sparrow species or major subspecies: history of its scientific description and naming, field identification, and range and geographic variation.

Did you know the pink-sided junco (dark-eyed junco subspecies) has Wyoming roots? A Smithsonian collecting trip, the South Pass Wagon Road expedition, made it to Fort Bridger, in the far southwest corner of what is now Wyoming, in the spring of 1858. Constantin Charles Drexler, assistant to the surgeon, collected a sparrow identified as an Oregon junco and shipped it back to Washington, D.C.

About 40 years later, experts determined it was the earliest collected specimen of pink-sided junco and Drexler, who went on many more collecting forays, lives on, famous forever on the internet.

Wright's feather-by-feather field identification comparisons will warm a birder's heart, as will the multiple photos. However, over half of each account is devoted to range and geographic variation. No map. No list of subspecies by name. To the uninitiated, including me, apparently, Wright's writing rambles. If you would become an expert on North American sparrows, you will have to study hard.

"Peterson Field Guide to Bird Sounds of Western North America" by Nathan Pieplow

Copyright 2019, Houghton Mifflin Harcourt.

It's here, the western counterpart of Nathan Pieplow's eastern book I reviewed in July 2017, https://cheyennebirdbanter.wordpress.com/2017/07/24/.

Each species gets a page with a small range map and a short description of habitat. The tiny painting of the male bird (and female if it looks different) is not going to help you with feather-splitting identification problems. It's just a faster way to identify the page you want if you are already familiar with the bird.

Each species' page has diagrams of the sounds it makes (spectrograms). They aren't too different from musical notation. The introduction will teach you how to read them. In addition to the standard index for a reference book or a field guide, there is an index of spectrograms. It works like a key, dividing bird sounds into seven categories; each of those are subdivided, and each subdivision lists possible birds.

Then you go online to www.PetersonBirdSounds.com to listen. I looked up one of my favorite spring migrants, the lazuli bunting. There are 15 recordings. Birds can have regional accents, so it was nice to see recordings from Colorado, including some made by Pieplow, a Coloradoan. If you've ever wanted to study birdsongs and other bird sounds, this is the field guide for you.

"A Season on the Wind, Inside the World of Spring Migration" by Kenn Kaufman

Copyright 2019, Houghton Mifflin Harcourt.

I referenced the advance reading copy of this book a couple months ago when discussing the coming development of the wind farm at Cheyenne's Belvoir Ranch. It gave me insights into the impact of wind energy on birds and bats.

The larger part of this book is about spring migration where birds and birdwatchers congregate in droves along the southwest shore of Lake Erie.

It's as much about the birds as it is the community of birders, beginning with those year-round regulars at the Black Swamp Bird Observatory like Kaufman and his wife, Kimberly Kaufman, the executive director, and the migrant birdwatchers who come from all over the world, some year after year.

Even if you know a lot about bird migration, this is worth a read just for the poetry of Kaufman's prose as he describes how falling in love with Kimberly brought him to northwestern Ohio, where he fell in love again – with the Black Swamp, a place pioneers avoided.

"Down the Mountain, The Life and Death of a Grizzly Bear" by Bryce Andrews

Copyright 2019, Houghton Mifflin Harcourt.

Are you familiar with the genre "creative nonfiction"? It means a book or other piece of writing is factual, but uses literary conventions like plot, character, scene, suspense. This is a suspenseful story. We already expect a death, based on the book's subtitle.

Rancher-writer-conservationist Andrews documents how a bear he refers to as Millie, an experienced mother with three cubs, gets in trouble in the Mission Valley of western Montana despite his efforts to protect her and other bears from their worst instincts.

Don't turn out the lights too soon after following Andrews into the maze of field corn where grizzlies like to gather on a dark night.

# 422 Cheyenne Big Day birders count 112 species

Sunday, June 23, 2019, Outdoors, page E2

No two Cheyenne Big Day Bird Counts at the height of spring migration have the exact same weather, people or bird list, which is why it is so exciting to see what happens.

This year, on May 18, we had decent weather. Last year we rescheduled because of a snowstorm – almost to be expected in mid-May lately. However, by afternoon, we had a couple showers of "graupel" – soft hail or snow pellets.

One of our best local birders, Greg Johnson, stayed home sick. Instead, we were joined by two excellent birders from out of town. Zach Hutchinson is the Audubon Rockies community naturalist in Casper. Part of his job is running five bird banding stations. In handling so many birds, he's learned obscure field marks on species we don't see often. If you shoot a bird with a digital camera, you can examine the photo closely for them.

The other visiting birder was E.J. Raynor. He came up from Fort Collins, Colorado, because he was our designated chaperone for birding the High Plains Grasslands Research Station. The south side of the station is now designated as the High Plains Arboretum and open to the public, but the area behind the houses is not. Normally we put in for a permit and this year we got E.J. instead.

He works for the Agricultural Research Service which operates the station. I thought he might be bored walking around with us, but his recent PhD is in ornithology, so I convinced him he should join us for as much of the day as possible, especially for the Wyoming Hereford Ranch part. People from all over the world visit it – including a Massachusetts tour guide and his 14 British birders a week before.

WHR put on a good show and E.J. and Zach were able to identify a female Rose-breasted Grosbeak, an eastern bird, which is nearly identical to a female black-headed grosbeak, a western bird.

We didn't get out to the station until early afternoon and then got graupeled and didn't find a lot of birds so I'm glad E.J. came early.

Counting as a group started at 6:30 a.m. at Lions Park. Surprisingly, we had people up at that hour who are new to birding. We hope they will join us again. I never get tired of seeing beginners get excited about birds.

By dusk, after Mark and I checked some of our favorite birding spots, the total bird list for the day looked like it might be about 90 species. But the next day we held a tally party at a local restaurant and the contributions of all 25 participants, including those who birded on their own, brought the total up to 112. Dennis Saville birded Little America, Chuck Seniawski birded F.E. Warren Air Force Base and Grant Frost covered some of the outer areas.

Now that most birders in Cheyenne use the global database eBird.org every day to document their sightings, the picture of spring migration is even more interesting than the single Big Day held each of the last 60 years. Migration ebbs and flows. Maybe we need to declare a Big Month and go birding every day in May.

## 2019 Cheyenne Big Day Bird Count, 112 Species

| | | | |
|---|---|---|---|
| Canada Goose | Ring-billed Gull | Western Kingbird | Mountain Bluebird |
| Blue-winged Teal | Caspian Tern | Eastern Kingbird | Swainson's Thrush |
| Cinnamon Teal | Double-crested Cormorant | Plumbeous Vireo | American Robin |
| Northern Shoveler | American White Pelican | Blue Jay | Gray Catbird |
| Gadwall | Great Blue Heron | Woodhouse's Scrub-Jay | European Starling |
| Mallard | Black-crowned Night-Heron | Black-billed Magpie | House Finch |
| Northern Pintail | Turkey Vulture | American Crow | Pine Siskin |
| Redhead | Osprey | Common Raven | Lesser Goldfinch |
| Lesser Scaup | Northern Harrier | Horned Lark | American Goldfinch |
| Common Goldeneye | Cooper's Hawk | Northern Rough- | Chestnut-collared Longspur |
| Common Merganser | Swainson's Hawk | winged Swallow | McCown's Longspur |
| Ruddy Duck | Red-tailed Hawk | Tree Swallow | Chipping Sparrow |
| Eared Grebe | Great Horned Owl | Violet-green Swallow | Clay-colored Sparrow |
| Western Grebe | Belted Kingfisher | Bank Swallow | Lark Sparrow |
| Rock Pigeon | Downy Woodpecker | Barn Swallow | White-crowned Sparrow |
| Eurasian Collared-Dove | Hairy Woodpecker | Cliff Swallow | Vesper Sparrow |
| Mourning Dove | Northern Flicker | Mountain Chickadee | Song Sparrow |
| Broad-tailed Hummingbird | American Kestrel | Red-breasted Nuthatch | Lincoln's Sparrow |
| American Coot | Western Wood-Pewee | White-breasted Nuthatch | Green-tailed Towhee |
| American Avocet | Least Flycatcher | Brown Creeper | Spotted Towhee |
| Killdeer | Dusky Flycatcher | House Wren | Yellow-headed Blackbird |
| Wilson's Phalarope | Say's Phoebe | Blue-gray Gnatcatcher | Western Meadowlark |
| Spotted Sandpiper | Cassin's Kingbird | Ruby-crowned Kinglet | Orchard Oriole |

| | | | |
|---|---|---|---|
| Bullock's Oriole | Worm-eating Warbler | Magnolia Warbler | Black-headed Grosbeak |
| Red-winged Blackbird | Northern Waterthrush | Yellow Warbler | Lazuli Bunting |
| Brown-headed Cowbird | Orange-crowned Warbler | Yellow-rumped Warbler | House Sparrow |
| Brewer's Blackbird | MacGillivray's Warbler | Wilson's Warbler | |
| Common Grackle | Common Yellowthroat | Western Tanager | |
| Great-tailed Grackle | American Redstart | Rose-breasted Grosbeak | |

May 2019, Cheyenne – High Plains Audubon Society "Flyer"

# 423 Giving away the ranch: What the Roundhouse Wind Energy Project application tells us

My friends and I are amateurs at tilting at windmills on behalf of wildlife and recreation. We are concerned about NextEra's Roundhouse Wind Energy Project slated to begin construction in August at the Belvoir Ranch, owned by the City of Cheyenne, and on adjoining private and state land.

The 120 499-foot-tall turbines will be scattered over 49,000 acres a few miles west of Cheyenne, from I-80 south to the state line and nearly from I-25 to Harriman Road (Wyoming Highway 218).

The friends I've met with are the representatives from The Nature Conservancy, Granite Canon Environmental Committee, Wyoming Pathways and Cheyenne Mountain Bike Association. I represent Cheyenne Audubon.

We've missed opportunities to voice our concerns due to inexperience but met several times with Roundhouse project director Ryan Fitzpatrick, Cheyenne native (I knew his 4th grade teacher) and Fort Collins resident. He's charming. But he works for NextEra. He told us to present our "asks" on behalf of wildlife and recreation but few were accommodated in the Roundhouse application

to the Wyoming Industrial Siting Council, http://deq. wyoming.gov/isd/application-permits/resources/ roundhouse-wind-energy-project/. The hearing is scheduled for June 13-14.

Audubon and the other groups are recommending relocating six turbines in the southwest corner, to benefit wildlife, recreation and safeguard adjacent protected areas: TNC's Big Hole conservation easement, Red Mountain Open Space and Soapstone Prairie Natural Area (the last two in Colorado). NextEra is relocating one, T8, at the city's request.

Birds, along with bats, are the wildlife killed directly by turbine blades. Hundreds of hours of wildlife studies for the application show WEST consultants found 34 bird species in the western part of the Belvoir and AECOM found 67 species in the southeastern private/state land area. AECOM included extra survey locations in riparian areas where songbirds hang out.

No studies were done in the northeastern part of the Belvoir since that was leased only a few months ago, in December 2018. Ironically, that's the part of the Belvoir where Audubon, TNC and

the University of Wyoming Biodiversity Institute found 24 species June 11 and 12 during the 2016 Bioblitz.

The eBird.org bird species list for adjacent Red Mountain and Soapstone Prairie combined is 144. Laramie County, where the Belvoir is located, has a list of 319 species. Why are the application's species numbers so low?

For every weekly survey of 60 minutes looking for eagles and other raptors, only an additional 10 minutes was devoted to songbirds. Eagles have major federal protection including substantial fines. Songbirds have less protection, even before the teeth were taken out of the Migratory Bird Treaty Act by the current federal administration.

Low species counts mean consultants are not giving us the picture of migration diversity that eBird.org records indicate, or the clouds of migrating birds from radar records found at BirdCast. org. I think WEST's estimate of 300 dead birds per year is low.

Methodology for counting dead birds after turbines begin operation was not ready at the time the application was submitted. The

bird surveys are not complete either. How can Audubon, an official party at the ISC hearing, make informed requests for changes?

Cheyenne Audubon will ask NextEra to have the project operator consult BirdCast.org and determine what nights clouds of migrating birds are thickest funneling up the Front Range and shut down the turbines for a few hours.

There's no room here to discuss the whole hawk and eagle situation, but I trust the U.S. Fish and Wildlife Service to take that on.

The 2008 Belvoir master plan, developed with public input, called for limited wind development. Without that same extensive public involvement, the city chose to let NextEra's turbines be spread across the entire ranch. The economic benefits will come at the expense of recreation and wildlife – even though those two, when handled properly, could provide economic benefits themselves in perpetuity, compared to the wind development's 30 years.

It would be so much easier to convince the powers that be of the economic value of birds and recreation if I had deep pockets.

2019

397

# 424 Participating at the Roundhouse hearing was an intense venture

Friday, July 5, 2019, Opinion, page A7

The Cheyenne-High Plains Audubon Society agrees clean energy is needed. However, wind energy is deadly for birds when they are struck by turbine blades.

Beginning last December, CHPAS discussed its concerns about the Roundhouse Wind Energy development with company, city and county officials. The 120-turbine windfarm will extend from Interstate 80 south to the Colorado state line and from I-25 west to Harriman Road.

The Wyoming Industrial Siting Council hearing for the approval of Roundhouse Wind Energy's application was held June 13 in a quasi-legal format.

Cheyenne-High Plains Audubon Society filed as a party, preparing a pre-hearing statement. The other parties were the Wyoming Department of Environmental Quality's Industrial Siting Division, Roundhouse and Laramie County, also acting on behalf of the city of Cheyenne.

We all presented our opening statements. Then the Roundhouse lawyer presented her expert witnesses, asking them leading questions. Then I, acting in the same capacity for CHPAS as the lawyer for Roundhouse, cross-examined her witnesses. One was a viewshed analysis expert from Los Angeles, the other a biologist from Western EcoSystems Technology, the Cheyenne consulting firm that does contract biological studies for wind energy companies across the country.

Then CHPAS presented our expert witness, Daly Edmunds, Audubon Rockies' policy and outreach director. Wind farm issues are a big part of her work. She is also a wildlife biologist with a master's degree from the University of Wyoming.

We were rushed getting our testimony in before the 5 p.m. cutoff for the first day because I was not available the next day. I asked permission to allow Mark Gorges to read our closing statement the next day, after the applicant had a chance to rebut all the conditions we asked for.

The seven council members chose not to debate our conditions. Some conditions were echoed by DEQ. But it was a hard sell, since Wyoming Game and Fish Department had already signed off on the application.

Here are the conditions we asked for:

1) Some of the recommended wildlife studies will be one and a half years away from completion when turbine-building starts in September. Complete the studies first to make better turbine placement decisions.

2) Do viewshed analysis from the south and share it with adjacent Colorado open space and natural area agencies.

3) Get a "take permit" to avoid expensive trouble with the U.S. Fish and Wildlife Service if dead eagles are found.

4) Use the Aircraft Detection Lighting System so tower lights, which can confuse night-migrating birds, will be turned on as little as possible. This was on DEQ's list as well.

5) Use weather radar to predict the best times to shut down turbines during bird migration.

6) Be transparent about the plans for and results of avian monitoring after the turbines start.

7) Relocate six of the southernmost turbine locations because of their impact on wildlife and the integrity of adjacent areas set aside for their conservation value.

The second half of the hearing dealt with county/city requests for economic impact funds from the state. The expected costs are from a couple hundred workers temporarily descending on Cheyenne requiring health and emergency services.

At the June CHPAS board meeting, members approved staying involved in the Roundhouse issue. The Roundhouse folks have a little mitigation money we could direct toward a study to benefit birds at this and other windfarms. There is a Technical Advisory Committee we need to keep track of. And we need to lobby to give Game and Fish's recommendations more legal standing so they can't be ignored.

It's too bad I don't watch courtroom dramas. The hearing would have been easier to navigate. But everyone – DEQ employees, the Roundhouse team, council members, hearing examiner, court reporter – was very supportive of CHPAS's participation. They rarely see the public as a party at these hearings. I just wish we could have had one or more conditions accepted on behalf of the birds.

# 425 Bird families expand in summer

Sunday, July 21, 2019, Outdoors, page E2

Early summer exploded with babies. In addition to our family adding the first baby of the new generation (do wild animals relate to their grand-offspring?), I noticed a lot of other baby activity.

Driving past Holliday Park at twilight at the end of June I caught a glimpse of what looked like three loose dogs. They were a mother racoon and two young scampering across the lawn.

Walking our dog around the field by our house, I saw a ground squirrel mother herd a youngster out of the street and back to the safety of the grass. There's also an explosion of baby rabbits in that field driving everyone's dogs crazy.

We have a pair of Swainson's hawks nesting in our neighborhood, and they are using the field as their grocery store. I'm not sure exactly where they are nesting, but I'm guessing it is one of the large spruce trees. Whenever I'm at the field, I catch a glimpse of at least one hunting. But I also glimpse them from my kitchen window soaring, meaning I can add them to my eBird. org yard list. The yard list is all the species I've seen from the window or while out in the yard. The Swainson's have put me at 99 species so far – over about 12 years.

When it warmed up, we spent more time in our backyard and I noticed other signs of family life. We always have a raucous community of tree squirrels, one generation indistinguishable from the next, chasing each other round and round in our big trees.

This year I've been hearing a mountain chickadee sing. No, not the "chick-a-dee-dee-dee" call—that's their alarm call—but a sweet three-note song (listen at https://www.allaboutbirds.org/).

I'm also learning the various phrases American goldfinches use while they spend the summer with us. We've left our nyger thistle seed feeder up for them (no, nyger thistle is not our noxious weed and it is treated not to sprout). They sometimes come as a group of four, including two males and two females, and sometimes a younger one.

The downy woodpeckers have been visiting as well. They go for one of those blocks of seed "glued" together that you buy at the store. You would think they would go for bugs hiding in the furrowed bark of the tree trunks. Maybe they do, in addition to the seed block.

The robins have been busy. I observed a youngster walking through my garden as it tried to imitate the foraging action of the nearby adult, but it finally resorted to begging to be fed.

Within the space of a couple days, I was contacted about two problem robins attempting to build nests on the tops of porch lights. Porch lights, because they usually provide a shelf-like surface under the safety of the roof overhang, are quite popular. But not everyone trying to use the adjacent door likes getting dive-bombed by the angry robin parents.

In the first situation, Deb, our former neighbor, said the robin was trying to build a nest on a porch light with a pyramidal top. The bird could not make her nest stick and all the materials from all her attempts slid off and accumulated on the porch floor. Providing another ledge nearby might not have worked for such a determined bird. Instead, Deb opted for screening off the top of the light. Hopefully Mama Robin found a better location in Deb's spruce trees.

Our current neighbor, Dorothy, texted me the next day, wondering what she and her family were going to do about being attacked by the robin which had built a nest on her (flat-topped) front porch light. Maybe avoid walking out the front door and walk out through the garage instead, I said. I asked her if she had a selfie stick so she could take pictures of the inside of the nest to show her two young boys.

Down at Lions Park a new colony of black-crowned night-herons has been established. Listen for them behind the conservatory. The colony at Holliday Park is still going strong.

In the far corner of Curt Gowdy State Park, I caught a glimpse of a bird family I hadn't seen together before. Way up on the nasty El Alto trail, I saw a brown songbird I couldn't identify readily. And then the parent came to feed it, a western tanager. The youngster has a long way to go before attaining either the look of its mother, if female, or if male, the bright yellow body with black and white wings and the orange head like its father.

# 426 Audubon Photography Awards feature Pinedale photographer

Sunday, August 11, 2019, Outdoors, page E2

Last month, a familiar name appeared on my screen, "Elizabeth Boehm."

I was reading an email from the National Audubon Society listing the winners of the 2019 Audubon Photography Awards.

I have never met Elizabeth in person. But she was one of the people who replied when I put out a request on the Wyobirds e-list for photos of the few bird species we didn't have for photographer Pete Arnold's and my book published last year, "Cheyenne Birds by the Month." She generously shared six images.

With my similar request on Wyobirds back in 2008 for "Birds by the Week" for the Wyoming Tribune Eagle, Pete supplied most of the 104 photos (the others were stock), and he contributed 93 for the book. Here's the small world connection: Pete is Elizabeth's neighbor whenever he and his wife visit his wife's childhood home in Pinedale.

Now here is the big world connection: Elizabeth won the 2019 Audubon Photography Awards in the professional category. To qualify as a professional, you must make a certain amount of money from photography the previous year.

A week later, Audubon magazine arrived and there, printed over a two-page spread, like the grand prize winner, was Elizabeth's winning photo: two male sage grouse fighting on an entirely white background of snow.

I decided it was time to get to know Elizabeth better and interviewed her by phone about her prize-winning photography. Elizabeth won the Wyoming Wildlife magazine grand prize a couple years ago and one year she was in the top 10 for the North American Nature Photography Association. Her photos have been published in Audubon magazine. "I was totally surprised," she said of her latest win.

More than 8,000 images were submitted by 2,253 U.S. and Canadian photographers. Categories included professional, amateur, youth (13-17 years old), Plants for Birds (bird and a plant native to the area photographed together) and the Fisher Prize (for originality and technical expertise).

Elizabeth started shooting landscapes and wildflowers 25 years ago, then started selling images 10 years later, adding wildlife to her subjects. Now she works her day job only two days a week.

Of her winning image she said, "I usually go out in the spring. I know the local leks. I like snow to clean up the background. The hard part of photographing fights is they are spontaneous. It's kind of a fast, quick thing."

The males fight in the pre-dawn light for the right to be the one that mates with all the willing females. "I set up the night before or in the middle of the night. It's better waiting and being patient," she said.

Elizabeth visits leks one or two times a week March through April. This past spring was too wet for driving the back roads. Even the grouse weren't on the leks until late. They don't like snow because there is nowhere to hide from the eagles that prey on them.

This winning photo is from three or four years ago. Elizabeth came across it while searching her files for another project and realized it could be special with a little work.

Audubon allows nothing other than cropping and a few kinds of lighting and color adjustments. At one point, Audubon requested Elizabeth's untouched RAW image. See the 2019 rules, and 2019's winning photos, at https://www.audubon.org/photoawards-entry. Her camera is a Canon EOS 6D with a Canon 500 mm EF f/4L IS USM lens. The photo was taken at 1/1500 second at f/5.6, ISO 800.

In September, National Audubon will finalize the schedule for the traveling exhibit of APA winners.

Elizabeth sells prints at the Art of the Winds, a 10-artist gallery on Pinedale's Main Street. You can also purchase images directly from her at http://elizabethboehm.com. She offers guided local birding tours and is also the organizer for the local Christmas Bird Count.

Photographers are a dime a dozen in the Yellowstone – Grand Teton neighborhood where Elizabeth shoots. She works hard to have her work stand out. She also donates her work to conservation causes like Pete's and my book, which is meant to get more people excited about local birds and birdwatching.

Look on the copyright page of "Cheyenne Birds by the Month" for the list of Elizabeth's contributions. You can find the book online through the University of Wyoming bookstore, the Wyoming Game and Fish store and Amazon, etc.

In Cheyenne it's at the Wyoming Game and Fish Department, the Cheyenne Depot Museum, Wyoming State Museum, Cheyenne Botanic Gardens, Riverbend Nursery, Cheyenne Pet Clinic, Cheyenne Regional Medical Center's Pink Boutique, Barnes and Noble, PBR Printing and out at Curt Gowdy State Park.

# 427 Nestling ID benefits from crowd sourced help

Sunday, September 15, 2019, Outdoors, page E2

Cheyenne resident Priscilla Gill emailed me a bird photo that her son, Matthew Gill, took Aug. 6. Could I identify the birds?

Digital technology is wonderful. Thirty years ago, I would get phone calls asking for ID help (and I still do) but it can be difficult to draw a mental picture. I must figure out how familiar with birds the callers are so I can interpret the size and color comparisons they make.

At least with an emailed photo, the ease of identifying the bird is only dependent on the clearness and how much of the bird is showing. In this case, the photo clearly showed two little nestlings so ungainly they were cute. They were black-skinned, but all a-prickle with yellow pin feathers and had large, lumpy black bills. They were nestled on top of a platform of sticks balanced high up on the pipe infrastructure at a well pad.

Those bills first made me think ravens. However, the nest was near Greeley, Colorado, where ravens are rarely seen.

Digital photos are easy to share. I forwarded the photo to Greg Johnson, my local go-to birder who enjoys ID challenges. But after a couple days without a reply,

I figured he was somewhere beyond internet contact, so I sent the photo on to Ted Floyd, Colorado birder and editor of the American Birding Association magazine.

He had no idea. No one has ever put together a field guide for nestlings. Julie Zickefoose comes close with her book, "Baby Birds: An Artist Looks into the Nest" (my review: https://cheyennebirdbanter.wordpress.com/2016/05/30/watching-one-bird-at-a-time/), where she sketched nestlings of 17 species at regular intervals.

Ted suggested I post the photo to the ABA's Facebook group, "What's this bird?"

Meanwhile, Greg was finally able to reply: mourning dove. They only have two young per nest, and they build stick nests.

By this time, I had joined the Facebook group and was starting to get replies. It's a little intimidating – there are 39,000 people in the group. There were 13 replies and 37 other people "liked" some of those replies, essentially voting on their ID choice.

I was surprised to see a reply from someone I knew, my Seattle birding friend, Acacia. Except for the person who suggested pelicans (based on the enormous

bills), the replies were split between mourning dove and rock pigeon. I was most confidant about the reply from the woman who had pigeons nest on her fire escape.

On reflection, "pigeon" seemed to make more sense, and Greg agreed. Pigeons are known for adapting to cities because the buildings remind them of cliffs they nest on in their native range in Europe and Asia. It seems odd to think of them nesting in the wild, but there's a flock around the cliffs on Table Mountain at the Woodhouse Public Access Area near Cheyenne. Mourning doves and Eurasian collared-doves, on the other hand, are more likely to hide their nests in trees.

But birds can sometimes adapt to what we humans present them with. Short of following the nestlings until they can be identified via adult plumage or comparing them to photos of nestlings that were then followed to adulthood, we can't say for sure which species they were.

Out there in the open, did these two make it to maturity? I wonder how easy it would be for hawks to pick off both the parents and young.

Here in Cheyenne at the

end of August, I've noticed the field by my house has gotten very quiet at ground level – virtually no squeaking ground squirrels anymore. However, many mornings I'm hearing the keening of the two young Swainson's hawks probably responsible for thinning that rodent population. The youngsters and parents sit on the power poles and watch as my friend Mary and I walk our dogs past.

The two kids have even been over to visit at Mark's and my house. One evening while out in the backyard, I happened to look up and see the two sitting on opposite ends of the old TV antenna that still sways atop its two-story tower. That gives new meaning to the term "hawk watching." They leave white calling card splats on the patio, so I know when I've missed one of their visits.

Another day, as I did backyard chores accompanied by the dog, one of them sat in one of our big green ash trees, sounding like it was crying its heart out – maybe it was filled with teenage angst, knowing how soon it needed to grow up and fly to the ancestral winter homeland in the Argentinian grasslands.

# 428 How 3 billion breeding birds disappeared in past 48 years

Sunday, October 13, 2019, Outdoors, page E2

"Decline of the North American avifauna" is the title of the report published online by the journal Science on Sept. 19.

The bird conservation groups I belong to summed it up as "3 billion birds lost."

In a nutshell (eggshell?), there are three billion (aka 29%) fewer breeding birds of 529 species in North America than in 1970.

The losses are spread across common birds such as western meadowlark, as well as less common birds in all biomes. While the grasslands, where we live, lost only 720 million breeding birds, that's 53% – the highest percentage of the biomes. And 74% of grassland species are declining. Easy-to-understand infographics are available at https://www.3billionbirds.org/.

Two categories of birds have increased in numbers: raptors and waterfowl. Their numbers were very low in 1970 due to pesticides and wetland degradation, respectively. Eliminating DDT and restoring wetlands, among other actions, allowed them to prosper.

The 11 U.S. and Canadian scientists crunched data from ongoing bird surveys including the North American Breeding Bird Survey, the Christmas Bird Count, the International Shorebird Survey and the Partners in Flight Avian Conservation Database.

Weather radar, which shows migrating birds simply as biomass, shows a 14% decrease from 2007 to 2017.

Two of the contributors to the study are scientists I've talked to and whose work I respect. Adriaan Dokter of Cornell Lab of Ornithology is working with me, Audubon Rockies and the Roundhouse developers. We want to see if weather radar can predict the best nights to shut down wind turbines for the safety of migratory birds passing through the wind farm they are building at the southwest corner of I-80 and I-25.

I've met Arvind Panjabi with Bird Conservancy of the Rockies headquartered in Fort Collins, Colorado, on several occasions. BCR does bird studies primarily in the west as well as educational programs.

How does the number of birds make a difference to you and me? Birds are the easiest animals to count and serve as indicators of ecological health. If bird numbers are down, we can presume other fauna numbers are out of whack too - either, for instance, too many insects devouring crops or too few predators keeping pest numbers down. Ecological changes affect our food, water and health.

The decline of common bird species is troubling because you would think they would be taking advantage of the decline of species less resilient to change. But even invasive species like European starling and house sparrow are declining.

The biggest reasons for avian population loss are habitat loss, agricultural intensification (no "weedy" areas left), coastal disturbance and human activities. Climate change amplifies all the problems.

A coalition including Audubon, American Bird Conservancy, Cornell Lab of Ornithology, Environment and Climate Change Canada, Bird Conservancy of the Rockies and Georgetown University have an action plan.

There are seven steps we can all take. The steps, with details, are at https://www.3billionbirds.org/. Most of them I've written about over the past 20 years so you can also search my archives, https://cheyennebirdbanter.wordpress.com/.

1. Make windows safer. Turn off lights at night inside and outside large buildings like the Herschler Building and the Cheyenne Botanic Gardens during migration. Break up the reflections of vegetation birds see in our home windows during the day.

2. Keep cats indoors. Work on the problem of feral cats. They are responsible for more than two-thirds of the 2.6 billion birds per year cats kill.

3. Use native plants. There are 63,000 square miles of lawn in the U.S. currently that are only attractive to birds if they have pests or weeds.

4. Avoid pesticides. They are toxic to birds and the insects they eat. Go organic. Support U.S. bill H.R. 1337, Saving America's Pollinators Act. Contact Wyoming's Representative Liz Cheney and ask that registration of neonicotinoids be suspended. Birds eating seeds with traces of neonics are not as successful surviving and breeding.

5. Drink shade-grown coffee. It helps 42 species of migratory North American birds and is economically beneficial to farmers.

6. Reduce plastic use. Even here, mid-continent rather than the ocean, plastic can be a problem for birds. Few companies are interested in recycling plastic anymore.

7. Do citizen science. Help count birds through volunteer surveys like eBird, Project FeederWatch (new count season begins Nov. 9), the Christmas Bird Count (Cheyenne's is Dec. 28), and if you are a good birder, take on a Breeding Bird Survey route next spring.

To aid grasslands in particular, support Audubon's conservation ranching initiative, https://www.audubon.org/conservation/ranching.

In a related Science article, Ken Rosenberg, the report's lead author, says, "I am not saying we can stop the decline of every bird species, but I am weirdly hopeful."

# 429 Alaskan bird behavior intrigues birdwatchers

Sunday, November 10, 2019, Outdoors, page E2

Our family lost its guide to Alaska in October. My husband Mark's brother Peter, a Catholic priest in southeast Alaska for 51 years, died at age 84.

Peter was an inveterate explorer, from his days growing up in the Bronx a block from Van Cortlandt Park – which is more than 300 acres larger than Central Park – to voluntarily relocating to Alaska. His extensive foreign travels with parishioners took him several places the last 20 years.

Whenever we visited, Peter was our tour guide: Ketchikan, Sitka, Juneau, Skagway, Haynes, Fairbanks, Denali, Anchorage, Homer (search "Alaska" at http://cheyennebirdbanter.wordpress.com). Like his father and three brothers, he was a fisherman and camped and hiked. But he also was interested in botany and Native cultures. He didn't just reside in Alaska. He knew the state's history, political and natural.

Peter became more interested in birds after he retired. It's hard to ignore them in southeast Alaska. For instance, Sitka Sound, opening onto the Pacific, has an abundance of gulls, ravens and bald eagles that mingle with Sitka townsfolk and summer cruise ship visitors.

Spend time among ravens that walk within ten feet of you unafraid and you will never mistake a crow for a raven again: enormous bills, bushy cowls of neck feathers, bouncy landings, deep croaking voices. And you know they are staring at you, calculating if you might share food.

One raven I met after the memorial mass for Peter accompanied me to the Sitka National Historical Park parking lot. It chose a dark blue car and fussed at its door, all the while looking over at me hoping I had a key to food inside.

I've been reading John Marzluff and Tony Angell's book, "Gifts of the Crow." Marzluff's study on the University of Washington campus revealed that crows remembered the faces of researchers that captured and banded them, so the birds mobbed them whenever they saw these researchers again. Luckily, the researchers wore masks. The original crows taught subsequent generations to recognize the masks.

Crows have many human-like behaviors because their brains operate in ways very similar to ours. The book is full of technical explanations. Crows especially, and the other corvids to some degree, jays, ravens and magpies, have developed a relationship with people.

The local indigenous people, the Tlingit, divide their clans into two groups, Raven and Eagle/Wolf (there's a north-south divide for the second group).

Because ravens and crows do not have bills strong enough to break the skin of other animals, they are known to lead predators, including hunters, to prey and then feast on the leftovers.

Perhaps the parking lot raven updated the tradition, finding park visitors have food.

I think all the other cars in the lot were white National Park Service vehicles because the visitor center was closed. And you know that the agency forbids feeding wildlife in its parks, so that's why the raven chose a blue car. And maybe it picks out people who aren't wearing park service uniforms.

Southeast Alaska is not particularly cold, but it is darker in winter than the lower 48, and much rainier, so everyone has enormous windows to maximize natural light. From Peter's rooms at the rectory, he had a panoramic view of Crescent Bay and its resident bald eagles.

Sitka's bald eagles are not as chummy as its ravens, but they have their favorite perches around town. One is a piling outside the marina breakwater. On our last day,

Mark and I walked the waterfront out far enough to look back at the rocky structure and I caught a glimpse of something in the distance swimming towards it.

Neither of us had binoculars, if you can believe it. Mark didn't have his camera with the zoom lens either. I have better than 20/20 distance vision but still, all I could tell was some brown animal was swimming. But it wasn't a consistent movement forward like the usual animal paddling. More like the jerkiness of the breaststroke.

Then there was a flash of white. Hmm, maybe a bald eagle? Have you seen any of the online videos of bald eagles catching fish too heavy to fly to land and instead swimming, using their wings like oars on a rowboat?

We waited and sure enough, the brown animal climbed onto the rocks and it became an eagle, white head and tail visible – but not what it beached. At least one raven flew over to inspect it. I wonder if Peter ever observed this behavior.

I don't think this will be our last trip to Alaska, now that two generations of our family reside nearby in Seattle. But we will have to find a new guide – or do more homework.

2019

# 430 Conservation ranching is for the birds – and for the cows

*Sunday, December 15, 2019, Outdoors, page E2*

You'll run across arguments saying our farmlands would be put to better use raising food crops for people instead of forage crops for cattle. Maybe so – back east.

But Wyoming's remaining rangeland, its prairie grasslands and shrublands, is not suited to raising crops. We don't have the water or the soils. But we do grow excellent native forage, originally for buffalo, now for cattle.

And what a great system it is – no fossil fuels required to harvest that forage – the animals do it for you! On top of that, good range management is good for birds.

However, grassland birds were identified as the group having declined the most in the past 48 years, according to https://www.3billionbirds.org/.

At a recent Cheyenne Audubon meeting, Dusty Downey, Audubon Rockies' lead for the organization's Conservation Ranching Initiative, explained part of the problem is grassland conversion. When ranchers can't make enough on cattle, they might try converting rangeland to cropland or to houses and other infrastructure. With hard work, cropland can be restored someday, but houses are a permanent conversion and wildlife suffers habitat loss.

Eighty-five percent of grasslands and sagebrush steppe is privately owned. So Dusty, raised on, and still living on, a ranch by Devils Tower, and his boss, Alison Holloran, a wildlife biologist, thought reaching out to ranchers about enhancing their operations could benefit both birds and cattle. Offering a financial incentive makes it attractive and might keep land in ranching.

National Audubon picked up the idea and made it a national program. The "Grazed on Audubon Certified Bird Friendly Land" logo can help ranchers get anywhere from 10 to 40 cents per pound more, depending on the market.

Conservation ranching is now popular in Dusty's Thunder Basin neighborhood where ranchers know him and his family. Through the program, ranchers learn techniques for maximizing production over the long term that also benefit birds and they get help finding funding for ranch improvements. With third-party certification, they earn the privilege of selling their meat at a premium price to people like me who value their commitment.

We also value meat free of hormones and antibiotics, so that is part of the certification. And we appreciate that cows eating grass produce less methane, part of the climate change problem, than if they eat corn.

Dusty said in the past 15 years, grass-fed beef sales have grown 400 percent, from $5 million a year to $2 billion.

Audubon-certified beef is available at Big Hollow Food Coop in Laramie, the Reed Ranch in Douglas, in Colorado, other western states and online. See https://www.audubon.org/where-buy-products-raised-audubon-certified-land#.

Grazing prairie looks simple. But grazing management is both art and science.

What does the vegetation need? How is it interacting with weather and grazers? Grassland vegetation needs grazing to stay healthy. Dusty cited a four-year study that showed an ungrazed pasture was not as productive or as diverse as one that had been grazed properly. Grazed plots showed five times more birds, two times more arthropods (food for chicks) and five times more dung beetles (the compost experts) than ungrazed plots.

Grazing grasslands down to bare ground like the buffalo did looks bad, but in the right context it allows highly nutritious plants to grow that can't compete otherwise. It also aids bird species that require bare ground or very short grass somewhere in their lifecycle, between courtship and fledging.

My experience with prairie plants in the Habitat Hero demonstration garden at the Cheyenne Botanic Gardens showed plants grazed down to ground level by rabbits rebounded the next spring.

But you can't let the rabbits in year-round or the same season year after year.

The gold standard when I was studying range management at the University of Wyoming was rest-rotation grazing. Now it's producing a changing mosaic of plants by adjusting grazing timing on a multi-year cycle for any given pasture, tailored to the plants there and the rancher's goals. Laramie County Conservation District helps local landowners figure it out.

For an elegant explanation of the dance between animal and prairie plant, read a recent blog post by Chris Helzer, https://prairieecologist.com/2019/11/13/what-does-habitat-look-like-on-a-ranch/. He is the director of science for The Nature Conservancy-Nebraska.

Chris talks about growing a shifting mosaic of plants that will be more resilient through drought and other extremes. He also said, "Chronic overgrazing can degrade plant communities and reduce habitat quality, but a well-managed ranch can foster healthy wildlife populations while optimizing livestock production."

Next time you meet a rancher, restaurant owner or grocery store manager, ask them if they've heard about Audubon-certified meat. Tell them it's good for birds – and cows.

# 2020

## 431 Be a Citizen Scientist in your backyard

Sunday, January 19, 2020, Outdoors, page E2

Along with the news last fall that there are 3 billion fewer birds in North America than in 1970, conservation organizations came out with a list of seven actions people can take.

No. 7 on the list is "Watch birds, share what you see." In fact, citizen science efforts, like the 120th annual Christmas Bird Count season that finished up Jan. 5, provided part of the data for the study that showed the bird decline.

There aren't enough scientists to collect data everywhere, so they depend on us informed lay people to help them.

There's another organized opportunity coming up for you to count birds Feb. 14-17: the Great Backyard Bird Count.

### GBBC history

Begun in 1998 by the National Audubon Society and Cornell Lab of Ornithology, the GBBC dates always coincide with Presidents' Day weekend. Scientists wanted to get a snapshot of where the birds are late winter, before spring migration begins.

The difference between this public participation bird count and the others at the time, is data reporting is entirely online. Some results are nearly real time on the website, like watching the participant map light up sporadically every few seconds as someone else hits "Send."

In 2002, Cornell started another online citizen science project, eBird, which collects data year-round from citizen scientists. In 2013, the GBBC was integrated with eBird. And now both have global participation from birdwatchers in 100 countries.

At the GBBC website you can find all kinds of interesting information about last year's count and prepare for this year.

### 2019 broke records

There were 209,944 checklists submitted in 2019. A checklist is the list of birds seen by one person or a group birding together. GBBC asks participants to bird for a minimum of 15 minutes and to not travel more than 5 miles for one checklist. Originally, the emphasis was on watching the birds in your backyard, but you can bird anywhere now.

There were 32 million birds counted, of 6,849 species. Columbia counted 1,095 species, the most of any country, even though only 1,046 checklists were submitted (there were 136,000 checklists for the U.S.). This time of year, a lot of our North American summer birds are in Columbia and other Central and South American countries.

The list of top 10 species most frequently reported starts with the cardinal, not native to Cheyenne, and the junco, common at our feeders, made second place. All the birds on this list were North American because the majority of 224,781 participants last year were from our continent. Birders in India are getting excited, though, and that might change someday.

California made the top of the list of states for most checklists submitted, 10,000. All the top 10 states were coastal, either Great Lakes or ocean. That's where the most people live.

Trends in North America showed up during the 2019 count, such as a high number of evening grosbeaks in the east. Canada had fewer finches because of a bad seed crop, and apparently the finches went south because there were higher numbers of finch species – red crossbills, common redpolls, and pine grosbeaks – in the northern states.

### You can prepare ahead

The GBBC website has links to other sites to help you identify birds (if you don't have a copy of my "Cheyenne Birds by the Month" already!):

--Merlin (also available as a free phone app) will ask you questions about the comparative size of the bird, color, activity, habitat, and give you a list of possibilities.

--All About Birds and the Audubon Bird Guide are both helpful.

--And if the weekend finds you in Central or South America, check out the link for Neotropical Birds Online.

### Take photos

Don't forget to take photos – there's a contest with these categories:

--Birds in their habitat
--Birds in action
--Birds in a group
--Composition—pleasing arrangement of all features
--People watching birds.

## How to get involved

Participate in the Great Backyard Bird Count Feb. 14-17

Count with CHPAS locally

Join Cheyenne–High Plains Audubon Society members from 10 a.m. to noon Feb. 15, for free at the Children's Village at the Cheyenne Botanic Gardens, 710 S. Lions Park Road. We'll bird a little around the park and then come back and enter our data. All ages are welcome.

And we have binoculars to share. Contact bgorges4@msn.com if you have questions.

If you are new to GBBC and want to participate on your own, participation is free. Instructions are at https://gbbc.birdcount.org/.

If you already eBird, submit checklists (15-minute minimum) to your account at http://ebird.org.

# 432

Sunday, February 16, 2020, Outdoors, page E2

## Wyobirds gets tech update and Wyoming Master Naturalists gets initial discussion

Technology drives changes in the birding community as it does for the rest of the world. We always wonder how hard it will be to adapt to the inevitable.

In January, the folks at Murie Audubon, the National Audubon Society chapter in Casper, announced that they would no longer pay the fees required for hosting the Wyobirds elist. There have been plenty of donations over the years to offset the $500 per year cost but, they reasoned, now that there is a no-cost alternative, why not spend the money on say, bird habitat protection or improvement? Also, the new option allows photos and the old one didn't.

But the new outlet for chatting about birds in Wyoming works a little differently, and everyone will have to get used to it. We've changed before. We had the Wyoming Bird Hotline until 2006 for publicizing rare bird alerts only. No one called in about their less than rare backyard birds, their birding questions and birding related events like they do now on Wyobirds.

The only problem with leaving the listserv is figuring out what to do with the digital archives. They may go back to 2004, the first time Wyobirds was mentioned in Cheyenne Audubon's newsletter.

Now the Wyoming birding community, and all the travelers interested in coming to see Wyoming birds, can subscribe to Wyobirds (no donations necessary) by going to Google Groups, groups. google.com, and searching for "Wyobirds." Follow the directions for how to join the group so that you can post and get emails when other group members post. I opted to get one email per day listing all the postings. That will be nice when spring migration begins and there are multiple posts each day.

Google Groups, a free service from Google, is one way the giant company gives back, and we might as well take advantage of it.

### Wyoming Master Naturalists

Wyoming is one of only five states that does not have a Master Naturalist program, however, it's in the discussion stage.

What is a Master Naturalist and what do they do? Jacelyn Downey, education programs manager for Audubon Rockies who is based near Gillette, explained at the

January Cheyenne Audubon meeting that programs are different in each state.

Most are like the Master Gardener program, offering training and certification. Master naturalists serve by taking on interpretive or educational roles or helping with conservation projects or collecting scientific data. The training requires a certain number of hours, and keeping up certification requires hours of continuing education and service. But it's not a chore if you love nature.

Master Gardeners is organized in the U.S. through the university extension program. Some Master Naturalist programs are too, as well as through state game and fish or parks departments or Audubon offices or other conservation organizations or partnerships of organizations and agencies.

Colorado has at least two programs, one through Denver Audubon and another in Ft. Collins to aid users of the city's extensive natural areas.

Dorothy Tuthill also spoke. She is associate director and education coordinator for the University of Wyoming Biodiversity Institute. She pointed out that several of its programs, like the

Moose Day surveys in which "community scientists" (another term for people participating in citizen science) gather data, are the kinds of activities a Master Naturalist program could aid.

Audubon and the institute already collaborate every year with other organizations and agencies on the annual Wyoming Bioblitz. It's one day during which scientists, volunteers, teachers, families and kids together gather data on flora and fauna in a designated area. This year's Bioblitz will be July 17-19 near Sheridan on the Quarter Circle A Ranch, the grounds of the Brinton Museum.

With a Wyoming Master Naturalist program, a trained corps of naturalists could be available to help agencies and organizations by visiting classrooms, leading hikes, giving programs and helping to plan and participating in projects and surveys.

Audubon chapter volunteers are already involved in these kinds of things: adult and child education, data collection on field trips and conservation projects. Many of us might broaden our nature expertise beyond birds and learn more about connecting people to nature.

But it would be nice to wear a badge that guarantees for the public that we know what we are talking about.

Just how a Wyoming Naturalist Program would be set up is being discussed right now. Maybe a Google Group needs to be formed. If you'd like to be in on the discussion, please contact Dorothy Tuthill at dtuthill@uwyo.

edu and Jacelyn Downey at jdowney@audubon.org.

# 433 High capacity water wells can negatively affect birds, other wildlife

Sunday, March 8, 2020, Outdoors, page E2

The relationship between groundwater and surface water is important to birds and other wildlife – and people.

Some surface water is merely runoff from rain and snow that hasn't yet soaked in and recharged the groundwater. Other surface water, like wetlands, is the result of high groundwater levels. Springs along a creek also depend on an adequate amount of groundwater.

Groundwater and surface water along streams and in wetlands grow vegetation wildlife depends on for shelter and food. Seventy percent of Wyoming's bird species require these wetter areas.

Precipitation can vary from year to year, but on average, it recharges the groundwater – the aquifer. Aquifers are geologically complicated, but mostly water flows through permeable layers much the same way surface water drains. In Laramie County, both surface and groundwater flow somewhat west to east.

If someone puts in a well and starts pumping, it will lower the water table – the top of the groundwater – for some distance from the well. If the water is for domestic use, it is filtered through a septic system and mostly returned to the groundwater. However, if it is used for irrigating lawns and gardens, much of it evaporates and is lost. If too many wells are sipping from the same aquifer, the water table drops, and people are forced to drill deeper wells.

Another side effect of the water table dropping is wetlands and streams dry up, affecting wildlife.

Wyoming has complex water laws for allocating surface water. The first person to homestead on a creek got the senior water rights. In a drought year, he might be the only one allowed to remove water from the creek.

Groundwater rights are not as clear-cut, as far as I can tell. More than 25 years ago, I remember being in eastern Laramie County putting on an Audubon presentation for the Young Farmers club. It was on the negative effects of human population growth. The farmers were already complaining then about the growing number of developments and the wells causing the water table to drop.

In 2015, the Laramie County Control Area Order was established in eastern Laramie County to keep an eye on the situation.

Before that, 2010 to 2014, the Natural Resources Conservation Service, under the Agricultural Water Enhancement Program, spent taxpayer funds to buy out 24 irrigation wells at $200,000 each within this same area, saving 1 billion gallons annually. The farmers could grow dryland wheat instead.

And now, in the same area, the Lerwick family is asking for a permit to drill eight high-capacity wells for maximum production of 1.5 billion gallons per year for agricultural purposes. We assume it's for irrigation, and that irrigation water will not be recharging the aquifer much. I don't get it. Permitting new high-capacity wells after paying to retire others in the same area makes no sense at all.

Neighboring farmers and ranchers are alarmed. Professional hydrologists can predict how it will negatively impact their water supplies. Creeks and wetlands, the few out there, will dry up and the neighbor's wells will have to be re-drilled.

Beginning March 4 and for 30 days, the state engineer is asking for comments about the effectiveness of the 2015 Laramie County Control Area Order which guides groundwater development in the area of the proposed wells. Call 307-777-6150 to find out exactly how to comment.

On March 18, the state engineer will hold a hearing regarding the Lerwick permits and will hear from the affected neighboring farmers and ranchers – 17 of them.

Is it fair for someone to get a new well permit that will cause all his neighbors the expense of drilling deeper? Instead, can a community, through governmental agencies, come to an agreement that an area is no longer suitable for irrigated agriculture?

The Ogallala Aquifer, of which this High Plains Aquifer is part, extends under parts of Wyoming, South Dakota, Nebraska, Kansas, Colorado, New Mexico and Texas. For several generations, farmers have been mining it. We can call it mining because more water is extracted than returned. It is not a sustainable situation for anyone.

Cheyenne – High Plains Audubon Society is speaking up for the birds and other wildlife, but it's doubtful wildlife will be considered much in the calculation of acre-feet and gallons per minute and other details of water rights. We already know that in the last 50 years grassland birds have lost the most population of any North American habitat type. Unsustainable mining of water in Laramie County, should the new high-capacity wells be permitted, won't help.

# 434 Flock of bird books arrives this spring

Sunday, April 5, 2020, Outdoors, page E2

Spring is when Houghton Mifflin Harcourt likes to send out bird books to review – completely forgetting that as spring migration gets going, birders have less time to read. Maybe we'll have more time to read this year. Luckily, birding in Wyoming, without Audubon field trips, is a solitary experience perfect for ensuring huge social distances.

I've suggested that we all get social and share our bird sightings on the Cheyenne-High Plains Audubon Society group Facebook page and through the Wyobirds Google Group. By posting sightings on eBird.org, everyone can "Explore" each other's Laramie County bird sightings.

Peterson Field Guide to Birds of North America, Roger Tory Peterson (and contributions from others), 2020, Houghton Mifflin Harcourt, 505 pages, $29.99.

This latest edition of the classic field guides follows the 2008 edition, the first to combine Peterson's eastern and western guides in one book. And now the birds of Hawaii have been added.

Peterson died in 1996, so additional paintings, range map editing, etc. are the work of stellar artists and ornithologists. Bird names are updated, now showing the four species of scrub-jays, except that I heard last month it was decided to drop the "scrub" from their names.

But to be a birder, one must regularly invest in the most up-to-date field guide.

White Feathers, The Nesting Lives of Tree Swallows, Bernd Heinrich, 2020, Houghton Mifflin Harcourt, 232 pages, $27.

If anyone can make eight springs of excruciatingly detailed observations interesting, Bernd Heinrich can. He wanted to know what purpose is served by tree swallows adding white feathers to their nests.

Every spring, hour after hour, he observed the comings and goings of pairs using his nest box and noted when they brought in white feathers to line (insulate?) and cover (hiding eggs from predators?) the nest inside the box.

Or, the white feathers might only advertise that a nesting cavity is taken.

Birdsong for the Curious Naturalist, Your Guide to Listening, Donald Kroodsma, 2020, Houghton Mifflin Harcourt, 198 pages, $27.

Here's where you can find out what a tree swallow sounds like when it starts singing an hour before sunrise.

In fact, you can skip this book and learn a lot by going to the associated free website, www.BirdsongForTheCurious.com. There are multiple songs each for most songbird species, as well as ideas for collecting your own data.

The book has chapters explaining topics such as:

"Why and How Birds Sing," "How a Bird Gets Its Song" and "How Songs Change over Space and Time."

Unflappable, Suzie Gilbert, 2020, https://www.suziegilbert.com/.

I read the first chapter for free online, and I think it will be a very entertaining novel. Here's the synopsis: "Wildlife rehabber Luna and Bald Eagle Mars are on a 2,300-mile road trip with her soon-to-be-ex-husband and authorities hot on their heels. What could possibly go wrong?"

Nature's Best Hope, A New Approach to Conservation That Starts in Your Yard, Douglas W. Tallamy, 2019, Timber Press, 255 pages, $29.95.

Tallamy first wrote "Bringing Nature Home" in 2007 where, as a professor who studies insects and ecology, he explains that it is important for all of us to plant native plants to benefit native wildlife.

Thirteen years later, Tallamy can cite a lot more research making his point: native plants support native insects which support other native wildlife (and support us). For instance, almost all songbird species, even if they are seed eaters the rest of the year, need to feed their young prodigious amounts of caterpillars plus other insects.

These caterpillars of native butterflies and moths can't eat just any old plant. They must chew on the leaves of the plants they evolved with – other leaves are inedible. Good news: rarely does the associated plant allow itself to be decimated.

Native bees, except for some generalists, also have a nearly one-on-one relationship with the native nectar and pollen-producing plants they've evolved with. You may see bees working flowers of introduced plants, but chances are they are the introduced European honeybees.

What's a concerned backyard naturalist to do? Become part of Tallamy's army of gardeners converting yards and wasted spaces of America into Homegrown National Park, http://www.bringingnaturehome.net/. A link there will take you to the National Wildlife Federation's Native Plant Finder, which lists our local natives based on our zip codes.

It's not necessary to vanquish every introduced plant, but we must add more natives. The best way is by replacing turf. Here in Cheyenne, the Board of Public Utilities is encouraging us to save water by replacing water-thirsty bluegrass with water-smart plantings. Plants native to our arid region (12-15 inches of precipitation annually) fit the bill perfectly – and they aid our native pollinators at the same time.

In next Sunday's Cheyenne Garden Gossip column, I will discuss exactly how to do that.

# 435 How to become a birdwatcher

Saturday, May 2, 2020, ToDo, page B2

I am living under the flight path of major construction. A Swainson's hawk is plucking cottonwood branches from one neighbor's tree and taking them over my house to another neighbor's tree to build a nest.

Lately, a gang of 60 or 70, puffed up and strutting around in shiny black feather jackets, shows up along our back wall – no motorcycles for them – they're common grackles. They even scare away the bully robin that keeps the house finches from the black oil sunflower seed we've put out.

A pair of northern flickers has been visiting the seed cake feeder. We know they are male and female – he has the red mustache. The black and white pair of downy woodpeckers are visiting regularly. The male has the red neck spots.

One small, yellow-breasted stranger shows up every day at the nyger thistle seed feeder. It's a female lesser goldfinch, not a regular species here. We recognize that her yellow, black and white feather scheme is arranged differently from the American goldfinch's.

I look forward to the springtime antics of birds in my backyard, but this year, millions of people are discovering them for the first time in their own yards and neighborhoods. Suddenly, it's cool to notice birds and nature. It's almost cool to be called a birdwatcher.

Would you like to be a birdwatcher, or a birder? Here's how.

## Step 1 – Notice birds

Watch for bird-like shapes in the trees and bushes and on lawns. Watch for movement. This time of year, birds are making a lot of noise and song. See if you can trace the song to the bird with his beak uplifted and open.

## Step 2 – Watch the birds for a while

Are they looking for food like the red-breasted nuthatches climbing tree trunks and branches?

Are they performing a mating ritual like the Eurasian collared-dove males that launch themselves from the top of a tree or utility pole, winging high only to sail down again in spirals?

Are they picking through the grass like common grackles do, looking for grubs to eat? Are they flying by with a beak full of long wispy dead grasses for nest building like the house sparrows do?

## Step 3 – Make notes about what you see

Or sketches, if you are inclined.

## Step 4 – Bird ID

But if you want to talk to other birdwatchers, you need to do a little studying.

You are in luck if you live in the Cheyenne area. In 2018, Pete Arnold and I put together a picture book of 104 of our most common birds, "Cheyenne Birds by the Month." You'd be surprised how many birds you probably already know. Go to www.yuccaroadpress.com/books to examine current purchasing options.

You can also go to www.allaboutbirds.org. You can

type in a bird name or queries like "birds with red breasts" (which covers all shades from pink and purple to orange and russet). If you click on "Get instant ID help," it will prompt you to download the free Merlin app. It will give you size comparison, color, behavior and habitat choices and then produce an illustrated list of possibilities – nearly as good as sending a photo to your local birder.

The best way to learn birds is to go birdwatching with someone who knows more than you. But since that probably isn't possible this spring, settle for a pair of binoculars and hone your eye for noticing field marks – the colors and shapes that distinguish one bird species' appearance from another's.

Keep in mind that even expert birders can't identify every bird – sometimes the light is bad and sometimes, and often for a species as variable as the red-tailed hawk, it doesn't look exactly like its picture in the field guides by Peterson, Kaufman or Sibley.

## Step 5 – Go where the birds are

In Wyoming, that is generally wherever there is water – and trees and shrubs. At least that's where you'll find the most bird species per hour of birding. But the grasslands are special. Drive down a rural road, like nearby Chalk Bluffs Road, and watch to see what birds flock along the shoulders and collect on the barbwire fence: meadowlarks, lark buntings, horned larks. Watch out for traffic.

## Step 6 – Invite the birds to visit you

Plant trees and shrubs and flowers and use no pesticides. Put out a bird bath, put out a feeder. Keep them clean. Keep cats indoors. I have more detailed advice on bringing birds to your backyard here: https://cheyennebirdbanter.wordpress.com/2018/11/01/basic-wild-bird-feeding/.

## Step 7 – Join other birdwatchers

Some of the nerdiest birders I know will say they prefer to bird alone, but they still join their local Audubon chapter. In Cheyenne, that's the Cheyenne-High Plains Audubon Society. People of all levels of birding expertise are welcome. Sign up for free email newsletters today and join when you are ready.

## Step 8 – Give back to the birds

People do not make life easy for birds. Our activities can affect birds directly and indirectly. Today, I read that the popular neonicotinoid pesticides affect birds' abilities to successfully migrate if they eat even a small amount of treated seed, or an insect that has eaten treated plant material.

Writing letters to lawmakers is one option, but so is planting native plants, and so is recording your bird observations through citizen or community science projects like www.eBird.org and taking part in other conservation activities.

## Step 9 – Call yourself a birdwatcher or a birder

You can do this as soon

2020

409

as you start Step 1, noticing birds. Not everyone does. Welcome to the world of birdwatching!

# 436   2020 Big Day Bird Count best in 18 years

Friday, June 5, 2020, ToDo, page B2

Cheyenne Audubon's 61st Big Day Bird Count May 16 was the best in 18 years: 141 species, with 39 people contributing observations. In those 18 years, the total number of bird species counted ranged from only 104 to 132.

Thinking about the decline in North American birds over the past 50 years, it isn't surprising that the average count for 1992-2002 is 147 species (range: 123 – 169) and the average count for 2009-2019 is 114 (range: 104 – 128).

In a way, I think the pandemic made a difference this year, plus a lucky break offset not being able to access F.E. Warren Air Force Base and part of the High Plains Grasslands Research Station.

The Cheyenne Big Day is held the third Saturday in May, as early as May 13 and as late as May 21, hopefully catching the peak of spring migration.

Sometimes migration runs late, as it apparently did in 1993 (record high total count 169 species), when wintering species like dark-eyed junco and Townsend's solitaire were counted – but we also aren't clear how far from the center of Cheyenne people were birding back then – some of our winter birds go only as far as the mountains 30 miles to the west.

Sometimes, like 1993, we get interesting shorebirds, usually heading north earlier than songbirds. Or, if the reservoirs are full, we don't have any "shore" and thus few shorebirds.

1993 and 2020 have some other interesting comparisons. Great-tailed grackles, birds of the southwest, were first reported breeding in Wyoming in 1998, and now their Cheyenne presence is spreading. Eurasian collared-doves, escaped from the caged bird trade and now nesting in our neighborhoods, were not recorded here before 1998.

But in 1993, we knew where to find burrowing owls. Now that location is full of houses.

The number of observers might matter, especially their expertise. Traditionally, we meet as a large group and hit the hotspots one at a time, Lions Park, Wyoming Hereford Ranch, the research station. The experienced birders might zero in on a vireo's chirp buried in the greenery while the bored novice birder notices American white pelicans flying overhead at the same time.

But this year might be proof that birding on our own (at least by household) as we did, ultimate physical distancing, could be more productive. All the birding hotspots were birded first thing in the morning, when birds are most active and most easily detected.

In addition, it was a magnificent spring migration day. While home for breakfast, lunch and dinner between outings, Mark and I observed a total of 23 species in our backyard, more than any of the days before or after May 16, more than any day in the last 30 years.

Now that we have lots of local birders reporting to eBird, it is easy to see May 16 was the best birding day of May 2020 in Cheyenne. However, the next day we found species we missed, the pelicans and the American redstart.

The thrill of seeing colorful migrants and welcoming back locally breeding birds was as wonderful as every year. But I missed the gathering of birders.

To see the 2020 species list broken out by location (Lions Park, Wyoming Hereford Ranch, High Plains Grasslands Research Station and others) and the comparison with 1993, go to www.cheyennebirdbanter. wordpress.com.

## Cheyenne Big Day Bird Count, May 16, 2020

| | | | |
|---|---|---|---|
| Canada Goose | Clark's Grebe | Ring-billed Gull | Downy Woodpecker |
| Wood Duck | Rock Pigeon (Feral Pigeon) | Forster's Tern | Hairy Woodpecker |
| Blue-winged Teal | Eurasian Collared-Dove | Double-crested Cormorant | Northern Flicker |
| Cinnamon Teal | White-winged Dove | Great Blue Heron | American Kestrel |
| Northern Shoveler | Mourning Dove | Black-crowned Night-Heron | Peregrine Falcon |
| Gadwall | Common Poorwill | Turkey Vulture | Prairie Falcon |
| American Wigeon | Chimney Swift | Osprey | Olive-sided Flycatcher |
| Mallard | Broad-tailed Hummingbird | Northern Harrier | Western Wood-Pewee |
| Redhead | Sora | Sharp-shinned Hawk | Least Flycatcher |
| Ring-necked Duck | American Coot | Cooper's Hawk | Gray Flycatcher |
| Lesser Scaup | American Avocet | Swainson's Hawk | Cordilleran Flycatcher |
| Common Merganser | Killdeer | Red-tailed Hawk | Say's Phoebe |
| Ruddy Duck | Baird's Sandpiper | Ferruginous Hawk | Ash-throated Flycatcher |
| Chukar | Wilson's Snipe | Eastern Screech-Owl | Great Crested Flycatcher |
| Pied-billed Grebe | Wilson's Phalarope | Great Horned Owl | Western Kingbird |
| Eared Grebe | Spotted Sandpiper | Belted Kingfisher | Eastern Kingbird |
| Western Grebe | Lesser Yellowlegs | Red-headed Woodpecker | Loggerhead Shrike |

| | | | |
|---|---|---|---|
| Blue Jay | European Starling | Lark Sparrow | Great-tailed Grackle |
| Black-billed Magpie | Gray Catbird | Lark Bunting | Yellow-breasted Chat |
| American Crow | Brown Thrasher | White-crowned Sparrow | Northern Waterthrush |
| Common Raven | Northern Mockingbird | White-throated Sparrow | Black-and-white Warbler |
| Mountain Chickadee | Eastern Bluebird | Vesper Sparrow | Orange-crowned Warbler |
| Horned Lark | Mountain Bluebird | Savannah Sparrow | MacGillivray's Warbler |
| Northern Rough-winged Swallow | Veery | Song Sparrow | Common Yellowthroat |
| Tree Swallow | Swainson's Thrush | Lincoln's Sparrow | Northern Parula |
| Violet-green Swallow | Hermit Thrush | Green-tailed Towhee | Yellow Warbler |
| Bank Swallow | American Robin | Spotted Towhee | Blackpoll Warbler |
| Barn Swallow | House Sparrow | Yellow-headed Blackbird | Palm Warbler |
| Cliff Swallow | House Finch | Western Meadowlark | Yellow-rumped Warbler |
| Ruby-crowned Kinglet | Red Crossbill | Orchard Oriole | Wilson's Warbler |
| Red-breasted Nuthatch | Pine Siskin | Bullock's Oriole | Western Tanager |
| White-breasted Nuthatch | Lesser Goldfinch | Baltimore Oriole | Rose-breasted Grosbeak |
| Blue-gray Gnatcatcher | American Goldfinch | Red-winged Blackbird | Black-headed Grosbeak |
| Rock Wren | Chestnut-collared Longspur | Brown-headed Cowbird | Lazuli Bunting |
| House Wren | Chipping Sparrow | Brewer's Blackbird | |
| | Clay-colored Sparrow | Common Grackle | |

# 437 Cheyenne Audubon tries a new field trip strategy

Friday, July 10, 2020, ToDo, page A8

The Cheyenne – High Plains Audubon Society has been adapting to pandemic life. We now Zoom for our board meetings, and our fall lectures will probably also be via Zoom.

Field trips are harder to adapt. Our field trip chairman, Grant Frost, suggested a survey of the Cheyenne Greenway birds in late April, and many of us signed up to individually bird a section. Our May Big Day Bird Count was arranged similarly. At the end of June, we tried "separate but simultaneous" at Curt Gowdy State Park – choosing different trails.

This time, there was some pairing up – but it is much easier to keep two arms' lengths away from one person than a group. However, the trails between the visitor center and Hidden Falls were practically a traffic jam of heavy-breathing bicyclists, reported the birders who headed that way. They had to continually step off the trail to allow bikes to pass.

One of our Laramie Audubon friends took the trail from Crystal Reservoir towards Granite Reservoir and met up with the many participants of a footrace.

Mark and I were lucky. We chose a trail with little shade, not very conducive to a summer stroll. But the trail passes along the lake shore and creek, through ponderosa pine parkland, grasslands (sad to say, much of it has gone over to cheatgrass in the last five years), mountain mahogany shrubland, cottonwood draws and across a cliff face in the stretch of about two miles.

We saw 29 species: gulls over the lake, a belted kingfisher along the creek, chickadee in the pines, meadowlarks in the grassland, green-tailed towhees in the shrubs, a lazuli bunting in the cottonwoods and rock wrens in the rocky cliff. The total for the morning, including what the other eight participants hiking in the forest saw, was 71 species.

While we could see the runners on the trail across the water, Mark and I met only two people on our trail, a friendly father and son on their bikes. So, it was a little disconcerting to come back to the trailhead three hours later and find in addition to the two vehicles there when we started, 10 more. One was the park ranger's truck, one from Colorado, one from Oregon and the rest from Laramie County, like us. They must have all gone the other way.

A normal Audubon field trip serves at least two purposes besides recreation. One is to find birds and to report them, now that there is a global database, www. eBird.org. But the other is to learn from each other. Our local bird experts are happy to share their knowledge with newcomers. Even the experts discuss with each other their favorite field marks for identifying obscure birds.

This time, we did have someone new to birding show up, and one of our members graciously allowed her to accompany her. As we finished our hikes, we reported back by the visitor center where we gathered with our lunches under a pine – spaced as required. There was general conversation about birds we'd seen and other topics dear to birdwatcher hearts. I almost canceled the Zoom tally party I'd suggested for the evening but decided to go ahead with it anyway.

Five of us signed on, including our new birder – now a new chapter member. I'd invited people to share photos from the day and showed landscape shots of where Mark and I hiked. Mark shared his shots of a yellow warbler and a mountain bluebird. Someone photographed a nest of house wrens, and Greg Johnson shared two photos we could use to compare the beaks of hairy and downy woodpeckers – the best field mark for telling them apart (the hairy's is proportionally longer).

2020

411

Then it occurred to me, maybe we should have a tally party via Zoom after more field trips and not just during pandemics. It could be a way for bird photographers to show off their pictures and for all of us to learn more about identifying the birds we see. It's a chance for birders to flock together, something we like to do as much as the birds.

Our next socially distant field trip will be July 18. We'll meet at the Pine Bluffs rest area to explore the natural area behind it and document what we find for the annual Audubon Rockies Wyoming Bioblitz. Check for details soon at www.cheyenneaudubon.wordpress.com.

# 438 Summertime is family time for birds

Sunday, August 2, 2020, ToDo, page B1

I asked one of our sons if he'd done any interesting birdwatching lately. He said no, it isn't as exciting as during spring migration.

I would disagree. Return migration starts up in mid-July. Migrating shorebirds were at the Wyoming Hereford Ranch Reservoir No. 1 then, while low water levels made their favorite mudflats.

One sure sign of impending autumn is the hummingbirds coming into town. My red beebalm's first flower was in bloom for about two days when it attracted the first broad-tailed hummingbird July 11. It was finished nesting in the high country and it, or other broadtails, made daily visits for 10 days. Then, a male rufous hummingbird returning from breeding – maybe in northwest Canada – came by. As the beebalm reached its peak, we had a hummingbird buzzing in every daylight hour or so, checking the flowers' recharge of nectar and mostly ignoring the hummingbird feeder.

However, birdwatching in my neighborhood in July and August is more about family drama.

Kids are naturally noisy, and the Swainson's hawks in the nest two yards down are no exception. One of the young took a tumble and landed on a branch several feet below the nest, which is set in the top of a spruce. It cried all day, but I think it climbed back up because it looked like there were two young back on the nest July 24.

One day, I thought one of the young Swainson's had fledged and was sitting in our tree. I could hear a slightly off rendition of the call, maybe like a young bird still practicing, but couldn't spot it at all. Later, I realized it was a blue jay doing imitations. And then there were the three blue jays flitting through our backyard that didn't sound like full-fledged blue jays. They weren't. Husband Mark's photo showed one still had puffy baby feathers on its rump.

July and August are when many plants bear fruit here. Whole extended families of robins strip our chokecherries even though I don't think the fruit is ripe yet.

In the neglected front yard around the corner, there's a wonderful crop of thistle. Usually, it's the American goldfinches helping themselves, but the other day there was a lesser goldfinch, which is not as common. Both are species that nest later than other songbirds because they are waiting to feed their young chewed up thistle seed instead of insects, like the other songbird parents.

If you keep your eyes open, you may see parents feeding young, even after they've fledged, like the yellow warblers we saw along Crow Creek. And, when you see five house wrens hanging around the same willow tree, you know they are siblings who haven't dispersed yet.

Young crows take longer to mature. One of the smarter species of birds, not everything they need to know is hard-wired in their brain. They must learn it. After my cleaning the other day, my dental hygienist and I peered out the window wondering just why the young crow was rolling a rock-like object around – sorry, didn't think to bring binocs to my appointment.

One surprise this summer has been the number of mourning doves. Within a few years of the first sighting of Eurasian collared-doves in Wyoming, here in Laramie County in 1998, we quit seeing mourning doves breeding in our neighborhood. But this summer, if we look closely at the doves on the wires, many have the mourning dove's pointy-tailed silhouette. Perhaps they've finally learned to compete with the collared-doves for nest sites.

For some species, their parental duties are already finished, and they are free to flock around Cheyenne with their pals. The other morning, I estimated there were 150 common grackles carrying on boisterously in treetops and on lawns. Eventually, they will head south.

If bird behavior interests you, read Jennifer Ackerman's new book, "The Bird Way: A New Look at How Birds Talk, Work, Play, Parent and Think." Ackerman writes, in a very readable way, about the latest science that is discovering that birds approach those five kinds of behaviors in myriad ways.

I flipped to the section on parenting. From egg shape to nest shape to who feeds the young and how they are protected, birds have evolved strategies to suit their environment.

But it isn't always an eons-long process. If they aren't successful with a nest in one location one year, they may move to a different location the next.

Or they knock people on the head if they suspect they've harmed their chicks, like the Australian magpie does.

And those birds can remember people for 20 years. Yikes.

Thankfully, our birds are easier to live with, especially when we preserve prairie habitat and enhance the city forest, letting them enrich our lives.

# 439 Migratory Bird Treaty Act back in full force

Saturday, September 5, 2020, ToDo, page B2

Canada is happy again. It was not happy in December 2017 when the solicitor of the U.S. Department of the Interior reinterpreted the U.S. Migratory Bird Treaty Act so that it allowed industry to accidentally kill birds without any penalty – including birds that spend summers in Canada.

In August this year, the National Audubon Society, American Bird Conservancy, other conservation organizations and eight states successfully sued to get that reinterpretation reversed by U.S. District Court Judge Valerie Caproni.

It was the hat-making industry that early on ran afoul (afowl?) of people who value birds. Wading birds that grow luxurious plumes during the breeding season and other birds were being slaughtered so that the feathers could adorn women's hats – and sometimes whole birds were stuffed and perched on women's heads.

In 1896, Harriet Hemenway and Minna B. Hall, no slaves to fashion, organized the Massachusetts Audubon Society for the Protection of Birds to save the birds from decimation. Ten years before, George Bird Grinnell organized a group he called the Audubon Society in New York City.

By 1898, 16 more state Audubon societies were formed, leading to the founding of the National Audubon Society for the Protection of Birds in 1905.

In 1916, the U.S. and Great Britain, on behalf of Canada, signed the Migratory Bird Treaty. In 1918, the U.S. enacted the Migratory Bird Treaty Act to implement it. In later years, with bipartisan support, the treaty and the act were expanded to include agreements with Mexico, Japan and what is now Russia.

The U.S. Fish and Wildlife Service is the agency setting the MBTA policies and enforcing them.

Unless permitted by regulation, there's a prohibition to "pursue, hunt, take, capture, kill, attempt to take, capture or kill, possess, offer for sale, sell, offer to purchase, purchase, deliver for shipment, ship, cause to be shipped, deliver for transportation, transport, cause to be transported, carry or cause to be carried by any means whatever, receive for shipment, transportation or carriage, or export, at any time, or in any manner, any migratory bird…or any part, nest, or egg of any such bird."

Technically, teachers should not display abandoned robins' nests or migratory bird feathers in their classrooms without license from the U.S. Fish and Wildlife Service.

Without the MBTA, BP (British Petroleum) would not have paid $100 million in penalties for killing an estimated one million birds in the Deepwater Horizon oil spill. The fine went to wetland and migratory bird conservation as compensation.

Without the full protection of the MBTA between December 2017 and this August, snowy owls were electrocuted in four states, oil spills happened in three states, and there were other examples of avoidable bird deaths that the U.S. Fish and Wildlife Service investigated but could not penalize anyone for.

The potential for hefty fines has led to industry innovation, such as covering oil field waste pits and protecting birds from electrocution. There's still work to be done. Recent numbers show up to 64 million birds per year are still killed by powerlines, seven million by communication towers, half to one million by oil waste pits, and oil spills still happen.

It's hard for industry leaders to understand why birds should matter more than their profits. Birds are not just pretty faces. They work. They perform ecological services, which means they do things like keep other species in balance that can become pests to humans otherwise.

According to the study published in September 2019, "Decline of the North American Avifauna," by Cornell Lab of Ornithology, American Bird Conservancy, Environment and Climate Change Canada, U.S. Geological Survey, Bird Conservatory of the Rockies, Smithsonian Migratory Bird Center and Georgetown University, North America has 3 billion fewer birds today than in 1970 due to loss of habitat and other human-caused problems. What if they were all still with us? Would one advantage be that we would need fewer chemical pesticides and have fewer of their side effects?

We have a lot in common with birds. Birds are more like us than we ever expected. They learn, they plot, they communicate – even with other species. Jennifer Ackerman's latest book, "The Bird Way," explores what scientists are learning.

Environmental protection regulations have taken a hit in the last four years. As people who breathe air and drink water, more of us should be more concerned. At least the bird protections regained by the recent MBTA verdict will help people as well, if somewhat indirectly.

Next, the National Audubon Society is going to court to defend the National Environmental Policy Act. The birds will appreciate that.

2020

# 440 Fall migration: some birds hit hazards, others find feeding bonanzas

Friday, October 9, 2020, ToDo, page A8

Mark and I were camping in the Cascades with our granddaughter and her parents, watching American dippers fly up and down the Sauk River when the unseasonably early snowstorm hit Cheyenne Sept. 8-9.

The first rumor we heard about a bird migration catastrophe was from my sister and brother-in-law in Albuquerque who mentioned a lot of dead birds had been found in New Mexico after that same storm.

Albuquerque dropped from a record high of 96 degrees for the date to a record low of 40 with high winds. Up on Sandia Crest, overlooking the city, there was snow. Dead bird reports started coming in including 300 carcasses at White Sands Missile Range and more in other parts of the region.

Within a week, the major bird conservation organizations and the national media were writing about it. The best account is at the American Birding Association website, written by Jenna McCullough, a graduate student at the University of New Mexico.

But there were also anecdotal reports of dead migratory birds in the west back in August, the possible culprit being the wildfires. Apparently, when smoke fills the air, the migratory birds leave, even if it is prematurely, so they may not have eaten enough to store enough fat for the journey.

Perhaps the birds can't find the food they need along the way because the smoke is obscuring it. And breathing smoke weakens them. Mark and I had a taste of that on our drive back across Washington, Oregon, Idaho and Utah. Cheyenne had even thicker smoke Sept. 26 when it was engulfed in the orange plume from the Mullen fire.

Not all birds migrate. The non-migratory birds know their territory's food sources well and are less likely to starve during rough times. However, many of the species that migrate are insect eaters, insectivores. They migrate south as far as Central and South America when it is too cold up north for insects.

There are birds that specialize in finding hibernating insects and their eggs in tree bark, like the brown creeper in winter in Cheyenne. But warblers and flycatchers want more lively insects, and swallows require flying insects.

No insects fly for a while after a freezing event. Even if the birds had stored fat for the journey, they would burn a lot trying to keep warm. Swallows are known to huddle together, and Jenna found crevices in the bank along the Rio Grande stuffed with emaciated, dead swallows.

## What radar tells us

Weather radar stations around the country can pick up the movements of migratory birds. At Cornell Lab of Ornithology, the BirdCast team, besides forecasting migratory activity, processes the data to show the amount and direction of migration over time in active maps.

The night of September 8-9, you can see a hole in migratory activity centered on New Mexico, Colorado, and western Texas, Kansas and Nebraska – no bird movement. And none in much of Wyoming, but apparently no one reported high numbers of dead birds here.

Conversely, on Sept. 21 as I walked the dog along the Henderson ditch in the morning, the vegetation was alive with small birds busily hunting. I checked BirdCast and sure enough, overnight there had been a major migration push from eastern Montana down through Cheyenne and down all along the Colorado Front Range, as well as in the Pacific Northwest and along the East Coast. The night flyers had landed and were having breakfast.

## Drought delight

One upside to the drought is the drawdown of reservoirs. When reservoirs are full, there are few migrating shorebirds stopping to feed. We went up to Bump Sullivan mid-September, and it was a cornucopia: Mudflats full of feeding birds stretched a hundred yards between shore and water.

We saw sandpipers: Baird's, least, pectoral, and spotted, and long-billed dowitchers plus great blue herons and sandhill cranes probing for invertebrates with their bills. On the water, hundreds of American white pelicans were feeding in long strings, shoulder to shoulder.

On Sept. 21, I also heard my first dark-eyed juncos in our neighborhood. While insectivorous birds abandon us for the winter, the seed-eating juncos join us. They aren't quite ready to be bird feeder regulars yet – plenty of seed in the wild for now – but they haven't forgotten how to find us. It would be interesting to do a banding study and find out which mountain ranges our juncos nest in.

# 441 Project FeederWatch brightens winter with backyard birds

Saturday, November 7, 2020, ToDo, page B2

Nov. 14 marks the beginning of Mark's and my 22nd season participating in Project FeederWatch. It's a community/citizen science winter bird count endeavor started by Cornell Lab of Ornithology and Birds Canada back in 1987.

It's open to people of any age and any expertise level who are willing to put up a feeder and count the birds that visit and report them 1 to 21 times during the 21-week season. This year's season ends April 9. Even if you don't participate, there's a wealth of free data, bird I.D. help and information about feeding birds available at www.feederwatch.org (and fun stuff like the participants' photo contests).

Here's how Mark and I do it. Every year, we update the description of our backyard – size doesn't change but how many trees and shrubs might. We describe our birdbath and three bird feeders: sunflower seed tube, nyjer thistle seed tube and the cage that holds a block of pressed-together seed.

For the two-day count period, we choose Saturday and Sunday each week, even now that Mark is retired. There must be a minimum of 5 days between counts, so we stick with the same days each week – it's easier to remember.

We could print out an official tally sheet for each week, but we just use a scrap sheet of paper on the kitchen table. All our feeders, and the ground under them, are visible from the kitchen window.

During the count, we are looking for the largest number that can be seen at one time of each species – at the feeders and in our bushes and trees. We estimate snow depth and amount of time we watch. We don't spend hours at the window. We spend less than one hour over the two days and just check as we walk by.

By Sunday evening, we can enter the count data online – including any comments on bird interactions and observations of disease – and upload bird photos. There's now a phone app for reporting counts too.

It's fun looking at our own data. CLO makes cool charts. I can see how the number of species and number of individuals changes during a season. I can compare all 21 seasons by species – back in 1999-2000, we were seeing goldfinches nearly every week, much less often in 2019-2000.

Our yard's landscaping has changed and matured. Over the 1999-2000 season, we saw 12 species total. Over 2019-20, it was 21 species, though one week only one bird, a junco, was seen during the two-day count period.

There were 20,000 participants last year, but only 27 in Wyoming, urban and rural. We could use more data to give scientists a more accurate view of our birds. Consider joining.

The participation fee of $18 ($15 for CLO members) funds nearly the entire endeavor, including mailing a research kit to first timers: instructions and bird i.d. poster. We all can opt for the calendar, 16-page annual report and a digital subscription to Living Bird, a 70-page, full-color quarterly magazine normally available for the minimum $39 CLO membership fee.

What will you see at your feeders? Here's the list of the top 25 species based on the percentage of Wyoming participants reporting them last season:

Eurasian collared-dove: 77
House finch: 74
House sparrow: 66
American goldfinch: 66
Dark-eyed junco: 66
Black-capped chickadee: 66
American robin: 59
European starling: 55
Northern flicker: 55
Red-breasted nuthatch: 55
Downy woodpecker: 48
Black-billed magpie: 44
Blue jay: 37
Mountain chickadee: 37
Red-winged blackbird: 33
American crow: 33
Pine siskin: 33
Rosy finch species: 25
Hairy woodpecker: 25
Common raven: 22
White-breasted nuthatch: 22
Common grackle: 22
Sharp-shinned hawk: 22
Wild turkey: 18
Song sparrow: 18

There's an irruption of pine siskins this year because there isn't a good seed crop in Canada. You may see more of them at your feeders.

Here in Cheyenne, we are unlikely to see wild turkeys or rosy finches, but the other species, and more, are all possible. If you go to Project FeederWatch's "Common Feeder Birds Interactive," set it for "Northwest" and "Black oil sunflower seed" and you'll find photos of most of our species. Click on each photo and discover what other kinds of food and feeders that species prefers.

CLO has the free Merlin phone app for identifying birds. You answer simple questions about location, size, color, behavior and habitat for your unknown bird and it shows you photos of possible birds.

For each species, CLO's All About Birds website, www.allaboutbirds.org, will give you multiple photos, sound recordings, range map, habitat, food, nesting, behavior information, conservation status, cool facts, backyard tips and their names in both Spanish and French.

I hope you'll join Project FeederWatch this winter with me and Mark. It is one of the things I like about winter.

# 442 First Cassin's finch visits Gorges backyard

Saturday, December 5, 2020, ToDo, page B2

By early November, our winter feeder birds are back.

House finches are the most abundant and show up every day. Juncos come when the weather's bad. This year, we are regularly hosting two red-breasted nuthatches and two mountain chickadees.

Occasionally, a downy woodpecker, flicker or collared-doves fly in. The goldfinches are unreliable, but their close cousins, the pine siskins, are showing up every day. That's unusual for them, but they are part of the flock pushed south this year due to a bad seed crop in the north.

I was gazing out the window at the birds busily flitting about the feeders and patio paving below, then realized I was seeing an odd bird in the mix.

House finches are the faded brown birds with faded stripes down their breasts. The males have pale pink heads that get redder in the spring. Pine siskins have stripes too but are smaller and their stripes are very dark - plus, on their wings they have a white bar and a flash of yellow.

The odd bird had the pine siskins' dark breast stripes, but it was the size of the house finch. It couldn't be dismissed as an aberrant house finch because there were light-colored markings on its head that house finches don't have, and the back of the top of its head was, well, kind of a pointy topknot. Time to get out the Sibley Field Guide to Birds: "Female Cassin's Finch," the 103rd species to fly over or into our yard.

The males have pink/red heads like house finches, but with the topknot being the brightest. Unlike the female, their breast stripes are very faint, fainter than the house finch's.

To get an overview of everything known about a bird species, I go online to Cornell Lab of Ornithology's "Birds of North America Online," but it doesn't exist anymore. It's been rolled into "Birds of the World" at www.birdsoftheworld.org, where my subscription is still good ($49 per year, or by the month or discounted for three years).

When I pulled up the Cassin's finch page, I was surprised to find a notation that I'd recorded this species in eBird seven times. Clicking on that link showed my two current observations, August last year in the Snowy Range, April 2014 and December 2013 in Hartville, June 2013 in a canyon in Washington State and July 2011 at Upper North Crow Reservoir.

This is a finch that breeds in coniferous forests of the Rocky Mountains, from just over the Canadian border to northern New Mexico and Arizona.

It can migrate altitudinally, spending the winter at lower elevations (Hartville, in Platte County, is at only 4,600 feet and Cheyenne at 6,100, compared to 10,000 feet in the Snowies) or latitudinally, flying as far south as central Mexico. Sometimes they just hang out if the seed crop is good. The one that visited us must have lost her flock.

Cassin's finch's closest relative is the purple finch, an eastern species, diverging from it genetically 3 million years ago. It diverged from the house finch 9 million years ago.

Ornithologists have classified Cassin's as a "cardueline" finch, a subfamily of finches of 184 species worldwide, including the Hawaiian honeycreeper species. In North America, it includes the redpoll, pine and evening grosbeaks, pine siskins, goldfinches, rosy finches, crossbills and our "rosefinches"—house, purple and Cassin's.

Besides sharing similar skull formation, cardueline species feed their young regurgitated seeds. Other perching birds feed theirs' insects. Cardueline species can grip a plant stem and extract seeds from flower heads. I see house finches and goldfinches do that in my wildflower garden all the time.

Sparrows wait until the seeds fall to the ground – I've never seen a junco, a species of sparrow, pluck seeds from plants or feeders, though one was experimenting last year.

I was curious if "cardueline" came from the same origins as "card" in cardinals, named for the religious figures in red robes, but red wouldn't hold for all the sub family species.

It's from "carduus," meaning wild thistle or artichoke. Artichoke is a giant thistle-type plant in the aster/sunflower family. And this makes perfect sense. These finches like to pluck seeds from flower heads, including thistle, coneflower, sunflower and aster.

I'm glad my cardueline finches can also pluck black oil sunflower seed out of our hanging tube feeder since it doesn't take long to clean out the seeds in our garden.

We look forward to hosting the birds during a winter we can't host people.

416 CHEYENNE BIRD BANTER

# 2021

## 443 December 2020 Southeastern Wyoming Christmas Bird Counts compared

Saturday, January 9, 2021, ToDo, page B2

There are many variables affecting the number of birds and bird species seen on the Christmas Bird Count. Weather is a big one. Dec. 19, the Cheyenne counters met up with strong winds that put a damper on small bird numbers. None of us were mean enough to shake them out of the bushes.

Count compiler Grant Frost and some of the other 13 participants were able to find a few of the missing species count week (three days before and after the count day) when the wind moderated.

A week later the weather was "spitty" with snow squalls, reported Jane Dorn, compiler for the Guernsey – Ft. Laramie Christmas Bird Count Dec. 27. Mark and I planned to drive up and help the five participants, but over the years we've had iffy weather like that turn into white-knuckle driving, so we stayed home.

Although both CBCs are in southeast Wyoming, Cheyenne is 80 miles south as the crow flies and, at 6063 feet, 1,700-1,800 feet higher than Guernsey and Ft. Laramie. The topography is different too.

As I read through Jane's list, I could imagine where the birds were. The bald eagles and ducks would have been on Greyrocks Reservoir, which was open – unlike Cheyenne's much smaller lakes, which were pretty much completely frozen.

The many robins and solitaires would be at Guernsey State Park, in the junipers and pines in the hills. Goldfinches, siskins and nuthatches would have congregated at feeders in Hartville, and the belted kingfisher would be somewhere along the North Platte River or the Laramie River, at Fort Laramie National Historic Site or at the Oregon Trail Ruts State Historic Site. Raptors could have been anywhere – there's a lot of unobstructed sky in the 15-mile diameter count circle.

The number of people, how long they are out counting and how much distance they cover, whether by human propulsion or vehicle, makes a difference. That's why, if you get into the scientific use of CBC data, the bird numbers are statistically shaped by these effort factors.

The lists for both counts are combined below, Guernsey-Ft. Laramie in italic numbers for species also seen in Cheyenne, and with names and numbers in italics for species not seen in Cheyenne. The abbreviation "cw" is for birds seen "count week."

The list starts out with one of the outstanding birds seen, the greater white-fronted goose (the forehead is white). Grant found it at Lions Park. It pays to examine every bird in a flock of Canada geese.

This individual was late in migrating from its Arctic breeding grounds. Since it is a nearly circumpolar arctic species, it would be interesting to see if any of them are found this late between breeding and wintering ranges – in the middle of Eurasia.

### Cheyenne CBC

Dec. 19, 2020; 33 species total plus 8 count week
*Guernsey-Ft. Laramie CBC; Dec. 27, 2020*
47 species total plus 3 count week

Greater White-fronted Goose  1
Cackling Goose 10, *48*
Canada Goose 1339, *3,387*
*American Wigeon 2*
Mallard 182, *441*
Domestic (White) Mallard 1
*Green-winged Teal 53*
Common Goldeneye 3, *1*
*Hooded Merganser 5*
*Common Merganser 213*
*Wild Turkey 75*
Rock Pigeon 145, *1013*
Eurasian Collared-Dove 81, *138*
Great Blue Heron 1, *1*
*Golden Eagle 1*
Northern Harrier cw
Sharp-shinned Hawk cw
Northern Goshawk cw
Bald Eagle 7
Red-tailed Hawk 4, *2*
Rough-legged Hawk 1, *2*
Great Horned Owl 1, *cw*
Belted Kingfisher 1, *3*
Downy Woodpecker 3, *1*
Hairy Woodpecker 1
Northern Flicker 8, *21*
American Kestrel 2, *5*
*Merlin 1*
*Prairie Falcon cw*
Northern Shrike 1
*Stellar's Jay 8*
Blue Jay 2, *22*
Black-billed Magpie 26, *14*
American Crow 90, *5*
Common Raven 7, *1*
*Horned Lark 15*
*Black-capped Chickadee 48*

| | | | |
|---|---|---|---|
| Mountain Chickadee 7, *13* | *Marsh Wren 1* | *Red Crossbill 2* | *Pink-sided - 19* |
| *Golden-crowned Kinglet cw* | European Starling 167, *181* | Pine Siskin 4, *33* | White-crowned Sparrow cw, *12* |
| Red-breasted Nuthatch 6, *11* | Townsend's Solitaire 3, *81* | American Goldfinch cw, *38* | Song Sparrow cw, *4* |
| White-breasted Nuthatch 1, *7* | American Robin *cw, 541* | American Tree Sparrow 9, *4* | *Red-winged Blackbird 23* |
| Pygmy Nuthatch 1 | House Sparrow 244, *9* | Dark-eyed Junco 30, *66* | |
| Brown Creeper 5 | House Finch 37, *60* | *Slate-colored - 9* | |
| *Canyon Wren 1* | Cassin's Finch cw | *Oregon - 5* | |

# 444

Saturday, January 30, 2021, ToDo, page B1

## Great Backyard Bird Count causes columnist to ponder diversity

The Great Backyard Bird Count is coming up Feb. 12-15. You can now take part by watching and reporting the birds you see at your bird feeders – or anywhere in the world, aka the real Great Backyard!

Now that the GBBC has gone global, it has a fresh website, https://www.birdcount.org/. Becca Rodomsky-Bish, with the Cornell Lab of Ornithology, charged with its redesign, wanted comments from a small group of reviewers and I was invited. I have in the past contacted CLO for information about their programs for these columns and I've taken part in the GBBC since nearly the beginning.

This is also the year that major environmental organizations are looking at their lack of diversity – both staff and outreach – because of incidents like Black birder Christian Cooper's experience in Central Park.

I think CLO's plan to invite GBBC participants around the world to submit photos of themselves and their families and friends birding during the event will do much to illustrate diversity.

Normally, birders talk about bird species diversity and how to protect

and improve it.

To measure human diversity in the local birding community, we can look at our local Audubon chapter. This is what we see: participants in events, members and board members are evenly split between male and female. In photos from the chapter's beginnings almost 50 years ago, it has always been like this. Human sexual orientation isn't as visible and hasn't come up during meetings and field trips.

We usually have a diversity in age, at least between 50 and 90 years old with the occasional younger outlier. Mark and I were unusual, bringing our kids along on field trips starting when they were infants. We've met teenagers occasionally who are into birds. But the lack of kids, I think, is more about how families choose to spend their limited time together. It's when the kids leave home that parents finally look for new activities. In the 39 years I've been involved in Audubon chapters, we've never run out of people in the upper age bracket.

A few years ago, the chapter established a grant program for education and conservation projects in Laramie, Goshen and Platte counties. We've had several

teachers successfully use our grants. Their students might be who will join when they are 50. But we could certainly use ideas and volunteers to help us reach more younger people.

Birding is adaptable for the disabled, though being able to see and/or hear a bird, however poorly, is rather necessary for birdwatching. No need to take a bird hike. A little black oil sunflower seed on the ground or in a feeder will help bring the birds in viewing range. You might start feeding the birds a couple weeks before the GBBC.

What about socio economic diversity?

Birdwatching at its most basic doesn't cost a thing. Birds are everywhere. You can check out a field guide from the public library. The CLO has many free resources online. I'm beginning to think of the internet as a public utility like water and everyone needs a device, a digital bucket, to capture some of the flow.

Old or cheap binoculars can be helpful, but not necessary for watching birds at a backyard feeder. Our local field trips are free, and except during pandemics, carpooling is often available.

I've talked to people at every socioeconomic level

who enjoy watching birds, whether it's the flock that comes every afternoon for their black oil sunflower seed handout or the flock that flew over their tour group in some exotic location. Some birdwatchers tune in to backyard bird behavior, some strive to add to their global bird life list.

Birds attract people from all walks of life. However, there is a higher percentage of wildlife biologists among birders than in ordinary social circles. I'm happy to say over the years there is an increase in the percentage that are women.

Our Audubon chapter is not as racially diverse as Cheyenne. I'm not sure how to change that. We advertise our existence and wait for people who have made a connection to birds and who want to meet other bird-happy people and learn from each other and share sightings and support the well-being of birds (and other wildlife and people).

Many birders point to a "spark bird," the bird they noticed and then wanted to find out more about, eventually finding more and more interesting birds – and finding they are all interesting birds.

Birds bring together all

---

418

CHEYENNE BIRD BANTER

sorts of people. Let's put on our binoculars as birdwatching badges, whatever quality they are, and find each other where the birds and birders gather. Maybe we'll see each other outside during the Great Backyard Bird Count.

# 445 Wildlife Conservation license plate: one way to give to Wyoming

Friday, March 5, 2021, ToDo, page A10

I've always wanted a vanity license plate – or what the Wyoming Department of Transportation calls a "Personalized Prestige Plate." It would be like a high-quality bumper sticker that doesn't leave residue when you pull it off.

Wyoming has several categories of special license plates: Radio Amateur, Pioneer, EMT, Disabled Vet, University of Wyoming, among others. But with our county plates starting with "2," I always thought it would be fun to have one that says, "2 BIRD." Turns out someone has that one already – I once parked next to them at the dentist's office.

Finally, Wyoming came out with a special plate that supports wildlife conservation. It features a mule deer buck on the far left, then our Wyoming bucking horse and rider silhouette in highway-sign yellow, followed by "WC" and four digits. The governor issued a challenge that 2,020 license plates be sold by the end of 2020, and the goal was barely met. That leaves less than 7,979 available, until they start using letters.

At dot.state.wy.us/wildlife plate, you'll find it costs $180, with $150 going to state wildlife conservation and $30 for the cost of the plate. It can be renewed each year for $50

in addition to your regular license fee. Because we'd barely touched our travel budget in 2020, thanks to the pandemic, Mark and I decided the WC plate would be a good investment for both our vehicles – and maybe an easy way to tell, when in parking lots, our blue Subaru from its many siblings.

The funds go to wildlife conservation, specifically the Wildlife Crossing project.

Currently, in Wyoming there is an average of 6,000 vehicle accidents per year involving large wildlife. We know where the favorite wildlife crossings are. Instead of being slaughtered, the animals can be funneled to wide bridges planted with native vegetation. These, as well as wildlife underpasses, make the highways 80-90% safer for both wildlife and people. See more numbers at wgfd.wyo.gov/wildlife-in-wyoming/migration/wildlife-crossing.

You can donate directly to the Wildlife Crossing project to pay for these bridges and underpasses rather than buy a plate.

Wyoming has a considerable number of anti-tax residents and legislators, so it is good to see support for this project, although because it also saves human lives, you would think the funds should come from the state

transportation budget.

Hunting and fishing licenses are another way we tax ourselves. No one complains because the funds go to the Wyoming Game and Fish Department. I don't think of poachers being anti-tax. I think of them as vandals.

According to an article in the March 2021 issue of Wyoming Wildlife, the Wyoming Game and Fish Department receives 85% of its revenue from hunting and fishing licenses and other fees, and federal taxes on firearms, fishing tackle and other outdoor equipment. The remaining 15% comes from grants for special projects. It does not receive any appropriated state funds.

Years ago, when Wyoming first offered lifetime fishing licenses, Mark and I bought them for ourselves and our kids. We thought of it as an investment in Game and Fish. And it makes it easy to go ice fishing on Jan. 1.

I wish Wyoming had a way for people who enjoy watching wildlife to contribute to Game and Fish for the well-being of nongame species – including birds.

There are huge cuts in the state budget starting this year due to the downturn in the oil, gas and coal industries, which paid the taxes that supported the state in

the past. The global economy is modernizing, and it is unlikely these industries will boom again as they have after previous busts.

Because we have no state income tax, there isn't an efficient way for Wyoming residents to contribute to the funding of other state entities, such as health and education. Having no income tax has been considered a selling point for getting people to relocate here. But when government services are diminished or cut altogether, not many people will want to come.

I suggest Wyoming start a 1% income tax everyone pays. Just like I'm proud to have a license plate that shows what I support, I would think all of us would be proud to support our state. To make it simple, we could all pay 1% of whatever amount our federal income tax is based on, before or after exemptions. I suppose you could prorate it for people who spend part of the year living elsewhere.

Millions of people contribute money to what they believe in. Why can't we residents have more ways to put money into Wyoming? Meanwhile, get your Wildlife Conservation license plates now!

# 446 Birds in the news: salmonella, predator aversion, wind turbines, song identification

Saturday, April 3, 2021, ToDo, page B2

I get bird news in so many more ways now besides mail: Facebook, podcasts, blogs, emails. And even from friends and the radio.

Kathy Jenkins asked if I'd heard the National Public Radio report on an outbreak of salmonella at bird feeders around the country. You can tell the bird victims because they will often sit quietly all fluffed up on a feeder perch when other birds have flown away. They are usually finches.

It's a disease passed around from bird to bird where they congregate at feeders. The cure, when you see sick birds, is as simple as taking down your feeder for a week and scrubbing it well with a solution of soapy water and a little bleach and rinsing it well before refilling.

There are a variety of other communicable bird diseases and cleaning of feeders every couple of weeks – and bird droppings in the vicinity – is good preventive maintenance and avoids having to suspend feeding because there are signs of disease.

On the other hand, painting stinky stuff around the nesting territories of endangered shorebirds is a good idea.

Researchers in New Zealand found that the enticing scent of chicken and other easily procured prey species mixed with petroleum jelly and slathered on rocks attracted predators. After a month of constant reapplication, the predators, ferrets and feral cats, learned that the smells offered no food rewards. They seemed to have moved on before the double-banded plovers, wrybills and South Island pied oystercatchers came in to nest.

Successful hatching doubled for the plovers and wrybills and tripled for oystercatchers. Keeping up this aversion training each season may lead to population increase over time instead of the current decreasing numbers.

The impact on birds of a proposed wind turbine project in Albany County was recently incorrectly compared by someone quoted for a Wyoming Public Radio story.

Wind energy proponents frequently cite the statistics that more birds are killed by cats than by wind turbines. The problem is that the kinds of birds killed by cats are more likely to be common birds in urban and suburban areas than the long-distance migrants like shorebirds (though they are also at risk on breeding grounds), raptors and warblers.

And because wind generation continues to increase and companies are not required to make public how many birds are killed, we only have their word for the comparison.

I still think we should fill current infrastructure with solar panels before littering the landscape with turbines, especially with their massive concrete pedestals, miles of underground cables and unrecyclable components.

I'd like to apologize to everyone who tried to attend the virtual Cheyenne Audubon meeting in March and was stymied by our human-caused technical error.

We hope to have the evening's guest speaker, Nathan Pieplow, visit Cheyenne later this spring for birding and a book signing.

Pieplow is the author of the Peterson Field Guide to the Bird Sounds of Western North America (and the eastern version). You can learn to hear an unfamiliar bird and look it up in his field guide, or at least narrow it down to a category of sound type and then compare with the bird sounds at https://academy.allaboutbirds.org/peterson-field-guide-to-bird-sounds/.

The field guide has spectrographs of bird sounds, very much like musical notation. The introduction gives you instructions on how to learn to "read" spectrographs. You can also use a phone app like Song Sleuth to record birds and see the spectrograph and get an identification suggestion.

Pieplow's March talk was on interpreting common bird sounds. Who knew that the sound of red-winged blackbirds in the spring in the cattails is actually a duet, the female joining in midway to declare "My mate is taken!"?

The more bird sounds are studied, the more variation is found. Brown thrashers can go off on a riff for over an hour and never repeat themselves.

A group of red-winged blackbird males in a marsh will use a series of call notes to keep in touch and apprise each other of danger, but another group 50 miles away uses a different set of calls.

Cowbird nestlings, hatched from eggs dropped in other bird species' nests, don't sound the same as the host nestlings, but get fed anyway.

We don't hear what birds hear because their hearing is better and more discriminatory. Kind of like the way they can see more "frames per second" than we can, they can hear more nuances than we can.

There is endless room for more research, including uploading your phone recordings of birds you hear to eBird.org. As Pieplow said, there are 10,000 bird languages – at least as many as there are bird species in the world.

# 447 Mullen Fire changes forest habitats

Saturday, May 1, 2021, ToDo, page B2

It isn't good, it isn't bad. We can't make moral judgements. It just is. This is the message Jesse McCarty had for us about the Mullen Fire.

McCarty is a wildlife biologist and on the natural resources staff of the Medicine Bow – Routt National Forest's Laramie Ranger District. The Mullen Fire started Sept. 17, 2020, on the forest in the Savage Run Wilderness Area. The source of ignition is still under investigation.

From there, firefighters were able to keep it from burning an area around Lake Owen critical to the safety of Cheyenne's water supply. But on Sept. 26, the wind pushed the fire down and around on a one-day, 30,000 acre-run to the east. That's a swath 6 miles wide and 8 miles long.

That was the day Cheyenne's skies turned orange, even though we were 70 miles downwind of the fire. That is the day that if you breathed that orange air, your lungs didn't feel right for a couple months afterwards.

(To see the extent of the fire, go to the website that tracks wildland fires, https://tinyurl.com/mullen-wildfire.)

The Cheyenne – High Plains Audubon Society invited McCarty to talk about what the effects of the fire were and will be on wildlife, especially birds, and what restoration work is planned.

This forest has been using certain bird species as indicators of habitat. Not all bird species specialize in a narrowly described habitat, but each species monitored is tied to a particular one. For instance, the Lincoln's sparrow is found around wet mountain meadows. As the meadow fills in with trees over time, there will be more forest species such as the brown creeper.

After a fire, the American three-toed woodpecker moves in. A species of the spruce-fir habitat, it is most numerous where insects are taking advantage of dying trees. When the flush of those insects is over and low growth is sprouting, another bird species will move in. On it goes until the spruce-fir forest is re-established and golden-crowned kinglets are at home again.

The Forest Service is continuing its bird surveys this summer. It also keeps an eye on threatened and endangered species and others in special, protective categories.

Field biologist Don Jones of Laramie asked an important question. In view of the warming climate (the forest was experiencing another drought year in 2020), will areas that were once spruce-fir come back, or will the vegetation of a drier climate prevail, like pine-juniper? Jones is young enough that he may see the answer in his lifetime.

The more than 55 people (not counting instances of more than one person per screen) around the state and beyond who were participating in the Zoom meeting were also concerned about other wildlife, such as the large mammals. McCarty said that there didn't appear to be large mammal carcasses in the wake of the fire. The new vegetative growth after the fire will attract big game.

The insect life will have taken a hit where it couldn't find moist places to hide, McCarty said, but there is not much fire science related to insects.

When McCarty visited the forest in December, he found green growth. Sometimes, he said, this is from the caches of seeds squirrels and other small animals make. Also, the heat of the fire will have opened the serotinous cones of lodgepole pine, releasing seed. Aspen growth is also stimulated by fire.

The spread of cheatgrass is a concern and so the forest is using applications of Rejuvra, an herbicide that keeps it from germinating. There will also be grass seeding and tree planting in critical areas such as steep slopes.

Burned areas in the Savage Run Wilderness Area will not be repaired – the definition of a wilderness area is that people do not interfere with ecological processes there.

For most of us in the audience, the Medicine Bow is our forest, and we want to know how we can volunteer to help it recover. This year, the forest is not allowing volunteers within the burn area, but you can find other volunteer needs by contacting Aaron Voos, aaron.voos@usda.gov.

As the summer recreation season gets started, we will find trails and campgrounds in the fire area that are closed. Please honor the forest's directives for your own safety until hazardous trees have been dropped and burnt slopes are stabilized.

And make sure you don't cause the next forest fire.

# 448 2021 Big Day brings in birds and birders

Friday, June 4, 2021, ToDo, page B2

It's a chicken or the egg conundrum. Which comes first, lots of birds or lots of birders?

It's true that the more birders there are out looking, the more birds are seen. But the way to get more birders out to look for them is for there to be more bird reports coming in. That piques interest and more birders go out looking instead of doing mundane household chores.

Mark, my husband, was out nearly every morning the first two weeks in May to one of several of his favorite hotspots: Wyoming Hereford Ranch, Lions Park (both are Wyoming Important Bird Areas), Laramie County Community College (the pond areas) or F.E. Warren Air Force Base (ponds there too).

When he came home, he'd give me a report on what

interesting migrants he'd seen and show me photos he'd taken before adding them to the checklists of birds he'd seen and entered through the eBird.org phone app. He'd tell me too who else he'd met, mostly birding friends, but sometimes visitors.

In the evening he liked to check eBird to see what sightings local birders had entered for the day. And he'd check birdcast.info to see if birds were going to be making a strong migratory push through our area overnight – and coming to earth here to rest and refuel in the morning.

Every year, for 60-plus years, the Cheyenne – High Plains Audubon Society designates a date for its Big Day Bird Count and hopes to hit the biggest migratory push. It's usually the third Saturday in May. Sometimes we've had icy storms and wonder if we should pick a later date. Sometimes eBird reports show that there just isn't a peak to the migration. We wonder, too, if climate change means we should move it up a week.

This year we had a good lead-up that encouraged more people to be out on our Big Day, May 15. We had a couple of sharper than average birders joining us too, Nathan Pieplow, author of the "Peterson Field Guide to Bird Sounds of Western North America," and his friend, Will Anderson.

Nathan signed books the evening before. It was going to be an outdoor event, but thank goodness the Hales family lent us one of the WHR barns as backup, because a good gully washer blew in.

Saturday morning was chilly and foggy, but the birds and birders were out. We weren't all in one big group, but we would get the scoop on cool birds from each other when we met up.

The next day, Mark started compiling the list of birds, looking at checklists on eBird for sightings in the Cheyenne vicinity.

At least 30 people submitted, or were included on, 74 checklists. I submitted a couple just for our bird feeders when we took a break at home.

It was one of the best Big Days in Cheyenne in a while: 136 species. And the warbler count was very good: 12 species.

Sunday, there were still a lot of migratory birds in town including 50 pine siskins under our thistle feeder for an hour.

But the show was over by Monday – both out in the field and at our now deserted feeder.

This year, migration seems to have peaked on the Saturday we picked, making it like Christmas in May.

## Cheyenne Big Day Bird Count, May 15, 2021

Compiled from 74 (51 unique) eBird lists. At least 30 people participated. 36 Species

Snow Goose
Canada Goose
Wood Duck
Blue-winged Teal
Cinnamon Teal
Northern Shoveler
Gadwall
Mallard
Northern Pintail
Green-winged Teal
Redhead
Lesser Scaup
Bufflehead
Ruddy Duck
Pied-billed Grebe
Horned Grebe
Eared Grebe
Western Grebe
Rock Pigeon
Eurasian Collared-Dove
Mourning Dove
American Coot
Black-necked Stilt
American Avocet
Semipalmated Plover
Killdeer
Marbled Godwit
Least Sandpiper
Long-billed Dowitcher
Wilson's Phalarope

Red-necked Phalarope
Spotted Sandpiper
Willet
Ring-billed Gull
California Gull
Double-crested Cormorant
American White Pelican
Great Blue Heron
Black-crowned Night-Heron
White-faced Ibis
Turkey Vulture
Northern Harrier
Sharp-shinned Hawk
Cooper's Hawk
Broad-winged Hawk
Swainson's Hawk
Red-tailed Hawk
Great-horned Owl
Burrowing Owl
Belted Kingfisher
Downy Woodpecker
Hairy Woodpecker
Northern Flicker
American Kestrel
Peregrine Falcon
Western Wood-Pewee
Willow Flycatcher
Least Flycatcher
Gray Flycatcher
Dusky Flycatcher

Say's Phoebe
Cassin's Kingbird
Western Kingbird
Eastern Kingbird
Plumbeous Vireo
Warbling Vireo
Loggerhead Shrike
Blue Jay
Black-billed Magpie
American Crow
Common Raven
Mountain Chickadee
Horned Lark
Northern Rough-winged Swallow
Tree Swallow
Violet-green Swallow
Bank Swallow
Barn Swallow
Cliff Swallow
Ruby-crowned Kinglet
Red-breasted Nuthatch
Blue-gray Gnatcatcher
Rock Wren
House Wren
European Starling
Gray Catbird
Brown Thrasher
Northern Mockingbird
Townsend's Solitaire
Veery
Swainson's Thrush

Hermit Thrush
American Robin
House Sparrow
House Finch
Pine Siskin
Lesser Goldfinch
American Goldfinch
Chipping Sparrow
Clay-colored Sparrow
Brewer's Sparrow
Lark Sparrow
White-crowned Sparrow
Vesper Sparrow
Savannah Sparrow
Song Sparrow
Lincoln's Sparrow
Green-tailed Towhee
Spotted Towhee
Yellow-headed Blackbird
Western Meadowlark
Orchard Oriole
Bullock's Oriole
Red-winged Blackbird
Brown-headed Cowbird
Brewer's Blackbird
Common Grackle
Great-tailed Grackle
Northern Waterthrush
Common Yellowthroat
Orange-crowned Warbler
Nashville Warbler
Virginia's Warbler

| MacGillivray's Warbler | Yellow-rumped Warbler | Rose-breasted Grosbeak | Indigo Bunting |
| Yellow Warbler | Townsend's Warbler | Black-headed Grosbeak | |
| Chestnut-sided Warbler | Wilson's Warbler | Blue Grosbeak | |
| Blackpoll Warbler | Western Tanager | Lazuli Bunting | |

# 449 Close encounters of the robin kind found in backyard

Saturday, July 10, 2021, ToDo, page B2

You could say that the robins in our backyard are benefitting from global warming this summer.

After 32 years managing without it, Mark and I had air conditioning installed, and the robins discovered it offered a good nesting location.

We normally can keep things comfortable by closing windows before the outside temperatures get hotter than the inside, plus the basement stays chilly. But with warmer and sometimes smokier summers, it seemed like the right time to invest in heat pump technology, referred to as a mini-split. It also provides heat and can be hooked up to a solar electric system someday.

The robins built their whole nest before we were aware. It is on top of the new conduit in the corner by the back door we don't use much. With the roof overhang, it is well protected.

I've heard about robins building nests on porch lights and attacking anyone who goes in and out the associated door. Gardening takes us back and forth below this nest location, but neither robin parent divebombed us or the dog. Mark even put up our 8-foot stepladder once to take a photo and there were no complaints.

Every time I glanced at the nest when a parent was on it, incubating the eggs, staring at me, I'd apologize for another disruption.

Finally, the day came when I noticed, looking out the kitchen window, that one of the parents was pausing on one of our fence posts with a big juicy, bright green caterpillar in its beak. There were many more treats for the nestlings, but caterpillars seemed the most popular.

It takes a lot of herbivorous prey to raise baby robins and I wondered what plant damage the robins were averting this summer. Gosh, it might have been the right year for growing cabbages. My last efforts were aborted by caterpillars.

By June 19, there was one large nestling left in the nest, almost filling it. By June 20, the nest was empty. I didn't see any speckle-breasted baby robins anywhere.

I went to the corner of the yard by the compost bins to re-pot houseplants. As I approached, a robin flew in, perching on a branch eye level with me. I stopped, and we looked each other in the eye. I murmured congratulations in case it was one of our parent robins. Then it flew to a new perch a few feet away and I turned, and we locked eyes again.

Most wild animals are interested in staying away from people unless we are handing out food. Otherwise, they don't encourage our attention because that is often dangerous.

The robin shifted position again, caught my eye, and then flew off around the upright junipers. I could hear again the quiet call it had been making on the other side of the bushes, plus another odd one. So, I circled the junipers and when I got to the point where I could see into the interior, there was the fledgling.

Unlike a killdeer which tries to draw you away from its nest, I felt like the robin had led me to the fledgling. Minutes later, the fledgling flashed away to another shrub, but I didn't go in pursuit.

Within a week, June 26, I saw a robin sitting on the nest again. Less than three feet away, a male house sparrow with a beak full of dry grass waited patiently for the robin to take a break. His mate waited behind him. I know we have a housing shortage in Cheyenne, but does the robin have a spare room, or what?

We still have a feeder hanging over the patio, under the clear corrugated plastic roof. It's one of those cage types that uses the blocks of seed that seem to be glued together. The red-breasted nuthatches visit it multiple times a day, pecking away.

A pair of these birds nested in a rotten stub on a tree across the street. We think these are the birds flying over our low house to our feeder. On June 25, I saw five nuthatches on the feeder, probably the whole family dining together. They are completely at home. In fact, as I walk back and forth doing chores, I sometimes remember to look up to where, two or three feet over my head, a nuthatch is completely unconcerned by my presence, or that I've stopped so close.

Maybe, like the geese in the park, they read body language and distinguish between danger and safety.

# 450 Neighborhood Swainson's hawks fledge three; fall migration underway

Saturday, August 14, 2021, ToDo, page B2

Just as they did last year, a pair of Swainson's hawks nested in the neighbors' spruce tree two houses down.

Thanks to some tree pruning in between, Mark and I had a perfect view of the nest from our bathroom window.

I'm sure the hawks were a little put out this spring to discover after their long migratory haul from Argentina that the field adjacent now sports a three-story apartment building under construction. But about a quarter mile away is the Greenway and the railroad right of way, still plenty of open space and tasty ground squirrels.

By July 7 we could see two fuzzy white heads in the nest. Nearly three weeks later, they were mostly brown. And then the youngsters started climbing out of the nest and onto the tree branches. That's when we realized there were three of them.

We think the day one of the juveniles left the nest for the first time was July 25. At 6 a.m., it was sitting on a bare branch just over our back wall, looking straight back at us through the kitchen window.

There were a few days the youngsters cried a lot for parental attention. One day they landed in our tree and then all three circled low over our block. It's become quieter, but they are still spending time in the neighborhood, sometimes on the nest tree.

It amazes me that a large hawk, best suited for flying grasslands in search of rodents (summer) and large insects (winter), would choose to nest in a residential neighborhood. I'm glad we can provide the big trees they require to successfully breed.

## Hummingbirds

The hummingbirds are a mystery this year. Their favorite red beebalm was halfway through blooming the last week in July and I hadn't seen them yet.

I checked my records on eBird.org and saw since 2013 they have arrived for a three-week stay starting the last week of July or the first week of August. My beebalm is blooming ahead of schedule and they may miss it. I caught a glimpse of one hummingbird July 30 as it flitted quickly over other flowers.

Maybe the red beebalm is early this year because of all our earlier hot weather and moisture. Maybe the broad-tailed hummingbirds are later because our mountains, where they nest, have been unusually full of nectar-filled flowers and they are staying longer.

Maybe we should all put up our hummingbird feeders anyway. Remember, use a little heat to dissolve 1 part white sugar in 4 parts water. Use no other sugar types, use no red dye, and replace any nectar that gets cloudy-looking.

## Weidensaul's new book

Mark and I are reading "A World on the Wing: The Global Odyssey of Migratory Birds" by Scott Weidensaul. A whole chapter is devoted to Swainson's hawks and unraveling the mysteries of their breeding and migration using new tracking technology.

The book also discusses the number of ways migrating birds are killed by human actions, directly and indirectly, that are preventable.

For instance, because many songbirds migrate at night, one of their navigational aids is starlight. Unfortunately, the glow from cities is attracting them and studies show more migrants in cities than there used to be. But when the small birds land in the mornings, do they find the trees and shrubbery full of insects they need to eat to recharge? Sometimes, they find well-lit skyscrapers and become disoriented, circling until exhausted, falling to the ground, discovered dead on the sidewalk in the morning.

City night light is detrimental to other life too, including plants and people without room-darkening shades. It increases with each porch and parking lot light left on. But it can also be decreased by one resident, one business owner and one municipality at a time.

For your home security lighting, see if you can use motion detection technology. You'll save money on your electric bill. For parking lot lights and streetlights, choose those that are hooded, lighting only what's below and not the sky. You'll save money, too.

Without our own astronomical observatory, like Flagstaff, Arizona, I don't think we will become an International Dark Sky City, asking Cheyennites to drive with only parking lights on, but it would be neat.

Fall migration has already begun. The Swainson's hawk family will head south sometime after the middle of September. Only six or eight weeks after fledging, the young Swainson's all over western North America make a journey of as much as 7,000 miles to the Argentine pampas. I imagine it looks something like Wyoming grasslands there. Safe travels, kids and parents.

CHEYENNE BIRD BANTER

# 451 Dry Creek restoration to improve hydrology, habitat

Saturday, September 4, 2021, ToDo, page B2

Jeff Geyer is fixing Cheyenne's Dry Creek.

First, how did it get its name? Jeff, Laramie County Conservation District water specialist, told me that unlike Crow Creek, our other stream that starts in the mountains, Dry Creek starts somewhere on the F.E. Warren Air Force Base. He said it never had much of a channel, with the water frequently spreading out in flat, temporarily marshy areas and percolating into the water table below as it flowed after a rain or snow event.

Fast-forward 160 years. The Greenway now follows Dry Creek as it crosses northern Cheyenne west to east, parallel to Dell Range Boulevard. At North College Drive it heads southeast to the new East Park and crosses under Interstate 80. It joins Crow Creek near where the sewage treatment plant is today on Campstool Road.

What's changed is the Dry Creek watershed, which drains two-thirds of Cheyenne. More land surfaces surrounding the creek have been paved and built on over the last 30 to 40 years as Cheyenne expands. Jeff says you can see the change on Google Earth (use the free Pro version you can download).

Snowmelt and rainfall aren't absorbed by pavement and roofs, so they run off into Dry Creek, making much higher flows. Higher flows are faster. Faster flows are straighter. Straighter flows have more energy to erode the soil. Between Campstool and I-80, that energy cut 5-foot-deep banks and sent good soil into Crow Creek where it gets deposited in the downstream reservoirs - not good for reservoirs, or the fish in Crow Creek.

In 2019, Jeff started to fix a small section of Dry Creek that will make a difference. The idea is to slow the creek down by increasing its sinuosity which will reduce the energy of the water. The water flow needs to look more like a traveling snake – looping to one side and then to the other, rather than a straight stick.

Mathematically, a straight stream has a sinuosity of 1 – the ratio of the distance the water travels is 1 to 1 with the length of the valley. Jeff would like to see a sinuosity of 1.2 or 1.4, meaning that in a 100 feet of valley length, the water would loop an extra 20 to 40 feet.

The banks of a sinuous stream will still erode a bit, but much of the dirt will be deposited in the next curve – slow moving streams can't carry as much soil suspended in the water.

While some earth work was required to reduce the 5-foot cutbank in places to give Dry Creek access to the flood plain during rainfall or snowmelt events, much of the sinuosity building is being done with willow stems, logs, posts and stakes.

At just the right location and angle in the stream bottom, Jeff and volunteers pounded in stakes in a line and then wove willow stems, forming a "Beaver Dam Analog." The willows were from a nearby location where they die back and new willows continue growing.

The woven willows are like snow fence that slows the wind, making the snow drop out into drifts. This structure slows water carrying dirt so the dirt will drop and form a bar where willows will grow, and their roots will stabilize the stream bed. There is already a nice stand of coyote willows in one spot.

Up on the flood plain are "Post Assisted Log Structures." Logs are pinned to the flood plain to make a rough passage that will also slow water down.

Long term, slower stream flow will allow more water around the creek to be absorbed and stored. That underground water flows like surface water and will eventually resurface in the creek, recharging it. Jeff is hoping for a little water to be always in Dry Creek – maybe it will need a new name.

Changing Dry Creek's hydrology, Jeff also expects to provide the moisture needed for more diverse vegetation for wildlife habitat. Mule deer and ermine have been seen. Cheyenne Audubon members have been making bird observations. Lorie Chesnut, a member, was instrumental in obtaining a $3,000 grant through the National Audubon Society's Western Water Network Grants this year that paid for the stakes and native plants.

As Jeff surveyed the conservation district-managed pasture that surrounds the first phase of the hydrology project (and a second phase that has just begun to the south), he frowned at all the 6-foot-tall mullein stalks and the other non-native weeds. Much more work will be required to transform the pasture into prairie more useful to ground-nesting birds and other wildlife, bringing it back to its formerly lush and flower-filled self.

# 452 Cheyenne birders search Pennsylvania and New York woodlands for eastern birds

Saturday, October 9, 2021, ToDo, page A12

Mark and I couldn't hear any birds over the sound of wind in the leaves. That's not unusual for Wyoming, but we were in Pennsylvania where the trees will grow a complete canopy without anyone planting them. Finding birds is dependent on hearing them, even more so than here.

We were at the Churchville Nature Center in Bucks County, my favorite place to bird when visiting my aunt. The goldenrod and purple asters were in full bloom in the little meadow and robins were picking fruit from all kinds of shrubs. But in the trees, it seemed birdless until we reached a little swale protected from the wind and suddenly there was a swarm of chickadees, titmice and warblers for a few minutes.

There were no birds to be seen on the reservoir. The waterbirds and shorebirds must have already tucked in for the coming storm, waiting for the afternoon's deluge.

We counted only 11 species altogether. For the Saturday morning bird walk before our visit, 19 local birders listed 64 species. Timing and experience make a big difference. I keep forgetting to look into hiring local bird guides when we travel.

In the Ithaca, New York, area, we had the help of our son, Bryan and his wife, Jessie, both avid birders. They have experience identifying birds we rarely see in Cheyenne, like black-throated green warbler. They pointed out the sound of a Carolina wren, unseen in the brush. They also pointed out that sometimes one-note calls in the trees are chipmunks or tree frogs.

The Finger Lakes region has a plethora of public land to explore and bird. We hiked the gorge at Watkins Glen State Park our first morning, as early as Jessie could get us on the road. It is black shale sculpted by water, dim and deep and deafening – no birds could be heard over the numerous waterfalls full of rain. The sun rarely reaches into the gorge at 9 a.m., but later the steep trail is crowded with people.

Have you heard of Finger Lakes National Forest? It's a scattering of parcels between Seneca and Cayuga lakes, tiny compared to any of the national forests in Wyoming, but then again, with all those trees in the way, the boundaries are not very noticeable. We hiked the Potomac trails where, in late September, fall color was just beginning to show.

Our second day of birding hikes began with the Dorothy McIlroy Bird Sanctuary northeast of Ithaca. A creek and wetlands attract a lot of birds to this property owned and managed by the Finger Lakes Land Trust. It commemorates a woman who had a significant role in the early days of the Cornell Lab of Ornithology. The shrub fen and peat swamp were bordered by hemlock trees, unusual for the immediate area, but all were old friends of mine from my central Wisconsin days.

Next, we hiked and birded nearby Bear Swamp State Forest Park. Didn't see any bears but found interesting mushrooms. Jessie found a red eft, the teenage stage of the eastern newt.

I've read that the overpopulation of deer has affected eastern forests, browsing the shrub and young tree understory layer of vegetation to the point that you can see quite a way between the tree trunks. It must negatively affect birds that specialize in that layer.

Where there was normal understory, I made a new friend, a small tree, striped maple, named for the vertical ridges on its stems. It is also known as moosewood. It's a favorite moose food and the name of my favorite Ithaca restaurant.

One stop we made between Philadelphia and Ithaca was to see the Rodale Institute, a proponent of organic gardening and farming beginning in 1947. Back in 1978, I contributed a story to their magazine, an interview with the designer of a safer bluebird house. Mark and I opted for the self-guided tour of the fields and greenhouses, which you can hear at their website.

Rodale is now a proponent of organic regenerative agriculture as well as planting for pollinators. However, they apparently haven't banned outdoor cats yet, so they aren't entirely bird-friendly. Ironically, in the shrubbery by the creek there were a lot of catbirds.

While we wistfully compared the unwanted extra precipitation the East has had lately with our western drought, we are still happy with our choice to live in Wyoming, where the horizon stretches much farther.

# 453 Fall reservoir birding is a leisurely affair, mostly black and white

Saturday, November 13, 2021, ToDo, page A12

Birds and birders are in a rush during the spring.

The birds are hurrying to get from their wintering grounds to their breeding grounds. But fall birding is as leisurely as that of the birds' migration south.

On the fourth Saturday in October, the Cheyenne Audubon field trip was to Fossil Creek Reservoir Natural Area, Fort Collins, Colorado, about 45 miles south of Cheyenne.

A reservoir during

426                    CHEYENNE BIRD BANTER

migration seasons is the avian equivalent of a truck stop, a crossroads with each species having its own itinerary. Birders are looking for the most interesting birds, the most exotic license plates.

Since ducks, geese and other water birds placidly rest or feed (unless a bald eagle passes by), every birder gets a chance to look through a spotting scope at them. We had five scopes on this trip.

We were dismayed to see the low water level. Much of the reservoir was mudflats with Fossil Creek trickling from pond to pond. Then we realized there were four kinds of shorebirds probing in the mud.

American avocets, shorebirds, waded in shallow water. These birds of the western Great Plains are ghostly white with black wings by the time they head south for a winter mostly on beaches, including those in southern U.S. and Mexico. In spring they have cinnamon-pink heads and necks.

No need for special optics to enjoy the many American white pelicans we saw, also white with black wing markings. With wingspans of 90 to 120 inches, they fly in lines, like geese, and sometimes spiral with thermals. Another bird of the Great Plains and Intermountain West, they head south to water that stays open so they can fish.

There were rafts of gulls, almost all ring-billed, also the most common gull around Cheyenne. It prefers to nest inland in the northern states and Canada and winter inland in the south and along the Pacific and Atlantic coasts.

We also found a lesser black-backed gull. In winter they are most common along the Atlantic and Gulf coasts, less common inland in the eastern half of the U.S. But the latest range map shows an influx into eastern Colorado. Perhaps the state tourism department invited them to make the trip from their summer homes located anywhere from Iceland to Siberia.

A raft of American coots, each bird the darkest slate gray accented with a bright white bill, was enjoying a day of rest in their migratory trip – or maybe not. Their range map (www.AllAboutBirds.org) shows some can be found year-round in a narrow strip along the east edge of the Rockies from Montana through Wyoming and Colorado.

Western grebes, dark gray from the top of their heads and down the back of their thin necks, but white from their chins to their breasts, were busy diving for small fish. They were stopping over, heading for the Pacific Coast, anywhere from Vancouver Island to Mexico. The range map shows them year-round inland in central Mexico too, but I don't know if that's a population of birds that doesn't migrate or if some northern birds join the locals.

Buffleheads, small black and white ducks, were bobbing around, playing a game of one-upmanship, furiously beating their wings, "standing" on their toes to look large and menacing, while raising their crests of white, then diving. They breed up in western Canada and think much of the U.S., including Cheyenne, is a lovely place to spend the winter.

There was a handful of lesser scaup, another black and white duck, but with a pale blue bill. Breeding from Alaska down to Wyoming, they head south either for the Pacific Coast or the southern states, or even the southernmost tip of Central America, or the Caribbean. Definitely not as cold tolerant as the buffleheads.

Common mergansers, the females sporting their shaggy red-feather crests, mixed with other, sleeker, redheaded ducks, including those known as redheads, plus a few canvasbacks, distinguishable by combined forehead and bill silhouettes forming straight diagonals.

In Wyoming, common mergansers may be seen year-round. Whether the same individuals stick around all year, or the ones from farther north move down for the winter, I don't know.

Redheads breed in Wyoming, but this western species likes to go at least as far south as New Mexico.

Canvasbacks breed in central Colorado and north into Alaska, but they head south for winter, some only as far as southern Colorado.

Finally, yes, there were Canada geese and mallards, the most recognizable waterbirds. You will see their permanent flocks and the winter ducks like buffleheads – and birdwatchers – around Cheyenne reservoirs if there's open water this winter.

# 454

Saturday, December 11, 2021, ToDo, page A12

## Bird feeding safety: clean feeders, cat fencing, glass obstruction

Winter is the most popular season for feeding birds.

Watching birds from your window is an entertaining and affordable, even educational hobby to lighten long winters. But please keep safety in mind as you apply these tips.

**Cleanliness**

Whether you choose a tube feeder, hopper feeder (looks like a little house), cage (for blocks of seed or suet) or platform feeder, make sure it is scrubbable.

The Cornell Lab of Ornithology recommends every two weeks taking feeders apart and brushing out all the detritus and washing them in a diluted bleach solution. You can use your dishwasher instead. Rinse feeders well and let dry thoroughly before refilling.

Wear gloves when handling dirty feeders or wash your hands afterwards.

Seed that gets wet can harbor mold and bird diseases. If you notice any finches with disfigured faces, it's time to take down all your feeders for a week to temporarily disburse (social

2021

427

distance) the flock while you get them clean.

The one best seed – most nutritious and most popular – for our local seedeaters is black oil sunflower seed. But unless you can afford to buy hull-less, you will have moldering hulls below the feeder. If you feed one of the bird seed mixes, there are a lot of seeds in it our birds won't eat, and they also end up making a kind of mat you'll want to rake up regularly. At our house we hang the feeders over the patio and sweep often.

Finches like nyjer ("thistle" that doesn't sprout) seed. It is very fine, requiring tube feeders with smaller holes or a fabric "sock." The hulls are tiny and blow away. If you put out suet, make sure the weather is cold to keep it from going rancid – or dripping.

### Window strikes

Birds have a hard time identifying glass. They see the reflection of sky and vegetation, smack into your window and die or are severely injured, becoming a snack for other animals. Or if two of your windows on opposite sides of your house line up, they may think they can fly through.

Your regular window

screens can break the reflection and soften the impact. There are other strategies and stickers that can be stuck to the outside of the glass (see https://abcbirds.org/glass-collisions/stop-birds-hitting-windows/).

The easy strategy is to place your feeders within three feet of your favorite bird-watching window—or even stick a suction-cup feeder on the window itself. That way, when the sharp-shinned hawk startles your flock, none of them will be moving fast enough to hurt themselves bumping into the window.

### Cats

Our cats love bird-feeding season. They sit on the windowsill for hours, entranced. But if you haven't made your felines into indoor cats yet like Lark and Lewis, please don't feed the birds.

What about the neighbors' cats? That's tricky. You might be able to convince neighbors that indoor cats are safer, healthier and more fun, and that they could then take up bird feeding like you.

Realistically, you are going to have to cat-proof your bird feeding station. While it is good to have cover, shrubs and trees, near your feeder so seed-eating birds can escape

hawks, you don't want it so close cats can pounce on birds feeding on the ground.

You might try encompassing the area under the feeder, where the birds feed on the ground, with a short fence – one you can step over. The idea is that while a cat can sneak up on a flock unobserved, having to leap the fence will give the birds the visual warning they need to escape.

### Water

Water is another way to attract birds – if you can keep your winter birdbath clean. It also has to stand up to freezing and thawing (unless you add a heater) and it needs to be easy to remove ice from or clean, like a flexible plastic trash can lid.

Birds should be able to reach the water when perched on the rim. If there is a sloping edge or sloping rock, birds will also be able to walk in for a bath.

Squirrels

Our fox squirrels are entertaining, but they can destroy birdfeeders and scarf down all your birdseed. We have a tube feeder that shuts down when any animal heavier than a finch sits on it.

Funnel-shaped barriers can be mounted on the pole below a feeder and/or placed

over the top of a feeder, especially one that is hanging. Our feeders hang from the underside of our patio roof.

You can also distract squirrels by feeding them peanuts nearby.

### Timing

Decide how much seed you can afford. Put seed out at the times of day you are most likely to enjoy watching your feeder. Being consistent will bring the most visitors, but if your seed isn't available, the flock will move on to one of their other regular daily stops.

### More information

The Feederwatch.org website is a fantastic free resource. You can find out what birds are seen in our area, each species' favorite foods and the best types of feeders for each.

The Project Feederwatch season runs early November into April. See feederwatch.org to join anytime and add your sightings.

The Christmas Bird Count has a feeder-watching component too. See cheyenneaudubon.org to find out how to take part for free in the local count Dec. 18.

# 455 Barb and Mark Gorges, Champions for bird conservation, Unsung Heroes special

Thursday, December 30, 2021, front page

By Rachel Girt

Receiving a request from a stranger for help identifying a bird is quite commonplace for Barb and Mark Gorges.

"About 50% of the time, they are describing

a Northern Flicker," Barb explained.

For over 32 years, the Gorgeses have shared their wealth of knowledge about biology, conservation and ecology through programs,

field trips, articles and books.

"People come to our programs or on a field trip with us and hopefully get hooked into birdwatching a little bit more," Barb said.

The couple's passion for

conservation and wildlife led them to join the Cheyenne-High Plains Audubon Society when they moved to Cheyenne in 1989. Together, they have held nearly every office existing in the chapter,

CHEYENNE BIRD BANTER

sometimes more than once.

Founded in 1974, the chapter promotes conservation and appreciation of birds through education and habitat stewardship in southeast Wyoming. Today, the chapter has approximately 148 members.

When it comes to working in the trenches, the Gorgeses are the first to volunteer, whether it is planning new gardens or caring for the Habitat Hero Demonstration gardens in the city, explained chapter member Lorie Chesnut in her nomination of the Gorgeses as Unsung Heroes.

Their enthusiasm for nature has been the motivating force for the chapter and other groups, Chesnut added.

They are the chapter's primary contact for persons interested in bird identification, behavior, conservation and where to bird in the region. The couple also recruits regionally and nationally known speakers for the chapter's conferences and programs, and organizes field trips.

"Barb and Mark have carried the chapter on their shoulders for a number of years, keeping us solvent, successful and vital as an organization," Chesnut wrote.

Chesnut called the Gorgeses cheerleaders for the chapter and Cheyenne's best emmisaries, attracting newcomers to the area.

When Chesnut and her husband decided to move to Wyoming in 2016, they contacted the local Audubon chapter. Barb responded to their query, patiently answering their questions

about living here. "Once we relocated to Cheyenne, Mark and Barb were wonderful hosts, making us feel at home and counseling us about how to make the most of our new Wyoming Adventure," Chesnut wrote.

Their experience was not unique, Chesnut said. Several people have moved to Cheyenne after talking to the Gorgeses and learning about the opportunities for hiking, birding and wildlife viewing here.

For the Gorgeses, their passion for the outdoors started early.

Barb had visions of spending her career outside after graduating with a degree in natural resource management from the University of Wisconsin-Stevens Point. A series of seasonal jobs led Barb to the West and eventually to Montana, where she met Mark.

Growing up in New York state, Mark said his interest in the outdoors started with family fishing and camping trips and being in the Boy Scouts.

After graduating college with a degree in biology, Mark served in the U.S. Navy for three years.

Mark then headed West to pursue a graduate degree in fish and wildlife management at Montana State University in Bozeman. After obtaining his degree, he worked in fisheries in Wyoming and Montana and the U.S. Army Corps of Engineers in Washington state.

The couple met in Miles City, Montana, where they both worked for the Bureau of Land Management. They

moved to Cheyenne 32 years ago, when Mark took the position of fisheries biologist at the state BLM office.

Barb pursued her interests in writing, teaching and quilting while raising their children. She taught quilting at Laramie County Community College for about 15 years and launched a small business making quilt labels.

Her first book, "Quilt Care, Construction and Use Advice" was based on columns she wrote for the Wyoming State Quilt Guild's newsletter.

While volunteering to write press releases for the local Audubon chapter, the Wyoming Tribune Eagle outdoors editor asked Barb to pen a column on birdwatching in 1999. She continues to do that and began writing a gardening column for the newspaper about 10 years ago.

She and local photographer Pete Arnold also collaborated on a newspaper column dedicated to a bird of the week for two years. Then, they transformed the columns into a book, "Cheyenne Birds by the Month: 104 Species of Southeastern Wyoming's Resident and Visiting Birds."

Most recently, Barb published her third book, "Cheyenne Garden Gossip: Locals Share Secrets for High Plains Gardening Success." The book details insights from more than 100 Cheyenne gardeners she interviewed.

Gardening is another aspect of birding that the Gorgeses encourage. Both are Laramie County Master Gardeners and serve on

the Cheyenne Habitat Hero Committee. Mark also volunteers regularly at the Paul Smith Children's Village at the Cheyenne Botanic Gardens. He does all sorts of horticultural and regular gardening work.

The couple's connection to Master Gardeners became a perfect fit with Audubon's cultivation of natural habitat for birds, Mark said. Audubon's Habitat Hero program provides resources to create bird habitats. By planting bird-friendly gardens with native plants, communities can help conserve birds.

The native flowers attract the pollinators, which attract the birds as a food source, Mark said.

The Gorgeses have helped secure National Audubon grants to create Habitat Hero Demonstration Gardens at the Cheyenne Board of Public Utilities and Cheyenne Botanic Gardens. Many additional volunteers have helped plant and care for the gardens.

Because of the Gorgeses' efforts, thousands of dollars have been awarded to the chapter, bringing beauty and knowledge to the residents of Laramie, Goshen and Platte counties, Chesnut added.

"Birdwatching is kind of a gateway to thinking about environmental issues and the impact of the choices we make," Barb explained.

Anyone interested in learning more about birds in the region or joining the chapter should visit cheyenneaudubon.org, Barb said.

# 2022

**456** Saturday, January 8, 2022, ToDo, page A10
## Ghosts of Christmas Bird Counts past visit local birdwatcher

Christmas 1989 was Mark's and my first Cheyenne Christmas Bird Count and the first time we met most of the people attending the tally party afterwards.

May Hanesworth, then in her mid-80s, was the count compiler and we met at her apartment. It was a scary place for me, the mother of two boys, ages 1 and 4, because it was filled with breakable figurines of birds. But May, a retired music teacher, was not concerned, and the boys and I often visited on Cheyenne – High Plains Audubon Society business after that. Later, I learned that she and her husband, Bob, secretary of the Cheyenne Frontier Days Committee for 25 years, had had a gracious home on 8th Avenue.

When May could no longer handle the job of compiler, Jane Dorn took over. She and her husband, Robert, are the authors of "Wyoming Birds." When Jane retired, they moved to Lingle and started the Guernsey – Fort Laramie CBC. We were planning to drive over to help this year, but both of our vehicles developed unroadworthy symptoms the day before.

Our next compiler was Greg Johnson, but he's moved on to bigger things – he's the CBC editor for our region. Grant Frost has the Cheyenne job now.

As the business end of the 1989 tally party got going, the Lebsack girls, maybe middle school age, took our boys into the kitchen to play. Their dad, Fred, was one of those birders who loved to geek out on subjects like the finer points of feather coloration. He died prematurely in 2011, and his widow, Judy, and his daughters now live in California.

The one person we had met previous to the tally party was Nels Sostrum, thanks to the regional Audubon director whom we knew through our old chapter in Miles City, Montana.

We always thought we would see more of Nels when he retired from the state, but he jumped straight into his other hobby, painting. And then he met Anne, and through her he acquired stepchildren and step-grandchildren. I hadn't seen him for a long time when his obituary showed up in the paper this fall. Nels's painting of Battle Creek in the Sierra Madre Range hangs on our wall and will always remind me of birding with him.

John Cornelison was at that tally party, too. After tallying the birds, the discussion turned to electing a new chapter president. John volunteered. Mark volunteered to be vice president and I, program chair. Later, we learned that John was the founding president back in 1974.

For many years, John and his wife, Joanne, invited the chapter to hold the tally party at the Westgate community building where there is a large living room with plenty of space for tables and chairs and laying out potluck contributions. Mark and I saw John and Joanne at an event last fall and were saddened to learn of John's death in December. The family asked that our chapter be one of two organizations receiving memorial donations.

I'm guessing that Jim and Carol Hecker were also at our first Cheyenne tally party. Jim was one of the pediatricians our boys saw at the Cheyenne Children's Clinic. He encouraged me to drop the "Dr." when addressing him outside the office and it took a while to do so.

When the boys were young, we had a CBC tradition to stop by mid-morning at their house to warm up, eat Christmas cookies and drink hot chocolate - and count the birds at their feeders. We still enjoy their hospitality.

In 1989, Mark and I were younger than the usual Audubon demographic. People with children spend most of their organized social time in kid-related groups. But there is something to be said for hanging out with people old enough to be your kids' grandparents and great-grandparents, especially when the real relatives live far away. And yes, we raised two sons who are assets to the Cheyenne CBC, whenever their families' Christmas travel schedules coincide with the date (not this year).

This Christmas, our toddler-aged granddaughter received her first pair of binoculars – from her maternal grandfather. This birding thing can be infectious!

Whatever your age and birding ability, look up www.

CheyenneAudubon.org. Join us for hybrid programs, field trips and other activities such as the 8th Annual Cheyenne Habitat Hero Workshop Jan. 29 at Laramie County Community College, also available virtually.

## Cheyenne Christmas Bird Count

Dec. 18, 2021, 15 participants, 36 species observed count day and 5 count week. List compiled by Grant Frost.

CW – count week: species observed on one of the three days before or after, but not on the count day.

Snow Goose CW
Cackling Goose 781
Canada Goose 1512
Northern Shoveler CW
Mallard 226
Northern Pintail 2
Ring-necked Duck CW
Common Goldeneye 19

Common Merganser CW
Rock Pigeon 230
Eurasian Collared-Dove 162
Wilson's Snipe 1
Cooper's Hawk CW
Northern Goshawk 1
Bald Eagle 2
Red-tailed Hawk 7

Rough-legged Hawk 4
Ferruginous Hawk 1
Eastern Screech-owl 1
Great Horned Owl 3
Belted Kingfisher 1
Downy Woodpecker 1
Northern Flicker 14
Northern Shrike 2
Blue Jay 11
Blacked-billed Magpie 53
American Crow 87
Common Raven 8
Mountain Chickadee 8

Horned Lark 13
Red-breasted Nuthatch 5
White-breasted Nuthatch 2
European Starling 233
Townsend's Solitaire 14
American Robin 8
House Sparrow 170
House Finch 75
American Tree Sparrow 26
Dark-eyed Junco 53
Song Sparrow 2
Red-winged Blackbird 24

# 457

Saturday, February 5, 2022, ToDo, page A10

# How to keep prairie birds, and us, safe

"Nurturing the Prairie" was the theme of this year's Cheyenne Habitat Heroes workshop held last month. For me, that includes the plants, animals and people.

Cheyenne sits in the middle of the shortgrass prairie so what we "townies" do matters as well.

Zach Hutchinson, workshop presenter and community science coordinator for Audubon Rockies, reminded us of the study showing North America has lost 2.9 billion birds, including 53% of grassland birds, since 1970. This means that for every 100 birds you could count along a certain distance of our county roads then, today you would only count 47.

One of the biggest causes is loss of habitat, including the conversion of undeveloped land into subdivisions, commercial property or cropland. Cheyenne is going through a terrific building phase. The landscaping in new high density residential neighborhoods will soon draw in birds, but not the

grassland birds. It is the ring of small-acreage landowners around the city who can make a difference.

First, what shape is the acreage in? Is it full of native prairie grasses and what range managers call forbs, which the rest of us call wildflowers? Or was it overgrazed and is now full of invasive weeds like toadflax and needs renewal?

Another workshop speaker, Aaron Maier, range ecologist for Audubon Rockies, talked at length about regenerative agriculture and how farmers are changing their practices, so they spend less on fertilizers and trips with the tractor yet sequester more carbon, capture more moisture and accumulate more beneficial soil microbes.

Aaron also talked about healthy grassland grazing practices benefiting wildlife as well, as laid out by the Audubon Conservation Ranching Initiative. Ranchers following Audubon's guidelines for best practices for land, wildlife and

livestock management are guaranteed premium prices for their product marked as "Audubon Certified."

But the small acreage owner is probably not going to be grazing cattle. In fact, without 30-36 acres and a seasonal rotation plan, they can't even graze one horse for one year (without supplemental feed) but must keep them much of the year in a corral to avoid making their entire property into a dust bowl.

Not to say that there aren't grassland birds that sometimes enjoy bare ground – after all, they evolved alongside the buffalo, famous for creating mosaics of bare ground in their migrations.

A lot of small acreage owners don't have livestock, but they do have cats and dogs that can be very detrimental to grassland birds. It's easy to see how, once you realize grassland birds nest on the ground.

Horned larks, western meadowlarks, vesper sparrows, savannah sparrows and

other grassland bird species have come up with various ruses and camouflages to avoid native predators. However, they haven't evolved yet to deal with what the American Bird Conservancy considers to be an invasive species: cats.

Cats kill more than a billion birds a year in the U.S. Zach pointed out that popular "trap, neuter and release" programs have a flaw – they allow cats to go back outside and kill more native birds and small mammals. It's a touchy subject. I admit to having been the owner of an indoor/outdoor cat up until 1990, when I started keeping my cat indoors. Four cats later, I'm a proponent of catios – screened outdoor areas – and taking leashed cats for walks.

Grassland birds nest sometime between April and July. That's a good time to keep dogs on a leash so they won't find and eat bird eggs. And it's an excellent time to abstain from mowing both the previous year's and current

2022

431

year's growth. If you value wildlife, mow only after consulting the professionals over at the Laramie County Conservation District.

However, you may want to forgo much vegetation around your house and outbuildings. The national Firewise program, firewise. org, has guidelines for protecting property from fire on the forest edges as well as in the grasslands.

And what can we townies do for grassland birds? Use less energy. Buy less new stuff. Every energy source I can think of has been detrimental to wildlife: harvesting whale oil, excavating peat, cutting firewood as well as producing the climate-changing fumes of coal, oil and natural gas and the toxic residue of nuclear, and building the cleaner but often habitat- and migration-disrupting installments of hydro, wind and solar power.

It seems as soon as we come up with energy saving changes – like families having fewer children and more efficient appliances, someone invents something like the new energy-intensive game of cryptocurrency mining. Don't mind me, I'm a trifle depressed after watching a new movie, the very dark comedy, "Don't Look Up."

But I plan to look up – spring bird migration will commence any day now.

# 458 Raptors entice birdwatchers to follow the "The Nunn Guy" in cold early start

Saturday, March 19, 2022, ToDo, page A10

A Cheyenne Audubon field trip in mid-February, starting at a frosty 8 a.m., usually attracts only a handful of diehards. But throw the word "raptor" into the publicity and suddenly there are 20-some people milling around in the parking lot at Lions Park, anxious to go see eagles, hawks, falcons and owls.

Or maybe it was the thought of travelling south to a balmier climate. Our destination, "Raptor Alley," starts in Nunn, Colorado, 30 miles south of Cheyenne. And it was balmy – 50 degrees, sunny, no wind and dry gravel roads.

We met our tour guide, Gary Lefko, "The Nunn Guy," at the Soaring V Fuels gas station/store. A seasoned trip leader knows how important it is to start a birding trip with empty bladders, especially in the nearly treeless farm fields of eastern Colorado.

Gary was also prepared with raptor identification handouts. Good thinking, because Mark and I discovered just before we left Cheyenne that many in the group considered themselves novice bird watchers.

Caravanning is not the ideal way to introduce people to birds. With carpooling, we pared down the number of vehicles to nine. When we joined Gary, he used handheld radios to tell our car what he was seeing, and then I texted a message to one person in each vehicle, such as "Red-tailed on the pole on the right up ahead."

Our end point was Pierce, Colorado, 5 miles south on U.S. Highway 85, but 30 miles as we shuttled back and forth along the county roads spaced on a 1-mile grid.

Gary later sent me his bird list from the trip, and even though Mark and I were only two cars behind him, he counted more raptors than we did:

Northern Harrier: 2
Bald Eagle: 2
Red-tailed Hawk: 6
Rough-legged Hawk: 4
Ferruginous Hawk: 3
Great Horned Owl: 4
American Kestrel: 2
Prairie Falcon: 3

We also documented rock pigeon, Eurasian collared-dove, black-billed magpie, horned lark, European starling and western meadowlark – 14 of them!

Gary frequently pulled over and jumped out of his trusty Subaru to train his spotting scope on a raptor in a lone treetop, on top of a utility pole or floating in the sky, giving everyone a chance to take a look. We may not have walked any miles, but we had plenty of exercise climbing in and out of our vehicles.

Raptor Alley is Gary's invention, and the genesis can be traced back to his wife giving him a bird feeder nearly 25 years ago. He bought 14 more feeders, but what hooked him and made him go buy binoculars and a field guide, was seven Monk parakeets visiting his feeders. The feral, bright green, tropical birds made themselves at home in Colorado Springs for a while.

Relocating to the outskirts of Nunn (current population 586) in 2002, Gary has now identified 135 bird species around his house. He's also just a couple miles from the western border of Pawnee National Grassland, a 30- by 60-mile tract administered by the U.S. Forest Service that is famous in international birding circles.

In some ways, Gary fits the stereotype of the birding loner, patrolling Weld County roads in search of avian rarities, but he also wants to spread the joy of birdwatching. When his mother told him years ago about the Florida birding trail, his first thought was, "Colorado needs one!"

Birding trails, routes like Raptor Alley, are mapped with notes about accessibility, conditions and birding highlights. Modern versions are on the internet and who better than Gary, an IT professional and web designer, to provide it? He started out with a five-county area he called the Great Pikes Peak Birding Trail. I have a t-shirt from that iteration.

It evolved into the Colorado Birding Trail, https://coloradobirdingtrail.com/, run by Colorado Parks and Wildlife. You can find "Raptor Alley" on the map, click on the link and get mile-by-mile directions and helpful hints like, "Be careful pulling

432     CHEYENNE BIRD BANTER

onto the shoulder of roads, as many are soft and you could get stuck."

Gary has identified 23 raptor species hanging out there in the winter. Why there? Good prey base – lots of rodents, and lots of perches for watching for them.

Along the way, Gary picked up graduate courses from Colorado State University in conservation communication and a certificate in non-profit administration. Gary's project for his certificate involved a whole new venture, setting up

the Friends of the Pawnee National Grassland, https://www.friendsofthepawnee-grassland.org/.

Part of that is an iNaturalist project to document the plants and wildlife, https://www.inaturalist.org/projects/birds-and-more-of-

the-pawnee-national-grassland. iNaturalist is global, community-based science, a perfect fit for a man with a personal mission to bring people to nature.

Thanks, Gary, for taking us to visit your birding "patch."

# 459 WGFD bird farm pheasants recruit hunters; sage grouse farming appeases developers

Saturday, April 9, 2022, ToDo, page A9

It's a matter of degrees when you are in charge of raising thousands of ring-necked pheasants.

Ben Milner, bird farm coordinator for Wyoming Game and Fish Department's Downar Bird Farm near Yoder, is also scrupulous about cleanliness, especially with the storm clouds of avian flu gathering on the eastern horizon.

Ben gave the Cheyenne – High Plains Audubon Society a tour in mid-March, before the eggs start rolling in. Before we could enter the facility, we had to step onto a soapy mat and squelch around a bit to kill any germs. Inside, it looked clean enough to perform surgery.

Each year, 18,000 pheasants are produced here and another 16,000 at Game and Fish's bird farm in Sheridan. Sheridan started in 1938 and Downar in 1963.

Each fall, Ben holds back 135 roosters and 1,350 hens for breeding, while the rest are released for hunting. The breeders make their home in nine acres of enormous pens secured against predators.

When the spring breeding season kicks in, each hen would normally stop laying after filling a nest with

12 to 15 eggs. But because employees go out every day to collect eggs, and hens have access to nutritious food, each averages 40 to 50.

The eggs are sorted, cleaned and stored in racks sized for pheasant eggs, smaller than chicken eggs, at 55 degrees, which suspends development of the embryos. When there are 6,700 eggs, they move to the giant incubator and 99.7 degrees. The racks tip every one to three hours to imitate the hen turning the eggs in her nest, keeping the embryos from sticking to the shells.

After 19 days, the eggs are placed in the hatcher, in chick-sized trays where they can hatch. After that, chicks move into brooder houses, where heaters set at 100 degrees substitute for brooding hens. They are soon pecking at waterers and feed.

After two weeks, the chicks are allowed to walk in and out of small outside pens and then eventually into the larger pens. These pens are so large that they are farmed. The crop is kochia – an invasive weed in everyone's garden, but it provides good cover and food in addition to the purchased feed.

The old brood stock is

released in May at Springer and Table Mountain Wildlife Habitat Management Areas as well as several walk-in areas.

Wyoming has not allowed raising exotic or native game animals privately, except exotic birds. At Downar, Game and Fish settled on ring-necked pheasants, natives of Asian jungles. Private bird farms order eggs and raise pheasants and other exotic gamebird species. Very few escape and reproduce because they are hunted by sportspeople and predatory animals.

Why does Game and Fish continue to produce an artificial population of pheasants, basically for put-and-take hunting? Ben sees pheasants as a way to introduce hunting to kids and adults, including women who have traditionally made up a small percentage of hunters. Game and Fish sponsors three kids-only hunt days each season on the Springer Wildlife Habitat Management Area and four in November at Glendo State Park to help recruit the next generation of sportsmen and women.

Historically, it was hunters who raised funds through licenses and tags and lobbied

for wildlife so that it wouldn't be extirpated by other interests such as farming, ranching, mining and energy extraction. So, thank those early hunters when you enjoy watching Wyoming wildlife.

Unfortunately, a few developers, alarmed by decreasing populations, think the bird farm method will make up for the loss of sage grouse habitat due to development. I'm discouraged that somehow influential people were able to convince the Wyoming legislature that this could be done by a private company.

Legislation gave Diamond Wings Upland Game Birds five years to give it a try, but this session they had to ask for and received another five, despite a large turnout against.

It turns out raising sage grouse is not like raising chickens – or pheasants.

First, there are no captive flocks to gather eggs from. Diamond Wings is allowed to steal up to 250 eggs per year from hens in the wild. So much for calling this captive "breeding." Sage grouse hens do not lay more eggs when they lose them, like the pheasants do. Plus, sage grouse chicks apparently

2022

433

need more instruction from the hens to succeed, unlike the pheasants.

Studies in Utah and Colorado concluded that captive breeding is not a viable way to increase sage grouse populations. Wildlife biologists say protecting sagebrush habitat is best. And what's good for sage grouse is good for other sagebrush-dependent wildlife.

People from many areas of expertise agreed on a Wyoming sage grouse management plan back in 2015 to keep them from being listed as threatened or endangered, avoiding a host of public land use restrictions.

For an update on sage grouse, please join Cheyenne Audubon April 19, 7 p.m., in the Cottonwood Room, Laramie County Library, 2200 Pioneer Ave. A Zoom link will be available at www.CheyenneAudubon.org close to the date.

# 460 How power production underlies bird problems and other bird news

Saturday, May 14, 2022, Outdoors (online version because newsprint version misplaced)

Last month, you may have read that a subsidiary of NextEra Energy will be paying a hefty fine for killing eagles at its wind developments, including Roundhouse, on the southwest edge of Cheyenne.

The company took a big gamble by not applying for an eagle "take" permit. The permit would have required expenditures, but now the company will have to spend money on remediation plus the fines.

Three years ago, I signed up to be party to the Wyoming Department of Environmental Quality's hearing on Roundhouse, representing Cheyenne Audubon. I still get the occasional registered letter with news about changes to the Roundhouse development plan.

I wasn't surprised to get a call from Ryan Fitzpatrick, my main Roundhouse contact. It sounds like the company may be installing a system to sense raptors approaching wind turbines so the turbines can be shut down before slicing an eagle. That's the system at the Top of the World wind farm. Ironically, bald eagles are increasing in number, however golden eagles are not faring as well.

Avian influenza reached Wyoming last month in poultry and wild birds. I've been following the story day by day through my favorite birding institutions, including the National Audubon Society and Cornell Lab of Ornithology. It seems to be travelling with migrating waterfowl and possibly affecting songbirds. It might be a good idea to put away the bird feeders for a while, instead of having to scrub them frequently. We usually take ours down for the summer anyway.

If you find any dead birds, report them to the Wyoming Game and Fish Department. Pick them up the same way you do dog droppings: grasp the bird with your hand gloved in a plastic bag. Then pull the bag carefully inside out over the bird and seal the bag shut.

I don't know if bird flu is transmissible to cats, but this would be a good time to start keeping your cats indoors so they won't eat dead birds.

Good news for sage grouse and the diverse group of people who in 2015 worked so hard to come to an agreement on the policies to protect primary habitat areas in Wyoming. The number of oil and gas parcels offered for lease has been reduced on critical sage grouse habitat for the Bureau of Land Management's next sale. There are scads of current leases that are not being drilled, so don't blame sage grouse for high gas prices.

But we really need to drop the conventional use of fossil fuels as soon as possible, and that isn't just because it is getting too warm for cute little pikas living on our mountain tops.

Let's consider the cost of air pollution to humans. It isn't healthy to breathe emissions from tailpipes and smokestacks or smoke from the increasingly frequent wildfires attributed to warming climate.

There's wildfire destruction itself. I saw concrete examples recently while driving to Louisville, Colorado. On one side of four-lane-wide Dillon Road, there is a very nice residential area that burned down to the concrete foundations that are now shaded by dead black trees. The houses on the other side of the street are safe, so far.

Mark's and my sons are doing their part for fighting climate change. Both drive electric cars. We plan to follow suit as soon as we need to replace a car.

Going electric is only going to help birds if the source of the power doesn't produce climate-warming pollution, slice them with turbine blades or cover grasslands and deserts with solar panels. To me, it looks like the most harmless alternative is solar panels on existing infrastructure. There's a million square feet of roof on our Lowe's distribution center. There's a nearly quarter-mile-long south-facing wall on the new eastside Microsoft installation. Could Cheyenne be forward thinking enough to write building codes that require buildings to produce power?

These are my daydreams this spring as I watch my first flock of white-crowned sparrows flit from shrub to shrub along Crow Creek on its way to the mountains to nest.

May 21 is the Cheyenne Big Day Bird Count, and once again, Cheyenne Audubon, www.CheyenneAudubon.org, will document the diversity of avian migration for the scientific record. Think about joining us. Someday, someone will examine our records, hopefully documenting increasing diversity here on out as we get a handle on our power problem.

434     CHEYENNE BIRD BANTER

# 461 Cheyenne Big Day Bird Count catches Arctic visitor

Saturday, June 4, 2022, ToDo, page A10

I'm sure our Cheyenne Big Day Bird Count compiler for Cheyenne Audubon, Grant Frost, was thinking to avoid cold, nasty weather when he picked May 21 instead of the 14 for the count, but it snowed the day before anyway. Our total of 125 species is not too shabby considering the weather was chilly, but luckily not windy.

We had several highlights during the bird count:

Red-throated loon juvenile was seen at Sloans Lake for several days before and on the count. It is considered rare in Wyoming, wintering on either coast and nesting in the Arctic.

Common loon juvenile, same place.

Broad-tailed hummingbird was trying to get nectar out of frozen crabapple blossoms at the Cheyenne Botanic Gardens.

Harris's sparrow may winter next door in Nebraska but is seldom seen here.

Red-headed woodpeckers showed up in two locations, including a pair in one.

Baltimore oriole, the eastern counterpart to our Bullock's, came by with a female.

No eagles were seen.

I came across the scan of a "Tribune Eagle" article about the 1982 Big Day, which was held a week earlier than this year's, where 40 people contributed to the count. The total number of species seen was nearly the same – 124.

The difference between species seen in 1982 and 2022 was 29, a figure close to how many were spotted in 2021, but not this year, 27. But if you look at eBird for the first three weeks of May this year in Laramie County, 185 species are listed. Some species passed through before our count day and may have still been present in smaller numbers, leading us to miss any sightings.

Aside from all the species name changes in the last 40 years, the most interesting assessment is in what species were not logged on the 1982 list but were in 2022:

--Cackling goose was split from Canada goose in 2004.

--Eurasian collared-dove was first observed in Wyoming here in Cheyenne in 1998.

--Great-tailed grackle in 2003 was my first Cheyenne observation.

--Common raven, though they have always been reliably seen starting about 10 or 15 miles west of town, my first Cheyenne observation wasn't until 2010.

The 1982 count lists five winter species we didn't see this count: bufflehead (duck), rough-legged hawk, northern shrike and at the time, what are now subspecies of dark-eyed junco listed as two species, Oregon junco and gray-headed junco. Maybe they migrated earlier this year thanks to weather or climate change.

Evening grosbeak made the 1982 list, but it is hard to find them anywhere these days. They are listed as a globally threatened species.

Black-bellied plover and mountain plover, grassland species recorded in 1982, rarely make our count anymore, but eBird has sightings recorded for April 2020 – when everyone was out birding more than usual.

Our Big Day count area is essentially the same as our Christmas Bird Count, a 7.5-mile diameter circle centered on the state Capitol building. There are more trees to attract birds than in 1982, or in 1956, when only 85 species were counted, according to early compiler May Hanesworth. But as the surrounding grasslands are built upon, mowed and invaded by free-roaming dogs and cats, the grassland birds will be harder to find.

## Cheyenne Big Day Bird Count, May 21, 2022

125 species, 19 participants

| | | | |
|---|---|---|---|
| Cackling Goose | Clark's Grebe | Common Loon | Willow Flycatcher |
| Canada Goose | Rock Pigeon | Red-throated Loon | Dusky Flycatcher |
| Wood Duck | Eurasian Collared-Dove | Double-crested Cormorant | Say's Phoebe |
| Blue-winged Teal | Mourning Dove | American White Pelican | Western Kingbird |
| Cinnamon Teal | Broad-tailed Hummingbird | Great Blue Heron | Eastern Kingbird |
| Northern Shoveler | American Coot | Black-crowned Night-Heron | Warbling Vireo |
| Gadwall | American Avocet | Turkey Vulture | Blue Jay |
| American Wigeon | Killdeer | Osprey | Black-billed Magpie |
| Mallard | Marbled Godwit | Northern Harrier | American Crow |
| Northern Pintail | Least Sandpiper | Cooper's Hawk | Common Raven |
| Green-winged Teal | Semipalmated Sandpiper | Swainson's Hawk | Black-capped Chickadee |
| Redhead | Western Sandpiper | Red-tailed Hawk | Mountain Chickadee |
| Ring-necked Duck | Wilson's Phalarope | Great Horned Owl | Horned Lark |
| Lesser Scaup | Red-necked Phalarope | Belted Kingfisher | Northern Rough-winged Swallow |
| Common Goldeneye | Spotted Sandpiper | Red-headed Woodpecker | |
| Ruddy Duck | Solitary Sandpiper | Downy Woodpecker | Tree Swallow |
| Pied-billed Grebe | Greater Yellowlegs | Northern Flicker | Violet-green Swallow |
| Eared Grebe | Willet | American Kestrel | Bank Swallow |
| Western Grebe | Ring-billed Gull | Olive-sided Flycatcher | Barn Swallow |
| | California Gull | Western Wood-Pewee | Cliff Swallow |

2022

| | | | |
|---|---|---|---|
| Ruby-crowned Kinglet | House Sparrow | Lincoln's Sparrow | Great-tailed Grackle |
| Red-breasted Nuthatch | House Finch | Green-tailed Towhee | Orange-crowned Warbler |
| Blue-gray Gnatcatcher | Pine Siskin | Spotted Towhee | MacGillivary's Warbler |
| Rock Wren | American Goldfinch | Yellow-breasted Chat | Common Yellowthroat |
| House Wren | Chipping Sparrow | Yellow-headed Blackbird | Yellow Warbler |
| European Starling | Lark Sparrow | Western Meadowlark | Yellow-rumped Warbler |
| Gray Catbird | Lark Bunting | Bullock's Oriole | Wilson's Warbler |
| Northern Mockingbird | White-crowned Sparrow | Baltimore Oriole | Western Tanager |
| Townsend's Solitaire | Harris's Sparrow | Red-winged Blackbird | Black-headed Grosbeak |
| Swainson's Thrush | Vesper Sparrow | Brown-headed Cowbird | Lazuli Bunting |
| Hermit Thrush | Savannah Sparrow | Brewer's Blackbird | Indigo Bunting |
| American Robin | Song Sparrow | Common Grackle | |

# 462 Fledge week observations entertain local birdwatcher

Saturday, July 2, 2022, ToDo, page A9

This summer, I miss the company of our dog of 16 years, Sally, while gardening in the backyard.

One day at the end of May, I stopped to take a flower photo, crouching down at eye level, a little glad that she wasn't around to photobomb my efforts. When I glanced off to the side, I realized I was also eye level with one of our robins, three feet away. He was watching me intently. I took his picture, and he didn't even blink.

At least once a spring, I hear from folks who are being strafed by furious robins every time they try to use their front or back door because the robins have built their nest on the porch light and are very territorial about anyone coming near.

This is the second year we've had robins nesting above our back door, and where I was crouching was only a few feet away from it. We more often use the attached garage's back door, but still, early June chores took me back and forth a lot, and the robins were courteous. I tried to return the favor, stopping whenever our paths were about to cross so

as not to delay their delivery of worm meat to the young in the nest.

Serendipitously, my garden digging coincided with the robins' hunt for food – I brought lots of tiny critters closer to the surface, and watering transplants brought out the worms. Mark saw the young minutes after they fledged, and I saw one just once. I hope the neighbor's cat didn't get them. Sally, the bird dog, would have tried. As of June 20, the robins were incubating another set of eggs.

The week before, two avian families arrived at our feeders. The only food out, besides the thistle in case a goldfinch comes by, is a chunk of suet-type stuff we stuck in one of those hanging cage feeders. A plain dark brown bird landed on the cage and stabbed at the brown stuff. There was something familiar about it, the bill, the shape – oh, baby starling! And then its two siblings and a parent showed up, and it was like watching a human family with small children visit the ice cream shop. A lot of shuffling and bumping and to-ing

and fro-ing.

Millions of starlings must go through this feeding performance every year – how else could there be millions of starlings out there to perform those "murmurations," clouds of birds performing sky-high arabesques captured on videos playing on the internet?

A few minutes later, a small, plump, light brown bird landed on the cage, fluttering its wings. Its parent quickly followed, a male house sparrow – they are the ones with the black goatees. He pecked the suet stuff and fed the slightly smaller bird. Soon, he was besieged by two more young, all three rapidly fluttering their wings, apparently the "feed me" signal. Nearly as amusing to watch was both families navigating our bird bath at the same time. Sparrow and starling shoulders bumped together.

The Swainson's hawk pair nested again in the neighbor's spruce tree. Every time I walk the neighborhood, I see at least one adult flying. We think the pile of sticks and whitewash in our driveway in May and early June was the

adults searching one of our overhanging silver maples for the perfect nesting materials, breaking off green sticks and dropping rejects.

I was concerned that the new apartment building in the field adjacent to our neighborhood would be a problem for the hawks' hunting, but this year, the church's gravel parking lot is home to a new colony of ground squirrels.

On June 16, one of the young hawks was trying to perch in our trees and was getting mobbed by blue jays.

The red-breasted nuthatches returned to the nesting cavity in the mountain ash tree across the street. The mountain chickadee seems to have nested somewhere else this year – but can sometimes be heard singing.

Before our shrubs leafed out, I saw a house wren checking out a tree cavity across the alley. Now he sings nonstop all day.

It's hard to make myself take a walk without Sally, but there are plenty of friends and neighbors I meet when I do, including the wild animals.

436     CHEYENNE BIRD BANTER

# 463   Merlin's "Sound ID" uncovers hidden birds
Friday, August 5, 2922, ToDo, page A10

Learning to identify birds by sight is simple – page through the field guide until you see a bird that matches or go birdwatching with someone who knows more than you.

One shortcut to the process is the Cornell Lab of Ornithology's free Merlin app.

You give it the bird's specs – relative size, color, behavior/habitat – and it gives you a short, illustrated list of possibilities. You can also give it a bird photo from your phone (including a photo of the screen on the back of a camera) and hit "Get Photo ID."

Learning to identify a bird by song or call is easy here on the edge of the Great Plains. Our most common birds vocalize while walking on lawns and prairie, sitting on bare branches and fence posts, swimming on water or soaring above. I can see robins chirping, crows cawing, house finches singing, collared-doves moaning and house sparrows cheeping.

It turns out I'm missing the birds that like to hide in vegetation but can still be heard. I've always thought that some winter I would sit down with a compilation of western bird song recordings and memorize them – hasn't happened in the last 30 years.

But now Merlin has a new feature, "Sound ID." It came out last summer as part of the free app, but it's this summer people are talking about it, even our Airbnb host, for whom it sounded like his gateway drug to birdwatching addiction.

The first step is to download the Merlin app for Android or iOS. Then open the menu (those three little lines stacked up) and choose Bird Packs. Install the one for "US: Rocky Mountains." This helps Merlin give you better choices. You can change it if you visit elsewhere.

Choose "Sound ID" from the home screen. Tap the microphone icon and hold out your phone towards the bird sound you hear. Closer is better, but start recording where you are first, in case moving closer scares the bird away. I found that Merlin doesn't hear everything I hear.

Merlin creates a spectrogram of what it hears, and it scrolls across the top of your screen. Eventually, it creates a list of the birds it is hearing, including a photo of each. Each time Merlin hears a species, it highlights the name so you can connect sound and name. Also, if you click on the bird, you'll get a list of other recorded sounds you can compare for that species, to double check Merlin's accuracy.

Early one morning recently, I stood on a corner in my neighborhood, recording and watching as half a dozen bird names filled my screen. But wait – great-tailed grackle? We have them in Cheyenne, usually at the country club and the air base, but I have not heard their loud, raucous calls on my side of town. How do I tell Merlin I heard common grackles instead? But I will still give every shiny blackbird's tail a closer look.

On the other hand, while I was hiking the Headquarters Trail at the end of July, Merlin told me I was hearing a warbling vireo. I hardly ever see them, so I have never perfected identifying them by sight, but now that musical warbling in trees along a creek will have me considering them when I hear it again.

And there's more. You can add these sound recordings to your eBird checklists. You can see if it's a bird already on your life list. Or Merlin will generate lists of birds where you plan to travel. It can sort them by most common at the top of the list. And for the most competitive birders, it can generate a list of birds they haven't seen in that area – their target species.

The Cornell Lab of Ornithology can tell you how all this magic happens. Mostly, it is from the crowd-sourced data from its community scientists all over the world – us birdwatchers.

Some 30 years ago, Beauford Thompson, a sixth-grade teacher at Davis Elementary School, told me we would have handheld devices that would help us do all kinds of things. I was imagining typing notes, maybe a digital day planner. Now, I use my smart phone for video calls, photographing and identifying flowers, reading books, tracking hikes, finding recipes and cafes, and counting birds.

Recording birds could become another time-eater, but learning bird songs and calls and contributing to the global avian knowledge is worthwhile. But let's not forget to sometimes go outside and enjoy the world empty-handed again.

# 464   How will the IRA affect birds?
Friday, September 2, 2022, ToDo, page A11

People affect birds – individual birds and whole populations – all the time.

Sometimes, we have a negative effect on birds, such as glass walls and bright lights that steer migrating birds to their deaths.

Sometimes we have a positive effect, such as growing windbreaks across the Great Plains that encouraged blue jays to follow the trees west. Or we create reservoirs to store water and the ducks and other water birds use them.

I've been looking at what the environmental organizations have to say about the new Inflation Reduction Act and how it will affect birds.

The IRA should help birds (and people) affected by climate change as it encourages actions for cleaner air. Clean air reduces climate change effects such as severe weather and the timing of

2022     437

seasonal changes.

Encouraging the switch to electric vehicles is good. Electricity can come from any source of fuel. If the source is a fossil fuel power plant, then pollution controls can be centrally located, rather than depending on vehicle owners to attend to maintenance. I don't know about you, but diesel fumes from the truck ahead of me at the stop light is something I won't miss.

It looks like the fossil fuel industries are losing out after spending the last 50 years fighting clean air regulations instead of finding technology to keep air clean.

Birds will certainly benefit from clean air, but I wonder how much that will be offset by the drawbacks of solar and wind energy production – the emphasis of the IRA.

Can we make smart changes?

If you remember, in 2019 I signed up to testify on behalf of Cheyenne Audubon at the Wyoming Department of Environmental Quality hearing on NextEra's Roundhouse wind development. It is located partly on the City of Cheyenne's Belvoir Ranch property.

What I discovered was that NextEra was required to show the impact of wind turbines and the numerous new roads on wildlife. Yes, the state asks for certain information, but it seems to me that a state or federal agency, rather than the company, should be performing the field investigations on the public's behalf, and doing so to a greater depth. And the results should be put in context with nearby developments, like the other windfarm adjacent to Roundhouse on the north side of Interstate 80.

Wyoming, famous for its wind, is slated to be covered with wind turbines. Our "big empty" is also slated to be covered with industrial solar developments. Solar will affect grassland birds, though it will be an army of graduate students who discover exactly how.

A new study on the effect of industrial solar fields on Wyoming's hoofed wildlife was recently examined by a Wyofile reporter, https://wyofile.com/report-industrial-solar-disrupts-big-game-movements/.

The study shows that the chain link fence required by

the National Electric Code kept migrating antelope out, essentially losing that amount of habitat.

Let's say you don't care about birds or other wildlife. Let's say you care more about Wyoming's economy.

Keep in mind that our second-largest economic sector is tourism.

During the pandemic years, tourists have discovered more of Wyoming than the Tetons and Yellowstone. They are finding our favorite local recreation areas. The tourists I talk to appreciate our wide-open natural spaces and wildlife the way most of us do. But I don't think thousands of acres of wind turbines and solar panels are going to enhance the views that tourists come here for, especially when they come from states that are already covered in industrial and agricultural development.

I still think smart, clean energy development is about integrating it with current infrastructure.

Currently, solar is more people-friendly, the source to concentrate on. No possibility of flying blades or deep vibration noises.

Think about the acres of

parking lots that could be roofed with solar panels. Think about the acres of roofs everywhere, especially the giant warehouses we have in Cheyenne. And Walmart's warehouse also has a lovely south-facing wall, as do the Microsoft data center buildings. Or maybe fill in the uninhabitable acres around wind turbines. The Germans are looking at solar canopies over their autobahns, https://www.rechargenews.com/transition/solar-panel-covered-autobahn-could-speed-german-energy-transition/2-1-854215.

Even our (electric) cars could have solar energy-collecting skins someday. You would go to the carwash to wash away dirt to improve your energy production. Although I suppose then no one would want to park in a solar-panel-roofed parking lot.

Yes, solar and wind have energy storage issues. But there are many brilliant minds in the world and the rewards of the marketplace to spur them on. Let's hope their solutions are bird-friendly, wildlife-friendly and at the very least, people-friendly.

# 465 Audubon volunteer reflects on 40 years

Friday, October 7, 2022, ToDo, page A11

In October 1982, 40 years ago this month, I attended my first Audubon meeting – as president.

My husband of one month, Mark, had been the first president the previous two years of a new chapter, the Rosebud Audubon Society. It covered Miles City and southeastern Montana.

Mark and his friends wanted to start an environmental club and Audubon appealed to them, especially birdwatching field trips.

Miles City, in 1980, had 9,600 people (2020 census: 8,300). The closest other incorporated town is Forsyth, 46 miles west, population 1,600 today. The closest big

city is Billings, 146 miles west, population 66,000 then and 117,000 today.

We were very creative in finding programs for our monthly meetings. We had natural resource professionals to call on from the offices in town: Custer County Conservation District, Montana Fish, Wildlife and Parks

Department, U.S. Bureau of Land Management, and from the U.S. Department of Agriculture, the Fort Keogh Research Station and the Natural Resource Conservation Service, as well as Miles Community College.

We could also borrow films from Audubon's Rocky Mountain Regional Office in

Boulder, Colorado. First, it was those big reels in metal cans, later VHS cassettes.

Our early newsletters were typed, duplicated, folded, stuffed into envelopes with address stickers and stamps applied. Later, we did without the envelope.

Mark took me on my first Christmas Bird Count two months later. We were covering some area near either the Tongue or Yellowstone rivers, which converge outside town. It was zero degrees. Because Mark, the wildlife biologist, was better at bird identification, I got to keep the list with pencil and paper. I couldn't manage them without taking off my mittens, and it was soon painful. A few years later, we took our older son along, using snowdrifts as diaper changing tables.

The chapter got involved in local projects – city parks, if I remember. I don't have the old newsletters, but we used to mail a copy each month to the state archives.

Then, in 1989, Mark took the fisheries biologist job with BLM's state office here in Cheyenne, a big city that presumably would have a big Audubon chapter where we could simply volunteer for a committee or two.

Cheyenne High-Plains Audubon Society was founded in 1974 and celebrated its 40th anniversary in 2014. But in 1989, it was at a low point. The founders were tired. We finally met them at the 1989 Christmas Bird Count tally party.

To revive the chapter, Mark volunteered to be vice president, I volunteered to be program chair, and 10 other people stepped in to fill other positions. In the 33 years we've been with Cheyenne Audubon, Mark has been president nine years and I seven.

Cheyenne is a great location for finding speakers for monthly programs. In addition to all the county, state and federal offices corresponding to the ones in Miles City, we are less than an hour's drive from the University of Wyoming and graduate students looking for audiences.

National Audubon's regional office was still in Boulder in the 1990s, and staff would come up to visit. Later, it was abolished, and we had our own state office in Casper. That's gone now, and we work with staff at the regional Audubon Rockies office in Fort Collins, Colorado.

We have members who can do travelogs of their nature-based trips. And there are staff from several other environmental organizations in town to speak about issues. We host speakers seven times a year.

I started writing this column in 1999, and thanks to Cheyenne Audubon, I've never lacked for topics beyond the birds in my backyard.

The chapter has grown. We now average 150 dues-paying members per year, with another 300 friends on our email mailing list (see www.CheyenneAudubon.org).

We've lobbied local, state and federal governments on environmental issues. We add our expertise to city park and conservation district plans. We offer educational and conservation grants. We invite the public to join us for our programs and monthly birdwatching field trips. We are planning our ninth Habitat Hero workshop in February.

After 40 years, newsletters are digital, programs can be offered in-person and virtually, field trip bird lists are entered on the eBird phone app, and grant money seems to be attracted to us. We work hard to get it spent on worthwhile projects that support our view that what is good for birds and other wildlife will be good for us too.

What hasn't changed is the need to speak up for the welfare of birds, other wildlife and people. New threats to our mutual health and safety seem to show up every day. But at least watching birds gives us mental health breaks. Those birds, and the people who love them, have taught me a lot these last 40 years.

# 466 Audubon Rockies' Hutchinson discusses community science

Friday, November 4, 2022, ToDo, page A10

Zach Hutchinson is Audubon Rockies' community science coordinator. He is currently located in Casper, although he plans to relocate to Cheyenne as soon as local real estate prices are realistic.

He spoke at Cheyenne Audubon's October meeting about community science, which started out being called "citizen science." The new name is more inclusive – you don't have to be a U.S.

citizen to participate – or be a college-educated scientist.

Zach said community science contributed to Bird Migration Explorer, https:// explorer.audubon.org/. I'll have more about this new endeavor in a future column.

Community science has also contributed data to the State of the Birds 2022 report, www.stateofthebirds.org/2022.

Most groups of birds, including our grassland birds, are still losing population, while others increased during the last couple decades. For instance, waterfowl increased because they benefitted from concentrated efforts by sporting groups, although you don't have to be a hunter to buy a Federal Duck Stamp to contribute.

This year's report highlights North American species that are at the "tipping

point," which means, after having lost 50% or more of their population since 1970, the report said, "These 70 species are on a trajectory to lose another 50% of their remnant populations in the next 50 years if nothing changes."

Thirteen of those tipping point species occur in Wyoming regularly, either as residents or migrants, some considered common and others uncommon on this scale: abundant, common, uncommon, rare.

I didn't include the species that are rare in our state in this list of 13:

--Greater Sage-Grouse
--Western Grebe
--Rufous Hummingbird
--Mountain Plover
--Long-billed Dowitcher
--Lesser Yellowlegs
--Red-headed Woodpecker
--Olive-sided Flycatcher
--Pinyon Jay
--Evening Grosbeak
--Black Rosy-Finch
--Chestnut-collared Longspur
--Bobolink

The primary causes of downward population trends are:

1. Habitat loss.
2. Cats (2.6 billion birds a year).
3. Windows, (624 million).
4. Vehicle collisions (214 million).
5. Industrial collisions, including wind turbines (64 million).

Zach went over the seven ways we can help birds:

1. Make windows safer day and night.
2. Keep cats indoors.
3. Reduce lawn, plant natives.
4. Avoid pesticides.
5. Drink shade-grown coffee.
6. Protect our planet from plastic (Think of waterbirds mistaking floating plastic for food.).
7. Watch birds, share what you see.

For more about each point, see www.birds.cornell.edu/home/seven-simple-actions-to-help-birds/.

"Watch birds, share what you see," means taking part in community science. Zach said this is how we find out about population trends, range expansion, and if there are losses, we can see where in the life cycle it happens so that action can be focused.

You've probably heard me talk about www.eBird.org before. Birdwatchers submit lists of birds they've seen, anywhere and anytime, using smart phones or computers.

I can delve into the data on the website and discover 272 species have been observed at the Wyoming Hereford Ranch headquarters, 216 at Lions Park and 151 at the High Plains Grasslands Research Station where the Cheyenne Arboretum is located.

The Christmas Bird Count is the most famous annual community science project, with this year's being the 123rd.

Two years ago, Zach said, 80,000 people took part, counted 2,355 species (worldwide), and traveled 500,000 miles on foot, by skis and by other means. Check https://cheyenneaudubon.org/ to find out about participating in the Cheyenne count in December.

The Great Backyard Bird Count, a snapshot of where birds are in late winter, celebrated its 24th anniversary last February. In 192 countries, 384,641 people participated and 7,099 species were counted on 359,479 checklists submitted. It's held over Presidents' Day weekend.

Zach runs bird banding stations every summer and people sign up to help (https://rockies.audubon.org/). Birds are caught in fine "mist nets" and then are measured and banded.

Bird banding provides data on demographics, productivity, recruitment (adding individuals to the population) and survival – when a bird previously banded is recaptured, or a band is recovered from a dead bird.

This year, 54 species were netted at Zach's stations. Usually, 500 new birds are banded but this summer it was only 340, probably because the drought has affected breeding and recruitment, Zach said.

Audubon Rockies launched a new community science project last summer on the Yampa River in Colorado. People on commercial float trips, including Zach, counted birds: 55 species and 732 individual birds. Stopping for a few minutes in a calm eddy in otherwise inaccessible places to count birds will add richness to the tourists' experiences and give science a new perspective.

There are other community science endeavors, such as iNaturalist, which is interested in plants as well as animals. Some have been very specific, such as The Lost Ladybug Project.

Consider becoming a community science participant in one or more ways.

# 467 Unusual birds "on the road" this fall in southeastern Wyoming

Friday, December 23, 2022, ToDo, page A10

On Nov. 9, a friend called to tell me she heard a story on KUWR, Wyoming's National Public Radio affiliate, about a Blackburnian warbler that blew across the Atlantic to an island off the southwest British coast, exciting birdwatchers.

It's ironic that this eastern North American bird was named by a German zoologist for an English naturalist, Anna Blackburne (1726-1793). She never saw a live specimen, but her name seems appropriate because the 5-inch-long male burns with a flaming orange throat and head on a body that is otherwise black and white.

We've had a few Blackburnians accidentally find their way to Wyoming. Under the Explore tab on eBird.org, you'll find that my husband, Mark Gorges, was the last to record one in Wyoming, a female, on May 28 at Wyoming Hereford Ranch.

Warblers typically eat insects, so the lost warbler Mark saw could find them in late May. Warblers leave the north in September and October, when cold weather limits their food supply.

However, beginning Nov. 11, Chuck Seniawski has had a pine warbler visiting his Cheyenne feeder nearly every day through Nov. 27, so far. This is another lost eastern North American

440                    CHEYENNE BIRD BANTER

species – and it is way late for an insect eater.

Pine warblers, according to Doug Faulkner's "Birds of Wyoming," published in 2010, are "vagrants." Their normal migration, breeding and winter ranges in the Eastern U.S. and southeastern Canada are nowhere near Wyoming.

However, Doug wrote, every fall there is at least one reported in Wyoming, usually between mid-August and mid-September. Doug's only winter report was a pine warbler that spent five days in December 1988 eating peanut butter at a feeder in Gillette.

Chuck says his pine warbler pecks at his sunflower feeders, hunts on the ground underneath and uses the birdbath. He isn't sure if the bird is eating seed bits or finding something else. When he posted a photo, Don Jones, eBird regional data reviewer in Laramie, who spent four years back East, agreed with his identification. Also, Chuck had just seen one in Central Park in New York City.

Pine warblers look a little like a female or a winter-plumage male American goldfinch, yellowish with dark wings with two white wingbars, so maybe we should all examine our feeder birds more closely.

Serious birders stake out reservoirs during fall migration, including the Laramie Plains Lakes. Jonathan Lautenbach was rewarded with being the first to record two king eiders, sea ducks, on Nov. 12 through 18 at Lake Hattie. He reported they were a female and a juvenile male, plain brown. The adult male, not seen, would be half white and half black, with a bright yellow-orange "bill-shield" on its forehead.

eBird shows these king eiders as the first to be recorded in Wyoming. Doug Faulkner does not list them at all in his 2010 book, which is a comprehensive review of bird sightings up until that point.

King eiders breed in the Arctic, across northern-most Canada. They winter around coastal Alaska and northeastern Canada, but there are frequent winter sightings in lower 48 states, most often coastal. They are also usually female and juvenile birds.

Cheyenne birder Grant Frost was probably checking Sloans Lake in Lions Park for interesting ducks and other waterbirds when he came across a small flock of bushtits Nov. 3 and again Nov. 27. "Peterson's Field Guide to Birds of North America," published in 2020, describes their habitat as brushy woodlands and pine-oak forests of the southwest.

But if you look closely at Peterson's range map, it shows this thin line of purple (meaning year-round resident) drawn up the Front Range of Colorado, practically pointing to Cheyenne. More bushtits may be in our future. Look for pale brown and gray, 4.5-inch-long birds building sack-like hanging nests.

Grant also found a blue-headed vireo at Lions Park on Nov. 1, and it was last seen there Nov. 3 by Vicki Herren.

Vireos are much like warblers – eating insects – but also fruit in winter. This species breeds across Canada, through New England and down through the Appalachians. It winters along the southeast coasts of the U.S.

It's possible that the birds from western Canada would head south through Wyoming to get to the Texas Gulf Coast. They are just hard to pick out from other vireos and warblers bouncing around in the trees.

Unusual bird observations submitted to eBird automatically get flagged. You are asked to write a description of your observation and submit a photo, if you can. Someone appointed by eBird for that area will decide whether your record becomes public.

These days, eBird and the Wyoming Bird Records Committee work together. Find out more about the committee at https://wybirdrecordscommittee.wordpress.com/.

# 2023

**468** Friday, January 6, 2023, ToDo, page A8
## Several remarkable observations from the Cheyenne Christmas Bird Count

Perhaps somewhere in the archives of Rocky Mountain National Park is my signature on a piece of paper from the cylinder on Hallett's Peak, proving I made it to the top in August 1973.

Short of birth, death and graduation records, most of us don't lead a permanently, well-documented life. But if you participate in a Christmas Bird Count, you can look yourself up online, at least back to 2005. More important are the number of birds counted, distances traveled and the weather conditions. That data goes back to 1900 (1974 for Cheyenne).

Explore the data at https://netapp.audubon.org/cbcobservation/. The address changes whenever the sponsor, the National Audubon Society, reorganizes its website.

This year was the 123rd Christmas Bird Count, straddling the year end of 2022-2023. The Cheyenne count was held Dec. 17, within a 7.5-mile-diameter count circle centered on the State Capitol.

The 20 participants together walked 26 miles, drove 76 miles and watched feeders for 15 hours.

Here is the list of 51 species and how many were seen of each, plus a few notes.

*Cackling Goose: 97*

These geese used to be lumped with Canada geese as four smaller subspecies, sometimes as small as a mallard, and are showing up more often.

*Canada Goose: 1,148*

These may be a mix of a non-migratory local flock and some migrating here when there's open water.

*Snow Goose: 1*

Oh no – is this species of goose thinking about wintering here too?

*Mallard: 354*
*Northern Shoveler: 8*
*Redhead: 1*
*Ring-necked Duck: 2*
*Green-winged Teal: 22*
*Common Goldeneye: 7*
*Gadwall: 2*
*Rock Dove (pigeon): 129*

There's a much larger flock in northeast Cheyenne that eluded us.

*Eurasian-collared Dove: 181*
*Wilsons's Snipe: 3*

They know where there's a spring providing open water.

*Northern Harrier: 5*
*Sharp-shinned Hawk: 2*
*Cooper's Hawk: 1*
*Bald Eagle: 4*
*Red-tailed Hawk: 12*
*Rough-legged Hawk: 4*
*Ferruginous Hawk: 2*
*Eastern-screech Owl: 1*
*Great-horned Owl: 2*

Good showing of raptors, including the merlin and kestrel listed below.

*Belted Kingfisher: 2*

Always a couple along Crow Creek.

*Downy Woodpecker: 5*
*Hairy Woodpecker: 1*
*Northern Flicker: 15*
*American Kestrel: 1*

Not all of them migrate farther south.

*Merlin: 1*
*Northern Shrike: 2*
*Blue Jay: 13*

This eastern bird continues to increase in numbers here.

*Black-billed Magpie: 80*

It should really be the state bird since it stays year round and cleans up carcasses.

*American Crow: 133*
*Common Raven: 30*

Lorie Chesnut videoed a flock of 25. Jane Dorn, who studied ravens for her master's degree, said young birds may flock, otherwise, ravens hang out in ones and twos. To tell them apart from crows, listen for the raven's croak compared to the crow's caw. Also, when flying, the raven's tail looks like the point of a diamond. The crow's looks like a half-circle fan. Crows are only 17.5 inches from beak tip to tail, ravens are 24 inches.

*Black-capped Chickadee: 14*

I need to be more careful in assuming all the chickadees I see are mountains and check for their white "eyebrows," which the black-cappeds don't have.

*Mountain Chickadee: 22*
*Horned Lark: 9*
*Red-breasted Nuthatch: 4*
*Brown Creeper: 2*

These are very hard to see. They are like a moving piece of bark on a tree trunk.

*European Starling: 444*
*Townsend's Solitaire: 10*

This relation of the robin is slenderer and is all gray. It likes to sit at the tip-top of trees, especially junipers, eating their berries.

*American Robin: 5*

Every year there are a few that winter here. We aren't sure if these birds spent the summer here or if these are birds that came from farther north.

---

442  CHEYENNE BIRD BANTER

*Cedar Waxwing: 6*

Waxwings only show up when they find fruit still on the tree or shrub, so seeing them is very lucky.

*House Sparrow: 432*
*House Finch: 119*
*American Goldfinch: 2*
*American*
*Tree Sparrow: 42*

In summer, small flocks of sparrows are often chipping sparrows. But they leave in fall and the tree sparrows come for the winter.

*Dark-eyed Junco: 59*
*Song Sparrow: 2*

They are almost always year round, by a creek.

*Bushtit: 10*

This is the flock our Christmas Bird Count compiler, Grant Frost, has been watching this fall. We are happy they stayed for their first count here. If they make it through the winter, they might decide to stay and make a state breeding record.

*Pine Warbler: 1*

This is the same bird that has been hanging out in Chuck Seniawski's back-yard this fall. Nice it could stick around and provide a count record.

*Golden-crowned Kinglet: count week, not observed day of the count*

Not an unusual bird in winter, but there are not many to be seen, plus they are tiny and not notice-able in the treetops where they hang out.

# 469 Habitat leasing to provide new tool for Wyoming conservation

Friday, February 3, 2023, ToDo, page A10

Bob Budd dropped the name of a new conservation tool during his book talk for the Cheyenne–High Plains Audubon Society last month. So, I asked Bob for more details a few days later.

The new habitat leasing program Bob mentioned is like conservation easements, but short term and with a lower price per acre.

First, let's look at what a conservation easement is. Wikipedia has an extensive definition showing it dates back to the 1950s or ear-lier, but to summarize, an interested landowner finds willing partners to pay him or her not to use their land for certain purposes.

Those certain purposes are most often development or subdivision, especially on farm and ranch land. Because the acreage can no longer be subdivided under a conserva-tion easement, the property loses the value associated with subdivision.

The consenting land-owner (it's always a volun-tary agreement) is paid for conserving the land. Who pays the landowner? In Bob's experience as executive director of the Wyoming Wildlife and Natural Re-source Trust, a state agency, in setting up some of these, it can be a mix of money from a non-profit organization like The Nature Conservancy, a farm or ranch organization, and government agencies, including the U.S. Depart-ment of Agriculture's Natural Resource and Conservation Service (NRCS).

The conservation ease-ment becomes part of the property, passed along to the next owner.

The landowner may recog-nize the conservation value of their land and seek the easement. The buyers of the easement confirm the conser-vation value and the financial value through an extensive appraisal process.

The land could be valuable for wildlife. Or maybe for other ecological services such as absorbing precipitation to decrease flooding down-stream, instead of increasing impervious pavement.

Around here, raising beef cattle doesn't pay as well as selling land for homesites. A rancher may be lucky enough to improve their bottom line by leasing some of their property for windfarms, solar farms or oil and gas development. Up until now, a conservation easement was one of few ways to be paid for providing wildlife habitat. But they are a hard sell be-cause they are forever.

Well, now we will have habitat leasing. The details are still being ironed out, but Wyoming will be piloting the concept this spring. Sim-ply, a landowner can sign a contract for a period of years, maybe 10 or 15, and receive an annual payment in return for maintaining their acreage to benefit habitat, migration routes and/or other ecologi-cal services.

It could be habitat for a species of concern, with an agency contacting the landowner to see if they want to sign up.

For this pilot program, Bob has been working with NRCS and the Wyoming Game and Fish Depart-ment. The habitat they are interested in is the seasonal migration corridors that have caught the public's attention.

The Wyoming Migration Initiative has identified routes that are used by deer and an-telope year after year as they move north from their win-tering grounds to higher-ele-vation breeding grounds and back again in the fall.

These routes have to pro-vide forage for the travelers, or they will fail to breed suc-cessfully. The animals don't seem to find new routes when there are obstacles.

So, protecting the his-toric migration routes with habitat leasing seems like a fair transaction for both the rancher and the wildlife.

Forty years ago, when I was in college studying natu-ral resources and then range management, environmen-talists and ranchers seemed to be on opposite sides of the fence, especially on topics like wolf reintroduction. But the wildlife folks and the ranchers have found they have much in common. Here in Wyoming, for instance, they have been successful in collaborating on how to help sage-grouse.

Bob has been there on the sage-grouse work. He's the ranch-raised kid who, as he explains in his new book, "Otters Dance," used to run

2023

443

feral through the willows, watching the birds and frogs.

In his previous job as the ranch manager for The Nature Conservancy's Red Canyon Ranch near Lander, Bob worked out ways to keep cows and wildlife happy simultaneously. He knew his stream restoration worked when a family of otters moved in.

The ranchers Bob knows are knowledgeable about the wildlife on their places and always interested in learning more.

I remember Bob saying one time when Mark and I visited Red Canyon Ranch that when the environmentalists visited, they wanted to see the cows and find out how many there were. Visiting ranchers wanted to see the endangered plants.

Bob, himself, is a special kind of person. Pick up his book at the Wyoming State Museum, Game and Fish headquarters, the Cheyenne Frontier Days museum or online and see what I mean. He can take the fence down between two opposing camps. With him riding herd on this habitat leasing pilot, I'm pretty sure it can be successful.

# 470 Birders get look behind the scenes, find more eBird perks

Friday, March 3, 2023, ToDo, page A7

eBird has come a long way since its debut in 2002.

As a means of collecting scientific bird data by offering birders a place to save their bird lists, Cornell Lab of Ornithology invented an ingenious bit of community (or citizen) science, and it just keeps getting better.

Anyone can go to eBird. org and sign up for free. The website, under the Help tab, has tutorials on how to enter your bird sightings.

Don Jones, University of Wyoming graduate student studying sagebrush songbirds, and Cheyenne Audubon's February guest speaker, said that for Wyoming, 15,000 different birdwatchers have submitted 200,000 checklists so far. Wyoming eBird data was recently added to the Wyoming Natural Diversity Database.

Globally, as of the 20th anniversary May 2022, 820,000 eBirders contributed 1.3 billion observations.

Since scientists are expected to use eBird data, there is a review process. Once, I received a polite email from the regional reviewer asking if I had indeed seen 49 black-crowned night-herons at Holliday Park, and if so, could I send more information.

When I explained that it was a breeding colony that has been there for years (and is still there, but a bit diminished as the park loses the big cottonwoods), my report was accepted. Today, I can look up that night-heron sighting on eBird and tell you it started May 29 at 7:45 a.m. and I saw 220 birds of 15 species while walking 1.5 miles in an hour.

Don, a volunteer eBird reviewer for 10 years, explained that if the reviewer doesn't think you have enough information to verify the entry, the entry can stay on your list, but it won't be publicly available. Don's been in that boat, especially when birding abroad when he's discovered he's made identification mistakes. But then he was able to fix them.

The globe is divided into review areas. We are in the Laramie/Goshen counties area. Volunteer reviewers familiar with bird life here, like Don, set a filter for each species, specifying which months it might be seen and maximum number seen at one time. The number is higher for a migratory species during migration months than during breeding months when birds spread out and become secretive.

Filters do change over time. Perhaps an invasive species like the Eurasian collared-dove has moved in or another species, like the dickcissel, is becoming rarer.

In the last few years, eBird has added new perks for birdwatchers. One is signing up for notices for birds you'd like to see.

For instance, I can generate a list of species I haven't seen in Laramie County but others have – target species. My 87 target species seem to be a lot of rarities – species unlikely to be seen here, but maybe common elsewhere. For instance, eBird has only three reports of prairie warbler, an eastern species, in Cheyenne, in 2000 and 2001. I have a much better chance of finding native burrowing owls last reported in 2022.

Once you know what birds you want to see, you can sign up for alerts. There are two kinds. Rare Bird Alerts are for species the American Birding Association considers rare for your area of interest. If I sign up for Needs Alerts for Laramie County, I'll be alerted whenever someone reports a species I haven't seen here yet.

Note: When eBird says "Laramie," they mean our county, not our neighboring town to the west.

eBird is handy for preparing for a birding trip to an area you aren't familiar with by showing where publicly accessible hotspots are and generating a list of species for you. You can see the latest observations.

You can even generate a multiple-choice species identification quiz for a location at a particular time of year, either with photos or bird sounds.

After your trip, you can pull together all the checklists you submitted and add notes and photos to make a "Trip Report" to save and share.

Under the Science tab are all sorts of wonderous interpretations of eBird data: Visualizations of bird abundance, abundance trends, migratory route animations plus improved range maps showing breeding, wintering and migration areas for each species.

There's the list of published studies using eBird data. There were 160

444        CHEYENNE BIRD BANTER

peer-reviewed publications in 2022, like this one: "Bai, J. P., Hou, D. Jin, J. Zhai, Y. Ma, and J. Zhao (2022) "Habitat Suitability Assessment of Black-Necked Crane (Grus nigricollis) in the Zoige Grassland Wetland Ecological Function Zone on the Eastern Tibetan Plateau." Diversity 14(7)."

It's incredible to think we birdwatchers, while having fun watching birds all over the world, with just a little extra effort, maybe using the mobile app, can contribute knowledge that helps birds.

For questions about eBird in Wyoming, contact Don at djones46@uwyo.edu.

# 471

Friday, April 7, 2023, ToDo, page A9

## McLean biography traces politics of passage of Migratory Bird Treaty Act

It's spring migration season.

Mid-March, the mountain bluebirds were back and could be seen at the High Plains Arboretum, where we hope they will find the new nest boxes put up by Rustin Rawlings.

Cheyenne Audubon's March 18 field trip to Lingle to see northern cardinals was a success. I never expected to see one in Wyoming. They even produced a breeding record for the state last summer.

We also checked out the reservoirs at the Springer Wildlife Habitat Management Unit, finding two trumpeter swans and sandhill cranes amid the usual ducks and geese. The following week, a storm of snow geese showed up.

Studies show there are fewer birds in North America than 50 years ago. But the losses would be much worse if not for the 1918 Migratory Bird Treaty Act.

Because "our" birds cross international boundaries, it is important that there is international law. And we have the persistence of one man to thank for it, George P. McLean (1857-1932), former governor of and U.S. senator from Connecticut.

McLean's great-great nephew, Will McLean Greeley, has written his biography, "A Connecticut Yankee Goes to Washington: Senator George P. McLean, Birdman of the Senate." An archivist by trade, Greeley inherited the story of a man who is a fascinating subject, with a well-documented life, too.

I admit, I jumped right to Chapter 8, "Saving the Birds," only skimming the previous chapters describing McLean's rise from homespun-wearing farm boy to governor. He was a champion of nature, especially birds.

In the late 1800s, wild birds were being harvested for feathers for women's hats (the impetus for the founding of the National Audubon Society). Chickens were less available than wild birds, including songbirds, for putting on dinner tables. States had hunting regulations, but not well enforced.

McLean's first efforts, as a Republican elected to the Senate in 1911, were for federal protection of migratory gamebirds. But as he acquired supporters, including hunters, gun and ammo manufacturers, the U.S. Agriculture Department and conservation groups like the National Audubon Society, he extended protection to songbirds and insectivorous birds.

In the era before chemical pesticides, protecting insect-eating birds protected crops. So, it made sense to add the Weeks-McLean bird protection bill to the massive ag appropriations bill. President Taft was so tired on the last day of his administration that he signed it without reading it.

The Weeks-McLean bill also defined bird hunting seasons, allowed federal laws to supersede state wildlife laws when more stringent, placed a five-year-ban on killing vulnerable species including whooping cranes, wood ducks and swans, and funded seven federal field agents and 172 local game wardens paid by the U.S. Agriculture Department.

Protecting birds should be an easy sell, but Missouri Sen. James Reed felt the need to oppose McLean at every stage, even though biographer Greeley discovered the two men and their families socialized outside work.

McLean next had to work with a Democratic president, Woodrow Wilson, and the distractions of World War I, the 1918 pandemic, plus opponents wanting the Supreme Court to judge whether the federal government could usurp the states' control of wildlife. Greeley points out that this was the Progressive Era, and McLean was one in the best sense of progressivism, including federal regulations we take for granted today, like labor laws.

Since international treaties are more impervious to Supreme Court decisions, McLean went after one next. In the final push, he had to let the president's man take credit for the 1918 Migratory Bird Treaty Act. Updated over the years but briefly tampered with during the 2017-2021 presidential term, it's recovered its full effectiveness.

Although George McLean became one of the wealthy elite, he was not miserly with his estate. He left substantial sums to all his nieces and nephews, great-nieces and nephews, employees and conservation groups like the Connecticut Audubon Society. His estate established the McLean Fund and the 4,400-acre McLean Game Refuge.

Historical and political biographies are not my usual literary fare, but I was intrigued by the bird connection. And I was rewarded with a riveting story in which, under the leadership of one man, America and the other treaty signers were convinced to do the right thing for birds.

Author Greeley found that McLean's rationale for

protecting birds was that he found them to be beautiful. So, get outside this spring and look for those beautiful birds. Or at least look out your window.

Then say a little thank you to George P. McLean that the robins weren't all baked into pies and that they are still patrolling your lawn.

# 472 Longspurs animate local shortgrass prairie

Friday, May 5, 2023, ToDo, page A7

Eastern Laramie County has no mountains, but it is not flat.

We were looking for birds north of Hillsdale (a town name indicating the varied topography), walking across the shortgrass prairie on a very fine morning (meaning no wind) in late April. We were surrounded by small birds popping up, circling us and then upon landing, becoming invisible.

A nearby windbreak was full of robins and red-winged blackbirds, but up the hill, where the grass was well-grazed, barely an inch tall, it was full of grassland birds like western meadowlarks and horned larks. And lots of longspurs.

Your field guide, if not brand new, will show them as McCown's longspurs. While John P. McCown was stationed with the U.S. Army in Texas along the Rio Grande, he collected several bird specimens. Presumably it was winter, when these longspurs are wintering there and in southern New Mexico and due south in Mexico.

McCown sent the specimens back east and the ornithologists determined his longspur was a new-to-them species. In 1851, they named it in honor of McCown.

However, by 2020, a closer examination of McCown's career showed that he'd served on the frontier with less than perfect integrity

and then joined the Confederate army. Altogether, he became someone the North American Classification Committee of the American Ornithological Society did not want to honor, and the bird's name was changed to thick-billed longspur. The committee is considering removing people's names from all bird names, which in North America would affect 150 species. You will need a new field guide when that happens.

But out on the prairie, the birds have no nametags, only their markings. The thick-billed longspur has a heavy, seed-cracking bill. It is closely related to sparrows and eats seeds all winter. However, in spring it eats insects and invertebrates and will feed them to its young.

As we walked, the longspurs kept popping up and circling us. Perhaps we were kicking up insects as we walked, just like cows or buffalo. It's also time for the males to do their aerial territorial mating display. They are marked with distinctive black bibs this time of year, and with their tails fanned out, white with dark center stripe and black lower edge, they are, after seeing so many, easy to separate from the more numerous horned larks, which have much blacker tails.

Over the last 50 years, the thick-billed longspur

population is down 94%, mostly due to changes in their habitat. They are on Wyoming Game and Fish Department's "Species of Greatest Conservation Need" list which includes 80 bird species: https://wgfd.wyo. gov/Habitat/Habitat-Plans.

Wyoming has about 27% of the world's thick-billed longspurs. Their breeding range is primarily eastern Wyoming, much of Montana and small extensions into Colorado, Nebraska, Saskatchewan and Alberta.

Game and Fish attributes population declines to prairie fragmentation by agriculture (plowing), urbanization (subdivisions) and fire suppression. Stressors include energy development (including wind energy), invasive species (like cheatgrass), off-road recreation, altered fire and grazing regimes (longspurs prefer heavily grazed areas), drought and climate change.

Maybe the academics can study how many houses per square mile can be built on the prairie before longspurs decamp. But not all homeowners take care of their property in the same way.

First, to protect ground-nesting birds like longspurs, meadowlarks and horned larks, people are keeping their dogs off the prairie, or at least on a leash, April through July.

Second, people who value grassland species of all kinds

refrain from mowing too often, especially April through July to protect the nesting birds, but also to reduce extreme fire risk.

It seems counterintuitive. If the shortgrass prairie grasses are repeatedly cut back (some people erroneously believe they need to mow more than once every couple years), the grasses begin to struggle. The less-shaded soil gets too hot, and heat-loving species move in, such as the more combustible, non-native cheatgrass.

Prairie grasses are so cool. They have deep roots so they can make a comeback from drought and grazing. Wanda Manley, who lives out on the prairie and has a master's degree in range management, told me that even after a (normal) grass fire, the growing point of each grass plant stays green and recovery is rapid. But where the prairie has been abused, fires are so hot, the soil burns, and recovery will take much longer.

If you are someone who owns a patch of prairie and a riding mower and who enjoys a reason to get out there on a nice day, why not leave the mower parked and grab your binoculars? Walk out, maybe to the top of one of the hills, and listen for the music of the longspurs, these small birds that have been visiting our prairie every spring for thousands of years.

446     CHEYENNE BIRD BANTER

# 473 Puffin paradise tour held along mid-coast Maine

Friday, June 9, 2023, ToDo, page A9

This year is the 50th anniversary of Project Puffin. The story is there were no Atlantic puffins left along the coast of Maine – too many egg collectors – and one man, Stephen Kress, dreamed up a way to successfully bring them back.

Kress was spending time on Hog Island, at the Audubon camp, when he became intrigued by the puffin problem. He arranged to bring pufflings back from Canada, where there were plenty. He improvised a frame filled with tin cans to imitate the burrows they were hatched in and carried them back where he had excavated burrows on Eastern Egg Rock, not far from the Audubon camp.

Puffins are supposed to return to the colony where they fledge, but when these birds were old enough to return to breed, they didn't seem to want to hang out at Eastern Egg Rock. Kress decided they needed to see that it was a popular place to live, so he set up puffin decoys and eventually established a thriving colony.

The lessons learned with Project Puffin have been applied to troubled seabird colonies all over the world. More than a dozen different kinds of seabird decoys are manufactured at Audubon's mainland property across from Hog Island. Made of long-lasting hand-painted molded plastic, they are shipped around the world.

Project Puffin, www. projectpuffin.audubon.org, is now part of Audubon's Seabird Institute. Mark and I were fortunate to be on a Holbrook Travel tour to Hog Island in late May to see the puffins and other seabirds and to hear a talk by Don Lyons, the institute's director.

Lyons talked about the evolution of technology for tracking birds, from biologging tags to geolocators. The new Motus Wildlife Tracking System makes use of fixed stations set up around the world to record birds wearing trackers as they pass by.

Lyons told us about the Aleutian terns that nest in Alaska and winter in Indonesia. Before tracking, people in Indonesia thought they were seeing common terns in the winter. Many terns in winter plumage look similar.

So much more is known now about where birds go, and where they might be heading into disaster, such as proposed off-shore windfarms. Not only can birds collide with wind turbine blades, but some are lost trying to avoid wind farm turbulence blocking their traditional migration route.

But it's not just migration. Many seabirds spend almost their entire lives aloft. When they breed in a colony, to feed their young they are often gone for several days, flying hundreds of miles to collect enough food from the sea.

The Seabird Institute can share information so that with a little tweaking, perhaps these long-distance fliers won't need restoration like the puffins.

This summer, four researchers are voluntarily stranded on Eastern Egg Rock. It has no trees, just a few observation blinds and one little shack for keeping electronics safe. When they saw our boat, they jumped up on the roof and waved. We did not go ashore but with binoculars we saw 45 puffins on the rock and around us swimming and flying.

We saw other breeding seabirds and shorebirds. That day's trip gave me four 'life birds' (species I'd never seen before): the Atlantic puffin, purple sandpiper, razorbill (the puffin's close relative) and roseate tern, my 690th life bird.

It was a beautiful sunny day, with water calmer than we had hoped for. And as always, it was fun meeting and travelling with other devoted birders and to make it to Hog Island, of which we'd heard so much as Audubon members the last 40-plus years. And the food was excellent, culminating in a Maine lobster dinner.

There were eight legs to our journey home: the pontoon ferry to the mainland, car ride to the bus stop in Damariscotta, two buses, the plane, a cart ride most of the nearly mile-long length of DIA's B terminal to make our Fort Collins bus connection, and the drive home.

I was disappointed that my smart phone battery seemed to be draining rapidly. I couldn't figure out where to plug it in on the plane so I could keep reading the book I downloaded. So, at nine percent I shut it down.

The next day, as I set my GPS app to measure how far I was walking, I realized the app had tracked my trip home.

There was the orange line leaving Hog Island, the winding road to the bus stop in Damariscotta, all the bridges over inlets, wetlands and rivers of mid-coast Maine. And then the plane banking over Boston and the Atlantic before setting a straight line across southern Ontario until it ended over Lake Michigan when I shut the phone off.

# 474 House sparrow effect demonstrated in backyard

Friday, July 7, 2023, ToDo, page A9

I blame the puppy.

When she joined our family last September, I thought it would be a good idea to feed the birds hulled sunflower seeds so that there wouldn't be hulls under the feeder for her to chew.

I noticed over the winter that we had more house sparrows visiting us than usual. Normally, few ever bothered to crack open our usual black oil sunflower seeds. They might look around under the feeder for scraps, but they don't usually sit on the feeder pulling out seeds.

Turns out they really like sunflower seeds – if they don't have to deal with the hulls. And with more hanging around, more of them thought about nesting here, but not for the first time.

Last year, after the robins fledged one batch of young and laid one or two eggs for the next, the house sparrows started building their nest on top. They completely covered the robin eggs with a hollow ball of dry grass stems. The robins left. We took the whole mess down. House sparrows are one of three non-native bird species in the U.S. that are legal to disturb or kill without a permit, the others being starlings and pigeons, probably for agricultural reasons originally.

This year we were happy to see the robins return to the ledge over our back door. It looked like the eggs hatched mid-May, about the time we left on a trip. The puppy left, too, so I didn't worry about her picking up any fledglings falling out of the nest.

When we came home, the nest was empty. Mark took it down so the house sparrows wouldn't take it over – the world does not need more house sparrows. Originally native to Europe and Asia, they have done a wonderful job of colonizing the globe, except for Antarctica.

I was surprised that the robins didn't want to nest on our ledge a second time. They seemed to have abandoned our yard. On the other hand, every time I stood at the kitchen sink and looked out the window, there would be a male house sparrow on the wire, sometimes with dry grass in his beak. He would fly off somewhere to the left, out of sight.

Late June, Mark and I were standing not far from the robins' favorite inner corner of our house exterior. We were discussing exactly where to set up a kennel to keep the puppy (and garden) safe when we aren't outside with her.

Barely 6 feet as the spider crawls from the robins' favorite ledge are various electrical boxes hanging on the wall. I noticed one had a long piece of dead grass sticking out from the bottom. We opened the box and found it half full of dry grass and a dozen feathers, and four tiny eggs. House sparrow eggs.

Dry grass and electrical connections are not a good combination. We cleaned out the box and put duct tape over the hole in the bottom where the wires, and house sparrows, entered it.

Within hours, we had robins in the yard again. They were our robins, the ones that don't spook or attack us when we cross paths in the backyard. Within two days there was most of a robin nest rebuilt on the ledge.

Some of the dry grass the robins collected, they pulled from the spots where puppy pee has killed the lawn, so we have the puppy to thank for making building materials so accessible.

For years I've heard that as cute as invasive house sparrows are, they steal nesting cavities from native birds like bluebirds, or the red-breasted nuthatches that nested across the street in a tree hollow the last two years. This year, house sparrows have it.

Otherwise, I've never seen a house sparrow nest in the wild, except for the old hollow trees in one corner of Lions Park. Mostly I've seen their messy nests sticking out of large commercial signs and other cavities of the human-built environment.

This recent experience shows that house sparrows can interfere with nesting of birds that don't use cavities, like robins.

On to the next mystery: Why am I seeing gulls in Cheyenne this summer? Usually, they don't get any closer to town than the landfill. There must be a new source of food around here, either trash or fish.

Oh, and I saw a hummingbird in Lions Park at the back of the amphitheater. It was inspecting the railings, maybe for spider webs for making a nest? It's about four weeks earlier than we see hummingbirds in town when they return from nesting in the mountains.

# 475 Bird Banter includes news from backyard and beyond

Friday, August 4, 2023, ToDo, page A9

Last month, I reported how quickly my robins moved into their favorite nest spot after the house sparrows' unfortunate nest was removed nearby.

Well, the robins abandoned their nest after too much activity under and over it when our lawn sprinkler system was being repaired. As soon as they left, the house sparrows took over their nest. They are constantly bringing in more grass to fix it and I think I heard young birds cheeping.

On July 26, I saw our first broad-tailed hummingbird this year in our red beebalm. She didn't like the way it was fenced in, so I think we'll take the fence down temporarily and keep the puppy corralled elsewhere. The same day, but earlier, I heard a broad-tailed hummingbird at 10,000 feet in the Snowy Range, so not all the hummers up there were finished with nesting.

In Riverton the week before, I stayed at the dorms at Central Wyoming College. Sleeping next to an open window, one morning I could hear the call of a goldfinch over and over – at 5 a.m. Was it a call of warning or distress?

Mark and I also did a little birding with friends up there who live next to Ocean Lake. Marta introduced us to her kingbirds, phoebes, and tree swallows, but the owls were not at home.

Many bird topics come up at Cheyenne Audubon board meetings. At our July meeting, our vice president was incredulous that the U.S. Forest Service has been doing prescribed burns in June and July to reduce underbrush in the Pole Mountain area west of Cheyenne. They've scorched hundreds of trees where birds may have been nesting at that time of year.

The chapter has funded a transmitter for one of the birds in a white-faced ibis migration study. Among the birds captured in Wyoming was a glossy ibis – quite far from home.

We've also been assisting Rustin Rawlings in collecting data for NestWatch on the eight bluebird boxes he put up at the High Plains Arboretum this spring. House wrens and tree swallows are enjoying them. Maybe the mountain bluebirds will show up next year when the boxes are more weathered.

National Audubon is holding a conference in November for training and inspiring community environmental activists. It will be in Estes Park, Colorado. Getting there could be tricky that time of year but it would be an incredible experience if you want to go.

Three board members met Cidney Handy, the new Audubon Rockies staff member based in Cheyenne. She is a range ecologist for Audubon's Conservation Ranching Initiative. She's looking for Wyoming ranchers who would like to enroll for free and who are interested in learning bird-friendly ranching practices that can bring them a premium price in the retail market.

Cheyenne Audubon board members briefly discussed the Bureau of Land Management's calls for public comments on conservation leasing and updating sage grouse policies.

The Bird Conservancy of the Rockies, headquartered in nearby Brighton, Colorado, has a lot of activities for all ages coming up. Check their website at https://www.birdconservancy.org/.

I subscribe to BirdLife International's email newsletters - https://www.birdlife.org/. It is a partnership of 120 national organizations in more than 115 countries. Bad news about disappearing bird species is balanced by stories of victories. Recently, the European Union's Parliament voted to pass the EU Nature Restoration Law. There were 140 amendments, but the final text was accepted by a margin of 36 votes out of a total of 648.

I'm not on Twitter, but I always thought it was cute – the chubby bluebird logo and calling posts "tweets." But the new owner is changing the platform's name to "X." It feels like he's targeting sweet little birds. Instead of tweeting, will people be "X-ing" things? Sounds like crossing them out.

On the other hand, ChangeX, Microsoft's community grants program, has been providing Cheyenne Audubon with thousands of dollars for the Native Prairie Island Program for which we partner with the Laramie County Conservation District. Homeowners can call LCCD to request native plant seeds and borrow the seeder to spread them on new septic fields and other areas that need restoration.

And that brings us back to our own backyards. What are you doing for the birds? Keeping your cats indoors or in a screened outdoor catio? Using motion-activated yard lights to keep nights darker? Using bird-friendly yard and garden practices (look up Habitat Hero online)?

Are you throwing out hummingbird nectar when it gets cloudy? Make sure you provide nectar that is 1 part regular white granulated sugar to 4 parts water. No dye. No other kinds of sugar. And think about planting more red tubular flowers like red beebalm, Monarda didyma.

2023

449

# 476 Crow loses life but aids researchers for years to come

Friday, September 8, 2023, ToDo, page A11

When you invite friends for dinner, you don't expect them to greet you and then say, "By the way, you have a dead crow on your front lawn."

After dinner I took a plastic trash bag with me and went out to investigate. It was still there. A sleek black pile of feathers.

Because Highly Pathogenic Avian Influenza (HPAI) is still going around, as well as West Nile Virus, I decided I didn't want to look closely at this bird. I used the poop bag technique, putting the bag over my hand, grasping the bird and then pulling the bag over it and tightly tying it shut. And I put the first bag into a second. Then, on the outside, I wrote the date.

If you know me, you know that dead birds go into the freezer. Next, I emailed Elizabeth Wommack, curator and collections manager at the University of Wyoming Museum of Vertebrates. She collects dead animals with spines, like birds, mammals, fish, reptiles and amphibians.

The day I was headed over to Laramie, I was packing the crow into a lunch-sized cooler with re-freezable blocks of ice when Mark said, "Take the other birds, too."

"What other

birds?" I asked.

He went to the freezer and brought back three small birds, each in their own plastic, sealable sandwich bag.

"There are no notes on these," I said. "Where did they come from? And when?"

"I don't know."

This is why I always write the information down, usually on a scrap of paper I stick in the bag with the bird. It's been at least four years since the last time we dropped off any dead birds and apparently, our dead bird memories don't go back even that far. They are most likely birds we found in our yard, perhaps window strike fatalities.

I found Elizabeth on campus at the Biodiversity Institute. In the intake room I filled out a simple form while she took the crow out to examine it.

The bird was molting and she pointed out the new feathers coming in and how worn the edges of the old feathers were. There were tiny ants around the crow's face. Their first target is the eyes, but I had frozen them before they could do any damage. When the bird is taxidermied, the stomach contents will be recorded. Sometimes crows eat inedible things like foil,

Elizabeth said.

She said it's a hard time of year for the adults now. Molting takes lots of energy and they also have their adult-sized children harassing them, begging for food. Plus, crows and other corvids like ravens and magpies are more susceptible to West Nile than other bird species.

The museum makes tissue samples available to researchers in the future who may determine the cause of death. The crow is a more valuable specimen than the other three birds because the location and date of collection and death are known.

One of the other birds was a Townsend's solitaire. In (a gloved) hand, you realize how slender it is compared to its cousin, the robin. The second was an orange-crowned warbler. Elizabeth gently pushed back the grayish, yellowish feathers on its head, and you could barely see a smudge of orange.

The third bird was much easier to i.d., a red crossbill. The crossed tips of the upper and lower parts of the bill are unmistakable.

But the best that can be said about these three birds is that they died in Cheyenne sometime before the day they were donated.

Elizabeth is glad to have more specimens, including game animals – or parts of them. Contact her, ewommack@uwyo.edu, to find out exactly which parts she needs.

Bird flu is still out there. On the U.S. Department of Agriculture's Animal and Plant Health Inspection Service, you can find a map showing how many cases have been detected by state and data for each case. As of the end of August, Wyoming had 137 cases reported while surrounding states ranged from 35 (Nebraska) to 231 (Colorado). Wyoming's most recent case listed was from Park County July 31.

There's more information from the U.S. Fish and Wildlife Service and the U.S. Centers for Disease Control and Prevention. Only one human case of bird flu has been found so far.

To report diseased wildlife, contact the Wyoming Game and Fish Department. Their website, https://wgfd.wyo.gov/, has more information about HPAI. Search "bird flu" and look for the news article, "Game and Fish asks public for help with HPAI."

# 477 Newcomers: Not all of Wyoming's birds have been here forever

Friday, October 6, 2023, ToDo, page A11

Just as many of us living here today didn't have family in Wyoming at the turn of the last century or earlier, many bird species didn't either.

I was perusing "Birds of Wyoming" by Douglas W. Faulkner, published in 2010, where I found records of first observations and population explosions of species in the state. Keep in mind that in the early days, there weren't many people recording bird species.

First, there are the familiar species introduced to North America from Europe, but it took some of them a while to make it to Wyoming.

Rock pigeons were introduced in eastern North America in the early 1600s and it's unknown when they made it here.

House sparrows were released for the first time in New York City in 1851, but it was the ones introduced in Salt Lake City that made it to Evanston in 1874. Birds released in Denver in 1877 expanded their population and were settled in Cheyenne by the 1890s.

First reports of European starlings in Wyoming were in Laramie and Laramie County in 1937. Starlings made Jackson Hole by 1941. Large counts, over 1000 birds, didn't become routine until 1990.

Eurasian collared-doves were first sighted in Florida in the 1980s, but the first Wyoming state record was in 1998 here in Cheyenne at the Wyoming Hereford Ranch. I remember being with the group there trying to distinguish them from ring-turtle doves, pet birds that escaped from time to time.

Sometimes bird species are native elsewhere in North America and eventually get to Wyoming: Chimney swift, 1924; blue jay, 1958 (dramatically increasing in the 1970s) and eastern bluebird, 1901, near Cheyenne but as of 2010 most breeding was seen along a bluebird trail in Crook County in northeastern Wyoming.

The indigo bunting, the eastern equivalent of our lazuli bunting, started infiltrating Crook County and was documented in 1949. Another was documented in southwestern Wyoming by 1959.

A blue grosbeak was first reported in 1962 in Torrington and is now considered a regular breeder in the state, especially along the North Platte.

There are species that were already in Wyoming when ornithologists started looking, but the distribution and numbers have increased. American white pelicans were up in Yellowstone National Park but colonized Pathfinder Reservoir near Casper by 1984. Double-crested cormorants were also up in Yellowstone and also flocked to the new reservoirs in the eastern part of the state.

The American crow barely showed up on Christmas Bird Counts around Wyoming in early years. Faulkner wrote, "Fewer than 4% of the CBCs reporting crows had totals exceeding 300 individuals from 1951-2004. Roughly around year 2000, many CBCs experienced up to a ninefold increase...." Certainly Cheyenne did and they are still here.

Common ravens were always seen a few miles west of town, but they have increased their numbers statewide and have been seen in Cheyenne sporadically the last 10 years.

Bobolinks, fancy-looking songbirds of the grasslands, were spotted this year in eastern Laramie County where Darrel and Marilyn Repshire's wetland pasture was undergoing restoration. Historically, they were common around Cheyenne, but more recently breeding birds were concentrated in irrigated meadows in northeast Wyoming.

The common grackle was documented in Wyoming in 1858 but was rare here in the southeast. Then it had a population explosion in the 1970s. The first record of the similar great-tailed grackle was 1989, near Cheyenne, now slowly spreading across the state.

Interestingly, one of our most common backyard birds, the house finch, was only present along Wyoming's southern border, including Cheyenne, in the first half of the twentieth century. Faulkner wrote that they expanded in the late 50s, with Casper's first nest record in 1984 and Jackson Hole's in the mid-90s.

Game birds from beyond North America are a group that often gets released for hunting. They don't often naturalize widely and so new birds are constantly released, like the popular ring-necked pheasants of China and East Asia. Sometimes you'll see escapees from bird farms like the chukar of Europe and Asia. Also from that region are gray partridges which sometimes can make a go of it around grain crops.

The wild turkey was introduced in Wyoming in western Platte County in 1935 in a trade with New Mexico for some sage-grouse.

The Canada goose, by the early 20th century, could be found in the river basins in Eastern Wyoming, but that wasn't enough for hunters and introductions of all kinds of subspecies were made.

Northern Bobwhite is only native to southeastern Wyoming. If you see it elsewhere in the state, those birds are from introduced populations.

Whether they immigrate with help from people or move in on their own, birds are constantly testing new locations for suitability—for living and raising young.

2023

451

# 478 Bird strikes, bird movements interest UW students

Friday, November 3, 2023, ToDo, page A8

Cheyenne Audubon's October meeting attendees heard from two University of Wyoming students about their bird studies, and about WYOBIRD, the Wyoming Bird Initiative for Resilience and Diversity.

Katie Shabron is an undergraduate who is already involved in bird studies, measuring the number of birds killed by colliding with windows on campus. Window collisions are the second worst human-caused hazard for birds in the U.S.—the first is loose cats.

This fall there was a terrible slaughter of migrating songbirds in one day caused by the perfect reflection of sky by the all-glass façade of the giant McCormick convention center on the edge of Lake Michigan in downtown Chicago. It's been a hazard for years.

On the UW campus, trees and building facades have been mapped and a phone app made available to students allows them to record instances of dead birds as well as no dead birds.

Katie said the first year's data didn't pick up many dead birds, creating more questions such as, are campus building layout and design not conducive to bird strikes? Are birds hitting windows but fluttering away and dying elsewhere?

Katie said the most effective defense against window strikes, if it's too late to install the special glass, is sheets of tiny dots that stick to the outside of windows. They are only visible from outside. Turning out lights at night, which UW does, and especially on tall buildings in big cities, reduces strikes, too.

PhD candidate Emily Shertzer is focused on tracking birds across their full annual cycle as she studies the effect of gas field development on birds near Pinedale.

Traditionally, bird studies have taken place during breeding season, when birds are returning to their breeding grounds and sticking around their nests. That was partly due to the tracking methods available, like banding.

Emily bands birds with the traditional metal leg band with the unique number that can be looked up through the national Bird Banding Laboratory if the bird is recaptured. Her birds also have three colored bands in unique combinations so that they can be identified by sight, without having to capture them.

Radio telemetry has been around for a while for large animals, but now that radio tags can be small enough for birds to wear, it's possible to track the bird's location every 30 seconds. It's much easier to figure out what kills birds on the breeding grounds where fledgling mortality is high.

Through the MOTUS system of stationary antenna towers being set up around the world by different entities, a bird's more extensive travels can be tracked as they fly by and ping an antenna. Each passing bird can be identified and the owner of the tracking device on it is notified.

Emily has set up a similar, but small system to track her study birds around their breeding territory after she finds a nest of a Brewer's sparrow, sage sparrow or sage thrasher. Being able to track the young birds means she can find them quickly if they die and discover the reason such as hail or other bad weather or predators. If birds can make it past the fledgling stage and all the predators and accidents waiting to happen, they might live seven or eight years.

The condition of the parent birds predicts the condition of the young and their ability to survive those early days. Apparently, human development within their breeding area does negatively affect fledgling survival.

Do these attached radio devices make a bird more likely to die? No, they don't seem to. The tags have to be less than 3% of the bird's weight.

Emily's subjects are migratory birds, and their routes can be traced by equipping them with geolocators. These don't send signals but instead, they record light levels, showing the timing of sunrise and sunset where the bird is. Turns out this is a way to tell the bird's migratory route and where it spends the winter. But you must recapture that bird when it returns in the spring to get the data. Good thing males are fairly faithful to their breeding site.

Finding out where birds die is a step towards improving conditions. Emily cited the example of two populations of one songbird species that breed hundreds of miles apart in southeastern U.S. One population was doing well, the other was rapidly declining. A study showed the population doing well spent the winter spread out in Central and northern South America. The declining group all spent the winter together in a comparatively small area where their habitat was rapidly being destroyed.

We need lots more ornithologists studying birds to understand their many characteristics and behaviors. UW's WYOBIRD program gives students more field experience and builds interest in bird studies. Check out the opportunities for involvement and support, https://wyobird.org/.

Emily is also looking for funding for her continuing studies. Contact her at eshertze@uwyo.edu.

# 479 Biologist attends meeting of flyway council protecting migratory birds

Friday, December 8, 2023, ToDo, page A9

Do you know what North American flyway you live in? I was born in the Atlantic, grew up in the Mississippi, live in the Central and visit my granddaughter in the Pacific.

Birds don't usually change flyways like that, but they can be blown off-course. And some prefer to migrate east-west, like harlequin ducks. They winter off the Pacific coast, from Alaska down to Oregon, and head east, inland, to breed, including in northwestern Wyoming.

Wyoming is split between the Central and Pacific flyways, right down the Continental Divide, and although he lives in Cheyenne, Wyoming Game and Fish Department biologist Grant Frost represented Wyoming at a meeting of the Pacific Flyway Council last year. He shared his experience at the November Cheyenne Audubon meeting.

The flyway councils are the people who gather all the data they can get their hands on, looking at academic studies as well as non-governmental and governmental wildlife data (including eBird more and more) in August or early September. There are also reports from Canada and Mexico and a second meeting in the spring.

The health of populations of migratory game birds like ducks, geese, mourning doves and sandhill cranes is considered. Non-migratory gamebirds like pheasants are not. Then the council conveys hunt limits for each species to the states which translate them into limits for hunting regulations that come out in fall.

While the councils can tell each state the maximum number of a species that can be killed, the states can choose to be more conservative.

The Pacific Flyway Council and the other councils were established in 1952. In 2006, the flyways established nongame bird technical committees. Grant was assigned to the one for the Pacific. He's been a bird nerd for a long time, but the meeting gave him new perspectives on birds and bird management.

With the Pacific Flyway extending from a corner of New Mexico on up to Alaska, there are a wide variety of concerns and cultural ways of relating to birds.

For instance, Utah has a swan hunting season, but not Wyoming.

Bald eagles were protected nationally in 1940, but back then Alaska offered a bounty on them because they were competing for salmon. Then Alaska became the 49th state in 1959 and began protecting bald eagles.

Sometimes, managing one species to increase their population has the secondary effect of depleting another species of concern. Grant mentioned the population of common ravens, arguably one of the smartest birds, which has become more successful by adopting the habit of shopping for groceries/roadkill along the highways. But more ravens mean more of them finding and eating the eggs of species of concern like sage grouse and desert tortoises.

Sometimes three species are involved: eagles fly over Caspian tern nesting colonies, causing the terns to flush. While their nests are unprotected, gulls gobble the eggs and young.

Then there are the Indigenous people who have always hunted birds and collected eggs, especially in a tough environment like Alaska. Grant heard gull eggs are particularly prized. Bird pelts go into traditional crafts.

The 2023 Alaska Subsistence Spring/Summer Migratory Bird Harvest was April 2 through August 31, when eggs are most plentiful, and feathers are brightest. Thirty species are listed: waterfowl, waterbirds, shorebirds, seabirds, cranes and owls. Owls? They are big, but here in Wyoming they are not considered game. Perhaps in Alaska they are not considered food either but harvested for feathers.

Grant also sat on the raptor subcommittee which was addressing the number of peregrine falcons being taken by falconers. Who knew we'd finally have this problem when for years we thought peregrines might become extinct? Then there are the protected golden eagles, much more numerous than bald eagles, killing sheep. And there's discussion about the effects of climate change.

Next year the Pacific Flyway Council will be meeting in Wyoming. The meeting Grant attended was in Juneau, Alaska.

There are new ways of studying migration every year and new knowledge gained from them, influencing the work of all the councils.

Recently I checked out a new book from the Laramie County Library, "Flight Paths: How a Passionate and Quirky Group of Pioneering Scientists Solved the Mystery of Bird Migration," by Rebecca Heisman. It's not too technical for most readers, especially people who read my bird columns.

I noticed a bookplate inside said that the book was added to the library's collection on the advice of a library patron. Thank you, whoever you are! I'll try to return the book as soon as I can so someone else can read it!

2023

453

# 2024

## 480 Latest Cheyenne Christmas Bird Count looks a bit different from 1956's

Friday, January 5, 2024, ToDo, page A9

Recently, Bob Dorn shared the results of the Cheyenne Christmas Bird Count from the December 1955-January 1956 count season, the 56th CBC (overall).

It's interesting to compare the differences over 68 years:

--Then, the 7.5-mile-diameter count circle was centered on the KFBC radio station when it was on East Lincolnway, where Channel 5 is now, Dave Montgomery told me. Now it's the Capitol.

--The percentage of open country was higher. Laramie County had only 60,000 people, today 100,000.

--The Cheyenne Audubon Club became the Cheyenne – High Plains Audubon Society in 1974.

--Married women now get to use their own first names. I first met Mrs. Robert Hanesworth, May, in 1989, when she was the Cheyenne count compiler.

--Lt. Col. Charles H. Snyder could have been with F.E. Warren Air Force Base, giving count participants access to that part of the count circle. Today we have retired Col. Charles Seniawski birding the base for us.

Seven bird names have changed:

--Some Canada geese are now cackling geese.

--Marsh hawk is now northern harrier.

--Red-shafted and yellow-shafted flickers are now northern flicker.

--American magpie is now black-billed magpie.

--Gray shrike is now northern shrike.

--Common starling is now European starling.

--The white-winged, Oregon and pink-sided juncos were combined with other juncos as the dark-eyed junco.

Participation has changed, too. We had 24 people help this time compared to only seven in 1956. So naturally we traveled more hours and more miles by foot and vehicle. And back in the 50s, apparently the hours put into watching bird feeders weren't separated.

For our Dec. 16 count we had similar weather – not too windy, no snow – but warmer, around 50 instead of 40.

As for the birds themselves, we counted more species and more geese, crows and starlings. Interestingly, we reported a greater variety of ducks, hawks and falcons, too.

But it's been a long time since we've seen evening grosbeaks. This may be the result of the decline in their population overall.

This year's highlights were the northern goshawk in Western Hills seen during count week (CW), the three days before and after count day, and the lone snow goose at Lions Park.

### Audubon Field Notes – 56th CBC

Published by the National Audubon Society in Collaboration with the U.S. Fish and Wildlife Service

56th Christmas Bird Count; Vol. 10, No.2; two dollars per copy; April 1956

436. Cheyenne, Wyo. (7 ½-mile radius centering from radio station KFBC on east edge of town; city parks and cemeteries 30%, open prairie, deciduous & evergreen trees 20%, prairie roadside 10%, open meadows, reservoirs and creek bottoms 40%).

Jan. 2, 8 a.m. to 4:30 p.m. Clear; temp. 31 degrees to 46 degrees; wind W, 15-35 m.p.h.; no snow. Seven observers in 3 parties. Total party-hour, 56 (6 on foot, 50 by car); total party-miles, 148 (8 on foot, 140 by car).

Canada Goose 12
Mallard 20
Rough-legged Hawk 5
Golden Eagle 1
Marsh Hawk 1
Red-shafted Flicker 13
Hairy Woodpecker 1
Downy Woodpecker 1
Horned Lark 2029 (4 Northern)
American Magpie 25
American Robin 3
Bohemian Waxwing 2
Gray Shrike 1
Common Starling 98
House Sparrow 138
Evening Grosbeak 25
House Finch 4
Pine Grosbeak 4
Pine Siskin 2
White-winged Junco 1
Oregon Junco 86
(pink-sided 84)
American Tree Sparrow 18
Lapland Longspur 22

Total, 23 species (2 additional subsp.), about 2512 individuals. (Observed in area count period: American Goldeneye, Ring-necked Pheasant, Mountain Chickadee, Mockingbird, Townsend's Solitaire, Common Redpoll, Slate-colored Junco).

Charles Brown, Mrs. Charles Brown, Mr. & Mrs. Robert Hanesworth, Wilhelmina Miller, Lt. Col. and Mrs. Charles H. Snyder (compiler) (Cheyenne Audubon Club).

## Cheyenne Christmas Bird Count – 124th CBC

Dec. 16, 2023; 24 participants

Feeder watch time: 11 hours, 36 minutes; Walking: 12 hours, 25 minutes, 19.56 miles; Driving: 4 hours 19 minutes, 87.3 miles

Compiler: Grant Frost

Cackling Goose 183
Canada Goose 1686
Snow Goose 1
Mallard 207
Northern Shoveler 18
Green-winged Teal 3
Lesser Scaup CW
Common Goldeneye 2
Rock Dove (pigeon) 479
Eurasian Collared-Dove 107
Mourning Dove 3

Ring-billed Gull CW
Golden Eagle 1
Northern Harrier 9
Sharp-shinned Hawk CW
Cooper's Hawk 1
American Goshawk CW
Bald Eagle 2
Red-tailed Hawk 6
Rough-legged Hawk CW
Ferruginous Hawk 1
Eastern Screech-Owl CW

Great Horned Owl 1
Belted Kingfisher 2
Downy Woodpecker 3
Hairy Woodpecker 1
Northern Flicker 21
American Kestrel 2
Merlin 2
Northern Shrike 1
Blue Jay 2
Black-billed Magpie 84
American Crow 108
Common Raven 12
Black-capped Chickadee 1
Mountain Chickadee 9
Horned Lark 270
White-breasted Nuthatch 1

Red-breasted Nuthatch 2
Brown Creeper 1
Winter Wren 1
European Starling 221
Townsend's Solitaire 5
American Robin 2
House Sparrow 205
House Finch 63
Pine Siskin 1
American Goldfinch 9
American Tree Sparrow 13
Chipping Sparrow 1
Dark-eyed Junco 61
White-crowned Sparrow 6
Song Sparrow 2
Red-winged Blackbird 50

# Acknowledgements

Writing Bird Banter columns for the Wyoming Tribune Eagle's readers for 25 years has been a privilege.

Outdoors section editor Bill Gruber's invitation to write a monthly (or twice a month for a while) column in 1999 came about because I was the one writing news releases about Audubon chapter events. They included my name and phone number so Bill started calling me when he needed the local angle on bird news. I'm forever in his debt for giving me this role in the community.

Bill lasted a couple more years before he left for a new career culminating in becoming a Florida state parks superintendent. The Outdoors section under Ty Stockton followed along pretty much the same. He came with me when I attended the meeting of the nearly all-male racing pigeon club. Ty now works for state government.

Cara Easton Baldwin, before she acquired her married name, was next. She was more of a skier and runner than hunter and fisher. I still have the wool ski socks she gave me for Christmas. She's gone on to work for the governor's office and for other publications and organizations.

Shauna Stephenson was the next Outdoors editor. She came with me on the mountain plover story to take the pictures, riding behind me on the 4-wheeler over the fields. You can recognize her era because of the many sub-headlines she used. She went on to work for Trout Unlimited and last I checked, she is a market gardener in Montana and serves on the board of the Mountain Mamas, an environmental group.

Kevin Wingert was more of a temporary Outdoors editor. However, he oversaw the "Bird of the Week" project that I did with photographer Pete Arnold from mid-2008 to mid-2010. I think he headed east when he left. Those Bird of the Week short essays and photos (not included here) became the book, "Cheyenne Birds by the Month," which has sold more than 1200 copies locally since 2018 (and more through print-on-demand – including 14 copies in Italy, inexplicably).

Around 2010, the Outdoors section was placed in the E section, the Features section, starting on page E2 in full color. Page E1 was reserved for colorful feature stories, sometimes my column, if the Features editor thought it was fun or important enough. I reported to Jodi Rogstad, the Features editor. We got along well. She sold the managing editor on the idea of me adding a monthly garden column in 2012. Often, she would shoot me column ideas, suggesting localizing a topic she had seen on the national wire service. She started editing "Dear Book" for me before she left to follow her husband back to their home state of Minnesota.

Ellen Fike stepped in for a little while. In the confusion, a couple columns did not appear in the WTE, only through Dave Lerner's digital news service. I've included them in this collection anyway.

At some point, after the WTE was bought by APG, it was announced columnists would no longer be paid even the current pittance. It took two more years for me to realize I could ask for a free subscription, and I did.

Niki Kottmann took over as Features editor in the fall of 2019, just in time for the pandemic. There were major upheavals at the WTE, cutting staff hours and pay. So, I asked Niki if she'd like to make a little extra finishing editing "Dear Book" and she did. Then she left for a better paying journalism job in the same town in California as her boyfriend was moving to.

Will Carpenter, my tenth editor, has just left the paper. He did not get to be Features editor. Since there was no one left to supervise except us freelancers, he was classified as the Features reporter. When he first arrived in fall 2021, Mark and I took him for a bird walk around Lions Park. He was a long way from his southern roots but game to learn about Wyoming and Cheyenne. Though he came straight from college, I appreciated his level of editing – and the weekly columns he wrote.

Of course, my columns do not get in the paper without the approval of the managing editor. When I started, she was Mary Woolsey, followed by Reed Eckhardt and currently Brian Martin.

For this iteration of Bird Banter, I'd like to thank graphic designer Chris Hoffmeister for her patience and great ideas. Elizabeth Sampson and Ursula Vigil, thank you for reading through and cleaning up typos and missing words and grammatical mistakes.

Once my picture started showing up alongside each column back in 1999, readers seeing me at the grocery store began to stop to tell me their bird stories. Doctors and lawyers abhor people asking for their professional advice in social situations, but I adore it. It's been a way to hear what birds are up to around town. And get invited to give bird talks to local clubs. These days readers are less likely to call and more likely to text or email me a bird photo with their question.

Besides editors and readers, I'd like to thank all the people sharing information, most of whom are named in the columns. From the members of the pigeon racing club to the staff of the University of Wyoming Biodiversity Institute and UW Department of Zoology and Physiology and staff members of government agencies and environmental non-profits, it's been a pleasure meeting you. In particular, Jane and Robert Dorn, authors of "Wyoming Birds," have always been readily available authorities.

And finally, Audubon: all the members of the Cheyenne – High Plains Audubon Society, the staff of Wyoming Audubon (now rolled into the regional office), staff of Audubon Rockies and the staff and members of the National Audubon Society, all of whom continue to battle for the birds: thank you.

# INDEX OF TOPICS

**Numbers correspond to article numbers, not page numbers.**

Audubon, 46, 465

Belvoir Ranch, 249
Big Day Bird Count, 5, 67, 111, 138, 167, 167a, 232, 245, 331, 371, 381, 382, 396, 411, 422, 436, 448, 461
Binoculars, 35, 213, 372
BioBlitz, 358, 385
Bird art, 105, 207
Bird atlas, 2
Bird banding, 39, 233
Bird Banter archives, 366
Bird baths, 83
Bird behavior, 345, 407, 425, 429
Bird checklist, 198
Bird conservation, 362, 428, 430, 445, 464, 469, 471, 475
Bird counts, 61
Bird diseases, 7
Bird education, 74, 78
Bird feeding, 12, 21, 53, 69, 283, 307, 325, 376, 416, 454
Bird festivals, 79, 290
Bird film, 149
Bird finding, 398
Bird flashcards, 131
Bird flu, 210
Bird gardening, 68, 72
Bird gifts, 223
Bird habitat, 77, 323, 324, 447, 451
Bird hotline, 166
Bird houses, 6, 69, 163
Bird i.d., 8, 33, 75, 81, 108, 109, 165, 236, 300, 427
Bird i.d.: ducks, 135, 227
Bird i.d.: sound, 289
Bird irruptions, 339
Bird list, 52
Bird names, 299
Bird nests, 115
Bird news, 460
Bird of the Week, 302
Bird people, 349
Bird photography, 377, 426
Bird quiz, 369
Bird records, 110
Bird rehab, 73
Bird reports, 65, 66
Bird rescue, 85
Bird safety, 244, 318, 329, 410, 457
Bird science, 446, 470, 478

Bird song, 30, 119
Bird taxonomy, 11
Bird trails, 458
Bird watching, 1, 9, 13, 23, 44
Bird species
  Atlantic Puffin, 473
  Bald Eagle, 129
  Bluebirds, 107
  Burrowing Owl, 412
  Cardinal, 125
  Cassin's finch, 442
  Cedar Waxwing, 226
  Chickadee, 164
  Condor, 188, 413
  Cranes, 19, 103, 133
  Crossbills, 230
  Crow, 322, 476
  Dipper, 144
  Doves, 189, 240
  Ducks, 179
  Eagles, 386
  Flammulated Owl, 200
  Geese, 337
  Goshawk, 241
  Grebes, 365
  Gulls, 181
  House Sparrow, 474
  Hummingbirds, 43, 197, 373
  Juncos, 14, 88
  Kinglet, 384
  Kite, 348
  Loons, 365
  Mountain Bluebird, 261
  Mountain Plover, 231
  Owls, 93, 191, 351, 356
  Pelican, 41, 314
  Peregrine, 330
  Pigeon, 98, 99, 100, 101
  Raptors, 390, 406, 458
  Red-bellied Woodpecker, 95
  Red-breasted Nuthatch, 203
  Ring-necked pheasant, 459
  Roadrunner, 309
  Robin, 327
  Rosy-finches, 234
  Sage-Grouse, 209, 222, 401, 402
  Sandhill Crane, 287, 352, 388
  Sharp-tailed Grouse, 193
  Shorebirds, 218
  Snowy Owl, 328

  Starling, 160
  Swifts, 334
  Thick-billed Longspur, 472
  Townsend's Solitaire, 90
  Trumpeter Swan, 153
  Turkey Vultures, 225
  Warblers, 136, 221
  Woodpeckers, 208, 216
Birdathon, 137
BirdCast, 420
Bird-friendly chocolate, 58
Birding, backyard, 26, 27, 132, 139, 146, 199, 215, 217, 251, 276, 333, 399
Birding, beginning, 158
Birding, by ear, 397
Birding, kids, 312, 347, 370
Birding Cheyenne, 212, 303, 306, 415
Birding ethics, 359
Birding hotspot, 96
Birding naked, 245
Birding publications, 154
Birding software, 25
Birding tours, 122
Birding trails, 151
Birding Wyoming, 37, 40, 50, 145
Birds & climate change, 343
Birds & energy, 243, 265
Birds & energy, wind, 353, 387
Birds & energy, wind, Roundhouse, 419, 420, 423, 424
Birds & water wells, 433
Birds, baby, 115, 228
Birds in novels, 305
Birds in winter, 239
Birds in Yellowstone, 259
Birdsong, 140
Birdwatchers, 59, 60
Birdwatching, 32, 112, 114, 383, 435, 453
Birdwatching, Merlin, 463
Birdy gifts, 124
Book reviews, 116, 157, 186, 195, 196, 201, 229, 242, 248, 252, 254, 256, 257, 260, 262, 264, 266, 279, 294, 297, 305, 308, 310, 321, 326, 332, 336, 378, 393, 406, 409, 415, 417, 421, 434
Breeding Bird Survey, 38

Cats indoors, 3, 28, 134, 194, 364, 391

Cheyenne Audubon, 360
Christmas Bird Count, 18, 20, 54, 55, 89, 91, 92, 127, 155, 156, 183, 206, 403, 443, 456, 468, 480
Citizen science, 389, 395, 466
Conservation stamp, 56
Cornell Lab of Ornithology, 282

eBird, 128, 291, 342, 392
Environmental education, 16
Exotic birds, 76

Field guides, 34, 62
Field trips, 437
Fledging, 70, 143
Flyways, 479

Great Backyard Bird Count, 22, 57, 94, 159, 286, 431, 444

Habitat Hero, 367
Habitat: backyard, 106
Habitat: prairie, 104
Hanesworth, May, 13
Hazard: windmills, 173
Health outdoors, 224
Hunting, 48
Hybrid car, 169

Important Bird Area, 121, 263
Interview, 455

Lefko, Gary, 458

Migration, 80, 118, 149, 176, 354, 479
Migration, fall, 10, 45, 148, 440, 450, 467
Migration, spring, 4, 24, 36, 161, 344, 371
Migratory Bird Treaty Act, 404, 405, 439, 471, 479
Moose, 204, 205

Nesting, 246, 296, 438, 449, 450, 462

Open space, 47

Pine beetle, 285

458      CHEYENNE BIRD BANTER

**Numbers correspond to article numbers, not page numbers.**

Project FeederWatch, 17, 51, 82, 120, 150, 180, 206, 350, 400, 441

———————

Quilting, 270, 288

———————

Ranching, conservation, 430
Range management, 130
Roadside Attraction, 247, 250, 253, 255, 267, 271, 272, 273, 274, 275, 277, 278, 281, 295, 298, 301, 311, 313, 315, 317, 319, 320
Rocky Mountain Bird Observatory, 162
Roundhouse wind, 419, 420, 423, 424

———————

Scouts, 29, 49
Scouts, outdoor cooking, 42

Skin cancer prevention, 211
Songbird brains, 379
Spark bird, 335
Spring snow, 64
Squirrels, 185, 187

———————

Taxonomy, 147
Thanksgiving Bird Count, 15, 84, 123, 152, 206
Travel
    Alaska, 174, 175, 235, 268, 269, 346, 429
    Albuquerque, 170
    Arizona, 214
    Balloon Fiesta, 293
    California, 238, 394
    Cape May, 177, 178, 338

Colorado, 361
Costa Rica, 414, 418
East, 63, 71
East Coast, 220
eBird, 304
Florida, 341, 368
Hawaii, 257, 258
Massachusetts, 237, 316
Midwest, 202
Mountains, 374
New York, 452
Pennsylvania, 452
Texas, 355, 363, 380
train, 280
Virginia, 375
World birding, 408

Wyoming, 171, 172, 184, 190, 219, 357

———————

University of Wyoming Vertebrate Museum, 476

———————

West Nile Virus, 142, 168
WGFD calendar, 86, 182
WGFD campaign, 87
Wildfire, 113, 126, 141,
Wildlife careers, 192
Wildlife safety, 292
Wildlife watching license, 340
Window hazards, 31
Winter, 102
Wyobirds, 97, 166, 432
Wyoming birds, 477

# INDEX OF PEOPLE, PLACES, BOOKS, AGENCIES, ORGANIZATIONS, ETC.

**Numbers correspond to article numbers, not page numbers.**

1838 Rendezvous Historic Site 320
2016 State of the Birds report 384
3BillionBirds.org 428

A Birder's Guide to Florida 341
A Birder's Guide to the
 Texas Coast 355
A Birder's Guide to Wyoming 398
A Guide to the Birds of East
 Africa novel 254
A Murder of Crows
 documentary film 322
A Naturalist at Large, The Best Essays
 of Bernd Heinrich 409
A Season on the Wind 419, 421
A World on the Wing: The Global
 Odyssey of Migratory Birds 450
Abernathy, Ian 358
Access to Wyoming's Wildlife 151, 184
Ackerman, Jennifer 393, 438
Acopian Center for Conservation
 Learning 225
Adams, Jean 190
Adirondack Mountains 220
Admundson, Marta and Larry 37
AECOM 423
Agate Fossil Beds National
 Monument 255
Agricultural Water
 Enhancement Program 433
Aimakapa Fishpond, Hawaii 258
Aircraft Detection
 Lighting System 424
Alaska Bird Observatory 174, 235
Alaska Marine Highway 346
Alaska Maritime National Wildlife
 Refuge 174, 268
Alaska Raptor Center 174, 177
Alaska Subsistence Spring/Summer
 Migratory Bird Harvest 479
Albuquerque 309, 335
Albuquerque International
 Balloon Fiesta 293
Alcova Reservoir 184, 188, 190
Alden, Peter 398
All Things Reconsidered, My Birding
 Adventures 223
AllAboutBirds.org 370, 384, 390, 391,
 400, 416, 435, 441
Allyn, Lela 185

American Bird Conservancy 194, 283,
 329, 410, 439, 454
American Birding Association
 154, 176, 190 223, 238, 299, 306,
 312, 336, 339, 393, 394, 398, 411,
 427, 428, 440
American Ornithologists' Union 11,
 147, 236, 239, 299
American Pigeon Racing Union 101
American Society of
 Mammologists 186
American Wind Energy
 Association 173
American Wind Wildlife Institute 387
Amtrak 280
Amundson, Marta and Larry 475
Anderson, Art 103, 119, 121, 179, 305
Anderson, Stan 80, 121
Anderson, Will 448
Andrews, Bryce 421
Angell, Tony 429
Angier, Natalie 338
ANKI app digital flashcards 414
Anzalduas County Park, Texas 380
Apache-Sitgreaves National Forest
 113, 126, 191
Appalachian Trail 375
Aransas National Wildlife Refuge 380
Arapaho National Wildlife
 Refuge Complex 357
Arapaho tribe 249
Arctic Autumn, a Journey to
 Season's Edge 326
Arenal Observatory Lodge,
 Costa Rica 418
Argo, Bill 202
Argo, Ellen and George 202
Arnold, Pete 302, 415, 417,
 426, 435, 455
Art of the Winds 426
Atlas missiles 249
Atlas of Birds, Mammals, Reptiles
 and Amphibians in Wyoming 110,
 125, 188, 236
Audubon at Home program 217
Audubon Awards, Laramie County
 School District 1 science fair 192
Audubon Center at Garden
 Creek 190, 233
Audubon Conservation Ranching
 Initiative 428, 457, 475

Audubon Minnesota 318
Audubon Photography Awards 426
Audubon Rockies 358, 360, 367, 385,
 386, 398, 422, 424, 428, 430, 457,
 465, 466, 475
Audubon Wyoming 130, 131, 137, 151,
 192, 243, 263, 342, 358, 401
Audubon Wyoming Community
 Naturalist 233
Audubon, John James 218
Auk, The 299
Ayres Natural Bridge 253

Baby Birds: An Artist Looks into the
 Nest 383, 427
Backstrom, Jan 415
Baer, Karl 402
Bakken, Bradley Hartman 197
Bald and Golden Eagle
 Protection Act 390
Baldwin, Cara Eastwood 366
Banks, Tim 390
Barber, Brian 360
Barnes and Noble 426
Barr Lake State Park,
 Colorado 176, 331
Basal cell carcinoma 211
Basset, Clint 367
Batbayar, Nyambayar 388
Battle Creek 184, 200
Bauldry, Vincent 107
Bayshore Summer, Finding Eden in a
 Most Unlikely Place 297
Baytown Nature Center 355
Bear River Migratory Bird
 Refuge, Utah 345
Bear Swamp State Forest Park,
 New York 452
Beason, Jason 334
Beaver Dam Analog 451
Beck, Jeff
Becoming an Outdoor Woman 204
Belvoir Ranch 249, 385, 419, 423, 464
Benkman, Craig 230
Bent, Arthur Cleveland 365, 384
Berquist, Francis 168
Berquist, Francis and Janice 182
Berquist, Greg 182
Best Backyard Bird report 146

Betz, Myrtyle Scharrer 341
Big Bend National Park 363
Big Hole 249
Big Horn Mountains 247
Bighorn Canyon National
 Recreation Area 247
Bighorn National Forest 247
Bildstein, Keith 225
Bioblitz 385, 423
Biological Conservation journal 395
Bird banding 466
Bird Banter 366
Bird Conservancy of the Rockies
 (See also Rocky Mountain Bird
 Observatory) 428, 439, 475
Bird flu 210, 476
Bird houses 6
Bird Migration Explorer 466
Bird observatory 192
Bird of Conservation Concern 412
Bird of the Week 302, 415
Bird Song Hero game 404
Bird song recordings 140
Bird Tape for windows 410
Bird Watcher's Digest 124, 154, 223,
 368, 396, 414, 418
Birdathon 137
BirdCast 31, 420, 440
Birder's World 124, 154
Birdhouse Network 163, 216
Birding magazine 306
Birding Without Borders 398, 408
BirdingPal.org 398
BirdLife International 382, 404, 475
Birds Canada 441
Birds of Grand Teton
 National Park 190
Birds of North America Online 199,
 203, 208, 221, 223, 231, 302, 309,
 327, 337, 348, 412, 442
Birds of the Great Plains 201
Birds of the Photo Ark 404, 408
Birds of Wyoming 294, 327, 339, 343,
 348, 349, 365, 467, 477
Bird-Safe Building Guidelines 318
Birdsong by the Seasons 264
Birdsong for the Curious
 Naturalist 434
BirdSource 154, 159

460  CHEYENNE BIRD BANTER

**Numbers correspond to article numbers, not page numbers.**

BirdTape (for windows) 329

BirdYourWorld.org 404

Birek, Jeff 360

Black Hills, Wyoming 208, 219

Black Mountain fire lookout 171

Black, Jack 305

Blackburne, Anna 467

Blog posts 366

Bly, Bart 231

Boehm, Elizabeth 415, 426

Borie oil field 249

Bosque del Apache National Wildlife Refuge 19

Boulder Imaging 387

Bowers, Jennifer 191

Box, C.J. 305

Boysen State Park 317

Brazos Bend State Park, Texas 355

Breeding Bird Survey 38, 128

Broadmeadow Brook Sanctuary, Massachusetts 237

Bronx, The 220

Brunton binoculars 213

Brush State Wildlife Area, Colorado 361

Bucks County, Pennsylvania 452

BudBurst.org 343

Budd, Bob 47, 130, 357, 401, 469

Bump Sullivan reservoir 440

Bureau of Indian Affairs 192

Bureau of Land Management 192, 243, 272, 281, 401, 402, 460, 475

Bureau of Reclamation 192

Burrowing Owl Conservation Network 412

Buteo Books 398

———————

Cade, Tom 330

Caladesi Island State Park, Florida 341

Camp Carlin 249

Camp Laramie Peak 141

Canadian Pigeon Racing Union 101

Canadian Wildlife Service 80, 118

Candey, Dave 238

Canon PowerShot camera 377

Cape Cod National Seashore 316

Cape May Bird Observatory 177, 178, 338

Cape May Point State Park 178

Cape May, New Jersey 177, 178, 338

Caproni, Valerie, U.S. District Court Judge 439

Carleton, Scott 197

Carling, Matt 342

Carrey, Steve 285

Castenada, Sue 194

Cat Clinic of Cheyenne 194

Cat fencing 194

Catio 475

Cats Indoors 3

Central Wyoming College 475

Cerovski, Andrea (see also Orabona) 11, 26, 38, 39, 108, 121, 194

Cerro de la Muerte, Costa Rica 418

Champions of the Flyway 382

ChangeX grants 475

Check, Kim 233

Checklist of North American Birds 299

Checklist of the Birds of Cheyenne, Wyoming and Vicinity 198

Cheney, Liz 405

Chesapeake Bay Bridge-Tunnel 375

Chesnut, Lorie 451, 455, 468

Cheyenne - High Plains Audubon Society 1, 4, 12, 75, 103, 133, 158, 179, 184, 191, 192, 212 219, 221, 222, 232, 238, 243, 285, 322, 329, 331, 336, 342, 344, 360, 362, 367, 371, 382, 396, 397, 398, 406, 407, 408, 411, 415, 424, 434, 437, 446, 448, 455, 459, 465, 469, 480

Cheyenne Animal Shelter 364

Cheyenne Audubon (see also Cheyenne - High Plains) 198, 379, 423, 430, 432, 436, 451, 453, 456, 460, 464, 466, 475, 478, 479

Cheyenne Audubon Club 480

Cheyenne Big Day Bird Count 436, 460

Cheyenne Bird Checklist 239, 369

Cheyenne Birds by the Month 417, 426, 435, 455

Cheyenne Board of Public Utilities 367, 455

Cheyenne Botanic Gardens 106, 212, 360, 367, 415, 417, 426, 455, 461

Cheyenne City Council 364

Cheyenne Compost Facility 323

Cheyenne Depot Museum 417, 426

Cheyenne Frontier Days Committee 456

Cheyenne Garden Gossip: Locals Share Secrets for High Plains Gardening Success 455

Cheyenne Greenway 143, 437, 450

Cheyenne Habitat Hero Committee 455

Cheyenne Habitat Hero Workshop 457

Cheyenne Heritage Quilters 288

Cheyenne Mountain Bike Association 423

Cheyenne Pet Clinic 115, 228, 292, 365, 410, 426

Cheyenne Regional Medical Center Pink Boutique 426

Cheyenne Skin Clinic 211

CheyenneBirdBanter.com 376, 416, 435

Chihuahuan Desert, Texas 363

Chindgren, Steve 242, 243, 266

Chisos Mountains, Texas 363

Chocolate, shade grown 58, 124

Christmas Bird Count 428, 456, 466

Chugach State Park, Alaska 268, 269

Chugwater 64

Church, Christopher DVM 365

Churchville Nature Center, Pennsylvania 452

Citizen science 389, 395

Citizen Scientist, Searching for Heroes and Hope in an Age of Exteinction 395

City of Cheyenne 423, 464

Clapp, Bob and Rhea and Leo 60

Cobirds 189, 331

Code of Birding Ethics 191

Coffee 4, 124

Collard III, Sneed B. 417

Colorado Birding Trail 458

Colorado Breeding Bird Atlas 2

Colorado Crane Conservation Coalition 388

Colorado Field Ornithologists 361

Colorado State University Cooperative Extension 216

Colorado Wildscapes, Bringing Conservation Home 217

Columbia University 220

Community Based Occupational Education 119

Community Cat Initiative 364

Community science 466

Condors in Canyon Country 413

Congressional Award 385

Conn, Laura 228

Conservation Ranching Initiative 430

Conservation Reserve Program 193

Contributions to Biodiversity Science Award 349

Contributions to Wyoming Biodiversity Conservation 349

Cordoba, Mario 414, 418

Cornelison, John 360, 456

Cornell Lab of Ornithology (other than bird counts) 106, 119, 128, 140, 154, 163, 216, 246, 282, 306, 307, 328, 362, 378, 389, 392, 395, 404, 420, 428, 439, 440, 441, 444, 460, 463, 470

Costa Rica 414, 418

Costa Rica Wildlife Foundation 418

Costopolous, Barbara 156

Coyote Hills, California 238

Crane Creek Graphics 207

Crane Trust Nature and Visitors Center 352

Crane, Eva 39

Creamer's Field Migratory Waterfowl Refuge 235

Creekmore, Terry 168

Crescentia Expeditions 414, 418

Crossley, Richard 218

Crossroads Park, Casper 190

Crow Creek 348, 451, 460

Culek, Bernie 231

Curry, Martin 357

Curt Gowdy State Park 144, 161, 202, 390, 425, 426, 437

———————

Danzenbaker, Jim 190, 213

Decline of the North American Avifauna study 428, 439

Denali National Park 235

Denver Audubon 432

Denver to Fort Laramie stage line 249

Dermer, Steve 98, 99, 100

Devils Tower National Monument 201

Diamond Wings Upland Game Birds 459

Diamond, Jared 378

Dickinson, Rachel 266

Digiscope 377

Doherty, Kevin 243

Doktor, Aadrian 420, 428

Domenici, Dominic 188

Dorn, Jane and Robert 8, 11, 20, 32, 43, 45, 55, 75, 80, 91, 95, 96, 110, 152, 156, 171, 188, 190, 198, 208, 294, 330, 349, 367, 371, 384, 398, 403, 408, 411, 415, 456, 468, 480

Dorothy McIlroy Bird Sanctuary, New York 452

Doughton, Sandi 378

Down the Mountain, The Life and Death of a Grizzly Bear 421

Downar Bird Farm 459

Downey, Dusty 430

Downey, Jacelyn 432

Drayson, Nicholas 254

Droll Yankee bird feeders 185

Dry Creek 451

Dry Lake 190

Dugway, Bureau of Land Management Recreation Site 281

Duke Energy 387, 419

Duncraft bird feeders 185

Dunn, John 237, 361

Dunne, Pete 218, 221, 262, 297, 326, 338, 406

Dvergsten, Cindy 44

Dykstra, Beth and Brian 126, 204, 214

———————

Eagle River, Wisconsin 202

Eagle take permit 460

East Park (see also Kiwanis Park) 451

Eastern Egg Rock, Maine 473

Eastern Shore Birding and Wildlife Festival 375

Eastern Shore of Virginia National Wildlife Refuge 375

Eastern Shoshone tribe 249

Easton, Beth 65

461

**Numbers correspond to article numbers, not page numbers.**

eBird 128, 238, 239, 291, 303, 304, 306, 337, 342, 343, 344, 346, 354, 355, 360, 363, 366, 371, 376, 380, 382, 384, 394, 395, 398, 408, 416, 420, 425, 437, 446, 448, 450, 466, 467, 470

eBird Mobile 392

Edmunds, Daly 424

Edness Kimball-Wilkins State Park 190, 246

Electric hybrid car 169

Elliot, Lang 223

Environment and Climate Change Canada 428, 439

Erwin, Kathleen 193

Esterbrook Campground 171

Everglades National Park 341

Exotic bird rescue 76

Expedition Island 271

---

F. E. Warren Air Force Base 170, 232, 245, 337, 392, 422, 448, 451

Falcon Fever 242

Falconer on the Edge, A Man, His Birds… 266

Farr, Robert, DVM 194, 228, 292

Fastest Thing on Wings, Rescuing Hummingbirds… 373

Fatal Light Awareness Program 31

Faulkner, Doug 118, 148, 181, 200, 236, 294, 299, 327, 339, 343, 349, 365, 467, 477

Felley, Dave 38, 65

Feral Cats and Their Management 364

Field Guide to the Birds of California 393, 294

Field Guide to the Birds of North America 223

Finding Birds Around the World 398

Finding the Birds of Jackson Hole 190

Finding Your Wings 242

Finger Lakes Land Trust, New York 452

Finger Lakes National Forest, New York 452

Fishing license, lifetime Wyoming 56

Fitzpatrick, John 57, 360, 362

Fitzpatrick, Ryan 423, 460

Five Springs Falls Campground 247

Flaming Gorge National Recreation Area at the Confluence 275

Flamm Fest 200

Flaspohler, Dave 221

Flicker, John 161

Flight Paths: How a Passionate and Quirky Group of Pioneering Scientists Solved the Mystery of Bird Migration 479

Flights Against the Sunset 252

Flippo, Mark 363

Floyd, Ted 176, 306, 336, 348, 360, 377, 411, 427

---

Flyaway: How a Wild Bird Rehabber Sought Adventure… 279

Forsberg, Michael 290

Fort Collins 181

Fort Collins Audubon Society 158

Fort Kearny State Recreation Area, Nebraska 290

Fort Laramie 403, 443

Fort Laramie National Historic Site 156, 443

Fort Robinson State Park 104, 250

Fort Steele 267

Fossil Creek Reservoir Natural Area, Colorado 181, 453

Fox squirrel 185, 187

FOY, First of Year species observations 343

Franzen, Jonathan 305

Freedom, novel 305

Friend Park Campground 171

Friends of the Belvoir 419

Frost, Grant 412, 422, 437, 443, 456, 461, 467, 468, 479

---

Gallagher, Tim 242

Gammon, Dave 164

Garden Creek 50

Gardener, Pete 20

Gauthreaux, Sidney 420

Gaven Hill 175

Gentile, Olivia 306

Georgetown University 428, 439

Gerhart, Bill 193

Geyer, Jeff 451

Gibbons, Euell 215

Gifts of the Crow 429

Gilbert, Suzie 279, 434

Gill, Matthew 427

Gill, Priscilla 427

Ginter, Jana 246

Girt, Rachel 455

Glendo Reservoir 112, 201

Global Big Day 382

Gloucester Harbor, Massachusetts 237

Goncalves, Jose 122

Good Birders Still Don't Wear White 393

Gooders, John 398

Gordon, Jeffrey A. 393

Gorges, Barb 455

Gorges, Bryan and Jessie 375, 392, 414, 452

Gorges, David 238

Gorges, Greg and Mary Jane 341

Gorges, Jeffrey and Madeleine 355, 380, 440

Gorges, Mark 415, 424, 448, 455, 456, 465, 467

Gorges, Mike 238

---

Gorges, Peter 71, 175, 235, 268, 269, 346, 429

Goshen Hole 184

GPS 188

Gramm fire 208

Granite Canon Environmental Committee 423

Grayrocks Reservoir 114, 156

Great Backyard Bird Count 466

Great Pikes Peak Birding Trail 151, 458

Great Texas Coastal Birding Trail 151, 355

Great Wisconsin Birding Trail 151

Greeley, Will McClean 471

Green Cay Wetlands, Florida 341

Green River 271

Grenier, Martin 358

Grey Reef reservoir 190

Greyrocks Reservoir 352, 443

Griggs, Pam 238

Grinnell, Eleanor 119

Growing Native Plants of the Rocky Mountain Area 367

Gruber, Bill 341, 366

Guernsey 156, 339, 403, 443

Guernsey State Park 156, 443

Gulls of the Americas 237

Gunn, Carolyn 334

Guttman, Burton 242

---

Habitat Hero 360, 367, 430, 455

Hales family 448

Hall, Minna B. 439

Handy, Cidney 475

Hanesworth, May 13, 347, 382, 456, 480

Hannibal, Mary Ellen 395

Hanson, Dave 66

Hargis, Bob 190

Hart, Dick 53

Hart, Helen 257

Hartville 339, 442, 443

Harvard Bridge, Massachusetts 220

Harvard Forest 237

Haupt, Lyanda Lynn 378

Hawaii Audubon Society 258

Hawaii The Big Island Revealed 260

Hawaii Tropical Botanical Garden 257

Hawaii, Big Island 257

Hawaiian Volcanoes National Park 257, 258

Hawaii's Birds 258

Hawk Mountain Sanctuary 225

Hawk Springs Reservoir 184

Hawks in Flight 338, 406

Headquarters Trail, Medicine Bow National Forest 463

Hecker, Jim and Carol 17, 119, 211, 456

Heinrich, Bernd 332, 383, 409, 434

Heisman, Rebecca 479

---

Helzer, Chris 430

Hemenway, Harriet 439

Hensel fire 171

Hewitt, Dawn 418

Hewston, John 84, 125, 152

Hicks, Martin 26

Hiester, Tom 387

High Island, Texas 396

High Plains Aquifer 433

High Plains Arboretum 422, 471, 475

High Plains Grasslands Research Station, USDA 212, 331, 371, 384, 396, 411, 422, 436, 466

Highly Pathogenic Avian Influenza 476

Hill, Kathy 131

Hillsdale, Laramie County 472

Hilo, Hawaii 257

Hines, Ann 190

Historic Jamestowne, Virginia 375

Hoffmeister, Chris 415

Hog Island, Maine 473

Holbrook Travel 473

Holliday Park 153, 232, 246, 276, 296, 303, 306, 314, 324, 337, 415, 425

Holloran, Alison (see also Lyon, Alison) 360, 430

Holloran, Matt 222

Holsinger, Connie 360, 367

Hosafros, Judith 182, 349

Hotel Bougainvillea, Costa Rica 418

Houghton Mifflin 218

Houston Arboretum and Nature Center 355

How to Be a Good Creature… 417

How to Raise a Wild Child 370

Howell, Steve 237, 297, 409

Hudson, Edna 186

Hummingbirds and Butterflies field guide 321

Hutchinson, Zach 398, 413, 422, 457, 466

Hutton Lake National Wildlife Refuge 179, 184, 357, 390, 411

---

I.P. Johnson Center for Birds and Biodiversity 282

Identiflight 387

Identify Yourself, the 50 Most Common Birding Challenges 196

Identifying and Feeding Birds 308

Important Bird Area 184, 263, 329

iNaturalist 458, 466

Inflation Reduction Act 464

Ingelfinger, Franz 30

International Crane Foundation 388

International Dark Sky City 450

International Ecotourism Society 380

International Federation of American Homing Pigeon Fanciers 101

International Shorebird Survey 428

**Numbers correspond to article numbers, not page numbers.**

International Union for Conservation of Nature 375
Interstate 80 249
Ipswich River Audubon Sanctuary 237
Ithaca, New York 452

---

Jackson, Hertha 202
Jaramillo, Alvaro 393
Jennings, Mary 76
Jensen, Merrill 347
John James Audubon Center at Mill Grove, Pennsylvania 338
Johnsgard, Paul 328
Johnson, Greg 173, 188, 198, 331, 339, 377, 403, 411, 414, 415, 427, 456
Jones, Don 446, 467, 470
Jones, Lynds, 382
Joppa Flats Education Center, Mass Audubon 316
Juneau Audubon Society 347

---

Kaloko-Honokohau National Historical Park 258
Karlson, Kevin 218
Kauffman, Matthew 417
Kaufman Field Guide to Advanced Birding 321
Kaufman Field Guide to Insects of North America 229
Kaufman, Kenn 83, 116, 189, 252, 321, 325, 338, 419, 421
Keen, Sam 242
Keffer, Ken 233
Keffer, Larry 233
Kenai Fjords National Park, Alaska 268, 269
Kenai Peninsula 268, 269
Kerlinger, Paul 149
Ketcham, Jeff 212
Keto, Ruth 142, 185
Keyhole State Park 201
Kimberling, Brian 343
King, Rose-Mary 190
Klataske, Ron 290
Knopf, Fritz 231
Knorr, Owen A. 334
Koeppel, Dan 224
Komar, Nick 361
Kona, Hawaii 257
Kozlowski, Steve 208
Kress, Stephen 473
Kroodsma, Donald 195, 264, 434
Kruse, Gere and Barbara 357

---

Lake Champlain, New York 220
Lake Hattie 236, 467
Lake Marie, Snowy Range Scenic Byway 298
Lakota tribe 249

Lander Art Center 358
Langston, Al 205
Lanham, Chuck 249
Laramie Audubon Society 179, 342, 357, 390, 413
Laramie County 424, 433
Laramie County Community College 198, 448, 455
Laramie County Conservation District 106, 367, 451, 457, 475
Laramie County Control Area Order, water 433
Laramie County Library 106
Laramie County Master Gardeners 367, 455
Laramie Plains lakes 236, 390, 411, 467
Laramie Range 179, 219
Laramie Ranger District 446
Laramie River Greenbelt Trail 190
Laramie Rivers Conservation District 357
Larkin, Jean 76
Last Child in the Woods 224
Lautenbach, Jonathan 467
Lawrence, Gloria and Jim 32, 97, 121, 166, 245
Layton, Lois and Frank 177
Lebsack, Fred 10, 13, 45, 65, 456
Lee, Bob 160
Lefko, Gary 148, 151, 458
Lehman, Paul 170
Leica binoculars 372
Lek 193
Leopold, Aldo 404
Lerwick family 433
Letters from an Elk Hunt by a Woman Homesteader 204
Letters from Eden, A Year at Home, in the Woods 223
Leukering, Tony 181, 236
Levad, Rich 200, 334
Life Everlasting, The Animal Way of Death 332
Life List, biography 306
Lilygren, Mike 372
Lincoln Highway 249
Lindzey, Fred and Stephanie 233
Link, Mary 105
Lions Park 5, 8, 17, 96, 121, 136, 143, 158, 179, 184, 190, 213, 227, 232, 245, 263, 329, 330, 337, 356, 384, 396, 411, 415, 425, 436, 448, 466, 467
Little America 422
Lives of North American Birds 325
Living on the Wind 420
Lockman, Dave 131
Lone Pine publications 201
Lone Tree Creek 385
Long Island Sound, Connecticut 220
Lopez, Gus 168
Louv, Richard 224

Loxahatchee National Wildlife Refuge, Florida 341
Luce, Bob 91
Lyon, Alison (see also Holloran) 107, 121, 172, 243
Lyon-Holloran, Alison 263
Lyons, Don 473

---

Macaulay Library 282
Madson, Chris 125, 182, 349
Maestas, Cade 372
Maguire, Annie 257
Maier, Aaron 457
Maley, James 342
Mammoth Hot Springs 219
Manley, Wanda 472
Mariposa Road, The First Butterfly Big Year 310
Martin, Steve 305
Martinez del Rio, Carlos 197
Marzluff, John 429
Masear, Terry 373
Mason, Joanne 65
Massachusetts Audubon 237, 316, 398
Massachusetts Audubon Society for the Protection of Birds 439
Master Gardeners 432
Mauna Kea 257, 258
Maven binoculars 372
McCartney, Blaine 352
McCarty, Jesse 446
McClean, George P. 471
McCown, John P. 472
McCreary, O.C. 330
McCullough, Jenna 440
McDonald, Dave 88, 172
McEneaney, Terry 117, 121, 190, 219
McGinis, Samuel M. 417
McNamee, Stan 160
McNicholas, Wayne 348
Mead, Matt, Wyoming Governor 401, 402
Meadowlark Audubon Society 342
Means, Marcela 192
Medicine Bow National Forest 40, 81, 145, 171, 204, 374, 397
Medicine Bow Peak 413
Medicine Bow Range 172, 179
Medicine Bow-Routt National Forest 285
Medicine Bow-Routt National Forest 285, 446
Medicine Wheel National Historic Landmark 247
Mendenhall Glacier 346
Menomonee River Parkway, Wisconsin 335
Merlin Bird I.D. app 392, 397, 404, 435, 441
Merlin Sound I.D. 463

Merrill, Nancy 388
Michelson, Chris 219
Migratory Bird Treaty Act 194, 353, 382, 390, 404, 405, 439, 471
Miles City, Montana 465
Milner, Ben 459
Miner's Cabin Trail, Snowy Range Scenic Byway 301
Modesitt, Larry 361
Moench, Belinda and Don 65
Molt in North American Birds 297
Monitoring for Avian Productivity 39, 128, 233, 466
Montgomery, Sy 417
Monticello, Virginia 375
Moon, Deb 101, 425
Moore, Teresa 324
Moose Day 432
Moose, Shiras 204, 205
Morgan, Wendy 207
MOTUS 478
Mount Auburn Cemetery, Massachusetts 316, 398
Mount Roberts, Alaska 346
Mullen Fire, Medicine Bow National Forest 440, 446
Munro, Bill 171
Murie Audubon Society 50, 166, 190, 246, 371, 398, 432
Murie, Olaus and Mardy 177, 235
Murphy, Karagh 379
Mustang: The Saga of the Wild Horse in the American West 256
My Garden Neighbors 186
Myer, Judy 352
Mylar Park 143
Mystic River, Connecticut 220

---

Na' ali' I Plantation Bed & Breakfast, Hawaii 257
Naked birding 245
National Audubon Society 46, 106, 128, 259, 263, 302, 318, 353, 362, 401, 404, 426, 428, 430, 432, 439, 451, 455, 460, 471
National Bird Banding Laboratory 478
National Electric Code 464
National Environmental Policy Act 439
National Geographic Field Guide to the Birds 361
National Geographic Society 223, 404
National Park Service 192
National Pigeon Association 101
National Renewable Energy Laboratory 387
National Wildlife Federation 106
National Wildlife Health Center 210
National Wildlife Refuge system 341
National Wind Technology Center 387

463

**Numbers correspond to article numbers, not page numbers.**

Native Prairie Island Program 475

Natural Resources Conservation Service 433, 469

Natural Resources Defense Council 386

Nature's Best Hope, A New Approach to Conservation… 434

Nebraska Crane Festival 352

Nebraska National Forest 250

Nebraska Prairie Partners program 231

Nelson, Del 343

Nelson, Mark 186, 292

NestWatch 246, 410, 475

New Birder's Field Guide to Birds of North America 359

Newport, Rhode Island 220

NextEra energy company 419, 423, 460, 464

Nici Self Museum, Centennial 313

Nordell, Cameron 407

North American Bluebird Society 6, 107, 261

North American Breeding Bird Survey 428

North American Nature Photography Association 426

North Crow Reservoir 41, 246

North Platte River 443

Northern Cheyenne tribe 249

Nunn Guy, The 458

Obmascik, Mark 305

Obrecht, Jeff 239

O'Brien, Michael 218

Ocean Lake 475

Ocean Lake Wildlife Habitat Management Area 190

Oehler, Dick and Judi 202

Ogallala Aquifer 433

Ogin Inc. 353

Ohme Gardens, Washington 345

One Wild Bird at a Time 383

Orabona, Andrea (see also Cerovski) 360

Oregon Trail Ruts State Historic Site 156, 253, 443

Osborn, Sophie A.H. 413

Otters Dance book 469

Overland Trail along Bitter Creek 277

Owls of North America and the Caribbean 378

Pacific Flyway Council 479

Panjabi, Arvind 176, 208, 428

Pannell, Karen and Fred 407

Paraiso Quetzal, Costa Rica 418

Parish, Chris 413

Parks, Karen, DVM 194

Partners in Flight Avian Conservation Database 428

Patchwork birding 306

Pathfinder Reservoir 184

Patrick, Sue 93

Patuxent Wildlife Research Center 63

Paul Smith Children's Village 455

Paulson, Deb 190

Pawling, Bailey 192

Pawnee National Grasslands 184, 262, 458

PBR Printing 415, 417, 426

Pelton Creek, Medicine Bow National Forest 204

Peregrine Fund 188, 330, 387, 413

Perky-Pet bird feeders 185

Perrin, Jacques 149

Pete Dunn's Essential Field Guide Companion 218

Peterson Field Guide to Bird Sounds of Western North America 421

Peterson Field Guide to Birds of North America, 2008 256, 297, 330

Peterson Field Guide to Birds of North America, 2020 434

Peterson Field Guide to the Bird Sounds of Eastern North America 393, 397, 446

Peterson Field Guide to Western Reptiles & Amphibians 417

Peterson Guide to Bird Identification-in 12 Steps 409

Peterson Reference Guide to Sparrows of North America 421

Peterson, Mark 361

Peterson, Roger Tory 223, 256, 297, 312, 338, 434

PetersonBirdSounds.com 421

Pieplow, Nathan 393, 413, 421, 446, 448

Pilot Butte Wild Horse Scenic Loop 272

Pine beetle 285

Pioneer Park, Fairbanks 235

Piskorski, Renee 105

Plum Island, Parker River National Wildlife Refuge 316

Plymouth, Massachusetts 220

Point of Rocks Stage Station State Historic Site 274

Point Reyes Beach North, California 394

Point Reyes Bird Observatory 162

Point Reyes National Seashore 394

Pole Mountain, Medicine Bow National Forest 119, 285, 475

Portage Glacier, Alaska 268

Post Assisted Log Structures 451

Potter, Kim 200, 334 .

Power Company of Wyoming 386

Prairie Fire newspaper 328

Prairie Ghost 231

Prairie Partners program 162, 231

Prairie Spring: A Journey into the Heart of a Season 262

PrairieEcologist.com 430

Pranty, Bill 341

Prather, Jonathan 379

Prius 169

Project FeederWatch 455

Project Puffin 473

Project WILD 16

Pryor Mountain Wild Horse Range 247

Puschel, Marion 335

Pyle, Bob 310

Quilt Care, Construction and Use Advice 455

Quilt Index 288

Quilt Wyoming 270

Quintana Neotropical Bird Sanctuary 355

Randall, Mike 131

Raptor Alley, Colorado 458

Raptor Recovery Nebraska 328

Rawhide coal power plant 419

Rawhide Reservoir 1

Rawlings, Rustin 471, 475

Ray Lake 190

Raynes, Bert 190

Raynor, E.J. 422

Reader Rendezvous 368, 380

Red Canyon Ranch 39, 358

Red Gulch Dinosaur Tracksite 247

Red Lion SWA, Colorado 361

Red Mountain Open Space, Colorado 385, 419, 423

Reed, L.A. 186

Renewable Energy Systems 387

Repshire, Darrel and Marilyn 477

Rio Grand Nature Center State Park 170

Rio Grande Valley Birding Festival 380

Riverbend Nursery 417, 426

Riverboat Discovery, Fairbanks 235

Rivers and Wildlife Celebration, Nebraska 290

Roadless areas 40

Robin nest 70, 71

Rocky Mountain Arsenal National Wildlife Refuge 406

Rocky Mountain Bird Observatory (see also Bird Conservancy of the Rockies) 148, 162, 176, 186, 200, 208, 231, 334, 360

Rocky Mountain Garden Survival Guide 367

Rocky Mountain National Park 112, 374, 468

Rocky Mountain Raptor Center 246

Rocky Mountain Regional Office, National Audubon Society 465

Rodale Institute 452

Rodeo-Chediski fire 113, 126

Rodomsky-Bish, Becca 444

Rodriguez, Roy 380

Rogstad, Jodi 366

Rogstad, Lucy 50

Romero, Dave 121

Rosenberg, Ken 428

Rothwell, Reg 26, 106, 292

Roundhouse Wind 428, 460, 464

Roundhouse Wind Energy Project 419, 423, 424

Rowe Audubon Sanctuary 103, 133, 287, 290, 352

Rudd, Bill 417

Rutledge, Brian 243, 401

Ryder, Ron 1, 11, 189

Sacramento National Wildlife Refuge 238

Safina, Carl 157

Sagebrush Ecosystem Initiative 401

Sage-Grouse Implementation Team 401

Sampson, Scott D. 370

Sand Creek 190

Sandia Crest 170, 234

Sapsucker Woods, New York 282

Saratoga Lake public access area 184

Sartore, Joel 404, 408

Sauk River, Washington 440

Savage Run Wilderness Area 446

Saville, Dennis 422

Schmeekle Reserve 202

Schmoker, Bill 176

Schuckert, Susie 202

Scott & Nix field guide publishers 336

Scott, Oliver 32, 96, 136, 151, 190, 312, 398

Scouts 29, 42, 44, 49, 104, 141, 171, 335

Sea Watch 178

Seabird Institute, National Audubon Society 473

Seagraves, Tom 212

Seldovia, Alaska 268

Selva Verde, Costa Rica 418

Seminoe Reservoir 184

Seniawski, Chuck and Sue 106, 119, 216, 348, 414, 422, 467, 468, 480

Senner, Stan 161

Shabron, Katie 478

Shea, Ruth 153

Shell Canyon 247

Shenandoah National Park 375

Sherman Mountains 219

Shertzer, Emily 478

Shirley Basin 48

Shorebird Guide, The 218

Sibley Field Guide to Birds 442

Sibley Guide to Bird Life and Behavior 251

Sibley Guide to Birds 241

**Numbers correspond to article numbers, not page numbers.**

Sibley, David Allen 116

Sierra Madre Range 184

Sightings: Extraordinary Encounters with Ordinary Birds 242

Silence of the Songbirds 242

Simkins, Velma 189

Simpson, Alan 47

Singing Life of Birds, The 195, 229

Sinks Canyon State Park 190

Sitka National Historical Park 346, 429

Sitka, Alaska 175

Skaggs, Kent 287

Skinner, Karin 73

Smith, Doug 259

Smith, Shane 27, 106

Smithsonian Field Guide to the Birds of North America 336

Smithsonian Migratory Bird Center 439

Snetsinger, Phoebe 306

Snowy Range 145, 172, 184, 204, 205, 234, 374, 442, 475

Snyder, Larry 231

Soapstone Prairie Natural Area, Colorado 385, 419, 423

Society for Range Management 130

Soda Lake 246

Songfinder 289

Songs of Insects, The 229

Songs of Wild Birds, The 223

Sostrum, Nels 456

South Gap Lake, Snowy Range 400

South Pass City State Historic Site 315

Southeastern Raptor Center 387

Space Coast of Florida 368

Species of Greatest Conservation Need, WGFD 472

Spiker, Mayor Jack 212

Springer-Bump Sullivan Wildlife Habitat Management Area 184, 471

Squirrel, fox 185, 187

St. Lazarius Island, Alaska 175

State Bath House, Hot Springs State Park 295

Stebbins, Robert C. 417

Stephenson, Shauna 366

Stephenson, Tom 393

Stewart, Elinore Pruitt 204

Stillman, Deanne 257

Stockton, Ty 167, 366

Strickland, Dale 387

Stroock Forum 47

Strycker, Noah 398, 404, 408, 411

Student Conservation Association 335

Stutchbury, Bridget 242

Styskal, Colby 192

Sullivan, Brian 409

Sunderman, Steve 141

Sunol Regional Wilderness, California 238

Surbrugg, Sandra 211

Sutton, Clay 178

Swarovski binoculars 372

Symchych, Catherine, 23

———————

Table Mountain Wildlife Habitat Management Area 184, 201

Tallamy, Douglas 434

Tensleep Preserve 77

Terra Foundation 367

Terry Bison Ranch, near 232

Tessman, Steve 160

Thayer Birding Software 25, 174, 223

The Big Year, a Tale of Man, Nature, and Fowl Obsession 305

The Bird Way: A New Look at How Birds Talk, Work, Play, Parent and Think 438, 439

The Birds of North America, Golden Press 330, 335

The Coolest Bird: A Natural History of the Black Swift 334

The Genius of Birds 393

The Guide to Walden Pond... 409

The Living Bird 378

The Nature Conservancy - Nebraska 430

The Nature Conservancy 77, 358, 375, 385, 388, 423, 469

The Warbler Guide 393

The Young Birder's Guide to Birds of North America 332

Thompson III, Bill 43, 85, 196, 223, 308, 359, 368, 396

Thompson, Beauford 131, 463

Thorson, Robert 409

Thunder Basin National Grassland 201

Thunderhead Mountain, Steamboat Springs 388

Timberman, Ann 357

To See Every Bird on Earth 224

Tongass National Forest 175

Tony Knowles Coastal Trail, Alaska 268

Toops, Connie 321

Top of the World wind farm 387, 460

Townsend, John Kirk 231

Trail Ridge Road, Colorado 374

Treasure Island public access area 184

Tree, Wayne 52, 146

Tronstad, Lusha 358

True, Diemer 402

Turner, John 47

Tuthill, Dorothy 432

Tweit, Susan 367

———————

U.S. Centers for Disease Control 210, 476

U.S. Fish and Wildlife Service 106, 188, 193, 337, 386, 401, 402, 412, 413, 439, 476

U.S. Fish and Wildlife Service Endangered Species List 258

U.S. Forest Service 192, 208, 285, 402

U.S. Forest Service 192, 208, 285, 402, 446

U.S. Geological Survey 439

U.S. Highway 30 249

U.S. National Arboretum 375

U.S.S. Badger 202

Ula, Sadrul, UW electrical engineering 265

Unflappable, novel 434

Union Pacific Railroad 249

University of California Davis Arboretum 394

University of Wisconsin - Stevens Point 107, 202

University of Wyoming 222, 328, 470

University of Wyoming Biodiversity Institute 342, 349, 358, 385, 407, 413, 423, 432

University of Wyoming Extension office 410

University of Wyoming Haub School of Environment and Natural Resources 349

University of Wyoming Institute for Environment and Natural Resources 47

University of Wyoming Museum of Vertebrates 395, 476

University of Wyoming Wildlife Restoration Ecology

University of Wyoming Zoology and Physiology Department 197, 230, 342, 379

Upper North Crow Reservoir 442

———————

Van Cortlandt Park, New York City 429

Vascular Plants of Wyoming 349

Vedauwoo Recreation Area 397

VerCauteren, Tammy 162

Vyn, Gerrit 378

———————

Waipi'o Valley, Hawaii 257, 258

Walgren, Bruce and Donna 146

Warblers & Woodpeckers... 417

Warren Livestock Company 249

Warren Peak 190, 219

Washington Park Arboretum, Washington 345

Watkins Glen State Park, New York 452

Weaver, Brendon 372

Weidensaul, Scott 80, 116, 149, 290, 378, 420, 450

Welch, Ken 197

Welty, Joel Carl 365

WEST (see also Western EcoSystems Technology) 423

West Nile Virus 142, 168, 476

Western EcoSystems Technology Inc. 173, 387, 424

Western Sky Design 415

Western Water Network Grant 451

What the Robin Knows 332

What's This Bird? 427

Wheatland Reservoir No. 3 143

White Feathers, The Nesting Lives of Tree Swallows 434

White Mountain Audubon Society 214

White Mountain Wild Horse Herd Management Area 272

White Sands Missile Range 440

White, Lisa 393

White, Scott 393

Why Don't Woodpeckers Get Headaches? 229

Wiggam, Mike 204

Wild Migrations, Atlas of Wyoming's Ungulates 417

Wildcat Audubonn Society 231

Wilderness Society 386

Wildlife Crossing project 445

Wildscaping 367

Wile, Darwin 190

Wilson Journal of Ornithology 334

Wilson, Owen 305

Wind Cave National Park 104

Wind River Canyon Scenic Byway 319

Winged Migration 149

Wingert, Kevin 366, 415

Wissner, Catherine 106, 217, 244

Witcosky, Jeff 208

Wommack, Elizabeth 395, 476

Wood, Chris 290

Woodhouse, Gay 212

Worcester Polytechnic Institute 237, 316

World Birding Centers, Texas 380

World Center for Birds of Prey 330

Worthman, Tina 415

Wright, Jeananne 415

Wright, Rick 421

WY Outside 398

Wyobio database 385

WYOBIRD 478

Wyobirds 97, 166, 202, 238, 328, 365, 415, 432

Wyofile 464

Wyoming Audubon Chapters Campout 246

Wyoming Audubon Society 360

Wyoming Bioblitz (see also Bioblitz) 358, 432

Wyoming Bird Checklist 33, 125, 140

Wyoming Bird Flashcards 74, 131, 147

Wyoming Bird Hotline 166, 432

**Numbers correspond to article numbers, not page numbers.**

Wyoming Bird Initiative for Resilience and Diversity 478

Wyoming Bird Records Committee 110, 125, 188, 236, 300, 467

Wyoming Bird Trail app 398

Wyoming Birding Bonanza 342

Wyoming Birds book 43, 294, 330, 349, 384, 398, 411

Wyoming Birds Flashcards 302

Wyoming Citizen Science Conference 395

Wyoming Conservation Stamp Art Competition 105

Wyoming Department of Environmental Quality 424, 460, 464

Wyoming Department of Health 168

Wyoming Department of Transportation 445

Wyoming Dinosaur Center 311

Wyoming Frontier Prison Museum 278

Wyoming Game and Fish Commission 74, 402

Wyoming Game and Fish Department 86, 87, 96, 105, 106, 110, 125, 131, 182, 186, 193, 205, 239, 249, 292, 302, 330, 340, 349, 357, 402, 410, 419, 424, 426, 445, 459, 460, 469, 472 476, 479

Wyoming Geographic Alliance 385

Wyoming Hereford Ranch 148, 158, 184, 190, 263, 329, 330, 344, 348, 360, 396, 411, 422, 436, 448, 466, 467, 477

Wyoming Hereford Ranch Reservoir No. 1 165, 232, 438

Wyoming Hunting and Fishing Heritage Expo 78

Wyoming Important Bird Area 360, 415, 448

Wyoming Industrial Siting Council 419, 423, 424

Wyoming Master Naturalists 432

Wyoming Migration Initiative 469

Wyoming Native Plant Society 171, 358

Wyoming Natural Diversity Database 358

Wyoming Outdoor Council 401

Wyoming Outdoor Hall of Fame 177

Wyoming Pathways 423

Wyoming Public Radio 446

Wyoming Quilt Project 288

Wyoming Raptor Initiative 407

Wyoming State Office Bureau of Land Management 401

Wyoming State Engineer 433

Wyoming State Museum 415, 417, 426

Wyoming State Parks 398

Wyoming State Quilt Guild 270, 455

Wyoming Territorial Prison State Historic Site 273

Wyoming Tribune Eagle 302, 366, 415, 419, 455

Wyoming Wildlife and Natural Resources Trust 357, 469

Wyoming Wildlife calendar 182

Wyoming Wildlife Consultants 222

Wyoming Wildlife Federation 401

Wyoming Wildlife magazine 125, 349, 426, 445

Wyoming Wildlife Viewing Tour Guide 151, 190

Wyoming Wildscape, How to Design, Plant, and Maintain… 217

Wyoming's Wildlife Worth the Watching 131

———

Yale University 220

Yampa Valley Crane Festival 388

Yarmouth Port, Massachusetts 316

Yarnold, David 353

Year of the Bird 404

Yellowstone National Park 117, 190, 219, 259

Yellowstone Wolf Project 259

Yesteryear I Lived in Paradise 341

Young Birder's Guide, The 248

Young Farmers club 433

Young, Jon 332

———

Zeiss binoculars 372

Zickefoose, Julie 196, 223, 248, 383, 427

Zinke, Ryan, U.S. Department of Interior 401

Zoom 437

Printed in the USA
CPSIA information can be obtained
at www.ICGtesting.com
CBHW080604180824
13253CB00057B/768